Crystal Plasticity (Volume II)

Crystal Plasticity (Volume II)

Editor

Wojciech Polkowski

MDPI • Basel • Beijing • Wuhan • Barcelona • Belgrade • Manchester • Tokyo • Cluj • Tianjin

Editor
Wojciech Polkowski
Krakow Institute of
Technology
Poland

Editorial Office
MDPI
St. Alban-Anlage 66
4052 Basel, Switzerland

This is a reprint of articles from the Special Issue published online in the open access journal *Crystals* (ISSN 2073-4352) (available at: https://www.mdpi.com/journal/crystals/special_issues/crystal_plasticity_2).

For citation purposes, cite each article independently as indicated on the article page online and as indicated below:

LastName, A.A.; LastName, B.B.; LastName, C.C. Article Title. *Journal Name* **Year**, *Volume Number*, Page Range.

ISBN 978-3-0365-6287-2 (Hbk)
ISBN 978-3-0365-6288-9 (PDF)

Contents

About the Editor

Wojciech Polkowski

Dr. Wojciech Polkowski (born 1985) graduated from the Faculty of Advanced Technologies and Chemistry of the Military University of Technology in Warsaw. Through 2013–2016 he was working at the Faculty as an academic teacher (Assistant Professor), where he was involved in organizing and conducting classes, as well as in planning and (co)supervising students' diploma works. After receiving a Ph.D. diploma, he left the Military University of Technology in Warsaw, and since October 2016 he has been employed in a permanent position as a Postdoctoral Researcher in the Centre for High-Temperature Studies of the Foundry Research Institute in Krakow. After a restructuration into the Łukasiewicz Research Network (April 2019) he has been taking on the role of the Deputy Head of the Centre for Materials Research and the Leader of High-Temperature Research Area at the Łukasiewicz–Krakow Institute of Technology. He is an author or co-author of 131 scientifically disseminated works including: 54 publications in journals from the Journal Citation Report list, 57 conference communications and publications in proceedings, and 3 authored book chapters. Regarding his scientific work and achievements, he has received three awards from the Rector of the Military University of Technology (including the Award for the Best Thesis of the Year, 2009). In 2013–2014, he was the laureate of the Scholarship for the Best Ph.D. students founded by the Marshal of Mazovia. In 2017 he was awarded the three-year Scholarship of the Ministry of Science and Higher Education for Young Outstanding Scientists.

Preface to "Crystal Plasticity (Volume II)"

Dear Colleagues,

The extensive response to our call for papers for the first Special Issue volume on "Crystal Plasticity" was very staggering. I am sure that it was also a surprising feat for 25 excellent articles to be published in such a short amount of time.

The articles are freely accessible to read through the following link: https://www.mdpi.com/journal/crystals/special_issues/Crystal_Plasticity.

Our previous efforts have provided us with a completely new collection of original state-of-the-art research papers on both theoretical and experimental aspects of plastic deformation. Indeed, the wide spectrum of submitted papers allowed us to merge the most important research areas of the crystal plasticity field, i.e., research on the theoretical modeling of dislocation mechanisms and lab-scale validation of materials' structural/mechanical responses to (semi-)industrial processing. Furthermore, both conventional (e.g., steels, nonferrous alloys) and novel (intermetallics, composites, high entropy alloys) materials were investigated. It was my honor to host well-recognized worldwide authorities as well as young researchers and post-docs taking the "next step" in their scientific careers. This versatility of contributing authors and topics provides more proof for the high interest of the scientific community in revealing materials' behaviors at the atomic scale to macroscale under external loadings.

After closing the first volume of Special Issue, we had the feeling that there was still much research to be conducted in the field of crystal plasticity, and thus much potential for publishing activities... Therefore, we had no doubts about announcing the second volume of a Special Issue on crystal plasticity. In this book, a collection of completely new 26 original works is presented for readers.

We hope that the second volume of our Special Issue will be interesting for the scientific and academic communities, and that it will bring much inspiration for future research activities in the field of crystal plasticity.

Wojciech Polkowski
Editor

crystals MDPI

Article

X-ray Line Profile Analysis of Austenitic Phase Transition and Morphology of Nickel-Free Fe-18Cr-18Mn Steel Powder Synthesized by Mechanical Alloying

Eliza Romanczuk-Ruszuk [1,*], Krzysztof Nowik [2] and Bogna Sztorch [3]

1 Institute of Biomedical Engineering, Faculty of Mechanical Engineering, Bialystok University of Technology, Wiejska 45C, 15-351 Bialystok, Poland
2 Institute of Mechanical Engineering, Faculty of Mechanical Engineering, Bialystok University of Technology, Wiejska 45C, 15-351 Bialystok, Poland
3 Centre for Advanced Technologies, Adam Mickiewicz University in Poznan, Uniwersytetu Poznanskiego 10, 61-614 Poznan, Poland
* Correspondence: e.romanczuk@pb.edu.pl

Abstract: In this study, microstructural evolution and phase transition of nickel-free Fe-18Cr-18Mn (wt. %) austenitic steel powders, induced by mechanical alloying, were investigated. X-ray diffraction, scanning electron microscopy, and microhardness testing techniques were used to observe the changes in the phase composition and particle size as functions of milling time. The first 30 h of mechanical alloying was performed in an argon atmosphere followed by nitrogen for up to 150 h. X-ray diffraction results revealed that the Fe-fcc phase started to form after 30 h of milling, and its fraction continued to increase with alloying time. However, even after 150 h of milling, weak Fe-bcc phase reflections were still detectable (~3.5 wt. %). Basic microstructure features of the multi-phase alloy were determined by X-ray profile analyses, using the whole powder pattern modeling approach to model anisotropic broadening of line profiles. It was demonstrated that the WPPM algorithm can be regarded as a powerful tool for characterizing microstructures even in more complicated multi-phase cases with overlapping reflections. Prolonging alloying time up to 150 h caused the evolution of the microstructure towards the nanocrystalline state with a mean domain size of 6 nm, accompanied by high densities of dislocations exceeding $10^{16}/m^2$. Deformation-induced hardening was manifested macroscopically by a corresponding increase in microhardness to 1068 $HV_{0.2}$. Additionally, diffraction data were processed by the modified Williamson–Hall method, which revealed similar trends of domain size evolutions, but yielded sizes twice as high compared to the WPPM method.

Keywords: nickel-free austenitic stainless steel; phase transition; mechanical alloying; X-ray diffraction; profile analysis; whole powder pattern modeling

Citation: Romanczuk-Ruszuk, E.; Nowik, K.; Sztorch, B. X-ray Line Profile Analysis of Austenitic Phase Transition and Morphology of Nickel-Free Fe-18Cr-18Mn Steel Powder Synthesized by Mechanical Alloying. *Crystals* **2022**, *12*, 1233. https://doi.org/10.3390/cryst12091233

Academic Editor: Wojciech Polkowski

Received: 31 July 2022
Accepted: 23 August 2022
Published: 1 September 2022

1. Introduction

Mechanical alloying (MA) is a popular method for obtaining new materials with a controlled structure. This method can be used for the synthesis of materials with equilibrium and non-equilibrium structures, and the preparation of supersaturated, metastable crystalline, quasi-crystalline, intermetallic, nanostructured, and amorphous alloys [1,2]. The advantage of MA involves the possibility of obtaining nanocrystalline solid solutions and the occurrences of low-temperature phase transformations [2]. During the mechanical alloying process, parameters such as milling time, milling atmosphere, ball-to-powder ratio, or the size of grinding balls are selected. The milling time is one of the most important variables that affect the purity, structure, and properties of the final powder product. Therefore, many publications focus on the analysis of the properties of powders depending on the milling time [1,3–5]. Another important process parameter that affects the oxidation and contamination of powders is the milling atmosphere. Generally, an inert atmosphere, such as argon or helium, is used. Nevertheless, when milling reactive powders, such as titanium, aluminum, iron, or iron alloys, a different protective atmosphere can be used to introduce gas from the atmosphere into the powders [1]. Mechanical alloying of iron alloy powders in a nitrogen atmosphere introduces nitrogen into the matrix in a solid–gas reaction. One interesting issue is the development of nickel-free austenitic stainless steel with nitrogen obtained by mechanical alloying in a nitrogen atmosphere and then consolidating these powders. Therefore, many studies have been carried out on nickel-free austenitic steel powders mechanically alloyed in a nitrogen atmosphere, which allows the introduction of nitrogen into the matrix. The use of nitrogen atmosphere in the mechanical alloying process leads to the transformation of ferrite into austenite. Additionally, manganese is added to these materials to increase the solubility of nitrogen [6,7].

The influence of the milling parameters and atmospheres on the Fe-α→Fe-γ phase transformation and amorphization process of the nickel-free stainless steel powder were analyzed by researchers [8–11]. However, there is no systematic analysis of the manganese (Mn) and nitrogen (N) influence and MA process parameters on the powder properties. Amini et al. [9] used elemental powders for the synthesis of Fe-18Cr-18Mn (wt. %) stainless steel under a nitrogen atmosphere. A fully fcc (Fe-γ) phase structure was achieved after 96 h of ball milling in a high-energy ball mill. Prolonging the ball milling process in this atmosphere up to 144 h resulted in an amorphous phase. Similar results, for Fe-18Cr-11Mn (wt. %) steel elemental powders mechanically alloyed in argon were reported by Haghir et al. [10]. A fully austenitic structure was achieved after 120 h of milling using a planetary high-energy ball mill (Retsch, PM100, Haan, Germany). When nitrogen atmosphere was used, a complete phase transformation occurred 20 h earlier. This emphasizes the nitrogen interstitial effect on the bcc to fcc phase transformation.

Tehrani et al. [12] reported an influence of Mn content on the Fe-α→Fe-γ phase transformation on two Fe-18Cr-7Mn and Fe-18Cr-8Mn (in wt. %) steel powder compositions mechanically alloyed in argon using a planetary high-energy ball mill (Retsch, PM100, Haan, Germany). In the higher manganese content alloy (8 wt. %), the fully austenitic structure was detected after 100 h of MA, while in the material with lower manganese content, even after 150 h of milling, the Fe-α→Fe-γ phase transformation was incomplete. This suggests that Mn content in the mechanically-alloyed austenitic steel should be higher than 7%.

The results in the literature show that the influence of manganese content from 6 to 12% on the properties of nickel-free stainless steel was usually investigated. The literature appraisal revealed that the influence of manganese content from 6 to 12% on the nickel-free stainless steel properties was usually investigated. In this study, a higher manganese content (18%) and a mixed atmosphere of mechanical alloying were used.

The main goal of this work was to analyze the phase transformation process and morphology of the Fe-18Cr-18Mn steel powder after mechanical alloying using argon followed by a nitrogen atmosphere.

2. Materials and Methods

2.1. Mechanical Alloying of Powders

Iron, chromium, and manganese elemental powders (average particle size ~45 μm, 99.5% purity) supplied by Alfa Aesar (Kandel, Germany) were mixed and mechanically alloyed to obtain a nominal composition of Fe-18Cr-18Mn (in wt. %). The MA process was conducted in a planetary, high-energy ball mill Pulverisette 6 (Fritsch, Amberg, Germany) at a rotation speed of 250 rpm. Stainless steel balls (12 mm in diameter) were charged into a 500 mL stainless steel bowl with a ball to powder ratio (BPR) of 10:1. First, mixing of the powders was performed for 0.5 h at 150 rpm under argon atmosphere. For the first 30 h, the milling process was conducted under a pure argon atmosphere (>99.999%); after that, pure nitrogen (>99.999%) was applied. After 150 h of MA, the process was finished.

2.2. Morphology and Microhardness of Powders

The morphology of the powders was examined using scanning electron microscopy (SEM, Hitachi 3000N, Tokyo, Japan) with an energy dispersive spectroscopy (EDS) (Hitachi, Tokyo, Japan). Semi-quantitative chemical analysis of the main elements in the powders was performed. Figure 1 shows the SEM image of the surface area of the particle taken for the EDS analysis.

Figure 1. Example SEM image of a particle surface area for the SEM-EDS analysis.

Vickers microhardness ($HV_{0.2}$) tests of the powders after different mechanical alloying times were performed using the PMT-3 tester (PMT Labs, Wah, Pakistan) under a load of 1.96 N (0.2 kg) for 15 s. For accurate results, at least ten indents were conducted on each sample and then the results were averaged and standard deviations determined.

2.3. X-ray Diffraction of Powders

A small amount of the MA powder was taken for the X-ray diffractometry (XRD) (Bruker, Karlsruhe, Germany) analysis after 30, 60, 90, 120, and 150 h of the milling process. To minimize the powder contamination, loading and unloading of the powder were performed in a glove box under a protection argon atmosphere. The phase structure was measured by means of X-ray diffractometry using Bruker D8 Advance equipped with Cu anode (λ = 0.15418 nm) radiation of 40 kV and 25 mA. For all samples, the angular range (2θ) of 20° to 100° with a step width of 0.01 and an acquisition time of 5 s was used. The instrumental broadening of peak profiles was determined using a corundum (Al_2O_3) standard (NBS SRM 1976b) and processed accordingly to the Caglioti et al. formula [13], following the procedures described widely in the literature (e.g., [14]).

It is widely known that traditional, "single-peak" methods of extracting microstructural data (e.g., Williamson–Hall method) do not take strain anisotropy effects into consideration [15], which are manifested in that the widths of the peaks do not increase monotonically with the diffraction angle θ [16]. For that reason, Ungár and Borbély developed an upgraded dislocation model of the mean square strain by incorporating the so-called average dislocation contrast factor $\overline{C}$, which is known as the *modified* Williamson–Hall plot (MWH) [17]. The strain contribution can be expressed in terms of dislocation properties, as demonstrated in Equation (1):

$$\Delta K = 0.9/d + \left(\pi A b^2/2\right)^{\frac{1}{2}} \rho^{\frac{1}{2}} K \overline{C}^{\frac{1}{2}} + O\left(K^2 \overline{C}\right) \quad (1)$$

where $K = 2\sin\theta/\lambda$, $\Delta K = 2\cos\theta(\Delta\theta)/\lambda$, θ, $\Delta\theta$, λ are the diffraction angle, FWHM of the reflection and the wavelength of X-rays. $\overline{C}$ is the average dislocation factor, where the average is made over the equally populated equivalent slip system. Other physical parameters in Equation (1) are: the dislocation density ρ, respectively, A is a constant determined by the outer cutoff radius of the strain field, R_e, and O stands for noninterpreted, higher-order terms. Finally, b is the Burgers vector equal to $b = a_{bcc}3^{\frac{1}{2}}/2$ or $b = a_{fcc}2^{\frac{1}{2}}/2$ for bcc and fcc crystal systems, respectively, where a is the lattice constant. The value of $\overline{C}$ depends only on the ratios of the material's elastic constants c_{11}, c_{12}, and c_{44}, which can be further reduced to two parameters—elastic anisotropy $A_i = 2c_{44}/(c_{11} - c_{12})$ and the ratio c_{12}/c_{44} [18]. In the case of cubic lattice systems, $\overline{C}$ is a linear function of the fourth-order invariant of the hkl Miller indices [19], as shown in Equation (2):

$$\overline{C} = \overline{C}_{h00}\left(1 - qH^2\right) \tag{2}$$

where:

$$H^2 = \frac{h^2k^2 + h^2l^2 + k^2l^2}{\left(h^2 + k^2 + l^2\right)^2} \tag{3}$$

Equation (2) shows that, under the hypothesis that the sample has a random texture, which is entirely justifiable for a milled powder or randomly oriented polycrystal, $\overline{C}$ can be evaluated if the values of q and $\overline{C}_{h00}$ are known. $\overline{C}_{h00}$ is the average contrast factor corresponding to the $h00$ reflection, whereas the value of q determines the edge/screw type of dislocations [20].

To implement the strain anisotropy model, contrast factor coefficients had to be calculated for both Fe-bcc and Fe-fcc phases. As the austenitic transformation is caused by the continuous dissolution of Mn in the Fe matrix, it was assumed that the ferritic phase occurring at the beginning of MA was pure Fe, while the austenitic phase had the nominal composition of Fe-18Cr-18Mn. The necessary single crystal c_{ij} elastic constants of pure ferrite (Fe-bcc) were adopted from the literature [21], whereas Fe-18Cr-18Mn alloy elastic properties were estimated by the relations provided by Razumovskiy et al. [22]. These values were used to calculate the average contrast factors toward the $h00$ direction for edge $\overline{C}^e_{h00}$ and screw the $\overline{C}^s_{h00}$ dislocations, considering the $\langle 111\rangle\{110\}$ and $\langle 110\rangle\{111\}$ primary slip systems for bcc and fcc metals, respectively, using the online program ANIZC [23]. Similarly, the extreme values of q, corresponding to pure edge q_e and pure screw q_s dislocation characters were obtained by equations elaborated by Ungár et al. [18]. After obtaining $\overline{C}_{h00}$ and q, the average contrast factor can be calculated. The trend of $\overline{C}$ as a function of the scattering vector modulus ($s = 2\sin\theta/\lambda$) is plotted in Figure 2. As presumed, $\overline{C}$ reaches its maximum towards the $h00$ direction, which is the soft crystallographic direction in both bcc and fcc Fe, along which the strain proceeds the most easily [24]. Relevant parameters needed to incorporate the dislocation model are summarized in Table 1. As can be noticed, the anisotropy of Fe-18Cr-18Mn austenitic alloy is considerably higher than in the case of pure bcc Fe, which is manifested by bigger deviations of A_i from unity.

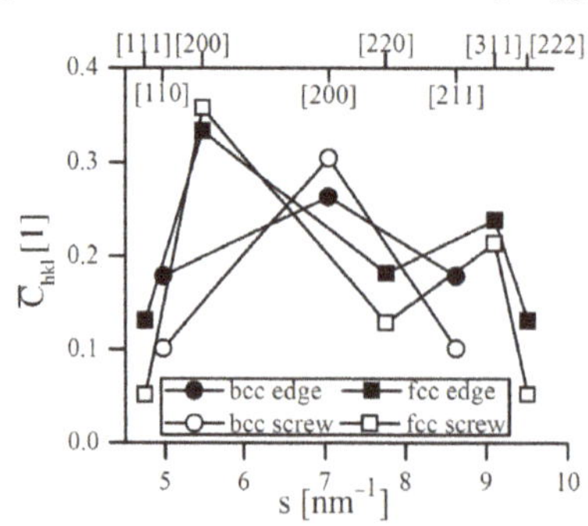

Figure 2. The average value of anisotropic contrast factor $\overline{C}$ versus the scattering vector s calculated for bcc and fcc microstructures of the Fe-18Cr-18Mn alloy, considering pure edge and pure screw dislocation characters.

Table 1. Elastic parameters required to implement the strain anisotropy model.

Parameter	Fe, bcc	Fe-18Cr-18Mn, fcc
c_{11}	230	219
c_{12}	134	153
c_{44}	116	152
A_i	2.417	4.606
$\overline{C}^e_{h00}$	0.403834	0.333565
$\overline{C}^s_{h00}$	0.361353	0.357697
q_e	0.721	1.821
q_s	2.597	2.561

Contrary to the classical methods of the line profile analysis (LPA), i.e., fitting peak profiles using arbitrarily defined, bell-shaped mathematical functions (top-down approach), the more sophisticated, bottom-up approach was recently developed. In the bottom-up approach, the entire diffractogram is modeled directly as a set of physical parameters affecting the peak's shape, width, and position. The size broadening calculation is based on the concept of column heights and the lognormal distribution of crystallite sizes is assumed (in contrast to the plain average obtained by classical LPA methods), which was largely verified in the case of highly deformed metals and finely dispersed powders [25]. Strain broadening was considered by the Krivoglaz–Wilkens theory of dislocations in distorted crystals [26].

The whole powder pattern modeling (WPPM) has been proposed as a universal technique for microstructure refinement. It provides detailed data on specimen microstructures by directly comparing model peak profiles with the entire experimental pattern, considering instrumental broadening and background [27,28]. Size and defect contributions are convoluted together with the instrumental component, and the pattern is then directly synthesized through Fourier transformation [29]. Despite the close analogy to the widely used Rietveld method, in WPPM, structural information is limited only to lattice parameters, while Rietveld refinement strictly relates integrated intensity to the structural model (atomic positions, occupancy, thermal factors, etc.) [28]. Another (but very similar to WPPM) approach is known as the convolutional multiple whole profile fitting (CWMP) [30], with the procedure of employing the instrumental profile being the only major difference. WPPM has been successfully utilized for investigating microstructural evolution in many ball-milled powders and other nanocrystalline materials [29]. Its algorithm has been implemented in a free and flexible software package, named PM2K [31], which was exploited in this study.

3. Results and Discussion

3.1. Morphology and Microhardness of the MA Powders

Figure 3 shows SEM images of the powder morphology as a function of milling time. The histogram of the average particle size is presented in Figure 4. It is clear that the time of the process impacts the powder morphology. As received, an iron powder is regular and smooth in shape (Figure 3a). The chromium powder is irregular and angular in shape (Figure 3b), whereas the manganese powder (Figure 3c) has a morphology similar to chromium, with a more regular and smoother surface. The first 30 h of MA revealed an irregular shape of the powder, with an average size of 165 ± 10 μm. After replacing argon with nitrogen and prolonging the milling time up to 60 h, an average particle size continues increasing up to 213 ± 11 μm. The next 30 h of milling (30 h in argon and 60 h in nitrogen) did not change the morphology of powder particles; however, the average size of particles slightly decreased to 200 ± 10 μm. After another 30 h of MA, significant particle refinement to 81 ± 8 μm was observed. Thus, at this stage of the MA process, after 30 h of milling in argon and 90 h of milling in nitrogen (total time of MA was 120 h), the powder fracturing took place. The MA process was performed up to a total of 150 h (30 h in argon and 120 h in nitrogen) and revealed bimodal distributions of particle, where approximately 75% of the powder had particle sizes of 51 ± 6 μm and 25% of powder had particle sizes of 10 ± 5 μm. This kind of bimodal powder distribution might have a favorable influence on the classical PM consolidation process, e.g., annealing, cold compaction, and sintering.

Figure 3. Morphology of the elemental powders: (**a**) Fe, (**b**), Cr and (**c**) Mn, and mechanically alloyed powder at different milling times, after: (**d**) 30 h, (**e**) 60 h, (**f**) 90 h, (**g**) 120 h, (**h**) 150 h.

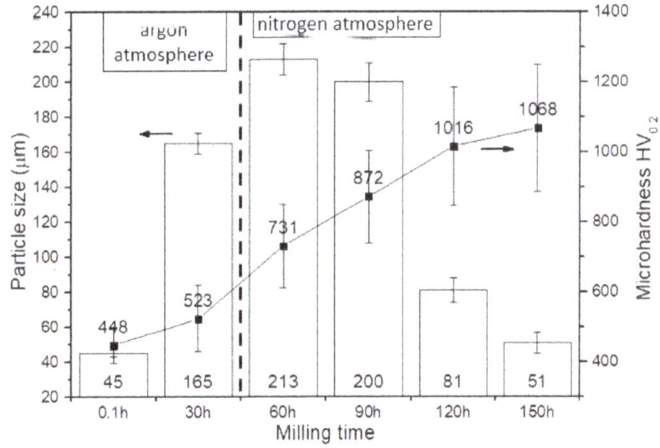

Figure 4. The particle size and microhardness of powders at different milling times.

The mechanical alloying process consists of the repetition of cold welding, fracturing, and rewelding of powder particles occurring in the milling jar by collisions of balls against each other and the walls of the jar. The cold welding leads to an increase in particle sizes, while the fracturing leads to the fragmentation of particles [32]. Analyzing changes in the morphology of particles as a function of milling time, it can be concluded that during the first 60 h of MA, cold welding dominates, increasing the average particle size. The subsequent 30 h of milling does not cause a significant alteration in the particle size. This means that the fracturing process, resulting in the particle size decrease, is in equilibrium with the cold welding process. After 120 h of MA, the fracturing phenomenon causes a continuous decrease in the particle size, which emphasizes domination of the fracturing over cold welding. After 150 h of ball milling, the average particle sizes were almost similar to the initial sizes of the particles used at the beginning of the process (but with the bimodal distribution). Similar trends were reported by Haghir et al. [10], whereas Cisneros et al. [33] and Duan et al. [34] observed that the initial sizes of the elemental powders of ~50 μm after mechanical alloying, up to 170 h in nitrogen, were reduced to 10 μm. The authors pointed out that first the powder particles flattened and then the prolonging of milling time became equiaxed.

Coarser particle sizes observed here, in comparison with the literature data [33,34], are due to the different milling devices and process parameters, including atmospheres. For instance, decreasing the ball to the powder weight ratio from 10:1 to 5:1 can cause a significant increase in the crystallite and particle size of the obtained powder [32]. Cisneros et al. [33] used an attritor ball mill with the ball to powder weight ratio of 30:1, the rotation speed of 300 rpm, and nitrogen. Duan et al. [34] used a planetary ball mill, for instance, with a rotation speed of 350 rpm, and with three different types of ball diameters. This explains the differences between the literature data and the results presented here. Moreover, the atmosphere used here differs from the literature cited. The powder obtained here is softer in comparison to the hardness measurements presented by the other authors. This suggests that nitrogen content in the bulk powder is lower and, therefore, the number density of hard nitride precipitations can be reduced. This is clearly observed especially when argon atmosphere is used.

Figure 4 depicts the microhardness variation of the as-milled powder as a function of the milling time. The average microhardness of the powder increased from 450 $\pm$ 35 $HV_{0.2}$ (after 0.5 h of powder blending) to 523 $\pm$ 45 $HV_{0.2}$ after the first 30 h of MA. Note that when the atmosphere changed from argon to nitrogen, the hardness of the powder abruptly increased from 523 $\pm$ 45 up to 731 $\pm$ 52 $HV_{0.2}$ within the same period of time. By further prolonging the milling time, an average hardness almost linearly increased, reaching 1068 ± 56 $HV_{0.2}$ after 150 h of milling.

The mechanical alloying process, due to collisions of the powder with the balls and jar wall, generated microstructure defects, which caused the hardening effect and a continuous increase in hardness. As expected, a greater increase in the hardness of the powder was measured after changing the milling atmosphere from argon to nitrogen. However, the final hardness measured in this work was about 5% lower in comparison with the literature data [7], where the nominal composition of the tested powder was Fe-18Cr-4Mn (in % wt.), a 2.5 times lower percent of manganese in comparison with this work [7]. It is worth noting that Salahinejad et al. [7] applied different mechanical alloying process parameters (e.g., the ball to powder weight ratio was 30:1) and nitrogen atmosphere through the whole period of MA. After 153 h of mechanical alloying, a fully amorphous phase was obtained. Moreover, the microhardness of powder after 99 h of milling, when the phase structure was crystalline, was 1070 HV similar to the value obtained in this work after 150 h of MA.

3.2. Chemical Composition of Powders

In this work, the nominal composition of the alloy (Fe-18Cr-18Mn-N) was selected based on the modified Schaeffler's diagram for austenitic stainless steel. The percentage of manganese used in this study was higher in comparison to the literature data [6,33,35–37]. Higher Mn content stabilizes γ-Fe phase and alters the final powder's properties. We calculated both the Ni and Cr equivalents assuming that the steel possessed 18% of Mn and 18% of Cr, to find out the minimum N content that ensured the austenitic structure. From these calculations, it followed that 0.75% of N content was enough to obtain the austenite phase structure at room temperature. This N content can also be diminished when the equivalent carbon in the powder is included. From the literature, it is known that 0.9 ÷ 1.2% of N and 0.02 ÷ 0.03% of C content in austenitic steel powder after 120–150 h of MA was measured [33,36].

Table 2 presents the changes in the main element composition of the steel powder as a function of the milling time. These results reveal that, with up to 90 h of MA, the Mn and Cr content continuously increase; however, after this time of milling, the concentrations of these elements are almost unaffected. Therefore, taking into account the optimization process, the expected composition of the alloy was achieved after 90 h of milling, which suggests that the process of MA should be interrupted. The MA process was continued for up to 150 h to find out whether—after such a long time of milling—an amorphous phase would occur. The content of nitrogen and carbon increased gradually with the increasing milling time. After 150 h of mechanical alloying, the nitrogen content was 0.9% and carbon was 0.03%; the content of N and C were similar to the data from the literature.

Table 2. SEM-EDS analysis of the iron, chromium, and manganese concentrations in the MA powders at different milling times.

Milling Time (h)	Element Composition (wt. %)				
	Fe	Cr	Mn	N	C
30	72.04 ± 3.46	12.28 ± 1.25	15.67 ± 0.89	0.12 ± 0.03	0.01 ± 0.01
60	67.08 ± 2.32	16.18 ± 0.86	16.74 ± 1.58	0.28 ± 0.05	0.015 ± 0.006
90	64.21 ± 3.86	17.86 ± 0.98	17.93 ± 2.03	0.35 ± 0.01	0.016 ± 0.003
120	64.33 ± 1.02	17.66 ± 1.53	18.01 ± 0.99	0.68 ± 0.08	0.023 ± 0.007
150	64.22 ± 2.29	17.92 ± 1.37	17.86 ± 1.24	0.93 ± 0.09	0.029 ± 0.002

3.3. X-ray Diffraction Analysis of the Powders

Several XRD patterns gathered at various stages of MA are shown in Figure 5a. As can be seen at the early stage of MA (5 h), the microstructure was purely bcc, and the main Fe-bcc phase was accompanied by several reflections of Mn-bcc. After 5 h of milling, no significant change in microstructure occurred as compared to the initial powder mixture (Figure 5c), besides broadening of reflections, and Mn peaks were still present. Eventually, after 30 h of MA, Mn reflections vanished, suggesting that it was finally dissolved in the Fe matrix. At this point, the Mn-induced phase transformation process was obvious, as the bcc peak intensities dropped drastically and fcc peaks became dominant (Figure 5a,c). After 60 h of MA, the bcc reflections were nearly elusive, with the exception of the most prominent [110] Fe-bcc peak. Prolonged milling caused a further increase in the intensity of the Fe-fcc phase. The Fe-bcc [110] peak trace, however, was still detectable even after 150 h of MA (Figure 5d).

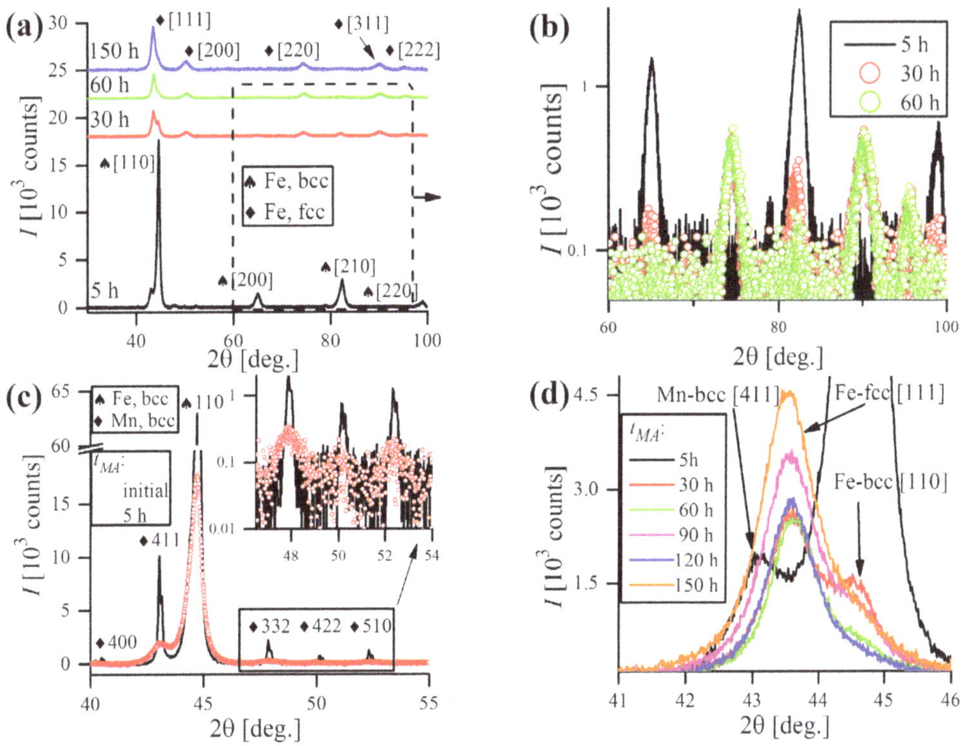

Figure 5. Microstructural evolution of Fe-18Cr-18Mn alloy: (**a**) overall comparison of XRD patterns after a certain MA time, (**b**) insight showing the vanishing of Fe-bcc peaks, (**c**) comparison of initial and after 5 h of MA patterns with emphasis on Mn-bcc reflections, (**d**) changes of the shape and intensity of most prominent Fe-bcc [110] and Fe-fcc [111] reflections during the MA course.

XRD patterns were fitted using the pseudo-Voigt function and the extracted data (peak 2θ position and FWHM) were processed in accordance with the MWH procedure, using the calculated contrast factor values for Fe-bcc and Fe-fcc phases, and regarding the instrumental broadening correction and proper background definition. The phase percentage was estimated by the ratio of integrated areas of peak profiles of Fe-bcc and Fe-fcc phases, respectively, as presented in Figure 6.

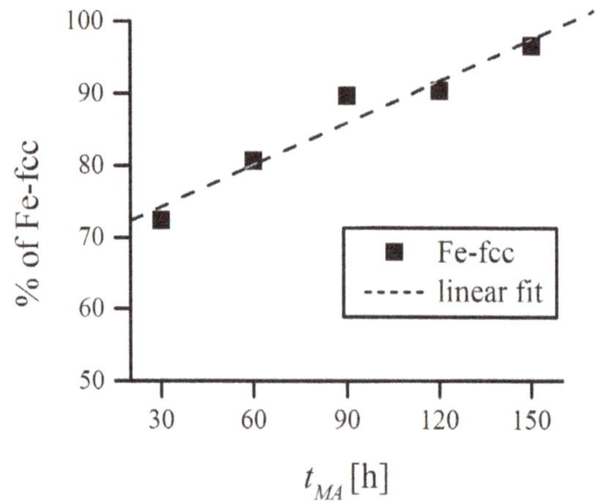

Figure 6. Phase percentage of Fe-fcc.

Extracting information from the experimental peak profiles, where multiple phases exist simultaneously, outlines the major problem concerning the traditional LPA methods—overlapping among peak profiles (especially within the region of Fe-bcc [110] and Fe-fcc [111] peaks). The most common solution to this difficulty is pattern decomposition—separating the individual contributions from different phases and subtracting the instrumental and background effects.

The main results of the MWH analysis are collated in Figure 7. The experimental values of q, which imply the edge/screw characters of dislocation, were evaluated using the methodology provided by Ungár et al. [18] (Figure 7a). The FWHM values were plotted according to the MWH relation (Figure 7b), where the intersection at $KC^{1/2} = 0$ gave the average crystallite size d. It is clear that the data points follow the linear trend in almost the exact manner and, therefore, were fitted using the linear function. It should be underlined that, due to the progressive bcc→fcc phase transformation over the MA time, the reliable determination of FWHM of Fe-bcc reflections was rather unobtainable after 30 h of MA, caused by the strong broadening and low intensity. Although after 30 h of MA, there were still three apparent Fe-bcc reflections present (Figure 5a), they were not enough to calculate d accurately using the MWH method.

As presented in Figure 7c, the domain size was reduced from ~20 nm to around 10 nm during the first 60 h of MA and saturates at this level until the end of milling. The slope of the MWH linear approximation is also plotted in Figure 7c, which can be related to the dislocation arrangement parameter M and the square root of dislocation density ρ. However, the M can be determined thoroughly only with the Fourier methods, such as WPPM. Consequently, the determination of ρ from the results obtained by MWH alone is improper and should be avoided [38]. Nevertheless, the inspection of the MWH slope provides some insight into the strain accumulation in the material, and it can be noticed that it mimics the trend of d in the opposite manner (Figure 7c). Finally, the dislocation character trend was plotted in the form of the edge dislocation fraction f_e (Figure 7d). In general, the dislocation structure is in favor of screw dislocations over the whole MA process, which might be caused by the rotational nature of the ball milling process, promoting the formation of twists in crystals [39]. The f_e value fluctuates in the 0.2–0.4 region and never exceeds the half edge–half screw line. This supports the other authors' conclusions that edge dislocations are unstable in fine Fe domains [24,40].

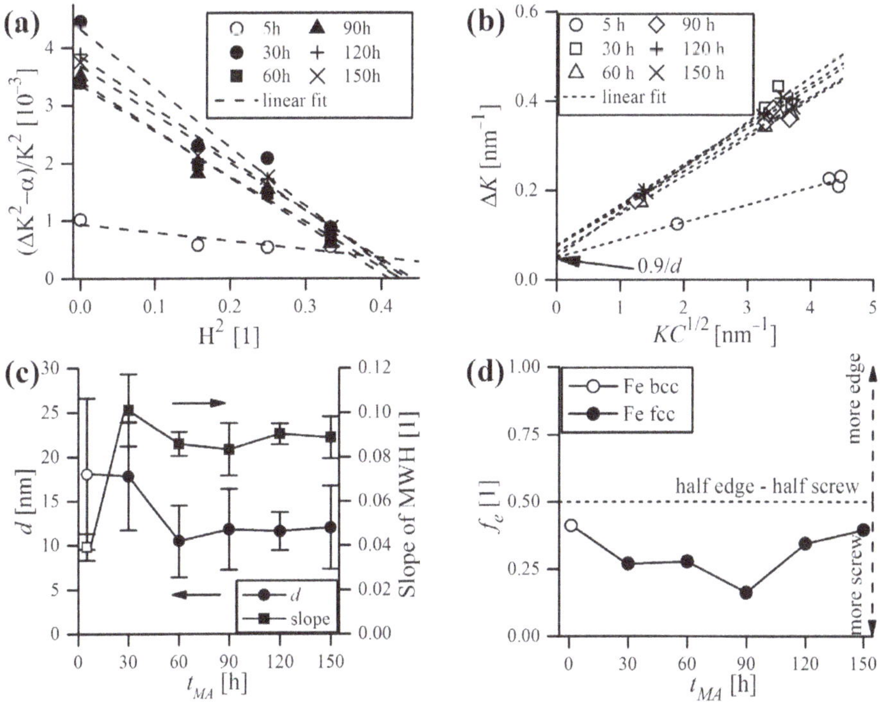

Figure 7. The principal result of the MWH analysis: (**a**) determination of experimental q values, (**b**) FWHM values plotted according to the MWH procedure, (**c**) evolution of crystallite size and MWH slope, (**d**) fraction of edge dislocations during MA.

The diffraction data were furtherly analyzed by the WPPM approach and the quality of fit can be perceived in Figure 8. Plots present the experimental data compared with model WPPM diffractograms, further decomposed into single analyzed phases (Mn-bcc, Fe-bcc, and Fe-fcc). A flat, nearly featureless residual line (the difference between experimental and calculated patterns) proves good quality of modeling. The graphical analysis is essential to determine the quality of fit and to ensure that the model is chemically plausible. The quality of WPPM fit, such as in other least-squares minimization methods, can also be defined by inspecting the discrepancy factors. In this study, the most straightforward discrepancy index, the weighted profile R-factor, R_{wp}, did not exceed 3.46%, which indicates a robust quality of fit, particularly considering the multiplicity of phases and overlapping peaks. Another marker, goodness of fit (GoF, related to the ratio between actual and expected R_{wp}) had a maximum of 1.52 (pattern after 5 h of MA) and a mean value of 1.21 (where unity indicates the flawless fit).

Unlike the MWH method, the pattern is not decomposed into individual line profiles, but the whole diffractogram is modeled by optimizing the value of several physical parameters. Figure 8a presents the pattern obtained after 5 h of MA, with a detailed view of multiple, minor Mn-bcc reflections, successfully refined using WPPM. Figure 8b–d magnify the most interesting, from the modeling viewpoint, 2θ region, where prominent Fe-bcc [110] and Fe-fcc [111] peaks extensively overlap. Additionally, inserts show the fit quality of the remaining reflections.

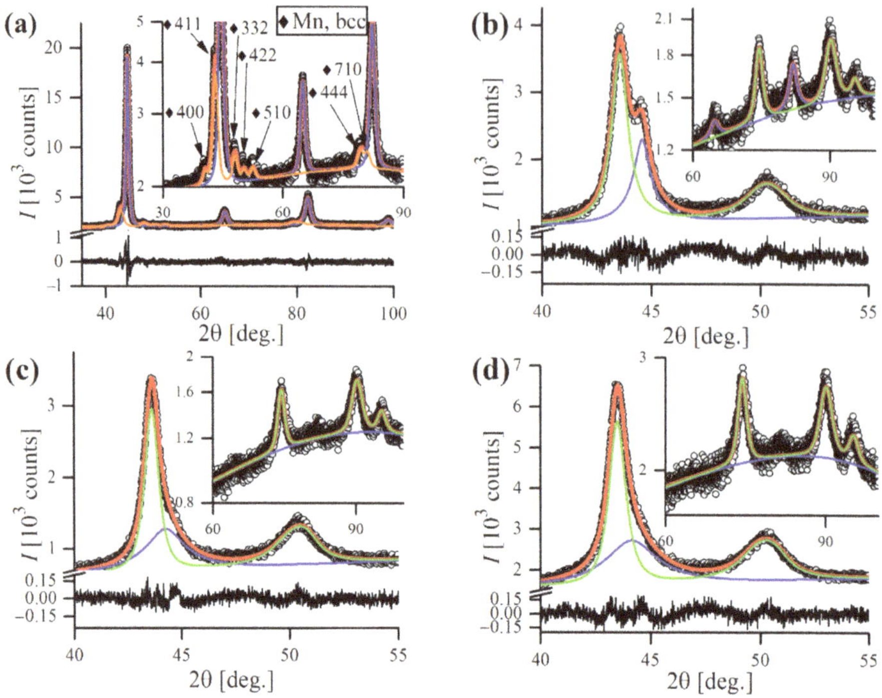

Figure 8. WPPM-refined experimental patterns gathered after: (**a**) 5 h of MA, (**b**) 30 h of MA, (**c**) 60 h of MA, (**d**) 150 h of MA. Legend: hollow dot—experimental data, red line—WPPM model, orange line—Mn-bcc phase, blue line—Fe-bcc phase, green line—Fe-fcc phase, black line—residual.

Basic results derived from WPPM refinement, in terms of size and strain evolution during MA, are revealed in Figure 9. It is a known fact that MA prompts the volume expansion of the unit cell, predominantly as a result of severe plastic deformation and persistent dissolution of alloying additives and various contaminants originating from the milling equipment into the matrix [28]. As depicted in Figure 9a, the lattice parameter a_{fcc} of the Fe-fcc phase increases constantly up to 90 h of MA and then exhibits a sharp rise and sets around this value.

The domain size d decreases with the MA time, following a similar trend as using the MWH method, and so does the lognormal standard deviation (Figure 9b,c). However, the d values obtained with the WPPM method are roughly half of that calculated by MWH. Both methods confirm that grain refinement is effective only during the beginning of MA (up to 30 h), with little reduction achieved afterward. It is worth mentioning that real materials have distributions of domain sizes, so the trends of the size distributions should also be followed, as demonstrated in Figure 10, and the corresponding values of the standard deviation of lognormal distribution are plotted in Figure 9c. It is clear that mechanical treatment causes the shift of the domain size distribution curve to lower values, the standard deviation being reduced accordingly, especially in the early stage of MA (Figure 10a). Eventually, a microstructure consisting of very fine (d = 6 nm) and narrowly distributed (σ = 3 nm) crystalline domains is established. Similar crystallite sizes were previously reported for other ball-milled fcc alloys (e.g., [41,42]).

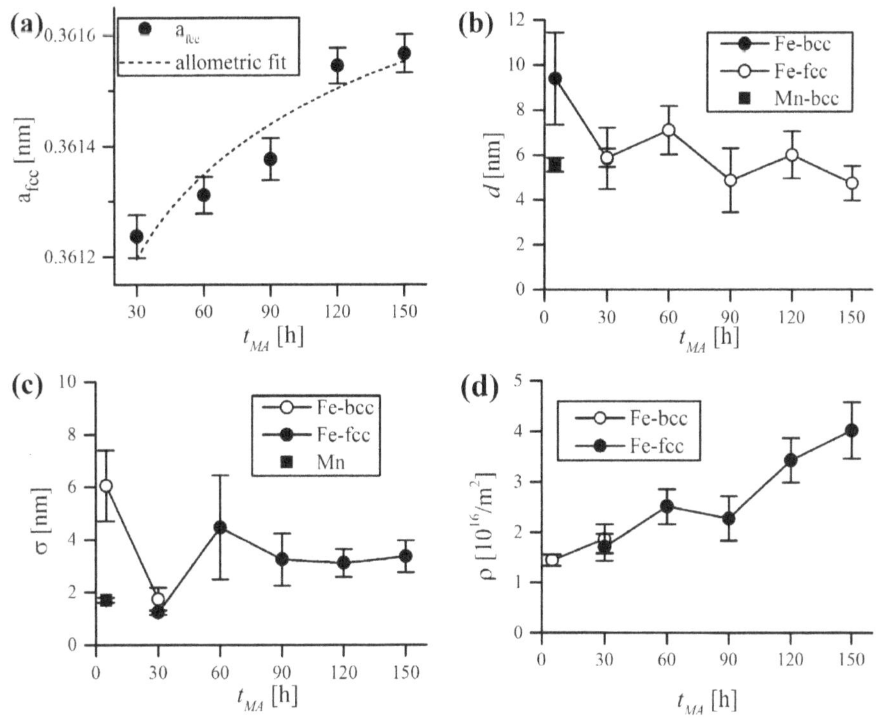

Figure 9. Basic microstructural data derived from WPPM refinement in terms of the progression of following the parameters during MA: (**a**) lattice parameter, (**b**) lognormal domain size, (**c**) lognormal standard deviation, (**d**) dislocation density. Error bars correspond to the estimated standard deviation of the WPPM fit.

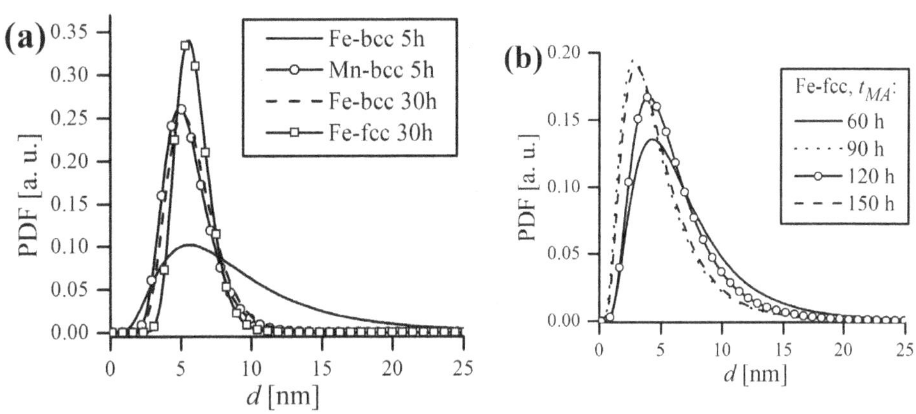

Figure 10. Lognormal domain size distributions of crystalline domains: (**a**) up to 30 h of MA, (**b**) after 30 h of MA, in terms of probability density function (PDF).

It is also clear that ball milling generates a high density of dislocations due to severe plastic deformation of ground material, which can be followed in Figure 9d. In general, the accumulation of defects in the Fe-fcc phase follows a similar trend to lattice expansion, with the same sharp rise after 120 h of MA. This is foreseeable, given the fact that the lattice inflation is caused not only by the dissolution of alloying additives but is directly caused by an increase in the dislocation density [43] and is further promoted by extrinsic dislocations pilling up at the grain boundaries [44]. Rawers and Cook suggested that the strain on the grain boundary could extend into the nanograin itself, expanding the lattice [45]. In this study, from the moment of appearing (30 h) up to 90 h of MA, the ρ of Fe-fcc phase oscillates around $2 \times 10^{16}/\text{m}^2$, then experiences a sharp rise and sets around $3.5 \times 10^{16}/\text{m}^2$, the effective outer cutoff radius R_e being 6.24 nm. The final value of R_e is around that of the mean crystallite size d (6.0 nm), and corresponds to an average of half the dislocation per crystalline domain (for $d = 6.0$ nm: $d/\left(\pi d^3/6\right) \sim 5.3 \times 10^{16}/\text{m}^2$), which is plausible. This is in line with HRTEM observations of ball-milled FeMo alloy by Rebuffi et al. [24], which concluded that dislocations are rather unlikely to be present in every single crystalline domain. As the accumulation of defects in powders, subjected to serious mechanical treatment, is the principal cause of microhardness growth during MA (Figure 4), ρ and HV can be related, which is demonstrated in Figure 11.

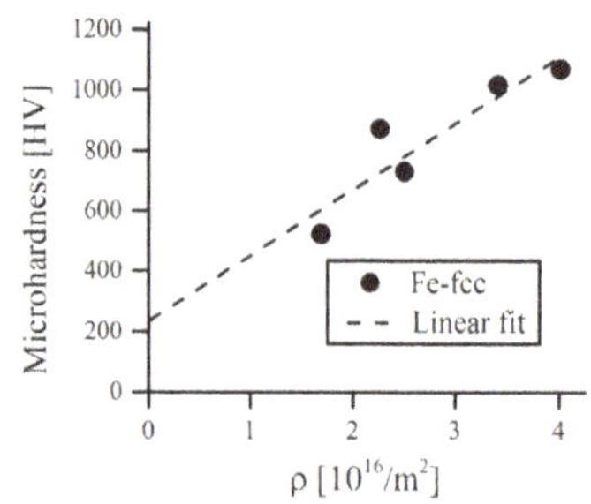

Figure 11. Relationship between dislocation density ρ and powder microhardness HV. Dashed line is the linear fit of the data.

Regarding the mutual dependence of ρ and R_e parameters in the Krivoglaz–Wilkens dislocation theory, Wilkens introduced the dimensionless parameter $M = R_e\rho^{-1} = R_e/l_d$ called the Wilkens parameter. As can be noticed, M is also directly related to the average dislocation distance l_d; therefore, it is used to determine the arrangements of dislocation arrangements in the strain fields [30]. The value of M much larger than unity ($M \gg 1$) indicates that the dislocations are very weakly correlated and randomly arranged (weak dipole character). In contrast, $M \leq 1$ signifies a distinctly correlated dislocation arrangement, accompanied by an intense screening of strain fields ($R_e < l_d$) (strong dipole character). In the present case, the decreasing value of M of the fcc phase to the value of ~1 with milling time (Figure 12) supports the hypothesis of gradual transformation from a dislocation cell structure with long-range strain fields into the nanocrystalline microstructure with a high density of dislocations, systematically distributed in grain boundaries.

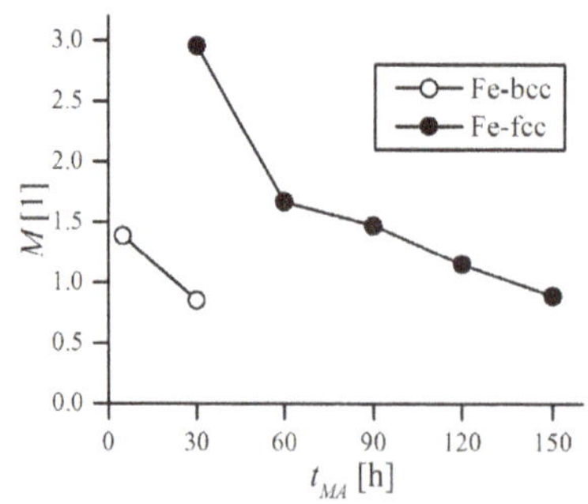

Figure 12. Trend of the Wilkens parameter M in the function of milling time.

The WPPM modeling quality was good (mean GoF = 1.21) and hassle-free in almost all cases, except the problems encountered during refining the pattern obtained after 30 h of MA. In that case, during refinement of the Fe-bcc phase, R_e constantly reached unrealistically small values (<1 nm), the lower limit of the continuum approach [46]. As a consequence, ρ rose to incredibly high levels a few times $10^{17}/m^2$, accompanied with very low M (~0.35). When the dislocation density in material becomes sufficiently high ($\gg 10^{16}/m^2$), the strain field parameters (ρ, R_e, M) can become uncertain in the sense that R_e constantly approaches 0; this is a known intrinsic problem with the Krivoglaz–Wilkens theory caused by the strong reciprocal correlation of R_e and ρ [47]. The simplest way to overcome this discrepancy is by imposing the arbitrary value of R_e and keeping it fixed during refinement. Here, we assumed that R_e was extended to the size of the entire crystalline domain ($R_e = d$, ~6 nm), and with this constraint, $\rho = 2.92 \times 10^{16}/m^2$ and $M = 0.85$ was obtained, which seemed perfectly reasonable.

4. Conclusions

The Fe-18Cr-18Mn-N nickel-free austenitic stainless steel powder was synthesized by mechanical alloying under argon for the first 30 h followed by nitrogen for up to 150 h. The formation of the nanocrystalline Fe-fcc phase involved the bcc to fcc phase transition induced by MA, which was studied via the analysis of powder diffraction data. After the first 30 h of MA in argon, an XRD examination revealed solubility of the main alloying elements (manganese and chromium) in the iron matrix and the mixture of fcc and bcc phase structure. Nitrogen incorporation into the powder intensified the phase transformation. The final morphology of the powder was globular but flattened, with a bimodal distribution. A greater increase in the hardness of the powder was measured after changing the milling atmosphere from argon to nitrogen.

The XRD line profile analysis was illustrated herein by means of a typical case of an experimental study of plastically deformed metal powders subjected to intense ball milling. It was demonstrated that the classic Williamson–Hall method can be relevantly enhanced by incorporating the anisotropic contrast factor, but there is no doubt that WPPM provides a far more detailed description of complex phenomena occurring during MA, and, in general, microstructure evolution. However, despite entirely different methodology of processing the XRD data, our comparison shows that results obtained by MWH and WPPM are coincident in terms of the domain size evolution trend, as both of them indicate that the final value of d halved during the MA cycle. The agreement between the grain size trend in WPPM and MWH certainly contributes to validating the calculations. However, each domain size value produced by MWH is generally twice higher compared to the results obtained by WPPM; however, both of these methods still indicate very fine, nanocrystalline sizes of crystallites (~10 nm or below). The source of the discrepancies in the d estimation might be numerous but is most likely caused by the fact that neither shape nor distribution of the domain size was taken into account in the MWH procedure. Thus, MWH should mostly be used for the preliminary analysis of diffraction lines and quantitative interpretation of its results should be discouraged.

Our simulations proved that the WPPM algorithm can be successfully adopted to quantify the microstructural parameters of more complex, multi-phase patterns when overlapping of line profiles occur. WPPM delivered detailed information on the evolution of the crystallite size and average defect density. Microstructural information attained by WPPM is useful in terms of optimizing and controlling the phase transformation during MA.

The SEM-EDS analysis of the powder revealed that manganese and chromium content continuously increases up to 90 h of MA; however, after this time, the concentration of these elements is almost unaffected, which suggests that the process of MA can be considered finished. Due to the gentle MA process parameters applied in this work in comparison to the literature, no amorphous phase of the steel powder was detected.

Author Contributions: Conceptualization, E.R.-R. and K.N.; methodology, E.R.-R., K.N. and B.S.; formal analysis, E.R.-R., K.N. and B.S.; investigation, E.R.-R. and K.N., data curation, K.N. and E.R.-R.; writing—original draft preparation, E.R.-R., K.N. and B.S.; writing—review and editing, E.R.-R. and B.S.; visualization, E.R.-R. and K.N.; supervision, E.R.-R. and B.S.; project administration, E.R.-R.; funding acquisition, E.R.-R. All authors have read and agreed to the published version of the manuscript.

Funding: This research was funded by the Institute of Biomedical Engineering, Faculty of Mechanical Engineering, Bialystok University of Technology, project no. WZ/WM-IIB/2/2020.

Acknowledgments: The authors are grateful to Zbigniew Oksiuta for the constructive suggestions during the planning and development of this research.

Conflicts of Interest: The authors declare no conflict of interest.

References

1. Amini, R.; Salahinejad, E.; Askari Bajestani, E.; Hadianfard, M. Microstructural and hardness evolution of mechanically alloyed Fe–Cr–Mn–N. *J. Alloys Compd.* **2010**, *497*, 369–372.
2. Dorofeev, G.A.; Lubnin, A.N.; Ulyanov, A.L.; Kamaeva, L.V.; Lad′yanov, V.I.; Pushkarev, E.S.; Shabashov, V.A. XRD characterization of mechanically alloyed high-nitrogen nanocrystalline Fe–Cr system. *Mater. Lett.* **2015**, *159*, 493–497. [CrossRef]
3. Amornpitoksuk, P.; Suwanboon, S. Correlation of milling time on formation of TiCoSb phase by mechanical alloying. *J. Alloys Compd.* **2008**, *462*, 267–270. [CrossRef]
4. Jain, A.; Agarwal, S.; Jain, I. Correlation between the milling time and hydrogen-storage properties of nanostructured ZrFeNi ternary alloy. *J. Alloys Compd.* **2009**, *480*, 325–328. [CrossRef]
5. Chaisan, W.; Yimnirun, R.; Ananta, S. Effect of vibro-milling time on phase formation and particle size of barium titanate nanopowders. *Ceram. Int.* **2009**, *35*, 173–176. [CrossRef]
6. Sumita, M.; Hanawa, T.; Teoh, S. Development of nitrogen-containing nickel-free austenitic stainless steels for metallic biomaterials—Review. *Mater. Sci. Eng. C* **2004**, *24*, 753–760. [CrossRef]
7. Salahinejad, E.; Amini, R.; Marasi, M.; Sritharan, T.; Hadianfard, M. The effect of nitrogen on the glass-forming ability and micro-hardness of Fe–Cr–Mn–N amorphous alloys prepared by mechanical alloying. *Mater. Chem. Phys.* **2009**, *118*, 71–75. [CrossRef]
8. Rawers, J.C.; Govier, D.; Doan, R. Nitrogen addition to iron powder by mechanical alloying. *Mater. Sci. Eng. A* **1996**, *220*, 162–167. [CrossRef]
9. Amini, R.; Hadianfard, M.; Salahinejad, E.; Marasi, M.; Sritharan, T. Microstructural phase evaluation of high-nitrogen Fe–Cr–Mn alloy powders synthesized by the mechanical alloying process. *J. Mater. Sci.* **2009**, *44*, 136–148. [CrossRef]
10. Haghir, T.; Abbasi, M.; Golozar, M.; Panjepour, M. Investigation of α to γ transformation in the production of a nanostructured high-nitrogen austenitic stainless steel powder via mechanical alloying. *Mater. Sci. Eng. A* **2009**, *507*, 144–148. [CrossRef]
11. Amini, R.; Salahinejad, E.; Hadianfard, M.; Marasi, M.; Sritharan, T. Characterization of Fe–Cr–Mn–N amorphous powders with a wide supercooled liquid region developed by mechanical alloying. *Mater. Sci. Eng. A* **2010**, *527*, 1135–1142. [CrossRef]
12. Tehrani, F.; Abbasi, M.; Golozar, M.; Panjepour, M. The effect of particle size of iron powder on α to γ transformation in the nanostructured high nitrogen Fe–Cr–Mn–Mo stainless steel produced by mechanical alloying. *Mater. Sci. Eng. A* **2011**, *538*, 3961–3966. [CrossRef]
13. Caglioti, G.; Paoletti, A.; Ricci, F.P. Choice of Collimator for a Crystal Spectrometer for Neutron Diffraction. *Nucl. Instr. Meth.* **1958**, *3*, 223–228. [CrossRef]
14. Scardi, P.; Leoni, M. Fourier modelling of the anisotropic line broadening of X-ray diffraction profiles due to line and plane lattice defects. *J. Appl. Crystallogr.* **1999**, *32*, 671–682. [CrossRef]
15. Scardi, P.; Leoni, M.; Delhez, R. Line broadening analysis using integral breadth methods: A critical review. *J. Appl. Crystallogr.* **2004**, *37*, 381–390. [CrossRef]
16. Ungár, T.; Révész, Á.; Borbély, A. Dislocations and Grain Size in Electrodeposited Nanocrystalline Ni Determined by the Modified Williamson-Hall and Warren-Averbach Procedures. *J. Appl. Crystallogr.* **1998**, *31*, 554–558. [CrossRef]
17. Ungár, T.; Borbély, A. The effect of dislocation contrast on X-ray line broadening: A new approach to line profile analysis. *Appl. Phys. Lett.* **1996**, *69*, 3173–3175. [CrossRef]
18. Ungár, T.; Dragomir, I.; Révész, Á.; Borbély, A. The contrast factors of dislocations in cubic crystals: The dislocation model of strain anisotropy in practice. *J. Appl. Crystallogr.* **1999**, *32*, 992–1002. [CrossRef]
19. Ungár, T.; Tichy, G. The Effect of Dislocation Contrast on X-Ray Line Profiles in Untextured Polycrystals. *Phys. Status Solidi A* **1999**, *171*, 425–434. [CrossRef]
20. Ustinov, A.I.; Polishchuk, S.S.; Demchenkov, S.A.; Petrushinets, L.V. Effect of microstructure of vacuum-deposited Fe100-xNix (30 < x < 39) foils with FCC structure on their mechanical properties. *J. Alloys Compd.* **2015**, *622*, 54–61.
21. Adams, J.J.; Agosta, D.S.; Leisure, R.G.; Ledbetter, H.M. Elastic constants of monocrystal iron from 3 to 500 K. *J. Appl. Phys.* **2006**, *100*, 113530. [CrossRef]

22. Razumovskiy, I.V.; Hahn, C.; Lukas, M.; Romaner, L. Ab Initio Study of Elastic and Mechanical Properties in FeCrMn Alloys. *Materials* **2019**, *12*, 1129. [CrossRef]
23. Available online: http://metal.elte.hu/anizc/ (accessed on 22 August 2022).
24. Rebuffi, L.; Troian, A.; Ciancio, R.; Carlino, E.; Amimi, A.; Leonardi, A.; Scardi, P. On the reliability of powder diffraction Line Profile Analysis of plastically deformed nanocrystalline systems. *Sci. Rep.* **2016**, *6*, 20712. [CrossRef] [PubMed]
25. Langford, J.I.; Louër, D.; Scardi, P. Effect of a crystallite size distribution on X-ray diffraction line profiles and whole-powder-pattern fitting. *J. Appl. Crystallogr.* **2000**, *33*, 964–974. [CrossRef]
26. Wilkens, M. The Determination of Density and Distribution of Dislocations in Deformed Single Crystals from Broadened X-ray Diffraction Profiles. *Phys. Stat. Sol.* **1970**, *2*, 359–370. [CrossRef]
27. Scardi, P.; Leoni, M. *In Diffraction Analysis of the Microstructure of Materials*; Mittemeijer, E.J., Scardi, P., Eds.; Springer: New York, NY, USA, 2004; pp. 51–91, ISBN 978-3-642-07352-6.
28. Scardi, P.; Leoni, M. Whole powder pattern modelling. *Acta Crystallogr. A* **2002**, *58*, 190–200. [CrossRef]
29. Leoni, M.; Scardi, P. Nanocrystalline domain size distributions from powder diffraction data. *J. Appl. Cryst.* **2004**, *37*, 629–634. [CrossRef]
30. Ribárik, G.; Jóni, B.; Ungár, T. The Convolutional Multiple Whole Profile (CMWP) Fitting Method, a Global Optimization Procedure for Microstructure Determination. *Crystals* **2020**, *10*, 623. [CrossRef]
31. Leoni, M.; Confente, T.; Scardi, P. PM2K: A flexible program implementing Whole Powder Pattern Modelling. *Z. Krist. Suppl.* **2006**, *23*, 249–254. [CrossRef]
32. Suryanarayana, C. Mechanical alloying and milling. *Prog. Mater. Sci.* **2001**, *46*, 1–184. [CrossRef]
33. Cisneros, M.M.; Valdes, E.; Vazquez, D.; Lopez, H.F.; Mancha, H.; Mendoza, G.; Mendez, M. Development of austenitic nanostructures in High-nitrogen steel powders processed by mechanical alloying. *Metall. Mater. Trans. A* **2002**, *33*, 2139–2144. [CrossRef]
34. Duan, C.; Chen, C.; Zhang, J.; Shen, Y.; Feng, X. Nitriding of Fe-18Cr-8Mn stainless steel powders by mechanical alloying method with dual nitrogen source. *Powder Technol.* **2016**, *294*, 330–337. [CrossRef]
35. Eliades, T.; Pratsinis, H.; Kletsas, D.; Eliades, G.; Makou, M. Characterization and cytotoxicity of ions released from stainless steel and nickel-titanium orthodontic alloys. *Am. J. Orthod. Dentofac. Orthop.* **2004**, *125*, 24–29. [CrossRef]
36. Salahinejad, E.; Amini, R.; Hadianfard, M.J. Structural evolution during mechanical alloying of stainless steels under nitrogen. *Powder Technol.* **2012**, *215*, 247–253. [CrossRef]
37. Ahmed, A.; Ghali, S.; Eissa, M.; El Badry, S. Influence of Partial Replacement of Nickel by Nitrogen on Microstructure and Mechanical Properties of Austenitic Stainless Steel. *J. Metall.* **2011**, *2011*, 639283. [CrossRef]
38. Borbély, A. The modified Williamson-Hall plot and dislocation density evaluation from diffraction peaks. *Scr. Mater.* **2022**, *217*, 114768. [CrossRef]
39. Kumari, S.; Singh, D.K.; Giri, P.K. Strain Anisotropy in Freestanding Germanium Nanoparticles Synthesized by Ball Milling. *J. Nanosci. Nanotechnol.* **2009**, *9*, 5231–5236. [CrossRef]
40. Swygenhoven, H.; Weertman, J.R. Grain boundaries and dislocations. *Science* **2002**, *296*, 66–67. [CrossRef]
41. Seelam, U.M.R.; Suryanarayana, C. Mechanically induced fcc phase formation in nanocrystalline hafnium. *J. Appl. Phys.* **2009**, *105*, 063524. [CrossRef]
42. Loudjani, N.; Benchiheub, M.; Bououdina, M. Structural, Thermal and Magnetic Properties of Nanocrystalline $Co_{80}Ni_{20}$ Alloy Prepared by Mechanical Alloying. *J. Supercond. Nov. Magn.* **2016**, *29*, 2717–2726. [CrossRef]
43. Mhadhbi, M.; Khitouni, M.; Escoda, L.; Suñol, J.J.; Dammak, M. Characterization of Mechanically Alloyed Nanocrystalline Fe(Al): Crystallite Size and Dislocation Density. *J. Nanomater.* **2010**, *2010*, 712407. [CrossRef]
44. Nazarov, A.A.; Romanov, A.E.; Valiev, R.Z. Disclination ensembles in ultrafine-grained materials produced by severe plastic deformation. *Scr. Mater.* **1996**, *34*, 729–734. [CrossRef]
45. Rawers, J.; Cook, D. Influence of attrition milling on nano-grain boundaries. *Nanostruct. Mater.* **1999**, *11*, 331–342. [CrossRef]
46. Seif, D.; Po, G.; Mrovec, M.; Lazar, M.; Elsässer, C.; Gumbsch, P. Atomistically enabled nonsingular anisotropic elastic representation of near-core dislocation stress fields in α-iron. *Phy. Rev. B* **2015**, *91*, 184102. [CrossRef]
47. Leonardi, A.; Scardi, P. Dislocation effects on the diffraction line profiles from nanocrystalline domains. *Metall. Mater. Trans. A* **2016**, *47*, 5722–5732. [CrossRef]

crystals

Article

Mechanical Properties of Low Carbon Alloy Steel with Consideration of Prior Fatigue and Plastic Damages

Qing Liu [1], Zhanzhan Tang [1,*], Xuan Yang [1], Zhixiang He [1], Hanyang Xue [1] and Hanqing Zhuge [2]

1 College of Civil Science and Engineering, Yangzhou University, Yangzhou 225127, China; mz120211015@stu.yzu.edu.cn (Q.L.); mz120210974@stu.yzu.edu.cn (X.Y.); mz120190743@yzu.edu.cn (Z.H.); dx120200087@yzu.edu.cn (H.X.)
2 College of Civil Engineering and Architecture, Zhejiang University of Science & Technology, Hangzhou 310023, China; 120043@zust.edu.cn
* Correspondence: tangzhanzhan@126.com

Abstract: Mechanical properties, including the fatigue behavior of metals, are usually determined from damage-free specimens, but it is not well known how these properties change with respect to prior damages; hence, the present work aims to understand the remaining mechanical properties of low carbon alloy steel Q345q with pre-damages. Low-cycle fatigue tests on the damage free specimens, tensile tests on the low-cycle fatigue damaged specimens, and fatigue tests on the plastic deformed specimens were carried out, respectively. The low-cycle fatigue life prediction formula was proposed. The influences of different kinds of pre-damages on the residual mechanical properties were analyzed. Results show that the stable hysteretic loops in the low-cycle fatigue tests are well-stacked. The material illustrates Masing behavior, and it has a good energy dissipation capacity. The ductility of the low-cycle fatigue-damaged materials decreases significantly in comparison with the undamaged ones. The low-cycle fatigue lives of Q345q steel are almost unaffected, so long as the pre-applied tensile strain is lower than 10%.

Keywords: plastic deformation; low-cycle fatigue; mechanical properties; pre-damage; low alloy steel; coupling damage

Citation: Liu, Q.; Tang, Z.; Yang, X.; He, Z.; Xue, H.; Zhuge, H. Mechanical Properties of Low Carbon Alloy Steel with Consideration of Prior Fatigue and Plastic Damages. *Crystals* **2022**, *12*, 967. https://doi.org/10.3390/cryst12070967

Academic Editor: Wojciech Polkowski

Received: 28 June 2022
Accepted: 7 July 2022
Published: 11 July 2022

1. Introduction

Low carbon alloy steels are widely used in the construction of buildings and bridges, energy storages such as fuel cells, and various mechanical equipment [1–3]. On the one hand, low-cycle fatigue (LCF) damage may be discovered in these structures during service due to seismic events and other extreme loading conditions. On the other hand, the materials in local regions on these steel structures and equipment may experience large plastic deformation caused by severe impact or other improper artificial operations. The coupling damage of alloy steel caused by LCF, and tensile plastic deformation is still unknown.

Fatigue of materials can be traditionally divided into high-cycle fatigue (HCF) and low-cycle fatigue scopes. Low-cycle fatigue involves bulk plasticity, whereas the deformation in high-cycle fatigue is in elastic range. A considerable number of studies on LCF of metals have been conducted to understand the fatigue life prediction and crack propagation rules [4–6]. Recently, Tsutsumi et al. [7] investigated the LCF behavior of butt-welded joints considering the inhomogeneous mechanical properties in the base metal, weld metal, and the heat-affected zones. They found that the effects of inhomogeneous material properties are very significant when evaluating the LCF of welded joints. Procházka et al. [8] studied the development of advanced techniques of the LCF test using miniature test samples. They reported that cyclic-strain and strain-life curves of the mini-sample geometries are almost of the same shape for rotor steels. Ho et al. [9] studied the LCF performance of gradient structured 316 austenitic steels under high strain amplitude loading conditions. They found that gradient structured steels exhibit inherently lower fatigue life than their coarse-grained counterparts. Huang et al. [10] conducted monotonic tensile and ultra-low-cycle-fatigue tests on Q235 and Q690 steels to investigate the fracture behavior under various loading conditions. It was found that the deformability of Q690 steel under monotonic loading and the fatigue life under LCF loading are lower than those of Q235 steel. Tong et al. [11] evaluated the LCF life of buckling-restrained braces (BRBs), and they proposed an evaluation method for BRBs based on the combination of the cumulative plastic deformation curves. Yoon et al. [12] studied the effect of multiple high-density pulsed electric currents on the LCF life of austenitic stainless steel. An improved LCF life in austenitic stainless steel was proposed and examined by the test results. Hua et al. [13] experimentally investigated the LCF behaviors of high strength steel Q690 exposed to different elevated temperatures. A fatigue model considering the elevated temperatures was proposed to describe the relation between the total strain amplitude and the load cycles to failure. Tang et al. [14] studied the effect of structural parameters on the LCF damage evolution of thin-walled steel bridge piers. Practical formulae to evaluate the LCF damage levels were proposed and validated by experiments. Bouazza et al. [15] conducted the LCF fragility of RC bridge piers and developed an LCF fragility curve based on the performance-based earthquake engineering methodology. These above-mentioned studies were devoted to the LCF behaviors, life prediction, and fracture models of metals. However, the coupling effects of LCF damages and other types of damages of these metals were not involved.

More than 30 years ago, Park et al. [16,17] proposed a practical damage index of concrete material and members, in which the combined effects of plastic deformation and plastic energy consumption caused by cyclic loads are considered. After that, many researchers tried to reveal the relation between the cyclic damage and the plastic deformation damage of metallic materials. Cadenas–Herrera et al. [18] studied the impact of the HCF damage on the fracture toughness parameters of aluminum alloy. They reported that the tensile static mechanical properties of the material are unaffected by the prior fatigue damage. Tang et al. [19] investigated the influence of pre-fatigue damages on the alloy steel material and steel members. It was found that the tensile mechanical properties of Q345 steel decrease significantly with consideration of the prior HCF damages. Paul et al. [20] conducted material experiments under pre-LCF, followed by tensile load conditions, to study the mechanical behavior and damage evolution during cyclic plastic deformation. It was found that the pre-ratcheting has a massive effect on the subsequent LCF life. López et al. [21] studied the effect of preceding cyclic loading on the tensile behavior of titanium alloy and found that pre-damage only affects the surface of the material, but not the fundamental mechanical properties. Moćko et al. [22] investigated the change of the strain distribution on the surface of the material caused by the pre-fatigue damages using the digital image correlation method. It was found that the pre-fatigue loads can change the formation of micro-damages in the material. Tang et al. [23] experimentally studied the combined effect of HCF and LCF damages and reported that the coupling effect of different types of damages is very significant, e.g., the residual life reducing to 0.6 of the original LCF life when the prior HCF damage reaches 0.4. Wang et al. [24] investigated the influence of pre-fatigue damage on the residual mechanical properties of P92 steel. It was found that the ultimate stress of the material considering pre-fatigue shows two stages; the first one is the initial rapid degradation stage and the second one is the linear decreasing stage. Jia et al. [25] studied the LCF and extreme LCF performance of the high strength steel Q690E, and an effective constitutive model was proposed. Based on the framework of damage mechanics, Peng et al. [26] investigated the effect of average strain on the LCF life of materials under different strain cycle ratios. The relationship between the maximum strain, strain amplitude, material properties, and LCF life of the material was discussed. The above-mentioned research mainly concentrated on the effect of prior HCF/LCF damages on the tensile properties of different kinds of metals. However, the change of the mechanical behavior is very much related to the material itself. Additionally, few studies have been conducted on the effect of tensile plastic deformation on the subsequent LCF behaviors.

In this research, strain-controlled fatigue tests were conducted under six different load amplitudes to study the LCF behavior of the alloy steel Q345q. Effects of the prior LCF damage on the remaining tensile properties were experimentally investigated. Moreover, the effects of the damage caused by tensile plastic deformation on the LCF behavior of the material were studied. Finally, predictive equations were proposed to determine the residual mechanical properties of the material.

2. Experimental Methods

The as-received material is low carbon alloy steel grade Q345q and its mechanical properties are similar to S355 and ASTM 50 steels [27]. Table 1 shows the chemical compositions in weight ratio of grade Q345q material provided by the factory of Baoshan Iron and Steel Co., Ltd. According to GB/T 228.1–2010 [28] and GB/T 15248–2008 [29], the shape and size of the test specimens were machined, as shown in Figure 1. All of the specimens were cut from the same plate with a nominal thickness of 10 mm. A total of 42 specimens were machined and tested for different purposes in this study. Four groups of specimens were prepared for the tensile test, the low-cycle fatigue test, the fatigue–tension test, and the tension–fatigue test, respectively.

Table 1. Chemical compositions of grade Q345q steel (wt%).

C	Si	Mn	P	S	Nb	V	Ti
≤0.18%	≤0.55%	0.9~1.70%	≤0.025%	≤0.02%	≤0.06%	≤0.08%	≤0.03%
Cr	**Ni**	**Cu**	**Mo**	**N**	**Als**	**Fe**	/
≤0.08%	≤0.50%	≤0.55%	≤0.20%	≤0.012%	≥0.015%	Balance	/

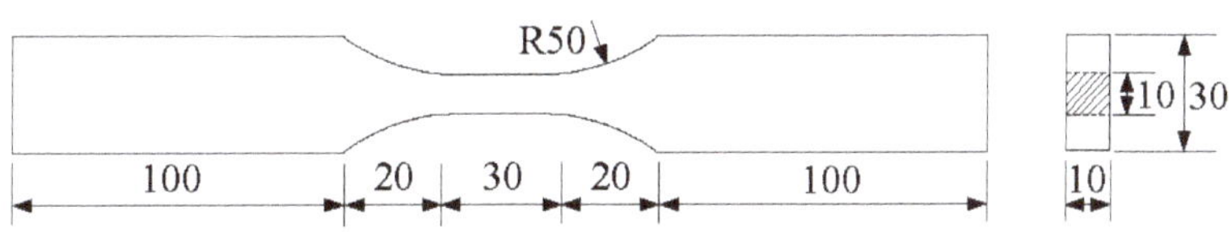

Figure 1. Geometric dimensions of specimens (mm).

Figure 2 shows the test machine with an installed specimen. The low-cycle fatigue test and the tensile test were performed on an INSTRON 8802 servo hydraulic machine in ambient air at room temperature. For the local deformation measurement, a clip-on extensometer with a gauge length of 25 mm was used and attached to the test specimens. The quasi-static tensile test was performed with a constant strain rate of 1 mm/min, while the cyclic load frequency for the low-cycle fatigue testing was kept to around 0.03 Hz. The low-cycle fatigue tests were conducted under fully reversed axial push and pull strain-controlled conditions. The test programs were designed as follows:

1. Tensile tests: to obtain the basic mechanical properties of the material such as the elastic modulus E_0, the yield stress σ_y, the ultimate stress σ_u, and the elongation at breakage δ_0. Static tensile tests were performed using 3 specimens.
2. Fatigue tests: to investigate the LCF behavior of the damage free material, a group of specimens were cyclically loaded up to fracture at selected strain amplitudes. The strain amplitude ($\Delta\varepsilon/2$) in the pure fatigue tests ranged from 1.0% to 3.0%. 18 specimens were used in the fatigue tests. The strain ratio $R = \varepsilon_{max}/\varepsilon_{min}$ was set as -1 for all fatigue load cases.
3. Tension–fatigue tests: to study the effect of large tensile plastic strain on the low-cycle fatigue behavior of the material, post-tension fatigue tests were carried out. First, the specimens were loaded with a very large deformation into the plastic stage. Afterwards, low-cycle fatigue loads were applied to the pre-damaged specimens.

9 specimens were used in the tension–fatigue tests. Triangular waveform was adopted for all cyclic loads.

4. Fatigue–tension tests: to study the effect of LCF damage on the tensile behavior of the material, pre-cyclic tests were performed to produce LCF damaged samples, and subsequent static tensile tests were carried out on these pre-damaged specimens. 12 specimens were used in these tests.

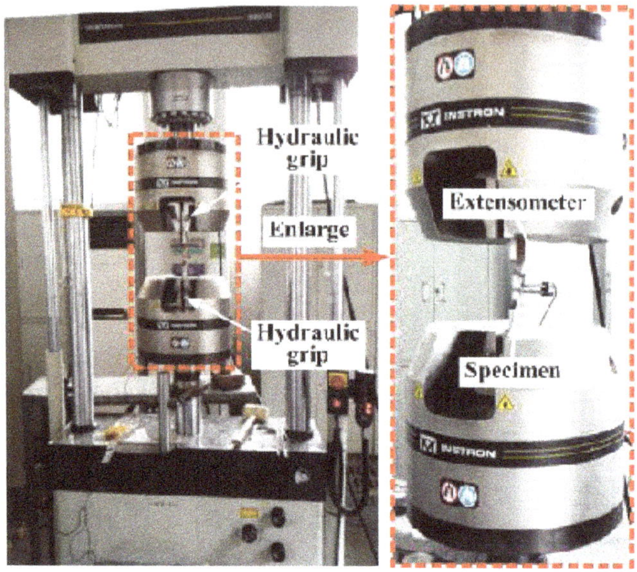

Figure 2. Test setup and instrumentation.

3. Results and Discussions

3.1. Tensile Tests

Three specimens were used in the monotonic tensile tests, in which two of them were used for repeated tests. The key parameters were derived from the stress–strain curves. Table 2 shows the monotonic tensile properties of the material, in which σ_y represent the yield stress, σ_u represents the ultimate stress, ε_u represents the ultimate strain corresponding to the ultimate stress, E_0 is the elastic modulus, and δ_0 denotes the elongation at breakage.

Table 2. Monotonic tensile properties of Q345 steel.

σ_y (MPa)	σ_u (MPa)	ε_u (%)	E_0 (GPa)	δ_0 (%)
389.6	529.2	26.0	204.2	39.5

3.2. Fatigue Tests

Constant-amplitude protocols were used for the LCF tests according to GBT 15248-2008 [29]. Six strain amplitude levels were designed, e.g., 1.0%, 1.5%, 1.7%, 2.0%, 2.5%, and 3.0%. 18 specimens were used in these tests. Figure 3 shows the hysteretic loops with different strain amplitudes in the fatigue tests. Table 3 lists the low-cycle fatigue lives of the specimens under different amplitudes of cyclic loads, in which N_f represents the fatigue life. As can be seen, low-cycle fatigue life decreases as the load amplitude increases in the constant-amplitude loading conditions.

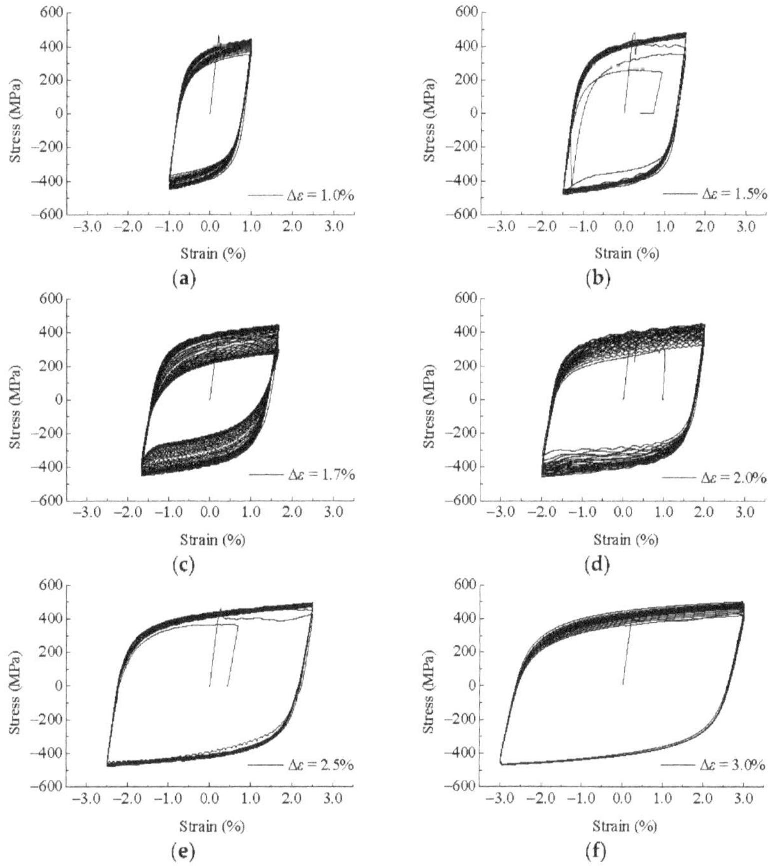

Figure 3. Hysteretic loops of low-cycle fatigue tests with strain amplitudes of: (**a**) 1.0%; (**b**) 1.5%; (**c**) 1.7%; (**d**) 2.0%; (**e**) 2.5%; (**f**) 3.0%.

Table 3. Low-cycle fatigue lives of Q345 steel.

No.	$\Delta\varepsilon/2$ (%)	N_f	Average Life
1		364	
2	1.0	430	405.0
3		421	
4		276	
5	1.5	312	304.7
6		326	
7		244	
8	1.7	220	234.0
9		238	
10		214	
11	2.0	198	194.7
12		172	
13		128	
14	2.5	140	139.0
15		149	
16		71	
17	3.0	65	77.7
18		97	

The total strain amplitude in LCF tests is composed of the elastic strain amplitude component and the plastic component as such:

$$\Delta\varepsilon/2 = \Delta\varepsilon_e/2 + \Delta\varepsilon_p/2 \quad (1)$$

where $\Delta\varepsilon_e/2$ is the elastic strain amplitude part, and $\Delta\varepsilon_p/2$ is the plastic strain amplitude part.

The relation between the LCF lives and the plastic strain amplitudes can be described by the Coffin–Manson equation:

$$\Delta\varepsilon_p/2 = \varepsilon'_f \cdot (2N_f)^c \quad (2)$$

where ε'_f is the fatigue ductility coefficient and c is the fatigue ductility exponent.

Figure 4 shows the fatigue lives of the test specimens and the fitting strain–fatigue life relation (see Equation (3)), which can be used to predict the LCF life of this kind of steel material.

$$\Delta\varepsilon_p/2 = 0.843 \cdot (2N_f)^{-0.64} \quad (3)$$

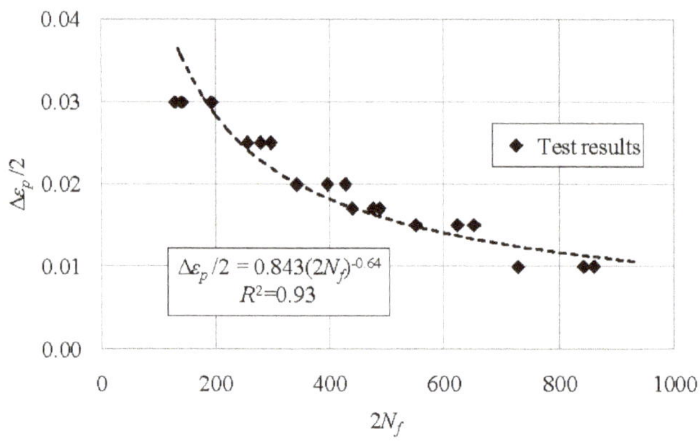

Figure 4. Fitting results based on the LCF tests.

Figure 5 shows the stress–strain loops at the $N_f/2$ load cycle under different strain amplitude conditions. It can be seen that the maximum stress in each case is basically on the coincident curves, and all the hysteretic loops are well-stacked and in a stable state. This means this kind of steel material illustrates Masing behavior, and it has a good energy dissipation capacity. Hysteretic curves can be fitted using the Ramberg–Osgood equation [30], and for LCF tests, the relation between the cyclic stress and strain can be expressed as:

$$\Delta\sigma/2 = K' \cdot (\Delta\varepsilon_p/2)^{n'} \quad (4)$$

where $\Delta\sigma/2$ is the stable stress amplitude, K' is the cyclic hardening coefficient, and n' is the cyclic hardening index. These stable values are obtained from the stress–strain loops at the $N_f/2$ load cycle as shown in Figure 5. The parameters in the Ramberg–Osgood equation can be calibrated via these experimental results. Parameter K' is 395.8 and n' is 0.118 for the low alloy steel Q345.

3.3. Fatigue-Tension Tests

The fatigue–tension tests were conducted to investigate the remaining mechanical properties of the specimens after being fatigue damaged. A constant strain amplitude of 1.7% was adopted in the fatigue–tension tests. Therefore, the corresponding fatigue life at this strain amplitude is 234.0 according to the previous pure fatigue tests. Eight different pre-fatigue levels were selected by applying different numbers of load loops (10, 20, 30, 40, 50, 60, 70, and 200) to the specimens. The pre-damage levels are represented using the ratio of the applied load cycles N and the corresponding fatigue life N_f. Figure 6 shows the stress–strain curves of the fatigue–tension tests, in which the static tensile curves are also illustrated for comparison. As can be seen, the ductility of the fatigue-damaged specimens decreases dramatically. When the specimen is severely fatigue-damaged, e.g., $N/N_f = 0.855$, both the strength and ductility degrade significantly.

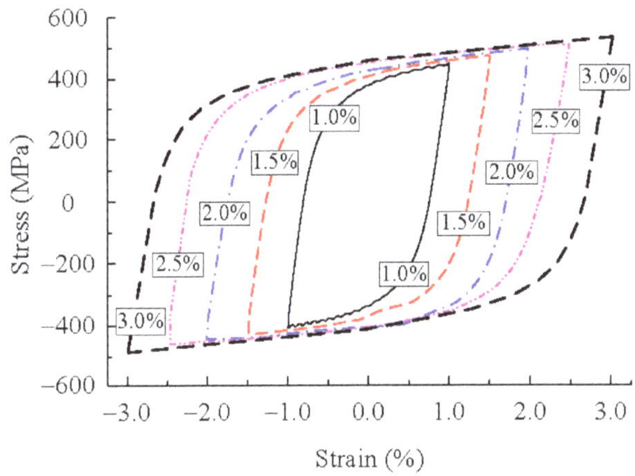

Figure 5. Stabilized cyclic stress–strain hysteretic loops in low-cycle fatigue tests.

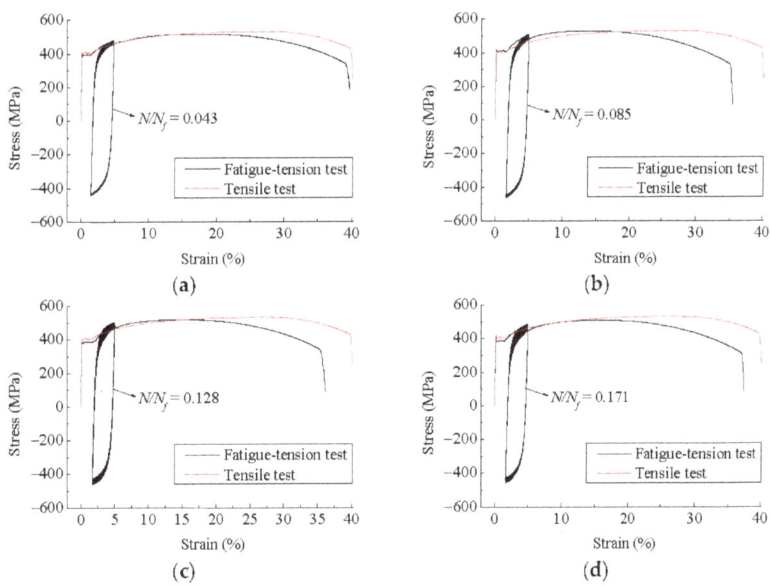

Figure 6. *Cont.*

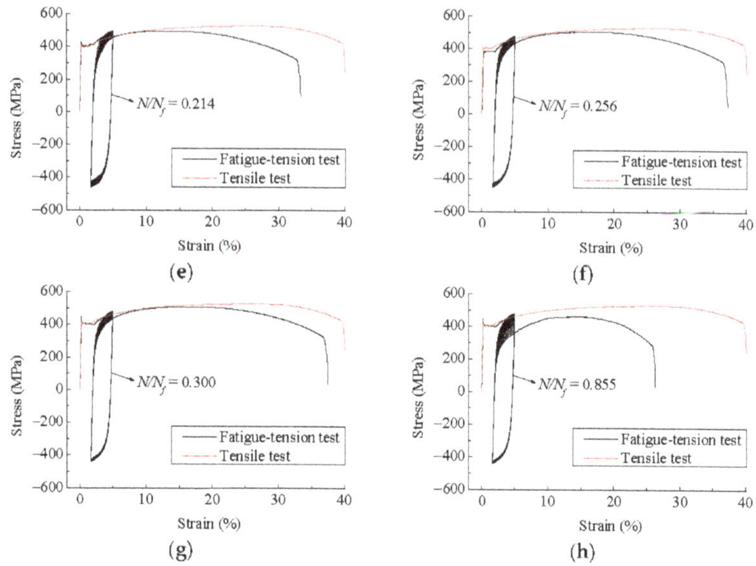

Figure 6. The stress–strain curves of fatigue-tension tests with strain amplitude of 1.70% and pre-fatigue damages of: (**a**) 0.043; (**b**) 0.085; (**c**) 0.128; (**d**) 0.171; (**e**) 0.214; (**f**) 0.256; (**g**) 0.300; (**h**) 0.855.

The fatigue–tension tests were also conducted with a constant strain amplitude of 2.5% to study the effects of the load amplitudes. The corresponding fatigue life with a strain amplitude of 2.5% is 139.0. Four different pre-fatigue levels were designed by applying 10, 20, 30, and 40 load cycles to the specimens. Figure 7 shows the stress–strain curves of the fatigue–tension tests with strain amplitude of 2.5%. As can be seen, the ultimate stress is almost unaffected by the pre-fatigue damage in these cases.

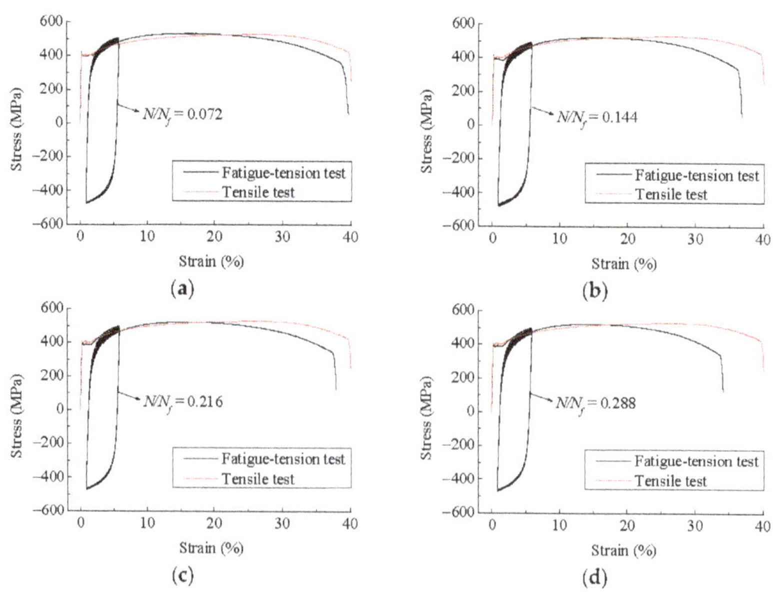

Figure 7. The stress–strain curves of fatigue–tension tests with a strain amplitude of 2.5% and pre-fatigue damages of: (**a**) 0.072; (**b**) 0.144; (**c**) 0.216; (**d**) 0.288.

Figure 8 shows the remaining mechanical properties of the specimens after being fatigue damaged, in which σ'_u represents the remaining ultimate stress, ε'_u represents the ultimate strain corresponding to the remaining ultimate stress, and δ denotes the elongation of the specimens with pre-fatigue damages. As can be seen in the figure, the ultimate strength and the elongation decrease as the pre-fatigue damages increase. The strain amplitudes of the cyclic loads have no significant effects on the deterioration tendency of the mechanical properties. Notable among the deterioration of the mechanical properties, the ultimate strain ε'_u corresponding to the ultimate stress decreases remarkably even if the pre-fatigue damage is slight, which indicates an earlier decrease of the tensile stress–strain curves (see Figures 6 and 7). The fitting equations of the test results are also given in these figures (see Equations (5)–(7)), which can be adopted for the prediction of the remaining mechanical properties of the fatigue-damaged materials. Note that for the fitting of the test results ε'_u, the result of the undamaged specimen was not used (Figure 8b).

$$\sigma'_u/\sigma_u = -0.151\cdot(N/N_f) + 1.000 \quad (5)$$

$$\varepsilon'_u/\varepsilon_u = -0.114\cdot(N/N_f) + 0.627 \quad (6)$$

$$\delta'_u/\delta_0 = -0.372\cdot(N/N_f) + 0.974 \quad (7)$$

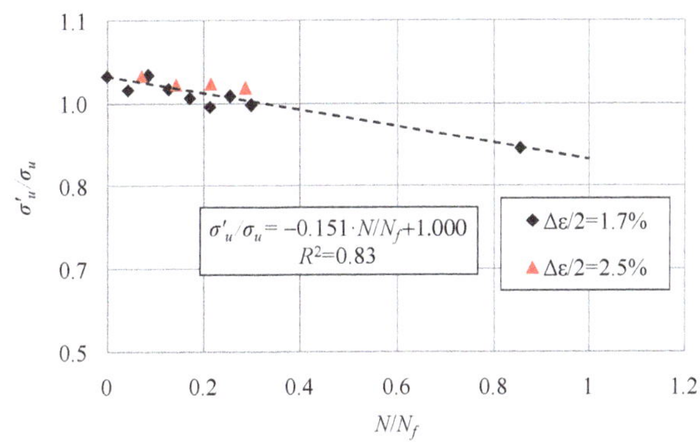

(a)

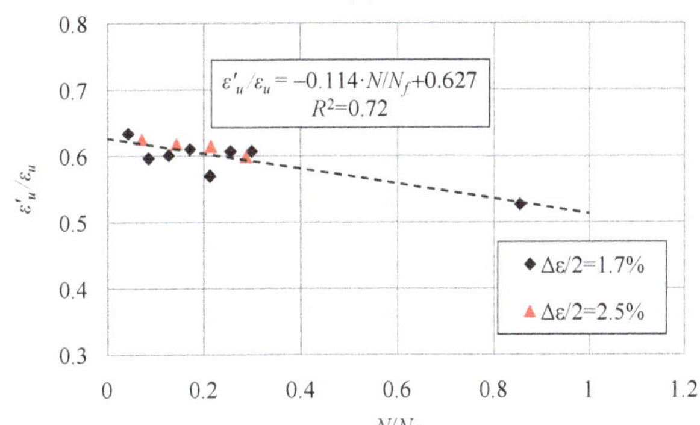

(b)

Figure 8. *Cont.*

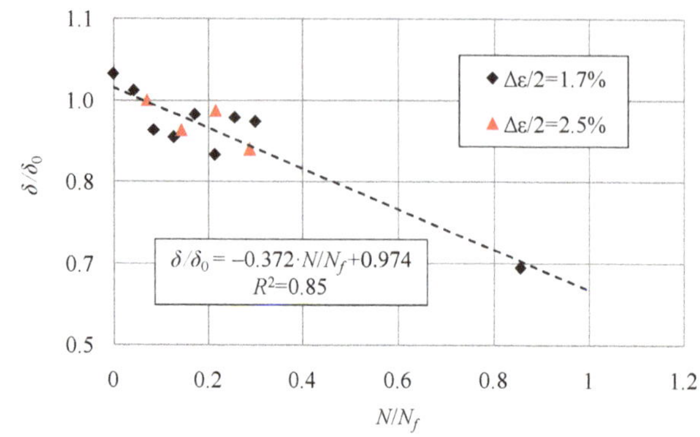

(c)

Figure 8. The remaining mechanical properties of the specimens after being fatigue damaged; (**a**) remaining ultimate stress; (**b**) strain corresponding to remaining ultimate stress; (**c**) remaining elongation at breakage.

3.4. Tension-Fatigue Tests

In this section, the effects of pre-damages caused by a large tensile plastic deformation on the LCF behavior of the low alloy steel are studied. Consequently, the tension–fatigue tests were performed, in which the specimens were firstly subjected to tensile strains of 3.35%, 4.20%, 5.03%, 5.86%, 6.70%, 7.54%, 8.38%, 9.20, and 10.10% to generate different pre-damage levels, and then subsequent low-cycle fatigue loads were applied to the same specimens. Figure 9 shows the stress–strain curves obtained from the tension–fatigue tests, in which constant strain amplitude of 1.7% was selected. Table 4 shows the remaining LCF lives of the specimens with consideration of the pre-damages caused by prior tensile deformation, in which N'_f represents the remaining fatigue life.

Table 4. Test results of the tension–fatigue tests.

No.	ε_t (%)	N'_f
1	3.35	185
2	4.20	221
3	5.03	192
4	5.86	180
5	6.70	190
6	7.54	247
7	8.38	201
8	9.20	229
9	10.10	238

Figure 10 shows the remaining LCF lives of the pre-damaged specimens, in which the fatigue life of the specimen without pre-damage is also indicated. As can be seen, the fatigue lives decrease due to the pre-damages induced by the tensile deformation in most cases. However, this kind of effect is not so notable. The remaining fatigue lives of the specimens with previous tensile deformation damages of 7.54% and 10.10% are even longer than that of the undamaged ones. Considering the large dispersion of the fatigue test results, it can be concluded that the fatigue lives are almost unaffected as long as the pre-applied tensile strain is lower than 10%.

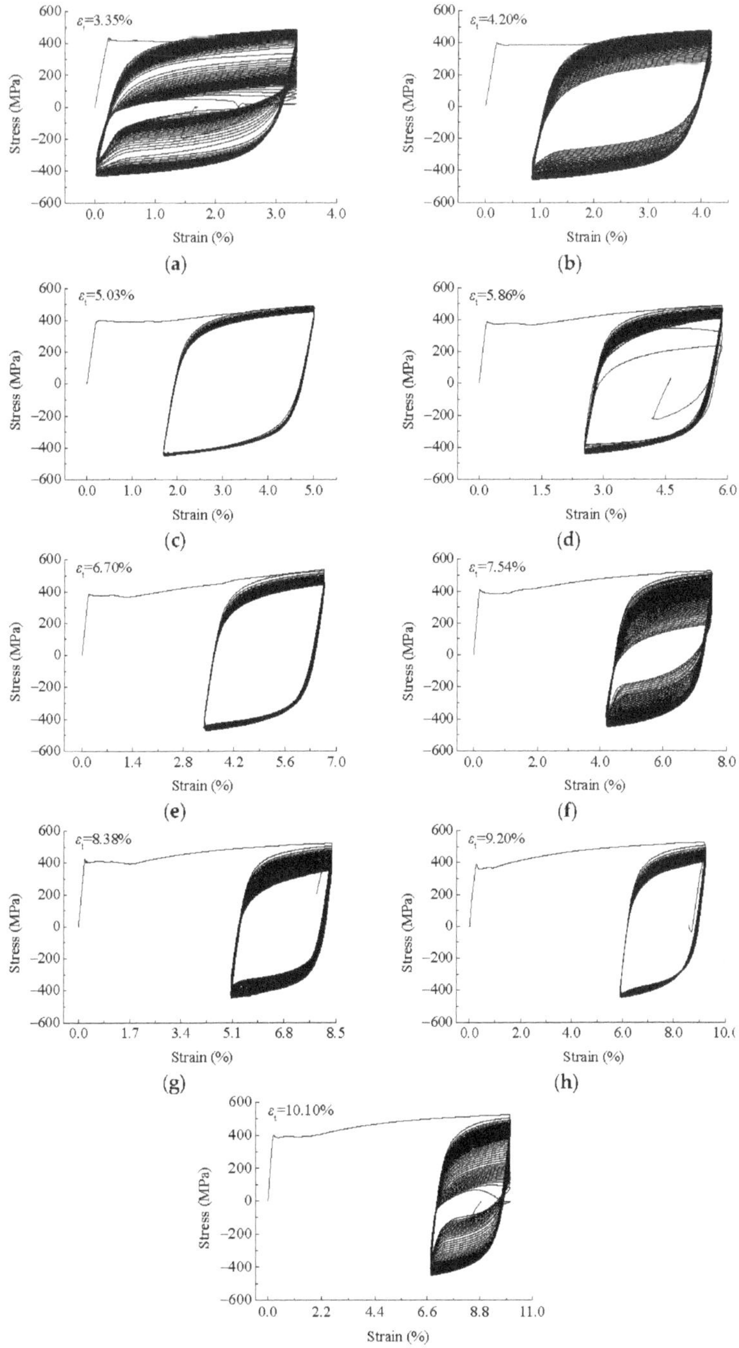

Figure 9. The stress–strain curves of tension–fatigue tests with a prior maximum tensile strain of: (**a**) 3.35%; (**b**) 4.20%; (**c**) 5.03%; (**d**) 5.86%; (**e**) 6.70%; (**f**) 7.54%; (**g**) 8.38%; (**h**) 9.20%; (**i**) 10.10%.

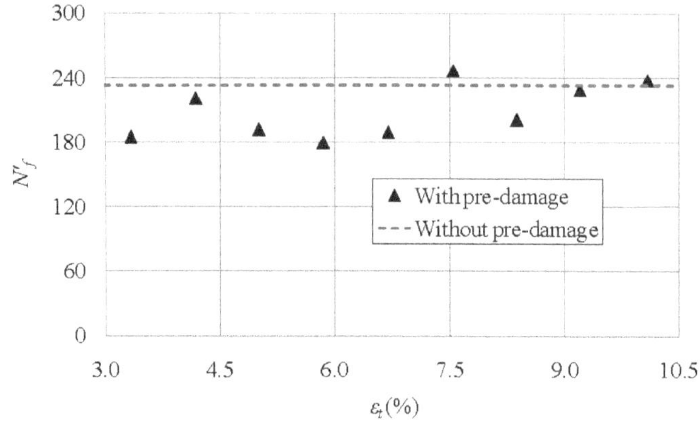

Figure 10. The remaining low-cycle fatigue lives considering prior tensile deformation damages.

4. Conclusions

The low-cycle fatigue behavior, the remaining mechanical properties with consideration of low-cycle fatigue damages, and the remaining fatigue lives considering the prior plastic deformation damages of low carbon alloy steel Q345q were experimentally studied herein. The following conclusions can be drawn, and they are limited to Q345q steel, which was adopted in the tests of this study.

1. The Manson–Coffin formula of Q345q steel was obtained based on the pure fatigue tests, which can be used to predict the low-cycle fatigue life of this kind of material. The stable hysteretic loops are well-stacked. The material illustrates Masing behavior, and it has a good energy dissipation capacity.
2. The ductility of the LCF-damaged materials decreases in comparison with the undamaged ones. When the specimen is severely fatigue-damaged, e.g., $N/N_f = 0.855$, both the strength and the ductility degrade significantly.
3. The strain amplitudes of the prior cyclic loads have no significant effects on the deterioration tendency of the remaining mechanical properties. The strain corresponding to the ultimate stress decreases remarkably, indicating an earlier decrease of the tensile stress–strain curves.
4. The obtained fitting equations of the remaining mechanical properties of the fatigue-damaged specimens can be adopted for the plastic behavior prediction of the LCF-damaged steel materials.
5. The low-cycle fatigue lives of Q345q steel are almost unaffected, as long as the pre-applied tensile strain is lower than 10%.

Author Contributions: Conceptualization, Z.T. and Z.H.; methodology, Z.T.; software, Q.L.; validation, Z.T. and Q.L.; formal analysis, Q.L. and X.Y.; investigation, H.X.; resources, Z.T.; data curation, Q.L. and Z.T.; writing—original draft preparation, Z.T. and Q.L.; writing—review and editing, Z.H. and H.Z.; visualization, Q.L.; supervision, Z.T.; project administration, Z.T.; funding acquisition, Z.T. All authors have read and agreed to the published version of the manuscript.

Funding: This research was funded by the Postgraduate Research & Practice Innovation Program of Jiangsu Province (SJCX22_1747, KYCX21_3225), the Qing Lan Project of Yangzhou University, and the National Natural Science Foundation of China (51708485).

Institutional Review Board Statement: Not applicable.

Informed Consent Statement: Not applicable.

Data Availability Statement: Data is contained within the article.

Acknowledgments: The corresponding author gratefully acknowledges the support of Yangzhou Zhongde Mining Machinery Co., LTD., Yangzhou, China for the manufacturing and testing work of some of the specimens.

Conflicts of Interest: The authors declare no conflict of interest.

References

1. Smith, B.H.; Szyniszewski, S.; Hajjar, J.F.; Schafer, B.W.; Arwade, S.R. Steel foam for structures: A review of applications, manufacturing and material properties. *J. Constr. Steel Res.* **2012**, *71*, 1–10. [CrossRef]
2. Stornelli, G.; Gaggiotti, M.; Mancini, S.; Napoli, G.; Rocchi, C.; Tirasso, C.; Di Schino, A. Recrystallization and grain growth of AISI 904L super-austenitic stainless steel: A multivariate regression approach. *Metals* **2022**, *12*, 200. [CrossRef]
3. Di Schino, A. Analysis of phase transformation in high strength low alloyed steels. *Metalurgija* **2017**, *56*, 349–352.
4. Guo, Y.; Fang, C.; Zheng, Y. Post-fire hysteretic and low-cycle fatigue behaviors of Q345 carbon steel. *J. Constr. Steel Res.* **2021**, *187*, 106991. [CrossRef]
5. Dong, Q.; Yang, P.; Xu, G.; Deng, J. Mechanisms and modeling of low cycle fatigue crack propagation in a pressure vessel steel Q345. *Int. J. Fatigue* **2016**, *89*, 2–10. [CrossRef]
6. Liao, F.; Wang, W.; Chen, Y. Parameter calibrations and application of micromechanical fracture models of structural steels. *Struct. Eng. Mech.* **2012**, *42*, 153–174. [CrossRef]
7. Tsutsumi, S.; Fincato, R.; Luo, P.; Sano, M.; Umeda, T.; Kinoshita, T.; Tagawa, T. Effects of weld geometry and HAZ property on low-cycle fatigue behavior of welded joint. *Int. J. Fatigue* **2022**, *156*, 106683. [CrossRef]
8. Procházka, R.; Džugan, J. Low cycle fatigue properties assessment for rotor steels with the use of miniaturized specimens. *Int. J. Fatigue* **2022**, *154*, 106555. [CrossRef]
9. Ho, H.S.; Lv, C.; Zhou, W.; Zhang, E. Low-cycle fatigue behavior of gradient structured austenitic stainless steels under high strain amplitude. *Fatigue Fract. Eng. Mater. Struct.* **2022**, *45*, 1818–1829. [CrossRef]
10. Huang, X.; Yuan, Y.; Zhao, J.; Wei, C. Comparative study on ultra-low-cycle-fatigue behavior of Q235 normal-steel and Q690 high-strength steel. *J. Constr. Steel Res.* **2022**, *194*, 107308. [CrossRef]
11. Tong, C.; Wu, J.; Hua, K.; Tian, H. Low-cycle fatigue life evaluation of buckling-restrained braces based on cumulative plastic deformation curves. *Adv. Struct. Eng.* **2022**, *25*, 336–354. [CrossRef]
12. Yoon, S.; Cui, Y.; Kimura, Y.; Gu, S.; Toku, Y.; Ju, Y. Improvement of low-cycle fatigue life of austenitic stainless steel by multiple high-density pulsed electric currents. *Int. J. Fatigue* **2022**, *156*, 106639. [CrossRef]
13. Hua, J.; Yang, Z.; Zhou, F.; Hai, L.; Wang, N.; Wang, F. Effects of exposure temperature on low–cycle fatigue properties of Q690 high–strength steel. *J. Constr. Steel Res.* **2022**, *190*, 107159. [CrossRef]
14. Tang, Z.Z.; Xue, H.Y.; Liu, H.; Zhang, W. Prediction of ultralow cycle fatigue damage of thin-walled steel bridge piers. *Adv. Steel Constr.* **2021**, *17*, 403–411.
15. Bouazza, H.; Djelil, M.; Matallah, M. On the relevance of incorporating bar slip, bar buckling and low-cycle fatigue effects in seismic fragility assessment of RC bridge piers. *Eng. Struct.* **2022**, *256*, 114032. [CrossRef]
16. Park, Y.J.; Ang, A.H. Mechanistic seismic damage model for reinforced concrete. *J. Struct. Eng.* **1985**, *111*, 722–739. [CrossRef]
17. Park, Y.J.; Ang, A.H.; Wen, Y.K. Damage-limiting aseismic design of buildings. *Earthq. Spectra* **1987**, *3*, 1–26. [CrossRef]
18. Cadenas-Herrera, P.; Amrouche, A.; Mesmacque, G.; Jozwiak, K.; Puchi-cabrera, E.S. Influence of residual fatigue damage on the fracture toughness parameters of an AA6082-T6 aluminium alloy. *Fatigue Fract. Eng. Mater. Struct.* **2010**, *33*, 54–65. [CrossRef]
19. Tang, Z.; Hu, X.; Jiang, J.; Xue, H.; Zhuge, H. Effects of pre-fatigue damages on ultimate strength of steel columns: From material to structure. *J. Constr. Steel Res.* **2022**, *195*, 107358. [CrossRef]
20. Paul, S.K.; Sivaprasad, S.; Dhar, S.; Tarafder, S. Cyclic plastic deformation and damage in 304LN stainless steel. *Mater. Sci. Eng. A* **2011**, *528*, 4873–4882. [CrossRef]
21. López, J.G.; Verleysen, P.; Baere, I.D.; Degrieck, J. Tensile properties of thin-sheet metals after cyclic damage. *Procedia Eng.* **2011**, *10*, 1961–1966. [CrossRef]
22. Moćko, W.; Brodecki, A.; Radziejewska, J. Effects of pre-fatigue on the strain localization during tensile test s of DP500 steel at low and high strain rates. *J. Strain Anal. Eng. Des.* **2015**, *50*, 349–351. [CrossRef]
23. Tang, Z.; Chen, Z.; He, Z.; Hu, X.; Xue, H.; Zhuge, H. Experimental and numerical study of combined high and low cycle fatigue performance of low alloy steel and engineering application. *Materials* **2021**, *14*, 3395. [CrossRef] [PubMed]
24. Wang, X.; Zhang, W.; Ni, J.; Zhang, T.; Gong, J.; Wahab, M.A. Quantitative description between pre-fatigue damage and residual tensile properties of P92 steel. *Mater. Sci. Eng. A* **2019**, *744*, 415–425. [CrossRef]
25. Jia, C.; Shao, Y.; Guo, L.; Liu, H. Cyclic behavior and constitutive model of high strength low alloy steel plate. *Eng. Struct.* **2020**, *217*, 110798. [CrossRef]
26. Peng, Y.; Liu, Y.; Li, H.; Xing, J. Research on low cycle fatigue life prediction considering average strain. *Mater. Res. Express* **2022**, *9*, 016521. [CrossRef]
27. *GB/T 714–2015*; General Administration of Quality Supervision, Inspection and Quarantine of the People's Republic of China, Structural Steel for Bridge. Standards Press of China: Beijing, China, 2015.

28. *GB/T 228.1–2010*; General Administration of Quality Supervision, Inspection and Quarantine of China, Metallic Materials-Tensile Testing-Part1: Methods of Test at Room Temperature. Standards Press of China: Beijing, China, 2011.
29. *GB/T 15248–2008*; General Administration of Quality Supervision, Inspection and Quarantine of China, The Test Method for Axial Loading Constant-Amplitude Low-Cycle Fatigue of Metallic Materials. Standards Press of China: Beijing, China, 2011.
30. Ramberg, W.; Osgood, W.R. Description of Stress-Strain Curves by Three Parameters. 1943. Available online: https://ntrs.nasa.gov/citations/19930081614 (accessed on 8 July 2022).

crystals MDPI

Article

Characterization of Microstructure and High Temperature Compressive Strength of Austenitic Stainless Steel (21-4N) through Powder Metallurgy Route

Arun Prasad Murali [1], Dharmalingam Ganesan [1], Sachin Salunkhe [1,*], Emad Abouel Nasr [2], João Paulo Davim [3] and Hussein Mohamed Abdelmoneam Hussein [4,5]

1 Department of Mechanical Engineering, Vel Tech Rangarajan Dr. Sagunthala R & D Institute of Science and Technology, Chennai 600062, India; arunprasadm@veltech.edu.in (A.P.M.); gdharmalingam@veltech.edu.in (D.G.)
2 Industrial Engineering Department, College of Engineering, King Saud University, Riyadh 11421, Saudi Arabia; eabdelghany@ksu.edu.sa
3 Department of Mechanical Engineering, Campus Universitário de Santiago, University of Aveiro, 3810-193 Aveiro, Portugal; pdavim@ua.pt
4 Mechanical Engineering Department, Faculty of Engineering and Technology, Future University in Egypt, New Cairo 11835, Egypt; hussein.hussein@fue.edu.eg
5 Mechanical Engineering Department, Faculty of Engineering, Helwan University, Cairo 11732, Egypt; hussein@h-eng.helwan.edu.eg
* Correspondence: drsalunkhesachin@veltech.edu.in

Citation: Murali, A.P.; Ganesan, D.; Salunkhe, S.; Abouel Nasr, E.; Davim, J.P.; Hussein, H.M.A. Characterization of Microstructure and High Temperature Compressive Strength of Austenitic Stainless Steel (21-4N) through Powder Metallurgy Route. *Crystals* **2022**, *12*, 923. https://doi.org/10.3390/cryst12070923

Academic Editor: Wojciech Polkowski

Received: 25 May 2022
Accepted: 23 June 2022
Published: 29 June 2022

Abstract: Exposure of the engine valve to high temperatures led to the degradation of the valve material due to microstructural instability and deteriorating mechanical properties. Performance enhancement and alteration in microstructures can be attained through the powder metallurgy route which is a viable method to produce near net shape components. In this current study, the development of austenitic stainless steel (21-4N) through the powder metallurgy route as an alternate material for engine valves was investigated. Mechanical alloying was carried out for the pre-alloyed mixtures and consolidated using vacuum hot pressing. Sintering parameters were fixed at 1200 °C, 50 MPa and at a vacuum level of 10-3 Torr. A scanning electron microscope was used to analyze the morphology of the milled powders. Densities for the hot pressed powders were compared with theoretical densities and found to be around 98–99%. Observations regarding grain size, the presence of austenitic grain, heterogeneous distribution of metal carbides and analysis of chemical composition along the metal matrix were determined using both optical and electron microscopes. X-ray diffraction was carried out for both the consolidated and powder samples. The hot pressed samples exhibited a hardness value of 410 ± 10 Hv. An isothermal compression test for the sintered samples was carried out at a temperature of 650 °C and strain rate of 0.001 s^{-1}. It is showed that the compressive strength of 1380 MPa. An analysis between the room temperature yield strength obtained from hardness measurement and the strengthening mechanism based on the microstructure was conducted. Grain size, dislocation and solid solution are the major strengthening mechanisms which strengthen the material. Overall, the development of valve steel material through the powder metallurgy route exhibited improved metallurgical and mechanical properties in comparison to the corresponding cast product.

Keywords: 21-4N valve steel; mechanical alloying; vacuum hot pressing; powder metallurgy

1. Introduction

Almost all engine valves consist of bi-metallic welded materials, which are composed of hardened steel stem and a superalloy. Superalloys previously used for valve manufacturing include Inconel 751, Pyromet 31 and Nimonic 80A which are resistant to heat, oxidation/corrosion and wear at elevated temperatures [1]. Due to the higher cost and processing requirements of nickel-based superalloys, valve manufacturers initiated the development of an alternate material consisting of titanium and austenitic stainless steels (21-4N). Due to the low density of titanium alloys, it had a good strength to weight ratio when compared to existing materials consisting of nickel superalloys. However, titanium alloys had poor room temperature ductility, low toughness properties and a high cost of production [2]. An alternate suitable material to improve strength, resistance to creep, corrosion and oxidation resistance at elevated temperatures is 21-4N austenitic heat resistant stainless steel [3]. It consists of a high chromium and manganese content which improve the material properties and provide structural stability. The high pressure and high operating temperature of automotive valves ranges between 800–950 °C, where the instability in strength was noted [4].

Zhang et al. analyzed the failure of valve heads due to the non-uniform distribution of temperature in the valve region, which resulted in an increased stress concentration and led to a crack in valve heads [5]. Similarly, a review of the failure of the engine valve conducted by Naresh et al. suggested that the principal reason for the failure of the valve is excessive heating leading to the degradation of valve material, fretting, microstructural instability and impact load [6]. If the engine valve is deemed poor, the output power decreases or can even lead to seizure. Overall, analyses regarding automotive failure have been conducted predominantly in engines [7].

Several researchers carried out the failure analysis of engine valves and interpreted the cause for its failures. Yu et al. investigated the failure of an exhaust valve made of 5Cr21Mn9Ni4N steel. It was stated that the reason for the failure is the disintegration of the austenitic structure due to the depletion of chromium from the grain structure. This led to impaired mechanical properties and subsequently fatigue failure [8]. Kum et al. found that the thermal deformation of valve heads leads to valve fracture. Moreover, they found that fractures occurred as a result of non-uniform stress distribution near the valve region [9].

Overall, it can be seen that materials produced through the cast route showed inhomogeneous grain growth throughout the product due to the uneven solidification rate during the casting process. To overcome these defects, alloys were developed through directional solidification to uplift the mechanical properties along the longitudinal directions. This will eliminate the weakening of the grain boundary and enhance high temperature mechanical strength. One such method to obtain high strength with fine grain structure is the powder metallurgy route. Near net shapes are obtained using the powder metallurgy process which reduces secondary machining processes such as milling, turning and drilling [10,11]. A better surface finish, close tolerance and post or secondary processing are needed for many parts in the service industry. With the introduction of powder metallurgy components, market review shows that 60% of components will only require one machining processes [12]. Powder consolidation techniques using vacuum hot pressing will eliminate unwanted and undesirable reactions within metal matrices due to the lower processing temperature [13]. Great flexibility during material design and selection can be achieved using powder metallurgy technique. Due to these advantages, materials developed using the powder metallurgy route exhibit superior properties in comparison to their counterparts developed through traditional methods such as the melting and casting route [14–16]. Components produced through the powder metallurgy route present fine microstructure and a homogeneous distribution of particles [17]. To consolidate metal powder particles in order to attain superior strength, hot pressing under vacuum conditions was carried out. During this process, both temperature and pressure were applied to the powder compacts. Vacuum condition was used to provide control over the atmosphere to prevent oxidation at a high temperature. These powder particles consist of a single phase before consolidation. It is attained by the process of mechanical alloying (MA). Without melting, the solid solution can be attained by violently deforming the powder mixtures under controlled conditions. This non-equilibrium process works based on energizing and quenching, whereby the alloyed materials are brought into the metastable state with the application of pressure and temperature. Moreover, studies related to hot compression for 21-4N were carried out previously using the casting route [18–20], whereas hot compression studies of 21-4N steel developed using the powder metallurgy route have yet to be reported.

The literature related to the comparison of mechanical and microstructural properties of similar materials with different processing routes are relatively scarce. Bartolomeu et al. studied the microstructure and mechanical properties of austenitic 316L stainless steel developed using both conventional casting and also the powder metallurgy route (hot pressing). It was reported that 316L stainless steel samples had a coarse grain structure in cast route whereas they possessed an equiaxed grain structure when developed using the hot pressing technique. A slight increase in hardness of 6% was reported for 316L developed through hot pressing. Yield and tensile strength increased by 33% and 50%, respectively, due to the equiaxed grain structure [14]. Garbacz et al. compared the mechanical and microstructural properties of platinum and rhodium (Pt-Rh) alloys manufactured using the isostatic pressing and casting route. When subjected to tensile testing, both the alloys exhibited an increase in tensile property by a factor of 2.5 when compared to alloys manufactured using the casting technique. The increase in tensile properties for Pt-Rh alloys developed using the powder metallurgy route can be attributed to grain size strengthening as a result of the fine grain size when compared with the coarser grain obtained with the casting route [15]. High entropy alloys consisting of Fe-Ni were developed by Larissa et al. using the cast and spark plasma sintering technique. Alloys developed using powder metallurgy possessed a nano-sized FCC metal matrix with a uniform and fine grained microstructure, whereas in contrast, cast alloys showed a coarse grained structure. Similarly, the mechanical properties were far superior when compared to its cast products [16]. Overall, the alloys developed through the powder metallurgy route showed improved mechanical properties when compared to cast products of a similar composition.

In the present work, we developed a suitable substitute for valve materials consisting of austenitic stainless steel (21-4N) using the powder metallurgy technique. It is very important that the microstructural stability of the material is evaluated using an optical micrograph (OM), scanning electron microscope (SEM) along with energy dispersive spectrum (EDS) and transmission electron microscopy (TEM) studies. The structural evolution of the powders was studied using SEM, and the X-ray diffraction (XRD) patterns were also evaluated for the hot pressed samples. The Vickers hardness (HV) for the consolidated samples was evaluated and compared with results obtained with various strengthening mechanisms.

2. Experimental Procedures

The chemical composition for the valve material of 21-4N austenitic stainless steel was developed with ferro-alloy powders of Fe-Ni, Fe-Mn and Fe-Cr which had a particle size of less than 25 μm. The desired chemical composition of 21-4N heat resistant steel is shown in Table 1.

Table 1. Individual elemental composition for 21-4N steel.

Alloy	Concentration of Elements in wt%					
	Cr	Ni	Mn	C	Si	Fe
21-4N	21	4	8.2	0.48	0.40	Rest

To convert the pre-alloyed mixtures into austenitic alloy, high energy planetary ball milling was used. MA was executed in a Fristch GmbH planetary ball mill with tungsten carbide balls and vials where the ball to powder ratio (BPR) was 10:1. Ball milling was carried out for 10 h at 300 rpm with toluene as the process control agent. The structural morphology of the milled powders at regular intervals were studied using SEM analysis. Milled powders were consolidated using vacuum hot pressing with a temperature, pressure and vacuum level of 1200 °C, 50 MPa and 10^{-3} Torr, respectively, for a period of 2 h. A schematic representation is shown in Figure 1 for vacuum hot pressing. After vacuum hot pressing, the dimensions of the consolidated sample obtained were 30 mm in diameter and 12 mm in height as shown in Figure 2.

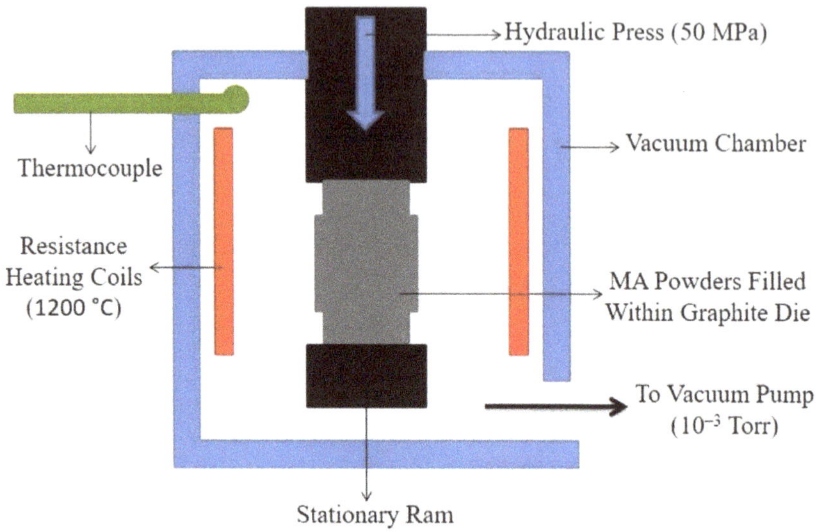

Figure 1. Schematic representation of vacuum hot pressing.

Figure 2. Schematic representation of vacuum hot pressed sample.

Vacuum hot pressed samples were analyzed for densification studies, microstructural examination and phase determination. Metallographic techniques were used to prepare the sample with ferric chloride as the etchant to reveal the microstructure. The phase distribution, shape and morphology of both powder and vacuum hot pressed samples were studied using SEM-EDS. INCA Xsight JOEL–JEM–2100 was used to carry out the transmission electron microscopy (TEM) study operated at 200 kV. The preparation of samples for TEM studies was initially carried out with Gatan model 656 for dimpling the specimens, followed by ion milling with Gatan model 691. Measurement of hardness for the hot pressed samples was evaluated at room temperature in a Vickers micro hardness tester using a FIE Model OMEGA hardness tester (Deckenpfronn, Germany). Indentation was carried out using a diamond intender at a load of 1 kg for a dwell time of 15 s. The bottom of the test specimens were filed flat before each testing. From each of the three hot pressed samples, 10 indentations were carried out and each of the indents were spaced 3 mm from each other. Zwick/Roell Z100 was used to carry out compression testing to evaluate the compression strength of the hot pressed samples as per ASTM E209, with a strain rate of

$0.001\ s^{-1}$ and at 650 °C. Specimens of cylindrical sizes with a height 10 mm and diameter 8 mm were subjected to uniaxial compression. During hot compression, the sintered samples were heated through resistance heating and the temperature measurements were conducted in thermocouples. To reduce the effect of friction during hot compression, the two ends of the sample were padded with a lubrication sheet consisting of graphite.

3. Results and Discussion

3.1. Structural Morphology of Milled Powders

Pre-alloyed mixtures of the powder particles consisting of ferro-alloys and traces of elemental powders that were blended inside a tungsten carbide vial are shown in Figure 3a. From the micrograph, it can be observed that the ferro-alloy particle mixtures consisted of various shapes such as irregular dentritic and flaky structures. Similarly, the morphology of the mixtures is shown in Figure 3b,c for a milling condition of 5 h and 10 h, respectively.

Figure 3. (**a**–**c**) Morphology of milled powders at regular milling interval.

From the SEM morphology, significant variations in size and morphology of the pre-alloyed powders during the entire milling time at regular intervals can be observed. The initial condition of the powder particles was soft and ductile. As the milling was carried out in wet conditions, these soft particles tended to agglomerate with each other during the early stages. With the continuation of milling, the agglomerated particles break down and flatten. It is evident that the milled powders are finer when compared to the initial condition after 5 h of milling. The formation of finer particles is due to the repeated fracturing and welding between the powder particles due to collisions between ball-powder-vial. Due to the high impact ball-powder-vial and shearing action between them, a high amount of energy was transferred to the milled powders. However, after 10 h of milling, the milled powders were found to be fragmented due to continuous work hardening. The surface properties revealed that the fracturing of the powder particles was dominant when compared to cold welding.

From Table 2, several studies carried out on the formation of austenitic stainless steel through MA are presented and it can be inferred that the formation of austenitic stainless steel depends on the milling parameters, namely starting materials, milling medium, BPR and milling time.

Table 2. Evolution of various austenitic grade stainless steels from different starting powders through MA.

Elemental/Pre-Alloyed Mixtures	Balls and Vial	BPR	Milling Speed (rpm)	Milling Time (h)	References
304, 316 and 310	Stainless Steel	5:1	300	50	[21]
Fe, Cr, Ni and W	Tungsten Carbide	10:1	300	25	[22]
Fe, Ni, Cr, Mo, Al and Mn	Stainless Steel	30:1	500	16	[23]
Fe, Cr, Ni, W and Ti	Stainless Steel	10:1	380	60	[24]
Fe, Cr, W, Fe-Ti and Fe-Y	Stainless Steel	8:1	225	100	[25]
Fe, Cr, Mn and Mo	Stainless Steel	25:1	300	60	[26]
Fe, Cr and Mn	Stainless Steel	25:1	300	100	[27]
Fe, Cr and Mn	Stainless Steel	10:1	400	144	[28]
Fe-Cr, Fe-Ni and Fe-Mn	Tungsten Carbide	10:1	300	10	Current study

When compared to stainless steel balls and vials, the kinetic energy of the moving particles within the vial is higher for tungsten carbide due to its higher density. If the austenitic stainless steel powders were to be developed from elemental powders, their evolution would require much more time compared to their development from pre-alloyed mixtures.

3.2. Evaluation of Density for the Hot Pressed Samples

Consolidation of the MA powders was carried out using hot pressing under a vacuum atmosphere. Density measurements for the vacuum hot pressed samples were performed using Archimedes principle. Theoretical density was calculated using rule of mixtures and compared with the hot pressed density as shown in Figure 4. The density attained for the hot pressed samples is averaged to be 7.67 g/cc which is 98.8% the density when compared to the theoretical density. A higher density was achieved due to MA, which fractured the powder particles to produce complicated shapes, which in turn increased the surface area of the powder particles required for hot pressing. Reduction in the size of the particles led to the formation of single compacts during hot pressing through grain boundary diffusion rather than lattice diffusion [29,30]. The high specific area of the powder particles initiated the surface diffusion mechanism during the initial stages of sintering. As the sintering progressed, grain boundary diffusion occurred and during later stages volume diffusion took place to produce highly dense compacts.

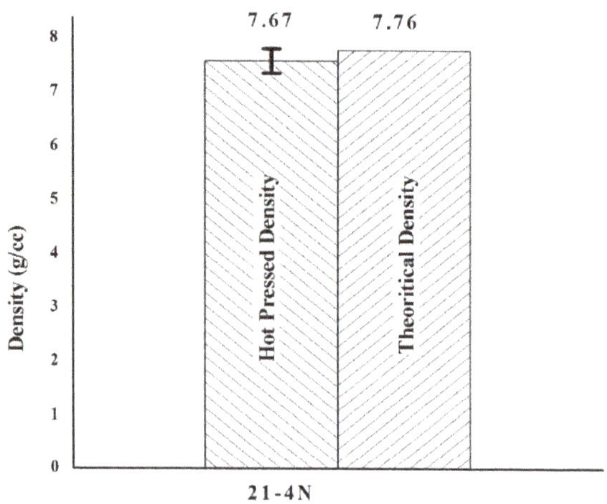

Figure 4. Density of hot pressed samples developed through powder metallurgy route.

3.3. Metallography Analysis

Figure 5a depicts the optical micrograph of 21-4N austenitic steel developed through vacuum hot pressing. The microstructure consists of equiaxed austenitic grains with grain boundary carbides distributed heterogeneously along the grain boundaries. The formation of grain boundary carbides is due to the slow cooling rate of 25 °C/min attained during vacuum hot pressing. The cooling rate plays an important role in the formation of metal carbides in an austenitic structure. This 21-4N alloy consists of low nickel and a high concentration of chromium in the metal matrix. Due to this, the grain boundary carbides consist of chromium-rich precipitates which are seen as dark phases within the micrograph. The light region represents austenitic grains. As the temperature for sintering is high (1200 °C), pores are not evident in the micrograph. Growth of grains did not occur at a faster pace at the beginning stages of sintering. However, at later stages of the sintering, cycle due to high temperatures, the available pores were rounded off. This can be interpreted in relation to a larger grain size in comparison to the crystallite size obtained as a result of MA. The sintering mechanism promotes the volume diffusion process through grain boundaries which act as vacancy sinks. The coalescence of pores occurs at high temperatures where grain growth is observed with grain boundaries as pinning points. Grain growth resulted in the rounding off of pores with little room for vacancies. Similarly, the density of the samples increased due to grain growth as well as volume diffusion.

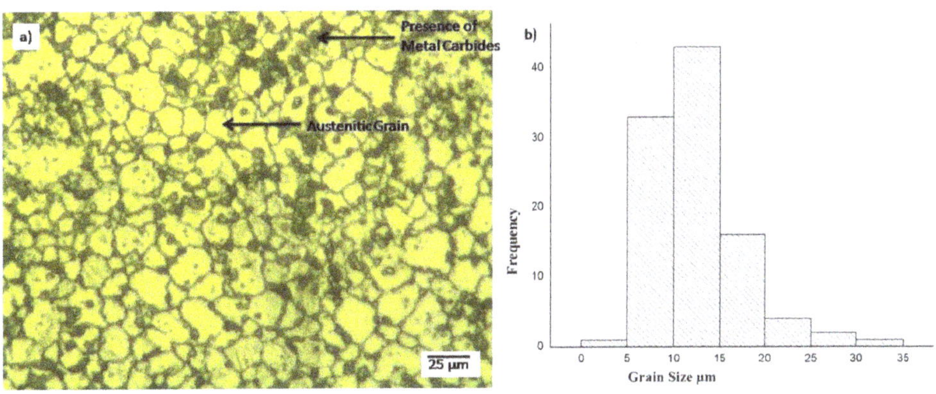

Figure 5. Optical micrograph of (**a**) 21-4N, (**b**) corresponding grain size.

The mean grain size of the austenitic grain was found to be around 7–12 µm as shown in Figure 5b. The grain size was measured using the intercept method as per ASTM E112. The size of the grains was much smaller compared to a similar cast product which contained 16 µm and 17 µm sized grains as reported by Ji et al. and Kumar et al., respectively [20,31]. The initial starting size of the powders followed by the MA inhibited the grain growth during sintering in order to obtain a fine grain size.

The precipitation of carbides along the grain boundaries can still be authenticated using a SEM-EDS analysis. Figure 6a,b show the SEM-EDS analysis of both the grain and grain boundaries of 21-4N austenitic stainless steel. The presence of carbide precipitates is due to the effect of alloying elements within the metal matrix which change the solubility of carbon in the austenitic matrix. As nickel and manganese are austenitic stabilizers, they enhance the precipitation of carbides by reducing their solubility within the metal matrix.

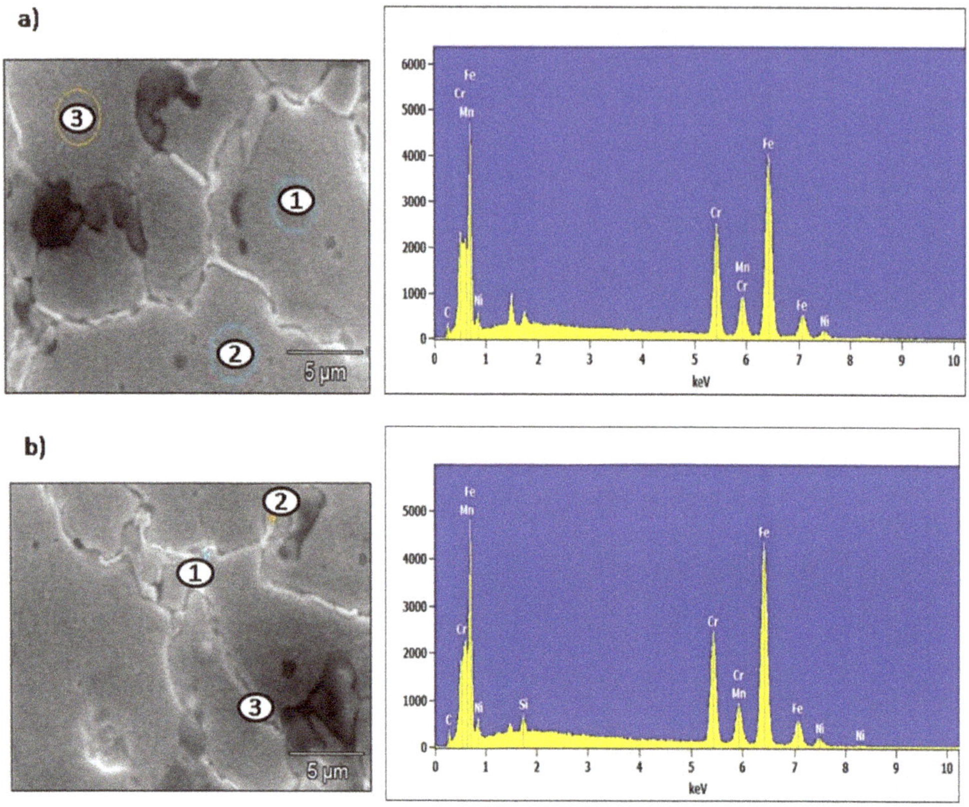

Figure 6. SEM-EDS analysis of 21-4N both at (**a**) grains and (**b**) grain boundaries.

In both the grains and grain boundaries, three sections of EDS measurement were selected and shown in Table 3. A variation in the chemical composition can be seen from the EDS analysis. It is evident that elemental migration took place from the grains to the grain boundary. From Figure 6a, concentration spots (1, 2 and 3) within the grain reveal the presence of an austenite stabilizing element such as nickel at an appropriate level. Similarly, at the grain boundary spot in Figure 6b, a higher concentration (1, 2 and 3) of chromium, carbon and manganese along with iron were found in excess in comparison to the required elemental composition. In comparison to similar studies conducted by Xu et al., the size of the carbide precipitates were not larger or continuously dispersed within the metal matrix [32]. Moreover, the development of the alloy through the powder metallurgy route led to the formation of fine grains and restricted the size of metal carbide precipitates.

Table 3. EDS concentration of elements both at grain boundaries and grains.

Location	Region	Concentration of Elements, wt%				
		Cr	Ni	Mn	C	Fe
Grain	1	19.9	3.8	8.5	0.25	Bal.
	2	19.2	3.6	8.3	0.30	Bal.
	3	18.7	4.1	8.7	0.27	Bal.
Grain Boundary	1	25.3	4.4	9.2	0.65	Bal.
	2	26.4	4.25	9.1	0.70	Bal.
	3	24.2	4.3	8.9	0.68	Bal.

The XRD analysis of the vacuum hot pressed steel is shown in Figure 7, and it can be confirmed that these metal carbides were mostly composed of $M_{23}C_6$ where M represents chromium, manganese and iron. The formation of carbides occurred mainly due to the higher C/Cr ratio. From the metallographic studies, it can be understood that $M_{23}C_6$ nucleates very easily at the beginning when compared to other types of carbides. Moreover, the structural shape of these metal carbides is of a globular or cellular type. These carbide precipitates are usually observed near the grain boundaries due to it being a favorable place for carbon, with an interstitial solid solution capability which rapidly diffuses near the grain boundaries. The intensity of austenite peaks was dominant when compared to $M_{23}C_6$ peaks. The formation of the $M_{23}C_6$ peak in hot pressed samples is due to slow cooling in the sintering cycle. During the sintering of MA powders at 1200 °C, the peaks of austenite became intense and sharper with diminished peaks of $M_{23}C_6$. The presence of $M_{23}C_6$ at the grain boundary improved the material resistance to heat and sliding of grains. Similarly, Figure 8 shows the XRD analysis of powder particles milled for 10 h. The austenitic peak is evident from the XRD analysis.

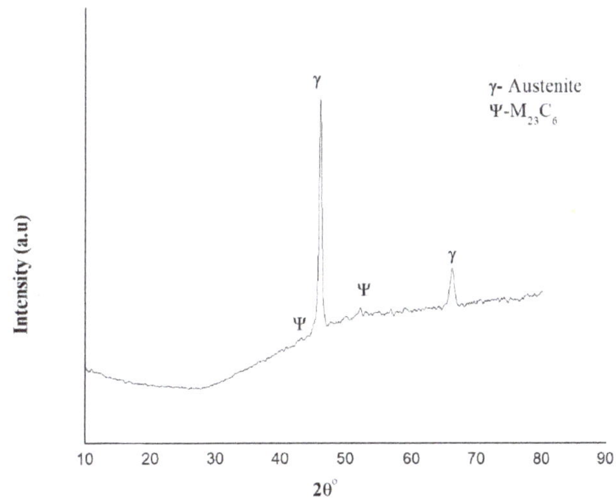

Figure 7. X-ray diffraction pattern for vacuum hot pressed sample.

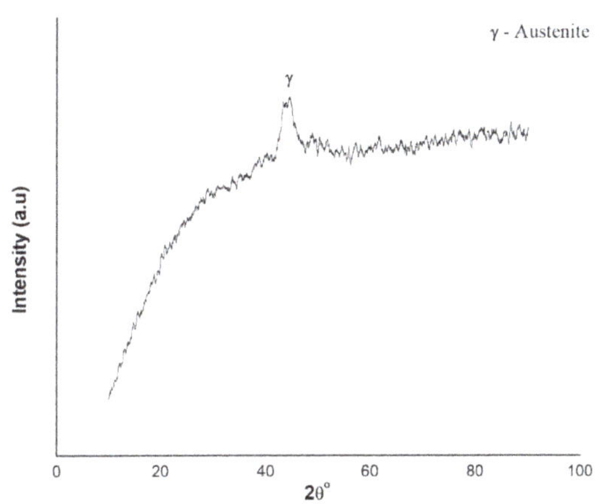

Figure 8. X-ray diffraction pattern for milled powder.

Figure 9a represents the TEM microstructure of the vacuum hot pressed steel. Within the austenitic grain, twins were available which indicates the presence of an austenitic matrix. Face centered cubic crystal alloys have a tendency to form twins within the microstructure when subjected to cold working and annealing as a result of low stacking fault energy. Due to low stacking faults, the energy required for twin formation is lower when compared to grain boundary formation. These kinds of fine austenitic twins are found as the material is subjected to cold working and annealing during ball milling and vacuum hot pressing, respectively. Few dislocations were also found within the grains due to hot pressing. The formation of nano crystalline grains along with dislocations is the outcome of severe plastic deformation. The dislocations are pinned at the grain boundaries. This is due to the presence of metal carbides which are present at the grain boundaries that resist dislocation due to the precipitate strengthening mechanism. These precipitates will impede movement or dislocation and resists plastic deformation. Due to MA, the powders consists of residual stresses which act as sites for nucleation [33]. During sintering, these powders nucleates and give rise to the particle stimulated nucleation phenomenon. Face centered cubic structures are found to have particle stimulated nucleation. As the holding time during sintering is 2 h, some grains nucleate near the grain boundaries. This is because a favorable place for grain nucleation is near the grain boundaries which are also the locations for rapid diffusion when compared to a lattice. From Figure 9a, few grains are seen in between the grain and grain boundary region. Grain sizes of the TEM samples were determined by the method of linear intercept. The mean grain size of 1 μm was obtained for the samples developed through vacuum hot pressing as shown in Figure 9b. Cast products with similar compositions had large precipitates of carbides along with σ-phases which were not visible near the grain boundary in the current study [3]. Due to the particle stimulated nucleation phenomenon, very few residual carbides were seen within the metal matrix along with the formation of grains and grain boundaries. Similarly, as the material was manufactured using MA and subsequently consolidated using hot pressing, σ-phases and other intermetallics were not visible as sintering took place under a vacuum atmosphere. The selected area diffraction (SAD) pattern, as shown in Figure 9c, indicates the formation of a nanocrystalline structure even after the consolidation of MA powders at 1200 °C. The rings represent the formation of nano crystallite structures of the metal matrix. Correspondingly, the outer ring represents the formation of face centered cubic structures, namely (111) and (222), respectively. Both the austenitic matrix and precipitates of the carbides co-exist within the metal matrix due to the fact that both have face centered cubic structures.

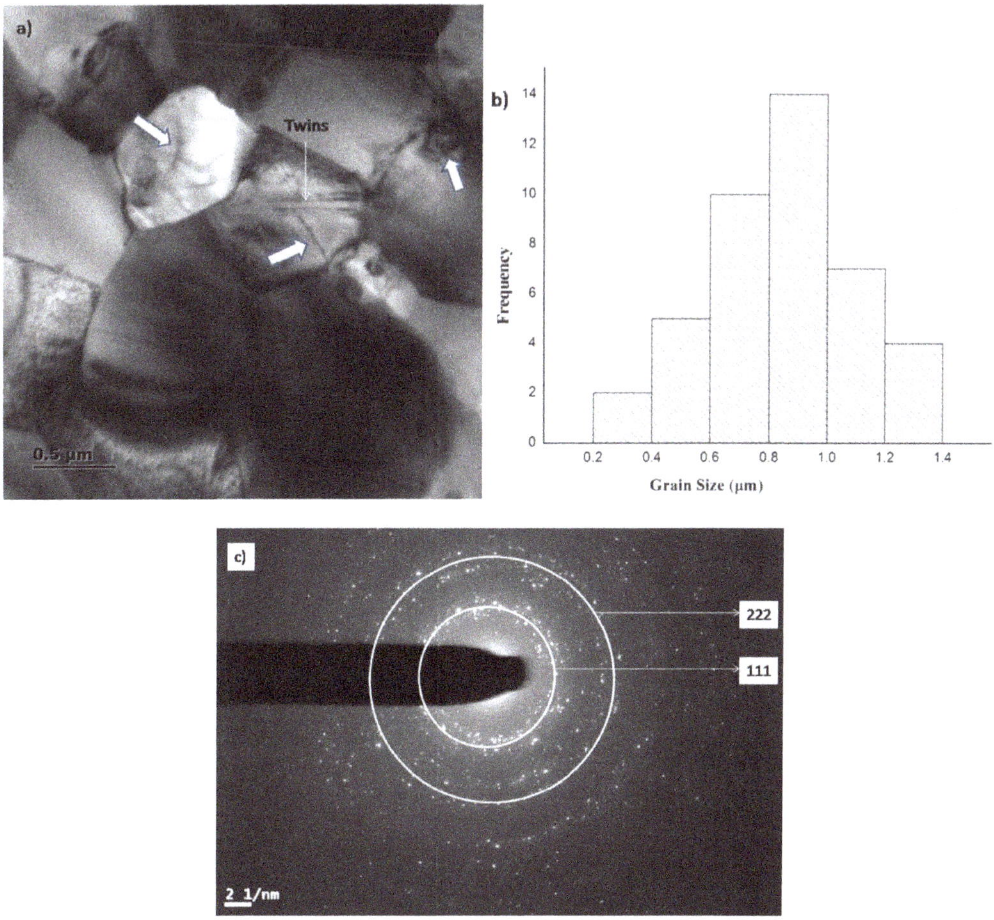

Figure 9. TEM micrograph of heat resistant austenitic steel of 21-4N (**a**) bright field image where arrow mark indicates dislocation lines, (**b**) distribution of grain size for the austenitic matrix, (**c**) corresponding SAD pattern.

3.4. Hardness Measurement

Measurement of hardness was carried out using Vickers microhardness. Table 4 shows the hardness value in comparison with other 21-4N heat-resistance steels manufactured through other processing routes. The hardness obtained through the powder metallurgy process was higher when compared to the hardness obtained through other processing techniques. The increase in hardness is due to the higher sintering temperature during the vacuum hot pressing technique [34]. Moreover, it can be seen that the presence of nano crystalline structures led to the formation of a fine-grained metal matrix which was evident from the microstructural analysis from Figure 9b which resulted in superior hardness for the alternate valve steel material developed using the powder metallurgy route. Similarly, the availability of precipitate strengthened metal carbides distributed heterogeneously (not as continuous structure) near the grain boundaries which had a significant effect on the increase in hardness. This acted as a strengthening mechanism in the formation of precipitates.

Table 4. Comparison of hardness values obtained for different valve steel made of 21-4N material.

Sl No.	Manufacturing Method	Hardness Value (Hv)	References
1	Casting	360 ± 15	[3]
2	Casting	320 ± 10	[35]
3	Casting	290 ± 10	[4]
4	Casting	375 ± 15	[36]
5	Hot rolled	343	[31]
6	Powder Metallurgy	410 ± 10	Present Study

3.5. Hot Compression Measurements

Figure 10 represents high temperature compression stress strain curves for 21-4N steel at 650 °C and 0.001 s^{-1}. Superior strength of 1340 MPa was reported for the material developed using the powder metallurgy technique. The flow stress hump that can be seen within the compressive stress strain graph is due to the formation of dynamic recrystallization which accumulates during the straining [37]. Figure 11a illustrates the micrograph of hot compressed samples at a temperature 650 °C and strain rate of 0.001 s^{-1}. Local bulging along with serrated irregular boundaries are indicated by arrow marks from Figure 11b. In the SEM micrograph, the serrated irregular boundaries are due to the movement of dislocations near the regions. Due to the fine microstructure, shearing of the particles is avoided, leaving behind deformations in the form of bulging. Similarly, near the serrated boundaries the structure consists of early stages of recrystallization. When compared to the compression test of the 21-4N cast product, steel developed through the powder metallurgy route showed higher compressive strength with a higher strain rate. Materials developed through cast products showed a decreasing trend in hot compressive strength. Huang et al., reported a maximum hot compressive stress value of 350 MPa with material developed through cast product [38]. Similar results of 350 MPa were reported by Li et al., and Ji et al., for 21-4N material developed through the cast process [19,20]. This can be attributed to the fact that inhomogeneous distribution of grain growth from the surface to inner layers occurred during uneven solidification rate. The development of the homogeneous and fine microstructure attained as a result of the powder metallurgy processing route showed substantial improvement in hot compressive values when compared to materials processed using the cast route.

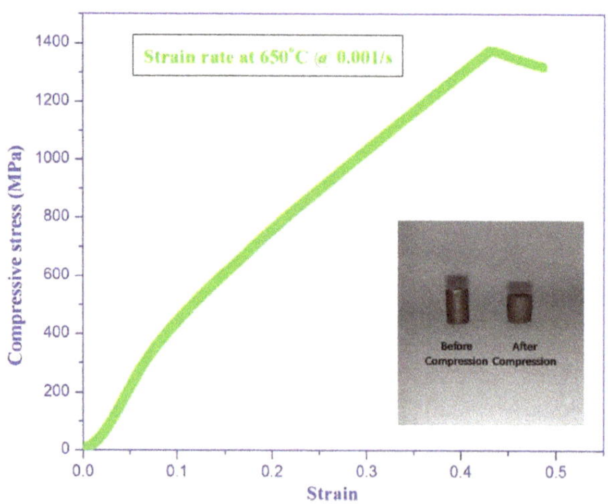

Figure 10. Hot compressive strength of the sintered sample, sample before and after compression (insert).

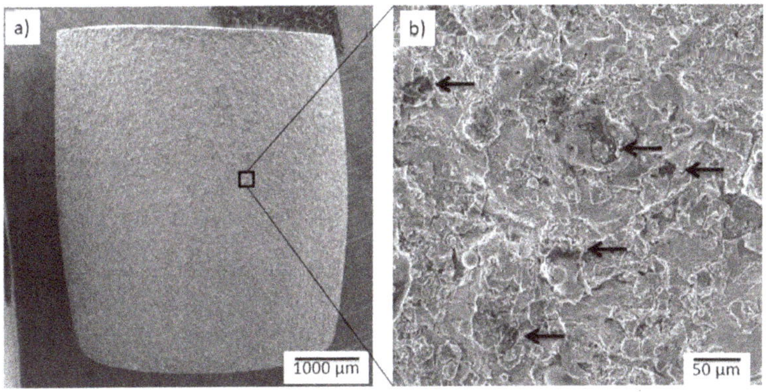

Figure 11. SEM micrograph of hot compressed sample (**a**) Overview of bulged sample (**b**) indication of serrated boundaries.

3.6. Corresponding Relationship between Mechanical and Microstructural Properties

The strength of the material at room temperatures was calculated and compared with the strengthening mechanism. As stated in Equation (1), the strength of the material at room temperatures was obtained with hardness measurements, where Hv is the value of the obtained hardness in terms of MPa and σ_y is the yield strength [39]. A multiplication factor of 9.8065 was used to convert the obtained hardness from Hv to MPa [40]. The value of yield stress as per Equation (1) is 1340 MPa.

$$\sigma_y = \frac{1}{3} H_v \quad (1)$$

For any alloy, the calculation of yield strength is also based on the combined contribution of the various strengthening mechanisms as stated in Equation (2) [39].

$$\sigma_y = \sigma_{ss} + \sigma_{gs} + \sigma_{dis} \quad (2)$$

where σ_{ss} is strengthening due to a solid solution, σ_{gs} is grain size strengthening and σ_{dis} strengthening due to dislocations. The effect of solid solution strengthening in an alloy is mainly due to both substitutional and interstitial types of elements present in the system. In the present study, substitutional elements consisted of chromium, nickel and manganese which were the major alloying elements. Meanwhile, the presence of interstitial or minor alloying elements based on their wt% was neglected, which in this case was carbon. The solvent iron which is strengthened due to substitutional elements is given by Equation (3) [41].

$$\sigma_{ss} = 0.00689 K X^n \quad (3)$$

where "K" is coefficient of strengthening and the corresponding values of "K" for the elements chromium, nickel and manganese are 1400, 6100 and 7000, respectively, "X" represents the elemental concentration in the present alloy in terms of atomic percentage and "n" is the constant valued 0.75 [42]. The effect of strengthening due of solid solution for the current study is found to be 447 MPa from Equation (3). The Hall Petch relation can be used to calculate the strengthening effect as a result of fine grain structure as per Equation (4).

$$\sigma_{gs} = \sigma_o + k d^{\frac{-1}{2}} \quad (4)$$

where "σ_o" is taken as 30 MPa for iron alloys which is known as friction stress, "k" is a constant and taken as 0.4 MN/m^2 for austenitic stainless steel grain sizes of less than 3 µm and "d" is the grain size (in meters) for the alloy [43]. Taking the average grain size of the samples from Figure 9b as 1 μm, the yield strength due to grain boundary strengthening is 430 MPa. The increase in strength due to dislocations present in the alloy is calculated using Equation (5) [44].

$$\sigma_{dis} = \alpha M G b \sqrt{\rho} \tag{5}$$

The efficiency of the hardening effect induced due to dislocation is denoted by "α" which as a constant ranges from 0.1 to 0.5, where the mean value is taken for calculation [42]. For a given FCC structure, "M" denotes Taylor's factor which is 3, "G" is termed as modulus of rigidity and for pure iron it is 83 GPa, "b" is burgers vector and for the FCC material it is 0.251 nm. Dislocation density for the vacuum hot pressed samples is expressed as "ρ" and is taken as 5×10^{14} m^{-2} [45]. From Equation (5), as a result of dislocation strengthening, the yield strength is 419 Mpa. From Equation (2) the overall yield strength obtained from the strengthening mechanisms, namely solid solution strengthening, grain size strengthening and dislocation strengthening can be found to be 1296 MPa. The obtained theoretical value as per Equation (1) is slightly overestimated by 3.39% which is in good agreement with the summation of all contributed strengthening mechanisms. The overestimated observation is due to either the summation or overestimation of individual strengthening contributions. Alloys consisting of solid solution strengthening lead to overestimation as a result of friction stress. The factor of friction stress is dependent on temperature, dislocation and slip system. Chauhan et al. reported that when comparing the calculated and measured values of yield stress, a small amount of overestimation for all alloys up to 53 Mpa occurs, whereas smaller values of 30 MPa were also reported by Li [42,46].

4. Conclusions

The current study reported on the development and evaluation of mechanical and microstructural properties of 21-4N austenitic steel through the powder metallurgy route. Parameters used for sintering gave rise to a balanced austenitic structure when compared to cast products with a similar composition. The 21-4N austenitic valve steel developed through MA to obtain a nano-crystalline structure led to the following conclusions.

During the initial stages of milling, pre-alloyed mixtures were soft and ductile. A prolonged milling time leads to the cold welding of particles and break down due to repeated fracturing. The morphology of the milled powders revealed that the fracturing of powder particles was dominant when compared to cold welding.

MA powders were consolidated using vacuum hot pressing which had a density of 98% in comparison to the theoretical density of the samples. High dense samples were obtained due to grain boundary diffusion and volume diffusion during sintering.

SEM-EDS studies revealed the presence of carbide precipitates near the grain boundaries. XRD analysis of the hot pressed samples confirmed the precipitates as $M_{23}C_6$, whereas these precipitates were not present in the powder particle. The formation of $M_{23}C_6$ carbide particle is due to slow cooling in the sintering cycle.

The TEM analysis revealed the presence of austenitic twins as materials were subjected to cold working and annealing. Dislocations were present within the metal matrix as an outcome of severe plastic deformation which is due to hot pressing. A fine grain structure was revealed during the TEM analysis with a mean grain size is of 1 μm.

When compared to other conventional processing techniques, the hardness value of 410 ± 10 Hv obtained through the powder metallurgy route was higher. The presence of a nano crystalline structure led to an increase in the hardness value. Evaluations of the strengthening mechanism clearly reveal that strengthening due to solid solution, grain size and dislocations are the dominant forces which increase the structural rigidity of the alloy.

Substantial improvements in the hot compression value of 1340 MPa were reported for 21-4N austenitic valve steel when compared to similar material developed through the cast product technique. The development of fine microstructure throughout the material resulted in the increase in hot compression values.

With the obtained results, 21-4N austenitic valve steel developed through the powder metallurgy route was demonstrated to have better strength and structural rigidity for the metal matrix which can be used for high temperature applications.

Author Contributions: Conceptualization and methodology, A.P.M. and D.G.; writing—original draft preparation, A.P.M. and D.G.; writing—review and editing, S.S., E.A.N., H.M.A.H. and J.P.D. All authors have read and agreed to the published version of the manuscript.

Funding: This research has received funding from King Saud University through Researchers Supporting Project number (RSP-2021/164), King Saud University, Riyadh, Saudi Arabia.

Institutional Review Board Statement: Not applicable.

Informed Consent Statement: Not applicable.

Data Availability Statement: Not applicable.

Acknowledgments: The authors extend their appreciation to King Saud University for funding this work through Researchers Supporting Project number (RSP-2021/164), King Saud University, Riyadh, Saudi Arabia. The authors would like to express their gratitude to R. Mariappan for his extended support through his valuable inputs during the course of the research work. TEM studies were carried out at PSG Institute of Advanced studies, India. We would like to extend our sincere gratitude to R. Rangarajan, Founder and Chancellor, Vel Tech Rangarajan Sagunthala R&D Institute of Science and Technology for providing facility in Metallurgical and Materials Laboratory to carry out this research work.

Conflicts of Interest: The authors declare no conflict of interest.

References

1. *ASM Handbook: Properties and Selection: Nonferrous Alloys and Special-Purpose Materials*; ASM International: Novelty, OH, USA, 1991; Volume 2.
2. Badami, M.; Marino, F. Fatigue tests of un-HIP'ed γ-TiAl engine valves for motorcycles. *Int. J. Fatigue* **2006**, *28*, 722–732. [CrossRef]
3. Choi, J.; Seok, C.S.; Park, S.; Kim, G. Effect of high-temperature degradation on microstructure evolution and mechanical properties of austenitic heat-resistant steel. *J. Mater. Res. Technol.* **2019**, *8*, 2011–2020. [CrossRef]
4. Witek, L. Failure and thermo-mechanical stress analysis of the exhaust valve of diesel engine. *Eng. Fail. Anal.* **2016**, *66*, 154–165. [CrossRef]
5. Zhang, Q.; Zuo, Z.; Liu, J. Failure analysis of a diesel engine cylinder head based on finite element method. *Eng. Fail. Anal.* **2013**, *34*, 51–58. [CrossRef]
6. Raghuwanshi, N.K.; Pandey, A.; Mandloi, R. Failure analysis of internal combustion engine valves: A review. *Int. J. Innov. Res. Sci. Eng. Technol.* **2012**, *1*, 173–181.
7. Hasan, M.R. Failure Investigation Report on Different Components of an Automotive Engine. *Int. J. Mech. Eng. Appl.* **2017**, *5*, 47–51.
8. Yu, Z.; Xu, X. Failure analysis and metallurgical investigation of diesel engine exhaust valves. *Eng. Fail. Anal.* **2006**, *13*, 673–682. [CrossRef]
9. Kum-Chul, O.; Sang-Woo, C.; Ji-Ho, K. A study of durability analysis methodology for engine valve considering head thermal deformation and dynamic behavior. In Proceedings of the SIMULIA Community Conference, Corp Providence, RI, USA, 19–22 May 2014.
10. Ivasishin, O.M.; Anokhin, V.M.; Demidik, A.N.; Savvakin, D.G. Cost-effective blended elemental powder metallurgy of titanium alloys for transportation application. In *Key Engineering Materials*; Trans Tech Publications: Freienbach, Switzerland, 2000.
11. Şap, E.; Usca, U.A.; Gupta, M.K.; Kuntoğlu, M. Tool wear and machinability investigations in dry turning of Cu/Mo-SiCp hybrid composites. *Int. J. Adv. Manuf. Technol.* **2021**, *114*, 379–396. [CrossRef]
12. M'Saoubi, R.; Czotscher, T.; Andersson, O.; Meyer, D. Machinability of powder metallurgy steels using PcBN inserts. *Procedia CIRP* **2014**, *14*, 83–88. [CrossRef]
13. Kurita, H.; Miyazaki, T.; Kawasaki, A.; Lu, Y.; Silvain, J.F. Interfacial microstructure of graphite flake reinforced aluminum matrix composites fabricated via hot pressing. *Compos. Part A Appl. Sci. Manuf.* **2015**, *73*, 125–131. [CrossRef]
14. Bartolomeu, F.; Buciumeanu, M.; Pinto, E.; Alves, N.; Carvalho, O.; Silva, F.S.; Miranda, G. 316L stainless steel mechanical and tribological behavior—A comparison between selective laser melting, hot pressing and conventional casting. *Addit. Manuf.* **2017**, *16*, 81–89. [CrossRef]
15. Garbacz, H.; Mizera, J.; Laskowski, Z.; Gierej, M. Microstructure and mechanical properties of a Pt–Rh alloy produced by powder metallurgy and subjected to plastic working. *Powder Technol.* **2011**, *208*, 488–490. [CrossRef]

16. Moravcikova-Gouvea, L.; Moravcik, I.; Omasta, M.; Veselý, J.; Cizek, J.; Minárik, P.; Cupera, J.; Záděra, A.; Jan, V.; Dlouhy, I. High-strength Al0. 2Co1. 5CrFeNi1. 5Ti high-entropy alloy produced by powder metallurgy and casting: A comparison of microstructures, mechanical and tribological properties. *Mater. Charact.* **2020**, *159*, 110046.
17. Schaffer, G. Powder processed aluminium alloys. *Mater. Forum* **2004**, *28*, 65–74.
18. Li, Y.; Ji, H.; Cai, Z.; Tang, X.; Li, Y.; Liu, J. Comparative study on constitutive models for 21-4N heat resistant steel during high temperature deformation. *Materials* **2019**, *12*, 893. [CrossRef]
19. Li, Y.; Ji, H.; Li, W.; Li, Y.; Pei, W.; Liu, J. Hot deformation characteristics—Constitutive equation and processing maps—of 21-4N heat-resistant steel. *Materials* **2019**, *12*, 89. [CrossRef]
20. Ji, H.; Huang, X.; Ma, C.; Pei, W.; Liu, J.; Wang, B. Predicting the microstructure of a valve head during the hot forging of steel 21-4N. *Metals* **2018**, *8*, 391. [CrossRef]
21. Wang, M.; Sun, H.; Zou, L.; Zhang, G.; Li, S.; Zhou, Z. Structural evolution of oxide dispersion strengthened austenitic powders during mechanical alloying and subsequent consolidation. *Powder Technol.* **2015**, *272*, 309–315. [CrossRef]
22. Susila, P.; Sturm, D.; Heilmaier, M.; Murty, B.S.; Subramanya Sarma, V. Microstructural studies on nanocrystalline oxide dispersion strengthened austenitic (Fe–18Cr–8Ni–2W–0.25Y2O3) alloy synthesized by high energy ball milling and vacuum hot pressing. *J. Mater. Sci.* **2010**, *45*, 4858–4865.
23. Phaniraj, M.P.; Kim, D.I.; Shim, J.H.; Cho, Y.W. Microstructure development in mechanically alloyed yttria dispersed austenitic steels. *Acta Mater.* **2009**, *57*, 1856–1864. [CrossRef]
24. Xu, Y.; Zhou, Z.; Li, M.; He, P. Fabrication and characterization of ODS austenitic steels. *J. Nucl. Mater.* **2011**, *417*, 283–285. [CrossRef]
25. Liu, D.H.; Yong, L.I.U.; Zhao, D.P.; Yan, W.; Fang, J.H.; Wen, Y.R.; Liu, Z.M. Effect of ball milling time on microstructures and mechanical properties of mechanically-alloyed iron-based materials. *Trans. Nonferrous Met. Soc. China* **2010**, *20*, 831–838. [CrossRef]
26. Tehrani, F.; Golozar, M.A.; Abbasi, M.H.; Panjepour, M. Characterization of nanostructured high nitrogen Fe–18Cr–xMn–4Mo austenitic stainless steel prepared by mechanical alloying. *Mater. Sci. Eng. A* **2012**, *534*, 203–208. [CrossRef]
27. Haghir, T.; Abbasi, M.H.; Golozar, M.A.; Panjepour, M. Investigation of α to γ transformation in the production of a nanostructured high-nitrogen austenitic stainless steel powder via mechanical alloying. *Mater. Sci. Eng. A* **2009**, *507*, 144–148. [CrossRef]
28. Amini, R.; Hadianfard, M.J.; Salahinejad, E.; Marasi, M.; Sritharan, T. Microstructural phase evaluation of high-nitrogen Fe–Cr–Mn alloy powders synthesized by the mechanical alloying process. *J. Mater. Sci.* **2009**, *44*, 136–148. [CrossRef]
29. Liu, D.; Xiong, Y.; Topping, T.D.; Zhou, Y.; Haines, C.; Paras, J.; Martin, D.; Kapoor, D.; Schoenung, J.M.; Lavernia, E.J. Spark plasma sintering of cryomilled nanocrystalline Al alloy-Part II: Influence of processing conditions on densification and properties. *Metall. Mater. Trans. A* **2012**, *43*, 340–350. [CrossRef]
30. Solay Anand, S.; Mohan, B. Effect of particle size, compaction pressure on density and mechanical properties of elemental 6061Al alloy through powder metallurgical process. *Int. J. Mater. Eng. Innov.* **2012**, *3*, 259–268. [CrossRef]
31. Kumar, N.; Arora, N.; Goel, S. Study on metallurgical and mechanical aspects of GMA welded nitronic steel under the influence of weld quenching. *J. Manuf. Processes* **2020**, *56*, 116–130. [CrossRef]
32. Xu, X.; Yu, Z.; Wang, J. Influences of microstructure on service properties of 5Cr21Mn9Ni4N heat resistant alloy. *Mater. Sci. Technol.* **2007**, *23*, 903–909. [CrossRef]
33. Weisser, M.A. Effect of Different Loading Conditions on the Accumulation of Residual Strain in a Creep Resistant 1% CrMoV Steel. Ph.D. Thesis, École Polytechnique Fédérale de Lausanne, Lausanne, Switzerland, 2013.
34. Ganesan, D.; Sellamuthu, P.; Prashanth, K.G. Vacuum Hot Pressing of Oxide Dispersion Strengthened Ferritic Stainless Steels: Effect of Al Addition on the Microstructure and Properties. *J. Manuf. Mater. Processing* **2020**, *4*, 93. [CrossRef]
35. Sato, K.; Saka, T.; Ohno, T.; Kageyama, K.; Sato, K.; Noda, T.; Okabe, M. *Development of Low-Nickel Superalloys for Exhaust Valves*; SAE Technical Paper; SAE: Warrendale, PA, USA, 1998.
36. Kumar, N.; Arora, N. Effect of solution treatment on slurry erosive wear performance of martensitic and nitrogen strengthened austenitic stainless steel. *Mater. Lett.* **2021**, *284*, 128932. [CrossRef]
37. Kunitskaya, I.; Spektor, Y.I.; Olshanetskii, V. Structural and kinetic features of dynamic recrystallization of alloyed austenite upon multipass hot deformation. *Met. Sci. Heat Treat.* **2012**, *53*, 498–502. [CrossRef]
38. Huang, X.; Wang, B.; Zang, Y.; Ji, H.; Guan, B.; Li, Y.; Tang, X. Constitutive relationships of 21-4 N heat-resistant steel for the hot forging process. *J. Mater. Res. Technol.* **2020**, *9*, 13575–13593. [CrossRef]
39. Srinivasan, D.; Corderman, R.; Subramanian, P. Strengthening mechanisms (via hardness analysis) in nanocrystalline NiCr with nanoscaled Y2O3 and Al2O3 dispersoids. *Mater. Sci. Eng. A* **2006**, *416*, 211–218. [CrossRef]
40. Broitman, E. Indentation hardness measurements at macro-, micro-, and nanoscale: A critical overview. *Tribol. Lett.* **2017**, *65*, 1–18. [CrossRef]
41. Lacy, C.; Gensamer, M. The tensile properties of alloyed ferrites. *Trans. AsM* **1944**, *32*, 88–110.
42. Chauhan, A.; Bergner, F.; Etienne, A.; Aktaa, J.; De Carlan, Y.; Heintze, C.; Litvinov, D.; Hernandez-Mayoral, M.; Oñorbe, E.; Radiguet, B.; et al. Microstructure characterization and strengthening mechanisms of oxide dispersion strengthened (ODS) Fe-9% Cr and Fe-14% Cr extruded bars. *J. Nucl. Mater.* **2017**, *495*, 6–19. [CrossRef]
43. Rajasekhara, S.; Ferreira, P.J.; Karjalainen, L.P.; Kyröläinen, A. Hall–Petch behavior in ultra-fine-grained AISI 301LN stainless steel. *Metall. Mater. Trans. A* **2007**, *38*, 1202–1210. [CrossRef]

44. Praud, M.; Mompiou, F.; Malaplate, J.; Caillard, D.; Garnier, J.; Steckmeyer, A.; Fournier, B. Study of the deformation mechanisms in a Fe–14% Cr ODS alloy. *J. Nucl. Mater.* **2012**, *428*, 90–97. [CrossRef]
45. Hin, C.; Wirth, B.D. Formation of Y_2O_3 nanoclusters in nano-structured ferritic alloys: Modeling of precipitation kinetics and yield strength. *J. Nucl. Mater.* **2010**, *402*, 30–37. [CrossRef]
46. Li, Q. Modeling the microstructure–mechanical property relationship for a 12Cr–2W–V–Mo–Ni power plant steel. *Mater. Sci. Eng. A* **2003**, *361*, 385–391. [CrossRef]

crystals

MDPI

Article

Understanding the Plastic Deformation of Gradient Interstitial Free (IF) Steel under Uniaxial Loading Using a Dislocation-Based Multiscale Approach †

Hao Lyu [1,*] and Annie Ruimi [2]

1 College of Transportation Engineering, Dalian Maritime University, Dalian 116026, China
2 Mechanical Engineering Program, Texas A&M University at Qatar, Doha P.O. Box 23874, Qatar; annie.ruimi@qatar.tamu.edu
* Correspondence: hao.lyu@dlmu.edu.cn
† Dedicated to the memory of Prof. Hussein M. Zbib, our mentor, colleague, and friend.

Abstract: Gradient interstitial free (IF) steels have been shown to exhibit a superior combination of strength and ductility due to their multiscale microstructures. The novelty of the work resides in the implementation of a modified slip transmission and a back-stress quantity induced by a long-range dislocation interaction in the dislocation-based multiscale model. This is an improvement over the model we previously proposed. Simulations are performed on IF specimens with gradient structures and with homogeneous structures. The macroscopic behavior of the samples under tension and compression is studied. The evolution of the microstructure such as dislocations, geometrically necessary dislocations (GNDs), and the effects of grain orientation is analyzed. Results show that with our enhanced model, the simulations can successfully reproduce the stress-strain curves obtained experimentally on gradient nano IF steel specimens under tension. The simulations also capture the tension-compression asymmetry (TCA) in specimens with homogeneous and gradient microstructures. The initial texture is found to have a significant effect on the TCA of specimens with gradient microstructures.

Keywords: multi-scale modeling; dislocations; gradient materials

Citation: Lyu, H.; Ruimi, A. Understanding the Plastic Deformation of Gradient Interstitial Free (IF) Steel under Uniaxial Loading Using a Dislocation-Based Multiscale Approach. *Crystals* **2022**, *12*, 889. https://doi.org/10.3390/cryst12070889

Academic Editor: Wojciech Polkowski

Received: 1 June 2022
Accepted: 19 June 2022
Published: 23 June 2022

1. Introduction

Gradient nanostructured materials are polycrystalline materials with microstructures containing grain sizes in the order of nanometers and exhibiting a gradient variation in grain size. Due to their outstanding combination of strength and ductility, these new kinds of materials have attracted tremendous interest over the few past decades and are ideal candidates for many practical applications [1–4]. Materials with different patterns of gradient microstructures have been successfully produced using severe plastic deformations (SPD) [5–14]. For instance, materials with nano-size grains on the surface and coarse grains in the center have been produced with the surface mechanical attrition treatment (SMAT) [2,15]. A 'bamboo' microstructure has been reported in [1,16] as successfully enhancing the mechanical properties of metals. (see for instance, [2,17] for their work on interstitial free (IF) steel, [18,19] for copper and [20] for magnesium). Other methods such as the high-pressure torsion method (HPT) have been employed to produce ultrafine-grained materials while maintaining their strength and ductility [9].

In experimental data on plastic deformations of various metals, evidence of a strong tension-compression asymmetry (TCA) behavior has been documented (i.e., see [21–24] for Mg alloys, [25,26] for body center cubic (BCC) alloys). The underlying mechanisms have been attributed to the activation of tension/compression twinning in hexagonal close-packed (HCP) alloys [27–29] or non-gliding forces contributing to the slip (aka non-Schmid effect) in BCC metals [30–32]. Recent experiments reported in [33] have also observed

TCA in gradient nano-structured copper as a consequence of the residual stress induced by surface mechanical grinding treatment. Similarly, TCA on iron-based metal has also been observed. For instance, in [34], a TCA behavior in dual-phase steel is reported while in [35], TCA in a 3D-printed stainless steel specimen is said to be induced by residual stress. TCA has also been observed for low-carbon steel and a correlation between the TCA and the loading direction has been demonstrated [36]. However, the underlying mechanisms leading to such asymmetry are not completely understood and the opportunity to propose an enhanced model which could explain how an optimal combination of strength and ductility can be obtained arose.

At the same time, a deeper understanding of the synergetic effects that multiple deformation mechanisms have at different scales is needed as is a thorough theoretical investigation of the underlying mechanisms responsible for the plastic behavior of materials with heterogeneous microstructures. Conventional or classical continuum plasticity (CCP) models describing plastic deformations of crystalline materials were popular in the last few decades. The basis for CCP relied on the assumption that by using an average grain size, an equivalent homogenized microstructure of strength depending on the average grain size (Hall–Petch relation) could be constructed [37,38]. However, these have been shown to lead to inaccurate predictions of the strength and ductility of the specimens with a high degree of spatial heterogeneity [39].

Subsequent improvements to CCP plasticity models based on gradient theories were proposed in the literature [40,41]. Among them, strain-gradient models assume that hardening results from the formation of geometrically necessary dislocations (GNDs) and statistically (or randomly) stored dislocations (SSDs). GNDs can accommodate the lattice curvature and act as barriers to gliding dislocations. SSDs arise as the result of plastic strains. They evolve from random trapping and are stored in the form of tangles, dipoles, etc. A summary of strain gradient plasticity models can be found in [42,43]. Stress-gradient plasticity models were introduced by Hirth in [44] to analyze pileup dislocations against grain boundaries and obstacles emitted from two sources. Using a continuum approach, Hirth [44] solved a singular integral equation with a kernel of Cauchy type on a finite interval. The idea was extended by Chakravarthy et al. and Liu et al. in [45,46] to capture how the flow stress depended on the obstacle spacing as in the Hall–Petch relation. In [47], Taheri et al. proposed a high-order stress gradient theory considering a general inhomogeneous state of stress. Stress and strain gradient models were first proposed independently but combining strain-gradient to stress-gradient models was effective in capturing the grain size effect in heterogeneous microstructures over a wide range of length scales [46]. This was done using a 2D Voronoi tessellation diagram, each Voronoi cell representing an individual grain.

In the present work, to consider the evolution of dislocations into the grain and the grain-grain interactions, we implement an enhanced version of the multiscale framework proposed by Zbib's group [39,48] based on a continuum dislocation dynamic model (CDD) coupled with a viscoplastic self-consistent (VPSC) model. In essence, in these models, CDD accounts for the evolution of dislocation density in polycrystals. The stress/strain state of individual grains is updated with this information and the plastic deformation of the entire sample subjected to external load can be predicted by the VPSC model. In [48], one of the essential features was to introduce the intrinsic length scale by means of the Nye's dislocation tensor. The effect of dislocations on flow stress was captured by incorporating GND and SSD densities into hardening laws or in the expression of the mean free path of gliding dislocations. This was shown to contribute to strain hardening [48,49]. In [39], implementing two material parameters (the grain size and the grain-size gradient) in the models lead to results closer to experimental data than when a Hall–Petch relation for homogeneous materials based on grain size only was used. In the dislocation-based theory by Taheri-Nassaj and Zbib [47], the dislocation pile-up against the grain boundary is shown to work as a strong barrier to stop the dislocation motion when the grain is under a homogeneous state of stress (Hall Petch effect), at the onset of the plastic deformation.

When the grain is subjected to an inhomogeneous state of stress, a stress gradient term that can capture the effect of dislocation pile up is introduced. In addition, the motion of the dislocations is affected by a long-range stress field (LRS, aka back-stress) caused by a group of dislocations in other grains or grain boundaries [50]. In the investigation of gradient nano-grained material, extra strengthening is attributed to the back-stress-induced GND pileups [14,41].

Regarding the role of grain boundaries in plastic deformations, they have been shown to play an important role in the dynamics of dislocations in polycrystalline materials. They can either act as barriers to the motion of dislocation [51–53] or emit dislocations across the grain boundaries [54]. Slip transfer through transparent, semi-transparent, and impenetrable grain boundaries are investigated. In a transparent grain boundary, slip transfer travels through directly without any strengthening effect. In contrast in [55], the motion of dislocation for an impenetrable grain boundary can be stopped by the resistance from the grain boundary and form pileups. Additionally, dislocations can cross the grain boundary by changing the Burgers vector and this process always includes absorption and nucleation of dislocations at the grain boundary. In the present work, we will employ and modify the dislocation flux model proposed by Hamid et al. in [55] to describe slip transmission across grain boundaries. The model is verified experimentally [55] with nano-indentation tests on two randomly selected adjacent grains and the results show that dislocation pileups between the indent and grain boundary propagate through the grain boundary.

In summary, the current dislocation-based multiscale approach proposed by Lyu et al. [39] can relate the macroscopic behavior of materials deforming plastically to their microstructures and it accounts for short-range interactions among dislocations (via the evolution equations), but not the effects of long-range interactions, including the long-range forces and associated back-stress. However, it is known that the magnitude of the back-stress is inversely proportional to the distance between a material point and the dislocations in neighboring grains, so the effect of the back-stress becomes more predominant with reducing the grain size and should be included. In this spirit, we improve on the model we previously proposed [39] by implementing a back-stress quantity in the simulations. The model for the back-stress is taken from [56,57] and it uses the Nye's dislocation tensor to describe the stress field arising from the continuous distribution of dislocations. Additional details are included in Section 2.2.

Section 2 provides the relevant details of the CDD-VPSC multiscale model as described in the work by [39,48]. Section 3 presents the results of the simulations as they pertain to the IF steel specimen. We show how the texture and gradient in texture affect the material's mechanical behavior. The models are implemented into a tension-compression loading scenario as we investigate if a TCA behavior resulting from various competing stresses can be captured. Results are compared with experimental data on TCA retrieved from the literature. A discussion of the result follows (Section 4) before we end with some conclusive statements (Section 5).

2. Materials and Methods

2.1. CDD-VPSC Model

As explained, the work presented here is based on the introduction of additional features in the multiscale theoretical framework we proposed earlier [48] and [39]. The multiscale framework relies on combining a continuum dislocation dynamic (CDD) with a visco-plastic self-consistent (VPSC) model, with the goal of understanding the grain size effect in materials with heterogeneous microstructures.

CDD-VPSC is a strain/stress gradient plasticity model which can capture the finite deformations of polycrystal materials represented by grains with certain orientations and volume fractions. At each time step, CDD tracks the evolution of the dislocation density in each grain under the stress state information provided by VPSC. This results in a critical resolved shear stress value and a dislocation density value (at this particular time) which is sent back to VPSC as it continues to iterate over the time domain until the average deviatoric stress of all grains equals the deviatoric stress of the effective medium. More details about the iterative process are contained in [58,59].

CDD and VPSC models are independent but they are connected via the Orowan relation, an expression introduced in the forties [60] relating the strain rate to the Burgers vector, the dislocation density, and the dislocation velocity. In the relation, dislocations refer to mobile dislocations. The dislocation velocity is a function of the critical resolved shear stress on a slip system needed to activate the dislocation motion. The critical shear stress is further decomposed into the sum of minimum stress to move a dislocation, a hardening term due to the dislocation-dislocation interaction, and a size-effect term describing the resistant effect arising from the dislocation pileups against the grain boundary. The hardening term is a function of statistically stored dislocation density, which is equal to the sum of mobile and immobile dislocation density. To capture the grain size effect, an intrinsic length scale was introduced within the geometrically necessary dislocation term considered in the mean free path of gliding dislocation (see Appendix A Equation (A3)) and a stress gradient term that accounts for the stress gradient aroused by non-uniform loading (see Appendix A Equation (A6)). Details of the formulation can be found in [39,48].

Briefly, in the CDD model [39], there are seven different dislocation-based mechanisms captured by seven distinct terms (see Equations (A1) and (A2) in Appendix A) which themselves differentiate the effects on mobile vs. immobile dislocations. The dislocation density on each slip system in one grain is assumed to be a scalar quantity. This is unlike the CDD model proposed by [61,62] in which the dislocation density is represented by a second-order tensor. In [39], the evolution of dislocation densities in each grain is represented by a set of nonlinear partial differential equations with seven coefficients obtained by fitting the evolution equation with discrete dislocation dynamics (DDD) and single-crystal tensile tests. In this work, we use the coefficients from [47,63].

In the VPSC model, each grain is treated as an ellipsoidal visco-plastic inclusion embedded in a homogenous effective matrix. The deformation of each grain is determined by rate-dependent crystal plasticity, which includes slip and twinning. VPSC model considers a dimensionless space with no spatial information of grains and domain size and is not intended to solve a boundary value problem (BVP). Using a self-consistent homogenization method, the macroscopic behavior of the polycrystal can be obtained by solving the stress equilibrium equation for the homogeneous medium. Details appear in Appendix B.

Our main contribution is the implementation of a back-stress and slip transmission quantity in the previous CDD-VPSC models we proposed in [39]. These two features are described in more detail below.

2.2. Back-Stress

In this work, we will use the back-stress model proposed by Akarapu and Zbib in [56,57], which uses the Nye's dislocation tensor to describe the stress field arising from the continuous distribution of dislocations. In the formulation, the shear stress field $\widetilde{\sigma}_{ij}$ is obtained by integrating Mura's formula (see [64]) as:

$$\widetilde{\sigma}_{ij}(\overline{x}) = C_{ijkl} \sum_{b=1}^{N'} \epsilon_{lnh} C_{pqmn} G_{kp,q}(\overline{x} - \overline{x}') A_{hm} V_b' \quad (1)$$

where ϵ_{lnh} is the permutation tensor, C is the stiffness matrix, $G_{kp,q}$ is the derivative of Green's functions, A_{hm} is the Nye's tensor at position $\overline{x}'$ and V_b' is the volume where the dislocations are homogenized (volume of a neighboring grain), N' is the total numbers of neighboring grains.

When the grain size was larger than 300 nm, we assumed isotropic conditions on the basis that the isotropic stress field in an array of dislocations converges to the anisotropic approximation when the distance between homogenized dislocations and the position of interest is greater than 80 Burgers vector [50]. The derivative of Green's functions can be approximated by [56]:

$$C_{pqmn}G_{kp,q}(\overline{x}-\overline{x}') = \frac{-1}{8\pi(1-\nu)}\left[(1-2\nu)\frac{\delta_{ni}(x_m - x'_m) + \delta_{im}(x_n - x'_n) + \delta_{mn}(x_i - x'_i)}{R^3} + 3\frac{(x_m - x'_m)(x_n - x'_n)(x_i - x'_i)}{R^5}\right] \quad (2)$$

where $R = \sqrt{(x_1 - x'_1)^2 + (x_2 - x'_2)^2 + (x_3 - x'_3)^2}$, ν is Poisson's ratio, δ Kronecker delta, $\overline{x}(x_1, x_2, x_3)$ refers to the position of interest and $\overline{x}'(x'_1, x'_2, x'_3)$ to the location where the dislocations are homogenized.

The stress field $\widetilde{\sigma}_{ij}(\overline{x})$ resolved on slip system α and τ^α is rewritten as the superposition of the resolved flow stress $\sigma_{ij}(\overline{x})$ and resolved back-stress $\widetilde{\sigma}_{ij}(\overline{x})$.

2.3. Slip Transmission

Three criteria are used to determine the occurrence of slip transfer: geometrical criteria [65], stress criteria [66], and dislocation density criteria [67].

In [29,65], the degree of coplanarity of slip systems is expressed by the mean of a parameter $m' = \left(\vec{n}_{in} \cdot \vec{n}_{out}\right)\left(\vec{d}_{in} \cdot \vec{d}_{out}\right)$ written as a function of the slip normal $\vec{n}$ and the slip direction $\vec{d}$. The parameter m' is a scalar number between 0 and 1. The upper bound 1 signifies that the grain boundary is transparent and the lower bound 0 means that the grain boundary is impenetrable. If the normal to the slip plane or the slip directions are in different directions, m' can take on a negative value. However, slip transfer cannot occur. A schematic of the slip transfer across grain boundary is depicted in Figure 1.

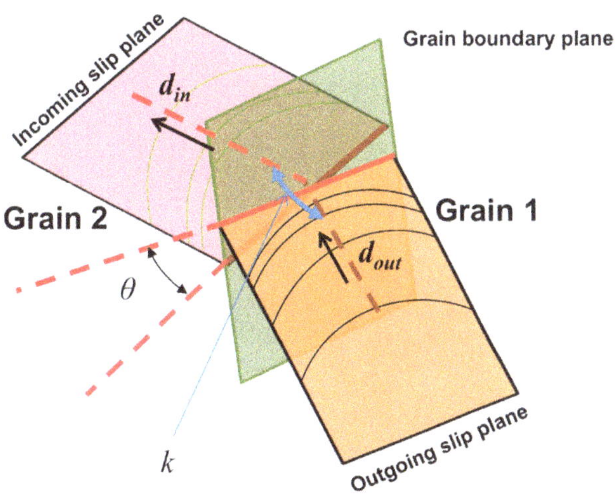

Figure 1. Schematic of slip transmission across grain boundary (θ is the angle between the normal direction of the incoming slip plane and the outgoing slip plane; k represents the angle between the incoming slip direction d_{in} and the outgoing slip direction d_{out}).

There are two additional requirements. First, the resistance from the grain boundary must be overcome. In this work as in [65], the resistance is expressed as

$$\tau_{GB} = (1 - m')\tau_S \quad (3)$$

where τ_S is the stress gradient term. τ_S can be approximated by $\frac{K}{\sqrt{L}}\left(1 + \frac{L}{4\overline{\tau}}|\nabla\overline{\tau}|\right)$, where K is Hall–Petch constant, L is grain size, and $\nabla\overline{\tau}$ is the gradient of effective stress. Secondly, the dislocation density in the outgoing slip system must be greater than in the incoming slip system. Once the above conditions are satisfied, the dislocation flux $\dot{\rho}^{\alpha}_{flux}$ between two grains can be measured using the following equation

$$\dot{\rho}^{\alpha}_{flux} = p\overline{v}^{\alpha} \cdot \nabla \rho^{\alpha}_{M} \cong p\overline{v}^{\alpha}_{g} \frac{\rho^{\alpha(in)}_{M} - \rho^{\alpha(out)}_{M}}{R} \tag{4}$$

where R is the distance between two neighboring grains, p is a parameter used by Shi and Zirky [67], which can be expressed as

$$p = \frac{m'_{\alpha\beta}\left(\frac{\tau^{\beta}_{out}}{\tau^{\beta}_{C}}\right)}{\sum_{i=1}^{N'}\left(m'_{\alpha i}\frac{\tau^{i}_{out}}{\tau^{i}_{C}}\right)} \tag{5}$$

in which $m'_{\alpha\beta}$ is the geometrical parameter between two neighboring grains mentioned above and N' is the number of all possible outgoing slip between two neighboring grains.

2.4. Implementation

To capture the strain/stress quantities and the dislocation density and to measure the strain and stress gradient as the result of the specimens deforming plastically, we use a spatial representative domain to represent the polycrystalline material. The domain is discretized into grains with spatial, size, and neighboring information using a 2D Voronoi tessellation diagram. Each Voronoi cell represents a grain with a stress state, from which the strain and dislocation densities are computed via the CDD-VPSC model. Then, the stress and strain fields (spatial) gradients can be numerically approximated using a moving least square method. Details on the technique can be found in [68].

We implemented the proposed models on a four-grain sample and verified that slip transmission and back-stress were captured accordingly. Figure 2 shows the stress-strain curves obtained from the simulations of the sample with—Σ5 grain boundary (see grain 1 and grain 3), which accounts for grain slip transmission and impenetrable grain boundary (between grain 1 and grain 3). As expected, transmission occurs from slip system 13 in grain 1 to slip system 24 in grain 3 with a transferred mobile dislocation density in the order of 10^{12} m^{-2}. The mobile dislocation density after slip transmission (red line) shows that the value in grain 1 is greater than for grain 3, after dislocation flux. Depending on the grain orientation, slip transmission may occur from grain 1 to grain 3 with further loading. The stress-strain curves also indicate a softening behavior. A possible explanation is that slip transmission induces more plasticity in comparison to the impenetrable boundary case because there is no resistance from dislocation pileups against grain boundaries.

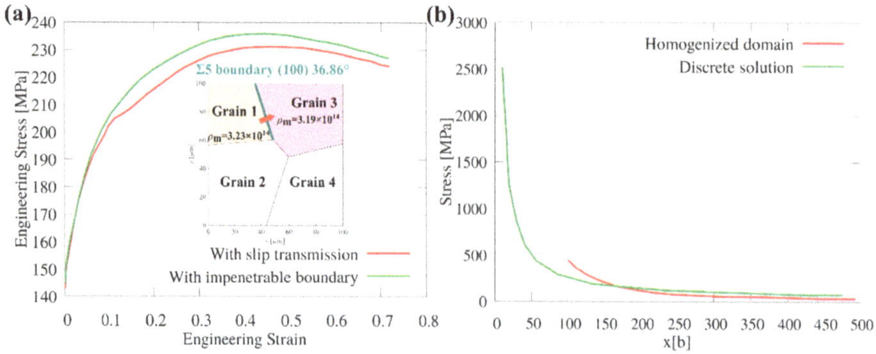

Figure 2. (**a**) Verification of the slip transmission model on a four-grain sample; (**b**) Verification of the back-stress model on a two-grain sample with a tilt grain boundary.

The implementation of the back-stress model was verified by a set-up of two grains with tilt boundary (an array of edge dislocations). Then the shear stress was evaluated and compared to the discrete solution of the stress field around an infinite edge dislocation. This is shown in Figure 2b. The results show that the solution for the shear stress converges when x is close to the source of the dislocation (x < 150b) but that the simulations closely reproduce the discrete solution otherwise.

2.5. Samples Generation

To generate samples representing specimens with homogenous and gradient structures, we first obtain 2D Voronoi tessellations using NEPER, a specialized software package for polycrystal generation [69]. Then, we employed a grain growth algorithm embedded in the NEPER code to obtain the desired gradient structure. Details of the procedure can be found in [70–72].

Figure 3 depicts samples with constructed gradient structures and cells size ranging from hundreds of nanometers to tens of microns. Using 35 μm as the average grain size, specimens with homogeneous structures and randomly assigned crystal orientations (i.e., homogeneous texture) were constructed (Figure 4a). The grain size was chosen to be the same as the one used by Wu et al. [2] in their experiments. The equivalent grain size distribution along the y axis for specimens with homogeneous structures and gradient structures are shown in Figure 3c,d, respectively.

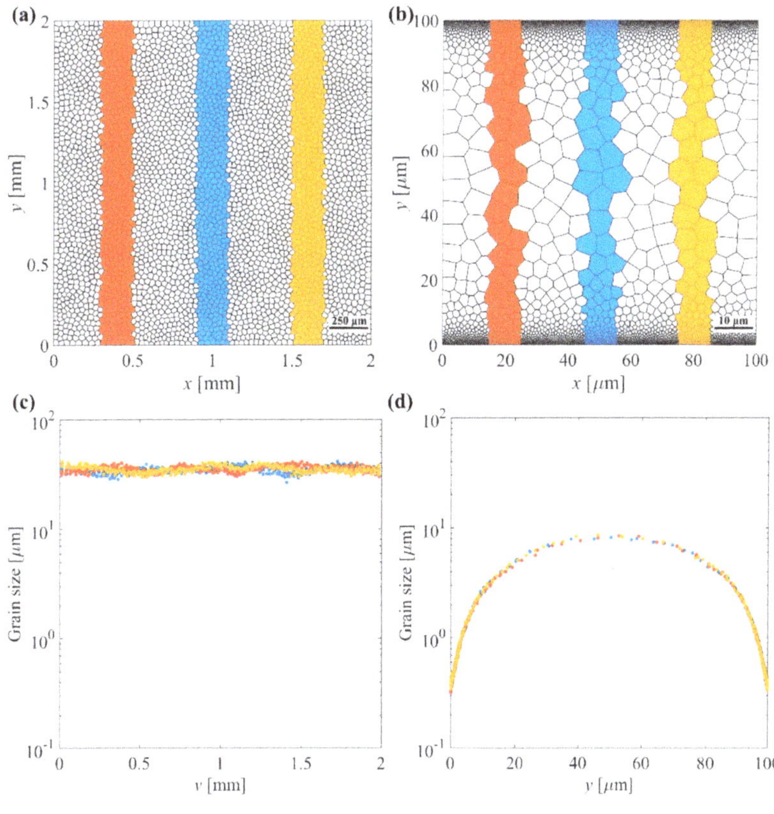

Figure 3. Constructed structures with 35 μm average grain size for specimens with (**a**) homogeneous structure and (**b**) gradient structure. Corresponding grain size distribution along the y axis for (**c**) homogeneous structure and (**d**) gradient structure.

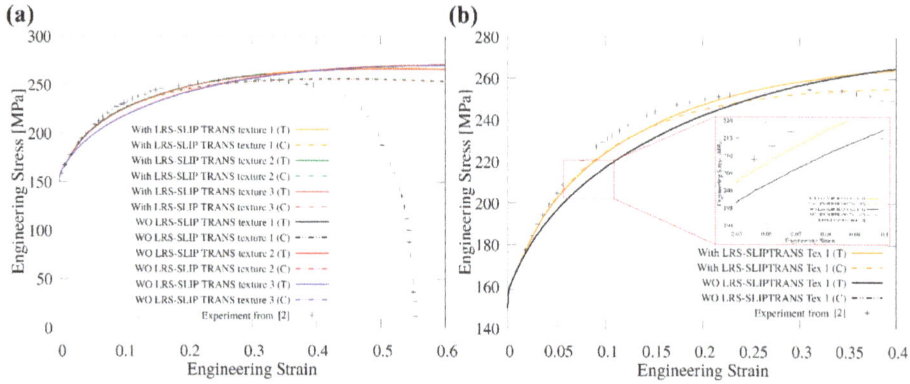

Figure 4. Predicted stress-strain curves of IF steel specimens with homogeneous structures. Results are shown for three initial textures. (**a**) Without (WO) and with back-stress (LRS) and slip transmission (SLIP TRANS) included; (**b**) Comparison of simulation results with and without LRS and SLIP TRANS. T denotes tension; C denotes compression.

3. Results

After implementing the back-stress and slip transmission in the multiscale CDD-VPSC model, simulations were performed on IF steel specimens with homogeneous microstructures with three different randomly assigned crystallographic orientations. The parameters used in the simulations were taken from [39] and are summarized in Table 1 below.

Table 1. Parameters used in the simulations.

Symbol	IF Steel (Unit)
c^* (Bailey–Hirsh hardening coefficient)	0.4
τ_0 (Internal friction) on [1 1 2]	11 MPa
τ_0 (Internal friction) on [1 1 0]	27.5 MPa
τ_0 (Internal friction) on [1 2 3]	25 MPa
C_{11} (Elasticity constant)	242 GPa
C_{12} (Elasticity constant)	150 GPa
C_{44} (Elasticity constant)	112 GPa
μ (Shear modulus)	80 GPa
K (Hall–Petch constant)	0.18 MPa/mm$^{-1/2}$
v_0 (Reference strain rate)	1×10^{-5} m/s
m (Strain rate sensitivity)	0.05
b (Magnitude of burger vector)	2.54 Å
R_c (Critical radius for annihilation coefficient)	15 b
q_1 q_2 q_3 q_4 q_5 q_6 q_7	0.02 1.0 0.002 0.002 0.018 0.001 0.1
Ω_{ij} (i = 1,48; j = 1,48) (Interaction matrix)	0.5

Using the neighboring information and the information about the grain orientation, the distribution of the misorientation angle was plotted (see Figure A1 in Appendix C). The distribution of misorientation angles in the specimens with three different textures is somehow identical.

Figure 4 shows the predicted stress-strain curves of the IF steel specimens with homogeneous structures in tension and compression. Results are shown for three textures and compared when back-stress (LRS) and slip transmission (SLIP TRANS) are not considered. Experimental data on tensile tests obtained by Wu [2] on IF steel specimens with homogenous structures and 35 μm average grain size is also indicated.

Figure 4a shows that the stress-strain curve obtained from the simulations closely captures the mechanical behavior observed experimentally by [2], regardless of the initial texture. Upon implementing the LRS and SLIP TRANS in the model, the TCA of the tensile strength can be observed at the instability point. Otherwise, TCA is barely noticeable for the three different cases (see Figure 4b). This also shows that implementing the back-stress and slip transmission yields a higher strain hardening rate at the beginning and results in a tension-compression asymmetry. Figure 4b shows that deactivating the back-stress and slip transmission leads to a much lower initial strain hardening but yields a higher tensile strength and higher ductility values than the ones obtained experimentally. TCA of flow stress can be observed at a strain of around 1.5%, which is measured by using

$$\mathrm{TCA} = \frac{2(|\sigma_C| - |\sigma_T|)}{(|\sigma_C| + |\sigma_T|)} \tag{6}$$

One can also find slightly higher flow stress under compression than under tension test before 1%. Higher flow stress under compression (positive TCA) was also captured by experimental tests of low-carbon steel [36] but with a much more pronounced asymmetry. Although the role of the initial texture and the microstructure on TCA was not completely understood by [36], they hinted that the loading direction was the reason for the behavior. In Figure 4a, the tensile strength is 266 MPa at 0.55 strain (approximately). This is 11 MPa larger than the strength obtained in a compression test at a 0.45 strain (approximately). These values refer to the homogeneous cases for which the initial random grain orientation has no effect. That explains why we have one tensile value and one compression value for the three different initial textures with different random grain orientations.

Because our multiscale model, as proposed, does not consider the reduction of the cross-sectional area, the simulations yielded a smaller slope in the true stress-strain curve and consequently, the onset of instability from the simulations is seen to appear later than in the experimental curves. In future work, we plan to consider the decrease in the cross-sectional area in the model to improve the accuracy. Similarly, the predicted stress-strain curves on IF steel specimens with gradient structures when back-stress and slip transmission are implemented in the model are shown in Figure 5. Using a zoomed view at the onset of instabilities for each of the three initial textures, we measured the tensile strength and the strain at the instability point. Then we calculated an average value of tensile strength and position of instability for the three textures. They are shown in Table 2.

The values show that tension-compression asymmetry (TCA) of flow-stress exists and depends on the texture. However, the tensile strength value varies by less than 5 MPa for the three textures. In addition, in comparison to specimens with homogenous structures, specimens with gradient structures have a higher strength and strain hardening.

When compared to experimental data obtained by Wu et al. [2] for tests on homogenous specimens, one can see that our simulation results can predict the stress-strain curves of the tensile test of IF steel with the same average grain size, in trend if not values. Even though the simulation values for specimens with gradient structures (Figure 5a) show that they exceed those recorded in the experiments, the trend of the stress-strain curves is similar to the macroscopic behavior of grain size gradient IF steel under the tensile test given in [2]. In addition, the grain size spatial distribution and the size of the specimens used in the simulations differ from the experimental samples. The strain-stress curves in Figure 5a (all textures combined) show that the tension-compression asymmetry (TCA) of stress is negligible at early strain stages and that the texture plays a role in the TCA of the flow stress. Texture increases strength and strain hardening, while it was inferred from Figure 3 that for homogenous structures, it did not. Figure 5b–d show that the specimens with texture 1 and texture 3 have a flow stress under compression that is consistently larger than the one in tension. In the specimens with texture 3, the flow stress in compression is always larger than that in tension. For the specimens with texture 2, the initial strain hardening rate in compression is higher than in tension, until approximately a 10 % strain after which the strain hardening decreases leading to a lower tensile strength. This agrees with what is shown in experiments in [36], where before 10% strain, higher flow stress in compression is recorded. The onset of instability for a gradient structure occurs earlier than in a homogeneous structure. Depending on the initial texture, the onset of instabilities in gradient structures varies from 0.2~0.25 (strain).

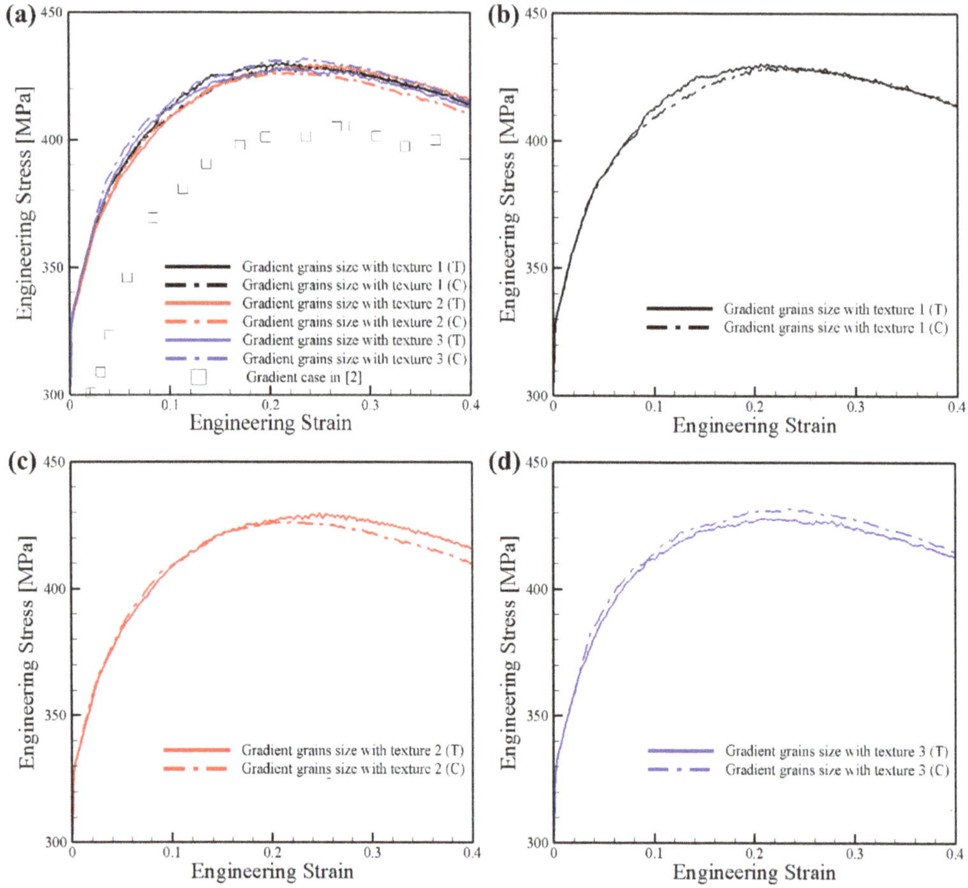

Figure 5. Predicted stress-strain curves of IF steel specimens with gradient structures in tension and compression after implementing back-stress and slip transmission in the model. (**a**) three textures combined; (**b**) texture 1; (**c**) texture 2; (**d**) texture 3.

Table 2. Average tensile strength and ductility for specimens with gradient structures in tension and compression.

		Measure 1 (Tensile Strength in MPa/Ductility)	Measure 2 (Tensile Strength in MPa/Ductility)	Measure 3 (Tensile Strength in MPa/Ductility)	Average Value (Tensile Strength in MPa/Ductility)
Texture 1	T	428/0.25	427/0.26	427/0.25	427/0.25
	C	430/0.22	429/0.22	429/0.21	429/0.22
Texture 2	T	428/0.25	428/0.26	429/0.26	428/0.26
	C	425/0.22	426/0.21	425/0.21	425/0.21
Texture 3	T	427/0.24	427/0.24	427/0.24	427/0.24
	C	430/0.24	431/0.23	431/0.23	431/0.23

Figure 6 shows the texture evolution of specimens with gradient structures in tension and compression. The texture evolution tracks the change of the misorientation angle with loading. Results are shown for two textures and compared to their respective initial texture. Results are recorded at 15 % strain. For both cases, the evolved texture shows the distribution of misorientation angle for a random texture without exhibiting a strong texture. In Figure 6a pertaining to specimens with texture 1, the evolution trend is similar in tension and compression. In specimens with texture 2 (Figure 6b), there is almost no change in the fraction of misorientation when the sample is under compression.

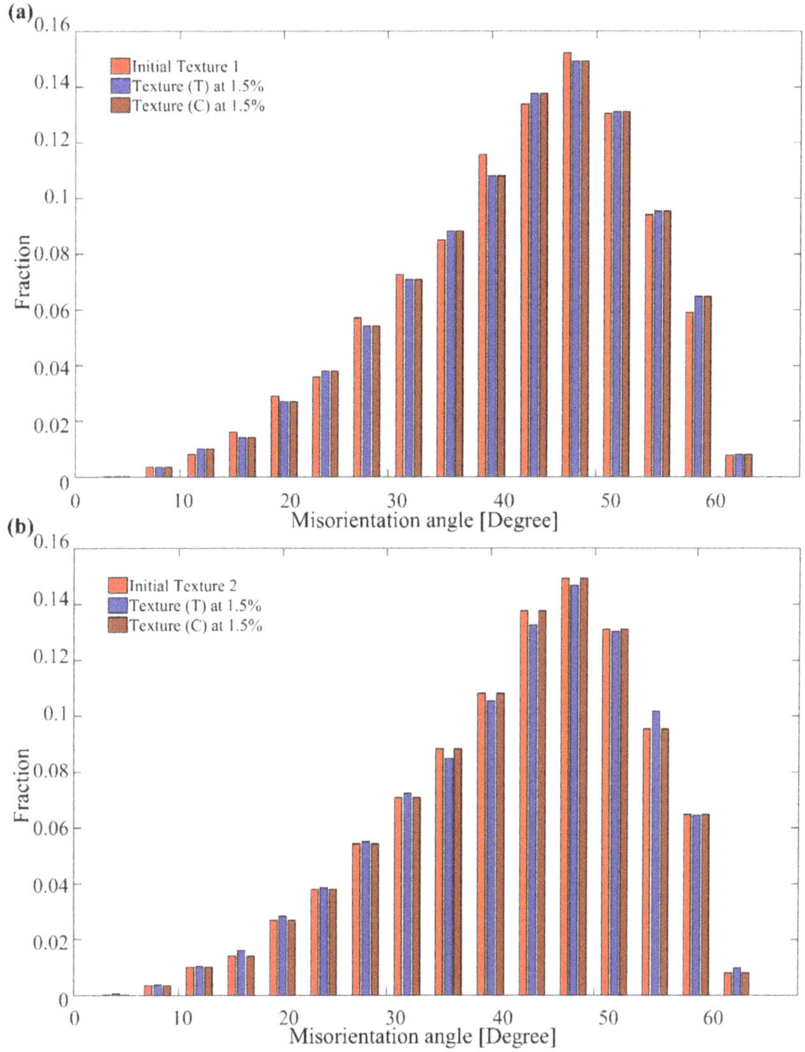

Figure 6. Comparison of misorientation angle distribution for specimens with gradient structures in tension and compression for two textures, (**a**) texture 1 and (**b**) texture 2. The corresponding initial texture is also shown. Results are shown for IF steel specimens at 15% strain.

Figure 7a shows the average mobile dislocation density vs. equivalent strain curves for IF steel specimens in tension and compression. The average mobile dislocation density is measured using the total dislocation density in the polycrystal sample divided by the total number of grains. Results are shown for three initial textures. Results on specimens with gradient structures are compared to the reference cases, for specimens with homogeneous structures. The curves show that at early strain stages, the mobile dislocation density increases much faster for the homogeneous cases than for the gradient structures. With further straining, the trend reverses, i.e., the average mobile dislocation density in specimens with gradient structures surpasses that of the homogeneous structures.

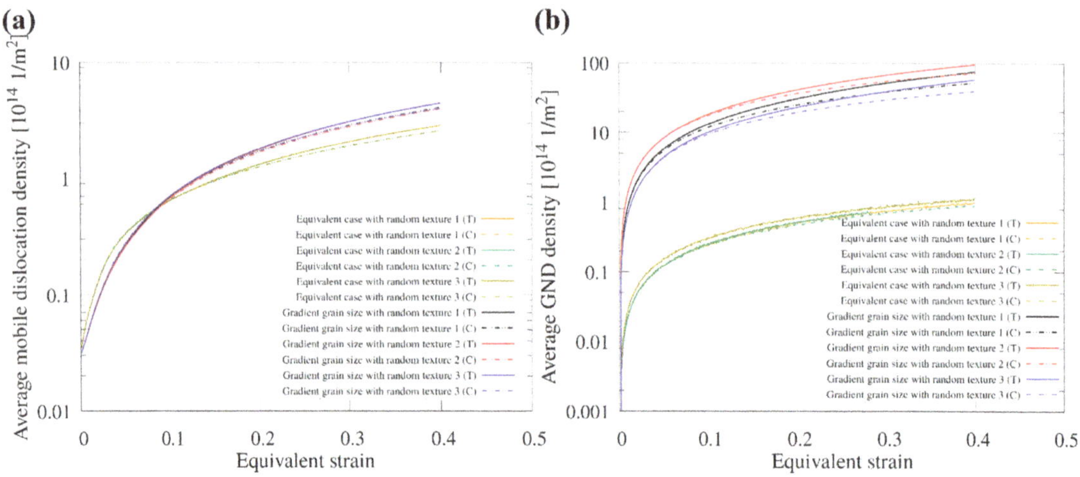

Figure 7. (**a**) Average mobile dislocation density vs. equivalent strain and (**b**) average GND density vs. equivalent strain, for IF specimens in tension and compression with three initial textures. Results are compared to samples with homogenous structures.

Figure 7a shows that there is a slight difference in the evolution of the average mobile dislocation density depending on if the specimens are in tension or compression for samples with homogeneous and gradient structures. However, the GND density for samples with gradient structures in tension is much higher than for specimens in compression (Figure 7b). This contrasts with the GND density of samples with homogeneous structures, which are similar in tension and compression.

Regarding the effect of initial texture, Figure 7a shows that the average mobile dislocation density does not depend on the initial texture, and that is true for both the homogeneous and the gradient cases. However, one can infer from the figure that the local plastic deformation and dislocation density evolution depend on the initial texture. This is due to the slip transmission and back-stress. This means that even though all polycrystal samples have the same total dislocation density, different textures will result in different local strain distributions and strain gradients. This can also be seen from the evolution of GND density shown in Figure 7b which shows that the average GND density vs. equivalent strain curves varies depending on the initial texture. Here, the assumption was that the initial GND density was 0 before it increased with straining. Additionally, the average GND densities for specimens with gradient structures are significantly higher than for those with homogeneous microstructures.

4. Discussion

We discuss the effects of implementing slip transmission and back-stress in the model as they pertain to IF steel specimens with homogeneous and gradient structures, in tension and compression.

4.1 Homogeneous vs. Gradient Structures

Figure 8 shows how the microstructure affects the GND density and the stress gradient quantities. Specimens with homogenous structures have a narrow unimodal grain size distribution. As a result, the critical resolved shear stress needed to activate the slip is the same for each grain and all grains experience the same amount of deformation simultaneously. Specimens with gradient structures have bimodal or multimodal grain size distributions, with large and small grains. The dislocation motion is more difficult to activate in small grains with a size of 100 nm compared to in large grains with a size of 10 µm. Small grains contribute more to the strengthening of the polycrystal materials and large grains endure more plastic deformation. In homogeneous structures, all grains are in equal-axial shape and have the same opportunities to activate dislocation motions. This leads to a relatively uniform deformation which further lowers the GND densities and rises the stress gradient due to dislocation pileups against the grain boundary. This is shown in Figure 8a. In this work, the stress gradient was approximated using the moving least square method (see Section 2.4). Considering that there is less activation of slips in the finer grain region, one would expect a lower average mobile dislocation density in the gradient structure at the early strain stage (5%).

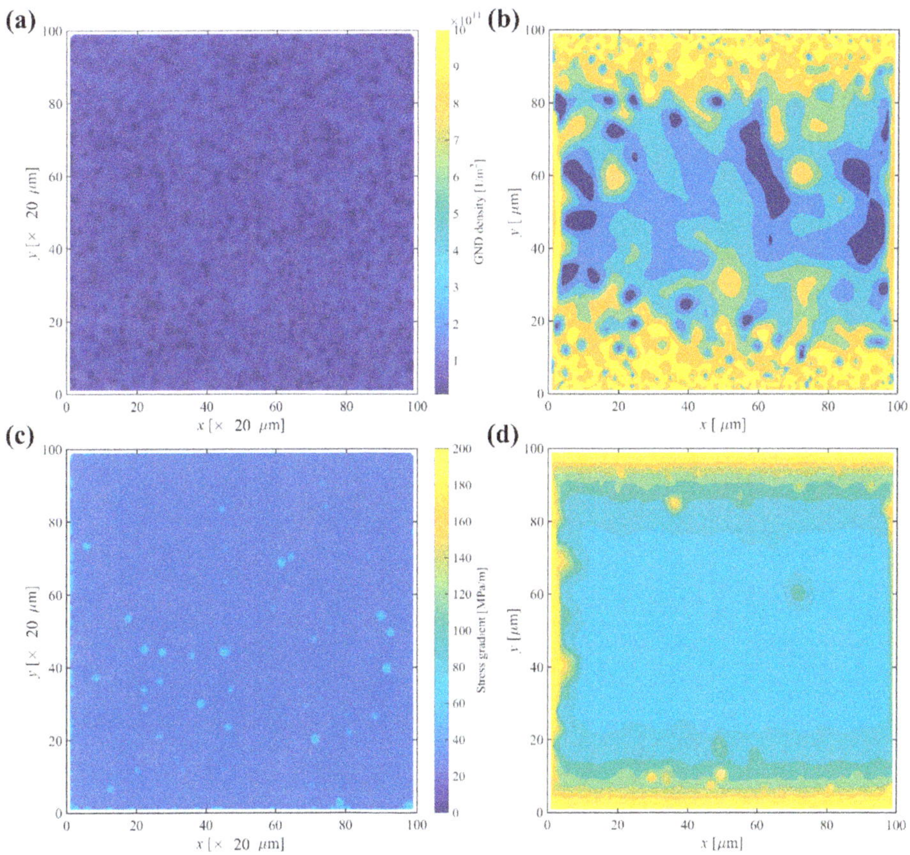

Figure 8. GND density in (**a**) homogenous structures and (**b**) gradient structures. Stress gradient in (**c**) homogenous structures and (**d**) gradient structures. The results are recorded on IF steel specimens at 5% strain.

As a result, the inhomogeneous deformation between small (hard) grains and large (soft) grains increases the accumulation of GNDs which explains the higher average GND density in Figure 7b. At the early strain stage, these GNDs burst which prevents the motion of dislocations significantly and strengthens the polycrystal material. When the GND rate starts dropping (≅5% strain), the average mobile dislocation density in gradient structures surpasses that of homogeneous structures. When the local stress state increases, more dislocation activities are triggered in the small grains. The stress gradient term also plays an important role in preventing slip activation. Figure 8d shows that the stress in gradient structures is almost three times higher than in homogeneous structures (Figure 8c). In gradient structures (Figure 8d), the stress gradient increases along the y-direction, not in a random fashion.

Figure 9 shows how the stress gradient varies between two strain stages. Results are shown for gradient structures with texture 2 in tension and compression. Four grains undergoing slip transmission have been chosen and labeled in Figure 9. As can be seen from Figure 8a, the stress gradient due to dislocation pileups can be significantly relieved by the slip transmission from grains A, B, and C to the neighboring grains. Further straining will decrease the stress gradient. Grain D experiences an increase in stress gradient, which makes the small grain even harder to deform and causes local stress concentrations. This is in contrast to what happens during compression: when the sign of the resolved back-stress is altered and superposed to the resolved stress afterward, the value of resolved shear stress is lower than the critical resolved shear stress, and slip transmission cannot be triggered. Thus, grains A, B, and C experience only a minimal increase in stress gradient.

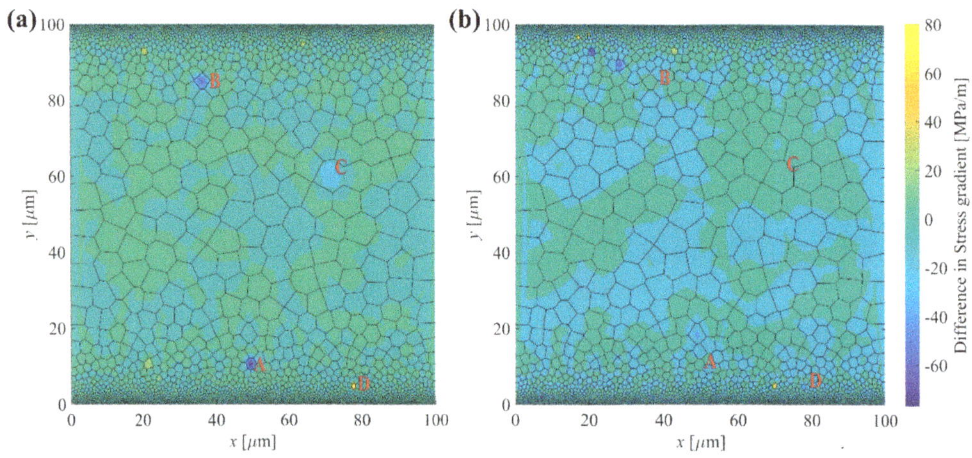

Figure 9. Difference in stress gradient at 5.0% and 5.1% strain for specimens with gradient structures (texture 2) under (**a**) tension and (**b**) compression. Points A to D (marked in red) are grains that endure significant stress gradient change and slip transmission.

We note that for smaller grains such as grain D, a higher stress gradient could be induced due to small grain size (see the top and bottom surface in Figure 8d). In such a case, the slip transmission can occur due to the back-stress and the local grain orientation. When more slip transmissions occur in the nanograin (NG) region, a reduction of macroscopic flow-stress can ensue. Thus, for a homogeneous structure with random texture, the TCA is independent of the initial texture and the local grain orientation. For a gradient structure with random texture, TCA is determined by the local grain orientation and the grain size.

4.2. Slip Transmission and Relation to Tensile Strength and Ductility

Table 3 shows the number of slip transmissions that occurred in the entire sample, in the transient region, and the nano-grain region for specimens with homogeneous and gradient structures, for three textures. Table 3 also lists the average values of the tensile strength and ductility for the various cases, recorded at 20% strain. As explained previously, these are average values taken over three measurements.

Table 3. Number of slips per region, tensile strength, and ductility in specimens with homogeneous and gradient structures, under tension and compression, for three initial textures. Results are recorded at 20 % strain.

			Number of Slip Transmission in the Entire Sample	Number of Slip Transmission in the Transient Region (y > 80 μm and y < 20 μm)	Number of Slip Transmission Which Occurred in the Nano Region (y > 90 μm and y < 10 μm)	Tensile Strength (MPa)	Ductility
Texture 1	Homogeneous	T	82,336	31,599	15,638	266	0.55
		C homogeneous structure	100,840	38,866	19,234	255	0.45
	Gradient	T	1564	435	19	427	0.25
		C homogeneous structure	1736	287	0	429	0.22
Texture 2	Homogeneous	T	79,896	31,193	15,754	266	0.55
		C homogeneous structure	110,350	43,863	23,049	255	0.45
	Gradient	T	1490	448	13	428	0.26
		C homogeneous structure	2036	583	3	425	0.21
Texture 3	Homogeneous	T	80,524	32,788	16,716	266	0.55
		C homogeneous structure	101,996	41,455	22,417	255	0.45
	Gradient	T	1978	464	55	427	0.24
		C homogeneous structure	2196	568	8	431	0.23

The reason why the sum of the second and third column values does not equal the value of the first column is that we did not list the number of slip transmissions in the coarse-grain region. In addition, since homogenous structures have no transient nor nano-grain regions, we used the number of slip transmissions in the relative region. For example, if the number of slip transmissions that occurred in the nano-region—y [0.95, 1] and [0, 0.05] μm for a gradient case, we compared the number of slip transmissions that occurred in the same area—y [0.95, 1] and [0, 0.05]μm for the homogeneous case.

4.3. Slip Transmission in Homogeneous vs. Gradient Structures

The values of Table 3 show that a significantly smaller number of slip transmissions occur in specimens with gradient structures than in specimens with homogeneous structures. Dislocations in small grains within the nano-region (region with grains of nano-size. i.e., for which y > 90 μm and y < 10 μm) can barely travel through the grain boundary, regardless of the misorientation. That is because the stress gradient (refer to Figure 8d) arising in the region of nano-size grains is at least twice as much greater than that in the region of coarse grains. Even when the geometrical factor is satisfied, dislocations still need to overcome the huge resistance from dislocation pileups and grain boundaries. Most slip transmissions occur in the center of the specimens and are locally dependent on the inhomogeneous deformation.

4.4. Slip Transmission in Compression vs. Tension

The values also show that more slip transmission occurs in compression than in tension. Slip transmission leads to the reduction of pileups and therefore reduces the barrier for dislocation motion causing further softening. This could explain why the tensile strength is higher than that in compression for a homogeneous structure. For the homogenous structure, ductility is lower in compression than in tension. A possible reason may be that slip transmission affects the total dislocation density indirectly, so more slip transmission does not necessarily result in a higher total mobile dislocation density (refer to Figure 7a, which shows the mobile dislocation density vs. strain). When looking at the number of slip transmissions that occurred during the entire loading process, we find that although more slip transmission occurs in a homogeneous structure under compression, the mobile dislocation density under tension is much higher, possibly because tension activates more slips.

In contrast, for specimens with gradient structures, the probability to trigger slip transmission varies with the region. In particular, slip transmission is more difficult in the nano region (y > 90 μm and y < 10 μm), which possesses very fine grains of size around 200 nm. For the gradient structures, the flow stress under compression is higher than under tension at the early hardening stage (less than 20%) (refer to Figure 5). The occurrence of TCA at the early stage could be attributed to less slip transmission in the nano region (see Table 3) and fewer activated slips due to grain size. However, subsequent deformation and onset of instability are complicated by the redistribution of strain/stress occurring at the later strain stage. As a result, there is not a clear correlation between tensile strength value and slip transmission. Yet, the number of slip transmissions in the nano-region affects the onset of instability in gradient structures. Less slip transmission in the nano-region results in earlier instability (low ductility).

4.5. Tension-Compression Asymmetry (TCA)

Tension-compression asymmetry in IF steel specimens have been evaluated in relation to their microstructure, texture, back-stress, and subsequent slip/slip transmission. In this work, and unlike in [59,60], we did not consider non-gliding forces, so we assumed that the non-glide slip was not responsible for TCA. With the bursting of GNDs due to non-uniform straining, the magnitude of the resolved-back-stress increases. Depending on the superposition of the resolved back-stress and the resolved stress, slips and slips transmission can be triggered.

Figure 10 shows the resolved back-stress in specimens with gradient structures under tension and compression. Results are recorded at 2.5 % strain. The values of the back-stress vary between ±200 MPa., while the critical shear stress (CRSS) is in the range of [90, 500] MPa. These values are the output of the simulations for slip system 16 for all grains at the 2% strain stage. Simulations track CRSS values on any slip system in any grain. This indicates that the resolved back-stress can aid or prevent the slip and the slip transmission, leading to a different strain hardening at the macroscopic scale.

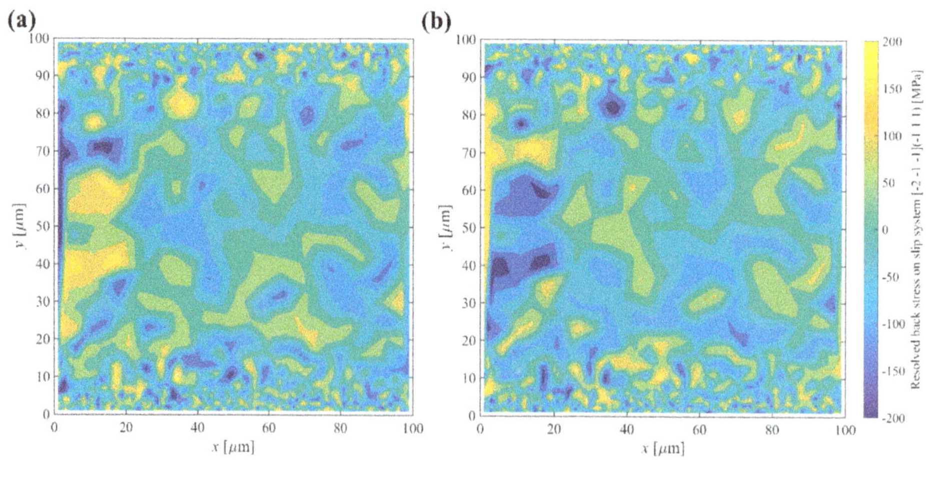

Figure 10. Resolved back-stress on slip 16 for specimens with gradient structures under (**a**) tension and (**b**) compression. Results are shown at 2.5% strain.

The values of tensile strength and ductility in Table 3 show that gradient structures exhibit much less tension-compression asymmetry than homogeneous structures. One reason may be that homogeneous structures have grain size in the order of 1μm so slips and slip transmissions occur more easily. Also, all equal axial grains have the same chance to deform, resulting in a more uniform deformation than for gradient structures. Thus, the magnitude of GND density and stress gradient in homogeneous specimens can be expected to be much lower than for a gradient specimen. Slip and slip transmission can easily be activated with low GND density and stress gradient. The critical resolved shear stress on slip system [-2-1-1] (-111) in homogeneous structures is around 70 MPa. This is much lower than the critical resolved back-stress (in order of 100 MPa) at approximately 0.1 % strain. Again, these values are retrieved as the output of our simulations, which tracks CRSS for any slip system in any grain. For homogeneous structures, the back-stress is negligible, initially. For gradient structures, the magnitude of the resolved shear stress (see Figure 10) is comparable to the critical resolved shear stress. The larger resolved back-stress is a result of non-uniform straining (formed GNDs) and the grain size, which can alter the activation of slips and slip transmissions depending on the sign.

With the results of the simulations, we were able to show that the local orientation played a key role in plastic deformation and macroscopic behavior of IF steel specimens. For samples with homogeneous structures, TCA does not depend on the initial texture if the texture has a random orientation. For samples with gradient structures, TCA is more sensitive to the local texture and particularly for the grains in the nano-region, even though the distribution of misorientation angles of three different samples are almost identical.

Producing a gradient microstructure using severe plastic deformation (SPD) such as SMAT will inevitably introduce a strong texture or grain orientation gradient into the texture of the sample [73]. To show this, we conducted another simulation using a texture gradient similar to the one reported in [74] for a low carbon steel specimen. Because the sample size differs from [74], we scaled the thickness of different regions with different fractions of the {110} textures The results are shown in Figure 11a and point that- an earlier onset of TCA occurs at the strain 5% and the value of TCA increases with further loading for the specimens with gradient grain size and gradient texture. This indicates that the local grain orientation in the nano layer and the gradient layer has a significant effect on the TCA behavior. Thus, a deeper investigation of the effect of local texture and texture gradient on deformation as well as a better understanding of the synergetic effects between the spatial distribution of grain size and grain orientation on the material macroscopic behavior will eventually lead to the development of new gradient materials.

Comparing the simulation results with the experimental data on low carbon steel by Koizumi et al. [36], Figure 11b shows that the predicted TCA value is lower and even in a more pronounced way, for the random texture. This can be attributed to the assumption of texture as well as the initial dislocation density and GND density.

Koizumi et al. have also shown that the loading direction (in the initial texture) could affect the magnitude and trend of the TCA in flow stress. Details of sample preparation are not given in [36], but residual stresses will likely exist in raw samples with no heat treatment. In the work by Chen et al. [35] on 3D printed stainless steel, there is a significant decrease in TCA when the sample is annealed. This could be explained by the release of residual stress in the homogeneous texture. Our simulations on specimens with gradient structures show that the TCA behavior is captured, and although the magnitude of TCA is much lower than the data recorded in experiments, it exhibits the same trend, i.e., a small hump at the initial strain stage followed by increasing TCA. The low initial TCA value predicted by the simulations could be attributed to a low dislocation capacity of the fine grains at the top and bottom surface. These fine grains can bare almost no plastic deformation and as a result, a stress concentration arises at the surface, regardless of the grain orientation. However, the strong texture in the transient and in the central region can lead to a very different deformation gradient, which could in turn affect the TCA.

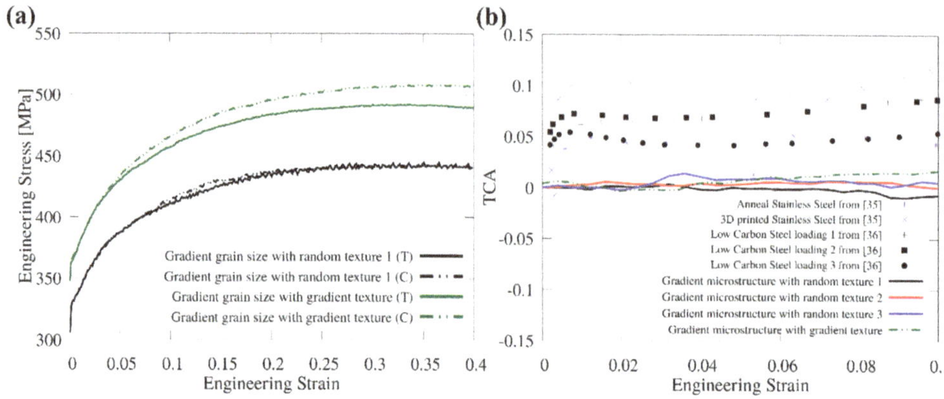

Figure 11. (**a**) Stress-strain curve of gradient grain size with random texture and gradient texture (**b**) TCA in flow stress vs. engineering strain and comparison with experimental data. Solid lines are simulation results; Dot plots in (**b**) are generated using experimental data from [35,36]. Blue dots are from tension/compression tests of 3D-printed stainless steel with and without heat treatment [35]; black data are experimental results of tension and compression tests by varying loading directions [36], 1-RD, 2–45°, 3-TR.

5. Conclusions

In this work, we improved on the multi-scale CDD-VPSC framework we previously proposed [39,68] by implementing a back-stress and slip transmission quantity in the constitutive model. With that, our simulations predicted the deformation of IF steel specimens under tension and compression, with three initial textures and in two specific grain-size regions (transient and nano). Comparisons between specimens with homogeneous and gradient structures were made. The following conclusions can be inferred from this study:

(1) Specimens with gradient texture cause deformation gradient along the grain size gradient direction. Plastic deformation occurs first in the coarse grains in the center region. Then it gradually expands to the transient and the nano region. Local inhomogeneous deformations were more easily induced for microstructures with a bi-or multi-modal grain size distribution and gradient size distribution along the y-direction.
(2) There is no clear correlation between slip transmission and tensile strength. More slip transmission in the nano region could explain the delay in the onset of instability.
(3) The back-stress quantity which arises from dislocations and slip transmission causes tension-compression asymmetry (TCA) behavior. TCA exists in specimens with homogeneous microstructure and with gradient microstructure. The simulations show that the predicted TCA values are lower than recorded experimentally. The initial texture of the specimens plays a predominant role in TCA.

Author Contributions: Methodology, A.R. and H.L.; validation, H.L.; formal analysis, H.L. and A.R.; writing—original draft preparation, H.L.; writing—review and editing, A.R.; visualization, H.L.; funding acquisition, H.L. All authors have read and agreed to the published version of the manuscript.

Funding: This research was funded by the National Natural Science Foundation of China, grant number 52101008, the Fundamental Research Funds for the Central Universities, grant number 3132022176, and the Start-up foundation of Xinghai talent plan-Dalian Maritime University, grant number 02502421.

Institutional Review Board Statement: Not applicable.

Informed Consent Statement: Not applicable.

Data Availability Statement: Not applicable.

Conflicts of Interest: The authors declare no conflict of interest.

Appendix A

The evolution of the dislocation density in the CDD model can be expressed by Equation (A1) for mobile dislocation ($\dot{\rho}_M^\alpha$) and by Equation (A2) for immobile dislocation ($\dot{\rho}_{IM}^\alpha$)

$$\begin{aligned}\dot{\rho}_M^\alpha = q_1\rho_M^\alpha \frac{\overline{v}_g^\alpha}{\tilde{l}_g^\alpha} - q_2 2R_c\left(\rho_M^\alpha\right)^2 v_g^\alpha - q_3\rho_M^\alpha \frac{\overline{v}_g^\alpha}{\tilde{l}_g^\alpha} + q_4\left(\frac{|\tau^\alpha|}{\tau_{cr}^\alpha}\right)^r \rho_{IM}^\alpha \frac{\overline{v}_g^\alpha}{\tilde{l}_g^\alpha} + \\ q_5\sum_{\beta=1}^N P^{\alpha\beta}\rho_M^\beta \frac{\overline{v}_g^\alpha}{\tilde{l}_g^\alpha} - q_6 R_c\rho_M^\alpha \rho_{IM}^\alpha \overline{v}_g^\alpha - q_7 R_c^3 \rho_M^\alpha \left(\rho_{IM}^\alpha\right)^2 \overline{v}_g^\alpha\end{aligned} \tag{A1}$$

$$\dot{\rho}_{IM}^\alpha = q_3\rho_M^\alpha \frac{\overline{v}_g^\alpha}{\tilde{l}_g^\alpha} - q_4\left(\frac{|\tau^\alpha|}{\tau_{cr}^\alpha}\right)^r \rho_{IM}^\alpha \frac{\overline{v}_g^\alpha}{\tilde{l}_g^\alpha} - q_6 R_c\rho_M^\alpha \rho_{IM}^\alpha \overline{v}_g^\alpha + q_7 R_c^3 \rho_M^\alpha \left(\rho_{IM}^\alpha\right)^2 \overline{v}_g^\alpha \tag{A2}$$

In Equations (A1) and (A2), $q_1 \sim q_7$ are coefficients obtained by fitting the evolution equation with discrete dislocation dynamics and single-crystal tensile tests. Terms 1 and 3 describe the increase of the mobile dislocation density due to multiplication and mobilization of immobile dislocation respectively. The annihilation of two mobile dislocations is captured in term 2. Mobile dislocation can also interact with other defects, leading to a reduction of mobile dislocations. The immobilization of moving dislocation, absorption of mobile dislocation by grain boundary, and dipole formation are considered in terms 3, 6, and 7. In this work, we used the coefficients from [47,63]).

$\tilde{l}_g^\alpha$ is the mean free path of mobile dislocation on slip system α, which can be expressed as

$$\tilde{l}_g^\alpha = \frac{c}{\sqrt{\sum_\beta W^{\alpha\beta}\left(\rho_M^\beta + \rho_{IM}^\beta + \left|\left|\rho_{\mathrm{GND}}^\beta\right|\right|\right)}} \tag{A3}$$

where c is a numerical constant in the order of 10, $W^{\alpha\beta}$ is a weight matrix similar to $\Omega^{\alpha\beta}$. $\left|\left|\rho_{\mathrm{GND}}^\beta\right|\right|$ is the effective density of GNDs on slip system β written as:

$$\left|\left|\rho_{GND}^\beta\right|\right| = \frac{1}{b}\sqrt{AA} \tag{A4}$$

where A is the Nye's tensor [74] that can be approximated by $A \approx \mathrm{curl}\,(-F^p) \cong \mathrm{curl}\,(-D^p)$, where $D^p = \sum_i^N \gamma^{(i)} s^{(i)} \otimes n^{(i)}$ [75–77], and where the derivative of D^p over space is approximated by a moving least square method.

τ_{cr}^α is the critical resolved shear stress needed to activate the dislocation motion. It can be decomposed as

$$\tau_{cr}^\alpha = \tau_0^\alpha + \tau_{\mathrm{H}}^\alpha + \tau_{\mathrm{S}}^\alpha \tag{A5}$$

where τ_0^α is an internal friction term that describes the minimum stress to move a dislocation, τ_{H}^α is a hardening term, which is dependent on current total dislocation density. The last term τ_{S}^α is a size-effect term, which describes the resistant effect arising from the dislocation pileups against the grain boundary. In the present work, we implement in our framework a simplified linear version of the stress gradient model such as

$$\tau_{\mathrm{S}}{}^\alpha = \frac{K}{\sqrt{L}}\left(1 + \frac{L'}{4\overline{\tau}}|\nabla\overline{\tau}|\right) \tag{A6}$$

where K is Hall–Petch constant, $\nabla\overline{\tau}$ is the spatial gradient of effective stress, L stands for the grain size, and L' is the average length of dislocation obstacles spacing. L' is also treated as the grain size.

R_c is the critical radius for reactions set to be 15 times the Burgers vector, r is 0.5, and $P^{\alpha\beta}$ is a cross-slip probability matrix describing the probability of screw dislocations cross-slip from slip system β to slip system α. When the Burgers vector on α-system and

β-system are parallel, cross slip is possible and can be determined stochastically using a Monte-Carlo analysis [77].

Appendix B

In the visco-plasticity self-consistent (VPSC) model, each grain is treated as an ellipsoidal visco-plastic inclusion embedded in a homogenous effective matrix. The deformation of each grain is determined by rate-dependent crystal plasticity including slip and twinning. The constitutive behavior at the local level is written as

$$\dot{\varepsilon}_{ij}(\overline{x}) = \dot{\gamma}_0 \sum_{\alpha=1}^{N} m_{ij}^{(\alpha)} \left(\frac{m_{kl}^{(\alpha)} \sigma_{kl}(\overline{x})}{\tau_{cr}^{(\alpha)}} \right)^n \tag{A7}$$

where $\dot{\gamma}_0$ is the reference strain rate, $m_{ij}^{(\alpha)}$ is the Schmid tensor associated with slip systems, n is a strain rate sensitivity exponent equal to $1/m$.

The linearized form of Equation (A9) I inside the domain of grain (r) is:

$$\dot{\varepsilon}_{ij}(\overline{x}) = M_{ijkl}^{(r)} \sigma_{kl}(\overline{x}) + \dot{\varepsilon}_{ij}^{0(r)} \tag{A8}$$

where $M_{ijkl}^{(r)}$ is the viscoplastic compliance and $\dot{\varepsilon}_{ij}^{0(r)}$ is the back-extrapolated term of grain (r). Similarly, the average strain rate is related to the stress in each grain via the expression

$$\dot{\varepsilon}_{ij} = M_{ijkl} \sigma_{kl} + \dot{\varepsilon}_{ij}^{0} \tag{A9}$$

At the polycrystal level, the relation between the macroscopic strain rate and the stress is

$$\dot{E}_{ij} = \overline{M}_{ijkl} \sum\nolimits_{kl} + \dot{E}_{ij}^{0} \tag{A10}$$

where $\overline{M}_{ijkl}$ and $\dot{E}_{ij}^{0}$ are respectively the macroscopic viscoplastic compliance and the back-extrapolated term. Invoking the concept of equivalent inclusion [63], the local constitutive behavior can be rewritten in terms of macroscopic compliance with the inhomogeneity included in an Eigenstrain rate term $\dot{\varepsilon}_{ij}^{*}$, as

$$\dot{\varepsilon}_{ij} = \overline{M}_{ijkl} \sigma_{kl} + \dot{E}_{ij}^{0} + \dot{\varepsilon}_{ij}^{0} \tag{A11}$$

Thus, the interaction of local and the macroscopic level can be expressed as

$$(\dot{\varepsilon}_{ij} - \dot{E}_{ij}) - \dot{\varepsilon}_{ij}^{0} = \overline{M}_{ijkl} \left(\sigma_{kl} - \sum\nolimits_{kl} \right) \tag{A12}$$

Appendix C

Additional supporting figures for the misorientation angle distribution for three initial textures.

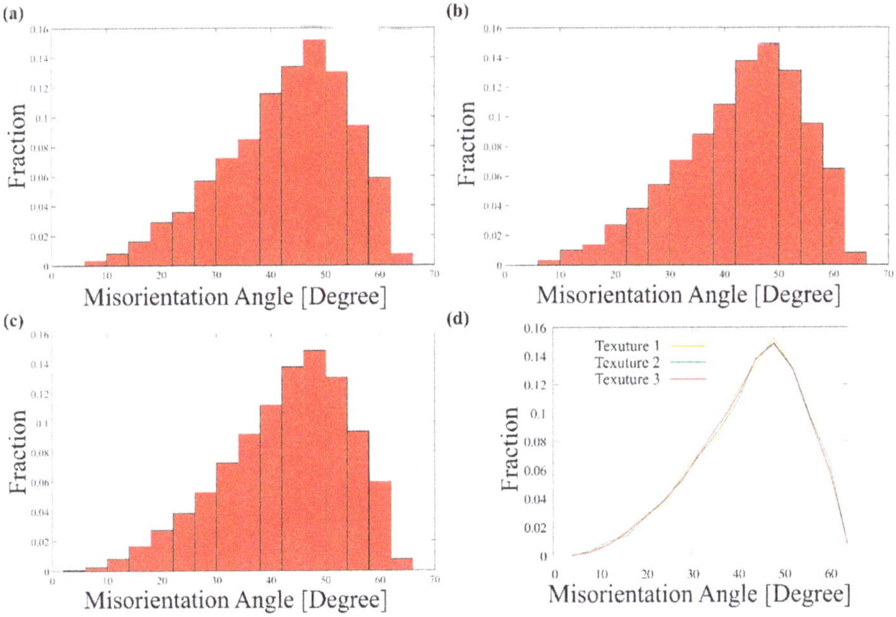

Figure A1. Effect of initial texture on distribution of misorientation angle (**a**) texture 1; (**b**) texture 2; (**c**) texture 3; (**d**) all combined.

References

1. Lu, K. Making strong nanomaterials ductile with gradients: Microstructures that increase metal crystallite size from nanoscale with surface depth are both strong and ductile. *Science* **2014**, *345*, 1455–1456. [CrossRef] [PubMed]
2. Wu, X.; Jiang, P.; Chen, L.; Yuan, F.; Zhu, Y.T. Extraordinary strain hardening by gradient structure. *Proc. Natl. Acad. Sci. USA* **2014**, *111*, 7197–7201. [CrossRef] [PubMed]
3. Shao, C.W.; Zhang, P.; Zhu, Y.K.; Zhang, Z.J.; Tian, Y.Z.; Zhang, Z.F. Simultaneous improvement of strength and plasticity: Additional work-hardening from gradient microstructure. *Acta Mater.* **2018**, *145*, 413–428. [CrossRef]
4. Wu, X.L.; Yang, M.X.; Yuan, F.P.; Chen, L.; Zhu, Y.T. Combining gradient structure and TRIP effect to produce austenite stainless steel with high strength and ductility. *Acta Mater.* **2016**, *112*, 337–346. [CrossRef]
5. Wu, Y.; Guelorget, B.; Sun, Z.; Déturche, R.; Retraint, D. Characterization of gradient properties generated by SMAT for a biomedical grade 316L stainless steel. *Mater. Charact.* **2019**, *155*, 109788. [CrossRef]
6. Wang, L.; Li, B.; Shi, Y.; Huang, G.; Song, W.; Li, S. Optimizing mechanical properties of gradient-structured low-carbon steel by manipulating grain size distribution. *Mater. Sci. Eng. A* **2019**, *743*, 309–313. [CrossRef]
7. Shi, Y.; Li, B.; Gao, F.; Wang, L.; Qin, F.; Liu, H.; Li, S. An outstanding synergy of high strength and ductility in gradient structured low-carbon steel. *Materialia* **2019**, *5*, 100181. [CrossRef]
8. Jamalian, M.; Hamid, M.; De Vincentis, N.; Buck, Q.; Field, D.P.; Zbib, H.M. Creation of heterogeneous microstructures in copper using high-pressure torsion to enhance mechanical properties. *Mater. Sci. Eng. A* **2019**, *756*, 142–148. [CrossRef]
9. Kang, J.Y.; Kim, J.G.; Park, H.W.; Kim, H.S. Multiscale architectured materials with composition and grain size gradients manufactured using high-pressure torsion. *Sci. Rep.* **2016**, *6*, 26590. [CrossRef]
10. Ma, E.; Zhu, T. Towards strength–ductility synergy through the design of heterogeneous nanostructures in metals. *Mater. Today* **2017**, *20*, 323–331. [CrossRef]
11. Li, J.; Cao, Y.; Gao, B.; Li, Y.; Zhu, Y. Superior strength and ductility of 316L stainless steel with heterogeneous lamella structure. *J. Mater. Sci.* **2018**, *53*, 10442–10456. [CrossRef]
12. Li, J.; Gao, B.; Huang, Z.; Zhou, H.; Mao, Q.; Li, Y. Design for strength-ductility synergy of 316L stainless steel with heterogeneous lamella structure through medium cold rolling and annealing. *Vacuum* **2018**, *157*, 128–135. [CrossRef]
13. Wang, P.; Xiang, Y.; Wang, X.; Liu, Z.; Qu, S.; Zhuang, Z. New insight for mechanical properties of metals processed by severe plastic deformation. *Int. J. Plast.* **2019**, *123*, 22–37. [CrossRef]
14. Wang, Y.F.; Wang, M.S.; Fang, X.T.; Guo, F.J.; Liu, H.Q.; Scattergood, R.O.; Huang, C.X.; Zhu, Y.T. Extra strengthening in a coarse/ultrafine grained laminate: Role of gradient interfaces. *Int. J. Plast.* **2019**, *123*, 196–207. [CrossRef]

15. Roland, T.; Retraint, D.; Lu, K.; Lu, J. Fatigue life improvement through surface nanostructuring of stainless steel by means of surface mechanical attrition treatment. *Scr. Mater.* **2006**, *54*, 1949–1954. [CrossRef]
16. Yang, M.; Pan, Y.; Yuan, F.; Zhu, Y.; Wu, X. Back stress strengthening and strain hardening in gradient structure. *Mater. Res. Lett.* **2016**, *4*, 145–151. [CrossRef]
17. Yang, X.; Ma, X.; Moering, J.; Zhou, H.; Wang, W.; Gong, Y.; Tao, J.; Zhu, Y.; Zhu, X. Influence of gradient structure volume fraction on the mechanical properties of pure copper. *Mater. Sci. Eng. A* **2015**, *645*, 280–285. [CrossRef]
18. Fang, T.H.; Li, W.L.; Tao, N.R.; Lu, K. Revealing extraordinary intrinsic tensile plasticity in gradient nano-grained copper. *Science* **2011**, *331*, 1587–1590. [CrossRef]
19. Yan, X.; Meng, X.; Luo, L.; Jing, Y.; Yi, G.; Lu, J.; Liu, Y. Mechanical behaviour of AZ31 magnesium alloy with the laminate and gradient structure. *Philos. Mag.* **2019**, *99*, 3059–3077. [CrossRef]
20. Kale, C.; Turnage, S.; Avery, D.Z.; El Kadiri, H.; Jordon, J.B.; Solanki, K.N. Towards dynamic tension-compression asymmetry and relative deformation mechanisms in magnesium. *Materialia* **2020**, *9*, 100543. [CrossRef]
21. Ning, J.L.; Xu, B.; Feng, Y.L.; Li, X.D.; Li, X.K.; Tong, W.P. Tension–Compression Yield Asymmetry Influenced by the Variable Deformation Modes in Gradient Structure Mg Alloys. *Acta Metall. Sin. (Engl. Lett.)* **2020**, *33*, 252–266. [CrossRef]
22. Lin, J.B.; Ren, W.J.; Wang, X.Y.; Ma, L.F. Tension–compression asymmetry in yield strength and hardening behaviour of as-extruded AZ31 alloy. *Mater. Sci. Technol.* **2016**, *32*, 1855–1860. [CrossRef]
23. Habib, S.A.; Khan, A.S.; Gnäupel-Herold, T.; Lloyd, J.T.; Schoenfeld, S.E. Anisotropy, tension-compression asymmetry and texture evolution of a rare-earth-containing magnesium alloy sheet, ZEK100, at different strain rates and temperatures: Experiments and modeling. *Int. J. Plast.* **2017**, *95*, 163–190. [CrossRef]
24. Kim, J.Y.; Greer, J.R. Tensile and compressive behavior of gold and molybdenum single crystals at the nano-scale. *Acta Mater.* **2009**, *57*, 5245–5253. [CrossRef]
25. Alkan, S.; Sehitoglu, H. Non-Schmid response of Fe3Al: The twin-antitwin slip asymmetry and non-glide shear stress effects. *Acta Mater.* **2017**, *125*, 550–566. [CrossRef]
26. Xing, Z.; Fan, H.; Tang, J.; Wang, B.; Kang, G. Molecular dynamics simulation on the cyclic deformation of magnesium single crystals. *Comput. Mater. Sci.* **2021**, *186*, 110003. [CrossRef]
27. Kurukuri, S.; Worswick, M.J.; Tari, D.G.; Mishra, R.K.; Carter, J.T. Rate sensitivity and tension-compression asymmetry in AZ31B magnesium alloy sheet. *Philos. Trans. R. Soc. A Math. Phys. Eng. Sci.* **2014**, *372*, 20130216. [CrossRef]
28. Ahmadikia, B.; Kumar, M.A.; Beyerlein, I.J. Effect of neighboring grain orientation on strain localization in slip bands in HCP materials. *Int. J. Plast.* **2021**, *144*, 103026. [CrossRef]
29. Wang, Z.Q.; Beyerlein, I.J. An atomistically-informed dislocation dynamics model for the plastic anisotropy and tension-compression asymmetry of BCC metals. *Int. J. Plast.* **2011**, *27*, 1471–1484. [CrossRef]
30. Cho, H.; Bronkhorst, C.A.; Mourad, H.M.; Mayeur, J.R.; Luscher, D.J. Anomalous plasticity of body-centered-cubic crystals with non-Schmid effect. *Int. J. Solids Struct.* **2018**, *139–140*, 138–149. [CrossRef]
31. Gröger, R.; Vitek, V. Impact of non-Schmid stress components present in the yield criterion for bcc metals on the activity of {110}<111> slip systems. *Comput. Mater. Sci.* **2019**, *159*, 297–305. [CrossRef]
32. Long, J.; Pan, Q.; Tao, N.; Lu, L. Residual stress induced tension-compression asymmetry of gradient nanograined copper. *Mater. Res. Lett.* **2018**, *6*, 456–461. [CrossRef]
33. Maeda, T.; Noma, N.; Kuwabara, T.; Barlat, F.; Korkolis, Y.P. Experimental Verification of the Tension-Compression Asymmetry of the Flow Stresses of a High Strength Steel Sheet. *Procedia Eng.* **2017**, *207*, 1976–1981. [CrossRef]
34. Chen, W.; Voisin, T.; Zhang, Y.; Florien, J.B.; Spadaccini, C.M.; McDowell, D.L.; Zhu, T.; Wang, Y.M. Microscale residual stresses in additively manufactured stainless steel. *Nat. Commun.* **2019**, *10*, 4338. [CrossRef]
35. Koizumi, T.; Kuroda, M. Evaluation of tension-compression asymmetry of a low-carbon steel sheet using a modified classical compression test method. *J. Phys. Conf. Ser.* **2018**, *1063*, 012167. [CrossRef]
36. Lehto, P.; Remes, H.; Saukkonen, T.; Hänninen, H.; Romanoff, J. Influence of grain size distribution on the Hall-Petch relationship of welded structural steel. *Mater. Sci. Eng. A* **2014**, *592*, 28–39. [CrossRef]
37. Li, J.; Weng, G.J.; Chen, S.; Wu, X. On strain hardening mechanism in gradient nanostructures. *Int. J. Plast.* **2017**, *88*, 89–107. [CrossRef]
38. Lyu, H.; Hamid, M.; Ruimi, A.; Zbib, H.M. Stress/strain gradient plasticity model for size effects in heterogeneous nano-microstructures. *Int. J. Plast.* **2017**, *97*, 46–63. [CrossRef]
39. Ma, A.; Roters, F.; Raabe, D. A dislocation density based constitutive model for crystal plasticity FEM including geometrically necessary dislocations. *Acta Mater.* **2006**, *54*, 2169–2179. [CrossRef]
40. Zhao, J.; Lu, X.; Yuan, F.; Kan, Q.; Qu, S.; Kang, G.; Zhang, X. Multiple mechanism based constitutive modeling of gradient nanograined material. *Int. J. Plast.* **2020**, *125*, 314–330. [CrossRef]
41. Xiao, X.; Chen, L.; Yu, L.; Duan, H. Modelling nano-indentation of ion-irradiated FCC single crystals by strain-gradient crystal plasticity theory. *Int. J. Plast.* **2019**, *116*, 216–231. [CrossRef]
42. Fleck, N.A.; Muller, G.M.; Ashby, M.F.; Hutchinson, J.W. Strain gradient plasticity: Theory and experiment. *Acta Metall. Mater.* **1994**, *42*, 475–487. [CrossRef]
43. Hirth, J.P. Dislocation pileups in the presence of stress gradients. *Philos. Mag.* **2006**, *86*, 3959–3963. [CrossRef]
44. Chakravarthy, S.S.; Curtin, W.A. Stress-gradient plasticity. *Proc. Natl. Acad. Sci. USA* **2011**, *108*, 15716–15720. [CrossRef]

45. Liu, D.; He, Y.; Zhang, B.; Shen, L. A continuum theory of stress gradient plasticity based on the dislocation pile-up model. *Acta Mater.* **2014**, *80*, 350–364. [CrossRef]
46. Taheri-Nassaj, N.; Zbib, H.M. On dislocation pileups and stress-gradient dependent plastic flow. *Int. J. Plast.* **2015**, *74*, 1–16. [CrossRef]
47. Lyu, H.; Ruimi, A.; Zbib, H.M. A dislocation-based model for deformation and size effect in multi-phase steels. *Int. J. Plast.* **2015**, *72*, 44–59. [CrossRef]
48. Lyu, H.; Taheri-Nassaj, N.; Zbib, H.M. A multiscale gradient-dependent plasticity model for size effects. *Philos. Mag.* **2016**, *96*, 1883–1908. [CrossRef]
49. Akarapu, S. *Dislocation Interactions with Interfaces*; Washington State University: Washington, DC, USA, 2009.
50. Lim, L.C.; Raj, R. Continuity of slip screw and mixed crystal dislocations across bicrystals of nickel at 573 K. *Acta Metall.* **1985**, *33*, 1577–1583. [CrossRef]
51. Zaefferer, S.; Kuo, J.C.; Zhao, Z.; Winning, M.; Raabe, D. On the influence of the grain boundary misorientation on the plastic deformation of aluminum bicrystals. *Acta Mater.* **2003**, *51*, 4719–4735. [CrossRef]
52. Priester, L. *Grain Boundaries: From Theory to Engineering*; Springer: Berlin/Heidelberg, Germany, 2013; Volume 172, ISBN 978-94-007-4968-9.
53. Kacher, J.; Eftink, B.P.; Cui, B.; Robertson, I.M. Dislocation interactions with grain boundaries. *Curr. Opin. Solid State Mater. Sci.* **2014**, *18*, 227–243. [CrossRef]
54. Hamid, M.; Lyu, H.; Schuessler, B.J.; Wo, P.C.; Zbib, H.M. Modeling and characterization of grain boundaries and slip transmission in dislocation density-based crystal plasticity. *Crystals* **2017**, *7*, 152. [CrossRef]
55. Akarapu, S.; Zbib, H. Dislocation interactions with tilt walls. *Int. J. Mech. Mater. Des.* **2008**, *4*, 399–406. [CrossRef]
56. Akarapu, S.; Zbib, H.; Hirth, J.P. Modeling and analysis of disconnections in tilt walls. *Scr. Mater.* **2008**, *59*, 265–267. [CrossRef]
57. Lebensohn, R. Modelling the role of local correlations in polycrystal plasticity using viscoplastic self-consistent schemes. *Model. Simul. Mater. Sci. Eng.* **1999**, *7*, 739–746. [CrossRef]
58. Lebensohn, R.A.; Tomé, C.N.; Maudlin, P.J. A selfconsistent formulation for the prediction of the anisotropic behavior of viscoplastic polycrystals with voids. *J. Mech. Phys. Solids* **2004**, *52*, 249–278. [CrossRef]
59. Orowan, E. Problems of plastic gliding. *Proc. Phys. Soc.* **1940**, *52*, 8–22. [CrossRef]
60. Hochrainer, T.; Sandfeld, S.; Zaiser, M.; Gumbsch, P. Continuum dislocation dynamics: Towards a physical theory of crystal plasticity. *J. Mech. Phys. Solids* **2014**, *63*, 167–178. [CrossRef]
61. Starkey, K.; Winther, G.; El-Azab, A. Theoretical development of continuum dislocation dynamics for finite-deformation crystal plasticity at the mesoscale. *J. Mech. Phys. Solids* **2020**, *139*, 103926. [CrossRef]
62. Li, D.; Zbib, H.; Sun, X.; Khaleel, M. Predicting plastic flow and irradiation hardening of iron single crystal with mechanism-based continuum dislocation dynamics. *Int. J. Plast.* **2014**, *52*, 3–17. [CrossRef]
63. Mura, T. Theory of Inclusions and Inhomogeneities. In *Micromechanics of Defects in Solids*; Springer: Dordrecht, The Netherlands, 1987; Volume 3, pp. 1–27.
64. Luster, J.; Morris, M.A. Compatibility of deformation in two-phase Ti-Al alloys: Dependence on microstructure and orientation relationships. *Metall. Mater. Trans. A* **1995**, *26*, 1745–1756. [CrossRef]
65. Bieler, T.R.; Eisenlohr, P.; Zhang, C.; Phukan, H.J.; Crimp, M.A. Grain boundaries and interfaces in slip transfer. *Curr. Opin. Solid State Mater. Sci.* **2014**, *18*, 212–226. [CrossRef]
66. Shi, J.; Zikry, M.A. Modeling of grain boundary transmission, emission, absorption and overall crystalline behavior in ω1, ω3, and ω17b bicrystals. *J. Mater. Res.* **2011**, *26*, 1676–1687. [CrossRef]
67. Lyu, H.; Ruimi, A.; Field, D.P.; Zbib, H.M. Plasticity in Materials with Heterogeneous Microstructures. *Metall. Mater. Trans. A Phys. Metall. Mater. Sci.* **2016**, *47*, 6608–6620. [CrossRef]
68. Quey, R.; Dawson, P.R.; Barbe, F. Large-scale 3D random polycrystals for the finite element method: Generation, meshing and remeshing. *Comput. Methods Appl. Mech. Eng.* **2011**, *200*, 1729–1745. [CrossRef]
69. Krill, C.E.; Chen, L.Q. Computer simulation of 3-D grain growth using a phase-field model. *Acta Mater.* **2002**, *50*, 3057–3073. [CrossRef]
70. Wakai, F.; Enomoto, N.; Ogawa, H. Three-dimensional microstructural evolution in ideal grain growth general statistics. *Acta Mater.* **2000**, *48*, 1297–1311. [CrossRef]
71. He, G.; Liu, F.; Huang, L.; Huang, Z.; Jiang, L. Controlling grain size via dynamic recrystallization in an advanced polycrystalline nickel base superalloy. *J. Alloys Compd.* **2017**, *701*, 909–919. [CrossRef]
72. Moering, J.; Ma, X.; Chen, G.; Miao, P.; Li, G.; Qian, G.; Mathaudhu, S.; Zhu, Y. The role of shear strain on texture and microstructural gradients in low carbon steel processed by Surface Mechanical Attrition Treatment. *Scr. Mater.* **2015**, *108*, 100–103. [CrossRef]
73. Nye, J.F. Some geometrical relations in dislocated crystals. *Acta Metall.* **1953**, *1*, 153–162. [CrossRef]
74. Shizawa, K.; Zbib, H.M. A thermodynamical theory of gradient elastoplasticity with dislocation density tensor. I: Fundamentals. *Int. J. Plast.* **1999**, *15*, 899–938. [CrossRef]
75. Shizawa, K.; Zbib, H.M. A Thermodynamical theory of plastic spin and internal stress with dislocation density tensor. *J. Eng. Mater. Technol. Trans. ASME* **1999**, *121*, 247–253. [CrossRef]

76. Das, S.; Hofmann, F.; Tarleton, E. Consistent determination of geometrically necessary dislocation density from simulations and experiments. *Int. J. Plast.* **2018**, *109*, 18–42. [CrossRef]
77. Rhee, M.; Zbib, H.M.; Hirth, J.P.; Huang, H.; De La Rubia, T. Models for long-/short-range interactions and cross slip in 3D dislocation simulation of BCC single crystals. *Model. Simul. Mater. Sci. Eng.* **1998**, *6*, 467–492. [CrossRef]

crystals

MDPI

Article

Unveiling the Mechanisms of High-Temperature 1/2[111] Screw Dislocation Glide in Iron–Carbon Alloys

Ivaylo Hristov Katzarov [1,2,*] and Ljudmil Borisov Drenchev [1]

[1] Institute of Metal Science, Equipment and Technologies with Hydro- and Aerodynamics Centre, Bulgarian Academy of Sciences, 67 Shipchenski Prohod Str., 1574 Sofia, Bulgaria; ljudmil.d@ims.bas.bg
[2] Department of Physics, King's College London, Strand, London WC2R 2LS, UK
* Correspondence: ivaylo.katsarov@kcl.ac.uk

Abstract: We have developed a self-consistent model for predicting the velocity of 1/2[111] screw dislocation in binary iron–carbon alloys gliding by a high-temperature Peierls mechanism. The methodology of modelling includes: (i) Kinetic Monte-Carlo (kMC) simulation of carbon segregation in the dislocation core and determination the total carbon occupancy of the core binding sites; (ii) Determination of kink-pair formation enthalpy of a screw dislocation in iron—carbon alloy; (iii) KMC simulation of carbon drag and determination of maximal dislocation velocity at which the atmosphere of carbon atoms can follow a moving screw dislocation; (iv) Self consistent calculation of the average velocity of screw dislocation in binary iron–carbon alloys gliding by a high-temperature kink-pair mechanism under a constant strain rate. We conduct a quantitative analysis of the conditions of stress and temperature at which screw dislocation glide in iron–carbon alloy is accomplished by a high-temperature kink-pair mechanism. We estimate the dislocation velocity at which the screw dislocation breaks away from the carbon cloud and thermally-activated smooth dislocation propagation is interrupted by sporadic bursts of dislocation activity.

Keywords: dislocations; diffusion; FeC alloy; dynamic strain aging

Citation: Katzarov, I.H.; Drenchev, L.B. Unveiling the Mechanisms of High-Temperature 1/2[111] Screw Dislocation Glide in Iron–Carbon Alloys. *Crystals* **2022**, *12*, 518. https://doi.org/10.3390/cryst12040518

Academic Editor: Wojciech Polkowski

Received: 19 March 2022
Accepted: 5 April 2022
Published: 8 April 2022

1. Introduction

Steel plasticity is strongly influenced by interactions between solute atoms, such as carbon, and dislocations [1]. It is commonly accepted that the segregation of solute atoms in the surroundings of dislocations, forming clouds of impurities known as Cottrell atmospheres, is the underlying atomistic mechanism behind steel hardening [2–4]. The atmosphere is known to pin dislocations and render them less mobile. As more carbon atoms segregate, the atmosphere grows around the dislocations and hinders dislocation motion. Higher stresses are then required to unpin the dislocation from the solutes. The hardening of a material that is aged for a certain period of time after undergoing plastic deformation is defined as static strain aging. In contrast to static strain aging, which occurs during the specimen rest time, another strain aging phenomenon, called dynamic strain aging takes place during the plastic deformation of a specimen. It is associated with the diffusion of impurities to a mobile dislocation temporarily arrested at obstacles.

Carbon-induced strengthening of bcc iron has been studied in [5] using molecular dynamics (MD) simulations and a stohastic model based on minimum-energy path calculations and transition state theory. Plastic deformation of a solid solution where dislocations pass a random distribution of atomic-size obstacles have been studied by MD simulations in [6]. Simulations of a single edge dislocation in Ni-Al alloy reveal that dislocations propagate through jerky avalanches, the size and duration of which are power-law distributed. Dislocation propagation above the depinning threshold has been characterized by the distributions of avalanche area and duration. The analysis of atomic-scale dislocation dynamics has revealed a number of signatures of criticality. Microscopic mechanisms of DSA in W-O interstitial solid solutions has been studied at atomic scale [7]. Simulations have been performed using a kinetic Monte Carlo model that accounts for kink-pair nucleation and solute diffusion. Solute cloud formation, solid solution strengthening and dislocation/solute coevolution leading to jerky flow have been revealed from the simulations. A kMC model of screw dislocation glide and solute diffusion in substitutional W-Re alloy has been developed in [8]. The model has been applied for studying the mechanisms governing dislocation kinetics. The intrinsic drawback of both MD and kMC simulations of dislocation/solute coevolution is that diffusion and kink-pair nucleation are rare events operating on different timescales.

Recent measurements by Caillard [9] in binary iron–carbon alloys with carbon concentrations of 1, 16 and 230 atomic parts per million (appm), reveal that the dynamic interaction between mobile dislocations and solute carbon atoms results in dynamic strain aging effects in the temperature range of 100–300 °C. Caillard found three regimes of behaviour: (i) the expected Peierls mechanism of kink pair formation and migration at low temperature; (ii) at intermediate temperatures carbon atoms become sufficiently mobile to reach and segregate in the core of nearest dislocation and immobialize it. Dislocations start to move in bursts—no dislocation motion is observed upon straining before a source is unlocked and emits many dislocations. This results in the dynamic strain aging effect characterized by avalanches of rapid dislocation glide corresponding to the observed jerky and serrated flow. (iii) Above about 200 °C, a new mechanism was discovered, namely viscous glide accomplished by a Peierls mechanism but with an activation energy almost twice that of the room temperature viscous flow. The transitions between these domains vary as a function of dislocation velocity, but so far a quantitative analysis and confirmation of observation with theory and modelling has been lacking.

A dislocation drag mechanism, by which dislocations can collect and transport carbon within their cores, takes place if the diffusion of carbon atoms and the motion of dislocations occur with rates in the same order of magnitude. It can be assumed that dislocations glide is accomplished by a high-temperature Peierls mechanism when carbon atoms in bcc-iron become mobile enough to follow gliding dislocations. To study the drag of carbon atoms by dislocations requires simulation techniques that capture carbon diffusion events, trapping and escaping from the core and the motion of dislocations simultaneously. Caillard [9] ascertained that dynamical strain aging is caused by the trapping of carbon by straight screw segments of 1/2[111] dislocations. For this reason we focus on the 1/2[111] screw dislocation in the present work. In [10] we have developed an atomistic kinetic Monte-Carlo (kMC) model describing carbon diffusion in the non-homogeneous stress field created by a 1/2[111] screw dislocation in bcc-iron, where the behaviour of individual atoms is explicitly taken into account. The kMC model employs information gathered from molecular statics simulations carried out in order to determine the activation energies required for carbon hops in the neighbourhood of the line defect [11]. The number of segregated carbon atoms forming a Cottrell atmosphere around a 1/2[111] screw dislocation, predicted by performing long time kMC simulations, has been validated against the carbon atmosphere visually identified by position-sensitive atom probe microscopy [12]. The kMC model allows us to study both the diffusing carbon residing in the dislocation core, and carbon atoms which move through the interstitial sites in dislocation surroundings. We have employed the kMC model to simulate carbon diffusion in bcc–iron leading to the formation of Cottrell atmosphere. The kMC approach also offers an atomistic view of the carbon drag mechanism. By setting the dislocation to glide with constant velocity v_{dis}, we can simulate the evolution of the carbon cloud around the moving linear effect. This approach allows us to estimate the maximal dislocation velocity $v_{\max}(T, C)$ at which the atmosphere of carbon atoms can follow a moving screw dislocation. When dislocation velocity is higher than $v_{\max}$, dislocations gradually break away from the carbon clouds. Carbon atoms trapped in the core can follow the dislocation if it glides slowly and viscously via a Peierls mechanism (the process of kink pair creation followed by kink migration). However, we do not consider kink pair formation and migration explicitly in the kMC model. We therefore turn now to the problem of predicting v_{dis} within the Peierls mechanism. v_{dis} is then the dislocation velocity averaged over the numerous acts of kink-pair creation and migration processes. The trapped C atoms strongly modify the kink pair formation enthalpy E_{kp}, which for a given resolved shear stress is a function of the solute concentration and the rate at which impurities are distributed among trap sites [13]. At lower solute atom mobility, the impurities remain behind in binding sites of higher potential energy as a dislocation segment moves between Peierls valleys, leading to a higher Peierls barrier and E_{kp}. The Peierls barrier becomes smaller as the rate at which solute atoms are distributed among the trap sites increases [13]. With increasing temperature, the number of solute atoms trapped in the core decreases, which also reduces the Peierls barrier [14]. Hence, as temperature increases, E_{kp} decreases as a consequence of the decreasing Peierls barrier. It could be expected that a screw dislocation can glide via a high-temperature Peierls mechanism if: (a) the kink-pair formation energy barrier, E_{kp}, becomes smaller as the mobility of carbon atoms trapped in the dislocation core increases at high temperature; (b) the carbon Cottrell atmosphere can follow the dislocation, that is, a dislocation's average velocity at a given carbon concentration and temperature is lower than $v_{\max}$. The transitions to high-temperature viscous glide accomplished by a Peierls mechanism vary as a function of dislocation velocity. The problem of predicting v_{dis} within the Peierls mechanism is related to predicting kink-pair formation enthalpy E_{kp} at a given carbon concentration and local shear stress τ. Here, we present a self consistent model that is able to predict the average velocity of screw dislocation in binary iron–carbon alloys gliding by a high-temperature kink-pair mechanism. The purpose of the present work is conducting a quantitative analysis of the conditions of stress and temperature at which a screw dislocation glide is accomplished by a high-temperature kink-pair mechanism and transition may occur between carbon drag and breakaway. We aim to understand whether the dislocation behaviour in the high temperature domain of DSA described in [9] is related to the occurrence of the "high-temperature Peierls mechanism". The structure of the paper is as follows. Section 2 presents our methodology of modelling: in Section 2.1, we examine the effect of trapped carbon on the motion of a straight dislocation between two adjacent Peierls valleys. Section 2.2 addresses the effect of carbon segregated in the dislocation core on kink-pair formation. In Section 2.3, we present a self consistent model for the determination of the average velocity of screw dislocation in binary iron–carbon alloys gliding by a high-temperature kink-pair mechanism under a constant strain rate. Sections 3 and 4 are our results and discussion sections. We conclude in Section 4.

2. Materials and Methods

2.1. Carbon Effect on the Dynamics of Straight Dislocation

Density functional theory (DFT) calculations have identified two core structures of the 1/2[111] screw dislocation in bcc Fe. The so called "easy core" (EC) is the stable configuration, and the metastable "hard core" (HC), which is very close in configuration to the "saddle point" core [15,16]. Tight-binding (TB) simulations of carbon interactions with the 1/2[111] screw dislocation found solute distribution to vary significantly between the easy and hard cores [17]. The binding sites of carbon, and their strengths, were determined by the TB calculations of carbon–screw dislocation interactions. These were performed using the Fe-C TB model of Paxton and Elsässer [18]. The binding energies of carbon to both the hard and easy cores, with the resulting distribution of carbon can be seen in Figure 1.

These binding energies agree well with experiments and atomistic/elastic calculations [16,19]. In agreement with DFT, the highest binding energy is found in the centre of the hard screw core [19]. As a dislocation moves from EC to HC to EC, the trap sites ahead of the dislocation line transform into HC sites before finally transforming into EC trap sites behind the dislocation line. These atomistic results provide data for determination of kink-pair formation energies as a function of stress and carbon concentration, C, using a line tension (LT) model of a screw dislocation [15]. A segment of a streight dislocation lying in its Peierls valley has to migrate towards or into the next Peierls valley so as to make an incipient or complete kink pair. Within the LT model, the dislocation is divided into segments of width b, the Burgers vector, along its length. The energy per unit length of dislocation is then described by the following line tension expression [15].

$$E = \sum_j E_j = \frac{1}{2}K\sum_j (x_j - x_{j+1})^2 + \sum_j E_P(x_j) - \sum_j \varepsilon_{1pq}\tau_{pr}b_r\xi_p x_j - \sum_k E_k^c(|x_j - x_C|) \quad (1)$$

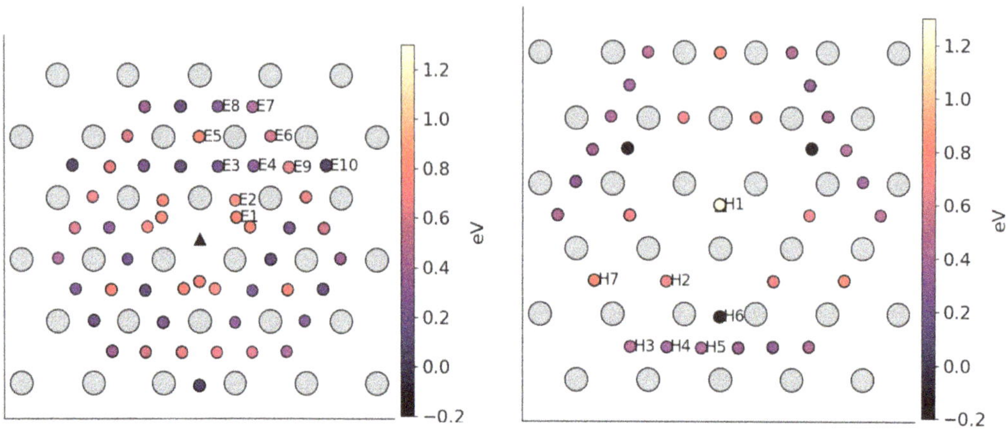

(**a**) (**b**)

Figure 1. The binding sites and binding energies of carbon segregated around: (**a**) easy 1/2[111] screw dislocation core; (**b**) hard 1/2[111] screw dislocation core.

The variable x_j describes the deviation of the jth segment from the dislocation's original position in the Peierls valley. The first term describes the energy penalty for two segments which have different amounts of deviation from the original Peierls valley towards the next and K is the associated "spring constant". The periodic Peierls energy landscape is described by an energy function, $E_P(x_j)$. The third term is the component perpendicular to the line sense ξ of the Peach–Kohler force arising from a local stress **ø** times the displacement of the jth segment. ε_{1pq} is the Levi–Civita symbol where 1-component is perpendicular to [111] direction. The final term expresses the energy associated with carbon atoms that are trapped at the binding sites in the dislocation core at a distance $|x_j - x_C|$ from the core, in which x_C is the position of the kth carbon atom relative to the elastic centre. Here, $0 < x < h$, where h is the period of the Peierls potential on the (110) plane. This term varies when a dislocation segment propagates between two adjacent Peierls valleys as a result of the variation in the binding energies $E_k^c(x)$ and carbon redistribution between trap sites. We parameterise $E_k^c(x)$ by fitting and interpolation of tigth-binding data [17]. We define

$$\chi_t = \sum_i \chi_i = const \quad (2)$$

as the total carbon occupancy of the core sites. We assume that χ_t is constant, that is, carbon will redistribute dynamically between trap sites during dislocation propagation between two Peierls valleys, but overall the dislocation will not absorb or reject carbon. We also only allow C to redistribute among traps within a plane perpendicular to the dislocation line, in view of the slower carbon pipe diffusivity [10]. We define the equilibrium occupation probability, χ_i^e of trap site i as [20],

$$\chi_i^e(x) = \frac{\chi_i e^{-E_i^c(x)/kT}}{\sum_j \chi_j e^{-E_j^c(x)/kT}} \tag{3}$$

Within the line tension model (1), we have a complete description of the energetics of the dislocation as a function of x and χ_t only in two limiting cases: (a) slow glide, in which traps are occupied according to (3), and (b) fast glide, in which all carbon atoms are fixed in the traps they occupy in the initial state before glide. Figure 2 shows potential energy profiles in both limiting cases for carbon concentration in the bulk C = 250 appm.

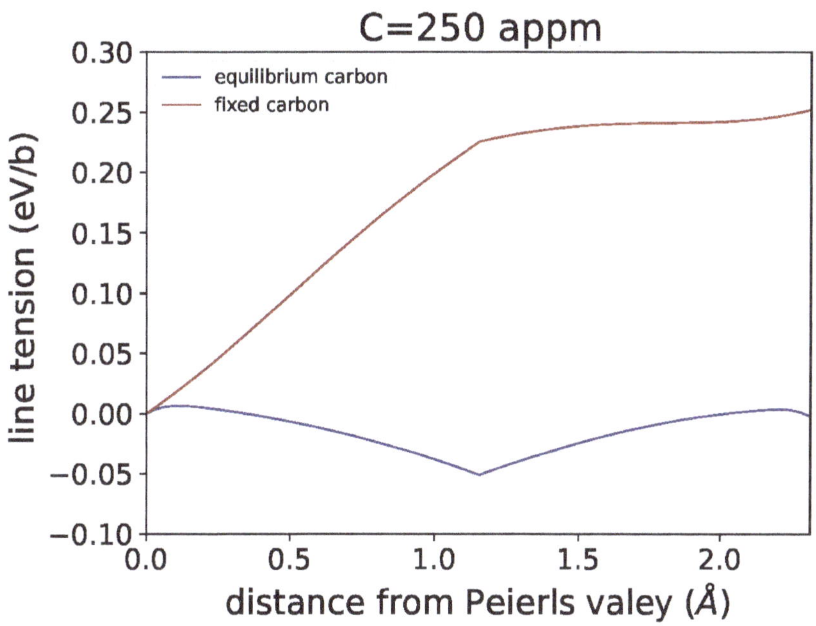

Figure 2. Potential energy of a straight 1/2[111] screw dislocation.

In the equilibrium limit, the effect of carbon trapped in deeper traps is strong enough that for carbon concentration in the bulk, C = 250 appm, the hard core is lower in energy than the easy core, and their roles are reversed. This result agrees well with DFT simulations by Ventelon [19]. When a carbon is placed in the vicinity of a relaxed easy dislocation core, a spontaneous reconstruction of the dislocation core occurs: from easy to hard. In the limit of rapid glide, carbon remains behind in sites of higher binding energy, leading to higher Peierls barrier. The actual dislocation energy profile between two Peierls valleys will be controlled by the dislocation velocity, v. Therefore, we need a theory that will predict the energy profile as a function of v. We may assume that the difference between the probability of the occupancy of trap i, $\chi_i(x)$, and its equilibrium value (3), $\chi_i^e(x)$ is the driving force for carbon redistribution between trap sites. For any of the binding sites, we find the following continuity equation,

$$\frac{\partial \chi_i(x,v)}{\partial t} = v\frac{\partial \chi_i(x,v)}{\partial x} = -\sum_{i \neq j}(\chi_i(x,v) - \chi_i^e(x))R \tag{4}$$

where $R = fexp(-E_i(x)/kT)\left(\chi_j^e(x) - \chi_j(x,v)\right)$ if $(\chi_i > \chi_i^e)$ and $\left(\chi_j < \chi_j^e\right)$; $R = fexp(-E_j(x)/kT)\left(\chi_j(x,v) - \chi_j^e(x)\right)$ if $(\chi_i < \chi_i^e)$ and $\left(\chi_j > \chi_j^e\right)$; f is an "attempt frequency" for carbon to escape from the ith trap. By solving (4), subject to the condition (2), we may determine the potential energy of the dislocation as a function of position between two Peierls valleys and velocity v.

2.2. Kink Pair Formation Enthalpy

Formation of a kink pair is a result of numerous random events in which a small segment of dislocation moves towards a neighbouring Peierls valley producing an "incipient" kink pair. The elastic attraction of the kinks leads to the annihilation of most of the "incipient" kink pairs. The formation of a stable kink pair is a result of increasing distance between kinks under the action of the applied shear stress. We do not consider these processes explicitly in our simulations. Trapped carbon strongly modifies the kink-pair formation energy, E_{kp}. E_{kp} is a function of v_{dis} since the distribution of carbon among binding sites will depend on how fast the dislocation is moving. Using the LT model, the energy, $E_j(C, x, v_{dis})$, of a dislocation segment of length b between two Peierls valleys, can be calculated. Then, using the "nudged elastic band" (NEB) method [21], we may calculate the kink pair activation energy, $E_{kp}(C, \tau, v_{dis})$ [14,15] for a given resolved shear stress, τ, and an assumed v_{dis}. However, dislocation velocity v_{dis} is a function of E_{kp}. We assume that v_{dis} is constant, and

$$v_{dis}(E_{kp}) = \frac{h}{t_r} \tag{5}$$

where we define an average relaxation time for kink-pair formation,

$$t_r = f_{kp}^{-1} e^{E_{kp}(C,\tau,v)/kT} \tag{6}$$

The frequency prefactor f_{kp} has been determined by comparing the velocity of a pure screw segment in bcc-Fe calculated from the kMC model of 1/2[111] screw dislocation with the experimentally estimated velocity [14]. We find $f_{kp} = 2.31 \times 10^9$ s^{-1}.

To find a self consistent solution of (5) and (6), and to determine E_{kp} at a given C and τ, we proceed with the following iterative process.

1. Assume an initial E_{kp}.
2. Calculate the corresponding v_{dis} using (5) and (6).
3. Determine the distribution of carbon from the continuity Equation (4), subject to (2); and calculate the segment energy, $E_j(C, x, v_{dis})$ from the LT model.
4. Calculate E_{kp} using the NEB and go to step 2.

This process is iterated until the E_{kp} calculated in step 4 is no longer changing to within some tolerance.

2.3. Self-Consistent Calculation of Dislocation Velocity in Binary Iron–Carbon Alloys

The velocity of a screw dislocation, steadily moving by the kink-pair mechanism, can be expressed as

$$v_{dis} = f_{kp} \frac{hL}{w} e^{-E_{kp}(\tau,C)/kT} \tag{7}$$

where w is the critical width of a stable kink pair and L is the dislocation length. Molecular dynamics simulations have identified the critical width of the 1/2 [111] screw dislocation as $w \approx 30b$ [22]. The increased probability of kink pair formation on all three [110] glide planes leads to an increased likelihood of kink pair collisions resulting in the formation of cross-kinks [13,14]. Equation (7) does not take into account the effects of cross-kink formation and self pinning on the dislocation mobility. In order to emulate experimental tests, which are performed under a constant strain rate, $\dot{\varepsilon}_0$, and temperature, T, we carry out strain rate-controlled simulations. The instantaneous shear stress resulting from a given $\dot{\varepsilon}_0$ is obtained as:

$$\tau(t) = 2\mu[t\dot{\varepsilon}_0 - \varepsilon_p(t)] \tag{8}$$

where ε_p is the total accumulated plastic strain and μ is the shear modulus

$$\varepsilon_p(t^{n+1}) = \varepsilon_p(t^n) + \delta\varepsilon_p(t^{n+1}) \tag{9}$$

t is the total simulation time and δt^{n+1} is the current timestep. The plastic strain update is obtained from the dislocation velocities calculated at each stress using Orowan's equation, as:

$$\delta\varepsilon_p(t^{n+1}) = \rho_d b v_{dis}(\tau)\delta t^{n+1} \tag{10}$$

where $v_{dis}(\tau)$ is the dislocation velocity, which depends on the resolved shear stress τ and carbon atoms trapped in the dislocation core; ρ_d is dislocation density. In order to determine self-consistently the velocity of screw dislocation in binary iron–carbon alloys gliding by a high-temperature Peierls mechanism, we proceed with the following iterative process.

1. Calculate the v_{dis} at the initial shear stress τ using (7) and determined by $E_{kp}(\tau, C)$.
2. From the dislocation velocity calculated at stress τ, we calculate the total accumulated plastic strain from (9) and (10).
3. The instantaneous shear stress τ resulting from the calculated accumulated plastic strain is obtained from (8).
4. Go to step 1

If dislocation breaks away from the carbon cloud, the steady dislocation propagation by the Peierls mechanism is replaced by a thermally-activated screw dislocation burst in carbon-free regions with a lower kink-pair formation energy interrupted by interactions with solute atoms. In the carbon-free regions, the kink-pair activation energy $E_{kp}(\tau, C)$ in (7) is replaced by the corresponding kink-pair formation enthalpy in bcc-Fe.

3. Results

The kMC treatment of carbon Cottrell atmosphere formation presented in this work considers a portion of the atmosphere contained inside a fixed volume parallelepipedal region $10 \times 10 \times 10$ nm^3, the rectangular cross section of which (10×10 nm^2) corresponds approximately to the surrounding dislocation core region of the atom probe experiments [12].

In the present study, the surroundings of the dislocation core (region with radius $R \approx 0.6$ nm in the direct neighbourhood of the dislocation centre [11]) are connected to an infinite carbon reservoir, such that no carbon depletion occurs in the matrix due to segregation [10]. The evolution of the number of carbon atoms segregated to form an atmosphere around a 1/2[111] screw dislocation at a background carbon concentration of 250 appm simulated by our kMC model is shown in Figure 3.

The kMC calculations allow us to predict the rate of formation and strength of carbon atmospheres.

The simulations show that at $T = 400$ K, an equilibrium carbon Cottrell atmosphere is formed after 100 s. At 500 K, the equilibrium carbon Cottrell atmosphere around a screw dislocation is formed even faster—0.3 s for 250 appm. At equilibrium, the average number of carbon atoms trapped in the screw dislocation core is around 5 C/nm at 400 K and 2.3 C/nm at 500 K.

From these kMC simulations, we determine the total carbon occupancy of the core binding sites χ_t.

The kMC treatment of the carbon drag mechanism presented in this section considers the evolution of the carbon cloud dragged by a mobile straight screw dislocation inside a fixed volume parallelepipedal region $60 \times 10 \times 10$ nm^3. At the starting time $t = 0$, the carbon atoms segregated to form an atmosphere around the 1/2[111] screw dislocation are at equilibrium with the background carbon concentration. We employ the kMC model to simulate the redistribution of carbon atoms when dislocation migrates in the (110) glide plane with constant velocity $v_{\rm dis}$ (Figure 4).

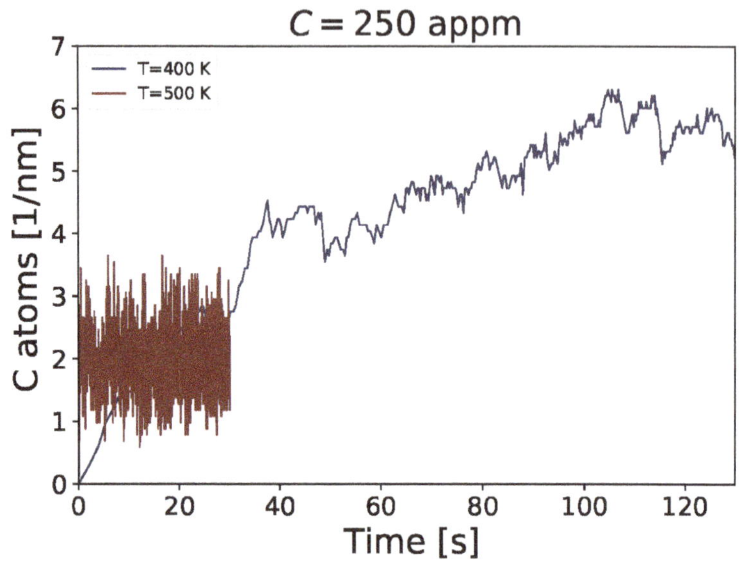

Figure 3. The number of carbon atoms per 1 nm trapped in the dislocation core at two different temperatures.

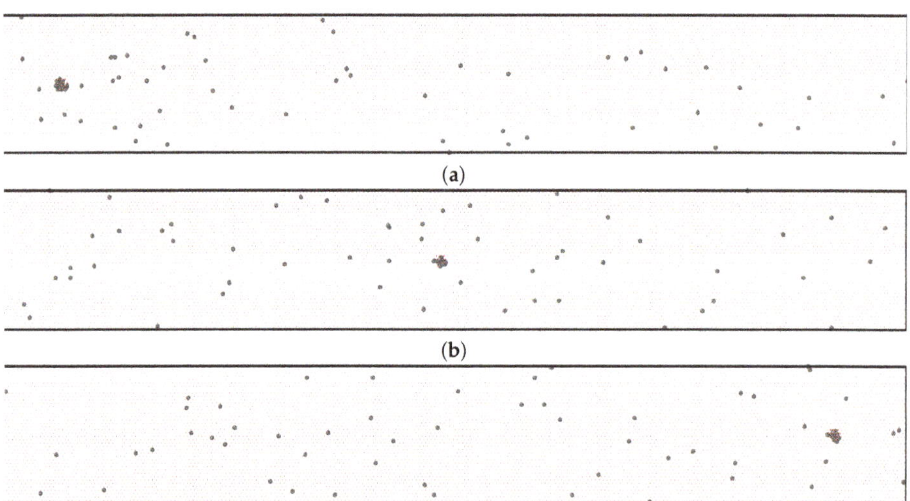

(**a**)

(**b**)

(**c**)

Figure 4. The motion of a screw dislocation dragging carbon, illustrated by a series of snapshots. KMC simulations are carried out at dislocation velocity v_{dis} = 0.1 nm/s and temperature 400 K; (**a**) t = 0 s; (**b**) t = 25 s; (**c**) t = 52 s.

The number of carbon atoms effectively dragged by the dislocation as a function of the dislocation travel distance is shown in Figure 5 for different dislocation velocities and temperatures of 400 K and 500 K. From the above simulations, we evaluate the maximal dislocation velocity $v_{\max}$ at which the atmosphere of carbon atoms can follow a moving screw dislocation (Figure 6).

Figure 7 shows the results of the self-consistent iterative procedure for the calculation of kink-pair formation enthalpy E_{kp} proposed in Section 2.2.

We may interpret Figure 7 in the following way. At high stress, E_{kp} is smaller because the process of dislocation segment transition to the adjacent Peierls valley is dominated by the applied stress.

At low stress, we observe a large E_{kp}, the largest being that of zero stress. As temperature increases, E_{kp} decreases as a consequence of the decreasing Peierls barrier resulting from the enhanced carbon mobility and lower C occupancy in the core. After determination of E_{kp} as a function of τ and C, we apply the self-consistent procedure described in Section 2.3 to determine the evolution of the 1/2[111] screw dislocation velocity during a strain rate-controlled simulation of dislocation motion in binary iron–carbon alloys. Dislocation density in our simulations is $\rho_d = 10^{13}$ m^{-2}. ρ_d sets the dislocation line length to a magnitude of approximately $L = (\rho_d)^{-1/2}$, which for $\rho_d = 10^{13}$ m^{-2}, gives $L = 1280b$. The simulations are performed under a constant strain rate, $\dot{\varepsilon}_0 = 10$ s^{-1}. Evolution of the velocity and stress with time during a strain rate-controlled simulation of dislocation motion at T = 400 K and T = 500 K are shown respectively in Figures 8 and 9.

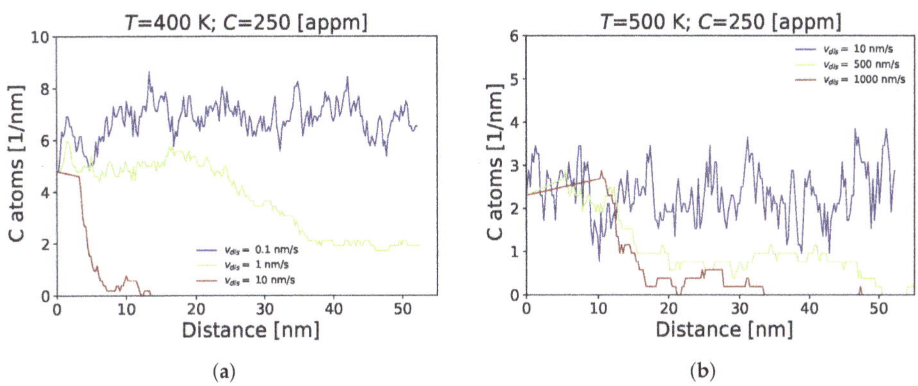

Figure 5. The evolution of the number of carbon atoms per 1 nm trapped in the core of a moving straight dislocation as a function of the travel distance; (**a**) T = 400 K; (**b**) T = 500 K.

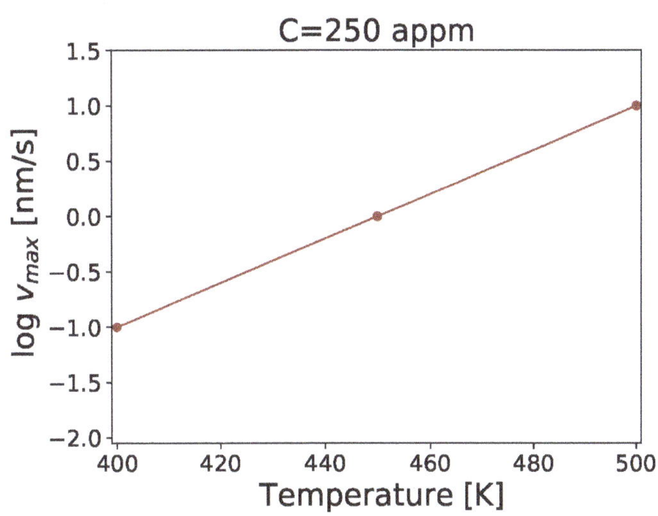

Figure 6. The maximal dislocation velocity $v_{\max}$ at which the atmosphere of carbon atoms can follow a moving screw dislocation at different temperatures.

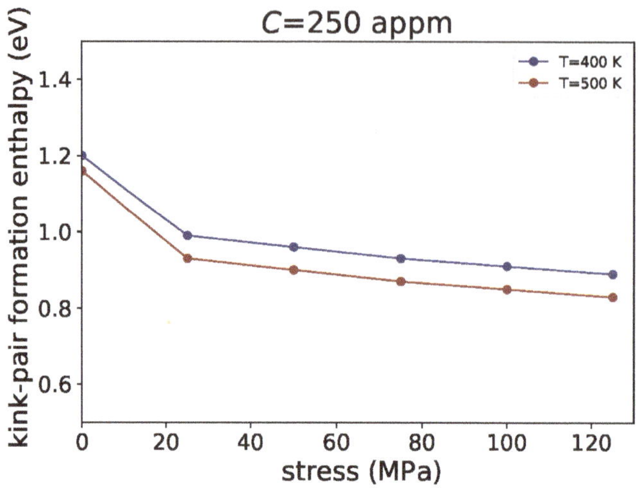

Figure 7. Kink pair formation enthalpy, E_{kp}, as a function of resolved shear stress, calculated by iterative solution of Equations (5) and (6).

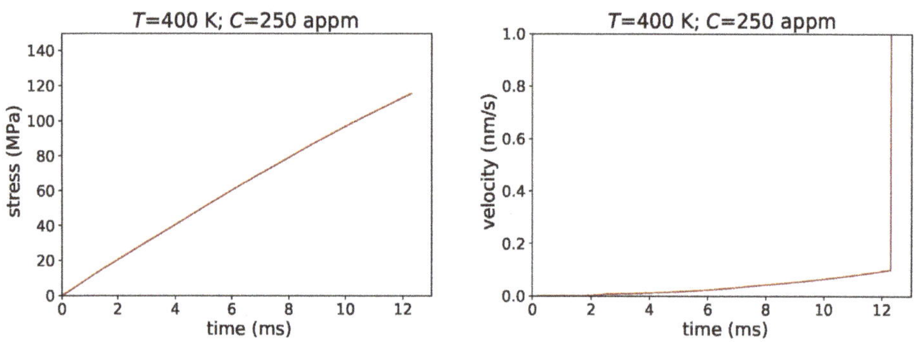

(**a**) (**b**)

Figure 8. Evolution of the stress and dislocation velocity with time during simulation of dislocation motion at T = 400 K; (**a**) stress; (**b**) velocity.

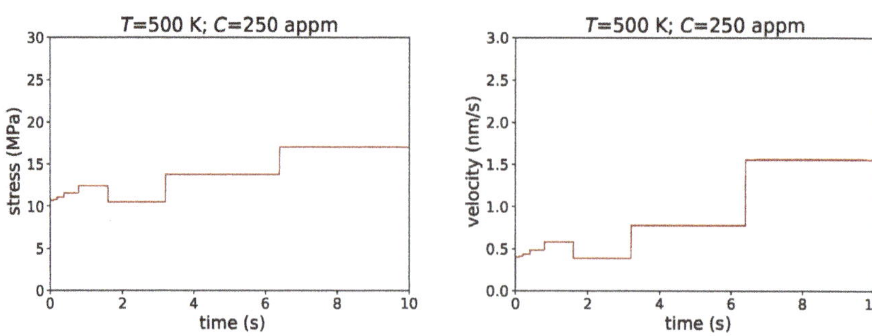

(**a**) (**b**)

Figure 9. Evolution of the stress and dislocation velocity with time during simulation of dislocation motion at T = 500 K; (**a**) stress; (**b**) velocity.

4. Discussion

We use our approach to simulate screw dislocation propagation by kink pair mechanism in an iron–carbon system at two temperatures—400 K and 500 K, which belong to two regimes of behaviour found by Caillard [9] (intermediate temperatures where dislocations start to move in bursts and high-temperatures above about 200 °C where dislocations move slowly and steadily by a possible Peierls mechanism). The experiments [9] show that the temperatures at which the screws move slowly, presumably by high-temperature kink-pair mechanism, are almost independent of the carbon concentration.

At 400 K, carbon atoms become sufficiently mobile to move to the nearest dislocations and hinder dislocation motion. The trapped solute atoms remain behind in binding sites of higher potential energy as a dislocation segment moves, leading to a higher Peierls barrier and kink-pair formation energy E_{kp} (Figure 7). Higher stresses are then required to increase kink-pair formation rate and dislocation velocity. The carbon atoms have sufficient time to follow the dislocation at lower dislocation velocities (Figure 5a). kMC simulations reveal that at 400 K and at background carbon concentrations of 250 appm, the maximal dislocation velocity $v_{\max}$ at which the atmosphere of carbon atoms can follow a moving screw dislocation is 0.1 nm/s. The accumulated plastic strain induced by dislocations gliding at low velocities $v_{dis} < v_{\max}$ can not efficiently relieve the shear stress resulting from the constant shear strain rate (Figure 8a). The rising shear stress increases the velocity of dislocation gliding by kink-pair mechanism. When v_{dis} exceeds $v_{\max}$, dislocations break away from the carbon clouds and start to move in carbon-free regions with a lower kink-pair formation energy interrupted by interactions with solute atoms. Plastic deformation of a solid solution where dislocations pass a random distribution of atomic-size obstacles have been studied in [6]. MD simulations of a single edge dislocation in a binary alloy reveal that dislocations propagate through jerky avalanches. In this work, we do not simulate dislocation motion after dislocation breaks away from the carbon cloud. We only consider the initial acceleration of dislocations in the carbon-free region. In the carbon-free region, we calculate the dislocation velocity using Equation (7), where the kink-pair activation energy $E_{kp}(\tau, C)$ is replaced by the corresponding kink-pair nucleation enthalpy $E_{kp}(\tau)$ in bcc-Fe which, for applied shear stress $\tau = 115$ MPa (Figure 8a), is $E_{kp}(\tau)$ = 0.35 eV [15]. Due to the lower kink-pair formation energy in bcc-Fe, after breaking away from the carbon cloud, the dislocation accelerates to a velocity which is six orders of magnitude higher then $v_{\max}$. The abrupt increase in the dislocation velocity relieves the stress, which results in decreasing v_{dis} below $v_{\max}$. After restoration of the kink-pair, the glide dislocation velocity gradually increases above $v_{\max}$, leading to repeated bursts of rapid dislocation glide. The studied thermally-activated smooth dislocation propagation is interrupted by abrupt sporadic bursts of screw dislocation activity.

At higher temperatures, the number of carbon atoms trapped in the core decreases, which reduces the barrier between two adjacent Peierls valleys. The Peierls barrier is additionally reduced as the rate at which carbon atoms are distributed among trap sites increases. Hence, as the temperature increases, the kink-pair formation enthalpy E_{kp} decreases as a consequence of the decreasing Peierls barrier (Figure 7). At higher temperatures, carbon has a sufficiently high mobility to keep up with faster dislocations (Figure 5b). Due to the higher $v_{\max}$ and lower E_{kp}, the accumulated plastic strain, induced by dislocations gliding by kink-pair mechanism, relieves the shear stress, resulting in relatively long periods of steady dislocation glide by kink-pair mechanism (Figure 9b). The periods of steady propagation are interrupted by more abrupt changes in stress and velocity due to the interactions between carbon and dislocations (Figure 9). The average dislocation velocity determined in the present simulation is 0.93 nm/s. The average velocity depends on the loading conditions. Figure 10 shows the evolution of the velocity with time during a strain rate-controlled simulation of dislocation motion for three different strain rates $\dot{\varepsilon}_0$. With an increase in the constant strain rate, the average dislocation velocities increase, and for $\dot{\varepsilon}_0$, shown in Figure 10, they are correspondingly 0.1 nm/s, 0.42 nm/s and 1.3 nm/s. The experimentally observed average dislocation velocity at C = 230 appm and T = 200 °C reported in [9] is 0.27 nm/s. Formation of cross-kinks resulting from kink collisions amount to self pinning points which reduce dislocation mobility. The present model does not take into account the effects of cross-kinks formation and self pinning on the dislocation mobility.

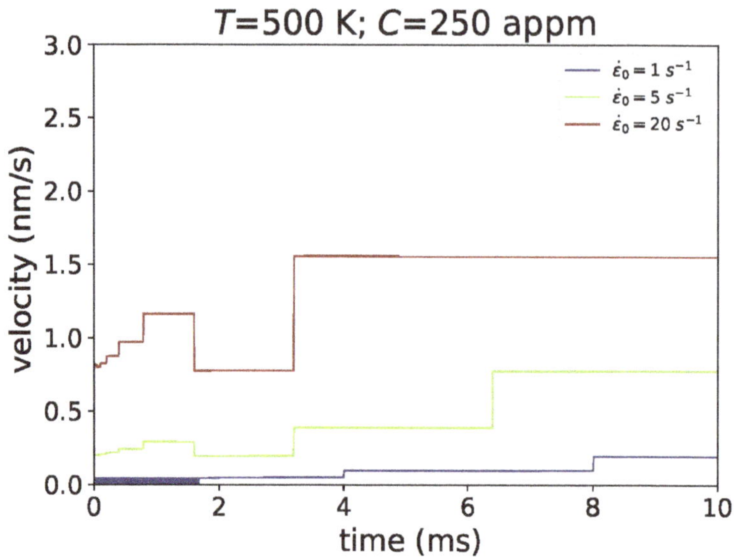

Figure 10. Evolution of the dislocation velocity with time during the simulation of dislocation motion at T = 500 K for three constant strain rates.

5. Conclusions

We have developed a self-consistent model for predicting the velocity of 1/2[111] screw dislocation in binary iron–carbon alloys gliding by a high-temperature Peierls mechanism. The methodology of modelling includes:

- KMC simulation of carbon segregation in dislocation core and determination the total carbon occupancy of the core binding sites.
- Evaluation of the effect of trapped carbon on the motion of a straight dislocation segment between two adjacent Peierls valleys.
- Determination of kink-pair formation enthalpy E_{kp} of a screw dislocation in iron—carbon alloy.
- KMC simulation of carbon drag and determination of maximal dislocation velocity $v_{\max}$ at which the atmosphere of carbon atoms can follow a moving screw dislocation.
- Self consistent calculation of average velocity of screw dislocation in binary iron–carbon alloys gliding by a high-temperature kink-pair mechanism under constant strain rate.

We conduct a quantitative analysis of the conditions of stress and temperature at which screw dislocation glide is accomplished by a high-temperature kink-pair mechanism. We conclude that several factors are responsible for the viscous dislocation glide observed above about 200 °C in iron—carbon alloy:

- At high temperatures, kink-pair formation enthalpy E_{kp} decreases as a consequence of the increased carbon mobility in the dislocation core and reduced number of segregated C atoms.
- The enhanced diffusivity of carbon both in the core region and in dislocation surroundings lead to higher maximal dislocation velocity $v_{\max}$ at which the atmosphere of carbon atoms can follow a moving screw dislocation.

In accordance with the experimental observations, the present simulations reveal that at high temperatures (T = 500 K), 1/2[111] screw dislocation glide in iron–carbon alloy is accomplished by a high-temperature kink-pair mechanism. At intermediate temperatures (T = 400 K), thermally-activated smooth dislocation propagation is interrupted by sporadic bursts of screw dislocation activity.

Author Contributions: Conceptualization, I.H.K. and L.B.D. All authors have read and agreed to the published version of the manuscript.

Funding: This research was funded in part by the Bulgarian Science Fund under the National Scientific Program "Petar Beron i NIE" (Project UMeLaMP), UKRI Grant EP/V001787/1 and by the European Regional Development Fund, within the Operational Programme "Science and Education for Smart Growth 2014–2020" under the Project CoE "National centre of mechatronics and clean technologies" BG05M20P001-1.001-0008-C01.

Institutional Review Board Statement: Not applicable.

Informed Consent Statement: Not applicable.

Data Availability Statement: Not applicable.

Conflicts of Interest: The authors declare no conflict of interest.

References

1. Caillard, D. An in situ study of hardening and softening of iron by carbon interstitials. *Acta Mater.* **2011**, *59*, 4974–4989. [CrossRef]
2. Cottrell, A.H.; Bilby, B.A. Dislocation theory of yielding and strain ageing of iron. *Proc. Phys. Soc. Sect. A* **1949**, *62*, 49–62. [CrossRef]
3. Clouet, E.; Garruchet, S.; Nguyen, H.; Perez, M.; Becquart, C.S. Dislocation interaction with C in α-Fe: A comparison between atomic simulations and elasticity theory. *Acta Mater.* **2008**, *56*, 3450–3460. [CrossRef]
4. Li, Y.; Choi, P.; Borchers, C.; Westerkamp, S.; Goto, S.; Raabe, D.; Kirchheim, R. Atomic-scale mechanisms of deformation-induced cementite decomposition in pearlite. *Acta Mater.* **2011**, *59*, 3965–3977. [CrossRef]
5. Allera, A.; Ribeiro, F.; Perez, M.; Rodney, D. Carbon-induced strengthening of bcc iron at the atomic scale. *Phys. Rev. Mater.* **2022**, *6*, 013608. [CrossRef]
6. Patinet, S.; Bonamy, D.; Proville, L. Atomic-scale avalanche along a dislocation in a random alloy. *Phys. Rev. B* **2011**, *84*, 174101. [CrossRef]
7. Zhao, Y.; Dezerald, L.; Pozuelo, M.; Zhou, X.; Marian, J. Simulating the mechanisms of serrated flow in interstitial alloys with atomic resolution over diffusive timescales. *Nat. Commun.* **2020**, *11*, 1–8. [CrossRef] [PubMed]
8. Zhao, Y.; Marian, J. Direct prediction of the solute softening -to-hardening transition in W-Re alloys using stochastic simulations of screw dislocation motion. *Model. Simul. Mater. Eng.* **2018**, *26*, 045002. [CrossRef]
9. Caillard, D. Dynamic strain ageing in iron alloys: The shielding effect of carbon. *Acta Mater.* **2016**, *112*, 273–284. [CrossRef]
10. Katzarov, I.H.; Drenchev, L.B.; Pashov, D.L.; Zarrouk, T.N.A.T.; Al-lahham, O.; Paxton, A.T. Dynamic strain aging and the rôle Dynamic strain aging and the rôle of the Cottrell atmosphere. 2022, *submitted*.
11. Nematollahi, G.A.; Grabowski, B.; Raabe, D.; Neugebauer, J. Multiscale description of carbon-supersaturated ferrite in severely drawn pearlitic wires. *Acta Mater.* **2016**, *111*, 321–334. [CrossRef]
12. Wilde, J.; Cerezo, A.; Smith, G.D.W. Three-dimensional atomic-scale mapping of a Cottrell atmosphere around a dislocation in iron. *Scr. Mater.* **2000**, *43*, 39–48. [CrossRef]
13. Gong, P.; Katzarov, I.H.; Nutter, J.; Paxton, A.T.; Rainforth, W.M. The influence of hydrogen on plasticity in pure iron–theory and experiment. *Sci. Rep.* **2020**, *10*, 10209. [CrossRef] [PubMed]
14. Katzarov, I.H.; Pashov, D.L.; Paxton, A.T. Hydrogen embrittlement I. Analysis of hydrogen-enhanced localized plasticity: Effect of hydrogen on the velocity of screw dislocations in α-Fe. *Phys. Rev. Mater.* **2017**, *1*, 033602. [CrossRef]
15. Itakura, M.; Kaburaki, H.; Yamaguchi, M.; Okita, T. The Effect of Hydrogen Atoms on the Screw Dislocation Mobility in Bcc Iron: A First-Principles Study. *Acta Mater.* **2013**, *61*, 18. [CrossRef]
16. Clouet, E.; Ventelon, L.; Willaime, F. Dislocation core energies and core fields from first principles. *Phys. Rev. Lett.* **2009**, *102*, 055502. [CrossRef] [PubMed]
17. Zarrouk, T. Atomistic Investigation of Dislocation-Assisted Carbon Migration in Iron. Ph.D. Thesis, King's College London, London, UK, 2022.
18. Paxton, A.T.; Elsässer, C. Analysis of a Carbon Dimer Bound to a Vacancy in Iron Using Density Functional Theory and a Tight Binding Model. *Phys. Rev. B* **2012**, *87*, 22. [CrossRef]
19. Ventelon, L.; Lüthi, B.; Clouet, E.; Proville, L.; Legrand, B.; Rodney, D.; Willaime, F. Dislocation core reconstruction induced by carbon segregation in bcc iron. *Phys. Rev. B* **2015**, *91*, 220102(R). [CrossRef]
20. Hirth, J.P.; Lothe, J. *Theory of Dislocations*, 1st ed.; McGraw-Hil Book Company: New York, NY, USA, 1968.

21. Henkelman, G.; Jónsson, H. Improved tangent estimate in the nudged elastic band method for finding minimum energy paths and saddle points. *J. Chem. Phys.* **2000**, *113*, 9978–9985. [CrossRef]
22. Itakura, M.; Kaburaki, H.; Yamaguchi, M. First-principles study on the mobility of screw dislocations in bcc iron. *Acta Mater.* **2012**, *60*, 3698–3710. [CrossRef]

crystals

MDPI

Article

The Cooling Rate and Residual Stresses in an AISI 310 Laser Weld: A Meso-Scale Approach

Edison A. Bonifaz * and Andrés S. Mena

Politécnico, Universidad San Francisco de Quito USFQ, Diego de Robles s/n y Vía Interoceánica, Quito 170901, Ecuador; amenac@alumni.usfq.edu.ec
* Correspondence: ebonifaz@usfq.edu.ec

Abstract: A three-dimensional coupled temperature-displacement finite element model was developed to generate values of temperature distribution, cooling rate, and residual stresses at the meso-scale level in a thick sheet AISI 310 laser welding test sample. High cooling rates (cooling time from liquid-to-solid temperatures) ranging from 960 °C/s to 2400 °C/s were observed when the calculations were made at the meso-scale level. These high cooling rates that arise during the formation of the weld pool originate the highest observed residual stresses that evolve throughout the weld during the entire heating and cooling cycles. An ABAQUS CAE meso model with dimensions of 10 × 5 × 1 mm (element size 100 μ) constructed from a global macro model of 40 × 10 × 75 mm (element size 1 mm) via the submodeling technique is presented in the present paper. In both analyses, macro and meso, the C3D8T thermally coupled brick, trilinear displacement and temperature elements were used. To mesh the entire plate with elements of regular size 100 × 100 × 100 μ, a total of 30 million elements are necessary. With the present approach, 1 macro mesh of 30 thousand elements (1 × 1 × 1 mm) and a meso mesh of 50 thousand elements (100 × 100 × 100 μ) were enough to simulate the weld problem at the meso-scale level.

Keywords: laser weld; heat input; cooling rate; residual stresses; fusion zone; meso scale; coupled temperature-displacement

Citation: Bonifaz, E.A.; Mena, A.S. The Cooling Rate and Residual Stresses in an AISI 310 Laser Weld: A Meso-Scale Approach. *Crystals* **2022**, *12*, 502. https://doi.org/10.3390/cryst12040502

Academic Editor: Bolv Xiao

Received: 9 March 2022
Accepted: 25 March 2022
Published: 6 April 2022

1. Introduction

Epitaxial solidification can be achieved if the welding process is adequately performed. It requires the solidification theory to control the columnar-to-equiaxed transition caused for the macroscopic heat input on the microstructure development. The analytical heat-flux solution methods presented in early approaches [1–8] use semi-empirically based computer programs with constant material properties. Consequently, several mathematical simplifications of the integral solution of the non-linear transient governing heat equation were made. As demonstrated in refs. [2–5], the heat input has a profound influence on the dendrite growth velocity, growth pattern, and melt pool geometry. The high peak temperatures and cooling rates (localized thermal cycling), grain structure, distortion, and reduced strength of a weld joint are produced by the intense heat input from fusion welding [9,10]. In the present work, the temperature-dependent material properties and forced convection due to Marangoni effects were incorporated into a three-dimensional coupled temperature-displacement finite element model.

The microstructure development and the evolution of the residual stresses are important aspects, both in the repair of gas turbine components and in relation to safety in the nuclear industry. The parameters that describe the heat input from the heat source are the most critical input data in the welding thermal analysis [9–11]. In regard to the residual stresses and distortions, the experimental optimization of the welding technique requires measurements and prototyping, which are expensive and time consuming. However, with the virtual prediction of residual stress evolution and microstructure development in the melt pool, the process can be optimized in the early stages of prototyping. Although these phenomena are difficult to simulate due to a lack of exact knowledge of welding conditions and material behavior, their prediction can provide substantial assistance in the accurate design and fabrication of welded structures. Accurately calculating the transient temperature field, that causes non-equilibrium phase formation in and around the welded joint, is the critical first step in creating a science base, not only for the design and analysis of welds, but also for the repair applications. The residual stresses that develop both in the heat-affected zone (HAZ) and the fusion zone (FZ) are harmful to the integrity and service-life of the welded part. In ref. [12], it is shown that residual stresses can cause problems, such as hydrogen-induced cracking, stress corrosion cracking, distortion, as well as initiating fracture and degrading the corrosion resistance of welded structures. To reduce them, pre- and post-heating is often used.

The accurate calculation and measurement of residual stresses at different scale levels still remains a major issue. The numerical prediction of thermal cycles is the prerequisite for subsequent calculations of residual stress distribution and microstructure development in welds [12–15]. Several key problems and issues remain to be addressed, especially at small scales. The main difficulties in the quantitative analysis of fusion welding are the scarcity of relevant data of properties at very high temperatures, i.e., much higher than the melting point, and the complexity of the physical processes. That is, the effects of fluid flow in the weld pool produced by the complex coupled transport phenomena are difficult to be addressed. Moreover, because of the extreme cooling conditions encountered during laser welding, the events that follow welding are far from equilibrium and, as a result, non-equilibrium phase formation in and around the welded joint occur. Therefore, the simulation of phase transformations (and the resulting microstructures), distortion and residual stress evolution in weldments remains a great challenge.

Physically based models used to simulate the entire thermal cycle of the welding process can provide a new understanding of the distortion evolution and microstructure development [14]. However, the physically based modeling of the entire thick plate and the melt pool as a whole, is a computationally formidable task. The lack of accurate stress–strain constitutive relations and the use of coarse grids create differences in the calculated and measured residual stresses. The uncertainties in the calculation procedures include the inaccuracies in the calculation of the thermal cycles as well as the approximations in the general frame for the mechanical model, especially when important solid-state transformations take place [12–15]. In previous works [10,16], it was well established that the cooling time from liquid-to-solid temperatures is longer at the fusion line and shorter at the weld centerline. As such, the cooling rate through the solidification temperature range increases and the dendrite arm spacing decreases from the fusion line to the centerline. In agreement with these previous numerical results, in ref. [16], it is also shown through weld micrographs that the solidification microstructure becomes finer from the fusion line to the centerline.

The modeling of forming processes involving the melting and solidification of metals, such as additive manufacturing (AM), casting or welding, all share some common features associated with those thermomechanical aspects that play an important role in the final material properties at the macroscale. The multiscale and Multiphysics phenomena involved, at different levels of significance, are imperative to be addressed when it comes to characterizing important aspects that perform a role in the final properties of the resulting parts. When running simulation studies with large assemblies, the submodeling feature allows us to refine the results for critical components without having to re-run the analysis for the entire assembly. In technical terms, submodeling helps transfer complex global loads from the entire structure to local sub-regions to obtain much more accurate results for the unknown field variables, such as temperature, stresses, and strains. Submodeling allows us to start the analysis process on a large assembly by running a "relatively coarse" parent analysis. This analysis may use coarser mesh than is really needed, might make simplified assumptions about contact and boundary conditions, or might neglect connection details.

The present research aims not only to verify that, at the microscale level, materials display strong size effects when the characteristic length-scale associated with non-uniform plastic deformation is on the order of microns, but also to clarify the link between cooling rates and solidification grain substructures. The prediction of crack initiation and propagation in weld samples is very important to identify direct energy deposition (DED) additive manufacturing (AM) process parameters for preserving the single-crystal nature of nickel-base superalloy gas turbine blades during repair procedures. The accurate prediction of the melt-pool shape and solidification parameters as a function of the processing parameters (including filler metal additions) is also important in the development of microstructural models for the design of functionally graded materials, using either laser additive manufacturing, or Wire + Arc additive manufacturing (WAAM).

2. The Model

The temperature $T(x; y; z; t)$ at any location $(x; y; z)$ and time (t) with respect to the moving heat source is calculated by solving the 3D transient nonlinear heat equation:

$$\frac{\partial}{\partial x}\left(k\frac{\partial T}{\partial x}\right)+\frac{\partial}{\partial y}\left(k\frac{\partial T}{\partial y}\right)+\frac{\partial}{\partial z}\left(k\frac{\partial T}{\partial z}\right)+\dot{Q}=\rho c_p\frac{\partial T}{\partial t} \tag{1}$$

Here, k, ρ and c_p are the temperature-dependent thermal conductivity, density and specific heat, respectively. T is temperature, t is time, and $\dot{Q}$ the internal heat-source term. The solution depends on the physical conditions existing at the *boundaries* of the workpiece and the conditions existing in the workpiece at some *initial time*. In this work, $\dot{Q}$ is 0, the initial temperature T_0 is 20 °C, and the latent heat was ignored. Convective and radiative boundary conditions for the heat exchange between the top surface of the workpiece and the surroundings beyond the laser heat source are expressed in the following equation:

$$-k\frac{\partial T}{\partial y}|_{top}+q(x,z,t)=h_t(T-T_s)+\sigma\varepsilon\left(T^4-T_s^4\right) \tag{2}$$

Here, h_t is the convection heat-transfer coefficient, T_s is the surrounding temperature, σ is the Stefan–Boltzman constant and ε is the surface emissivity. Values of ε = 0.7 and h_t = 242 Wm^{-2} K^{-1} to include forced convection in the area directly beneath the nozzle of the laser heat source were used. Convection heat loss at the workpiece top surface and all the other weldment surfaces were also accounted for using a value of h_t = 10 Wm^{-2}K^{-1}. The heat input $q(x,z,t)$ from the heat source to the workpiece is represented by a moving Gaussian power density distribution (Figure 1a and Equation (3)) applied over the top surface of the specimen during a period of time that depends on the welding speed (v). The characteristic heat distribution parameter, C, was selected based on previous experimental measurements of the fusion zone and heat-affected zone widths [17]. The mentioned parameter is very important because the shape and distribution of the heat input from the heat source greatly depend on its value. A distribution parameter C = 4.5 mm was used in all the simulations. The Gaussian distribution model is usually used in simulations of welding processes with high power densities, i.e., arc, laser, or electron beam welding. With the help of this heat source model, in which energy is distributed according to a Gaussian profile, it is possible to map deep penetration, while maintaining a small width, characteristic of "keyhole" welding techniques [18].

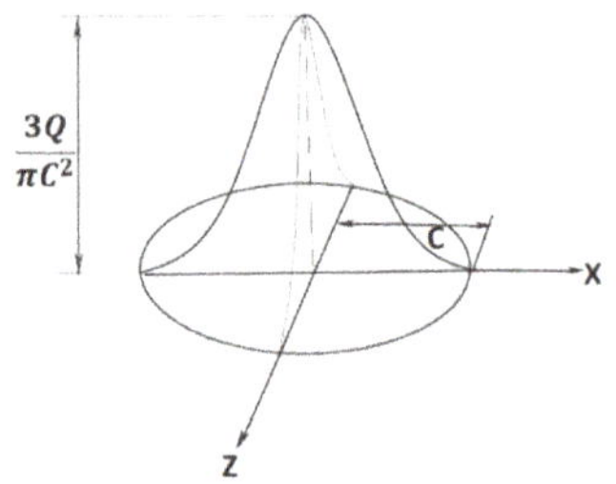

(a)

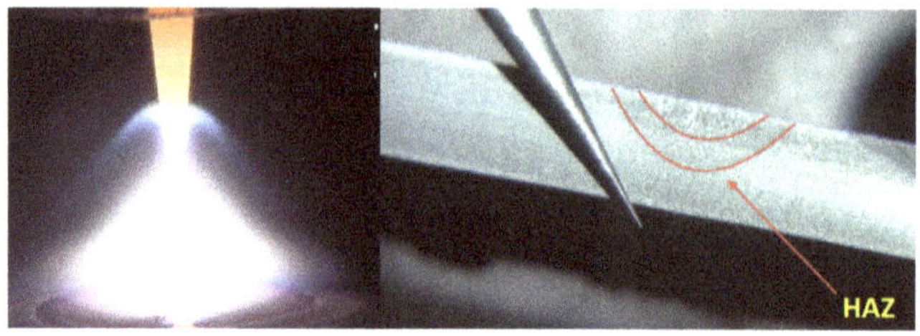

(b)

Figure 1. (**a**) Gaussian surface heat source model. (**b**) Tungsten arc profile and fusion zone and heat-affected zone (HAZ) widths.

The Gaussian power density distribution is described as follows:

$$q(x,z,t) = \frac{3Q}{\pi C^2} exp^{\{-3[(z-vt)^2+x^2]/C^2\}} \tag{3}$$

$Q = \eta_{th}Vi$, where η_{th} is the thermal efficiency, V is the voltage and i is the electric current. The thermal efficiency n_{th} is defined as the ratio of the power required for melting the volume of metal in the fusion zone P_{th} and the emitted laser beam power P_E [19], and is described by:

$$n_{th} = \frac{P_{th}}{P_E} = \frac{A\ .\rho\ .\ \text{v}\ .\ (c_p\ .\ (T_M - T_A) + h_M)}{P_E} \tag{4}$$

where A is the area of the cross-section of the molten pool, ρ the density, v the welding speed, c_p the specific heat capacity, T_M the melting temperature, T_A the ambient temperature, and h_M the latent heat for melting. Using Equation (4), it should be considered that the energy for heating the material outside of the molten pool as well as the energy for overheating the molten pool are assumed to be energy losses. Equation (4) can also be written as:

$$Heat\ input = \frac{n_{th}\ .P_E}{\text{v}} \quad \left[\frac{J}{mm}\right] \tag{5}$$

A thermal efficiency of η_{th} = 0.5 and a heat input of 120 J/mm were used in all the simulations. The ABAQUS [20] user subroutines FILM and DFLUX were written to account for the convection, radiation and heat-input distribution. Of a similar manner, the dependent elastic modulus and yield strength (0.2% offset) for AISI 310 austenitic steel obtained from ref. [21] and plotted in Figure 2d,e, respectively, were included into the ABAQUS material module. It is important to note that the AISI 310 steel can be welded relatively easily to other weldable steels, and also to the HAYNES® 230® alloy. Material property values at very high temperatures (i.e., above the liquidus temperature) are difficult to obtain, and most of the information found in the literature is related to fits, assumptions, simplifications, and effective properties used to simulate phase transformations and stirring of the weld pool. Therefore, we prefer to feed the model with the available data documented in Figure 2, and allow the code's material module adjust the values itself based on the most appropriate trend.

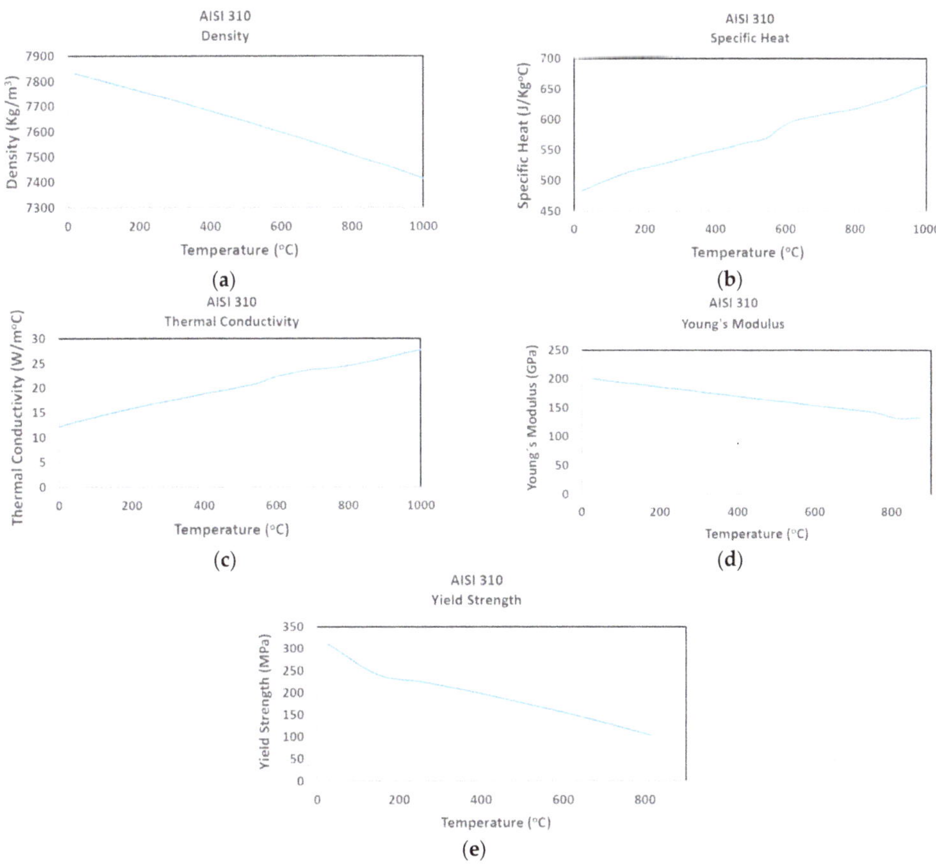

Figure 2. Temperature-dependent material properties for AISI 310 austenitic steel: (**a**) density; (**b**) specific heat; (**c**) thermal conductivity; (**d**) Young's modulus; and (**e**) yield strength.

The mechanical model framework used to represent the physical problem is presented in refs. [22,23] as follows:

The components of the total strain-rate tensor $\dot{\varepsilon}_{ij}$ are presented by the sum of the elastic $\dot{\varepsilon}^{e}_{ij}$ and plastic $\dot{\varepsilon}^{p}_{ij}$ components:

$$\dot{\varepsilon}_{ij} = \dot{\varepsilon}^{e}_{ij} + \dot{\varepsilon}^{p}_{ij} \tag{6}$$

The elastic part obeys Hooke's law and reads as follows:

$$\dot{\varepsilon}^{e}_{ij} = C_{ijkl}\dot{\sigma}_{kl}\,, \qquad C_{ijkl} = \frac{1+\nu}{E}\left(\delta_{ik}\delta_{jl} - \frac{\nu}{1+\nu}\delta_{ij}\delta_{kl}\right), \tag{7}$$

where C_{ijkl} is the elastic stiffness matrix, σ_{ij} is the stress tensor and a dot denotes the differentiation with respect to time t, E is Young's modulus, ν is the Poisson's ratio, and δ_{ij} is Kronecker's delta. The Levy–Mises equation accounts for the plastic part of the strain-rate tensor as follows:

$$\dot{\varepsilon}^{p}_{ij} = \frac{3}{2}\frac{\dot{p}}{q_{vm}}\sigma'_{ij} \tag{8}$$

Here, σ'_{ij} is the deviatoric stress tensor; the quantities

$$\dot{p} = \left(\frac{2}{3}\dot{\varepsilon}^{p}_{ij}\dot{\varepsilon}^{p}_{ij}\right)^{1/2} \tag{9}$$

and

$$q_{vm} = \sqrt{\frac{3}{2}\sigma'_{ij}\sigma'_{ij}} \tag{10}$$

are the equivalent von Mises plastic strain rate and stress, respectively. The substitution of the above equations into the equilibrium Equation (11)

$$\frac{\partial \sigma_{ji}}{\partial x_j} = 0 \tag{11}$$

leads to the governing equation to be solved by means of the finite element method. The elastic-plastic problem is a boundary-value problem that requires essential boundary conditions at each boundary point. Burnet [24] has shown that, for a well-posed problem, that is, to prevent the rigid-body motion of the entire structure, enough displacements (BCs) must be specified. In this work, to represent the fixation of the specimen to the base plate, appropriate mechanical boundary conditions were applied using the ABAQUS-BCs module. The eight corners of the thick sheet (the macro model) were firmly fixed with the ABAQUS option *ENCASTRE, and the ABAQUS option* ZASYMM was used to prevent rigid-body motion in the sub-model. Of this manner, loading from the master study was transferred as temperature or/and displacement values to the boundary between the surfaces included in the sub-model, and those neglected.

In a coupled temperature-displacement finite element model, the total strain increment at a material point from t → t + Δt can be expressed as:

$$\Delta\varepsilon_{ij} = \Delta\varepsilon^{e}_{ij} + \Delta\varepsilon^{p}_{ij} + \Delta\varepsilon^{th}_{ij} \tag{12}$$

Here, the total strain increment derived from the classical incremental plasticity theory can be accounted for an additive decomposition of the elastic, plastic and thermal strain component increments. On the other hand, a fairly general formulation adopted by most of the commercial finite element codes is the von Mises yield function defined by

$$f = q_{vm} - \sigma_y = \sqrt{\frac{3}{2}\sigma'_{ij}\sigma'_{ij}} - \sigma_y \tag{13}$$

Here, σ_y is the yield strength at temperature T, and q_{vm} is the von Mises stress. The yield criterion is given by

$f < 0$: Elastic deformation

$f = 0$: Plastic deformation

The use of the von Mises equivalent quantities implies the plastic isotropy (ideal plasticity) of the material. The C3D8T ABAQUS element type and appropriate built-in-type boundary conditions (thermal and mechanical) were used in the macro and meso models. From there, a transient nonlinear coupled temperature-displacement finite element model was performed to determine the evolution of temperature, cooling rate and residual stresses at the meso-scale level in a thick sheet AISI 310 laser welding test sample. The calculations of the distortions, stresses, and strains are computationally intensive, and they take significantly more computer time than the calculation of the transient temperature field.

The numerical simulations presented in the next section correspond to an autogeneous laser weld test (Figure 3) with a heat input of 120 J/mm (emitted laser beam power

P_E = 1500 Watts, heat source speed v = 6.25 mm/s and thermal efficiency n_{th} = 0.5) deposited on an AISI 310 austenitic steel macro mesh.

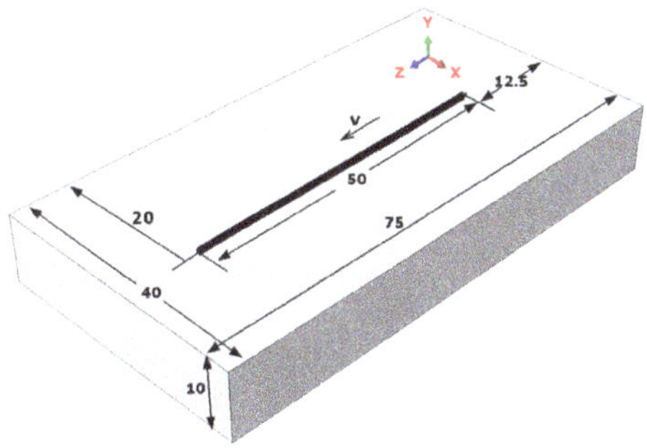

Figure 3. Schematic of the autogeneous laser weld test showing the overall dimensions, the direction of the welding speed, and the x, y, and z directions. Units in mm.

3. Results and Discussions

The multiscale simulation of microstructure and distortion evolution into the melt-pool is an important issue. It is, at the same time, a heavy and costly task, giving way to the use of different approaches traditionally used in multidomain approaches, such as level set or phase field. This may include, for example, the influence of cooling rates on the microstructure (formation of different phases, and grain texture), and resulting macro-scale properties. This also encompasses the modeling of heat sources combining fluid flow, solid mechanics, gas or plasma interaction, in order to predict the occurrence of various defects (for instance, porosity, segregation, and cracks) and evaluate their influence on local properties and part behavior. Since differences in the measured and calculated residual stresses are attributed to the absence of accurate stress–strain constitutive relations, uncertainties in the calculations of thermal cycles and the use of coarse grids, in this work, an efficient model order reduction approach is addressed. The numerical results obtained along three different paths located in the fusion zone were calculated in coarse and fine meshes. Figure 4 shows an autogenous laser weld deposited on a macro mesh of 40 × 10 × 75 mm, composed of 30,000 C3D8T elements of size 1 × 1 × 1 mm, and 3 thermal cycles calculated at points N1, N2 and N3 located ouside the fusion zone. It is observed that the cooling rates decrease as the analyzed locations are further away from the fusion zone.

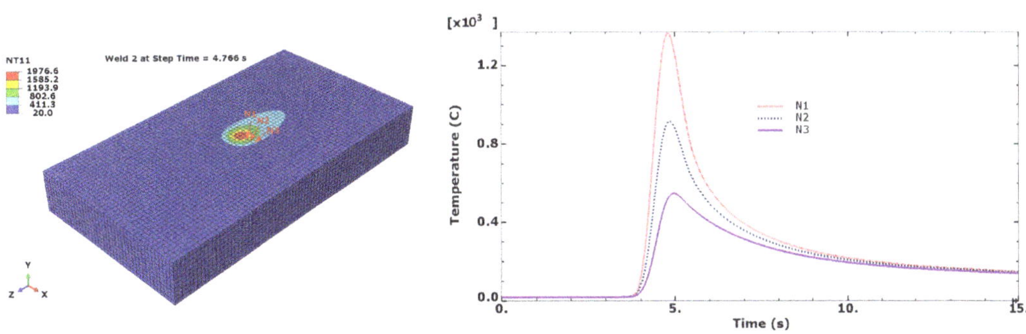

(a) (b)

Figure 4. (**a**) An autogenous laser weld performed in a macro mesh of 40 × 10 × 75 mm and (**b**) thermal cycles at documented N1, N2 and N3 locations. Element size 1 mm.

Figure 5 shows the rough Gaussian profiles of temperature vs. true distance along the documented paths.

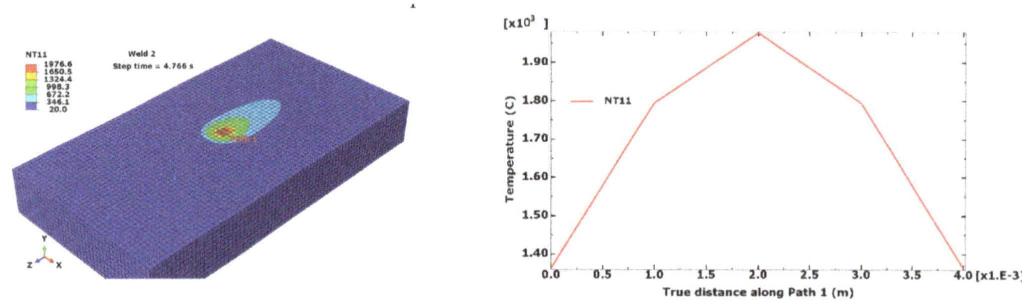

(**a**)

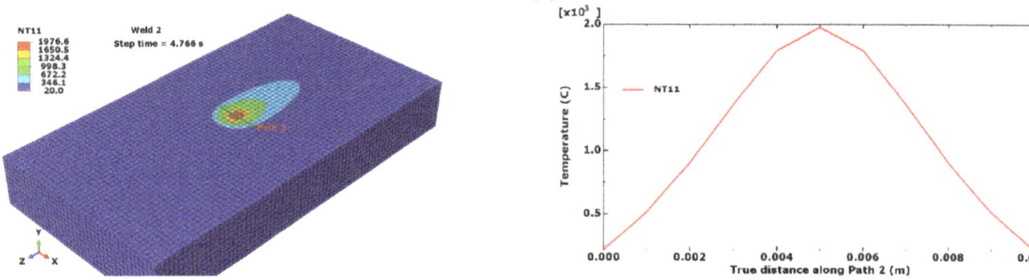

(**b**)

Figure 5. (**a**) Temperature distribution along Path 1 of 4 mm length and (**b**) temperature distribution along Path 2 of 10 mm length. Values obtained at the step time 4.766 s. Element size = 1 mm.

To obtain smoother Gaussian profiles, a fine meso grid was constructed from the global macro model via the submodeling technique. The domain of the meso model is located symmetrically at the top in the middle of the plate (see Figure 6a). In this figure, a description of the meso model with dimensions of 10 × 5 × 1 mm constructed from a global macro model of dimensions 40 × 10 × 75 mm (element size = 1 mm) via the submodeling technique using the Shape → Cut → Extrude options available in ABAQUS CAE is presented. A meso mesh of 50,000 C3D8T elements (element size = 100 μ) was generated. As expected, smoother Gaussian profiles of the temperature distribution along Paths 1 and 2 were obtained and are plotted in Figure 6b,c, respectively. The temperature–distance curves were obtained at the step time of 4.756 s.

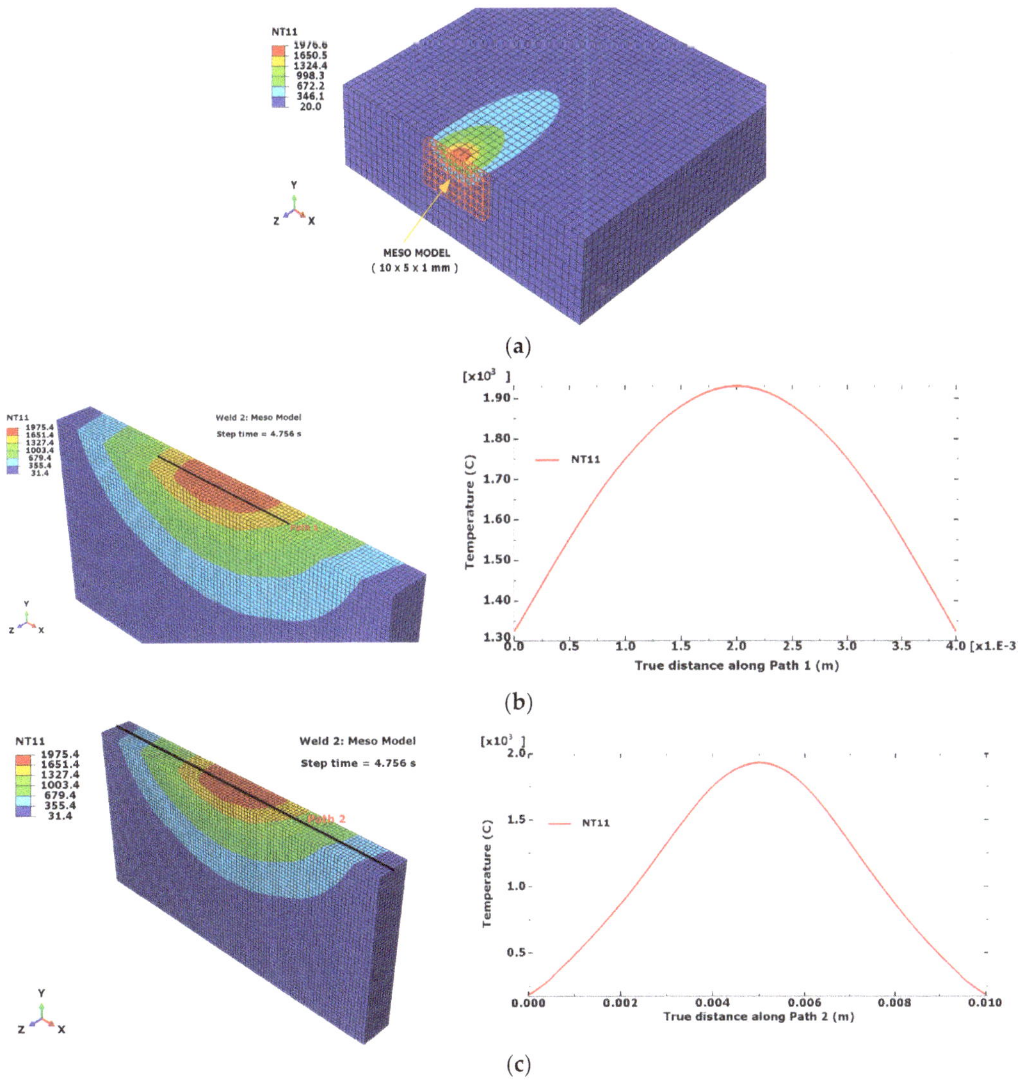

Figure 6. (**a**) Description of the meso model (10 × 5 × 1 mm) constructed from a macro model (40 × 10 × 75 mm) using the Shape → Cut → Extrude options available in ABAQUS CAE; (**b**) temperature distribution along Path 1; and (**c**) temperature distribution along Path 2. Values obtained at the step time 4.756 s. Element size = 100 μ.

Figure 7 shows a meso weld-pool cross-section, thermal cycles and cooling rates at six reported locations. The liquidus (T_L) and solidus (T_S) temperatures are also shown. For AISI 310 steel, the liquidus and the solidus temperatures are 1402 °C and 1354 °C, respectively [25]. Figure 7c reveals high cooling rates (through the solidification temperature range) in the points located at the meso fusion zone ranging from 960 °C/s at point D6 to 2400 °C/s at point D1.

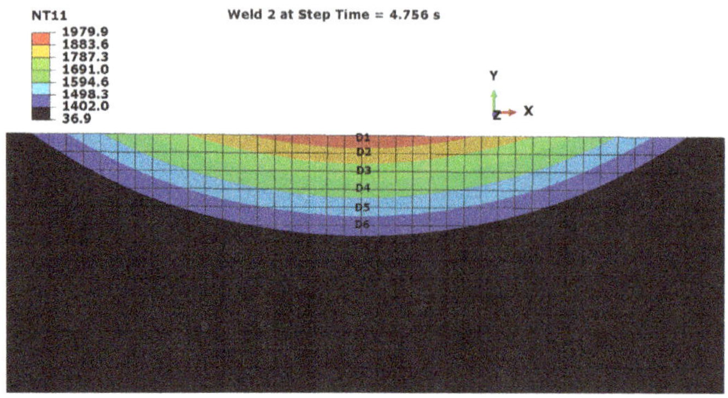

(a)

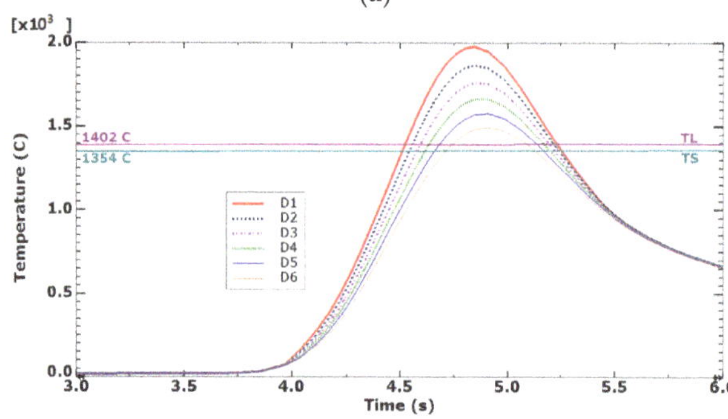

(b)

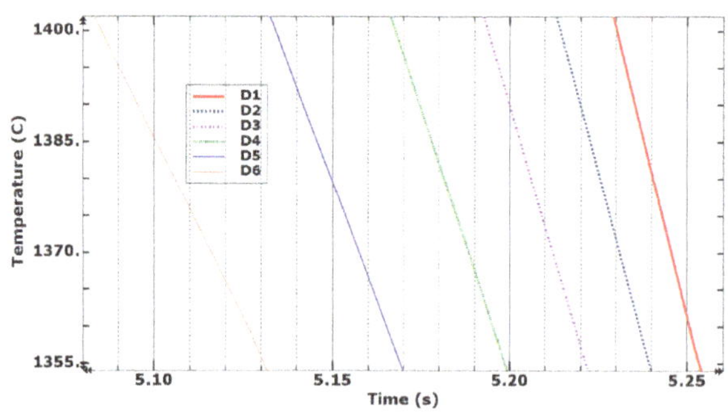

(c)

Figure 7. (**a**) Cross-section of the meso weld pool showing six locations (D1–D6) in Path 3; (**b**) thermal cycles (and the liquidus T_L and solidus T_S temperatures) at the documented locations; and (**c**) cooling rates (through the solidification temperature range) at the selected six locations.

Figure 8 shows the meso-level residual stresses (normal, shear and mises) at a cooling time of 60 s along Paths 1 and 2 (see the paths in Figures 5 and 6). The maximum observed value of 520 MPa corresponds to the normal stress S33 that follows the direction of the heat-source speed (longitudinal axes Z or 3). These high values explain the evolution of the distortion of the fusion zone due to the welding process. Macro residual stresses are commonly calculated with the finite element method using macro grids. However, rough residual stress profiles are commonly plotted due to the scarcity of numerical points. In the proposed meso model, as a fine mesh is used, smoothed profiles of residual stresses can be better graphed, as shown in Figure 8.

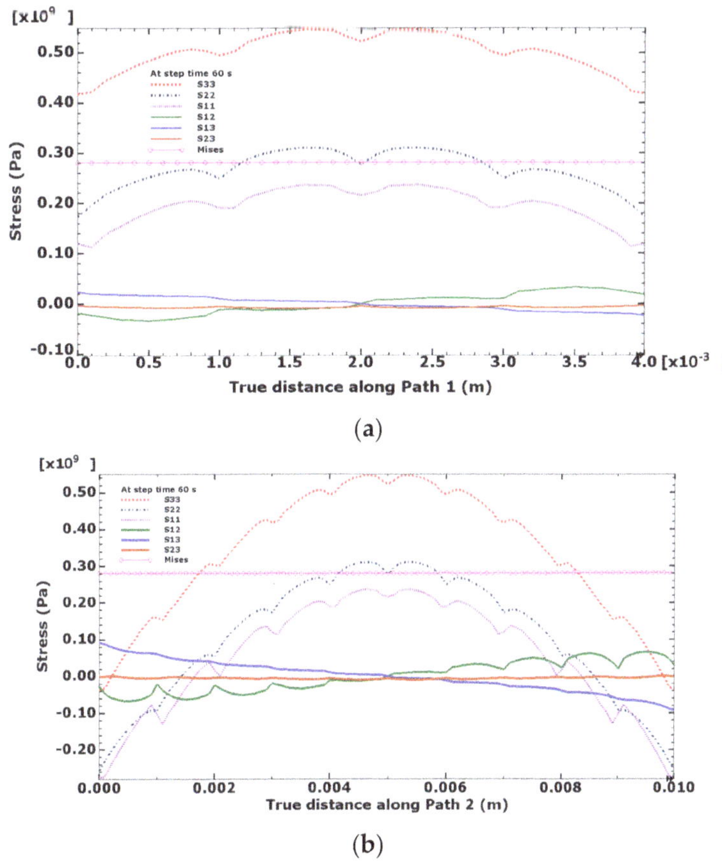

Figure 8. Meso-level residual stresses at cooling time 60 s (**a**) along Path 1 and (**b**) Path 2.

The present contribution is in line with the multiscale framework that covers the thermal and mechanical failure analysis of extended microstructural regions. The submodeling technique (a flexible strategy used to calculate plastic strains, temperature gradients, and residual stresses to any number of levels) was used to mesh a local portion of the part with a refined mesh, based on the interpolation of the solution from the initial coarse macro global model. In a future work, the presented meso-level residual stresses will be validated by a comparison to the experimental results performed under the same welding and boundary conditions. In addition, more numerical and experimental work will be performed to compare the best welding parameter combination to minimize residual stresses and the formation of stray and equiaxed grains. Depending on the experimental technique and magnification used, it is observed that dendrite secondary arm spacing belongs to the micro scale, cooling rate belongs to the macro scale, and residual stresses belong to the macro, meso or micro scale. Therefore, to find new expressions that link, for example, the grain substructure with the cooling rate that is imposed on the welding during solidification, the experimental and numerical work should be conducted at the same scale. At the present time, the connection between the scales is yet unclear [26]. An interesting overview of the state of the art in laser welding simulation can be found in refs. [27–29].

Figure 9a describes Path 3 (see also Figure 7a), which is used to calculate the stresses vs. true distance along Path 3 shown in Figure 9b. The stress evolution at location D1 during the first 10 s of the welding process is shown in Figure 9c. The point D1 located at the top surface is the starting point of Path 3 (coordinate 0.00), and the point D6 located at 0.5 mm from D1 in the Y (2) direction is the end point of Path 3.

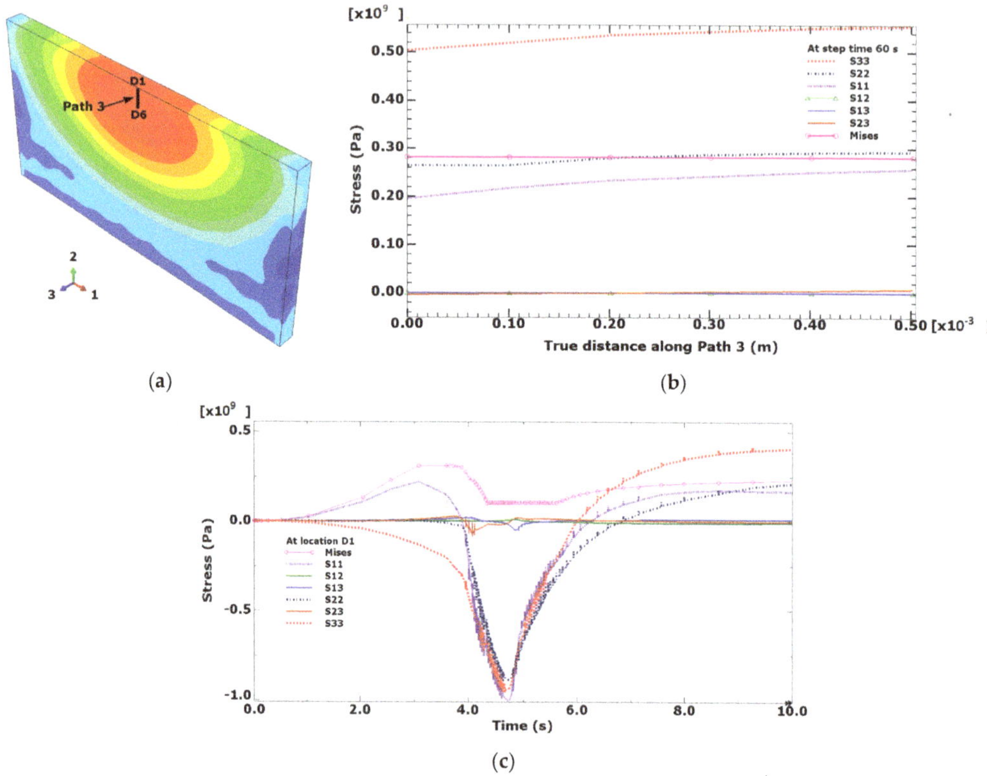

Figure 9. (**a**) Description of Path 3, (**b**) stress vs. true distance along Path 3 at step time 60 s, and (**c**) stress evolution at location D1.

After the process has started, it takes 4.756 s for the heat source to be positioned just above the meso-model domain located at the central part of the plate. At that time, maximum negative values of stresses appear (see Figure 9c). Figure 9c also shows the evolution of the stresses that change signs (from negative-to-positive values) as the cooling time increases.

The evolution of the meso-normal stresses S33 is shown in contour graphs in Figure 10. These stress contours evolve from 100% negative values at step time 3 s, to almost 100% positive values at step time 60 s. At this step time, the plate temperature has already cooled down to approximately 70 °C. It therefore means that only a few more seconds are needed to finish the cooling cycle, and thus the complete conversion to positive values. It is important to note that, in order to mesh the 40 × 10 × 75 mm plate with elements of a 0.1 × 0.1 × 0.1 mm size, a total of 30 million elements is necessary. With the present approach, one macro mesh of 30,000 elements (1 × 1 × 1 mm) and a meso mesh of 50,000 (0.1 × 0.1 × 0.1 mm) elements were enough to simulate the weld problem at the meso-scale level. Multiscale modeling is based on fundamental physical principles and experimental data. The goal is to predict the behavior and performance of complex materials across all relevant time-and-length scales. The challenge is tremendous due to the fact that, at the macro scale (centimeters), stresses arising from temperature gradients may be the controlling elements of the materials' performance. At the microscale (tens of micrometers), defects, such as dislocations controlling the mechanical behavior, occur, while large collections of such defects, including grain boundaries and other microstructural elements, govern the mesoscopic properties (hundreds of micrometers). The net outcome of these interactions governs the continuum behavior that can be described as a constitutive law. To analyze the thermal and mechanical phenomena on the micro scale, the proposed meso-model based on the interpolation of the solution from an initial relatively coarse macro global model has to be considered as the global model for a subsequent micro submodel. In this manner, the cooling rate–microstructure relationship can be calculated at the same scale level.

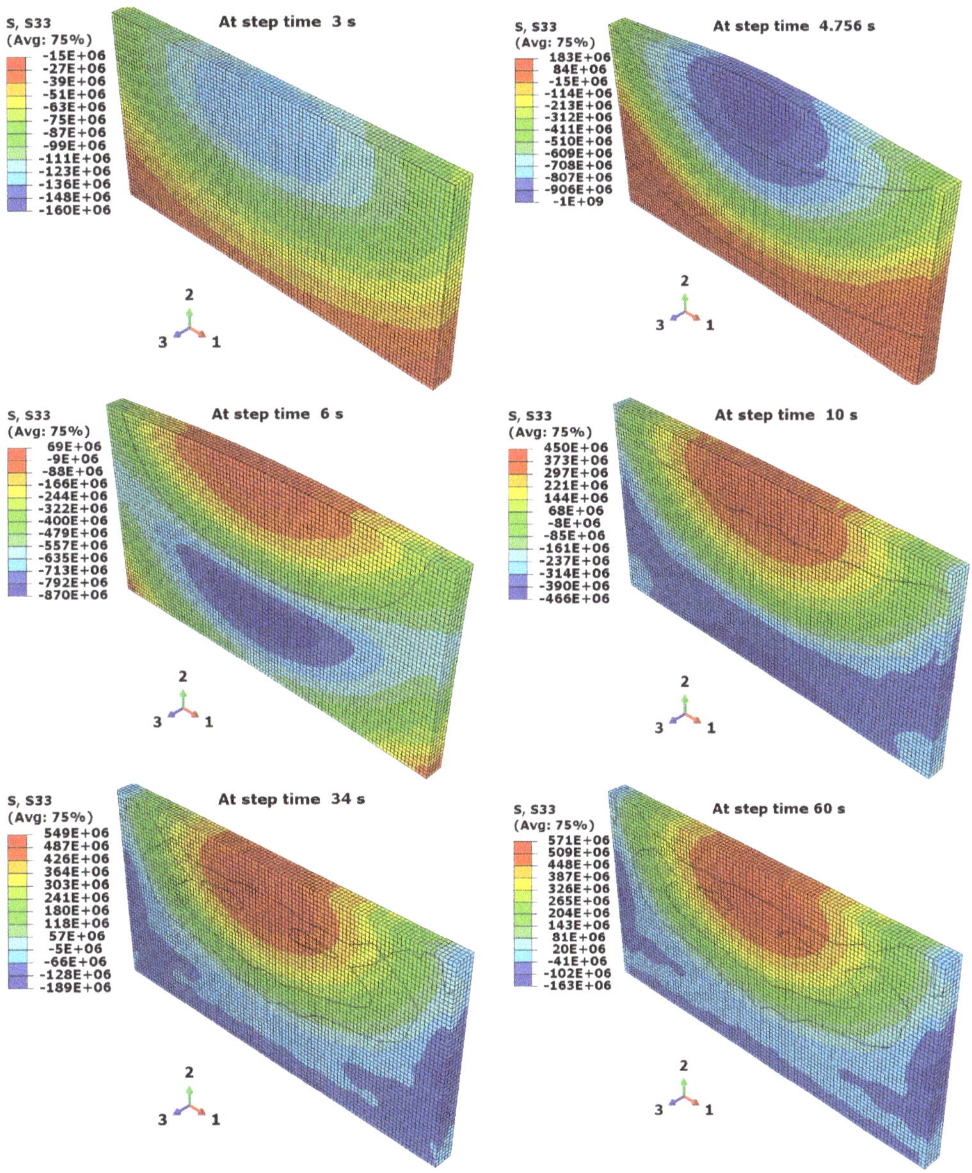

Figure 10. Evolution of the meso-level residual stresses (S33) at documented step times.

4. Conclusions

1. A new three-dimensional coupled temperature-displacement finite element meso model constructed using appropriate ABAQUS CAE submodeling options is proposed.
2. High cooling rates (through the solidification temperature range) ranging from 960 °C/s to 2400 °C/s were observed when the calculations were made at the mesoscale level. These high cooling rates that arise during the formation of the weld pool originate the highest observed residual stresses that evolve throughout the weld during the entire heating and cooling cycles.
3. To mesh the entire plate with elements of a regular size of $100 \times 100 \times 100$ μ, a total of 30 million elements are necessary. With the present approach, one macro mesh of 30 thousand elements ($1 \times 1 \times 1$ mm) and a meso mesh of 50 thousand elements ($100 \times 100 \times 100$ μ) were enough to simulate the weld problem at the meso-scale level.

Author Contributions: Investigation, E.A.B. and A.S.M.; Writing—original draft, E.A.B. All authors have read and agreed to the published version of the manuscript.

Funding: This research received no external funding.

Institutional Review Board Statement: Not applicable.

Informed Consent Statement: Not applicable.

Data Availability Statement: Not applicable.

Acknowledgments: E.A.B. would like to thank to the Universidad San Francisco de Quito for financial support.

Conflicts of Interest: The author declares no conflict of interest.

References

1. Fernandes de Lima, M.S.; Goia, F.A.; Riva, R.; Espírito Santo, A.M. Laser Surface Remelting and Hardening of an Automotive Shaft Sing a High-power Fiber Laser. *Mater. Res.* **2007**, *10*, 461–467. [CrossRef]
2. Liu, W.; Dupont, J. Effects of melt-pool geometry on crystal growth and microstructure development in laser surface-melted superalloy single crystals. *Acta Mater.* **2004**, *52*, 4833–4847. [CrossRef]
3. Park, J.; Babu, S.; Vitek, J.; Kenik, E.; David, S. Stray grain formation in single crystal Ni-base superalloy welds. *J. Appl. Phys.* **2003**, *94*, 4203. [CrossRef]
4. Scholz, F.; Cevik, M.; Hallensleben, P.; Thome, P.; Eggeler, G.; Frenzel, J. A 3D Analysis of Dendritic Solidification and Mosaicity in Ni-Based Single Crystal Superalloys. *Materials* **2021**, *14*, 4904. [CrossRef]
5. Vitek, J. The effect of welding conditions on stray grain formation in single crystal welds—Theoretical analysis. *Acta Mater.* **2005**, *53*, 53–67. [CrossRef]
6. Norman, A.; Ducharme, R.; Mackwood, A.; Kapadia, P.; Prangnell, P. Application of thermal modelling to laser beam welding of aluminium alloys. *Sci. Technol. Weld. Join.* **1998**, *3*, 260–266. [CrossRef]
7. Mokadem, S.; Bezencon, C.; Hauert, A.; Jacot, A.; Kurz, W. Laser Repair of Superalloy Single Crystals with Varying Substrate Orientations. *Metall. Mater. Trans. A* **2007**, *38A*, 1500–1510. [CrossRef]
8. Churchman, C.; Bonifaz, E.A.; Richards, N.L. Comparison of Single Crystal Ni Based Superalloy Repair by Gas Tungsten Arc and Electron Beam Processes. *Mater. Sci. Technol.* **2011**, *27*, 811–817. [CrossRef]
9. Bonifaz, E.A. Finite Element Analysis of heat flow in single pass arc welds. *Weld. J.* **2000**, *79*, 121s–125s.
10. Bonifaz, E.A.; Richards, N. Modeling Cast In-738 Superalloy Gas-Tungsten-Arc- Welds. *Acta Mater.* **2009**, *57*, 1785–1794. [CrossRef]
11. Tekriwal, P.; Mazumder, J. Finite element analysis of three dimensional heat transfer in GMA welding. *Weld. J.* **1988**, *67*, 150-s–156-s.
12. Debroy, T.; David, S.A. Physical processes in fusion welding. *Rev. Mod. Phys.* **1995**, *67*, 85–112. [CrossRef]
13. Brown, S.B.; Song, H. Implications of Three-Dimensional Numerical Simulations of Welding of Large Structures. *Weld. J.* **1992**, *71*, 55-s–62-s.
14. Thiessen, R.G.; Richardson, I.M. A physically based model for microstructure development in a macroscopic heat-affected zone: Grain growth and recrystallization. *Metall. Mater. Trans. B* **2006**, *37B*, 655–663. [CrossRef]
15. Kou, S. *Welding Metallurgy*, 2nd ed.; Wiley: New York, NY, USA, 2003.
16. Bonifaz, E.A.; Conde, J.M.; Czekanski, A. Determination of secondary dendrite arm spacing for IN-738LC gas-tungsten-arc-welds. *J. Multiscale Model.* **2019**, *10*, 1850012. [CrossRef]
17. Bonifaz, E.A. Thermo-mechanical analysis in SAE-AISI 1524 carbon steel gas tungsten arc welds. *Int. J. Comput. Mater. Sci. Surf. Eng.* **2018**, *7*, 269–287.

18. Kik, T. Heat source models in numerical simulations of laser welding. *Materials* **2020**, *13*, 2653. [CrossRef]
19. Ganser, A.; Pieper, J.; Liebl, S.; Zaeh, M.F. Numerical simulation of the thermal efficiency during laser deep penetration welding. *Phys. Procedia* **2016**, *83*, 1377–1386. [CrossRef]
20. ABAQUS. *Finite Element Software, version 6.14.2*; Simulia ABAQUS®Inc.: Waltham, MA, USA, 2018. Available online: https://www.3ds.com/es/productos-y-servicios/simulia/productos/abaqus/ (accessed on 10 February 2021).
21. The Engineering ToolBox. Resources, Tools and Basic Information for Engineering and Design of Technical Applications. Available online: https://www.engineeringtoolbox.com/stainless-steel-310-properties-d_2167.html (accessed on 20 October 2021).
22. Dunne, F.; Petrinic, N. *Introduction to Computational Plasticity*; Oxford University Press: Oxford, UK, 2005.
23. Bonifaz, E.A.; Richards, N.L. The plastic deformation of non-homogeneous polycrystals. *Int. J. Plast.* **2008**, *24*, 289–301. [CrossRef]
24. Burnett, D.S. *Finite Element Analysis: From Concepts to Applications*; Addison-Wesley Publishing Company: Boston, MA, USA, 1988.
25. Folkhard, E. *Welding Metallurgy of Stainless Steels*, 1st ed.; Springer: New York, NY, USA, 1988.
26. Bonifaz, E.A.; Watanabe, I. Anisotropic Multiscale Modelling in SAE-AISI 1524 Gas Tungsten Arc Welded Joints. *Crystals* **2021**, *11*, 245. [CrossRef]
27. Winczek, J. The influence of the heat source model selection on mapping of heat affected zones during surfacing by welding. *J. Appl. Math. Comput. Mech.* **2016**, *15*, 167–178. [CrossRef]
28. Dal, M.; Fabbro, R. An overview of the state of art in laser welding simulation. *Opt. Laser Technol.* **2016**, *78*, 2–14. [CrossRef]
29. Feng, Y.; Gao, X.; Zhang, Y.; Peng, C.; Gui, X.; Sun, Y.; Xiao, X. Simulation and experiment for dynamics of laser welding keyhole and molten pool at different penetration status. *Int. J. Adv. Manuf. Technol.* **2021**, *112*, 2301–2312. [CrossRef]

crystals

MDPI

Article

Micromechanical Effect of Martensite Attributes on Forming Limits of Dual-Phase Steels Investigated by Crystal Plasticity-Based Numerical Simulations

Tarek Hussein [1], Muhammad Umar [1,2,*], Faisal Qayyum [1], Sergey Guk [1] and Ulrich Prahl [1]

[1] Institute of Metal Forming, Technische Universität Bergakademie Freiberg, 09599 Freiberg, Germany; tarekelsesi22@gmail.com (T.H.); faisal.qayyum@imf.tu-freiberg.de (F.Q.); sergey.guk@imf.tu-freiberg.de (S.G.); ulrich.prahl@imf.tu-freiberg.de (U.P.)

[2] Department of Mechanical Engineering, Khwaja Fareed University of Engineering and Information Technology, Rahim Yar Khan 64200, Punjab, Pakistan

* Correspondence: umar.matrix@gmail.com

Abstract: This study analyses the effect of martensite grain size and its volume fraction in dual-phase (DP) steel on (1) the formability limit, (2) average global behavior under different loading conditions, and (3) damage initiation. The virtual RVEs (Representative Volume Elements) were constructed using DREAM.3D software with a variation of microstructural attributes. The numerical simulations were carried out using DAMASK, which evaluates the polycrystalline material point behavior and solves versatile constitutive equations using a spectral solver. The simulations were post-processed to obtain global and local stress, strain, and damage evolution in constructed RVEs. The global results were processed to obtain FLDs according to Keeler-Brazier (K-B) and Marciniak and Kuczynski (M-K) criteria. In this work, the capability of microstructure-based numerical simulations to analyze the FLDs has been established successfully. From Forming Limit Diagrams (FLDs), it was observed that formability changes by changing the strain hardening coefficients (n-values), the martensite fraction, and martensite grain sizes of DP steels. The improved formability was observed with lower martensite fraction, i.e., 17%, decreased martensite grain size, i.e., 2.6 µm, and higher strain hardening coefficient. The M-K approach shows the better capability to predict the formability by various loading conditions and clarifies the necking marginal zone of FLD. The damage propagation is also strongly affected by the loading conditions. The current study would be a good guide for designers during the manufacturing and selecting of appropriate DP steels based on the service loading conditions.

Keywords: dual-phase steel; forming limit diagrams; crystal plasticity; DAMASK; M-K approach; Keeler-Brazier approach

Citation: Hussein, T.; Umar, M.; Qayyum, F.; Guk, S.; Prahl, U. Micromechanical Effect of Martensite Attributes on Forming Limits of Dual-Phase Steels Investigated by Crystal Plasticity-Based Numerical Simulations. *Crystals* **2022**, *12*, 155. https://doi.org/10.3390/cryst12020155

Academic Editor: Wojciech Polkowski

Received: 20 December 2021
Accepted: 17 January 2022
Published: 21 January 2022

1. Introduction

Dual-Phase (DP) steel, due to its higher energy absorption capacity and reduced weight, is used in the automotive industry to achieve simultaneous high strength and elongation goals [1–3]. Hard phase martensite laths embedded in the softer ferrite matrix are responsible for reinforcing the solid aggregate, while ductility is incorporated by the matrix [4]. Dual-phase steel is a suitable example of a multi-phase material because of the significant difference in the mechanical properties of its phases, and it has widespread usage in the automotive industry. This peculiar combination of hard and soft phases imparts desirable properties in the material, i.e., low 0.2% proof stress and a high work-hardening coefficient (n-value). Generally, during cold forming processes of drawing and stretching, higher n-values exhibit uniform global formability by avoiding local thinning [5]. However, practical aspects often reveal unexpected necking and local failure in complex forming during bending or flanging processes. Due to the heterogeneous microstructure of soft ferrite and hard martensite, local straining causes unpredicted local abnormalities. The heterogeneity affects the micro-scale attributes of materials. Consequently, it influences the component scale's material properties, particularly the material damage behavior [6]. Therefore, it is of the utmost importance to investigate the relationship between the phases' heterogeneity and their microstructural attributes, especially martensite and ferrite fractions and their grain sizes [7].

Material formability is a fundamental mechanical property for vehicle bodies and structural members in the automotive sector. Therefore, accurate predictions of its indicators and reliable utilization are needed. For this purpose, FLDs are commonly used to estimate sheet metal formability. Its purpose is to predict safe, necking, and failure zones by adopting major and minor forming limit strain inside a diagram [8]. Experimentally, the Nakajima test detects these forming limit strains, which is used by performing a punch–die method on specimens with varying dimensions under different loading conditions. The test shows a higher accuracy for most materials, but it is expensive, slow, and relies on a complex specimen-shaping process. Therefore, FLDs are plotted using some well-established numerical models [9]. Tasan et al. [10] carried out nanoindentation experiments and concluded that the numerical damage could not be applied with better accuracy to the martensite phase because the indents are the same size or bigger than martensite grains. While there is a considerable difference in the indent and average grain size for ferrite, the identified parameters in their study could predict the damage in the ferrite phase [11].

Analytical models commonly applied to predict and plot FLDs by adopting major and minor limit strains are Swift, Chow–Hosford, Keeler–Brazier, and Marciniak–Kuczynski [11]. These approaches have been analyzed and compared in depth in research work by Basak and G Béres [12,13]. These approaches depend on some variables to develop the mathematical model and corresponding limit strains, i.e., Swift models require strain ratios and strain hardening coefficient values. Chow equations depend on the anisotropy coefficients, and for the Keeler–Brazier model, the thickness value and strain hardening coefficient are primarily important. The simulations in this study were performed using two approaches, i.e., the Keeler–Brazier, and M-K approaches, because of their efficiency, popularity, and simplicity in plotting FLDs of high-strength steels recommended by Duancheng Ma, Basak, and Béres. Kuang-Hua Chang et al. [14] used damage percentage detection to show the marginal zone of an FLD, where the area under this zone is safe, while the area above it is a damage zone.

Virtual modeling and simulation tools for the sheet metal processes play an essential role in predicting mechanical behavior and have become an essential and inevitable part of each industry [15,16]. These tools introduce full behavior prediction models of the metals, starting from their production to the heat treatment and testing processes. It helps engineers and designers to reduce the resource-consuming experiments and push the overall economic and technical aspects forward. A significant amount of work is being carried out toward developing, validating, and implementing numerical models to improve the accuracy of the simulation results. The numerical simulations method is an intelligent tool used to solve complex equations with different variables, called constitutive equations. These equations and variables represent the physical situations of the materials, e.g., deformation mechanisms, and help analyze the phenomena of mechanical deformation, damage, and failure. It can also predict crucial mechanical thresholds for high-end application materials during melting, casting, forming, and machining [17].

Hutchinson [18] built a model for FCC crystal, which was later expanded for BCC and HCP crystal structure and applied to DAMASK (Düsseldorf Advanced Materials Simulation Kit). Michel et al. [19] stated that the heterogeneity of ferrite and martensite phases in DP steels and their elastic stiffness coefficients play an essential role in the convergence behavior and stability schemes. After that, Diehl et al. [20] used some FFT assumptions and developed the stiffness coefficients applied later in the DAMASK framework. [7] The phenomenological crystal plasticity model is employed as plasticity law, which assumes that plastic deformation occurs on a slip system when the resolved shear stress exceeds a critical value. The resolved critical shear stress depends on the amount of the applied stress and can be calculated from a relation known as Schmid's law in Equation (1) [19,21].

$$\dot{\gamma}^{\alpha} = \dot{\gamma}_0 \left| \frac{\tau^{\alpha}}{S^{\alpha}} \right|^{n} \text{sgn}(\tau^{\alpha}) \quad (1)$$

Analytical techniques can merely solve partial differential equations used in these constitutive models more straightforwardly, while numerical methods are essential for complex forms. Many efforts have been made to reach a framework connecting these boundary problems with physical phenomena [18]. Many numerical methods, i.e., Finite Element Method (FEM), Finite Volume Method (FVM), spectral method, and the Fast Fourier Transform (FFT)-based (Crystal Plasticity Finite Element Method) CPFEM method are usually used [22,23]. The difference between FE and SP methods lies in their homogenization technique and consequent time saving, as Shanthraj et al. [24] reported. The analytical FLD strain calculation models must be solved numerically with a high-efficiency solver to construct FLDs [25].

A representative volume element (RVE) was constructed for this purpose as a virtual sample to express the properties and microstructure of the material in the sample. In this study, spectral solver methods were applied using Fast Fourier Transform (FFT) to solve the boundary value problem for mechanical equilibrium and damage phase field. An FFT-based spectral solver shows higher efficiency and faster computational time over the domain.

The elastic and plastic material properties are governed by the crystal plasticity-based constitutive model, which is extensively used to study the deformation in crystalline materials. Besides considering microstructural parameters, it also considers some fitting parameters to compensate for the influence of some complex phenomena happening during the processing route of a specific grade of the material [25,26]. The simulations presented in this work were performed using DAMASK, which is available as free and open-source software [25]. It aims to simulate the material using crystal plasticity principles within a finite strain framework for continuum mechanical considerations by modeling the material point (Fourier point) inside the constructed mesh. The plasticity laws implemented in DAMASK are represented by isotropic plasticity, phenomenological crystal plasticity, or dislocation density-based crystal plasticity. The phenomenological crystal plasticity models were used in these simulations of polycrystal models. Furthermore, as deformation of polycrystalline aggregate strongly depends on the respective orientations of grains inside the lattice, consideration of Euler angles and rotation matrices was taken during the implementation of the model [20,25].

In this work, the establishment of a crystal plasticity-based approach to evaluate the forming limits of multi-phase steels by numerical simulation has been carried out successfully. A plastic instability approach and a high-efficiency numerical solver were employed to investigate multi-phase materials' formability by adopting major and minor strains in FLDs. Specifically, the microstructural attributes of martensite, i.e., grain size and phase fraction under different loading conditions, were studied. Furthermore, the damage initiation and propagation were determined by a CPFFT-based spectral solver using a phenomenological model for plastic deformation and Hooke's law for elasticity. The conclusions of this study can help the sheet metal industries as a guide for the designers to process material in an improved way.

2. Methodology

The numerical simulation modeling approach is a systematic process that uses well-defined pre-processing steps, running a set of simulations, post-processing, and visualization of the results, as shown in Figure 1. Firstly, RVEs were constructed by varying martensite volume fraction and martensite grain size by using DREAM.3D [27]. Next, the DAMASK adopts individual grains details to run numerical simulations using given load and geometry configuration files. Constitutive equations and microstructural attributes for ferrite and martensite phases were adopted from Tasan's work [10]. Ductile damage and degradation parameters already implemented by Shanthraj et al. [24] were applied to the ferrite phase within a pre-developed model by Roters and Tasan et al. [20,25]. Finally, the values of local and global results were extracted after simulations and visualized by ParaView and other trend-plotting tools.

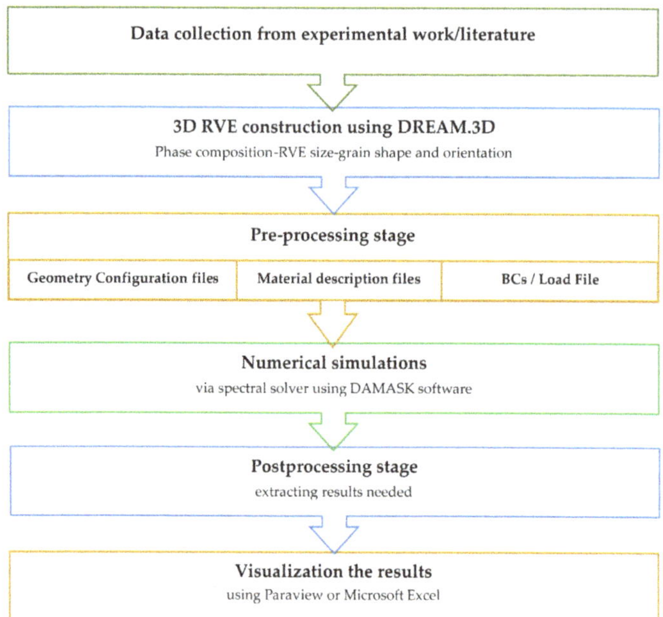

Figure 1. A simple flow chart showing the methodology adopted and data flow in this study.

2.1. Data Collection

The microstructural attribute values, e.g., martensite fractions and grain sizes, were taken from Tasan et al. [10], as shown in Table 1. The study was carried out for DP600 with the variation of martensite attributes (fractions 1.2%, martensite grain size from 1.0 to 4.3 µm) and a considerable ferrite grain size difference within 2.2 to 14.5 µm.

Table 1. The chemical composition and microstructural characteristics of DP 600 used in the current study. Reprinted with permission from Ref. [10] Copyright 2014 Elsevier.

Steel	Martensite (%)	Ferrite Grain Size (µm)	Martensite Grain Sizes (µm)
DP600	17.2	8.4 ± 6.1	2.7 ± 1.6
	18.4	4.9 ± 1.9	1.7 ± 1.1

2.2. RVE Construction

From DP steel values in Table 1, four RVE models (A–D) were constructed using DREAM.3D with variations in RVE dimensions, phase fractions, martensite grain size, and spatial distribution (refer to Figure 2). However, in all the five RVE models, the same hardness properties of martensite and ferrite grain size were considered, as shown in Table 2.

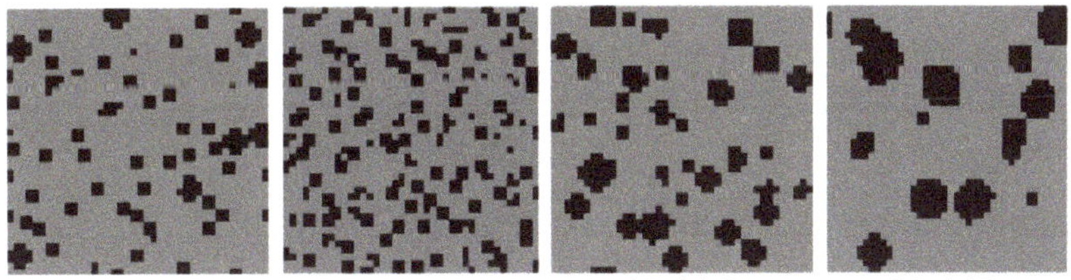

Figure 2. Top surfaces of 3D RVE models of the dual-phase steels, where the black particles represent the martensite phase inside the grey ferrite matrix.

Table 2. The projected RVEs used in the study with ferrite grain size 8.4 ± 6.1 µm for all the RVEs.

Parameter	Model A	Model B	Model C	Model D
Synthetic volume size (voxels)	40 × 40 × 10	40 × 40 × 10	40 × 40 × 10	40 × 40 × 10
Martensite grain size (µm)	2.7 ± 1.6	2.7 ± 1.6	4 ± 2	6 ± 2
Martensite volume fraction	17%	18%	17%	17%
Ferrite volume fraction	83%	82%	83%	83%

Suitable statistical distributions, cubic crystal structure, and ellipsoid grain shapes closely mimic the actual microstructure of DP Steel. In addition, different dimensions of synthetic volume were adopted to check the effect of RVE size on the global stress–strain curves and FLDs. Qayyum et al. [28] have shown a more detailed framework for the RVE generation using the DREAM.3D pipeline in their work. If interested, the readers are encouraged to refer to their work for further details.

2.3. Pre-Processing Stage

2.3.1. Material Properties

The ferrite and martensite phases in DP steel have some common elastic–viscoplastic properties, which help build a more manageable material file framework, despite variations in their mechanical behavior and properties. The material parameters and damage values of both phases were adopted from Qayyum et al. [7], wherein already developed and validated models from the framework of DAMASK [25] were adopted. The elastic coefficients, initial and saturated shear resistances of slip systems and fitting parameters from the already published literature [10,29] were used as presented in Table 3. Regarding damage, the already developed models from Roters et al. [25] were incorporated and adopted in the material configuration files of respective RVEs. The critical plastic strain value ϵ_{crit} for the ferrite phase was taken as 0.5 [10].

2.3.2. Boundary/Loading Conditions

Different loading conditions were used in the simulation process with four strain states on each RVE model along y-directions while controlling x-directions and freeing the z-directions (refer to Figure 3).

The periodic boundary conditions were stated as:

In the uniaxial tension state:

$$\dot{\mathbf{F}} = \begin{bmatrix} * & 0 & 0 \\ 0 & 1 & 0 \\ 0 & 0 & * \end{bmatrix} \times 10^{-3}\,\mathrm{s}^{-1} \qquad \& \qquad \mathbf{P} = \begin{bmatrix} 0 & * & * \\ * & * & * \\ * & * & 0 \end{bmatrix} \mathrm{Pa} \tag{2}$$

Table 3. Physical and fitting parameter values were adopted from the published literature for ferrite [29], reprinted with permission from Ref. [10] Copyright 2014 Elsevier, and martensite [7].

Parameter Definition	Symbol	Ferrite Attributes	Martensite Attributes
Lattice crystal structure	lattice structure	bcc	bcc
First elastic stiffness constant with normal strain	C11	233.3 GPa	417.4 GPa
Second elastic stiffness constant with normal strain	C12	135.5 GPa	242.4 GPa
First elastic stiffness constant with shear strain	C44	118.0 GPa	211.1 GPa
Shear strain rate	γ^{α}	$10^{-3}/s$	$10^{-3}/s$
Initial shear resistance on [111]	So [111]	95 MPa	405.8 MPa
Saturation shear resistance on [111]	S∞ [111]	222 MPa	872.9 MPa
Initial shear resistance on [112]	So [112]	97 MPa	456.7 MPa
Saturation shear resistance on [112]	S∞ [112]	412 MPa	971.2 MPa
Slip hardening parameter	h_0	1000 MPa	563.0 MPa
Interaction hardening parameter	hα,β	1	1
Stress exponent	n	20	20
Curve fitting parameter	w	2	2
Damage parameters			
Interface energy	g_0	$1.0\ \mathrm{J\ m^{-2}}$	-
Characteristic length	l_0	1.5 μm	-
Damage mobility	M	$0.01\ \mathrm{s^{-1}}$	-
Damage diffusion	D	1.0	-
Critical plastic strain	ϵ_{crit}	0.5	-
Damage rate sensitivity	P	10	-

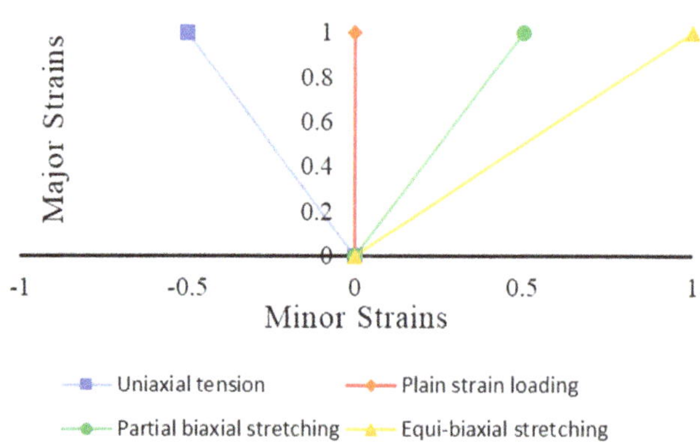

Figure 3. Projected loading states to conclude the forming limit diagram.

In the plane strain state:

$$\dot{\mathbf{F}} = \begin{bmatrix} 0 & 0 & 0 \\ 0 & 1 & 0 \\ 0 & 0 & * \end{bmatrix} \times 10^{-3}\,\mathrm{s}^{-1} \qquad \& \qquad \mathbf{P} = \begin{bmatrix} * & * & * \\ * & * & * \\ * & * & 0 \end{bmatrix} \mathrm{Pa} \tag{3}$$

In the partial biaxial stretching state:

$$\dot{\mathbf{F}} = \begin{bmatrix} 0.5 & 0 & 0 \\ 0 & 1 & 0 \\ 0 & 0 & * \end{bmatrix} \times 10^{-3}\,\mathrm{s}^{-1} \qquad \& \qquad \mathbf{P} = \begin{bmatrix} * & * & * \\ * & * & * \\ * & * & 0 \end{bmatrix} \mathrm{Pa} \tag{4}$$

In the equi-biaxial stretching state:

$$\dot{\mathbf{F}} = \begin{bmatrix} 1 & 0 & 0 \\ 0 & 1 & 0 \\ 0 & 0 & * \end{bmatrix} \times 10^{-3}\,\mathrm{s}^{-1} \qquad \& \qquad \mathbf{P} = \begin{bmatrix} * & * & * \\ * & * & * \\ * & * & 0 \end{bmatrix} \mathrm{Pa} \tag{5}$$

where $\dot{\mathbf{F}}$ and $\mathbf{P}$ are the rate of the deformation gradient and first Piola–Kirchhoff stress tensors, respectively. The coefficients denoted by '$*$' express the stated complimentary conditions. Using these conditions, the four strain states were applied in the y-direction, with a $1 \times 10^{-3}\,\mathrm{s}^{-1}$ iso-static strain rate, as shown in Figure 3. For readers not familiar with the modeling strategy, a brief model background is provided in the work of Qayyum et al. [7] and Duancheng Ma et al. [30].

2.4. *Evaluation of FLDs by M-K and K-B Approaches*

2.4.1. M-K Approach

In this approach, the engineering stress and strain values were extracted by applying customized subroutines. Then engineering stress–strain curves of the four loading conditions for each RVE were plotted. According to Duancheng Ma [30], Drucker's stability criterion claims that the forming limit occurs at the maximum stress (localized necking point) on the engineering stress–strain curve.

2.4.2. Keeler-Brazier Approach

In this approach, the true stress and strain values were extracted from applying already customized subroutines, and then the true stress–strain curve was plotted. As per the Keeler–Brazier model detailed in appendix A, the major strain values (ε_1) depend on t (RVE thickness), n (strain hardening coefficient of material), and ε_2 (minor strain values detected from the tensor matrix at an increment, corresponding to the maximum stress on the stress and strain curve). Therefore, the major and minor strain values were available for the four loading conditions. These and other extended details about post-processing and an activity flow chart of M-K and K-B approaches are given in Appendix A.

3. Results

Simulations of multi-phase DP steel were processed for 3D RVEs with damage consideration, wherein four models were virtually constructed to detect the variation in FLDs' behavior by varying microstructural attributes. By simulating different synthetic volumes of RVEs with the same ferrite grain sizes, the varying behavior of engineering stress–strain curves and FLDs was observed.

3.1. *Effect of Martensite Fractions on Stress–Strain Curve and FLDs*

The numerical simulations were carried out for models A and B as per Figure 2, and the difference in martensite volume fractions, as shown in Table 2. The behavior of stress–strain curves of uniaxial, plane strain, biaxial, and equi-biaxial loading is shown in Figure 4. By increasing martensite volume fractions, the yield and ultimate strengths increased while the values of corresponding engineering strains decreased. The stress values of uniaxial loading were almost like those of plane strain loading conditions but much lower than biaxial and equi-biaxial loading conditions. By calculating areas under engineering flow curves for the four loading conditions, the plastic work per unit volume increased by decreasing the martensite fraction in DP steels. The plastic work performed by the biaxial loading case ranged between 37 and 39 $\mathrm{MJ/m^3}$, while uniaxial and plane strain loading cases ranged from 28 to 36 $\mathrm{MJ/m^3}$.

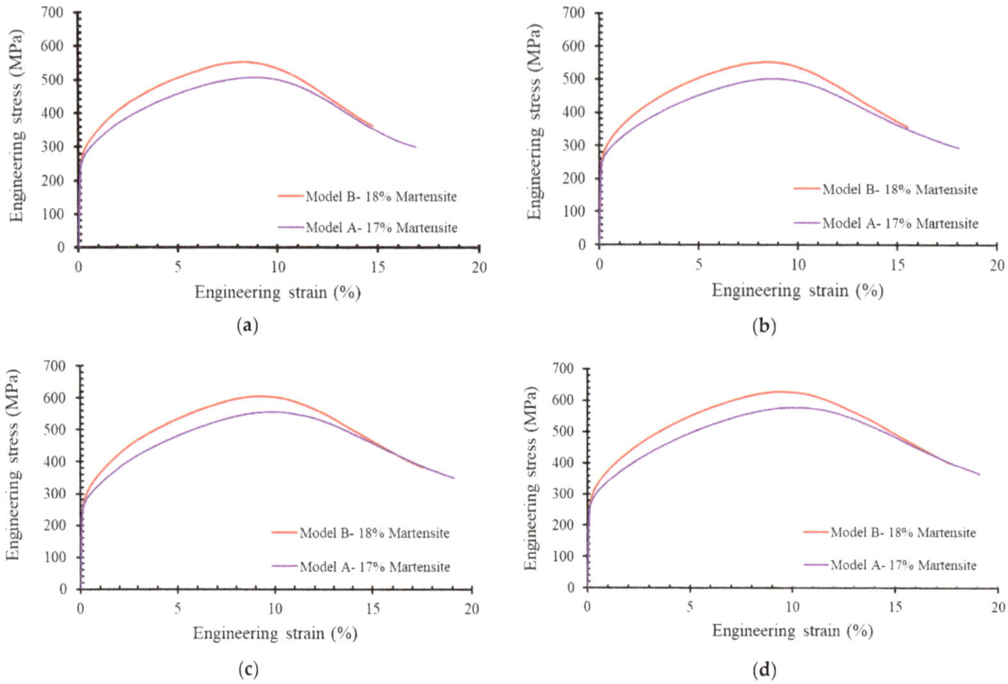

Figure 4. Comparison of stress–strain curves of different martensite volume fractions in (**a**) uniaxial tension, (**b**) plane strain tension, (**c**) partial biaxial tension, and (**d**) equi-biaxial loading.

Concerning FLDs, the results are more significant, as shown in Figure 5. It was observed that the increase of martensite volume fractions decreased the values of major and minor strains of FLDs slightly in the M-K approach. However, in the case of the K-B approach, the formability difference was not observed.

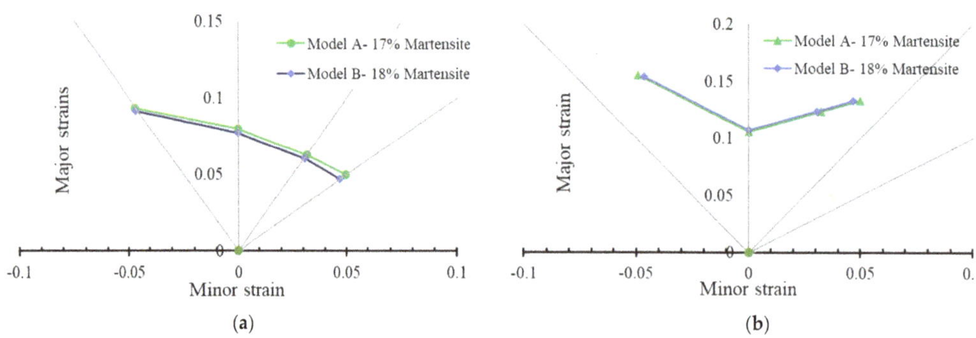

Figure 5. The FLDs in martensite fraction differences in the case of (**a**) M-K approach and (**b**) K-B approach.

By plotting the logarithmic true stresses and strains of models (A and B), both variables were directly proportional to each other; when the values of n-values were detected from the slope of the graph. It was found that by decreasing the martensite volume fractions, the values of n-values increased, as shown in Figure 6 and Table 4. The trend is similar to what is already reported in the literature [25].

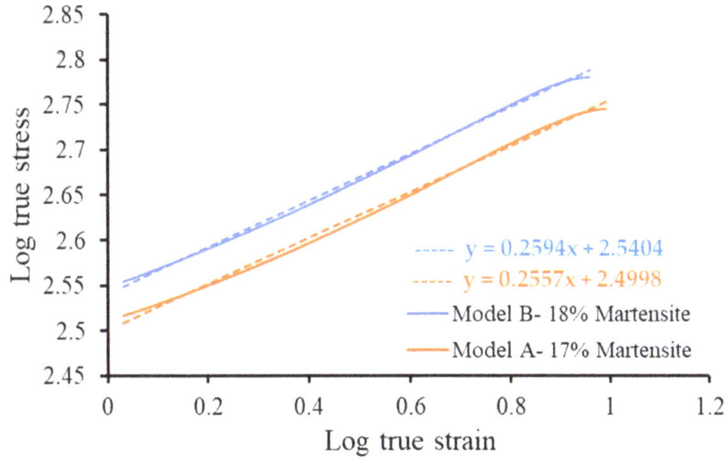

Figure 6. Comparison of strain hardening exponent (n) with corresponding martensite volume fraction, i.e., 18% and 17%.

Table 4. N-values versus martensite fractions of DP steels.

n-Values	Martensite Fractions
0.2594	18%
0.2557	17%

3.2. *Effect of Martensite Grain Size on Stress–Strain Curve and FLDs*

By simulations of models (A, C, and D) as per Figure 2, the influence of the difference in martensite grain sizes was plotted as stress–strain curves of the four loading conditions in Figure 7. The stress–strain curves do not significantly affect the variation of martensite grain size in uniaxial loading. Contrarily, it was observed that the maximum stress of medium and small martensite grain size was almost the same, but the RVEs with bigger martensite sizes failed at lower strain values.

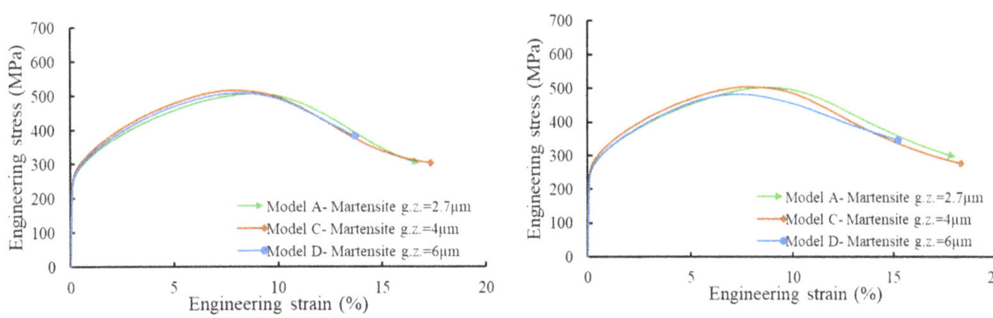

(a) (b)

Figure 7. *Cont.*

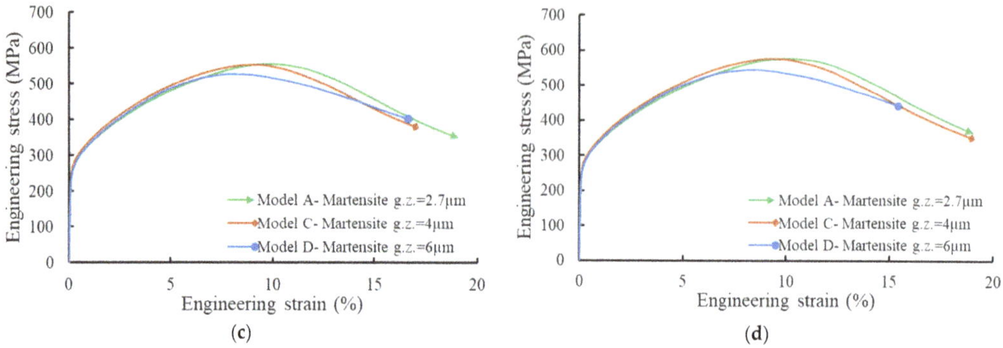

Figure 7. Comparison of stress–strain curves of martensite grain sizes in (**a**) uniaxial tension, (**b**) plane strain tension, (**c**) partial biaxial tension, and (**d**) equi-biaxial loading.

By increasing martensite grain sizes, the limiting major and minor strains decreased in the case of the M-K approach with some exceptions in uniaxial loading cases, as shown in FLDs in Figure 8. On the other hand, in the case of the K-B approach, the variation in martensite grain sizes did not have a remarkable effect on formability.

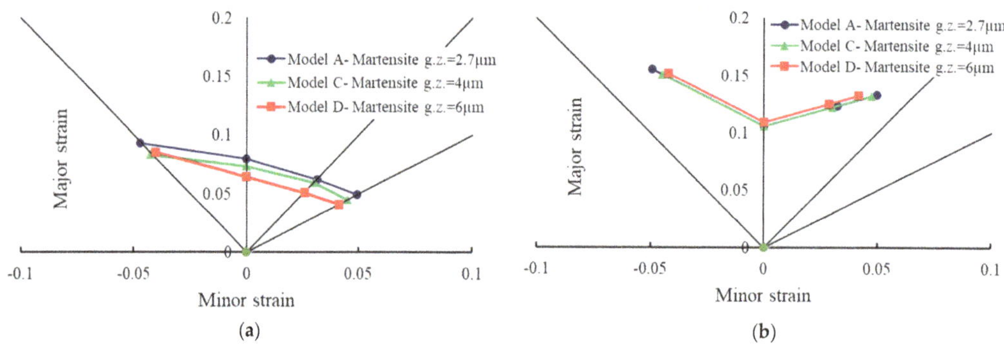

Figure 8. The FLDs in martensite grain sizes in the case of (**a**) M-K approach and (**b**) K-B approach.

3.3. Necking Band

In the case of the M-K approach, as shown in Figure 9, the FLD band was observed for models A and B, where the green line is the necking start at 0% of material degradation, the blue line is localized necking point, and the red line represents the fracture of the RVE at 20% of material degradation. The area under the green line is a safe zone, and above the blue line is the damage zone, where the necking starts.

While in the case of K-B, by using models A and D, the values of major strains at the plane strain loadings were almost the same when the damage values were changed, while the values of the major and minor strains in other loading conditions changed. In addition, the marginal zone of the FLD was not observed for either model, as shown in Figure 10.

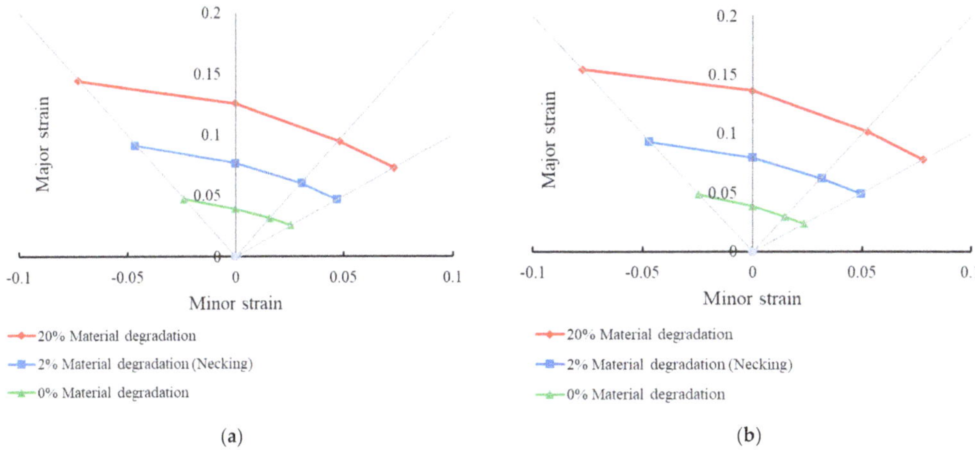

Figure 9. FLDs are showing marginal zones of (**a**) model A and (**b**) of model B.

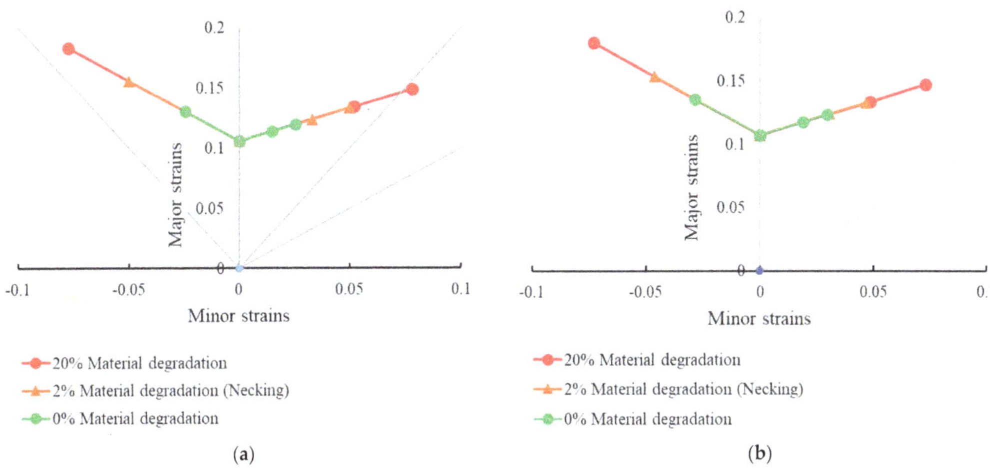

Figure 10. The FLDs marginal zones for (**a**) martensite grain sizes = 2.7 µm and for (**b**) martensite grain sizes = 6 µm.

3.4. Local Damage Evolution

Damage initiation in models A and C were analyzed on the top surfaces of 3D RVEs, as shown in Figures 11 and 12, respectively. The frames were extracted during post-processing of the simulation results to compare local behaviors of the RVEs at the necking point, which showed 20% and 30% of global material degradation, respectively. The local damage field is presented with the help of values ranging from 1.0 to 0.0, i.e., undamaged to fully damaged state, respectively.

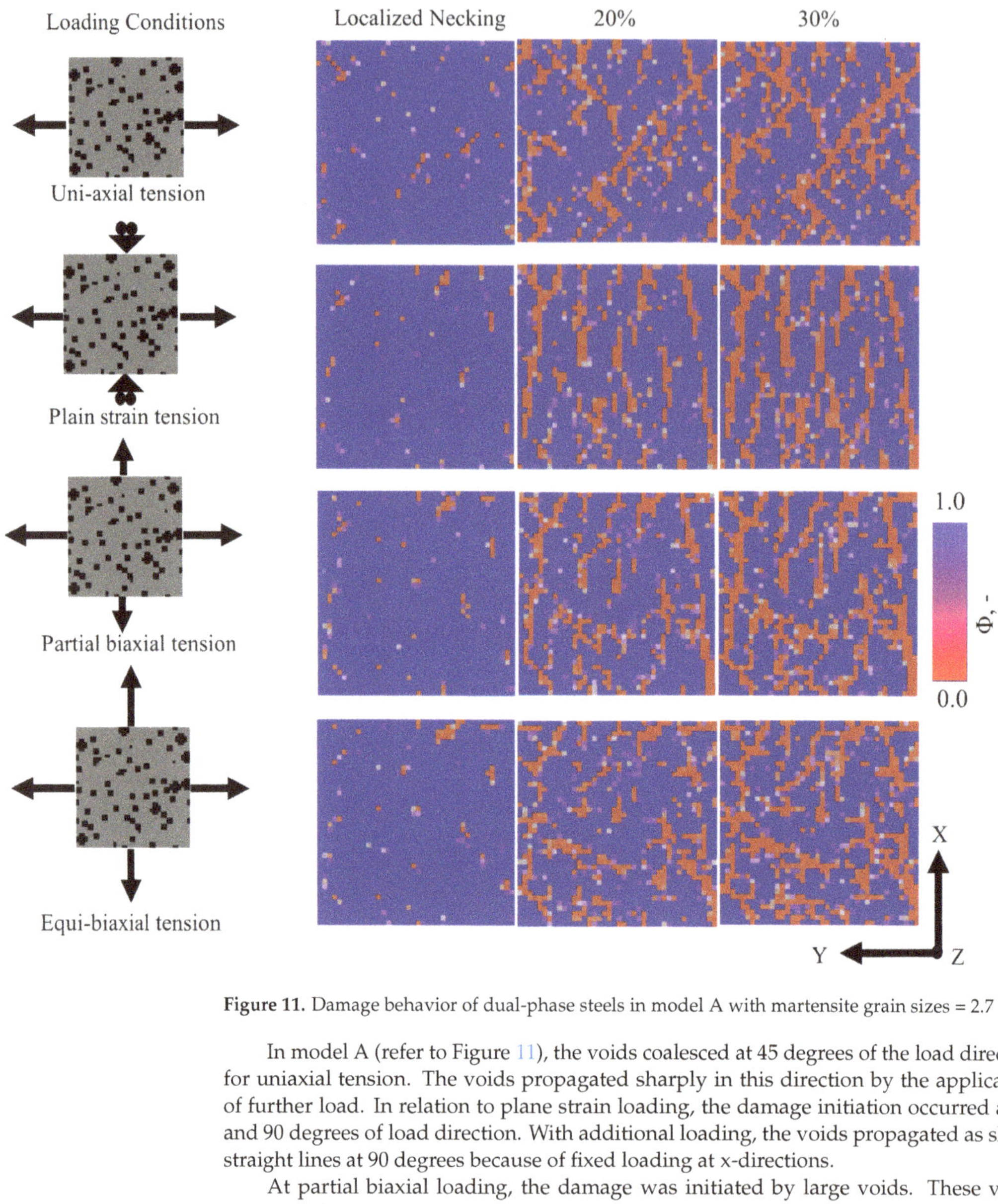

Figure 11. Damage behavior of dual-phase steels in model A with martensite grain sizes = 2.7 μm.

In model A (refer to Figure 11), the voids coalesced at 45 degrees of the load direction for uniaxial tension. The voids propagated sharply in this direction by the application of further load. In relation to plane strain loading, the damage initiation occurred at 45 and 90 degrees of load direction. With additional loading, the voids propagated as sharp straight lines at 90 degrees because of fixed loading at x-directions.

At partial biaxial loading, the damage was initiated by large voids. These voids propagated in 45 and 90 degrees to load directions. In contrast, in the equiaxial loading case, the damage propagated in different and random directions, such as 0, 45, and 90 degrees to load directions (refer to Figure 11). In the case of model C (refer to Figure 12), the local damage evolution did not differ much from model A, especially in the uniaxial case. However, it was observed in the plane strain case that the larger martensite grains constraint the damage behavior, and the damage tended to form around the martensite grains. The damage propagation also behaved in the same manner in the case of partial-biaxial and equi-biaxial loadings. The damage was initiated at the inter-granular level but continued growing trans-granularly with increased matrix degradation. In addition, stress relaxation near the damage areas affected the damage propagation and strain localization.

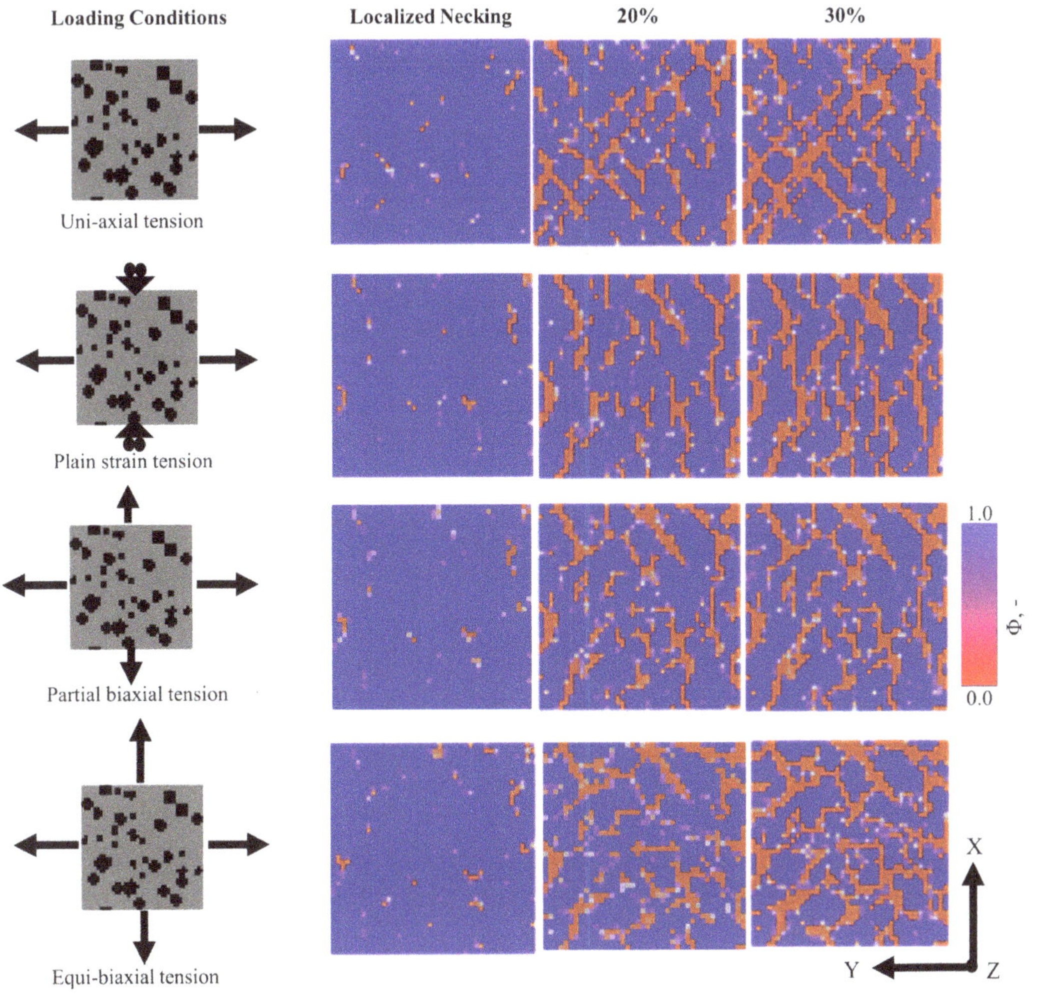

Figure 12. Damage behavior of dual-phase steels in model C with martensite grain sizes = 4 μm.

4. Discussion

Most analytical models used for plotting FLDs detect the forming limits at the necking point of materials, but calculating major and minor limit strains by numerical simulations is still required. In this study, these limiting strain values were detected from logarithmic strain tensor components, where the maximum value inside the tensor matrix represents the major strain, and minimum value is minor strain. Logarithmic strain is widely used in numerical simulations and is the best indicator for true strain, as its ability to count the strain values at every time interval and gives us an exact number when deformation occurs by a series of increments [31]. The strain tensor of this matrix represents an imaginary square unit, where minor strain is the minimum logarithmic change value in the unit square, and major strain is the maximum logarithmic change in this unit.

The increasing value of ultimate tensile strength by increasing phase fraction of the martensite is because of the enhanced reinforcement by uniformly distributed hard martensite particles [32]. On the other hand, the reduced capability of the material to undergo global strains with more martensite fraction is due to the brittle nature of martensite and the comparatively less force shifted to the ferrite grains. Furthermore, the martensite's decreased grain size provokes the surrounding ferrite's plastic deformation; therefore, the DP steel sample with a small martensite grain size showed comparatively better formability [33].

The behavior of FLDs agrees with the results of Duancheng Ma [30], where it applies on the right side of FLD with and without periodic boundary conditions, as shown in Figure 5 for the M-K approach. On the other hand, in the K-B approach, the simulations, as shown in Figure 5b, agree with Basak [8], where results are presented for DP steels 600, 800, and 1000, and G Béres [12], who worked on aluminum alloy Al 2008-T4. In Figure 5a, the FLDs with a relative difference in trends influenced by martensite fraction compared to those in Figure 5b, where no significant difference was observed, establish the argument that the M-K approach corresponds better to the crystal plasticity-based modeling approach. Furthermore, this effect was observed to be more pronounced in the case of using the M-K approach to evaluate the influence of variation in the martensite grain size, as shown in Figure 8.

Keeler–Brazier's equations did not illustrate the difference of formability by changing martensite grain sizes. Furthermore, they did not show a significant necking band, as the damage values at plane strain loading conditions were not altered, as shown in Figures 8 and 10. On the other hand, the damage values' dependency on engineering stress–strain curves can clarify the 0.2% proof stress points, onset of necking, and fracture for the M-K approach.

In addition, n-values were calculated for both models using Equation (6).

$$\sigma_t = k\ \varepsilon_t^n \tag{6}$$

where σ_t is the true stress, k is a constant, ε_t is the true strain, and n is the strain hardening coefficient. These n-values can keep their higher strength values in the pre-necking zones, reducing the risk of local strain accumulations and uniformly distributing the strain over the whole domain, improving materials' formability. True stress–strain curves of DP steels and the plastic work performed by biaxial and uniaxial loading using crystal plasticity models show a good agreement with Equations (7) and (8) used during experimental work.

$$\sigma_t = \sigma_e\ (1 + \varepsilon_e) \tag{7}$$

$$\varepsilon_t = \ln\ (1 + \varepsilon_e) \tag{8}$$

where σ_t is the true stress, ε_t is the true strain, σ_e is the engineering stress, and ε_e is the true strain. These equations are based on the ISO 16842 standard for studying biaxial tensile testing on sheet metals [34].

The local damage behavior of the DP steel samples upon varying loading conditions showed different outcomes, which can help understand the corresponding effect of microstructure. This is also affected by the morphology of the martensite particles, their orientation, aggregation, and presence in small islands. The initiation and propagation of local damage at the martensite–ferrite grain interface are because of the decohesion of the comparatively weak point in the aggregate [35,36]. The direction of the damage propagation in the case of tensile loading in ductile ferrite matrix, i.e., at 45 degrees, is caused by the shear bands [29]. A further extensive study is needed to understand this effect by considering the contributing factors.

5. Conclusions

The dependence of microstructural attributes on the formability limit of DP steels was analyzed. Specifically, the averaged global behavior and the effect of different loading conditions on damage initiation were investigated. Several 3D RVEs with varying microstructural attributes were simulated using a crystal plasticity-based numerical simulation model called DAMASK. The global and local stress, strain, and damage evolution of various RVEs revealed the internal phenomena during deformation. The global results were processed to obtain FLDs according to K B and M-K criteria. Appealing outcomes were observed, which can be summarised by the following points.

1. DAMASK can model FLDs of multi-phase materials with different loading conditions. Therefore, it can be used to study the effect of microstructural attributes on the formability of crystalline materials.
2. There are some limitations in the K-B model; firstly, the equations did not introduce a significant difference in FLDs by using different martensite grain sizes. Secondly, they did not possess the traditional marginal zones of FLDs. Contrarily, the M-K approach was proved to have good efficacy and agreement with the previous study of Duancheng Ma [7]. Moreover, it can clarify differences in mechanical behavior influenced by varying grain sizes of martensite in FLDs. Consequently, it shows comparatively safe necking and damage zones for different loading conditions.
3. In plotting FLDs, it was found that the lower the martensite fractions in DP steels, the better the formability. The precipitates of martensite act as obstacles, which restricts the slip deformation. Regarding martensite grain size in DP steels, the formability generally improves by decreasing martensite grain size. The higher strain hardening coefficient values (n-values), the better the formability because of comparatively low martensite phase fractions, which is a hard phase.
4. Plastic work and strength values of biaxial loading cases are higher than those in uniaxial and plane strain loading. At the same time, the plastic work values increase by decreasing martensite fractions in DP steels.
5. The difference in martensite grain sizes and loading conditions strongly affects the damage initiation and propagation behaviors of the RVEs, which could serve as a good guide on how to avoid damage propagation in the future.

Author Contributions: Conceptualization, T.H. and F.Q.; methodology, T.H.; software, T.H., F.Q. and M.U.; validation, T.H., F.Q. and S.G.; formal analysis, M.U.; investigation, T.H. and F.Q.; resources, F.Q. and S.G.; data curation, T.H.; writing—original draft preparation, T.H., M.U. and F.Q.; writing—review and editing, F.Q., M.U. and S.G.; visualization, T.H.; supervision, F.Q. and S.G.; project administration, S.G. and U.P.; funding acquisition, T.H. and U.P. All authors have read and agreed to the published version of the manuscript.

Funding: This research received no external funding.

Institutional Review Board Statement: Not applicable.

Informed Consent Statement: Not applicable.

Data Availability Statement: The simulation data are not publicly available but can be shared upon request.

Acknowledgments: The authors acknowledge the DAAD Faculty Development for Candidates (Balochistan), 2016 (57245990)-HRDI-UESTP's/UET's funding scheme in cooperation with the Higher Education Commission of Pakistan (HEC) for sponsoring the stay of Faisal Qayyum at IMF TU Freiberg. This work was conducted with the DFG-funded collaborative research group TRIP Matrix Composites (SFB 799). The authors gratefully acknowledge the German Research Foundation (DFG) for the financial support of SFB 799. Furthermore, Freunde und Förderer der TU Bergakademie Freiberg e.V. is acknowledged for providing financial assistance to Muhammad Umar. The authors also acknowledge the support of Martin Diehl and Franz Roters (MPIE, Düsseldorf) for their help regarding the functionality of DAMASK. Finally, the competent authorities at Khwaja Fareed University of Engineering and Information Technology, (KFUEIT) Rahim Yar Khan, Pakistan, and TU BAF Germany are greatly acknowledged for providing research exchange opportunities to Muhammad Umar at the Institute of Metal Forming TU BAF, Germany, under a memorandum of understanding (MoU).

Conflicts of Interest: The authors declare no conflict of interest.

Nomenclature

Acronym	
Symbol	*Description*
CPFEM	Crystal plasticity finite element method
DAMASK	Düsseldorf Advanced Materials Simulation Kit
DP	Dual-phase steels
FEM	Finite Element Method
FFT	Fast Fourier Transform
FLD	Forming Limit Diagram
K-B equations	Keeler–Brazier equations
M-K approach	Marciniak and Kuczynski approach
RVE	Representative volume element
SP	Spectral method
g.z.	Grain Size (μm)

Appendix A.

Appendix A.1. Running Simulation via CP Spectral Solver

The numerical simulations were processed by incorporating ductile damage and recording more increments after the damage initiation than those without damage in each RVE. During numerical processing by the spectral method, each grid point inside the mesh of the RVE acts as a computation point and represents individually defined deformation mechanisms, phase fractions, grain orientation, and homogenization schemes. Each increment after damage initiation records the material degradation behavior in detail. Once the damage is initiated, the numerical processing slows down and becomes intensively computed, and crashes after specific material degradation occurs for the RVE. The simulations and the load increments in each loading condition in this work were processed till converging.

Appendix A.2. Postprocessing Stage

The completed numerical simulations were post-processed using customized subroutines on DAMASK [37]. FLDs were plotted by adopting major and minor strains. The following two approaches are generally accepted, and commonly employed techniques based on some equations and engineering stress and strain curves to adopt the major and minor forming limit strains as forming limit criteria.

Appendix A.2.1. M-K Approach

For the M-K approach, after plotting engineering stress and strain curves, the maximum stress (localized necking point) was detected, and the strain tensors were recorded along with the increments. Major and minor values were detected from the strain tensor matrix at an increment, corresponding to this maximum stress in the stress–strain curve, as shown in the following process chart (see Figure A1).

Appendix A.2.2. K-B Approach

The Keeler-Brazier model depends on the major true strain value FLD0, true on the thickness of the virtual samples, and the strain hardening coefficient when the minor true strain equals zero [13]. It presents Equations (A1)–(A3) to predict the FLDs following the activities, as shown in Figure A2, where major true strain ε_1 is dependent on values of thickness t and strain hardening coefficient n.

For $\varepsilon_2 < 0$ (uniaxial case)

$$\varepsilon_1 = \mathrm{FLD}_{0,\ \mathrm{true}} - \varepsilon_2 \tag{A1}$$

For $\varepsilon_2 > 0$ (biaxial case)

$$\varepsilon_1 = \ln[0.6 \times (\exp(\varepsilon_2) - 1) + \exp(\mathrm{FLD}_{0,\ \mathrm{true}})] \tag{A2}$$

where,

$$FLD_{0,\, true} = \ln[1 + (0.233 + 0.413xt) \cdot n/t] \tag{A3}$$

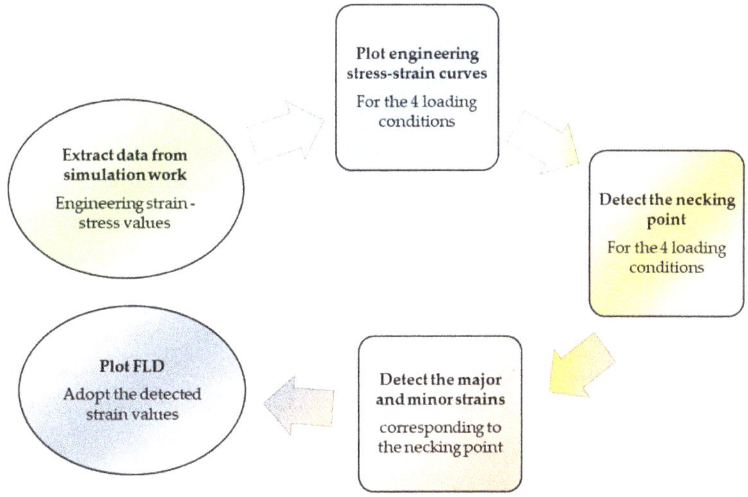

Figure A1. Process chart of plotting FLD via M-K approach.

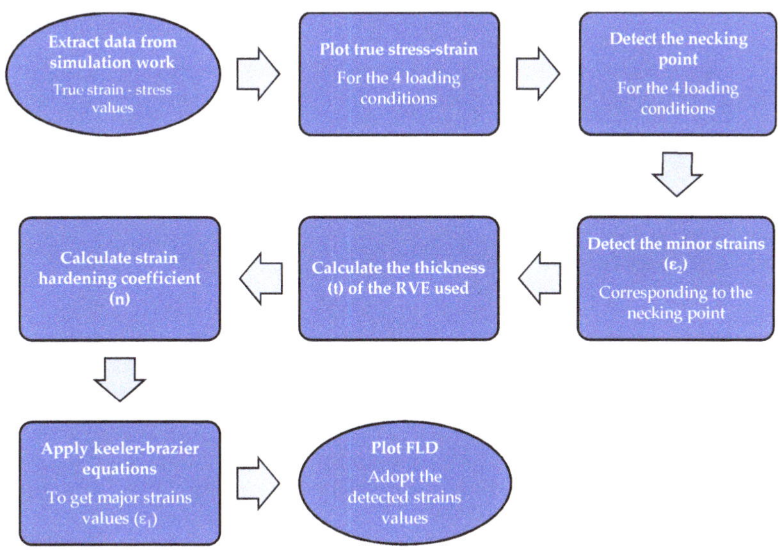

Figure A2. Process chart of plotting FLD via Keeler-Brazier (K-B) approach.

References

1. Qayyum, F.; Guk, S.; Prahl, U. Studying the Damage Evolution and the Micro-Mechanical Response of X8CrMnNi16-6-6 TRIP Steel Matrix and 10% Zirconia Particle Composite Using a Calibrated Physics and Crystal-Plasticity-Based Numerical Simulation Model. *Crystals* **2021**, *11*, 759. [CrossRef]
2. Umar, M.; Qayyum, F.; Farooq, M.U.; Khan, L.A.; Guk, S.; Prahl, U. Analyzing the cementite particle size and distribution in heterogeneous microstructure of C45EC steel using crystal plasticity based DAMASK code. In Proceedings of the 2021 International Bhurban Conference on Applied Sciences and Technologies (IBCAST), Islamabad, Pakistan, 12–16 January 2021; p. 6. [CrossRef]

3. Qayyum, F.; Shah, M.; Ali, S.; Ali, U. Bulge hydroforming of tube by rubber mandrel without axial feed: Experiment and Numerical Simulation. In Proceedings of the First International Symposium on Automotive and Manufacturing Engineering (SAME), Islamabad, Pakistan, 11 November 2015; Volume 1, Number 7. pp. 19–26.
4. Zaeimi, M.; Basti, A.; Alitavoli, M. Effect of martensite volume fraction on forming limit diagrams of dual-phase steel. *J. Mater. Eng. Perform.* **2015**, *24*, 1781–1789. [CrossRef]
5. Mohrbacher, H. Advanced metallurgical concepts for DP steels with improved formability and damage resistance. In Proceedings of the International Symposium on New Developments in Advanced High-Strength Sheet Steels, Vail, CO, USA, 23–27 June 2013.
6. Xiong, Z.; Kostryzhev, A.G.; Zhao, Y.; Pereloma, E.V. Microstructure evolution during the production of dual phase and transformation induced plasticity steels using modified strip casting simulated in the laboratory. *Metals* **2019**, *9*, 449. [CrossRef]
7. Qayyum, F.; Umar, M.; Guk, S.; Schmidtchen, M.; Kawalla, R.; Prahl, U. Effect of the 3rd dimension within the representative volume element (Rve) on damage initiation and propagation during full-phase numerical simulations of single and multi-phase steels. *Materials* **2021**, *14*, 42. [CrossRef] [PubMed]
8. Arrieux, R.; Bedrin, C.; Boivin, M. Determination of an Intrinsic Forming Limit Stress Diagram for isotropic metal sheets. In Proceedings of the 12th Biennial Congress of the IDDRG, Margherita, Italy, 24–28 May1982; pp. 61–71.
9. Akinribide, O.J.; Mekgwe, G.N.; Ajibola, O.O.; Obadele, B.A.; Akinwamide, S.O.; Olubambi, P.A. Studies on Mechanical properties of graphite reinforced Ti (Cx, N1-x) using nanoindentation techniques. *Procedia Manuf.* **2019**, *30*, 604–610. [CrossRef]
10. Tasan, C.C.; Hoefnagels, J.P.M.; Diehl, M.; Yan, D.; Roters, F.; Raabe, D. Strain localization and damage in dual phase steels investigated by coupled in-situ deformation experiments and crystal plasticity simulations. *Int. J. Plast.* **2014**, *63*, 198–210. [CrossRef]
11. Ma, D. Assessment of full field crystal plasticity finite element method for forming limit diagram prediction. *arXiv* **2018**, arXiv:1810.05742.
12. Béres, G.; Weltsch, Z.; Tisza, M. Comparative study on theoretical and experimental evaluation of forming limit diagrams. In *Proceedings of the IOP Conference Series: Materials Science and Engineering*; IOP Publishing: Bristol, UK, 2018; Volume 426, p. 12005.
13. Basak, S.; Bandyopadhyay, K.; Panda, S.K.; Saha, P. Prediction of formability of bi-axial pre-strained dual phase steel sheets using stress-based forming limit diagram. In *Advances in Material Forming and Joining*; Springer: Berlin/Heidelberg, Germany, 2015; pp. 167–192.
14. Chang, K.-H. *Chapter 13—Sheet Metal Forming Simulation*; Chang, K.-H.B.T.-D., Ed.; Academic Press: Boston, MA, USA, 2015; pp. 685–741. ISBN 978-0-12-382038-9.
15. Hussain, N.; Qayyum, F.; Pasha, R.A.; Shah, M. Development of multi-physics numerical simulation model to investigate thermo-mechanical fatigue crack propagation in an autofrettaged gun barrel. *Def. Technol.* **2020**, *17*, 1579–1591. [CrossRef]
16. Qayyum, F.; Guk, S.; Kawalla, R.; Prahl, U. On attempting to create a virtual laboratory for application-oriented microstructural optimization of multi-phase materials. *Appl. Sci.* **2021**, *11*, 1506. [CrossRef]
17. Bruschi, S.; Altan, T.; Banabic, D.; Bariani, P.F.; Brosius, A.; Cao, J.; Ghiotti, A.; Khraisheh, M.; Merklein, M.; Tekkaya, A.E. Testing and modelling of material behaviour and formability in sheet metal forming. *CIRP Ann.* **2014**, *63*, 727–749. [CrossRef]
18. Hutchinson, J.W. Bounds and Self-Consistent Estimates for Creep of Polycrystalline Materials. *Proc. R. Soc. A Math. Phys. Eng. Sci.* **1976**, *348*, 101–127. [CrossRef]
19. Michel, J.C.; Moulinec, H.; Suquet, P. A computational scheme for linear and non-linear composites with arbitrary phase contrast. *Int. J. Numer. Methods Eng.* **2001**, *52*, 139–160. [CrossRef]
20. Diehl, M.; An, D.; Shanthraj, P.; Zaefferer, S.; Roters, F.; Raabe, D. Crystal plasticity study on stress and strain partitioning in a measured 3D dual phase steel microstructure. *Phys. Mesomech.* **2017**, *20*, 311–323. [CrossRef]
21. McHugh, P.; Asaro, J.; Shih, C. Texture development and strain hardening in rate dependent polycrystals. *Acta Metall. Mat.* **1993**, *41*, 1461–1477. [CrossRef]
22. Umar, M.; Qayyum, F.; Farooq, M.U.; Guk, S.; Kirschner, M.; Korpala, G.; Prahl, U. Exploring the Structure–Property Relationship in Spheroidized C45EC Steel Using Full Phase Crystal Plasticity Numerical Simulations. *Steel Res. Int.* **2021**, 93. [CrossRef]
23. Umar, M.; Qayyum, F.; Farooq, M.U.; Khan, L.A.; Guk, S.; Prahl, U. Investigating the Effect of Cementite Particle Size and Distribution on Local Stress and Strain Evolution in Spheroidized Medium Carbon Steels using Crystal Plasticity-Based Numerical Simulations. *Steel Res. Int.* **2020**, *92*. [CrossRef]
24. Shanthraj, P.; Svendsen, B.; Sharma, L.; Roters, F.; Raabe, D. Elasto-viscoplastic phase field modelling of anisotropic cleavage fracture. *J. Mech. Phys. Solids* **2017**, *99*, 19–34. [CrossRef]
25. Roters, F.; Diehl, M.; Shanthraj, P.; Eisenlohr, P.; Reuber, C.; Wong, S.L.; Maiti, T.; Ebrahimi, A.; Hochrainer, T.; Fabritius, H.O.; et al. DAMASK—The Düsseldorf Advanced Material Simulation Kit for modeling multi-physics crystal plasticity, thermal, and damage phenomena from the single crystal up to the component scale. *Comput. Mater. Sci.* **2019**, *158*, 420–478. [CrossRef]
26. Zhou, R.; Roy, A.; Silberschmidt, V.V. A crystal-plasticity model of extruded AM30 magnesium alloy. *Comput. Mater. Sci.* **2019**, *170*, 109140. [CrossRef]
27. Groeber, M.A.; Jackson, M.A. Dream.3D: A digital representation enviroment for the analysis of microstructure in 3D. *Integr. Mater. Manuf. Innov.* **2014**, *3*, 5. [CrossRef]
28. Qayyum, F.; Chaudhry, A.A.; Guk, S.; Schmidtchen, M.; Kawalla, R.; Prahl, U. Effect of 3d representative volume element (Rve) thickness on stress and strain partitioning in crystal plasticity simulations of multi-phase materials. *Crystals* **2020**, *10*, 944. [CrossRef]

29. Umar, M.; Qayyum, F.; Farooq, M.U.; Guk, S.; Prahl, U. Qualitative Investigation of Damage Initiation at Meso-Scale in Spheroidized C45EC Steels by Using Crystal Plasticity-Based Numerical Simulations. *J. Compos. Sci.* **2021**, *5*, 222. [CrossRef]
30. Ma, D.; Eisenlohr, P.; Epler, E.; Volkert, C.A.; Shanthraj, P.; Diehl, M.; Roters, F.; Raabe, D. Crystal plasticity study of monocrystalline stochastic honeycombs under in-plane compression. *Acta Mater.* **2016**, *103*, 796–808. [CrossRef]
31. Rees, D.W.A. *Basic Engineering Plasticity an Introduction with Engineering and Manufacturing Applications—STRESS ANALYSIS*; Rees, D.W.A.B.T.-B.E.P., Ed.; Butterworth-Heinemann: Oxford, UK, 2006; pp. 1–32, ISBN 978-0-7506-8025-7.
32. Diehl, M.; Shanthraj, P.; Eisenlohr, P.; Roters, F. Neighborhood influences on stress and strain partitioning in dual-phase microstructures: An investigation on synthetic polycrystals with a robust spectral-based numerical method. *Meccanica* **2016**, *51*, 429–441. [CrossRef]
33. Tian, C.; Kirchlechner, C. The fracture toughness of martensite islands in dual-phase DP800 steel. *J. Mater. Res.* **2021**, *36*, 2495–2504. [CrossRef]
34. ISO Metallic Materials—Sheet and Strip—Biaxial Tensile Testing Method Using a Cruciform Test Piece. Available online: https://www.iso.org/standard/82087.html (accessed on 14 December 2021).
35. Tasan, C.C.; Diehl, M.; Yan, D.; Bechtold, M.; Roters, F.; Schemmann, L.; Zheng, C.; Peranio, N.; Ponge, D.; Koyama, M.; et al. An Overview of Dual-Phase Steels: Advances in Microstructure-Oriented Processing and Micromechanically Guided Design. *Annu. Rev. Mater. Res.* **2015**, *45*, 391–431. [CrossRef]
36. Ullah, M.; Wu, C.S.; Qayyum, F. Prediction of crack tip plasticity induced due to variation in solidification rate of weld pool and its effect on fatigue crack propagation rate (FCPR). *J. Mech. Sci. Technol.* **2018**, *32*, 3625–3635. [CrossRef]
37. Roters, F. DAMASK—The Düsseldorf Advanced Material Simulation Kit, Installation and Documentation. Available online: https://damask.mpie.de/ (accessed on 15 December 2021).

crystals

MDPI

Article

Twinning Behavior in Cold-Rolling Ultra-Thin Grain-Oriented Silicon Steel

Bo Zhang [1], Li Meng [1,*], Guang Ma [2], Ning Zhang [1], Guobao Li [3], Kun Liu [4] and Sheng Zhong [5]

1 Metallurgical Technology Institute, Central Iron and Steel Research Institute, Beijing 100081, China; zhangbo530925@163.com (B.Z.); zhangning@cisri.com.cn (N.Z.)
2 State Key Laboratory of Advanced Power Transmission Technology, Global Energy Interconnection Research Institute Co., Ltd., Beijing 102211, China; maguang@geiri.sgcc.com.cn
3 Baoshan Iron and Steel Co., Ltd., Shanghai 201900, China; ligb@baosteel.com
4 China Three Gorges Corporation, Beijing 100038, China; liu_kun@ctg.com.cn
5 ABB Corporate Research Center, 940 Main Campus Dr, Suite 200, Raleigh, NC 27606, USA; szhongud@gmail.com
* Correspondence: mengl@cisri.com.cn or li_meng@126.com

Abstract: Twinning behaviors in grains during cold rolling have been systematically studied in preparing ultra-thin grain-oriented silicon steel (UTGO) using a commercial glassless grain-oriented silicon steel as raw material. It is found that the twinning system with the maximum Schmid factor and shear mechanical work would be activated. The area fraction of twins increased with the cold rolling reduction. The orientations of twins mainly appeared to be α-fiber (<110>//RD), most of which were {001}<110> orientation. Analysis via combining deformation orientation simulation and twinning orientation calculation suggested that {001}<110> oriented twinning occurred at 40–50% rolling reduction. The simulation also confirmed more {100} <011> oriented twins would be produced in the cold rolling process and their orientation also showed less deviation from ideal {001}<110> orientation when a raw material with a higher content of exact Goss oriented grains was used.

Keywords: ultra-thin grain-oriented silicon steel (UTGO); cold rolling; twinning; Goss; {100}<011>

Citation: Zhang, B.; Meng, L.; Ma, G.; Zhang, N.; Li, G.; Liu, K.; Zhong, S. Twinning Behavior in Cold-Rolling Ultra-Thin Grain-Oriented Silicon Steel. *Crystals* **2021**, *11*, 187. https://doi.org/10.3390/cryst11020187

Academic Editors: Ronald W. Armstrong and Wojciech Polkowski

Received: 22 January 2021
Accepted: 12 February 2021
Published: 14 February 2021

1. Introduction

Ultra-thin grain-oriented silicon steel (UTGO steel, thickness $\leq$ 0.10 mm) is an important magnetic material mainly used for manufacturing intermediate and high-frequency transformers [1–4], thanks to its ability to increase core power while reducing core loss and volume. Although the manufacturing route for grain-oriented silicon steel has been developed for decades, it is still difficult to produce ultra-thin products using a conventional process which is based on secondary recrystallization. The difficulty is due to acceleration of the inhibitor coarsening during recrystallization and poor control over Goss orientation under large rolling reduction [5–9]. At present, the most prevalent production method to prepare ultra-thin grain-oriented silicon steel is to use commercial grain-oriented silicon steel sheets as starting material, then cold-rolling the steel sheets to the desired thickness followed by annealing processes [4,10,11].

In recent years, many studies have reported the formation of deformation twinning in silicon steel. Shi et al. [12] and Xie et al. [13] discovered that deformation twinning occurred in Fe-6.5% Si alloy in a medium temperature tensile and compress test, and that this twinning promoted the plastic deformation of the alloy. Dunn et al. [14] demonstrated that both slip and twinning were activated during cold rolling in Fe-3.25 wt.% Si alloy, and {001}<110> oriented twins could be formed in Goss single crystal at an early deformation stage. Rusakov et al. [15] studied the features of twinning in cold-deformed Goss single crystal in an Fe-3% Si-0.5% Cu alloy and found that twins with near {001}<110> orientation were formed at 5% reduction and the twinning orientation did not change during the subsequent deformation. Dorner et al. [16] reported that the area fraction of {001}<110> oriented twins increased with an increase of deformation reduction and reached the maximum at 61% deformation reduction when studying the evolutions of crystallographic orientations of cold-rolled Goss single crystals in Fe-3% Si alloy. Even with extensive studies, there are still disputes on how the twins in BCC structured grain-oriented silicon steels are formed and evolved during cold rolling.

It is known that deformation twinning has a pronounced grain orientation dependence in FCC [17–19] and HCP [20,21] metals or alloys, and similar effects of initial grain orientation on deformation twinning are also reported in BCC structure [22,23]. Fu et al. [22] found that twins tend to occur in grains with a tensile orientation near the <001> corner and a compressive orientation near the <101>-<111> line, and this twinning activation is closely related to their corresponding Schmid factor, respectively. In the production of ultra-thin grain-oriented silicon steel, it is preferred that the starting material has a strong Goss texture, so that grains with exact Goss in a certain extent in the resultant UTGO steel will occupy the great majority of area. The deviation of grain orientation from the exact Goss will affect the subsequent orientation transition routes [23]. However, the effect of deviation degree on the twinning behavior remains unclear. In this study, the twinning behaviors during cold rolling ultra-thin grain-oriented silicon steel have been systematically analyzed, and special attention has been focused on the influence of initial Goss orientation deviation on the twinning behavior. The results will help to comprehensively understand the cold rolling process and to provide a theoretical basis for preparing ultra-thin grain-oriented silicon steel.

2. Experimental Procedure

A 0.35 mm-thick commercial glassless grain-oriented silicon steel plate without a magnesium silicate layer was used as raw material to prepare the ultra-thin silicon steel. Its magnetic properties were B_8 = 1.89 T, $P_{1.7/50}$ = 1.12 W/kg, the average grain size was about 30 mm and the chemical compositions (mass fraction, %) were C: 0.0058, Si: 3, Mn: 0.0088, S: 0.0003, Al: 0.005, N: 0.001, Sn: 0.1. After being pickled with hydrochloric acid to distinguish the original grain boundaries, the sample plates of 300 mm (along rolling direction, RD) × 40 mm (along transverse direction, TD) were cut into two parts, as shown in Figure 1. The parts marked by the red block diagram are 10 mm (RD) × 5 mm (TD) and were used to confirm the initial orientation of the marked grains. The analysis was done using a Zeiss GeminiSEM500 field emission scanning electron microscope (SEM) equipped with an EBSD system and the accelerating voltage was 20 KV when measured. EBSD data was post-processed by the HKL CHANNEL5 software. The part marked by the green block is 300 mm (RD, rolling direction) ×30 mm (TD, transverse direction) and was cold rolled to 0.21–0.075 mm thick to investigate the twinning behavior. The microstructures and microtextures at different reductions were characterized and evaluated with a Confocal laser scanning microscope (CLSM) and EBSD technique, respectively. For the EBSD measurements, the samples were machined and electropolished with a 5% perchloric acid/alcohol solution to remove the surface strain layer. In the EBSD data, the tolerance angles for the orientations are set as 15°.

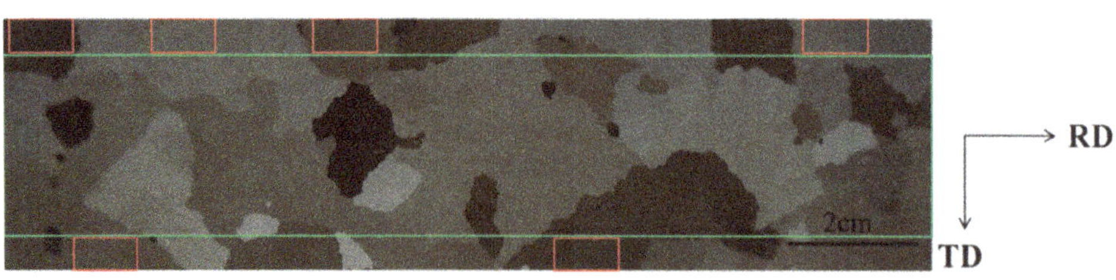

Figure 1. Schematic diagram of sample cutting.

3. Results and Discussion

3.1. Deformation Twinning Microstructure

The microstructures of the 0.075 mm rolled sheet were characterized by CLSM and are shown in Figure 2, and the textures measured by EBSD are presented in Figure 3. Parallel bands with serrated edge and different angles toward RD direction were found. The measurement shows the parallel bands are {001}<110> orientation and they are in <111> /60° relationship with the {111}<112> oriented matrix.

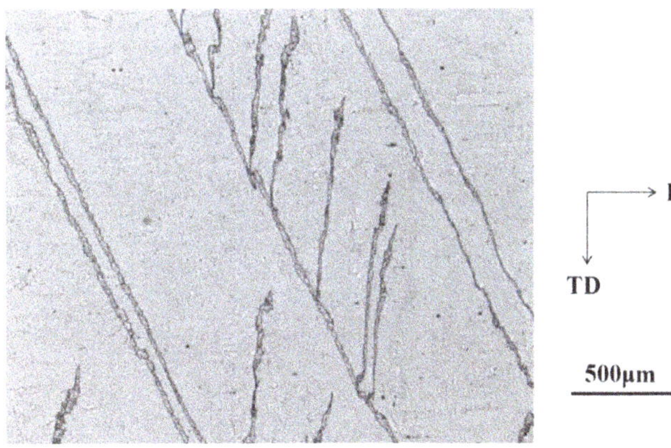

Figure 2. SEM micrographs in deformation twinning area of 0.075 mm sheet in rolling plane.

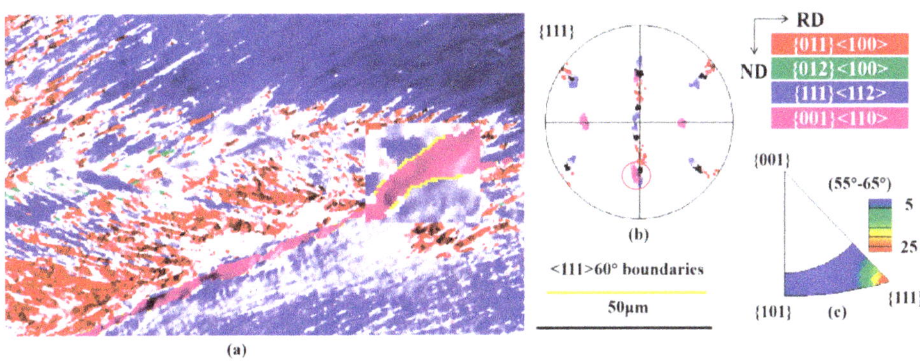

Figure 3. EBSD measurement of 0.075 mm sheet in transverse plane with step size = 0.2 μm: (**a**) EBSD orientation image mapping; (**b**) {111} pole figure; (**c**) rotation axes distribution for misorientation angle of 55–65° in (**a**).

3.2. Twinning Behavior in Goss-Oriented Grains during Cold Rolling

In this paper, the twinning Schmid factor and shear mechanical work of cold rolled Goss oriented grains at different strains are calculated and the effect of grain orientation on the selection of the twinning system is discussed [24–26]. The evolution of the twinning behavior during cold rolling in ultra-thin grain-oriented silicon steel is investigated by theoretical calculation and is verified via experiments.

The microstructures of the grains with a 3° angle deviated from exact Goss orientation at different reductions are depicted in Figure 4. The microstructure and microtextures at 78% reduction measured with EBSD are shown in Figure 5. With the increase of deformation reduction, the area fraction of twins increases. When the deformation reduction reaches 78%, the edge serration of the twins becomes more prominent. Twins with three different deviation angles from exact {001} <110> orientation are observed in Figure 5a, and it is suggested that they were formed at different deformation stages during cold-rolling.

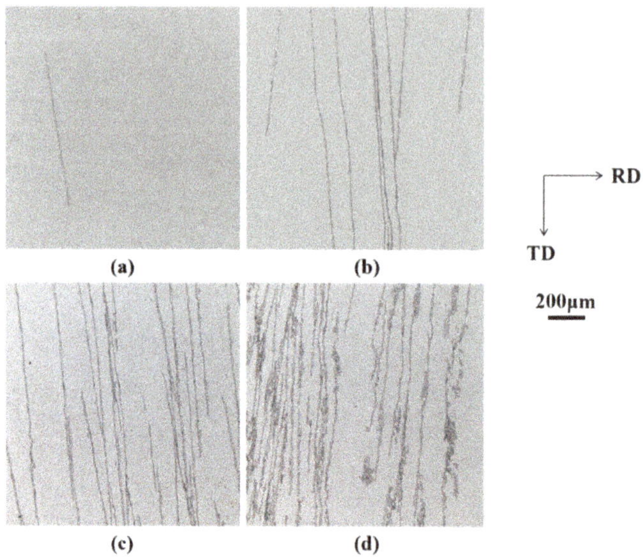

Figure 4. The morphology of deformation twinning within Goss-oriented grains at different reductions in rolling plane. (**a**) 32%; (**b**) 41%; (**c**) 53%; (**d**) 78% reduction.

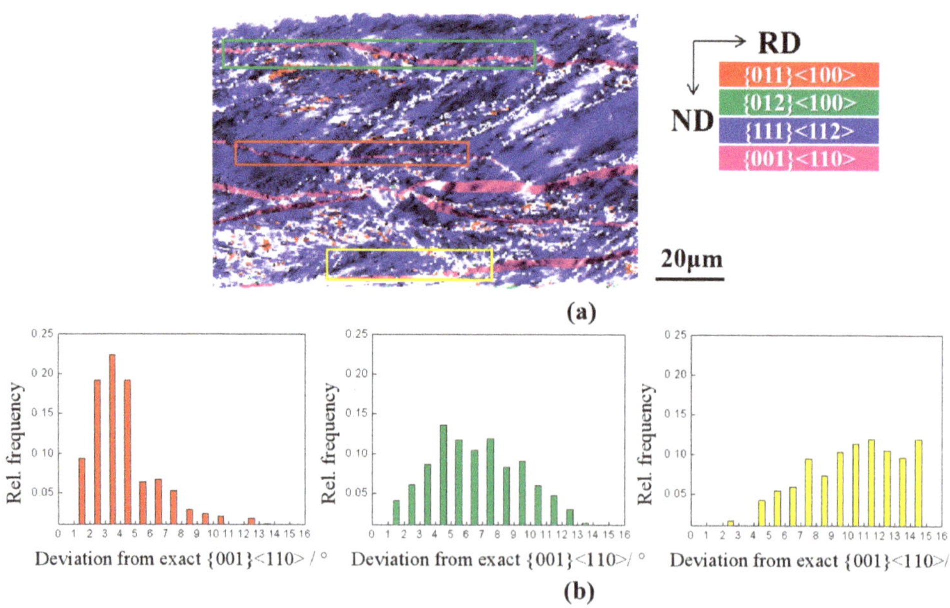

Figure 5. EBSD measurement in deformation twinning area of exact Goss orientated grains with a reduction of 78% in transverse plane with step size = 0.3 μm: (**a**) orientation image mapping; (**b**) the distribution of the deviation degree between twinning orientation and exact Goss orientation in the area marked by different color block diagram in (**a**).

According to Humbert's observation [24], the deformation twinning in BCC alloy can be obtained when every two adjoining (112) planes are displaced toward [111] direction by $\sqrt{3}a/6$. By these successive displacements, the BCC lattice is sheared by $\sqrt{2}/4$.

In the reference frame Y_T based on the corresponding twinning system, this plane strain is expressed by matrix E_T.

$$\text{Ey} = \begin{bmatrix} 0 & 0 & \frac{\sqrt{2}}{4} \\ 0 & 0 & 0 \\ \frac{\sqrt{2}}{4} & 0 & 0 \end{bmatrix} \tag{1}$$

In the reference frame based on deformed matrix orientation, the twinning strain matrix, E, is expressed as

$$E = N \cdot M \cdot E_T \cdot M^{-1} \cdot N^{-1} \tag{2}$$

Herein, M and N are corresponding transformation matrices of the twinning strain matrix from the reference frame based on corresponding twinning system to sample coordinate system and crystal coordinate system based on deformed matrix orientation successively, and the corresponding mechanical work of the twinning shear is calculated as

$$W = \frac{1}{2}(\varepsilon_{11} \cdot E_{11} - \varepsilon_{33} \cdot E_{33}) = \frac{1}{2}\varepsilon_{11}(E_{11} - E_{33}) \tag{3}$$

Herein, the rolling stress is simplified as equivalent force in the direction of ND and RD, that is, $\varepsilon 11 = \varepsilon 33$.

The twinning forming stages can be deduced based on the orientation relationship between twinning and deformed matrix. On the other hand, the selection of the twinning system can be analyzed based on the orientation of the deformed matrix when twinning occurs. The VPSC (visco-plasitic self-consistent) model has been applied to predict the orientation rotation path of Goss oriented grain during cold rolling and the result is shown in Figure 6.

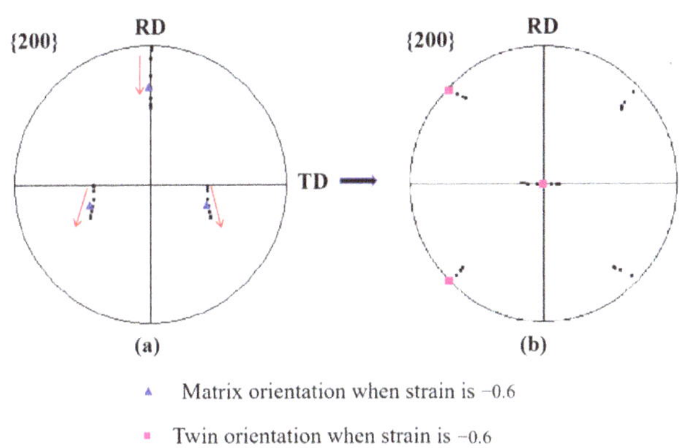

Figure 6. The orientation rotation path of Goss orientation: (**a**) corresponding twinning orientation; (**b**) during cold rolling, strain value = −0.6.

The calculation results suggest that during cold rolling, Goss orientation rotates to {111}<112> orientation along the route indicated by red arrows in Figure 6a, which is consistent with the literature [27]. In the calculated deformed matrix, the twinning orientations can be predicted based on the maximum Schmid factor and shear mechanical work of the twinning systems, as shown in Figure 6b. The orientation of the twinning forming during the cold rolling of Goss oriented grains is always α-fiber, close to {001}<110> orientation but with different degree of deviation. This is in good agreement with the experimental results, suggesting that twinning can occur in Goss oriented grains across a cold rolling process. The twinning orientation of exactly {001}<110> is achieved when the strain is about –0.6 and the deformed matrix orientation is (18° 48° 153°). Since most twins have orientations of exact or close to {001}<110>, we predict that twins mainly formed at this strain.

In a matrix of (18°, 48°, 153°) orientation at strain equal to –0.6, the Schmid factor and shear mechanical work of the twinning systems and their corresponding twinning orientations were estimated and are listed in Table 1. It is obvious that the Schmid factor and shear mechanical work of the twelve twinning systems increase at the same time; in other words, the maximum Schmid factor and shear mechanical work will correspond to the same twinning system, which leads to the twinning orientation of (136° 90° 90°). Similarly, the same phenomenon applies to the deformed orientations at different stains, and the calculated twinning orientations are always close to or near {001}<100>. The results confirm the assumption that the activated twinning system during cold rolling of Goss orientation grain always has the maximum value of Schmid factor and shear mechanical work. The greater the Schmid factor and shear mechanical work, the easier a twinning system would occur.

Table 1. The Schmid factor, shear mechanical work of twins in the deformed matrix with orientation of (18° 48° 153°).

Twinning System	Schmidt Factor	Mechanical Work	Twinning Orientation
$(112)[11\bar{1}]$	0.48	$0.167\varepsilon_{33}$	(46° 90° 0°)
$(121)[1\bar{1}1]$	0.26	$0.093\varepsilon_{33}$	(90° 141° 45°)
$(211)[1\bar{1}\bar{1}]$	0.36	$0.130\varepsilon_{33}$	(106° 116° 8°)
$(\bar{1}21)[\bar{1}1\bar{1}]$	0.26	$0.093\varepsilon_{33}$	(37° 116° 30°)
$(1\bar{1}2)[\bar{1}11]$	0.14	$0.046\varepsilon_{33}$	(122° 27° 105°)
$(11\bar{2})[111]$	0.52	$0.176\varepsilon_{33}$	(105° 64° 83°)
$(\bar{1}21)[\bar{1}\bar{1}1]$	0.46	$0.167\varepsilon_{33}$	(42° 180° 86°)
$(1\bar{2}1)[\bar{1}\bar{1}\bar{1}]$	0.14	$0.046\varepsilon_{33}$	(162° 84° 154°)
$(12\bar{1})[1\bar{1}\bar{1}]$	0.50	$0.176\varepsilon_{33}$	(161° 97° 116°)
$(\bar{2}11)[\bar{1}\bar{1}\bar{1}]$	0.38	$0.130\varepsilon_{33}$	(120° 153° 14°)
$(2\bar{1}1)[11\bar{1}]$	0.94	$0.333\varepsilon_{33}$	(136° 90° 90°)
$(21\bar{1})[1\bar{1}1]$	0.52	$0.185\varepsilon_{33}$	(141° 117° 60°)

3.3. *Effect of Goss Orientation Accuracy of Initial Grain on Twinning Behavior*

The majority of grains in the raw material have orientations close to Goss with 3°–10° deviation. The microstructures of 3° and 10° deviated Goss grains at 78% reduction were characterized by metalloscopy in the ND plane and by EBSD in the TD plane. The typical images are shown in Figures 7 and 8, respectively. Compared to 10° deviated Goss oriented grains, there is much higher density of twins in 3° deviated Goss oriented grains and the twinning orientations are more closely aligned to the exact {001}<110> orientation.

The crystal rotation routes of 10° deviated Goss orientation (0° 35° 0°) during cold rolling were simulated using VPSC model. Meanwhile, the Schmid factor and shear mechanical work as well as twinning orientation of the twinning systems were calculated in matrices with various orientations at different strain values. The results are shown in Figure 9 and Table 2. During cold rolling, the crystal rotation routes of grains with deviated Goss orientation are similar to that of exact Goss grains, showing deviated {111}<112> orientation at 78% rolling reduction. Moreover, under the same strain, the twining Schmid factor and shear mechanical work in deviated grains are both lower than that of exact Goss oriented grains. In a deviated Goss grain, the twins generally have a deviation angle with the exact {001}<110> orientation. The relationship of this deviation angle with {001}<110> orientation and strain is analyzed and the simulated results are presented in Figure 10. At the same strain, the twins in the initial deviated Goss grains have twinning orientation further away from the exact {001} <110> orientation. Since twinning systems are critical for achieving the final microstructures in ultra-thin grain-oriented silicon steel, the deviation degree of initial Goss oriented grain, therefore, would influence the evolution of the microstructure and microtextures of recrystallized steel. This influence is subject to future study.

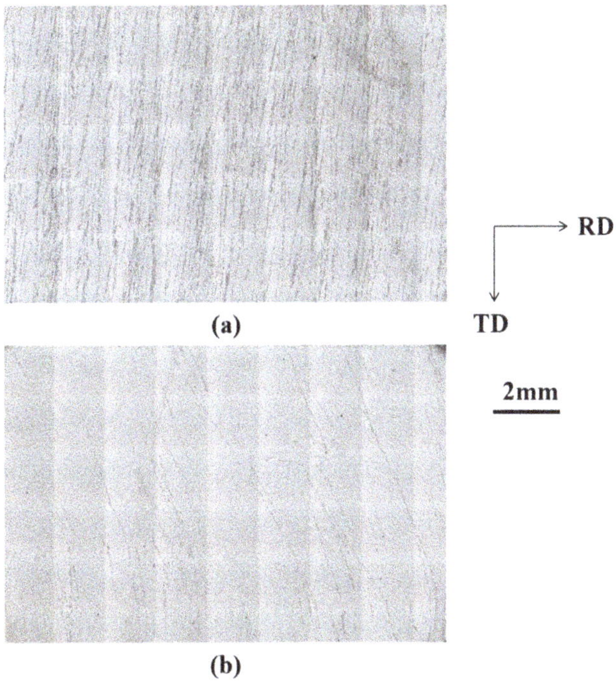

Figure 7. The effect of the deviation degree of the initial Goss orientation on twinning distribution in rolling plane: (**a**) 3° deviated Goss oriented grain; (**b**) 10° deviated Goss oriented grain.

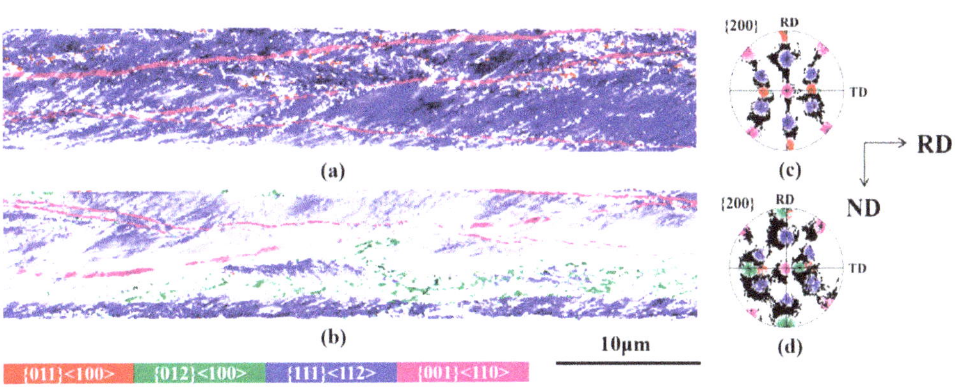

Figure 8. The effect of deviation angle of Goss oriented grain on twinning orientation in the lateral plane: (**a**) 3° deviated Goss oriented grain; (**b**) 10° deviated Goss oriented grain; EBSD measurement step size = 0.5 μm.

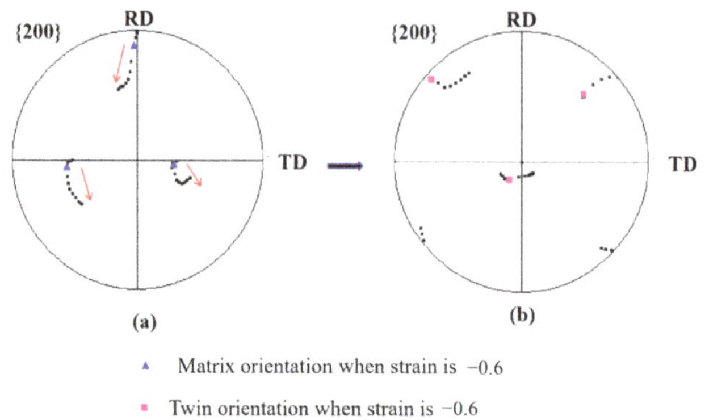

Figure 9. The orientation rotation path of (0° 35° 0°) orientation; (**a**) and corresponding twinning orientation (**b**) during cold rolling, strain = −0.6.

Table 2. Schmid factor and shear mechanical work of (0° 45° 0°) orientation and (0° 35° 0°) orientation at different strain in cold rolling (MSF and MMW represent maximum Schmid factor and shear mechanical work, respectively).

Strain	MSF of (0° 45° 0°)	MMW of (0° 45° 0°)	MSF of (0° 35° 0°)	MMW of (0° 35° 0°)
0	0.951	$0.336\varepsilon_{33}$	0.935	$0.329\varepsilon_{33}$
−0.2	0.962	$0.340\varepsilon_{33}$	0.938	$0.331\varepsilon_{33}$
−0.4	0.989	$0.350\varepsilon_{33}$	0.948	$0.336\varepsilon_{33}$
−0.6	0.988	$0.350\varepsilon_{33}$	0.965	$0.341\varepsilon_{33}$
−0.8	0.929	$0.329\varepsilon_{33}$	0.952	$0.337\varepsilon_{33}$
−1.0	0.845	$0.298\varepsilon_{33}$	0.887	$0.313\varepsilon_{33}$
−1.2	0.756	$0.269\varepsilon_{33}$	0.823	$0.288\varepsilon_{33}$
−1.4	0.714	$0.254\varepsilon_{33}$	0.776	$0.263\varepsilon_{33}$
−1.6	0.710	$0.251\varepsilon_{33}$	0.739	$0.242\varepsilon_{33}$

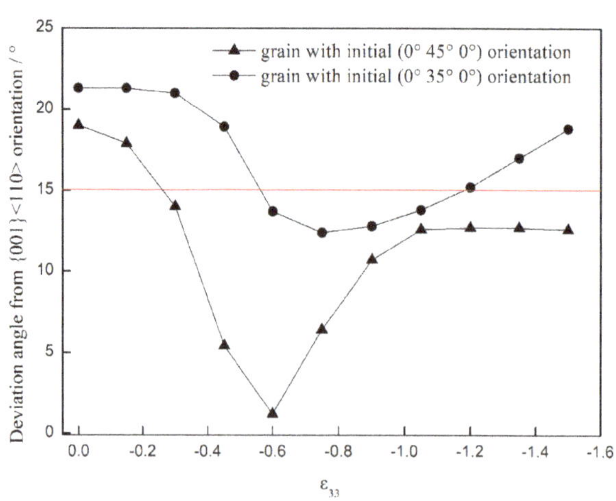

Figure 10. The effect of initial grain orientation on the deviation angle of twinning orientation from exact {001}<110> orientation during cold rolling.

4. Conclusions

The actual activated {112}<111> twinning systems are shown to have the maximum Schmid factor and shear mechanical work. The greater Schmid factor and shear mechanical work a twinning system has achieved, the easier it is to be activated.

Twinning can occur at different stages of rolling deformation in Goss oriented grain during cold rolling. The twinning area fraction increases with the increase of rolling reduction, and the twinning orientations are mainly located at α-fiber and are mostly around {001}<110> component. Based on VPSC simulation results, exact {001}<110> oriented twinning is achieved when the rolling reduction is 40–50%.

In the preparation of ultra-thin grain-oriented silicon steel, when an initial material has more grains with exact Goss orientation, more twins are observed in the deformed grains after cold-rolling and the twinning orientation is closer to exact {001}<110>.

Author Contributions: Conceptualization, L.M.; Data curation, B.Z.; Methodology, L.M.; Project administration, L.M.; Resources, G.M. and K.L.; Supervision, N.Z. and G.L.; Writing—original draft, B.Z.; Writing—review & editing, N.Z. and S.Z. All authors have read and agreed to the published version of the manuscript.

Funding: National Key Research and Development Program of China: 2016YFB0300301, 2017YFB0903901; Shanxi Provincial Science and Technology Major Special Project: 20191102004.

Acknowledgments: The authors gratefully acknowledge the funding of this work by the National Key Research and Development Program of China (Grant Nos. 2016YFB0300301 and 2017YFB0903901) and the Shanxi Provincial Science and Technology Major Special Project (20191102004).

Conflicts of Interest: The authors declare no conflict of interest.

References

1. Lobanov, M.L.; Redikul'Tsev, A.A.; Rusakov, G.M.; Kagan, I.V.; Pervushina, O.V. Effect of the grain orientation in the material used for the preparation of an ultrathin electrical steel on its texture and magnetic properties. *Phys. Met. Metallogr.* **2011**, *111*, 479–486. [CrossRef]
2. Moses, A.J. Energy efficient electrical steels: Magnetic performance prediction and optimization. *Scripta Mater.* **2012**, *67*, 560–565. [CrossRef]
3. Gao, X.H.; Qi, K.M.; Qiu, C.L. Magnetic properties of grain oriented ultra-thin silicon steel sheets processed by conventional rolling and cross shear rolling. *Mater. Sci. Eng. A* **2006**, *430*, 138–141.
4. Wang, Y.P.; An, L.Z.; Song, H.Y.; Wang, G.D.; Liu, H.T. Dependence of recrystallization behavior on initial Goss orientation in ultra-thin grain-oriented silicon steels. *J. Magn. Magn. Mater.* **2020**, *499*, 166290. [CrossRef]
5. Liu, H.T.; Yao, S.J.; Sun, Y.; Gao, F.; Song, H.Y.; Liu, G.H.; Li, L.; Geng, D.P.; Liu, Z.Y.; Wang, G.D. Evolution of microstructure, texture and inhibitor along the processing route for grain-oriented electrical steels using strip casting. *Mater. Charact.* **2015**, *106*, 273–282. [CrossRef]
6. Wang, Y.P.; Liu, H.T.; Song, H.Y.; Wang, G.D. Ultra-thin grain-oriented silicon steel sheet fabricated by a novel way: Twin-roll strip casting and two-stage cold rolling. *J. Magn. Magn. Mater.* **2018**, *452*, 288–296. [CrossRef]
7. Song, H.Y.; Liu, H.T.; Wang, Y.P.; Wang, G.D. Microstructure and texture evolution of ultra-thin grain-oriented silicon steel sheet fabricated using strip casting and three-stage cold rolling method. *J. Magn. Magn. Mater.* **2017**, *426*, 32–39. [CrossRef]
8. Nakashima, S.; Takashima, K.; Harase, J. Effect of thickness on secondary recrystallization of Fe-3%Si. *Acta Metall. Mater.* **1994**, *42*, 539–547. [CrossRef]
9. Harase, J.; Shimizu, R.; Takahashi, N. Effect of cold rolling reduction on the secondary recrystallization in 3%Si-Fe alloy. *J. Jpn. Inst. Met.* **1990**, *54*, 381–387. [CrossRef]
10. Arai, K.I.; Ishiyama, K. Rolled texture and magnetic properties of 3% silicon steel. *J. Appl. Phys.* **1988**, *64*, 5352–5354. [CrossRef]
11. Liang, R.Y.; Yang, P.; Mao, W.M. Effect of initial Goss texture sharpness on texture evolution and magnetic properties of Ultra-thin Grain-oriented electrical steel. *Acta Metall. Sin.* **2017**, *30*, 895–906. [CrossRef]
12. Shi, X.J.; Liang, Y.F.; Liu, B.B.; Ding, Z.Y.; Zhang, B.; Ye, F. Deformation twinning characteristics in hot-rolled Fe-6.5wt%Si alloy with different degree of order. *Mater. Sci. Eng. A* **2019**, *762*, 138095. [CrossRef]
13. Xie, J.X.; Fu, H.D.; Zhang, Z.H.; Jiang, Y.B. Deformation twinning in an Fe-6.5 wt.% Si alloy with columnar grains during intermediate temperature compression. *Mater. Sci. Eng. A* **2012**, *538*, 315–319. [CrossRef]
14. Dunn, C.G. Cold-rolled and primary recrystallization textures in cold-rolled single crystals of silicon iron. *Acta Metall.* **1954**, *2*, 173–183. [CrossRef]
15. Rusakov, G.M.; Lobanov, M.L.; Redikul'tsev, A.A.; Kagan, I.V. Specific features of cold deformation of a (110)[001] single crystal of an Fe–3% Si–0.5% Cu alloy related to twinning. *Phys. Met. Metallogr.* **2011**, *111*, 530–536. [CrossRef]
16. Dorner, D.; Zaefferer, S.; Raabe, D. Retention of the Goss orientation between microbands during cold rolling of an Fe3%Si single crystal. *Acta Mater.* **2007**, *55*, 2519–2530. [CrossRef]
17. Yang, P.; Xie, Q.; Meng, L.; Ding, H.; Tang, Z. Dependence of deformation twinning on grain orientation in a high manganese steel. *Scripta Mater.* **2006**, *55*, 629–631. [CrossRef]

18. Hong, C.S.; Tao, N.R.; Lu, K.; Huang, X. Grain orientation dependence of deformation twinning in pure Cu subjected to dynamic plastic deformation. *Scripta Mater.* **2009**, *61*, 289–292. [CrossRef]
19. Jo, S.Y.; Han, J.; Kang, J.H.; Kang, S.; Lee, S.; Lee, Y.K. Relationship between grain size and ductile-to-brittle transition at room temperature in Fe-18Mn-0.6C-1.5Si twinning-induced plasticity steel. *J. Alloy Compd.* **2015**, *627*, 374–382. [CrossRef]
20. Zhang, Z.; Wang, M.P.; Jiang, N.; Li, S.M. Orientation analyses for twinning behavior in small-strain hot-rolling process of twin-roll cast AZ31B sheet. *Mater. Sci. Eng. A* **2010**, *527*, 6467–6473. [CrossRef]
21. Jiang, M.G.; Yan, H.; Chen, R.S. Twinning, recrystallization and texture development during multi-directional impact forging in an AZ61 Mg alloy. *J. Alloy Compd.* **2015**, *650*, 399–409. [CrossRef]
22. Fu, H.D.; Zhang, Z.H.; Jiang, Y.B.; Xie, J.X. Applying the grain orientation dependence of deformation twinning to improve the deformation properties of an Fe-6.5 wt%Si alloy. *J. Alloy Compd.* **2016**, *689*, 307–312. [CrossRef]
23. Xie, J.X.; Fu, H.D.; Zhang, Z.H.; Jiang, Y.B. Deformation twinning feature and its effects on significant enhancement of tensile ductility in columnar-grained Fe-6.5wt.%Si alloy at intermediate temperatures. *Intermetallics* **2012**, *23*, 20–26. [CrossRef]
24. Humbert, M.; Petit, B.; Bolle, B.; Gey, N. Analysis of the γ-ε-α' variant selection induced by 10% plastic deformation in 304 stainless steel at −60 °C. *Mater. Sci. Eng. A* **2007**, *454–455*, 508–517. [CrossRef]
25. Wang, L.N.; Yang, P.; Mao, W.M. Analysis of martensitic transformation during tension of high manganese TRIP steel at high strain rates. *Acta Metall. Sin.* **2016**, *52*, 1045–1052.
26. Wang, L.N.; Yang, P.; Jin, T.; Mao, W.M. Different mechanisms of ε-M and α'-M variant selection and the influencing factors of ε-M reversion during dynamic tension in TRIP steel. *Acta Metall. Sin. Eng. Lett.* **2016**, *26*, 1863–1870. [CrossRef]
27. Walter, J.L.; Hibbard, W.R. Texture of cold-rolled and recrystallized crystal of silicon-iron. *Trans Metall. Soc. AIME* **1958**, *212*, 731–737.

crystals

MDPI

Article

Numerical Investigation into the Influence of Grain Orientation Distribution on the Local and Global Elastic-Plastic Behaviour of Polycrystalline Nickel-Based Superalloy INC-738 LC

Benedikt Engel *, Mark Huth and Christopher Hyde

Gas Turbine and Transmissions Research Centre (G2TRC), University of Nottingham, Nottingham NG7 2RD, UK; ezymah@exmail.nottingham.ac.uk (M.H.); christopher.hyde@nottingham.ac.uk (C.H.)
* Correspondence: benedikt.engel@nottingham.ac.uk

Abstract: Polycrystalline nickel-based superalloys tend to have large grains within component areas where high loads are dominant during operation. Due to these large grains, caused by the manufacturing and cooling process, the orientation of each grain becomes highly important, since it influences the elastic and plastic behaviour of the material. With the usage of the open source codes NEPER and FEPX, polycrystalline models of Inconel 738 LC were generated and their elastic and crystal plasticity behaviour simulated in dependence of different orientation distributions under uniaxial loading. Orientation distributions close to the [100] direction showed the lowest Young's moduli as well as the highest elastic strains before yielding, as expected. Orientations close to the [$\bar{5}$89] direction, showed the lowest elastic strains and therefore first plastic deformation under strain loading due to the highest shear stress in the slip systems caused by the interaction of Young's modulus and the Schmid factor.

Keywords: nickel-based superalloy; anisotropy; slip system modelling; Schmid factor; Young's moduli

Citation: Engel, B.; Huth, M.; Hyde, C. Numerical Investigation into the Influence of Grain Orientation Distribution on the Local and Global Elastic-Plastic Behaviour of Polycrystalline Nickel-Based Superalloy INC-738 LC. *Crystals* **2022**, *12*, 100. https://doi.org/10.3390/cryst12010100

Academic Editor: Wojciech Polkowski

Received: 24 December 2021
Accepted: 11 January 2022
Published: 13 January 2022

1. Introduction

Stationary gas turbines, as used in gas power plants, can be started from cold to a maximum output in less than 30 min [1]. Due to this short start-up time, gas power plants are able to react to shortages in the electricity supply caused by volatile feed in of renewables energies and can stabilize the electricity grid in times of high demand. Due to frequent start-up, shut down and load changes, the components in the hot gas section, especially the turbine blades and vanes undergo high mechanical and thermal loads, which has to be considered in the design process. Current design paradigms are based on deterministic lifetime approaches, which lead to conservative designs and inefficient component geometries caused by high safety factors. In recent years, these deterministic approaches were successfully replaced by probabilistic approaches, which are based on the statistical distribution of factors, which affects the high temperature fatigue life of the considered materials [2–4].

Polycrystalline nickel-based superalloys are widely used in the rear hot gas section as blades and vanes. Due to the casting manufacturing process and the component geometries, large grains with diameters up to 3–4 mm can form in areas where high loads appear during operations. Experimental strain controlled high temperature fatigue tests of nickel-base superalloys with large grains, i.e., low grain numbers in the gauge sections, show significantly large scatter in the lifetime as well as in the mechanical properties. As shown by several authors, the scatter in fatigue life and mechanical properties can be mainly attributed to two characteristics, the local grain orientation and the low amount of grains. Nickel-based superalloys show a high elastic anisotropy factors up to 2.7 [5]; therefore, the local orientation distribution of the grains within the gauge section have a significant influence on the elastic response of the component. This leads, in combination with the low amount of grains, to a non-averaging of mechanical properties as known for isotropic materials. For example, some austenitic steels show higher elastic anisotropies [6,7], but due to significantly smaller grains known for stainless steels, mechanical properties are averaging over the high amount of grains, and the macroscopic material behaviour can be classified as isotropic.

Furthermore, the local grain orientation influences the local plastic behaviour. If a critical shear stress within the slip systems of the grain is exceeded, dislocations start moving within the slip systems, which represents plastic deformation. For a face-centred crystal structure (fcc), such as nickel, slip mainly occurs preferably in one of the 12 slip systems of the type {111}[110], which are composed of four closed packed planes together with three close packed directions. As proposed by Schmid [8], the resulting shear stress within a slip system can be calculated by the product of the cosines of slip system direction and slip system plane (the so-called Schmid factor m) related to the loading direction and the applied normal load. Numerical investigations showed for 100,000 randomly orientated fcc single crystals, the minimum possible Schmid factor under normal loading is 0.28, which occurs only with a probability of 0.032%. The highest and also most frequently occurring Schmid factor is 0.5, which occurs with a probability of 16.4%. On average, the Schmid factor is calculated to 0.45 ± 0.041 for a single grain under uniaxial loading [9]. Seibel, Schmitz et al. [10,11] could significantly decrease the occurring scatter in a total strain Wöhler diagram for polycrystalline nickel-based superalloy Rene80, by determining the Schmid factor of the crack initiating grains using electron backscatter diffraction (EBSD), estimating the resulting shear stress and representing it over the lifetime. Further experimental and analytical investigations showed that with the usage of the product of the Young's moduli (E) and the Schmid factor (m) the so-called $E{\cdot}m$ value, predictions can be made in which grains plastic deformations concentrates. These grains tends to form persistent slip bands and provide crack initiations as shown in [12]. With the usage of polycrystalline finite element models and calculations of resulting shear stresses within the slip system, it could be proven that different orientation distribution within the materials can lead to improvements of the mechanical properties [13]. For smooth specimen made of random orientated coarse grain nickel-based superalloy Rene80 Young's moduli were in general higher compared to smooth specimen made from the same material batch but with a determined crystallographic texture (resulting from the casting process). In addition to lower Young's moduli and therefore lower stresses for tests under total strain control (compared to the random orientated material), the texture of the grains also leads to a shift of the Schmid factor distribution to lower values. This leads to lower resulting shear stresses in the slip systems and therefore to a delay in crack initiation and therefore longer lifetimes. Based on these findings, a probabilistic lifetime models for both variants of the material could be developed with a great accordance to observed experimental results. The same material batch, as discussed before, was investigated in [14] but for notched specimen. Contrary to all expectations, the material with the crystallographic texture didn't show any improvements in lifetime or mechanical properties. This can be attributed to the interaction of varying stress states along the notch with the crystallographic texture, which lead to high stiffness along the notch area, which increases the local stresses and shear stresses [15].

All above mentioned numerical researches were conducted using anisotropic elastic material models were calculated up to the point where the elastic resulting shear stress within the slip systems reaches a critical value and plastic deformation starts. The following paper investigates how local grain orientation also affects the plastic behaviour of polycrystalline nickel-based superalloys by numerical simulation. Specifically, the influence of different orientation distributions on the local and global onset of plasticity as well as the interaction of grains will be investigated.

2. Materials and Methods

2.1. Material

The investigated material is the polycrystalline nickel-based superalloy IN-738 LC, which has the composition shown in Table 1.

Table 1. Composition of IN-738LC [16].

Element	Ni	Cr	Co	Ti	Al	W	Mo	Ta	C	Nb
Wt.%	Bal.	16	8.58	3.42	3.36	2.67	1.81	1.9	0.105	0.9

This material is a precipitation strengthened nickel-based superalloy with around 59 vol.% of cuboid shaped γ' embedded in the γ matrix with grain sizes up to 3 mm in dependence of the manufacturing process and specimen design [17,18].

2.2. Material Modelling

Finite element simulations were carried out for polycrystalline models by using the open source codes NEPER (Version 4, by Romain Quey, MINES Saint-Étienne) [19,20] and FEPX (Version 1.1.2 by Paul Dawson, Cornell University) [21]. Neper is able to generate complex geometric components made of polycrystals by using three dimensional Voronoi tessellation and allows for a convenient meshing procedure. The generated model file can be exported as an input file for commercial finite element solvers or passed on to FEPX. FEPX is a finite element solver that allows for elastic and plastic calculations within polycrystals under the assumption of crystallographic characteristics within the material (i.e., slip system, anisotropy effects) to be performed. The main advantage of FEPX is the large range of possible material properties that can be considered. For example, FEPX includes non-linear kinematic hardening, rate-dependent slip within the slip systems, the evolution of state variables for crystal lattice orientation and slip system strength. In the following the main equations, which are important in this work, will be discussed where further details can be found in [21].

The elastic regime of the material in dependence of its orientation will be described by using a generalized Hooke's law for cubic crystals. The stress tensor $\underline{C}$ can be calculated by using the stiffness tensor $\underline{C}$, with the material constants C_{11}, C_{12} and C_{44}, multiplied by the elastic strain ε according to Equation (1).

$$\sigma = \underline{C}\varepsilon \tag{1}$$

Grains within polycrystalline materials are characterized by different grain orientations. In order to calculate the stiffness for a specific grain, the stiffness tensor must be rotated according to the grain orientation related to the loading direction. For plastic deformation, the kinematics of slip are given as a power law function by the following Equation (2).

$$\dot{\gamma}^{\alpha} = \dot{\gamma}_0 \left(\frac{|\tau^{\alpha}|}{g^{\alpha}} \right)^{\frac{1}{m}} sgn(\tau^{\alpha}) \tag{2}$$

With $\dot{\gamma}_0$ the fixed-state strain rate scaling coefficient, τ^{α} is the resolved shear stress of the αth slip system, m is the rate sensitivity coefficient and g^{α} is the current slip system strength or resistance to shear [22]. Note that in FEPX g^{α} is equal for the same family of slip systems. Since g^{α} describes the slip system strength, it can be further varied by using the modified Voce form, to account for cyclic hardening, i.e., slip system strength evolution (Equation (3)).

$$\dot{g^{\alpha}} = h_0 \left(\frac{g_s(\dot{\gamma}) - g^{\alpha}}{g_s(\dot{\gamma}) - g_0} \right)^{n} g_0 \tag{3}$$

h_0 is the fixed-state hardening rate scaling coefficient, g_0 is the initial slip system strength (which is considered equal for all slip systems) and n is the non-linear Voce hardening exponent. $g_s(\dot{\gamma})$ describes the slip systems saturation strength and is given by a power law according to Equation (4).

$$g_s(\dot{\gamma}) = g_{s0}\left(\frac{\dot{\gamma}}{\dot{\gamma}_{s0}}\right)^{m'} \tag{4}$$

g_{s0} is the initial slip system saturation strength, m' is the saturation strength rate scaling exponent and $\dot{\gamma}_{s0}$ is the initial saturation slip system shear rate. $\dot{\gamma}$ is the sum of the individual $\dot{\gamma}^{\alpha}$ values. Cyclic hardening [23] is not considered within the present paper.

2.3. Material Parameters

The elastic constants for IN-738 LC [16] were linearly interpolated between the given temperatures of 800 °C and 898 °C to calculate the elastic constants at 850 °C, as given in Table 2.

Table 2. Elastic constants of IN-738 LC at 850 °C.

C_{11} (GPa)	C_{12} (GPa)	C_{44} (GPa)
225.83	161.45	98.79

In order to model the plastic behaviour, IN-738 LC experimental tensile test data (for an assumed random orientation distribution) at 850 °C [24] were fitted and optimized against a model with sufficient amounts of grains for a strain rate of 0.001 s^{-1}. Figure 1 shows the experimental stress-strain data as well as the corresponding fitted elastic plastic material model.

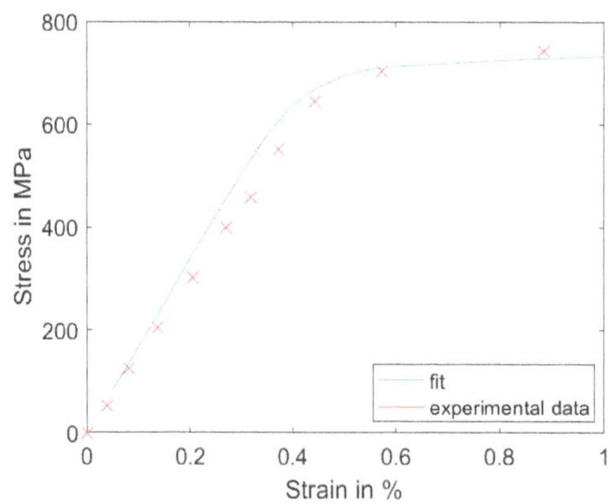

Figure 1. Experimental stress-strain data for IN-738 LC at 850 °C [24] and corresponding fitted elastic plastic polycrystalline model.

Plastic parameters, as shown in Table 3, where derived from the fitting and are used in the following simulations.

Table 3. Fitted plastic constants of IN-738 LC at 850 °C.

$\dot{\gamma}_0$ (s^{-1})	m	h_0 (MPa)	g_0 (MPa)	n'	g_1 (MPa)	$\dot{\gamma}_s$ (s^{-1})	m'
1.0	0.05	200	210	1	330	5×10^{10}	5×10^{-3}

2.4. Simulations

Material behaviour simulations were carried out for a cuboid with a 7 × 7 mm^2 cross section and 18 mm in length. 49 grains were generated using NEPER, which leads to an average grain size of 3.25 mm in diameter. Figure 2a represents the simulated specimen, within the visualization tool of NEPER.

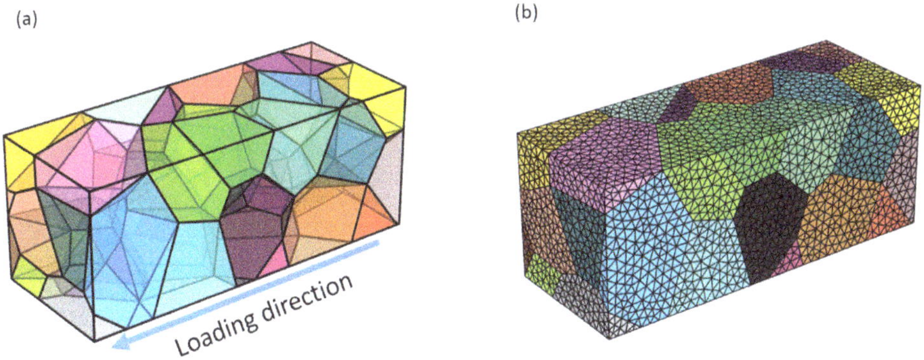

Figure 2. (**a**) Polycrystalline model with 49 grains generated by NEPER and (**b**) meshed equivalent for finite element analysis.

Meshing was conducted using the meshing tool integrated in NEPER, by using a relative cell length value (rcl) of 0.4, which leads to approximately 65000 10-node tetrahedral elements (C3D10), which gives a good compromise between calculation time and accuracy (see Figure 2b). A displacement load in loading direction (see Figure 2a) was applied on one face to ensure a uniaxial loading whereas the other face was locked for rotation and translation. The maximum displacement was set to 0.18 mm, which is the equivalent of 1% of total strain to allow for distinct plastic deformation. In order to investigate the influence of various orientation distribution, sets of rotational matrices, i.e., Euler angles, were generated and assigned to the grains.

In order to explain the reasons why the following orientation distributions were chosen, the $E{\cdot}m$ model for a single grain will be explained [9,12]. The main assumption of this model is that the resulting shear stress in a slip system of a single grain under uniaxial strain control is only dependent upon the Young's modulus, E, and Schmid factor, m, of the grain, both of which are defined by the local grain orientation. Under the assumption of an elastic behaviour, the resulting shear stress can be calculated using Equation (5). Note that this equation is only valid up to the point where the resulting shear stress, τ_{res},is equal to the critical shear stress, τ_{crit}, which is the point at which plastic deformation begins.

$$\tau_{res} = \varepsilon_{a,t}{\cdot}E{\cdot}m \tag{5}$$

For a constant total strain, $\varepsilon_{a,t}$, grains with a high value of $E{\cdot}m$ reach yield earlier compared to grains with low $E{\cdot}m$ values. The pole figures shown in Figure 3 represent the chosen orientation distributions (see Figure 3a) as well as the value of at 850 °C in dependence of the grain orientation, illustrated as an inverse pole figure (see Figure 3b).

Whereas the orientations close to [001], [011] and [111] directions will be named the same, orientations at (1) will be considered stiff orientations (close to the [$\bar{5}$89], since they show a maximum in the $E{\cdot}m$ model (see Figure 3b), whereas (2) will be considered soft, due to the correspondingly low $E{\cdot}m$ values. Furthermore, six random orientation distribution, were generated in accordance to the Haar measure to ensure statistically random grain orientations [4]. One model is divided in the middle in a half stiff and half soft grain orientation distribution in order to investigate how they interact; this will be considered a mix model.

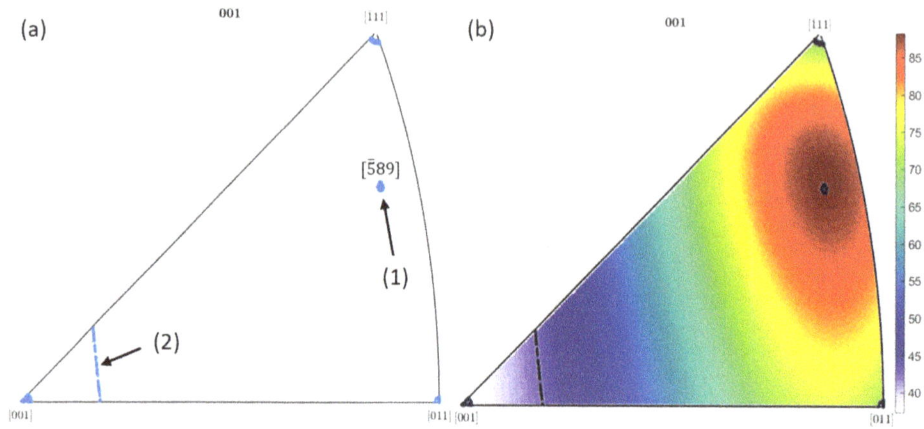

Figure 3. (**a**) Inverse pole figure of chosen grain orientation distributions and (**b**) $E{\cdot}m$ values at 850 °C.

3. Results

Figure 4 shows the macroscopic stress-strain curves, i.e., the mechanical response of the whole model (see Figure 2) for a uniaxial load in dependence of the different orientation distributions. The yield point is indicated with **x** for each curve whereby it was defined as the point at which the simulated elastic stress (based on the Hooke's law) differs by more than 5% from the simulated elastic-plastic results.

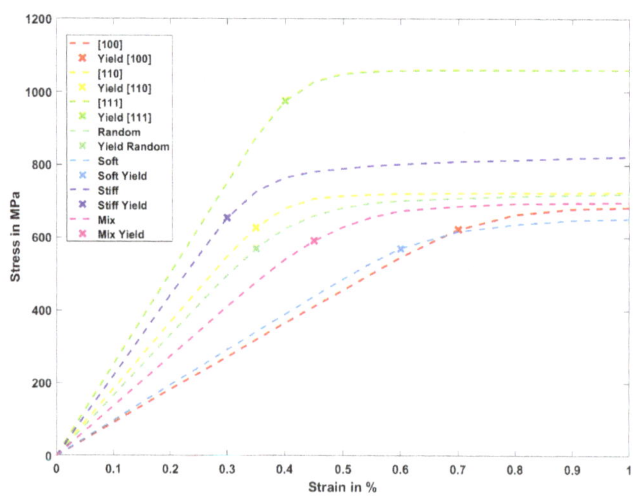

Figure 4. Simulated stress-strain curves for different grain orientation distributions for an INC-738 LC at 850 °C.

The overall stress-strain curves comparison shown in Figure 4 shows a significant scatter within the elastic regions. The detailed results for each curve is listed below:

- For the [100] orientation distribution the lowest Young's moduli of 91.19 GPa can be found. The yield point is reached at 622 MPa with an elastic strain of 0.7%.
- For the soft orientation distribution, which slightly differs from the [100] orientation distribution (see Figure 3), the Young's moduli is marginal higher with 97.58 GPa. However, yield is distinct lower and reached at 568 MPa and a related elastic strain of 0.6%.
- Highest Young's moduli is reached for the [111] orientation distribution with 250.45 GPa and yield at 973 MPa with an elastic strain of 0.4%.
- The stiff orientation distribution reaches yield first, i.e., shows lowest elastic strain compared to all other investigated orientation distributions. Besides a quite high Young's modulus of 220.37 GPa, yield is reached at 0.3% with a yield stress of 655 MPa.
- The [110] orientation distribution as well as the random orientation distribution show same elastic strains at the yield point with 0.35%. But due to slightly higher Young's modulus for the [110] orientation distribution with 182.67 GPa, the yield stress is also slightly higher with 625 MPa.
- In order to consider the influence of random orientation distributions and the resulting scatter, six simulations were statistically evaluated, where the average stress-strain curve is shown in Figure 4. For the random orientation distribution, the Young's modulus is 163.8 ± 4.8 GPa and yield stress is reached at 561.63 ± 15.17 MPa.
- The mix orientation distribution shows a Young's modulus of 136.37 GPa and yields at 590 MPa and 0.45% total strain.

Figure 5 shows the local elastic strain in the loading direction as well as the equivalent plastic strain for a simulated specimen with a random grain orientation at 1% total strain (turquoise curve, see Figure 4).

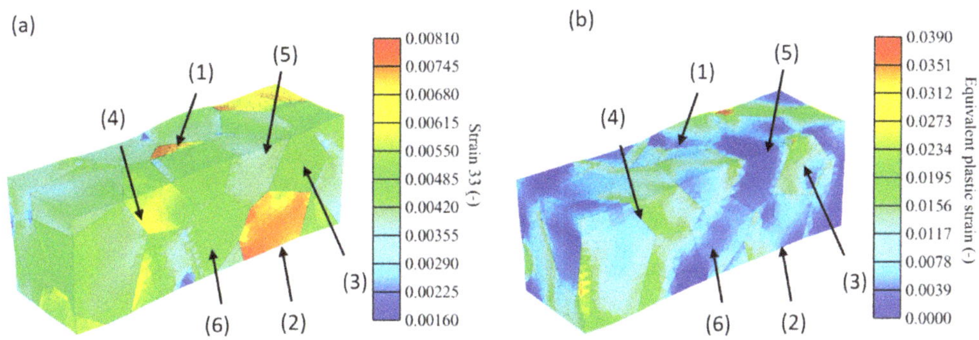

Figure 5. (**a**) Elastic strain S33 in loading direction and (**b**) equivalent plastic strain for a randomly orientated specimen.

The elastic strain distributions in loading direction are inhomogeneous, where some grains show significantly higher elastic strains compared to others. Interestingly, these high elastic strain grains show low equivalent plastic strains as seen at grain (1) and (2) in Figure 5b. Higher plastic strains have been observed in grains that show intermediate elastic strains, as demonstrated by grain (3). Grain (4) shows both, high local elastic strain as well as high equivalent plastic strain, whereas grain (5) and (6) see intermediate elastic strains, but almost negligible equivalent plastic strain. Figure 6 shows the distribution and development of equivalent plastic strain for the specimen shown in Figure 5, at 0.2%, 0.4%, 0.6% and 0.8% of total strain.

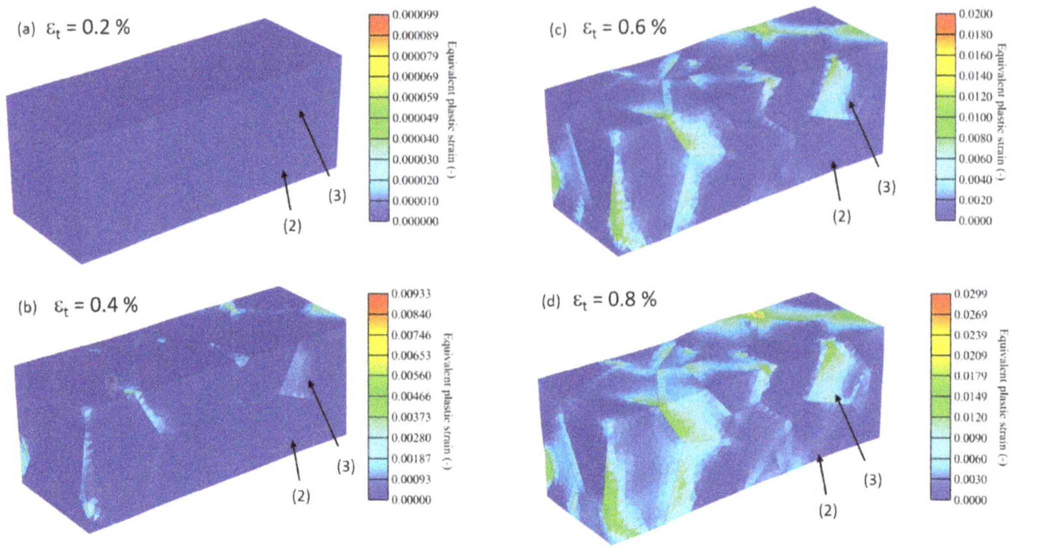

Figure 6. Distribution of equivalent plastic strain under (**a**) 0.2%, (**b**) 0.4%, (**c**) 0.6% and (**d**) 0.8% total strain for the randomly orientated grain distribution.

For a total strain of 0.2%, no plastic strain within the whole material can be determined. At a total strain of 0.4% some plastic deformation can be seen to occur at some grain boundaries. At this strain, the previously discussed grain (2) displays no plastic deformation, whereas grain (3) already does. Plastic deformation continues to increase as the total strain increases to 0.6% and grain (2) still shows only minor plastic deformation. For a total strain of 0.8% grain (2) show slightly increased plastic deformation whereas for grain (3) plastic strain is significantly increased.

Figure 7 shows the distribution of elastic strain in the loading direction and the equivalent plastic strain for a simulated specimen with the soft orientation distribution at 1% of total strain (light blue curve, see Figure 4).

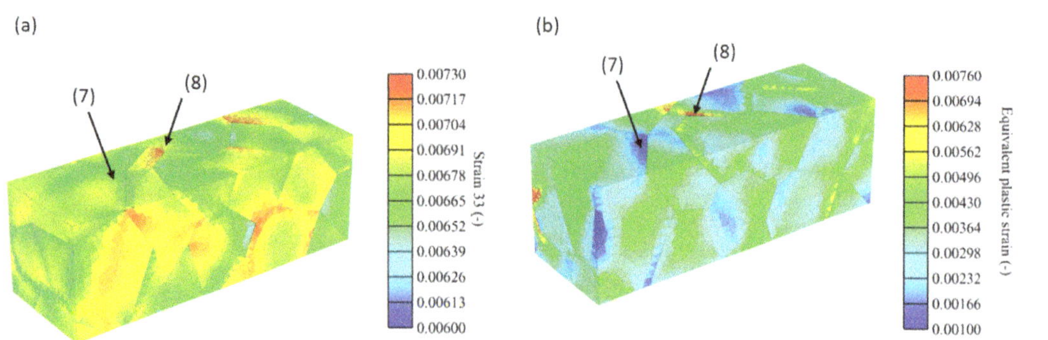

Figure 7. (**a**) Elastic strain S33 in loading direction and (**b**) equivalent plastic strain, for the soft orientated specimen.

The results shown in Figure 7 for a specimen with soft orientation distribution, show a much more homogeneous distribution of high elastic strains within the material (see scale range), compared with the previous shown specimen. This leads to an approximately uniform distribution of equivalent plastic strains with small areas of slightly increased equivalent plastic strain (see grain (8)) or slightly decreased equivalent plastic strain (see grain (7)). The small deviations arise from similar but not equal grain orientations (see pole figure in Figure 3a).

A contrary behaviour can be seen for the stiff orientation distribution in Figure 8 (see blue curve in Figure 4). The elastic strains are in general much smaller compared to the soft orientation distribution and high values are occasionally located at grain boundaries as the grains (9) shows.

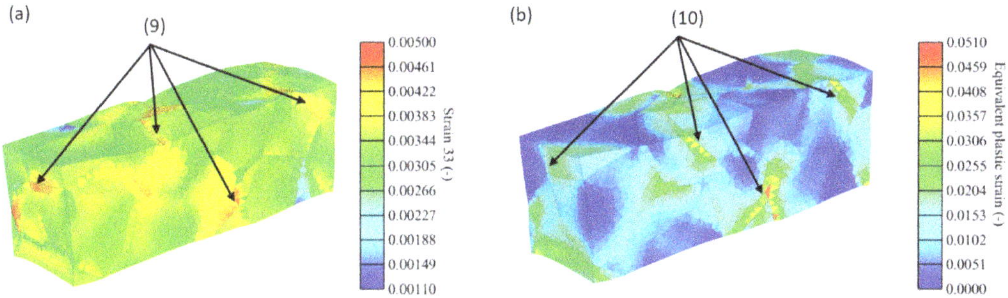

Figure 8. (**a**) Strain S33 and (**b**) equivalent plastic strain in loading direction for the stiff orientated specimen.

According to Figure 8b and the grains indicated with (10), it appears that highest plastic deformations mostly occur close to grain boundaries and decline into the grain. Areas which show high elastic strains also show high equivalent plastic strains. As the simulation also shows, high multi-axial deformation is added to the stiff specimen during a simulated uniaxial test.

Figure 9 shows the results for the mix specimen under a total strain of 1%. It should be noted here that the first 25 grains have a stiff orientation distribution whereas the last 24 show a soft orientation distribution (if counted from the right).

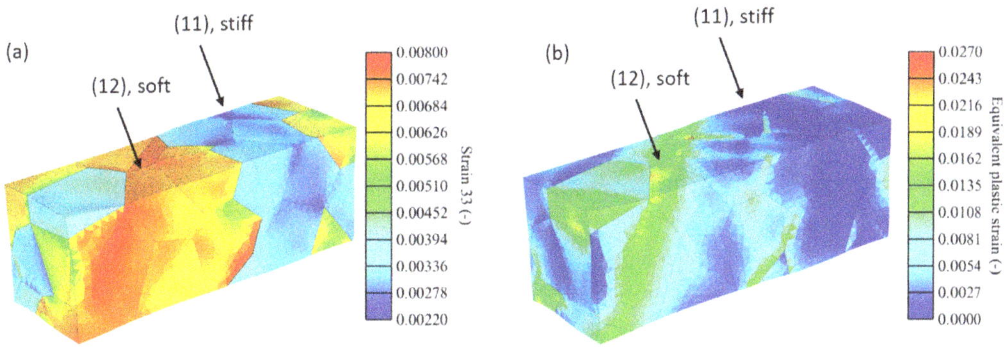

Figure 9. Elastic strain S33 in loading direction (**a**) and equivalent plastic strain (**b**) for the mix orientated specimen.

The elastic strains in the soft part of the specimen (12) are significantly higher compared to the stiff orientated part (11), which is caused by the grain orientation and their associated low Young's moduli. However, the soft part shows significantly larger equivalent plastic strains as compared to the stiff part.

4. Discussion

The stress-strain curves, shown in Figure 4, represent what was already assumed in the $E{\cdot}m$ model. Even if the [100] and the soft orientation distribution show similar Young's moduli (see Figure 10b), their yielding behaviour is different. This is due to the case that the [100] close orientations show average Schmid factors of 0.412 ± 0.0019, whereas the soft orientation distribution shows distinct higher average value for the Schmid factor of 0.455 ± 0.0036, as Figure 10a represents. The slightly increased Young's moduli together with higher Schmid factors for the soft orientation distribution leads to average $E{\cdot}m$ values of 43.57 ± 0.023 and thus leads to earlier plastic deformation at the same strains as for the [100] orientation distribution with average $E{\cdot}m$ values of 37.65 ± 0.1834.

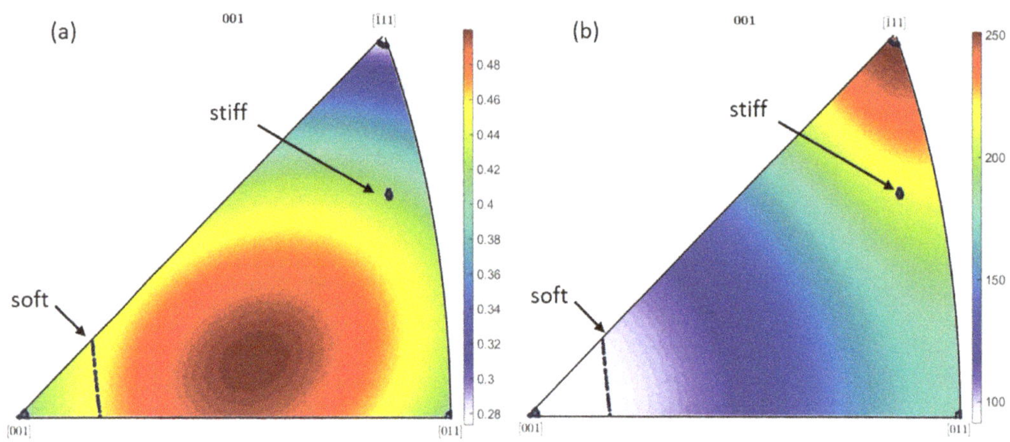

Figure 10. Inverse pole plot of the Schmid factor (**a**), inverse pole plot of the Young's moduli of IN-738 LC at 850 °C (**b**) and the chosen orientation distributions.

Comparing the stiff and the [111] orientations distributions also shows an interesting behaviour. For the [111] orientation the space diagonal of a unit cell of the investigated nickel-based superalloy aligns with the loading direction and shows a global maximum in Young's moduli with about 250 GPa. Despite the fact that the stiff orientation has a lower Young's moduli of 220 GPa it yields at significantly lower strains (see blue curve, Figure 4). Also in this case the Schmid factor distribution is the important factor. For orientations close to the [111] direction, Schmid factor shows a global minimum with average Schmid factors of 0.284 ± 0.0015, whereas the stiff orientation distribution shows average Schmid factors of 0.421 ± 0.0016. However, if both Young's moduli and Schmid factor are considered in combination within the $E{\cdot}m$ model, it becomes apparent that the stiff orientation distribution leads to a global maximum in $E{\cdot}m$ with 88.09 ± 0.038 as shown in Figure 3.

Thus, it follows that the stiff orientation distribution globally shows lowest elastic strain, i.e., start to plastic deform as first, compared to other orientation distributions, although this distribution doesn't show maxima in Schmid factor neither in Young's modulus. First as a combination in the $E{\cdot}m$ model, it becomes obvious why the stiff orientation distribution shows the explained behaviour. A comparison of the stiff orientation distribution and its Schmid factor distribution with 0.421 ± 0.0016 and the soft orientation distribution and its Schmid factor distribution with 0.455 ± 0.0036 also reveal the fact that a higher Schmid factor doesn't necessarily lead to higher shear stresses, as it could possibly be assumed. Moreover, the Young's moduli has also to be considered to make reliable statements about the occurring shear stress within the slip systems.

Since the individual orientations for the distributions [100], [110], [111], stiff and soft are all very close to each other, no large scatter in local mechanical properties were expected and proven by the finite element results. However, for the random orientation distribution, a statistical analysis has been carried out because of a large occurring scatter in local and global mechanical properties (see Figure 5). The following Figure 11 shows the distribution of the random grain orientations for all five simulated specimen within an inverse $E{\cdot}m$ pole plot; additionally, the grains (1)–(6) from Figure 5 are highlighted.

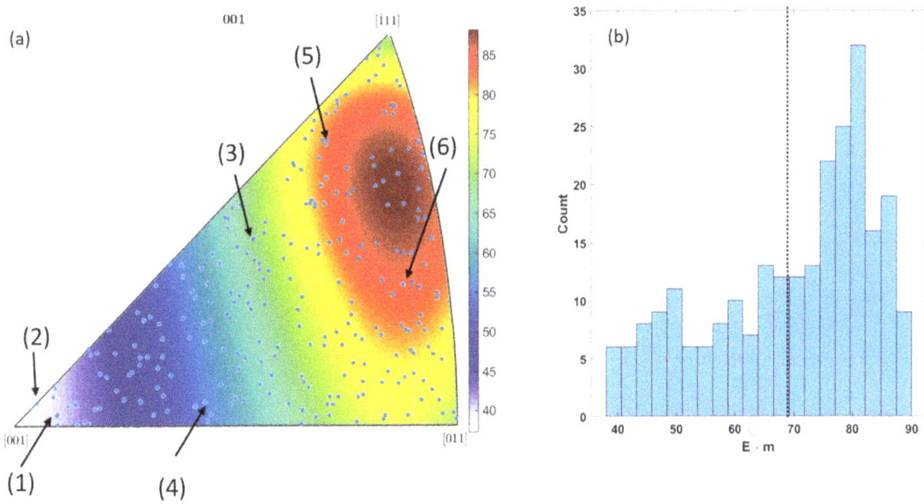

Figure 11. Distribution of random orientations in the inverse $E{\cdot}m$ plot (**a**) and its histogram (**b**).

It can be clearly seen at the inverse pole plot that the orientations are randomly distributed, whereas on average the $E{\cdot}m$ value is calculated to 69.57 ± 13.89 (see Figure 11b). The histogram also shows a shift of the distribution to higher $E{\cdot}m$ values, which underline the fact that for a random orientation, a high $E{\cdot}m$ value (> 70) is more likely than a low one. It should be noted for the random specimen, since the orientations and therefore the mechanical properties differ over a wide range, the strain distributions also are not homogeneously distributed compared to the specimen where orientation just differs slightly. As grain (2) in Figure 5 shows, large elastic strains in loading direction evolve, but only minor plastic strains are generated. This is caused by its orientation, as Figure 11a shows. The orientation is close to [100] and allows due to lowest $E{\cdot}m$ values high elastic strains before plastic deformations starts at a total strain of 0.6%, as the stress-strain curve in Figure 4 shows. In comparison grain (3) has an intermediate $E{\cdot}m$ value, which leads to lower elastic strains and more equivalent plastic strains. For the mentioned cases of grain (2) and (3) in Figure 5, the $E{\cdot}m$ approach works well to estimate the onset of plasticity, which is due to the case that both grains show total strains of same magnitude until yield is reached as the following Figure 12a shows.

But within the random orientated specimen, grains appear that show remarkably low total strain, and therefore there is nearly no plastic deformation as grain (5) indicates. Even with an $E{\cdot}m$ value of 82.16, the amount of strain that is generated within the grain is not sufficient enough to reach yield and to generate plastic deformation, even at a global total strain of 1%. The same behaviour can be found for grain (6), which has a calculated $E{\cdot}m$ value of 83.5 but only show low equivalent plastic deformation. It appears that the grains (2) and (3), both with much lower $E{\cdot}m$ value, generate much more local equivalent total strain as the connected grains (5) and (6). A similar behaviour can be observed in the mix specimen. The stiff part with an average $E{\cdot}m$ value of 88.1 ± 0.038 shows distinct lower equivalent total strains and therefore lower equivalent plastic strains compared to the soft part with an average $E{\cdot}m$ of 43.57 ± 0.023. Due to the higher local strains, the soft part of the specimen generates the majority of the plastic deformation. These results lead to the hypothesis that if grains with a low $E{\cdot}m$ value are interacting with grains with high $E{\cdot}m$ value, the softer grains will see higher local total strains and therefore more potential plastic deformation compared to stiff grains. This behaviour could already be observed in [9,12] during fatigue testing, where cracks either initiate in grains with high $E{\cdot}m$ value or in grains with a low $E{\cdot}m$ value, which are interacting and surrounded with grains with a high $E{\cdot}m$.

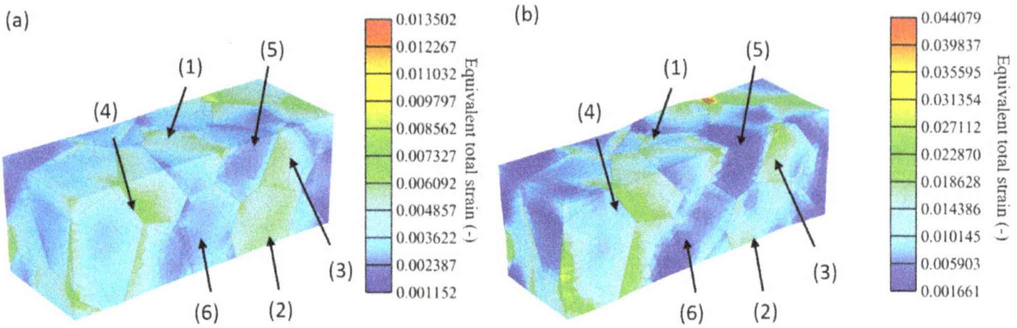

Figure 12. Equivalent total strain for (**a**) 0.4% global applied total strain and (**b**) 1% global applied total strain.

5. Conclusions

The presented work could clearly demonstrate by means of numerical simulations the influences of different grain orientation distributions on the local and global mechanical behaviour of IN-738 LC. It could be shown that grain orientations close to the [100] direction provide lowest Young's moduli and additionally highest elastic strains before yielding. It could be also shown that orientation distributions close to the [111] direction generates highest Young's moduli, but interestingly not the lowest elastic strains before yielding. This could be attributed to lowest Schmid factors associated with the [111] direction. Using the $E{\cdot}m$ model, which considers both Young's moduli and Schmid factor in dependence of the orientation, it could be found that orientations close to the $[\bar{5}89]$ direction show a global maximum in the $E{\cdot}m$ model and therefore lowest elastic strains before yielding. Since the $E{\cdot}m$ model is only able to predict the onset of plasticity if a uniform and homogeneous strain distribution is considered, random grain orientation distribution were simulated. Due to a high range of possible mechanical properties of the grains caused by elastic and plastic anisotropies, the total strain distribution is highly inhomogeneous. For regions in the material where the total strain is evenly distributed, the $E{\cdot}m$ allows for the indication of which grains show plastic deformation first. But, if grains with high and low $E{\cdot}m$ value interact, most of the total strain is taken by grains with low $E{\cdot}m$ value (due to low stiffness), which therefore show more plastic deformation compared to surrounding grains with a high $E{\cdot}m$ value.

The presented results were derived from modeling a specific nickel-based superalloy. But the determined elastic plastic behaviour in dependence of the grain orientation, i.e., the $E{\cdot}m$ value, can be applied to all cubic fcc metals as long the plastic deformation occurs within the {111}[110] slip systems. Extensions of the model to bcc metals as well as other slip systems are also possible. The shown approach is particularly interesting because almost all metallic materials show an anisotropic elastic behaviour on the microscale. Future investigations should focus on the application of the shown procedure to more complex geometries and multi-axial stress states.

Author Contributions: B.E. wrote the present paper and did the conceptualization and methodology, C.H. and B.E. supervised the numerical modelling conducted by M.H., C.H. and M.H. did the review and editing. All authors have read and agreed to the published version of the manuscript.

Funding: No funding.

Institutional Review Board Statement: Not applicable.

Informed Consent Statement: Not applicable.

Data Availability Statement: Not applicable.

Conflicts of Interest: The authors declare no conflict of interest.

References

1. Walsh, P.P.; Fletcher, P. *Gas Turbine Performance*; Blackwell Science Ltd.: Oxford, UK, 2004.
2. Mäde, L.; Schmitz, S.; Gottschalk, H.; Beck, T. Combined notch and size effect modeling in a local probabilistic approach for LCF. *Comput. Mater. Sci.* **2018**, *142*, 377–388. [CrossRef]
3. Mäde, L.; Kumar, K.; Schmitz, S.; Gundavarapu, S.; Beck, T. Evaluation of component-similar rotor steel specimens with a local probabilistic approach for LCF. *Fatigue Fract. Eng. Mater. Struct.* **2020**, *43*, 932–946. [CrossRef]
4. Moch, N. From Microscopic Models of Damage Accumulation to the Probability of Failure of Gas Turbines. Ph.D Thesis, Universitätsbibliothek Wuppertal, Wuppertal, Germany, 2019.
5. Rösler, J.; Harders, H.; Bäker, M. *Mechanisches Verhalten der Werkstoffe*; Springer Fachmedien Wiesbaden: Wiesbaden, Germany, 2012; ISBN 978-3-8348-1818-8.
6. Dou, Y.; Luo, H.; Zhang, J. Elastic Properties of $FeCr_{20}Ni_8X_n$ (X = Mo, Nb, Ta, Ti, V, W and Zr) Austenitic Stainless Steels: A First Principles Study. *Metals* **2019**, *9*, 145. [CrossRef]
7. Hasanzadeh, A.; Hamedani, A.; Alahyarizadeh, G.; Minuchehr, A.; Aghaei, M. The role of chromium and nickel on the thermal and mechanical properties of FeNiCr austenitic stainless steels under high pressure and temperature: A molecular dynamics study. *Mol. Simul.* **2019**, *45*, 672–684. [CrossRef]
8. Schmid, E.; Boas, W. *Plasticity of Crystals with Special Reference to Metals*; F.A. Hughes: Burnsville, NC, USA, 1950.
9. Engel, B. Einfluss der lokalen Kornorientierung und der Korngröße auf das Verformungs- und Ermüdungsverhalten von Nickelbasis Superlegierungen. Ph.D Thesis, Technische Universität Kaiserslautern, Kaiserslautern, Germany, 2019.
10. Seibel, T. Einfluss der Probengröße und der Kornorientierung auf die Lebensdauer einer polykristallinen Ni-Basislegierung bei LCF-Beanspruchung. Ph.D Thesis, Forschungszentrum Jülich, Jülich, Germany, 2014.
11. Schmitz, S.; Seibel, T.; Beck, T.; Rollmann, G.; Krause, R.; Gottschalk, H. A probabilistic model for LCF. *Comput. Mater. Sci.* **2013**, *79*, 584–590. [CrossRef]
12. Engel, B.; Beck, T.; Moch, N.; Gottschalk, H.; Schmitz, S. Effect of local anisotropy on fatigue crack initiation in a coarse grained nickel-base superalloy. *MATEC Web Conf.* **2018**, *165*, 4004. [CrossRef]
13. Engel, B.; Mäde, L.; Lion, P.; Moch, N.; Gottschalk, H.; Beck, T. Probabilistic Modeling of Slip System-Based Shear Stresses and Fatigue Behavior of Coarse-Grained Ni-Base Superalloy Considering Local Grain Anisotropy and Grain Orientation. *Metals* **2019**, *9*, 813. [CrossRef]
14. Engel, B.; Beck, T.; Schmitz, S. (Eds.) High temperatue Low Cylce Fatigue of the Ni-base Superallor René80. In Proceedings of the Eighth International Conference on Low Cycle Fatigue (LCF8), Dresden, Germany, 27–29 June 2017.
15. Engel, B.; Ohneseit, S.; Mäde, L.; Beck, T. Influence of Grain Orientation Distribution on the High Temperature Fatigue Behaviour of Notched Specimen Made of Polycrystalline Nickel-Base Superalloy. *Metals* **2021**, *11*, 731. [CrossRef]
16. Bayerlein, U.; Sockel, H. Determination of single crystal elastic constants from ds- and dr-ni-based superalloys by a new regression method between 20 °C and 1200 °C. *Superalloys* **1992**, 695–704. Available online: https://www.tms.org/Superalloys/10.7449/1992/Superalloys_1992_695_704.pdf (accessed on 1 December 2021).
17. Tobias, J.; Chlupova, A.; Petrenec, M.; Polak, J. (Eds.) Low Cycle Fatigue and Analysis of the Cyclic Stress-Strain Response in Superalloy Inconel 738LC. In Proceedings of the 18th International Conference of Engineering Mechanics, Svratka, Czech, 14–17 May 2012.
18. Šmíd, M.; Petrenec, M.; Polák, J.; Obrtlík, K.; Chlupová, A. Analysis of the Effective and Internal Cyclic Stress Components in the Inconel Superalloy Fatigued at Elevated Temperature. *AMR* **2011**, *278*, 393–398. [CrossRef]
19. Quey, R.; Dawson, P.R.; Barbe, F. Large-scale 3D random polycrystals for the finite element method: Generation, meshing and remeshing. *Comput. Methods Appl. Mech. Eng.* **2011**, *200*, 1729–1745. [CrossRef]
20. Quey, R.; Renversade, L. Optimal polyhedral description of 3D polycrystals: Method and application to statistical and synchrotron X-ray diffraction data. *Comput. Methods Appl. Mech. Eng.* **2018**, *330*, 308–333. [CrossRef]
21. Dawson, P.R.; Boyce, D.E. FEpX—Finite Element Polycrystals: Theory, Finite Element Formulation, Numerical Implementation and Illustrative Examples. 2015. Available online: http://arxiv.org/pdf/1504.03296v1 (accessed on 1 December 2021).
22. Kasemer, M.; Falkinger, G.; Roters, F. A numerical study of the influence of crystal plasticity modeling parameters on the plastic anisotropy of rolled aluminum sheet. *Model. Simul. Mater. Sci. Eng.* **2020**, *28*, 85005. [CrossRef]

23. Turkmen, H.S.; Miller, M.P.; Dawson, P.R.; Moosbrugger, J.C. A Slip-Based Model for Strength Evolution During Cyclic Loading. *J. Eng. Mater. Technol.* **2004**, *126*, 329–338. [CrossRef]
24. Lin, R.C.; Betten, J.; Brocks, W. Modeling of finite strain viscoplasticity based on the logarithmic corotational description. *Arch. Appl. Mech.* **2006**, *75*, 693–708. [CrossRef]

crystals

MDPI

Article

Microstructural Changes in Ni-Al-Cr-Based Heat-Resistant Alloy with Re Addition

Nina A. Koneva [1], Elena L. Nikonenko [1], Alisa V. Nikonenko [2] and Natalya A. Popova [1,*]

[1] Department of Physics, Tomsk State University of Architecture and Building, 2, Solyanaya Sq., 634003 Tomsk, Russia; koneva@tsuab.ru (N.A.K.); vilatomsk@mail.ru (E.L.N.)

[2] Department of Physics, Tomsk State University of Control Systems and Radioelectronics, 40, prospect Lenina, 634050 Tomsk, Russia; aliska-nik@mail.ru

* Correspondence: natalya-popova-44@mail.ru

Abstract: This paper presents scanning and transmission electron microscope investigations of the structure, phase composition, and morphology of a heat-resistant alloy modified by thermal treatment and additionally alloyed by rhenium. The rhenium alloy was obtained by using the directional crystallization technique. The structural investigations were carried out for two states of the alloy, i.e., (1) original (after the directional crystallization); (2) after the directional crystallization with 1150 °C annealing for 1 h and 1100 °C annealing for 480 h. It is shown that fcc-based γ- and γ'-phases are primary in all states of the alloy. The γ'-phase has an $L1_2$ structure, while γ-phase is a disordered phase. It was found that after directed crystallization, the volume fraction of the γ' phase is ~85%, the fraction of the γ-phase is less than 10%. Annealing leads to an increase in the γ'- phase up to 90%, the proportion of the γ-phase practically does not change. Rhenium is a phase-formation element. The investigations show that high-temperature annealing modifies the structural and phase conditions of the heat-resistant alloy.

Keywords: aluminum; nickel; superalloys; γ'- and γ-phases; close-packed phases; rhenium

Citation: Koneva, N.A.; Nikonenko, E.L.; Nikonenko, A.V.; Popova, N.A. Microstructural Changes in Ni-Al-Cr-Based Heat-Resistant Alloy with Re Addition. *Crystals* **2021**, *11*, 89. https://doi.org/10.3390/cryst11020089

Academic Editor: Wojciech Polkowski

Received: 29 December 2020
Accepted: 18 January 2021
Published: 21 January 2021

1. Introduction

The global political processes of the last century have triggered the emergence of materials used in the manufacture of the engine-related parts of jet aircrafts, i.e., superalloys. Further enhancement of the properties of materials operating under high temperatures has enabled them to increase their operating temperatures and operating voltages by improving the production processes and changing the chemical compositions of superalloys. Progress in the development of such alloys has made it possible to create modern jet engines with an even higher ratio of the developed thrust to the dead weight of the engine. Superalloys play a vital role in industrial gas turbines, coal refineries, and other installations where high temperatures and highly corrosive environments operate [1–3].

The success of modern technology is largely driven by the creation and application of metallic materials with the necessary service properties. The level of the requirements for these materials is constantly growing due to the new problems that arise when creating new equipment, particularly when operating under extreme conditions, such as high velocities, high temperatures, etc. The search for metallic materials that can operate in these extreme conditions is constantly ongoing [4]. One of the promising directions is the development of superalloys containing intermetallic phases. Some examples of these superalloys are those based on a mixture of γ'- and γ-phases [1–6], where the γ-phase is a disordered fcc solid solution based on, e.g., nickel and aluminum, while the γ'-phase (in this case, Ni_3Al) is an ordered structure with an $L1_2$ superstructure [7–9]. Currently, superalloys are often created on the basis of an alloy of nickel and aluminum alloyed with various refractory elements. In these modern superalloys, the γ'-phase is the primary one. For this reason, it is largely responsible for the formation of the superalloy properties.

With the development and creation of superalloys, the proportion of the γ-phase in the superalloy decreases, while the proportion of the γ'-phase increases (up to 90% or more). In real nickel-based superalloys, the composition of which has multiple components, along with Ni and Al, there are atoms of other elements, such as Ti, Cr, Co, Mo, W, Ta, Nb, Hf, and Re [1,2,10–16]. Alloying with Re increases the operating temperature and entails the appearance of topologically close-packed phases (TCP-phases) inside the material, which are called Frank–Kasper phases. These phases can have a complex effect on the structure and properties of superalloys [17–20].

It has also been established that Re is mainly localized in the γ-phase [21–23]. When concentrated in solid γ-solution, Re reduces the velocity of diffusion processes under working temperatures, and thus increases thermal stability of γ-matrix. Thus, with an increase in Re content, the long-term strength of heat-resistant alloys grows [21]. Apart from the increase in strength properties, a larger content of Re also results in increased numbers and sizes of γ-phase interlayers, which significantly reduce the coarsening kinetics of γ'-phase quasi-cuboids [21,22]. Moreover, Refs. [21–23] revealed inhomogeneities in the spatial distribution of Re: Areas enriched with Re are localized close to the border of the γ-/γ'-phases on the side of the γ-phase [21,22].

Thus, superalloys, their structures, and their mechanical properties under high temperatures and under low temperatures, as well as their stability in operating conditions, represent one of the greatest achievements in the physics of alloys nowadays. Their creation and the need to obtain high properties require studying the structure and properties of solid solutions, ordered phases, and intermetallic compounds, constructing numerous equilibrium diagrams, solving high-temperature strength problems, studying creep resistance, and solving numerous other problems. Since the morphology of the γ'-phase can have a decisive influence on the mechanical properties of the superalloy, it seems important to study the structures of various superalloys in detail using structural research methods, which, along with the study of the phase composition of multicomponent superalloys, allow detailed study of the morphology of the γ'-phase. The studies carried out within this work are aimed at solving some of the above-mentioned problems.

The aim of this work was to qualitatively and quantitatively study the structure, phase composition, and phase morphology of a complexly alloyed heat-resistant alloy additionally alloyed with Re both in the initial state and after heat treatment.

2. Materials and Methods

The study of the structure of the alloy was carried out in two states: (1) the initial state—after directional crystallization (DC); (2) DC + annealing at a temperature of T = 1150 °C for 1 h + annealing at the temperature of T = 1100 °C for 480 h. All states of the alloy had a monocrystalline structure with the [001] orientation.

The basic elements of the alloy were: Ni: ~70 at.%, Al: ~17 at.%, and Cr: ~5 at.%. The alloying elements were Mo: 3 at.%, W: 1.2 at.%, Ti: 1.3 at.%, Co: 2 at.%, C: 0.1 at.%, and Re: 0.4 at.%. The chemical composition was determined by X-ray fluorescence analysis.

Three methods were used in the present work: (1) transmission electron microscopy (TEM) (transmission electron microscope EM-125), (2) scanning electron microscopy (SEM) (electron microscope QUANTA 200 3D FEI), and 3) a method of X-ray diffraction analysis (DRON-7 diffractometer with 20–920 range for angles).

Using the SEM method, the following structural parameters were identified: the shapes and sizes of large particles of the γ' phase ($\geq$1 μm) and secondary phases, their relative positions, and their volume fractions. TEM images of the fine structures of the materials were used: firstly, to classify the structures; secondly, to identify phases and determine the size and volume fraction of the main γ- and γ'-phases and precipitates of the secondary phases ($\leq$1 μm) [24], as well as their localization sites; thirdly, to determine the scalar dislocation density (ρ). In the case of X-ray structural analysis, the images of samples were taken at room temperature on an X-ray diffraction meter. The positions of the diffraction maxima that were clearly related to the unit cell size were used to calculate the crystal lattice parameters of the present phases.

The average sizes of structural elements (phases) and the scalar dislocation density were measured from the corresponding micrographs using the secant method [25]. The data obtained were processed statistically. Phase analysis was carried out on the basis of data obtained via transmission electron microscopy from the interpretation of the corresponding microdiffraction patterns, observations in the bright and dark fields of high resolution, and the results of the X-ray structural analysis [26].

3. Results and Discussion

3.1. Phase Composition

The electron microscopy method showed the presence of a number of phases in different states of the alloy, the qualitative and quantitative composition of which depends on the heat treatment (Table 1).

Table 1. Phase composition and quantitative properties of phases.

Phases	Crystal Lattice Type	Spatial Group	Parameter of Crystal Lattice, nm	Volume Fraction of Phases (±1), %	
				After DC	After DC + Annealing
γ'	Cubic	Pm3m	a = 0.3568–0.3575	85.6	90.0
γ	Cubic	Fm3m	a = 0.3569	8.0	9.0
β	Cubic	$Pm\bar{3}m$	a = 0.288	5.0	0
χ	Cubic	$I\bar{4}3m$	a = 0.957–0.960	1.4	0
σ	Tetragon	$P4_2/mnm$	a = 0.910–0.960 c = 0.475–0.499	0	1.0

The phases observed in a superalloy can be classified into primary and secondary phases. This classification is based on the volume fractions of the phases, their roles in the alloy, and stationary or single presence. As seen in Table 1, the primary phases are γ'- and γ-phases. They form the basic structure of the alloy in almost all superalloys [1,2]. The γ'-phase is an ordered solid solution with the $L1_2$ superstructure (based on Ni_3Al); the γ-phase is an fcc-disordered solid solution based on nickel [1,2]. In the alloy under study, these phases are present in the form of γ'-phase quasi-cuboids [20] separated by γ-phase interlayers. The rest of the phases are secondary. Different secondary phases are present depending on the state of the alloy. The types and volume fractions of all present phases regardless of their state are given in Table 1. The deviation in determining the volume fractions of phases is ±1%.

As seen in Table 1, the volume fraction of the γ-phase is kept the same irrespective of the alloy state.

The secondary phases are phases β, χ, and σ (Table 1). The TEM studies showed that, under the symbol "β-phase", two phases can be present in the alloy: (1) β-NiAl with a crystal lattice parameter of a = 0.288 nm, and (2) AlRe with the same crystal lattice parameter of a = 0.288 nm. Both of them have a CsCl structure, and both can transform into tetragonal phases of the $L1_0$ type [27]. These phases are mutually soluble, they can form solid solutions, and their presence on the sides of the Ni–Al and Al–Re triangle of the Ni–A–Re triple diagram means [28] that they can form a larger area by connecting these positions. Both phases (AlRe and β-NiAl) can be called the β-phase.

The next two phases, σ and χ (see Table 1), refer to topologically close-packed phases or, in other words, to Frank–Kasper phases [17–20]. Their appearance is due to the presence of Re in the alloy. Indeed, the interaction of Re with Mo and W leads to the appearance of the σ and χ phases: The σ-phase is formed as a result of solidification at high temperatures, and the χ-phase is formed as a result of decomposition at lower temperatures or as a result of crystallization. In addition, the σ-phase arises as a result of the interaction of Re with Co,

Cr, Ni, and Al, and the χ-phase is formed due to the interaction with Mo and W, as well as with Co and Al.

3.2. *Phase Morphology*

It is known that the structure of superalloys is a mixture of (γ + γ′)-phases. This mixture has a morphologically correct crystallographic orientation and forms a quasi-monocrystalline structure. The presence of active phase-forming metals in the alloy complicates the structure of the superalloy and causes the appearance of areas with a damaged γ-/γ′-structure in them. In the investigated alloy, the phase-forming metal is mainly Re. The presence of Re contributes to the formation of a part of the β-phase, as well as to the complete formation of the σ- and χ-phases. The formation of these four phases seriously disturbs the structure of γ/γ′ quasi-cuboids [20]. Research has shown that these phases are shaped like needles or plates. Since Re does not uniformly fill the entire volume of the alloy and is only present in local areas, then, in all states of the alloy, only a part of the volume of γ/γ′ quasi-cuboids was damaged.

The conducted studies enabled presentation of the schemes of the alloy samples under study. They are given in Figure 1.

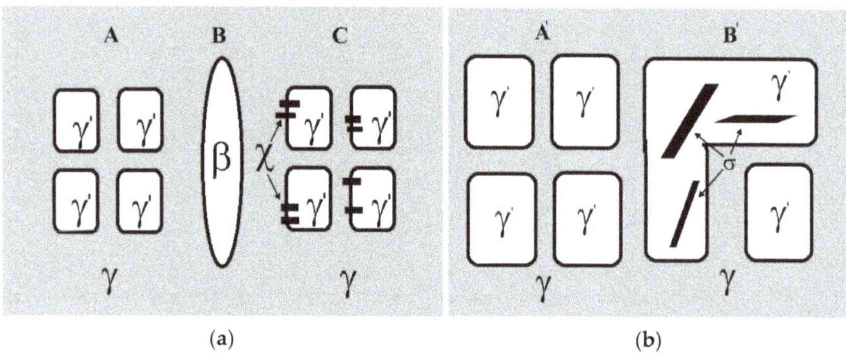

Figure 1. Schematic representation of the structure of the alloy: (**a**) the structure of the alloy in the initial state (after directional crystallization (DC)); (**b**) structure of the alloy after DC and annealing.

Let us comment on these figures.

A schematic representation of the structure of the alloy after DC (initial state) is shown in Figure 1a. It is clearly seen that the alloy contains three different morphological phase states, i.e., "A", "B", and "C". The volume fractions of these states, their phase compositions, and the sizes of the quasi-cuboids of the γ′-phase are given in Table 2.

Table 2. Structural phase states of the alloy and their quantitative data after DC.

The State	Volume Fraction of the State, %	Phase Composition	The Size of Quasi-Cuboids, nm	Particle Size of χ-Phase, nm
A	65	γ′ + γ	320 × 440	-
B	5	β	-	-
C	30	γ′ + γ + χ	320 × 440	30 × 240

It is clearly seen from the diagram in Figure 1a that state "A" is an ideal structure (γ + γ′) with slightly anisotropic distortion. Its TEM image is shown in Figure 2, which, along with the bright-field image (Figure 2a), indicates a microdiffraction pattern (Figure 2b). Additionally shown are the dark-field images obtained in the superstructure reflection [012] of the γ′-phase (Figure 2c) and in the reflection $[\bar{3}11]$, which, due to the parallelism of the crystal lattices of the γ′- and γ-phases [1,2], is simultaneously the main reflection for the γ'- and γ-phases (Figure 2d). It is these reflections that make it possible to separate the γ'- and γ-phases: the first reflection is the quasi-cuboids of the γ'-phase, and the second is the interlayers of the γ-phase. That is why, in Figure 2c, only the structure of the quasi-cuboids of the γ'-phase is visible, and in Figure 2d, both the structure of the quasi-cuboids of the γ'-phase and the interlayer of the γ-phase are visible.

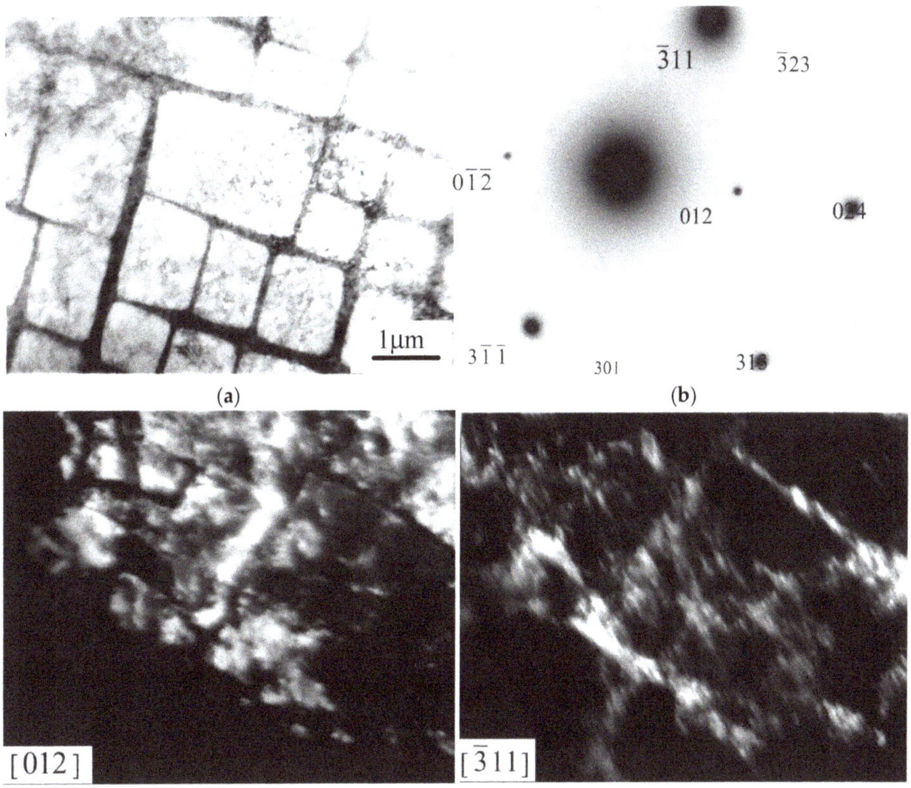

Figure 2. Electron microscopic image of a quasi-cuboid structure (γ' + γ) in the alloy after DC (structural phase state "A"): (**a**) bright-field image; (**b**) indicated microdiffraction pattern; (**c**,**d**) dark-field images obtained, respectively, in the superstructure reflection [012] and the main reflection [$\bar{3}$11].

Along with quasi-cuboids (γ + γ'), state "C" contains interlayers of the χ-phase. It was found that, in the "C" state, the quasi-cuboids of the γ'-phase are also anisotropic, and the volume fraction of the γ-phase is close to 5%. It was found that, first, the χ-phase interlayers can be localized in both the γ'- and γ-phases and, secondly, the crystallographic orientation of the χ-phase interlayers, like the orientation of the γ-phase interlayers, is parallel to the cubic directions of γ'-phases. The proof of the second statement is Figure 3, which shows an image of the structural phase state "C". Figure 3 shows the parallelism of the χ-phase interlayers and the γ-phase interlayers. Both phases have cubic crystal lattices; however, the differences in the parameters of the crystal lattices make them have some crystallographic misorientation.

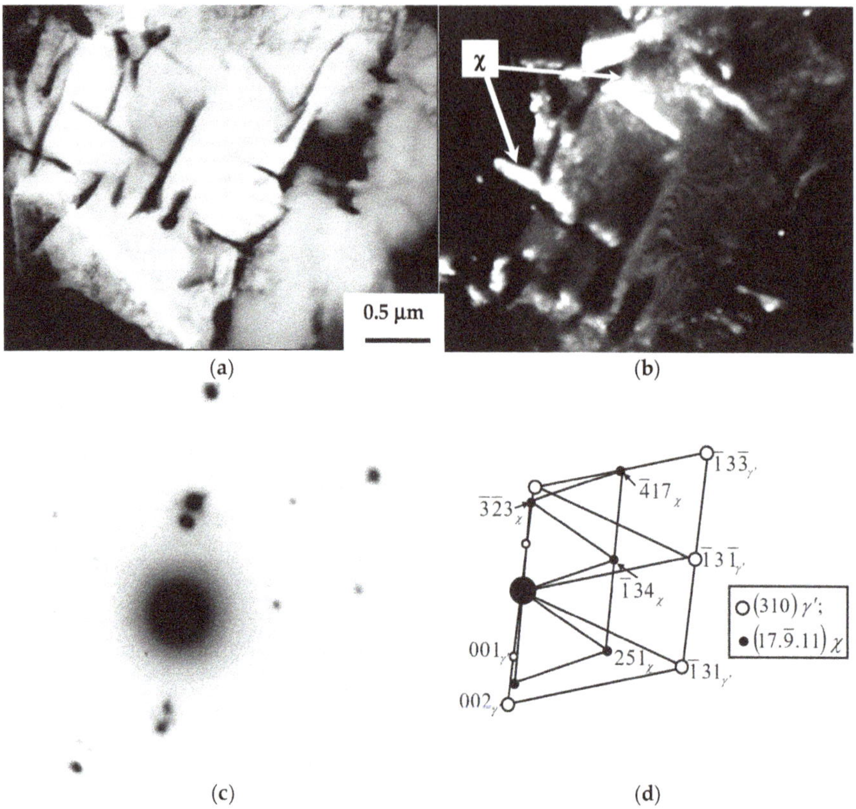

Figure 3. Electron microscopic image of the structural phase state "C" in the alloy after DC: (**a**) bright-field image; (**b**) dark-field image obtained in the reflection $[\bar{3}\bar{2}3]_{\chi}$; (**c**) microdiffraction pattern; (**d**) its indicated scheme.

The point is that the parameters of the crystal lattices of these phases are noticeably different (a_{γ} = 0.3569 nm; a_{χ} = 0.9570 nm). Figure 3c shows that the [002] direction, which is simultaneously the main direction for the γ′- and γ-phases, is almost parallel to the $[32\bar{3}]\chi$ direction. In other words, despite the difference in the crystal lattice parameters of these phases, the tendency toward a minimum of elastic energy led to the formation of a crystallographic orientation with the above-mentioned orientation ratio and parallelism of the χ- and γ-phase interlayers. In this case, there are no elastic stress fields. This is indicated by the absence of the structure of bending–torsion extinction contours in the electron microscopic images [20]. Thus, the γ- and γ′-phases are in almost coherent and stress-free contact with the χ-phase.

State "B" is distinguished by a completely damaged structure of quasi-cuboids. The entire volume of section "B" is occupied by a β-phase.

Consider the contact between the β-phase and the γ-/γ′-phase block (Figure 4).

The boundary between the block and the β-phase runs along the [011] direction of the γ′-phase. The orientation ratio of the β-phase and γ-/γ′-phases is [110]β II [111]γ′. This is a typical ratio for the mutual phase transformation fcc → bcc, for example, in steels [19]. Significantly, the extinction contours of bending–torsion are not observed near the β–γ/γ′ boundary (see Figure 4a). This means that the β-, γ-, and γ′-phases are coherently connected. The microdiffraction pattern shown in Figure 4b characterizes the structure of the β-phase. The presence of diffuse strands connecting reflections may indicate a nonstoichiometric composition against the formula indicated above, as well as the presence of a micro-phase separation in it. That is why this phase disappears after annealing.

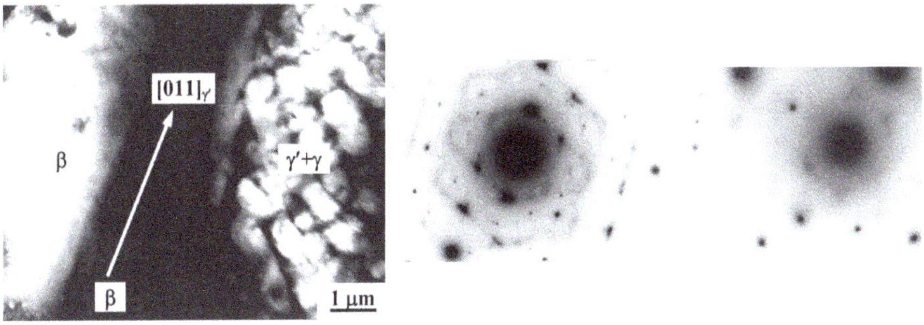

(a) (b) (c)

Figure 4. Electron microscopic image of the fine structure in the sample after DC: (**a**) junction of structural phase states "A" (γ' + γ) and "B" (β); (**b**) microdiffraction pattern obtained from state "B" (there are only reflections of the β-phase); (**c**) microdiffraction pattern obtained from state "A" (only reflections of the γ-/γ'-mixture are present).

The integral image of the alloy structure after DC, which was obtained with the SEM method at a lower magnification, is shown in Figure 5a. As seen in Figure 5a, most of the alloy's volume is represented by quasi-cuboids (solid and damaged), the contrast from which does not differ due to a relatively small increase. The separation by Re is clearly visible due to the precipitation of the β-phase, which is predominantly stabilized by this chemical element. Particles of the β-phase are also shown in Figure 5a.

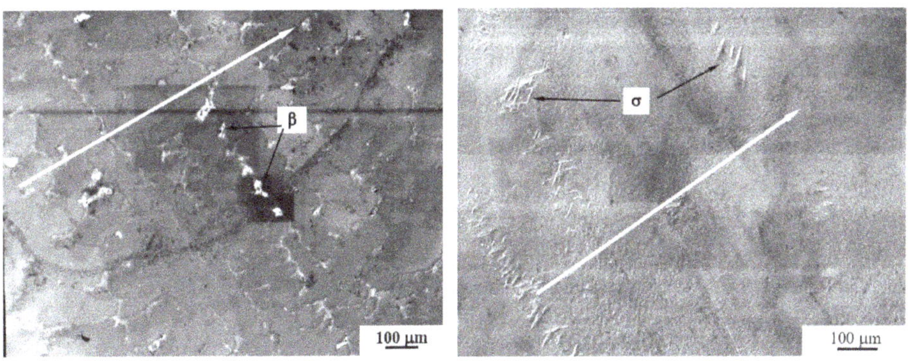

(a) (b)

Figure 5. Scanning electron microscopy (SEM) images of the alloy structure: (**a**) image of the alloy after DC; (**b**) image of the alloy after DC and annealing. The black arrows in (**a**) mark examples of particles of the β-phase, and those in (**b**) mark the σ-phase; white arrows show the direction of periodicity.

Thus, concluding the description of the morphology and phase composition of the alloy after DC, we note that 95% of its volume consists of quasi-cuboids of γ-/γ'-phases; 65% of these are ideal quasi-cuboids and 30% are quasi-cuboids containing χ-phase interlayers. Finally, 5% of the volume is occupied by the β-phase, the basis of which is the bcc-ordered ternary $NiAl_2Re$ phase.

A schematic representation of the structure of the alloy after DC and annealing is shown in Figure 1b. As can be seen, the alloy consists of ideal quasi-cuboids—the volume fraction of which is 90%—and quasi-cuboids distorted by the presence of the σ-phase, the volume fraction of which is 10%. Let us designate the corresponding structural phase states as "A′" and "B′", the quantitative data of which are given in Table 3.

Table 3. Structural phase states and quantitative data of the alloy after DC and annealing.

States	Volume Fraction of States, %	Phase Composition	The Size of Quasi-Cuboids,nm	The Size of σ-Phase Particles, nm
A′	90	$\gamma' + \gamma$	1440 × 1720	-
B′	10	$\gamma' + \gamma + \sigma$	1440 × 1720	290 × 4050

Electron microscopic images of the states "A′" and "B′", respectively, are shown in Figures 6 and 7.

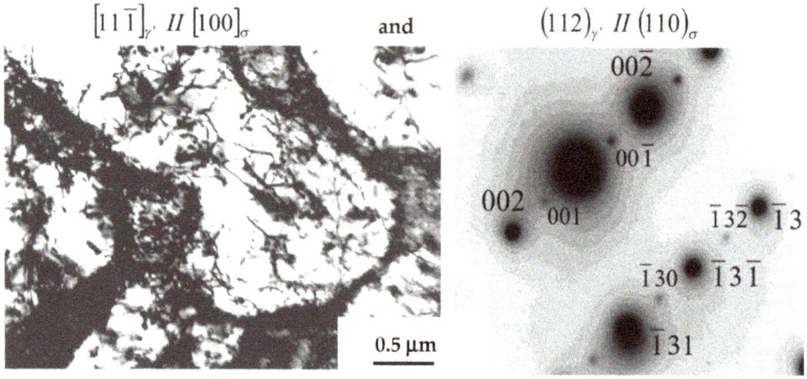

(a) (b)

Figure 6. Electron microscopic image of a quasi-cuboid structure ($\gamma' + \gamma$) in the alloy after DC and annealing (structural phase state "A′"): (**a**) bright-field image; (**b**) in the indicated microdiffraction pattern, there are reflections belonging to the (310) plane of the γ'-phase.

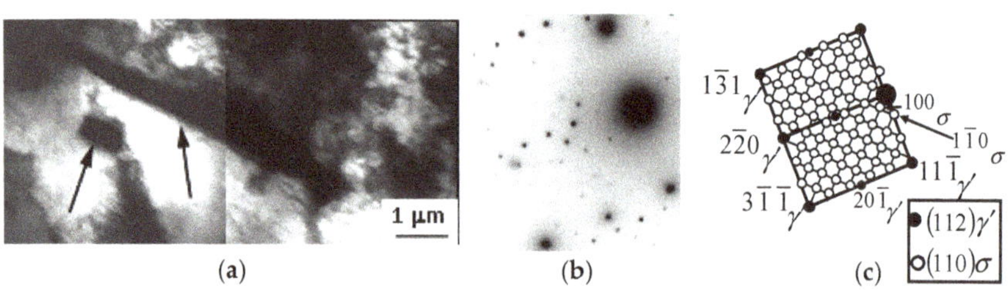

(a) (b) (c)

Figure 7. Electron microscopic image of the structural phase state "B′" in the alloy after DC and subsequent annealing: (**a**) bright-field image (arrows indicate particles of the σ-phase); (**b**) microdiffraction pattern; (**c**) its diagram.

Figure 6 shows a TEM image of a quasi-cuboid structure (γ'+γ) that corresponds to the "A′" state (Figure 1b). This figure clearly shows the γ'-phase cuboids separated by γ-phase interlayers.

Figure 7 shows the particles of the σ-phase, which have a plate-like shape and are located along the [011] and [001] directions of the γ'-phase. The diffraction analysis in Figure 7 shows that the crystal lattices of the γ'- and σ-phases have coinciding directions and planes.

The contact between the γ'- and σ-phases occurs without noticeable distortions. This is evidenced by the absence of bending–torsion extinction contours in the structure. As can be seen from Figure 7, the structure of quasi-cuboids is damaged in the places of formation of the σ-phase.

If, after DC, the ideal quasi-cuboids account for 65% of the alloy's volume, then in the state after DC and subsequent annealing, they already make up 90%.

Figure 5b shows an SEM image of the alloy's structure after DC and annealing. The local distribution of particles of the σ-phase is clearly seen, and it alternates with the ideal structure of γ-/γ'-phases.

Therefore, the analysis of Figure 5a,b indicates the presence of periodicity in the arrangement of spatial structures (structural phase states "A", "B", "C", "A′", "B′"). Similar results were obtained in [22], where an inhomogeneous distribution of phases containing a large amount of Re was theoretically proved. It was noted that this non-uniform distribution has a wave-like nature. It has been stated that this inhomogeneity is caused by the heterogeneous Re distribution. In this work, this fact was confirmed experimentally and was called the "periodicity". The direction of periodicity is indicated by a long arrow in the images of the structure (Figure 5). The ideal state of the γ-/γ'-mixture and the strongly damaged one (γ' + γ + secondary phases) alternate along it. The average width of each layer is given in Table 4.

Table 4. Quantitative data of the periodicity of the structure of the superalloy.

The State of the Alloy	The Width of Interlayers of Ideal Structure (γ' + γ), µm	The Width of Interlayers with Secondary Phases, µm	The Length of Periodicity, µm
After DC	120	10	130
After DC and annealing	420	25	445

In this case, in the alloy after DC, the main period length is occupied by an ideal structure (γ' + γ) that is 120 µm wide, the structure with secondary phases is very local—10 µm. In turn, in the structure of the alloy after DC and subsequent annealing, the width of ideal interlayers is, on average, about 3.5 times greater than the corresponding width of the alloy after DC, and the width of interlayers with secondary phases is ~2.5 times greater. There is no doubt that the periodic structure of the superalloy with a suitable orientation should strengthen the material. In other words, there are reserves at the mesoscale for strengthening the superalloy material.

3.3. *Crystallographic Correspondence of the Primary and Secondary Phases in All States of the Superalloy and the Problem of the Long-Range Stress Field*

The complex phase composition of the superalloy under study is not its only feature. An important structural property is the concentration layer separation, which leads to a periodic structure and localization of secondary phases in the (γ' + γ) mixture. This is clearly seen in Figure 5. A detailed simultaneous electron microscopic and diffraction study of the structure of areas with secondary phases localized in the γ'- and γ-phases showed the absence of extinction contours and, thus, the absence of noticeable internal stresses. This behavior is due to the good crystallographic agreement between the crystal lattices of γ/γ' and the secondary phases, which is organized during the separation of these phases. The corresponding orientation ratios are presented in Table 5. Such good crystallographic agreement suggests that alloying with Re, which brings a hardening effect, should not cause complications associated with internal stresses. This also means that the positive effect of Re alloying can be exploited in superalloys.

Table 5. Orientation ratios between the crystal lattice of the γ'-phase and the crystal lattices of the secondary phases in various states of the alloy.

State of the Alloy	Secondary Phase		
	β-Phase	χ-Phase	σ-Phase
After DC	$[110]_\beta II[111]_{\gamma'}$	$[32\overline{3}]_\chi II[001]_{\gamma'}$	–
After DC and annealing	–	–	$[100]_\sigma II[11\overline{1}]_{\gamma'}$

3.4. Sizes of Quasi-Cuboids, Secondary Phases, and Dislocation Structure

As noted above, in alloys both after DC and after DC and subsequent annealing, there are areas with ideal (undamaged) quasi-cuboids of the γ'-phase. A schematic representation of such quasi-cuboids and their sizes are shown in Figure 8 and Table 6, respectively. As can be seen in Table 6, the sizes of quasi-cuboids in the alloy after DC are the smallest.

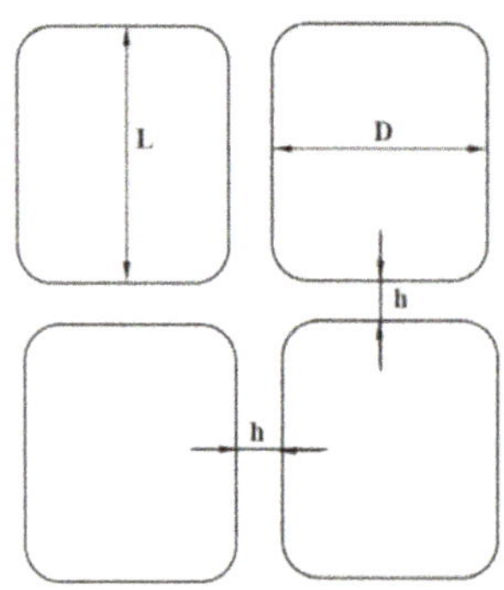

Figure 8. Schematic representation of quasi-cuboids: D—transverse; L—longitudinal size of γ'-phase quasi-cuboids; h is the width of the γ-phase interlayers.

Table 6. Quantitative properties of quasi-cuboids of the γ/γ' mixture.

The State of the Alloy	The Sizes of Quasi-Cuboids of γ'-Phase		The Width of γ-Phase Interlayers, h, nm	Scalar Dislocation Density, $\rho \cdot 10^{-10}$, cm^{-2}	
	Lateral Size, D, μm	Longitudinal Size, L, μm		in γ'-Phase, $\rho_{\gamma''}$	in γ-Phase, ρ_γ
After DC	0.32	0.44	80	3.8	8.8
After DC+annealing	1.44	1.72	170	3.0	8.3

In the alloy after DC and subsequent annealing, the dimensions of the quasi-cuboids of the γ'-phase (D and L), as well as the width of the interlayers (h), increase, and the dimensions (D) grow faster than h. The fact that the dimensions change disproportionately is due to two factors. The first is the different volume fraction of the γ- and γ'-phases. Secondly, there is a different volume fraction of ideal quasi-cuboids in each state of the alloy. Thirdly, the thickness of the γ-phase interlayers is nanometric, and the dimensions of the quasi-cuboids of the γ'-phase in the state after DC are also nanometric; in the sample after annealing, they are a few micrometers. In turn, the volume fraction of the distorted γ/γ' structure is due to the formation of various secondary phases (β, χ, σ) in it and the kinetics of this process.

The particle sizes of the secondary phases (see Tables 2 and 3) range from nanometric to microns. This means that a significant role in the formation of the superalloy properties will be played by particles periodically located in groups in the bulk of the superalloy material and belonging to the areas where the damaged γ/γ' mixture and secondary phases are located.

3.5. Scalar Dislocation Density (ρ)

The average values of the scalar dislocation densities in the γ'- and γ-phases are given in Table 6. As follows from Table 6, in the γ-phase, the value of ρ is much larger than in the γ'-phase. Annealing the alloy at a high temperature decreases the ρ value in both phases. This indicates a high degree of dislocation pinning in both phases of the superalloy. This is especially true for super-dislocations in the ordered γ'-phase.

The dislocation structure in the γ'-phase is cellular-meshy. In the γ-phase, it is not possible to classify the structure, since it is fixed in narrow channels.

3.6. Phase Transformations during Alloy Annealing after DC

As stated above, after high-temperature annealing of the DC alloy, a phase transformation occurs. Let us compare the phase composition and the volume fraction of the alloy phases in these two states (Table 7).

Table 7. Quantitative phase compositions of alloys.

The State of the Alloy	The Volume Fraction of Ideal γ'-Phase, %	The Volume Fraction of Damaged Mixture γ/γ', %	The Volume Fraction of β-Phase, %	Phases before and after Transformation
After DC	65	30	-	β+χ
After DC and annealing	90	10	5	σ

It is quite obvious that the phase transformation is complex. It was found that it occurs, first, in the interlayers of the γ-phase of the imperfect part of the structure; secondly, inside γ'-quasi-cuboids of the same part of the structure; thirdly, in the free space between areas with quasi-cuboids. A part of the area of the damaged quasi-cuboids and the β-phase transforms into an ideal zone of quasi-cuboids. The following transformation occurs: $[(\gamma' + \gamma)$ partially $+ (\beta + \gamma)$ completely$] \rightarrow (\gamma + \gamma' + \sigma)$. This transformation is of a diffusion nature and is connected with mass transfer. It is mainly localized in the areas of the β-phase and the damaged γ/γ' mixture. The volume fraction of the γ-phase is generally preserved, and the γ'-phase even slightly increases. Apparently, this occurs due to the decomposition of the β-phase. In this case, some Al atoms pass into the σ-phase. Re atoms from the β-phase also pass into the σ-phase. The atoms of Re, Co, Mo, W, and Al leave the χ phase to form the σ phase. Cr atoms are also added there. The diagrams of the equilibrium of Re with Mo, W, and other elements [25] show that at a close volume fraction of the σ- and χ-phases, the reaction is at quasi-equilibrium, and some of the elements from the β-phase also appear in the σ-phase.

4. Conclusions

The phase composition, morphology of the phases, and their localization in the bulk of the Ni-Al-Cr-Me superalloy doped with Re were studied with TEM and SEM methods. The volume fractions of the phases and the particle sizes of the phases were measured after directional crystallization (DC) and subsequent annealing.

During the research, the formation of refractory phases was established. These phases are β-, σ-, and χ-phases. The formation of these phases leads to hardening of the superalloy for the following reasons: First, the resulting phases are ordered; secondly, their morphological features, χ-, β-, and σ-phases, are shaped like needles or plates and contribute to barrier deceleration of dislocations; thirdly, the formation of phases β, σ, and χ led to the fact that the alloy became multiphase; fourthly, these phases, being refractory, increase the working temperature of the alloy.

The periodicity in the arrangement of spatial structures was established, where the ideal state of the γ/γ' mixture is interchanged with the strongly damaged one

(γ' + γ + secondary phases). There is no doubt that the periodic structure of the superalloy with a suitable orientation will help to strengthen the alloy. In other words, there are reserves at the mesoscale level for strengthening the superalloy material.

Author Contributions: N.A.K. directed the general research program; N.A.K., E.L.N., and N.A.P. conducted research and discussed the results; A.V.N. oversaw the software. All co-authors have contributed to the manuscript. All authors have read and agreed to the published version of the manuscript.

Funding: The research was carried out within the state assignment of the Ministry of Science and Higher Education of the Russian Federation (theme No. FEMN-2020-0004).

Conflicts of Interest: The authors declare no conflict of interest.

References

1. Stoloff, N.S.; Hagel, W.C. *Superalloys II*; Sims, C.T., Ed.; Wiley-Interscience Publication John Wiley & Sons Publisher: Hoboken, NJ, USA, 1987; p. 615.
2. Sims, C.T.; Stoloff, N.S.; Hagel, W.C. *Superalloys II: High-Temperature Materials for Aerospace and Industrial Power*; Metallurgiya Publisher: Moscow, Russia, 1995; Volume 1, p. 385.
3. Reed, R. *The Superalloys: Fundamental and Applications*; Cambridge University Press: Cambridge, UK, 2006; p. 363.
4. Hausmann, D.; Solís, C.; Freund, L.P.; Volz, N.; Heinemann, A.; Göken, M.; Gilles, R.; Neumeier, S. Enhancing the High-Temperature Strength of a Co-Base Superalloy by Optimizing the γ/γ' Microstructure. *Metals* **2020**, *10*, 321. [CrossRef]
5. Smith, T.; Esser, B.; Antolin, N. Phase transformation strengthening of high-temperature superalloys. *Nat. Commun.* **2016**, *7*, 1–7. [CrossRef] [PubMed]
6. Parsa, A.B.; Wollgramm, P.; Buck, H.; Somsen, C.; Kostka, A.; Povstugar, I.; Choi, P.; Raabe, D.; Dlouhy, A.; Müller, J. Advanced Scale Bridging Microstructure Analysis of Single Crystal Ni- Base Superalloys. *Adv. Eng. Mater.* **2014**, *17*, 216–230. [CrossRef]
7. Viswanathan, G.; Shi, R.; Genc, A.; Vorontsov, V.; Kovarik, L.; Rae, C. Segregation at stacking faults within the γ' phase of two Ni-base superalloys following intermediate temperature creep. *Scr. Mater.* **2015**, *94*, 5–8. [CrossRef]
8. Agudo Jácome, L.; Nörtershäuser, P.; Somsen, C.; Dlouhý, A.; Eggeler, G. On the nature of γ' phase cutting and its effect on high temperature and low stress creep anisotropy of Nibase single crystal superalloys. *Acta Mater.* **2014**, *69*, 246–264. [CrossRef]
9. Krutz, N.; Shen, C.; Karadge, M.; Egan, A.J.; Bennett, J.R.; Hanlon, T.; Mills, M.J. An Approach Toward Understanding Unstable Gamma Prime Precipitate Evolution and Its Effect on Properties. In *Superalloys 2020, The Minerals, Metals & Materials Series*; Tin, S., Hardy, M., Clews, J., Cormier, J., Feng, Q., Marcin, J., O'Brien, C., Suzuki, A., Eds.; Springer: Beijing, China, 2020; pp. 691–701.
10. Kuznetsov, V.P.; Lesnikov, V.P.; Popov, N.A.; Vasil'ev, A.S.; Popova, E.N. Phase Transformations in Single-Crystal Refractory Nickel Alloy with Tantalum, Rhenium and Ruthenium Additions. *Metal Sci. Heat Treat.* **2018**, *60*, 100–105. [CrossRef]
11. Kozlov, E.V.; Nikonenko, E.L.; Popova, N.A.; Koneva, N.A. Structure and composition of higherrhenium-content superalloy based on La-alloyed Ni–Al–Cr. *AIP Conf. Proc.* **2015**, *1683*, 020101.
12. Pollock, T.M. Alloy design for aircraft engines. *Nat. Mater.* **2016**, *15*, 809–815. [CrossRef] [PubMed]
13. Almirall, N.; Wells, P.; Yamamoto, T. Precipitation and hardening in irradiated low alloy steels with a wide range of Ni and Mn compositions. *Acta Mater.* **2019**, *179*, 119–128. [CrossRef]
14. Sun, W.; Zhu, Y.; Marceau, R. Precipitation strengthening of aluminum alloys by room-temperature cyclic plasticity. *Science* **2019**, *2019*, 972–975. [CrossRef] [PubMed]
15. Wang, Z.; Zhang, L.; Li, W. High throughput experiment assisted discovery of new Ni-base superalloys. *Scr. Mater.* **2020**, *178*, 134–138. [CrossRef]
16. Barba, D.; Alabort, E.; Pedrazzini, S.; Collins, D.M.; Wilkinson, A.J.; Bagot, P.A.J.; Moody, M.P.; Atkinson, C.; Jérusalem, A.; Reed, R.C. On the microtwinning mechanism in a single crystal superalloy. *Acta Mater.* **2017**, *135*, 314–329. [CrossRef]
17. Seiser, B.; Drautz, R.; Pettifor, D. TCP phase predictions in Ni-based superalloys: Structure maps revisited. *Acta Mater.* **2011**, *59*, 749–763. [CrossRef]
18. Rae, C.M.; Reed, R.C. The precipitation of topologically close-packed phases in rhenium-containing superalloys. *Acta Mater.* **2001**, *49*, 4113–4125. [CrossRef]
19. Kolobov, Y.R.; Kablov, E.N.; Kozlov, E.V. *Structure and Properties of Intermetallic Materials with Nano-Phase Hardening*; MISiS: Moscow, Russia, 2008; p. 328.
20. Kozlov, E.V.; Smirnov, A.N.; Nikonenko, E.L.; Popova, N.A.; Koneva, N.A. *Phase Morphology and Transformations due to Thermal Treatment of Ni-Al-Cri Ni-Al-Co-based Superalloys. Scale and Concentration Effects*; Innovatsionnoe mashinostroenie: Moscow, Russia, 2016; p. 175.
21. Blavette, D.; Caron, P.; Khan, T. An atom probe investigation of the role of rhenium additions in improving creep resistance of Ni-base superalloys. *Scr. Metall.* **1986**, *20*, 1395–1400. [CrossRef]
22. Mottura, A.; Reed, R.C. What is the role of rhenium in single crystal superalloys? *MATEC Web Conf.* **2014**, *14*, 1–6. [CrossRef]
23. Kozlov, E.V.; Nikonenko, E.L.; Koneva, N.A.; Popova, N.A. Effect of Re Content on Structure and Phase Composition of Ni-Al Alloys. *Bull. Russ. Acad. Sci. Phys.* **2005**, *69*, 1116–1120.

24. Kozlov, E.V.; Popova, N.A.; Kabanina, O.V.; Klimashin, S.I.; Gromov, V.E. *The Evolution of Phases State, Defect Structure, Internal Stresses and Redistribution of Carbon and Temper of Cast Structural Steel*; Izdatel'stvo SibGIU: Novokuznetsk, Russia, 2007; p. 177.
25. Saltykov, S.A. *Stereometric Metallography*; Metallurgy: Moscow, Russia, 1976; p. 271.
26. Hirsch, P.; Howie, A.; Nicholson, R. *Electron Microscopy of Thin Crystals*; Mir: Moscow, Russia, 1968; p. 574.
27. Lyakishev, N.P. *Diagrams of Binary Metal Systems*; Mechanical engineering: Moscow, Russia, 1996–2000; pp. 1–3.
28. Cornish, L.A.; Witcomb, M.J. A metallographic study of the Al–Ni–Re phase diagram. *J. Alloys Comp.* **1999**, *291*, 145–166. [CrossRef]

crystals

MDPI

Article

Numerical Investigation of Plastic Strain Homogeneity during Equal-Channel Angular Pressing of a Cu-Zr Alloy

Jittraporn Wongsa-Ngam [1,*], Nitikorn Noraphaiphipaksa [2,3], Chaosuan Kanchanomai [2] and Terence G. Langdon [4,*]

1 Department of Mechanical Engineering, School of Engineering, King Mongkut's Institute of Technology Ladkrabang, Bangkok 10520, Thailand
2 Department of Mechanical Engineering, Faculty of Engineering, Thammasat University, Pathumthani 12120, Thailand; noraphaiphipaksa@gmail.com (N.N.); kchao@engr.tu.ac.th (C.K.)
3 Nitikorn Research Partner Co., Ltd., Lumlukka 12150, Thailand
4 Materials Research Group, Department of Mechanical Engineering, University of Southampton, Southampton SO17 1BJ, UK
* Correspondence: jittraporn.wo@kmitl.ac.th (J.W.-N.); langdon@usc.edu (T.G.L.)

Abstract: A three-dimensional finite element method (3D FEM) simulation was carried out using ABAQUS/Explicit software to simulate multi-pass processing by equal-channel angular pressing (ECAP) of a circular cross-sectional workpiece of a Cu-Zr alloy. The effective plastic strain distribution, the strain homogeneity and the occurrence of a steady-state zone in the workpiece were investigated during ECAP processing for up to eight passes. The simulation results show that a strain inhomogeneity was developed in ECAP after one pass due to the formation of a corner gap in the outer corner of the die. The calculations show that the average effective plastic strain and the degree of homogeneity both increase with the number of ECAP passes. Based on the coefficient of variance, a steady-state zone was identified in the middle section of the ECAP workpiece, and this was numerically evaluated as extending over a length of approximately 40 mm along the longitudinal axis for the Cu-Zr alloy.

Keywords: ECAP; copper alloy; strain homogeneity; 3D FEM

Citation: Wongsa-Ngam, J.; Noraphaiphipaksa, N.; Kanchanomai, C.; Langdon, T.G. Numerical Investigation of Plastic Strain Homogeneity during Equal-Channel Angular Pressing of a Cu-Zr Alloy. *Crystals* **2021**, *11*, 1505. https://doi.org/10.3390/cryst11121505

Academic Editor: Wojciech Polkowski

Received: 7 November 2021
Accepted: 1 December 2021
Published: 3 December 2021

1. Introduction

The fabrication of bulk ultrafine-grained (UFG) materials has been widely investigated over the last two decades due to the potential to produce metals with superior mechanical and physical properties. Several techniques are available for fabricating these materials, but major emphasis has been placed on the use of equal-channel angular pressing (ECAP), in which a sample is pressed through a die constrained within an internal channel that is bent through a sharp angle [1]. In this procedure, a shear strain is introduced as the sample passes through the bend in the channel, and this is a very effective severe plastic deformation (SPD) processing method for producing bulk UFG materials for use in engineering applications. In ECAP processing, the cross-section of the sample remains unchanged so that the pressing may be repeated for multiple passes in order to achieve the required strain level.

The equivalent plastic strain, ε, introduced in a single pass through the die is given by a relationship of the form [2]:

$$\varepsilon = \frac{1}{\sqrt{3}}\left(2\cot\left(\frac{\Phi}{2}+\frac{\Psi}{2}\right)+\Psi\,\mathrm{cosec}\left(\frac{\Phi}{2}+\frac{\Psi}{2}\right)\right), \tag{1}$$

where Φ is the angle subtended by the two parts of the channel, and Ψ is the outer arc of curvature at the point of intersection of the two channels. The relationship in Equation (1) is therefore only a function of the die geometry, and it provides a very useful and simple procedure for estimating the average strain introduced during ECAP processing.

Nevertheless, an understanding of the local strain distribution during ECAP processing is also important since the strain imposed during processing relates directly to the internal microstructure, and, ultimately, it characterizes the overall material properties.

The finite element method (FEM) is a powerful tool that can be used to understand the deformation behavior of a material during the ECAP process. Earlier works on FEM simulations of the ECAP process were carried out by using two-dimensional (2D) simulations [3–9]. In these 2D finite element models, plane strain conditions were generally assumed in order to calculate the effects of the die geometries, the processing conditions and the material properties on the deformation behavior and the inhomogeneity of the materials. However, 2D simulation is used for a billet with a square cross-section, and it is not generally applicable to the round cross-sectional workpieces that are used in ECAP when employing a solid die [10–12]. Recently, three-dimensional (3D) FEM simulations were effectively performed in an analysis of deformation behavior during the first pass of the ECAP process [10,11,13–15]. Nevertheless, there are only limited reports related to multi-pass ECAP processing using 3D FEM for circular cross-sectional workpieces, and all of these reports are directed at the deformation of aluminum and aluminum alloys [16–18].

Because of this deficiency, the present investigation was initiated in order to use a 3D FEM simulation to investigate the deformation behavior and homogeneity evolution of the multi-pass ECAP processing, up to eight passes, of a copper alloy of Cu-0.1 wt.% Zr with billets with circular cross-sections.

2. The Principles of Finite Element Simulations

The commercial software Abaqus/Explicit version 2016 [19] was used to simulate the multi-pass ECAP processing of the Cu-0.1 wt.% Zr alloy. The workpiece and the die geometry were modeled according to descriptions in earlier experiments conducted at University of Southern California (USC), USA [20,21]. Specifically, the workpiece was in the form of a cylindrical billet with a diameter of 10 mm and a length of 70 mm, and it was processed using a rigid solid die with a channel angle (Φ) of 110° and an outer corner angle (Ψ) of 20°. These angles were therefore used in the simulation.

Figure 1a shows a schematic representation of an ECAP assembly. For convenience in discussion, a local coordinate system was set as x, y and z axes. The plane normal to the x, y and z axes are henceforth designated the X-plane, Y-plane and Z-plane, respectively. Four points (1–4) were marked on the cross-section of the mid-length of the billet, and these points were used to monitor the accumulated equivalent strain during the ECAP processing as shown in Figure 1b. Consecutive passes of the ECAP processing were modeled using an interconnected multi-channel die corresponding to the equivalent of processing route B_C, in which the billet is rotated by 90° around the longitudinal axis in the same sense between each pass [22] up to a total of 8 passes as illustrated in Figure 1c.

In the simulation, the overall behavior of the billet is taken as an elastic–plastic material. This model can describe the deformation behaviour of a material under severe plastic deformation during ECAP processing. To determine the material properties, the Cu-0.1 wt.% Zr billet was annealed as in earlier experimental work [20,21], and then the tensile testing was conducted. The stress–strain curve and the material properties used for the analysis are shown in Figure 2 and Table 1, respectively. The flow stress curve of the present material was determined until the maximum strain of 0.4 and the strain-hardening exponent was 0.68. However, during the ECAP process, the cumulative strain is expected to increase continuously with the number of passes. To numerically evaluate the cumulative strain beyond the limit of the flow stress curve, the flow stress is linear extrapolated from the end of the curve using a tangent line with a strain-hardening exponent of 0.68. The billet material was modeled with C3D8R (eight-node linear brick element). The die and punch were modeled as a rigid surface, and all simulations were performed with a pressing speed of 3 mm/s. The value of the friction coefficient between the die and the billet was assumed to be 0.1; this value is recommended when processing using MoS_2. An arbitrary Lagrangian–Eulerian (ALE) adaptive remeshing and mass scaling was used for all simulations to prevent failure of the mesh due to large deformation and also to reduce the total computation time.

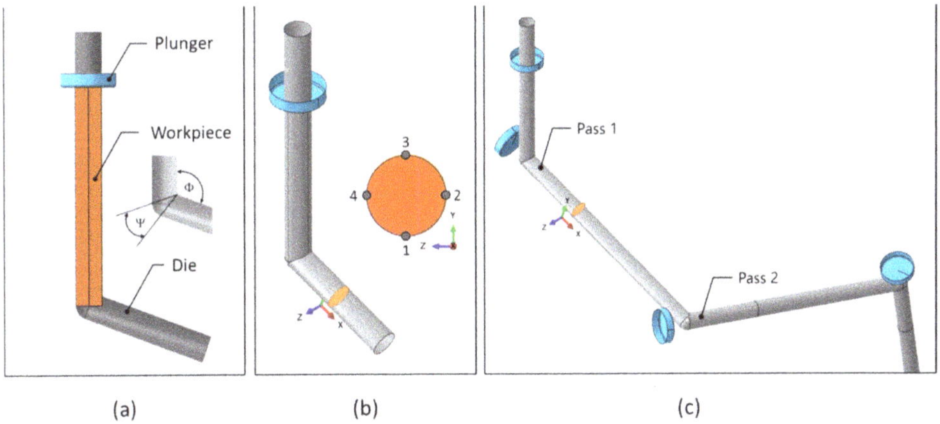

Figure 1. Schematic illustration of the ECAP process: (**a**) an ECAP assembly, (**b**) four points marked on the mid-length of the billet and (**c**) an interconnected multi-channel die.

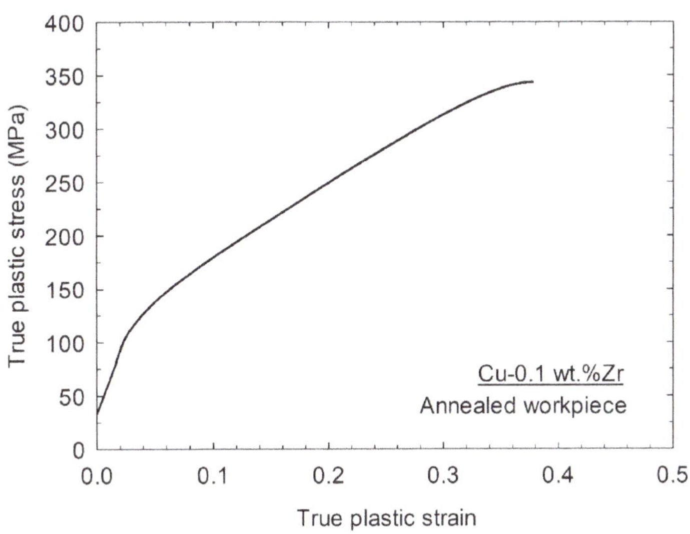

Figure 2. Experimental true plastic stress–strain curve for the Cu-0.1% Zr alloy.

Table 1. Material properties of an annealed specimen of Cu-0.1 wt.% Zr.

Properties	Value
Elastic modulus	110 GPa
Poisson's ratio	0.3
Density	8960 kg/m^3
Yield strength	33.9 MPa
Strain-hardening exponent (n)	0.68
Strength coefficient (K)	659 MPa

Before performing the simulation, a convergence test was carried out to assess the mesh sensitivity. Seven different numbers of elements, namely 1056, 1280, 1408, 2200, 3420, 5092 and 7680, were used to calculate the total strain energy. The relationship between the total strain energy and the numbers of elements was then plotted as shown in Figure 3. Based on the result in Figure 3, a mesh consisting of 2200 elements, with element sizes of 0.2–2.0 mm, was chosen in all simulations.

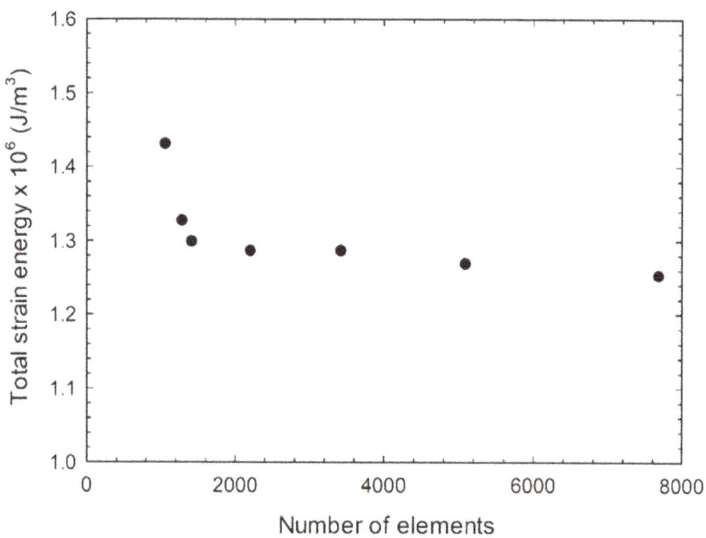

Figure 3. Convergence test for different numbers of elements.

3. Results and Discussion

3.1. *Equivalent Plastic Strain*

The equivalent plastic strain during the pressing stroke was monitored through the four points located at the cross-sections of the mid-length of the workpiece as illustrated in Figure 1b. Figure 4 shows the variation in the equivalent plastic strain tracing through points 1 to 4 located on the cross-section of the mid-length of the billet as it passes through the die for one pass. As a general overview observation, the deformation behavior of the four points is similar to the deformation commencing after approximately 40% of the pressing stroke. Thereafter, the equivalent plastic strain increases rapidly when the billet is pressed through the intersection or deformation zone of the ECAP channel, corresponding to 50% of the stroke, and, finally, it remains steady during the subsequent ECAP process.

On close inspection, there is a difference in the magnitude of the equivalent plastic strain at different points after one pass of ECAP. Thus, shortly after entering the intersection of the ECAP die, corresponding to approximately between 44% and 50% of the stroke, point 1 and point 3 located near the outer corner and inner corner of the die, respectively, have higher strains than those of points 2 and 4. This is because points 1 and 3 are deformed before points 2 and 4, but the local strains change with further pressing whereby each point reaches a peak value of strain and remains approximately constant after approximately 60% of the stroke. It is found that the highest equivalent plastic strain of ~1.04 occurs at point 3 located near the inner channel angle; the lowest of ~0.58 is at point 1 located near the outer corner of the die; and the strains at points 2 and 4 are identical at ~0.75, where this value lies between points 1 and 3. This result demonstrates the deformation inhomogeneity that exists in the billet after one pass of ECAP. Similar trends were observed in earlier research [23–25], in which deformation begins to take place at points near the outer and inner curves of the die, but the point near the outer curve has a lower deformation rate because it necessarily travels a longer distance.

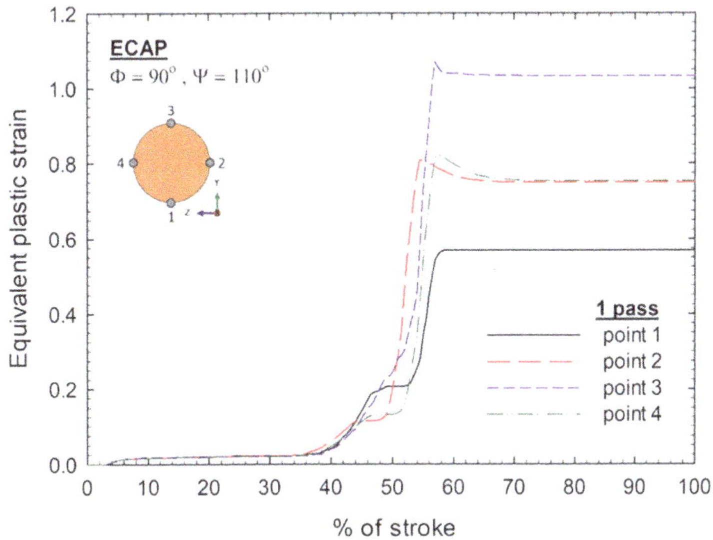

Figure 4. Equivalent plastic strain variation tracing through points 1, 2, 3 and 4 located on the cross-section of the mid-length of the billet during ECAP processing through one pass.

The equivalent plastic strain distribution contours on the transverse cross-section (X-plane) and longitudinal section (Z-plane) through ECAP for one pass are plotted in Figure 5a,b, respectively. Figure 5a shows that the strain distribution varied along the vertical direction, in which the strain values at the bottom are lower than those at the top of the billet, while the strain distribution along the horizontal direction is reasonably uniform. The values of the equivalent plastic strain at the top, middle and bottom are consistent with the local strain at points 1–4 as observed during ECAP through one pass depicted in Figure 4. In addition, the strain distributions on the transverse cross-section are consistent with the microhardness distributions measured on the Cu-Zr alloy in an earlier study [20]. The low effective plastic strain at the bottom corresponds to an area of low microhardness, which verifies the results from the FE simulation.

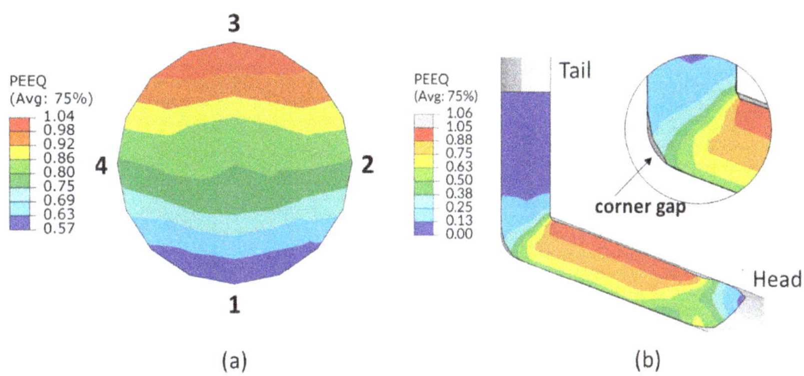

Figure 5. Equivalent plastic strain distribution contours for (**a**) transverse cross-section and (**b**) longitudinal direction through ECAP for one pass.

In practice, the lower values of the equivalent plastic strain at the bottom of the workpiece in Figure 5a are associated with the formation of a corner gap between the die and the workpiece at the outer corner as observed in Figure 5b. As the corner gap is formed during pressing, the workpiece is no longer in contact with the die wall, and this leads to a lower degree of deformation and, consequently, to a lower imposed strain and a lower hardness.

The presence of a corner gap is usually found during the ECAP processing of strain-hardening materials. Thus, the formation of a corner gap at the outer corner of the ECAP channel was reported through experimental observations [5,12,26] and numerical modelling [11,12,17,24,27–29]. In a study based on using pure aluminum as a model material [26], an examination of the formation of the corner gap between a strain-hardening material and a quasi-perfect plastic material led to the conclusion that a larger corner gap is formed in the material with the higher strain-hardening rate. It was also pointed out that the less deformation at the outer corner was due to a bending effect more than a shearing effect [28]. Another study examined the equivalent plastic strain rate in the plastic deformation zone and showed that the strain rate at the bottom surface was lower than that at the top surface, and, in addition, the strain rate at the bottom surface decreased with increasing angle Ψ [17].

The evolution of the accumulative equivalent plastic strains for these four points in consecutive ECAP processing of up to eight passes was monitored and plotted as shown in Figure 6. Since each point was rotated by 90° according to processing route B_C, the position of each point has a change of 90° in the channel leading to a change in the strain path and strain increment in consecutive passes. It is therefore obvious that the equivalent plastic strain of each point continued to increase with the increasing number of passes, but the increment in each point is different depending on the position of the point in the die for a given pass number. For example, after two passes, the strain increments of points 1, 2, 3 and 4 were 0.63, 0.87, 0.80 and 0.55, respectively. This means that the highest increment was found in point 2 because this point was rotated to the inner curve of the die. Conversely, the lowest increment occurred in point 4 because it was moved to the outer curve of the die. Similar results were also reported in a study of 6061 Al alloy in a circular cross-sectional ECAP process with a die angle (Φ) of 90° using route B_C and a finite volume method (FVM) simulation [30].

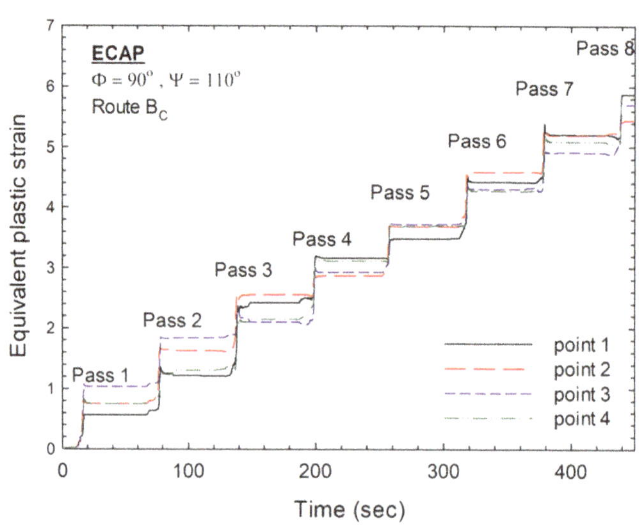

Figure 6. Equivalent plastic strain variation tracing through points 1, 2, 3 and 4 located in the cross-section of the mid-length of the billet during ECAP processing up to a total of 8 passes.

The average equivalent plastic strain (ε^p_{ave}) from the FE simulation was calculated by taking the average of the strain values over the cross-section at the mid-length of the billet, which were obtained by the following equation:

$$\varepsilon^p_{ave} = \frac{1}{n}\sum\nolimits_{i=1}^{n} \varepsilon^p_i, \tag{2}$$

where n is the number of nodes in the cross-section, and ε^p_i is the equivalent plastic strains at node i. The average strain per pass was given earlier as Equation (1). Therefore, the average equivalent plastic strain obtained by simulation can be plotted as a function of the number of passes and compared directly to the analytical model dictated by Equation (1) using a die geometry of Φ = 110° and Ψ = 20° as shown in Figure 7. It is readily apparent that the average equivalent plastic strains from the simulation are very close to the analytical values calculated from Equation (1). This agreement validates the results obtained by the simulation. Figure 7 also shows that the average equivalent plastic strain increases with the increase in the pass number.

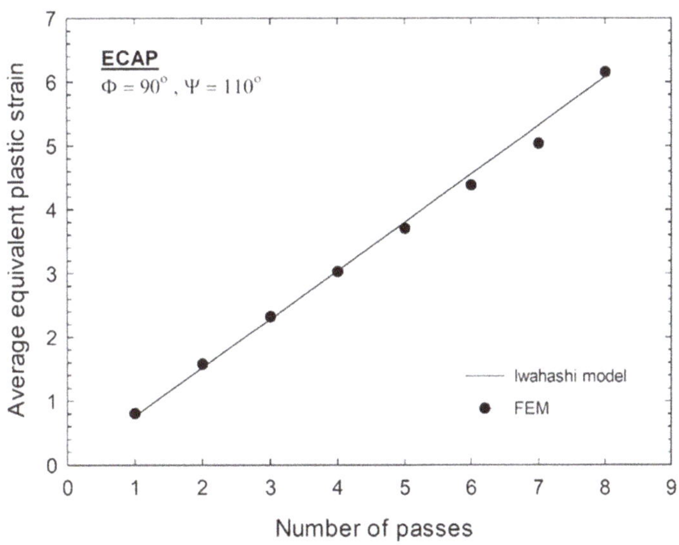

Figure 7. Average FEM simulation and the calculated equivalent plastic strains using Equation (1) from [2].

3.2. *Effect of Number of Passes on the Strain Homogeneity*

In addition to the magnitude of the accumulated strain, the strain homogeneity is also important in the design of the ECAP process. In practice, the effect of ECAP passes on the overall strain homogeneity can be assessed directly by measuring the degree of inhomogeneity and noting that lower values indicate better homogeneity.

The degree of inhomogeneity can be quantified by two different methods: using an inhomogeneity index (C_i) or a coefficient of variance (CV_{ε^p}). The inhomogeneity index (C_i) is based on the difference between two extreme strain values as defined by

$$C_i = \left(\frac{\varepsilon^p_{max} - \varepsilon^p_{min}}{\varepsilon^p_{ave}}\right), \tag{3}$$

where ε^p_{max}, ε^p_{min} and ε^p_{ave} denote, respectively, the maximum, minimum and average of the equivalent plastic strain. By contrast, the coefficient of variance (CV_{ε^p}) uses the standard deviation, which depends on the distribution of strain considering the value of the strain at all nodes in the section as defined by

$$CV_{\varepsilon^p} = \frac{Std\ \varepsilon^p}{\varepsilon^p_{ave}}, \tag{4}$$

where $Std\ \varepsilon^p$ is the standard deviation of the equivalent plastic strain. This latter value measures the amount of dispersion of the equivalent plastic strain at each node around the average strain as defined by

$$Std\ \varepsilon^p = \left(\frac{1}{n}\sum\nolimits_{i=1}^{n}\left(\varepsilon^p_i - \varepsilon^p_{ave}\right)^2\right)^{1/2}, \tag{5}$$

The strain distribution homogeneity for different passes during ECAP was determined using these two methods, where the inhomogeneity index (C_i) and the coefficient of variance (CV_{ε^p}) were estimated across the transverse section in the mid-length of the billet as shown in Figure 8. It is apparent that the values of the inhomogeneity index, C_i, are higher than the coefficient of variance, CV_{ε^p}, for all conditions, and this is consistent with earlier reports [15,29]. This difference arises because C_i is based on the difference between the maximum and the minimum values, whereas CV_{ε^p} is based on the distribution of strain for all nodes in the section. However, both C_i and CV_{ε^p} give the same tendency whereby their values decrease as the pass number increases, and this is especially true in the earlier stage of deformation as in passes 1–3. However, after four passes, the degree of homogeneity exhibits no significant change because the strains at all nodes on the cross-section are taken into consideration for the coefficient of variance. Therefore, this was selected to represent the strain homogeneity of ECAP.

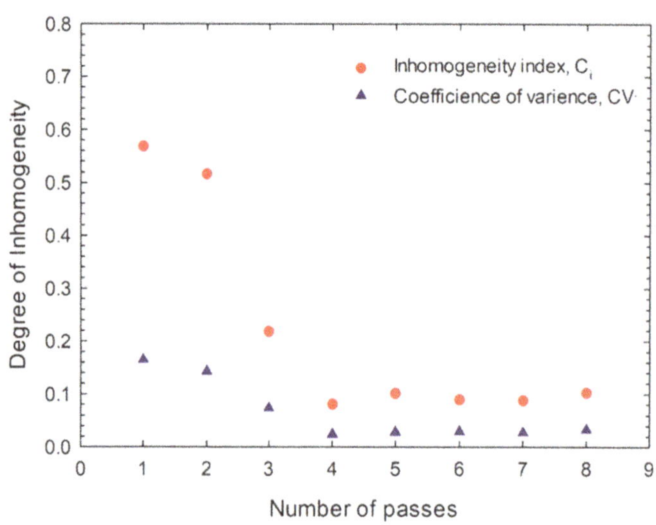

Figure 8. Inhomogeneity in equivalent plastic strain indicated by the inhomogeneity index and by the coefficient of variance.

In processing by ECAP, a sample experiences intense plasticity as it is pressed through the region of the intersection of the two parts of the channel, where this may be denoted as the plastic deformation zone (PDZ). The evolution of strain homogeneity with the increment of the number of passes can be explained by the strain distribution within the PDZ at the intersection of the die channel as shown in Figure 9. The results demonstrate that the variation in the equivalent plastic strain distribution in the PDZ decreases as the pass number increases as, for example, by 0.13 to 0.88 for one pass and 5.10 to 5.70 for eight passes, and this results in the development of a reasonable degree of deformation homogeneity in consecutive passes.

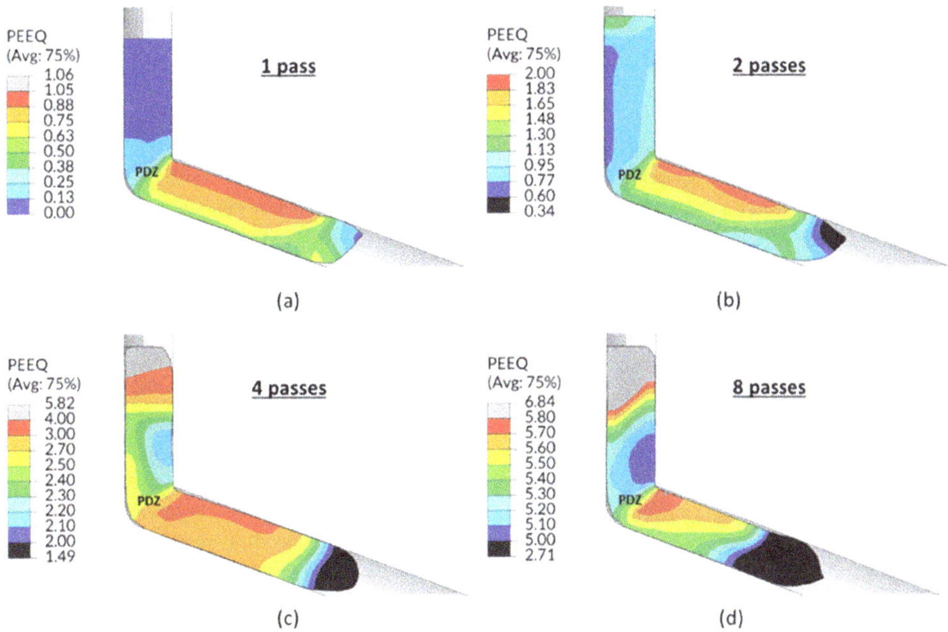

Figure 9. Equivalent plastic strain distribution contours of the workpiece during ECAP processing at 60% of stroke: (**a**) 1 pass, (**b**) 2 passes, (**c**) 4 passes and (**d**) 8 passes.

3.3. The Steady-State Zone

The steady-state zone is the region along the billet axis where strain is relatively uniform. To define the steady-state zone in an ECAP workpiece, the transverse cross-sections were cut perpendicular to the pressing direction throughout the length of the workpiece as shown in Figure 10. Then, the average equivalent plastic strains and the coefficients of variance were calculated over the cross-sectional surface and plotted over the length of the workpiece through 1, 2, 4 and 8 passes of ECAP as shown in Figure 11a,b, respectively.

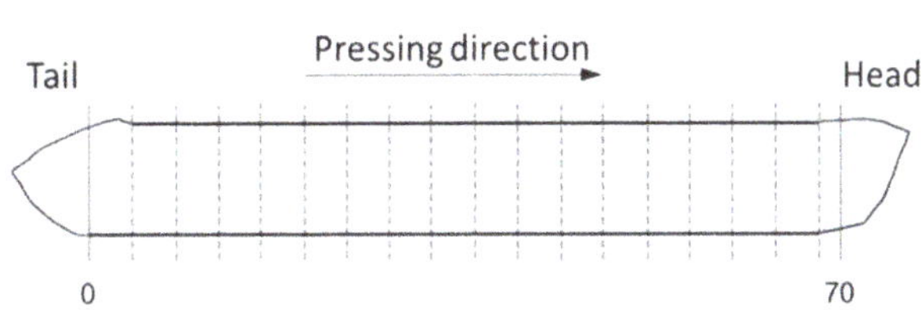

Figure 10. Schematic illustration showing the cross-sections in the analysis for the steady-state zone.

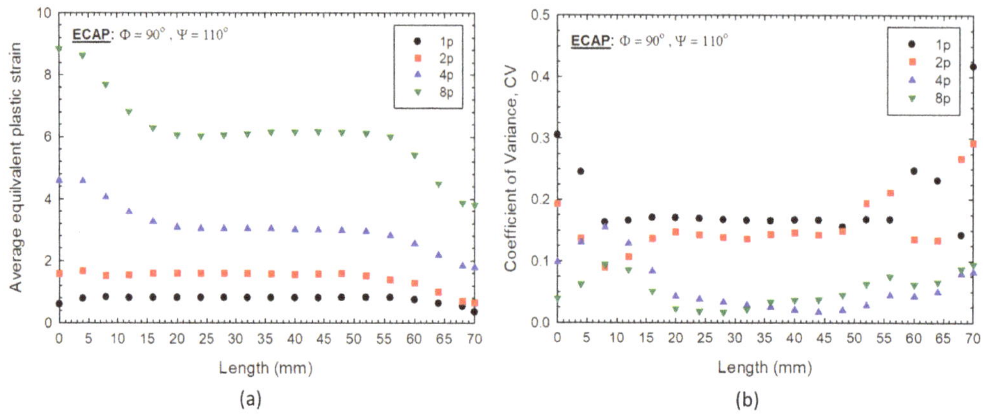

Figure 11. The plots of (**a**) average equivalent plastic strains and (**b**) coefficients of variance on different cross-sections over the length of the workpiece for various numbers of passes.

Figure 11a shows that the average equivalent plastic strain increases with the increasing number of passes for all sections along the length of the workpiece. Moreover, during ECAP through one pass and two passes, the values of the average equivalent plastic strain are relatively constant along the length and decrease slightly near the front of the workpiece due to the lack of any constraint at the exit of the ECAP die. In consecutive passes with route B_C, the maximum value of the average equivalent plastic strain lies at the back of the workpiece close to the load application point, and then it decreases with the distance away from the application point until it reaches a certain value and remains steady in the middle portion of the workpiece over a length of about 40 mm. It then becomes lower again near the front of the workpiece as is evident in the workpiece after ECAP for four and eight passes.

Figure 11b represents the degree of strain homogeneity in each cross-section over the length of the billet measured using the coefficient of variance, CV_{ε^p}, where a lower value indicates a better degree of homogeneity. These results show that the coefficient of variance decreases as the number of passes increases, where the CV_{ε^p} values for the workpieces after ECAP for four and eight passes are lower than 0.1. It is apparent that higher values of CV_{ε^p} exist at the front and back of the workpiece within distances of about 15 mm from either end for all conditions.

Finally, Figure 12 shows the equivalent plastic strain distribution contours along the length of the workpiece through ECAP after different numbers of passes. An inspection shows that there is a significant variation in equivalent plastic strain from the bottom to the top of the workpiece after ECAP for one and two passes as shown in Figure 12a,b, but this variation is gradually eliminated with the increasing number of ECAP passes. From the results plotted in Figures 11 and 12, it is concluded that the workpiece may be conveniently divided into three regions corresponding to the head, the intermediate steady-state zone and the tail. The steady-state zone is the region of reasonable homogeneity located between the non-uniformly deformed head at the front of the workpiece and the tail at the back of the workpiece. This is evident in Figure 12d, where the steady-state zone is clearly marked.

In this study, the length of the steady-state zone of the Cu-0.1 wt% Zr alloy after ECAP through eight passes is ~40 mm in the middle portion of the billet between the head and the tail. Minimizing the lengths of the non-uniform regions in the head and tail is of technological importance in order to maximize the length of the useful homogeneous portion of the ECAP sample. It was reported earlier that, if the angle Ψ is maintained constant, the non-uniform part in the head is longer in a strain-hardening material than that in a perfectly plastic material [8]. Furthermore, if the ECAP facility has a capability of applying back pressure, this will contribute to the uniformity of the metal flow during the

ECAP operation because the application of back pressure increases the homogeneity and leads to a filling of the outer corner of the die so that the deformation zone becomes closer to that of a localized shear band [31–33].

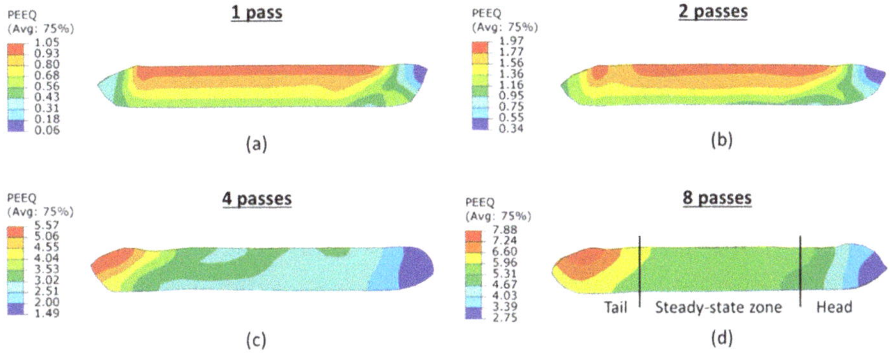

Figure 12. Equivalent plastic strain distribution contours along the length of the workpiece through ECAP for (**a**) 1 pass, (**b**) 2 passes, (**c**) 4 passes and (**d**) 8 passes.

The goal in the design of the ECAP process is to maximize the magnitude of the strain, to minimize the inhomogeneity and to minimize the force that is needed. In order to achieve these objectives, it is important to obtain an understanding of the effect of various parameters, including the die geometry, the material properties, the processing conditions and their interactions on the overall deformation behavior. The present investigation shows that the use of 3D FEM simulation provides an important contribution toward achieving a much improved understanding of the ECAP deformation process.

4. Conclusions

1. A 3D FEM simulation was successfully used to investigate the deformation behavior of a circular cross-sectional workpiece of a Cu-Zr alloy during ECAP processing, with die angles of $\Phi = 110^\circ$ and $\Psi = 20^\circ$, up to a total of eight passes using processing route B_C.
2. The results show that strain inhomogeneity is present in the early stages of ECAP processing, in which a lower equivalent plastic strain occurs on the outer corner of the workpiece due to the formation of a corner gap in a representative strain-hardening material. However, this strain inhomogeneity decreases with the number of passes and becomes reasonably uniform after four passes of ECAP.
3. The average equivalent plastic strains obtained from the cross-section in the mid-length of the workpiece after various passes is in very good agreement with the basic analytical model for ECAP, and this serves to verify the validity of the simulation results. The values of the average equivalent plastic strain tend to increase continuously with the increasing number of passes. In addition, the strain distributions on transverse cross-sections are in a good agreement with the microhardness values measured on the Cu-Zr alloy.
4. The effect of pass number on the strain homogeneity was evaluated using the inhomogeneity index (C_i) and the coefficient of variance (CV_{ε^p}), where a lower value means a higher homogeneity in the strain distribution. Both C_i and CV_{ε^p} significantly decrease with up to four passes of ECAP, and, thereafter, they remain reasonably constant so that there is a potential in ECAP processing for achieving strain homogeneity after processing through a significant number of passes.
5. Based on the coefficient of variance, it is concluded that, after pressing, the workpiece may be conveniently divided into three separate regions corresponding to the head,

steady-state zone and the tail. In this study, the steady-state zone extends over approximately 40 mm in length along the longitudinal axis of the workpiece.

Author Contributions: Methodology, J.W.-N., N.N., C.K. and T.G.L.; validation, J.W.-N., N.N. and C.K.; investigation, J.W.-N., N.N. and C.K.; resources, J.W.-N., N.N., C.K. and T.G.L.; data curation, J.W.-N., N.N. and C.K.; writing—original draft preparation, J.W.-N., N.N. and C.K.; writing—review and editing J.W.-N., N.N., C.K. and T.G.L.; supervision, J.W.-N., N.N., C.K. and T.G.L.; funding acquisition, J.W.-N., C.K. and T.G.L. All authors have read and agreed to the published version of the manuscript.

Funding: The work was supported by the European Research Council under ERC Grant Agreement No. 267464-SPDMETALS, the KMITL Research Fund under Project Number KREF01590 and Thammasat Postdoctoral Fellowship under Project Number TUPD12/2564.

Data Availability Statement: The raw and processed data generated during this study will be made available upon reasonable request.

Acknowledgments: We appreciate the assistance of the staff in the Mechanical Engineering Laboratory in KMITL.

Conflicts of Interest: The authors declare no conflict of interest.

References

1. Valiev, R.Z.; Langdon, T.G. Principles of equal-channel angular pressing as a processing tool for grain refinement. *Prog. Mater. Sci.* **2006**, *51*, 881–981. [CrossRef]
2. Iwahashi, Y.; Wang, J.; Horita, Z.; Nemoto, M.; Langdon, T.G. Principle of equal-channel angular pressing for the processing of ultra-fine grained materials. *Scr. Mater.* **1996**, *35*, 143–146. [CrossRef]
3. Li, S.; Bourke, M.A.M.; Beyerlein, I.J.; Alexander, D.J.; Clausen, B. Finite element analysis of the plastic deformation zone and working load in equal channel angular extrusion. *Mater. Sci. Eng. A* **2004**, *382*, 217–236. [CrossRef]
4. Figueiredo, R.B.; Pinheiro, I.P.; Aguilar, M.T.P. The finite element analysis of equal channel angular pressing (ECAP) considering the strain path dependence of the work hardening of metals. *J. Mater.Process. Technol.* **2006**, *180*, 30–36. [CrossRef]
5. Wei, W.; Nagasekhar, A.V.; Chen, G.; Tick-Hon, Y.; Wei, K.X. Origin of inhomogenous behavior during equal channel angular pressing. *Scr. Mater.* **2006**, *54*, 1865–1869. [CrossRef]
6. Figueiredo, R.B.; Cetlin, P.R.; Langdon, T.G. The processing of difficult-to-work alloys by ECAP with an emphasis on magnesium alloys. *Acta Mater.* **2007**, *55*, 4769–4779. [CrossRef]
7. Wei, W.; Zhang, W.; Wei, K.X.; Zhong, Y.; Cheng, G.; Hu, J. Finite element analysis of deformation behavior in continuous ECAP process. *Mater. Sci. Eng. A* **2009**, *516*, 111–118. [CrossRef]
8. Balasundar, I.; Raghu, T. Effect of friction model in numerical analysis of equal channel angular pressing process. *Mater. Des.* **2010**, *31*, 449–457. [CrossRef]
9. Yoon, S.C.; Jeong, H.-G.; Lee, S.; Kim, H.S. Analysis of plastic deformation behavior during back pressure equal channel angular pressing by the finite element method. *Comp. Mater. Sci.* **2013**, *77*, 202–207. [CrossRef]
10. Suo, T.; Li, Y.; Guo, Y.; Liu, Y. The simulation of deformation distribution during ECAP using 3D finite element method. *Mater. Sci. Eng. A* **2006**, *432*, 269–274. [CrossRef]
11. Basavaraj, V.P.; Chakkingal, U.; Kumar, T.S.P. Study of channel angle influence on material flow and strain inhomogeneity in equal channel angular pressing using 3D finite element simulation. *J. Mater.Process. Technol.* **2009**, *209*, 89–95. [CrossRef]
12. Mahallawy, N.E.; Shehata, F.A.; Hameed, M.A.E.; Aal, M.I.A.E. 3D FEM simulations for the homogeneity of plastic deformation in Al–Cu alloys during ECAP. *Mater. Sci. Eng. A* **2010**, *527*, 1404–1410. [CrossRef]
13. Su, C.W.; Lu, L.; Lai, M.O. 3D finite element analysis on strain uniformity during ECAP process. *Mater. Sci. Technol.* **2007**, *23*, 727–735. [CrossRef]
14. Leo, P.; Cerri, E.; De Marco, P.P.; Roven, H.J. Properties and deformation behaviour of severe plastic deformed aluminium alloys. *J. Mater.Process. Technol.* **2007**, *182*, 207–214. [CrossRef]
15. Basavaraj, V.P.; Chakkingal, U.; Kumar, T.S.P. Effect of geometric parameters on strain, strain inhomogeneity and peak pressure in equal channel angular pressing—A study based on 3D finite element analysis. *J. Manufac. Process.* **2015**, *17*, 88–97.
16. Xu, S.; Zhao, G.; Luan, Y.; Guan, Y. Numerical studies on processing routes and deformation mechanism of multi-pass equal channel angular pressing processes. *J. Mater.Process. Technol.* **2006**, *176*, 251–259. [CrossRef]
17. Cerri, P.; De Marco, P.; Leo, P. FEM and metallurgical analysis of modified 6082 aluminium alloys processed by multipass ECAP: Influence of material properties and different process settings on induced plastic strain. *J. Mater. Process. Technol.* **2009**, *209*, 1550–1564. [CrossRef]
18. Djavanroodi, F.; Omranpour, B.; Ebrahimi, M.; Sedighi, M. Designing of ECAP parameters based on strain distribution uniformity. *Prog. Nat. Sci. Mater. Int.* **2012**, *22*, 452–460. [CrossRef]
19. *ABAQUS User's Manual*; ABAQUS Inc.: Palo Alto, CA, USA, 2016.

20. Wongsa-Ngam, J.; Kawasaki, M.; Langdon, T.G. The development of hardness homogeneity in a Cu-Zr alloy processed by equal-channel angular pressing. *Mater. Sci. Eng. A* **2012**, *556*, 526–532. [CrossRef]
21. Li, J.; Wongsa-Ngam, J.; Xu, J.; Shan, D.; Guo, B.; Langdon, T.G. Wear resistance of an ultrafine-grained Cu-Zr alloy processed by equal-channel angular pressing. *Wear* **2015**, *326*, 10–19. [CrossRef]
22. Furukawa, M.; Iwahashi, Y.; Horita, Z.; Nemoto, M.; Langdon, T.G. The shearing characteristics associated with equal-channel angular pressing. *Mater. Sci. Eng. A* **1998**, *257*, 328–332. [CrossRef]
23. Kim, J.K.; Kim, W.J. Analysis of deformation behavior in 3D during equal channel angular extrusion. *J. Mater. Process. Technol.* **2006**, *176*, 260–267. [CrossRef]
24. Aal, M.I.A.E. 3D FEM simulations and experimental validation of plastic deformation of pure aluminum deformed by ECAP and combination of ECAP and direct extrusion. *Trans. Nonferrous Met. Soc. China* **2017**, *27*, 1338–1352. [CrossRef]
25. Priel, E.; Mittelman, B.; Trabelsi, N.; Cohen, Y.; Koptiar, Y.; Padan, R. A computational investigation of equal channel angular pressing of molybdenum validated by experiments. *J. Mater. Process. Technol.* **2019**, *264*, 469–485. [CrossRef]
26. Shan, A.; Moon, I.-G.; Ko, H.-S.; Park, J.-W. Direct observation of shear deformation during equal channel angular pressing of pure aluminum. *Scr. Mater.* **1999**, *41*, 353–357. [CrossRef]
27. Kim, H.S.; Seo, M.H.; Hong, S.I. On the die corner gap formation in equal channel angular pressing. *Mater. Sci. Eng. A* **2000**, *291*, 86–90. [CrossRef]
28. Luis-Pérez, C.J.; Luri-Irigoyen, R. Gastón-Ochoa, Finite element modelling of an Al–Mn alloy by equal channel angular extrusion (ECAE). D. *J. Mater. Process. Technol.* **2004**, *153*, 846–852. [CrossRef]
29. Agwa, M.A.; Ali, M.N.; Al-Shorbagy, A. Optimum processing parameters for equal channel angular pressing. *Mech. Mater.* **2016**, *100*, 1–11. [CrossRef]
30. Kim, W.J.; Namgung, J.C.; Kim, J.K. Analysis of strain uniformity during multi-pressing in equal channel angular extrusion. *Scr. Mater.* **2005**, *53*, 293–298. [CrossRef]
31. Lapovok, R.Y. The role of back-pressure in equal channel angular extrusion. *J. Mater. Sci.* **2005**, *40*, 341–346. [CrossRef]
32. Xu, C.; Xia, K.; Langdon, T.G. The role of back pressure in the processing of pure aluminum by equal-channel angular pressing. *Acta Mater.* **2007**, *55*, 2351–2360. [CrossRef]
33. Xu, C.; Xia, K.; Langdon, T.G. Processing of a magnesium alloy by equal-channel angular pressing using a back-pressure. *Mater. Sci. Eng. A* **2009**, *527*, 205–211. [CrossRef]

crystals

MDPI

Article

Strain Rate and Temperature Effects on Tensile Properties of Polycrystalline Cu_6Sn_5 by Molecular Dynamic Simulation

Wei Huang [1,2], Kailin Pan [1,2,*], Jian Zhang [1] and Yubing Gong [1]

[1] Engineering Research Center of Electronic Information Materials and Devices, Ministry of Education, Guilin University of Electronic Technology, Guilin 541004, China; huang0773@guet.edu.cn (W.H.); guet.zhang722@gmail.com (J.Z.); gybcome@163.com (Y.G.)

[2] School of Mechanical and Electrical Engineering, Guilin University of Electronic Technology, Guilin 541004, China

* Correspondence: pankl@guet.edu.cn

Abstract: Intermetallic compounds (IMCs) are essential in the soldering of electronic products and are composed mainly of Cu_6Sn_5 and Cu_3Sn. They must maintain reliable mechanical and electrical connections. As they are usually only a few microns thick, and it is difficult to study their mechanical properties by traditional methods. In this study, a 100 Å × 100 Å × 100 Å polycrystal with 10 grains was created by Atomsk through Voronoi tessellation based on a Cu_6Sn_5 unit cell. The effects of the temperature and strain rate on the tensile properties of the polycrystalline Cu_6Sn_5 were analyzed based on MEAM potential function using a molecular dynamics (MD) method. The results show that Young's modulus and ultimate tensile strength (UTS) of the polycrystalline Cu_6Sn_5 decrease approximately linearly with an increase in temperature. At high strain rates (0.001–100 ps^{-1}), Young's modulus and UTS of the Cu_6Sn_5 are logarithmic with respect to the strain rate, and both increase with an increase in strain rate. In addition, at low strain rates (0.00001–0.0005 ps^{-1}), the UTS has a quadratic increase as the strain rate increases.

Keywords: strain rate; temperature; polycrystalline Cu_6Sn_5; molecular dynamic; tensile properties

Citation: Huang, W.; Pan, K.; Zhang, J.; Gong, Y. Strain Rate and Temperature Effects on Tensile Properties of Polycrystalline Cu_6Sn_5 by Molecular Dynamic Simulation. *Crystals* **2021**, *11*, 1415. https://doi.org/10.3390/cryst11111415

Academic Editors: Wojciech Polkowski and Pavel Lukáč

Received: 26 October 2021
Accepted: 17 November 2021
Published: 19 November 2021

1. Introduction

With the integration and miniaturization of electronic products, electronic packaging and assembly technologies are progressing toward achieving high density and small size. Therefore, the size of solder joints has dropped sharply from hundreds of microns to tens of microns, or even a few microns [1]. Intermetallic compounds (IMCs) are formed between the solder joint and pad during soldering because of interfacial reactions. At present, the most commonly used solders in electronic products are the Sn–Ag–Cu (SAC) and Sn–Cu (SC) lead-free solders [2,3], and the primary composites of IMCs are Cu_6Sn_5 and Cu_3Sn [3]. IMCs are a prerequisite for reliable connection for electronic products [4]. Therefore, the mechanical properties of IMCs significantly affect those of the entire solder joints [3].

As the thickness of the IMC is only a few microns, it is difficult to obtain its mechanical properties using conventional methods. There are usually two methods to study the mechanical properties of IMC: experiments and simulation. In the experimental method, specific samples are usually required, and nanoindentations are adopted to test the hardness and Young's modulus of the IMC [5–11]. However, the results tested by nanoindentation fluctuate within a certain range because it is difficult to obtain a pure IMC to ensure the consistency of samples. In the simulation method, first-principles and MD methods are usually used. First-principles based on quantum mechanics can be used to obtain relatively accurate physical parameters of a material [12,13]. However, the first-principles method requires massive computing resources; therefore, only a system with just a few atoms can be calculated. Molecular dynamics (MDs) based on classical mechanic theories can ease this problem. With the development of computer technology, MD simulation can be used to simulate atomic systems at the micro and nanoscales. IMC properties such as thermodynamic, mechanical, and diffusion properties can be obtained by MD simulation [14–19]. The crystal size and strain rate can influence the tensile properties of a single crystal Ni_3Sn_4 via MD simulation [16,19]. Studies have shown that the strain rate affects the tensile properties of the materials, and the UTS of the material increases with the increase in the strain rate [20–25]. The results of previous studies were primarily based on experiments. However, because the thickness of IMCs is only a few microns, it is difficult to study their tensile properties by traditional experimental methods.

The previous studies on the mechanical properties of diffusion properties of IMCs were all based on monocrystals. However, most of the IMCs in solder joints are polycrystals, and their mechanical properties differ from those of monocrystals. In this study, the mechanical properties of a polycrystalline Cu_6Sn_5 were studied using the MD method, considering the effects of strain rate and temperature on the tensile properties. The temperature range in this study is within 250 to 500 K (melting point of Cu_6Sn_5 is 688 K [5]), and the strain rate is 0.00001 to 100 ps^{-1}. In this study, the strain rate exceeding 0.001 ps^{-1} represents a high strain rate, and from 0.00001 to 0.001 ps^{-1} represents a low strain rate. The tensile properties in both situations are discussed separately in Section 3.

2. Methodology

2.1. Structure of Cu_6Sn_5 Unit Cell

We performed MD simulations using a large-scale atomic/molecular massively parallel simulator (LAMMPS) [26] with the modified embedded atom method (MEAM) potential proposed by Baskes [27].

The Cu_6Sn_5 unit cell in this study is an η′ phase belonging to a monoclinic crystal system with the C2/c space group [28]. In Figure 1, the gray and blue balls represent Sn and Cu atoms, respectively [12,28].

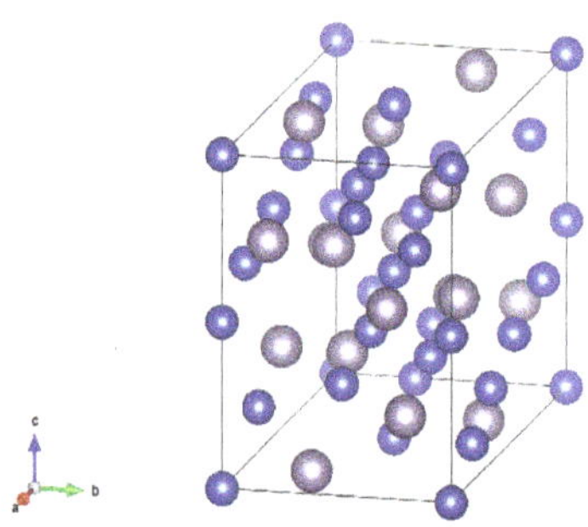

Figure 1. Cu_6Sn_5 (η′ phase) cell structure [12].

2.2. Modified Embedded Atom Method (MEAM) Potential

In the MEAM theory, the total energy of the system consists of two parts: the energy embedding an atom in the background electron density and a pair interaction, which is described by Equations (1) and (2) [27,29].

$$E_{tot} = \sum\nolimits_i [F_i(\overline{\rho}_i) + \frac{1}{2}\sum_{j(\neq i)} \phi_{ij}(R_{ij})] \quad (1)$$

$$F_i(\overline{\rho}_i) = AE_c \frac{\overline{\rho}_i}{\rho_{i0}} \ln(\frac{\overline{\rho}_i}{\rho_{i0}}) \quad (2)$$

where F_i is the embedding energy of the atom i, Φ_{ij} is the interaction potential between atoms i and j, and R_{ij} is the distance between atoms i and j. The parameters in the MEAM potential and their implications can be obtained from the literature [27,29]. For pure metals, the parameters in the MEAM potential function mainly include the cohesive energy E_c, equilibrium nearest-neighbor distance r_0, exponential decay factor for the universal energy function α, scaling factor for the embedding energy, the exponential decay factors for the atomic densities β (i, i = 0, 1, 2, 3), the weighting factors for the atomic densities t (i, i = 1, 2, 3), and the density scaling parameter ρ. For alloys, in addition to the above parameters, interaction parameters between the two pure metals could also be included, such as the binding energy between two different metal atoms, equilibrium nearest-neighbor distance between two types of atoms, and exponential decay factor for the universal energy function of the two atoms. Table 1 list the primary parameters in the potential according to references [29]. There is a many-body screening function in the MEAM, and the screening parameter C is adopted to determine the screening extent. C_{min} and C_{max} represent the minimum and maximum values of the screening boundaries, respectively [29]. Values of the C_{min} and C_{max} in this study are shown in Table 2 according to reference [29].

Table 1. Parameters of Cu_6Sn_5 MEAM potential.

Elements	E_c (eV)	A	r_0 (Å)	α	$\beta^{(0)}$	$\beta^{(1)}$	$\beta^{(2)}$	$\beta^{(3)}$	$t^{(1)}$	$t^{(2)}$	$t^{(3)}$	ρ_0
Cu	3.4	1.07	2.657	5.11	3.634	2.20	6.00	2.20	3.14	2.49	2.95	1
Sn	3.84	1	3.176	6.20	6.20	6.00	6.00	6.00	12.5	8.0	−0.38	1
Cu_6Sn_5	4.03		2.907	5.38								

Table 2. Screen parameters of the MEAM.

	(Cu, Cu, Cu)	(Cu, Cu, Sn)	(Cu, Sn, Cu)	(Sn, Sn, Sn)	(Sn, Cu, Sn)	(Sn, Sn, Cu)
C_{min}	0.8	0.8	0.8	1.29	0.8	0.8
C_{max}	2.8	2.8	2.8	4.43	2.8	2.8

3. Results and Discussion

3.1. Details of the MD Simulation

3.1.1. Monocrystalline and Polycrystalline Structures of Cu_6Sn_5

Based on a Cu_6Sn_5 unit cell, polycrystalline Cu_6Sn_5 can be created by Atomsk through Voronoi tessellation [30]. The box size and number of the grain seeds can be determined when creating polycrystals. For example, a 100 Å × 100 Å × 100 Å polycrystal with 10 grains is shown in Figure 2. Different colors represent different grain identifications (IDs).

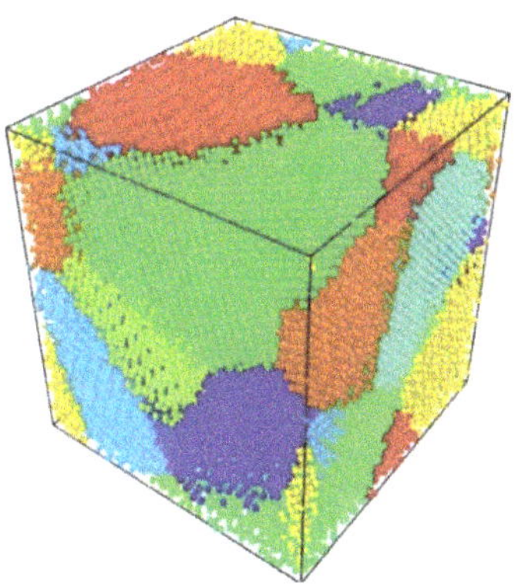

Figure 2. Polycrystalline Cu_6Sn_5 of 100 Å × 100 Å × 100 Å with 10 grains.

3.1.2. Validation of the MEAM Potential

Table 3 lists the stiffness constants calculated via the large-scale atomic/molecular massively parallel simulator (LAMMPS) with the MEAM potential in our study and by the first-principles method in reference [31]. Table 4 lists the elastic moduli of polycrystals, including bulk modulus (B), shear modulus (G), and Young's modulus (E), obtained via the

Voigt–Reuss–Hill (VRH) methods. Though some of the stiffness constants obtained from both methods differ, the value of the elastic moduli of the polycrystalline Cu_6Sn_5 is within the calculated result by Ghosh [32] and Lee [31]. Our MD simulations were performed based on this MEAM potential.

Table 3. Stiffness constants of monocrystalline Cu_6Sn_5.

Stiffness Constants (GPa)	C_{11}	C_{22}	C_{33}	C_{12}	C_{13}	C_{23}	C_{44}	C_{55}	C_{66}
Our MD	144.52	160.15	133.29	52.25	55.10	68.07	40.91	46.26	43.67
Lee et al. [31]	156.4	165.2	155.8	62.2	69.4	60.6	42.3	51.9	48.0

Table 4. Average elastic constants of polycrystalline Cu_6Sn_5 with VRH method.

Methods	*B* (GPa)	*G* (GPa)	*E* (GPa)
Our MD	87.25	43.40	111.67
Lee et al. [31]	95.61	46.23	119.435
Ghosh et al. [32]	84.6	37.0	96.9

3.1.3. Simulation Setting

By using the proposed MD simulation, the stress–strain relations of the polycrystals were studied by stretching the polycrystals at different temperatures and strain rates. The polycrystal was fully relaxed under NVE, NVT, and NPT ensembles alternately to make the system reach target setting temperatures and pressures and was stretched under the NPT ensemble. To access the effects of the strain rate, the polycrystals were stretched at stain rates of 0.00001 to 100 ps^{-1}. In addition, the temperature effect over a temperature range of 250–500 K with intervals of 50 K was considered in our study.

3.2. *Isotropic Analysis of Tensile Properties of Polycrystals*

To determine the isotropic characteristics of the polycrystalline Cu_6Sn_5, it was stretched along the x-, y-, and z-axis, respectively, at 300 K with a strain rate of 1 ps^{-1}. Figure 3 shows the stress–strain response. Stress_x, Stress_y, and Stress_z represent stretching along the x-, y-, and z-axis, respectively. The three curves almost overlapped before the polycrystalline Cu_6Sn_5 cracked, specifically in the linear elastic deformation phase. This indicates that the tensile properties of the polycrystalline Cu_6Sn_5 along the x, y, and z directions are very similar to each other. Therefore, the polycrystalline Cu_6Sn_5 in our study is isotropic, which is consistent with the real polycrystal materials [33].

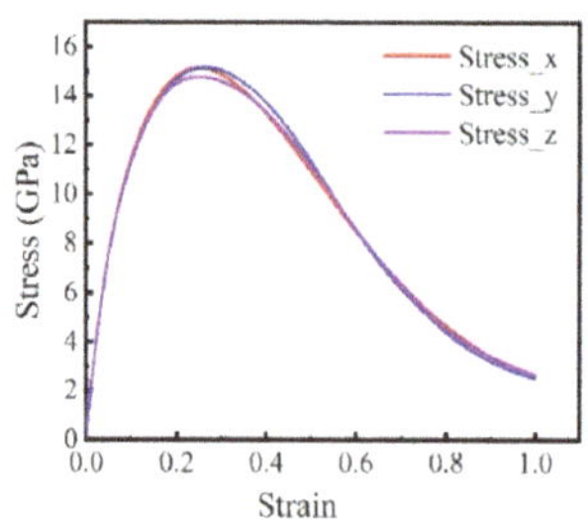

Figure 3. Stress–strains curves when stretching along x-, y-, and z-axis.

3.3. *Temperature Effect on Tensile Properties of Polycrystalline Cu6Sn5*

The stress–strain curves of the polycrystalline Cu_6Sn_5 at strain rates ranging from 0.001 to 100 ps^{-1} were recorded in the temperature range of 250 K to 500 K, as illustrated in Figure 4. The changing trends of all stress–strain curves are similar in the temperature range of 250–500 K, with the strain rate ranging from 0.001 to 100 ps^{-1}. The stretching deformation shifted from the elastic stage to the plastic stage and then cracked. There is no yield phase and plastic deformation in this tensile process, which indicates that the elastic deformation is prominent, and the failure of the polycrystalline Cu6Sn5 at strain rates from 0.001 to 100 ps^{-1} is due to brittle fracture. It was also observed that the stress–strain curve at the tensile strain rate of 100 ps^{-1} was very close to that at the strain rate of 10 ps^{-1}. This indicates that for a stain rate exceeding 10 ps^{-1}, its effects on the stress–strain response are negligible. Research shows that the failure modes of materials can be affected by strain rate [4]. It was found that elastic deformation is dominant before UTS and exhibits a brittle failure when the polycrystal is stretched, with the strain rate ranging from 0.01 to 100 ps^{-1}, as shown in Figure 4a–e. Plastic deformation occurs when the strain rate is less than 0.001 ps^{-1} despite the dominance of elastic deformation, as shown in Figure 4f. More information about the ductile performance under a low strain can be found in Section 3.5. The polycrystal exhibits obvious ductile deformation with strain rates lower than 0.001 ps^{-1}. Therefore, 0.001 ps^{-1} was chosen as a boundary. Therefore, in this study, strain rates ranging from 0.001 to 100 ps^{-1} are defined as high strain rate stretching, and those less than 0.001 ps^{-1} are considered as low strain rate stretching.

According to the generalized Hooke's Law, during linear elastic deformation, the stress (σ) and strain (ε) satisfy Equation (3), where E is Young's modulus. Therefore, the slope of the linear part of the stress–strain curve is recorded as Young's modulus.

$$E = \frac{\sigma}{\varepsilon} \tag{3}$$

The calculated results of Young's modulus according to Figure 4 are shown in Figure 5. It is observed that Young's modulus decreases approximately linearly with increasing temperature at different strain rates. Young's modulus decreases from 136.15 to 125.15 GPa, 137.29 to 126.90 GPa, 155.86 to 147.71 GPa, 180.26 to 174.33 GPa, 182.99 to 177.89 GPa, and 183.04 to 178.02 GPa when the temperature increases from 250 to 500 K at strain rates of 0.001, 0.01, 0.1, 1, 10, and 100 ps^{-1}, respectively. The decreasing rates of Young's modulus are 44.03, 41.57, 32.61, 23.71, 20.37, and 20.07 MPa/K, respectively. Young's modulus is the tensile modulus, which represents the ability of a material to resist elastic tensile deformation. Thus, the ability of the polycrystalline Cu6Sn5 to resist tensile deformation at a lower temperature is greater than that at a higher temperature.

The UTSs at different temperatures are shown in Figure 6. It is observed that the UTS decreases with an increase in temperature when the strain rate is kept constant. It decreases from 8.91 to 8.53 GPa, 10.90 to 10.53 GPa, 12.96 to 12.22 GPa, 15.29 to 14.46 GPa, 16.58 to 16.00 GPa, and 16.65 to 16.13 GPa when temperature increasing from 250 to 500 K at strain rates of 0.001, 0.01, 0.1, 1, 10, and 100 ps^{-1}, respectively. Moreover, it is observed that most strains at corresponding UTSs are nearly the same, while the strain rates are maintained constant, which indicates that temperature has little effect on the strain of the UTS during stretching at a strain rate ranging from 0.001 to 1 ps^{-1}, as shown in Figure 4.

From what has been discussed above, the temperature can affect Young's modulus and UTS during stretching, and they both decrease with increasing temperature. This is because both the initial kinetic energy and potential energy of the relaxed system at a high temperature are higher than those at a low temperature, as shown in Figure 7. The potential energy represents the interaction between atoms; thus, the stability of a system increases with a decrease in the potential. On the other hand, kinetic energy represents the random thermal motion between atoms. With an increase in temperature, the atomic motion is more intense, thus leading to an increase in the kinetic energy and, consequently, a decrease in atomic interactions. As the total energy is the sum of the potential and kinetic energy, the initial total energy of the system also increases as the temperature increases. The total energy of the system at the UTS and the energy absorbed during the stretching process (the total energy at UTS minus the initial total energy) are shown in Figures 8 and 9, respectively. The total energy of the system at the UTS increases with the increase in temperature, while the absorbed energy is almost constant. The change in the total energy is due to the initial energy differences, while the energy absorbed by the system is almost unaffected by the temperature at a constant strain rate stretching.

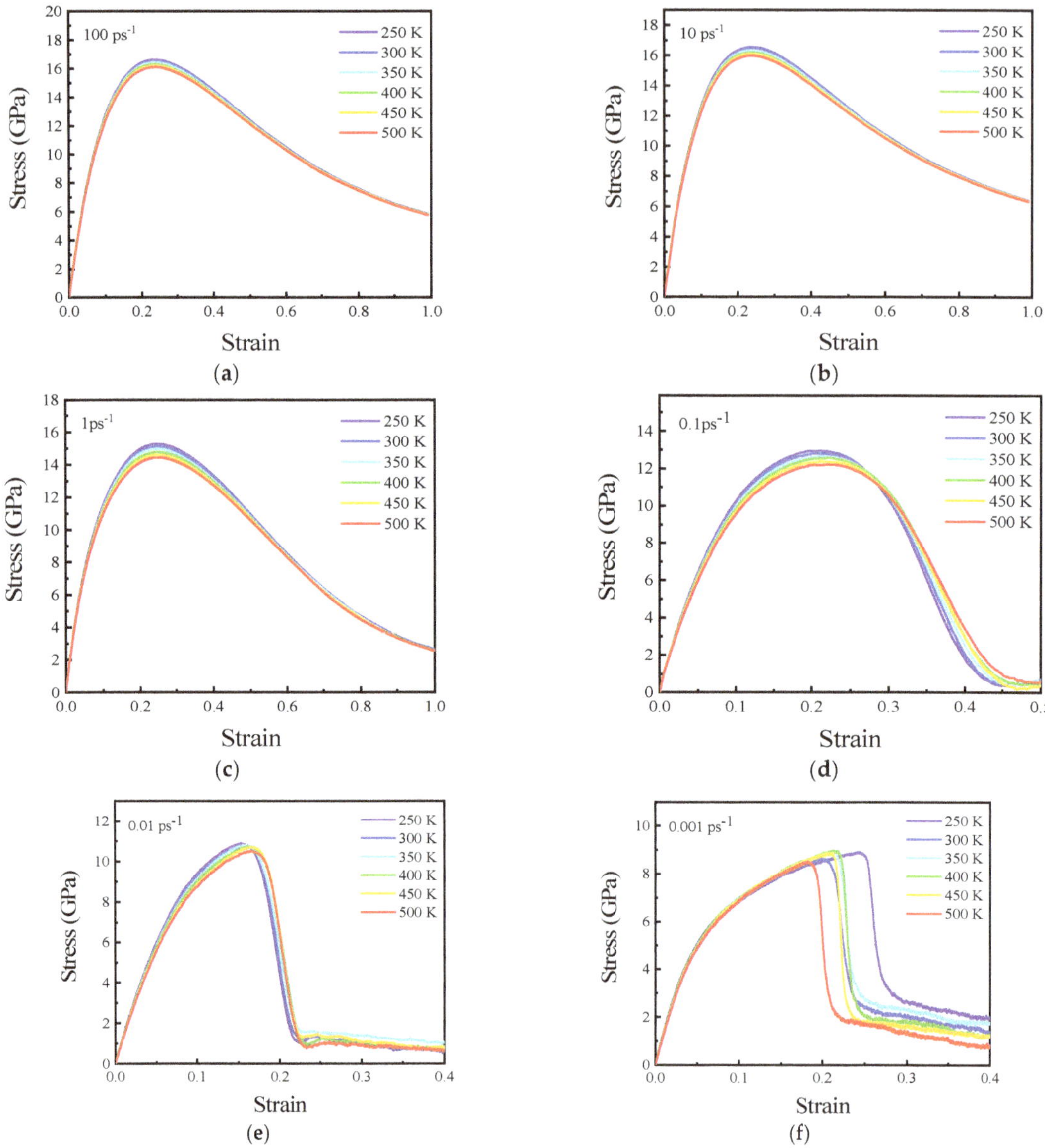

Figure 4. Stress-strain curves of polycrystalline Cu_6Sn_5 at temperatures of (**a**) 100 ps^{-1}, (**b**) 10 ps^{-1}, (**c**) 1 ps^{-1}, (**d**) 0.1 ps^{-1}, (**e**) 0.01 ps^{-1}, and (**f**) 0.001 ps^{-1}.

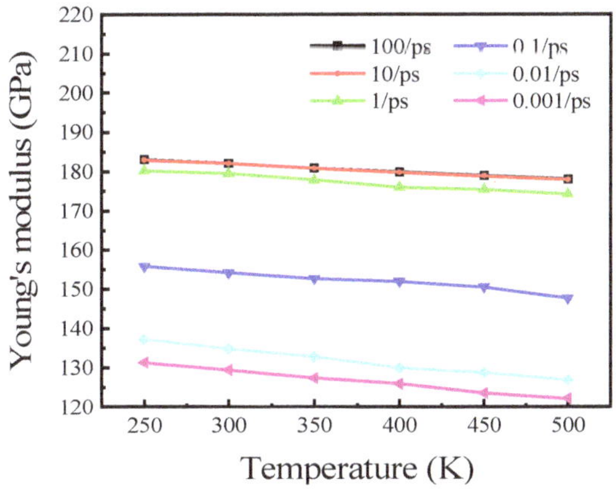

Figure 5. Young's modulus at different temperatures.

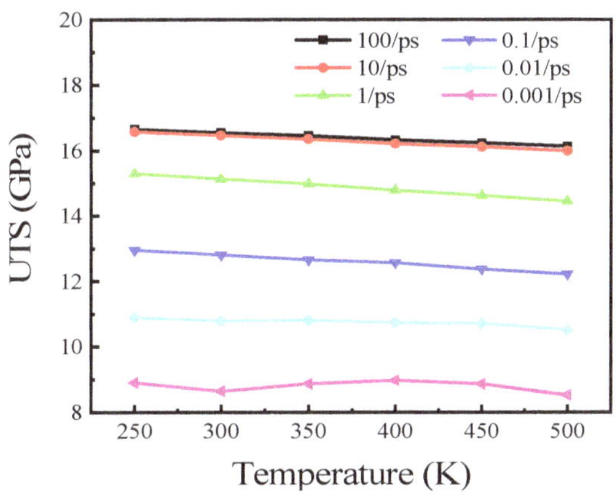

Figure 6. UTS at different temperatures.

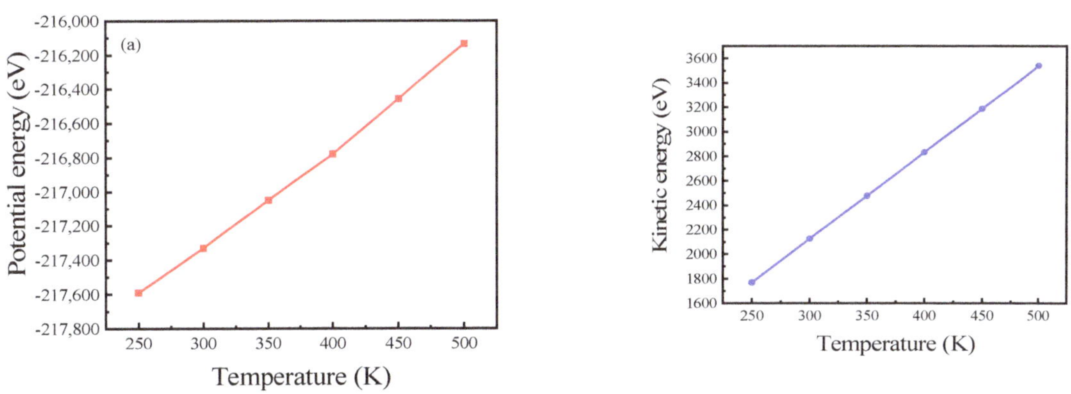

(**a**) (**b**)

Figure 7. Initial energy of the polycrystalline Cu_6Sn_5 of (**a**) potential energy and (**b**) kinetic energy.

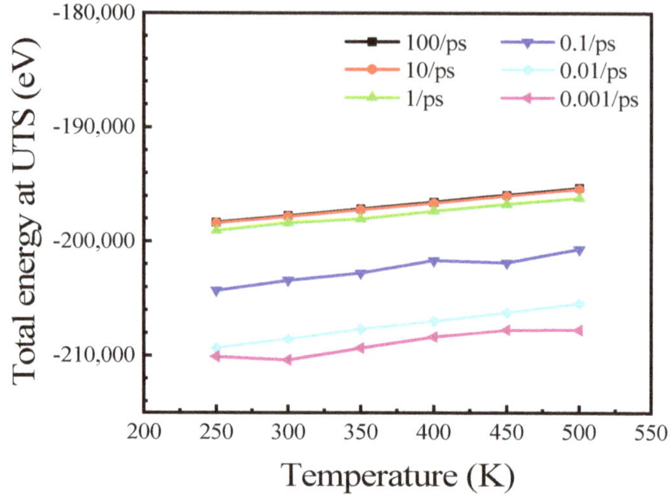

Figure 8. Total energy at UTS with temperature increasing.

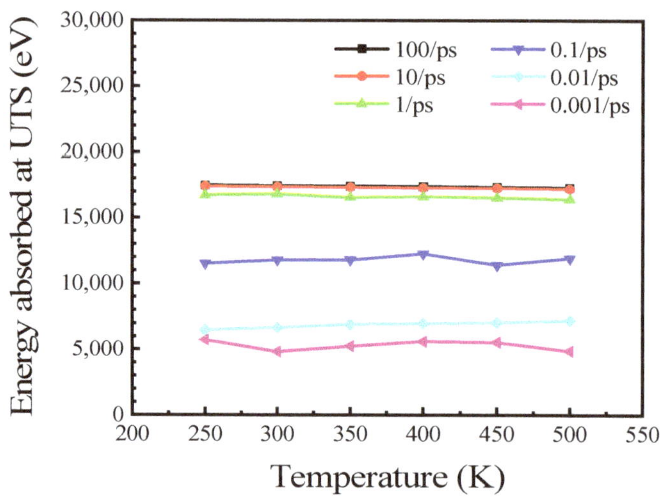

Figure 9. Energy absorbed at UTS with temperature increasing.

The above reasons account for the decrease in Young's modulus and UTS when the temperature increases. Both Young's modulus and UTS decrease approximately linearly with increasing temperature, but the linearity at higher strain rates (0.01 to 100 ps^{-1}) is better than that at lower strain rates (0.001 ps^{-1}), as shown in Figures 5 and 6. According to the previous description, 0.001 ps^{-1} is the critical strain rate between low and high strain rates. We infer that it may be a ductile–brittle mixed failure when stretching at this strain rate. The energy values absorbed at UTS with different temperatures are shown in Figure 9. It can be found that the absorbed energy exhibits inconsistency according to the temperature variation when stretching with 0.001/ps. Therefore, it can be concluded that the above reasons lead to the inconsistency of UTS with temperature change. This means that as the strain rate decreases, the polycrystalline Cu_6Sn_5 is subjected to more complex stresses when stretching. The strain rate effects on tensile properties of polycrystalline will be discussed in the next section.

3.4. High Strain Rate Effects on Tensile Properties of Polycrystalline Cu_6Sn_5

Figure 10 shows the stress–strain curves with strain rate changing from 0.001 to 100 ps^{-1} at (a) 250 K, (b) 300 K, (c) 350 K, (d) 400 K, (e) 450 K, and (f) 500 K. The strain rates changing from 0.001 to 100 ps^{-1} are denoted as high strain rates in this study. With an increase in the strain rate, the stress at the corresponding strain before the UTS is larger than that at a lower strain rate, and the plastic deformation is more significant at lower strain rates. Despite this, elastic deformation is predominant at high strain rate stretching. When elastic deformation is dominant in the stretching process (strain rate from 0.001 to 100 ps^{-1}), the corresponding strain of the UTS increases with the increase in the strain rate, which indicates that the tensile strength of the IMC at a high strain rate is higher than that at a lower strain rate. However, the stress-train response at the strain rate of 100 ps^{-1} is very close to that at the strain rate of 10 ps^{-1}. This means that when the strain rate is sufficiently large, the effect of the stress–strain response is almost negligible in this study.

Young's modulus and UTS were also calculated based on the stress–strain curves in Figure 10. The results are shown in Figures 11 and 12. Logarithmic coordinates were chosen for abscissa in both figures and the curves reveal the existence of logarithmic relationships between both Young's modulus and UTS to the strain rate. Moreover, they both increase with increasing strain rates. The mechanical properties of Cu_3Sn also exhibit a similar logarithmic trend [22].

We can presume that due to fast stretching at high strain rates, dislocations among grain boundaries occur later; thus, the effect of the dislocations is negligible. In this case, the forces at all positions of the polycrystalline Cu_6Sn_5 are almost the same. Therefore, the ability of the polycrystal to resist tensile deformation is enhanced, thus leading to an increase in Young's modulus and UTS.

To further evaluate the influence of the strain rate on the deformation of the polycrystalline Cu_6Sn_5, polycrystals with 0.25 and 0.5 strains at 300 K are shown in Figure 13. All the grains are uniformly deformed when stretched at the strain rate of 1 ps^{-1}, and the defects between the grain boundaries, even the elongation up to 0.5, are negligible, as shown in Figure 13a,b. This is because the polycrystals are destroyed before the dislocations between the grain boundaries occur owing to the speed of the tensile velocity. However, the deformation characteristics change exponentially with decreasing strain rate. When the strain rate is reduced to 0.001 ps^{-1}, there is adequate time for the grain boundaries to dislocate, which leads to defects in the polycrystal, as shown in Figure 13c,d. In addition, with a decrease in the strain rate, the dislocations are more significant among the grain boundaries at the same strain, which results in a decrease in the UTS of the material, as shown in Figure 12.

The radial distribution function (RDF) is the most common mathematical language used to describe the microstructure of liquid and amorphous materials. It denotes the ratio of the local density of a molecule to the bulk density at a distance of r around a central atom. In the binary intermetallic compounds system, the RDF for atoms α and β can be calculated as Equation (4) [16].

$$g(r) = \frac{V}{N_\alpha N_\beta}\left\langle \sum_{i=1}^{N_\alpha} \frac{n_{i\beta}(r)}{4\pi r^2 \Delta r} \right\rangle \tag{4}$$

where V represents the volume of the system and $n(r)$ the number of particles, which can be found in the shell from r to $r + \Delta r$ [16].

The calculated RDFs at different strain rates when the strain is 0.1 are shown in Figure 14. The peak value of the RDFs varies with the strain rates and then gradually approaches 1 with increasing r. It can be seen from the local enlargement that the RDF decreases with an increasing strain rate. This is because an increase in the strain rate causes more disorder in the atomic structure, which increases amorphization and hinders the formation of slip planes [34,35]. As there is no dislocation surface, the UTS improves [16]. The analysis results here are consistent with those shown in Figure 11. The energy absorbed by the system with different strain rates at the UTS is shown in Figure 15. As the strain rate increases to a larger value, more energy is absorbed by the system when reaching the UTS. Figure 10 shows that the corresponding strain value at the UTS changes over a small range; thus, the increased stress causes more energy to be absorbed.

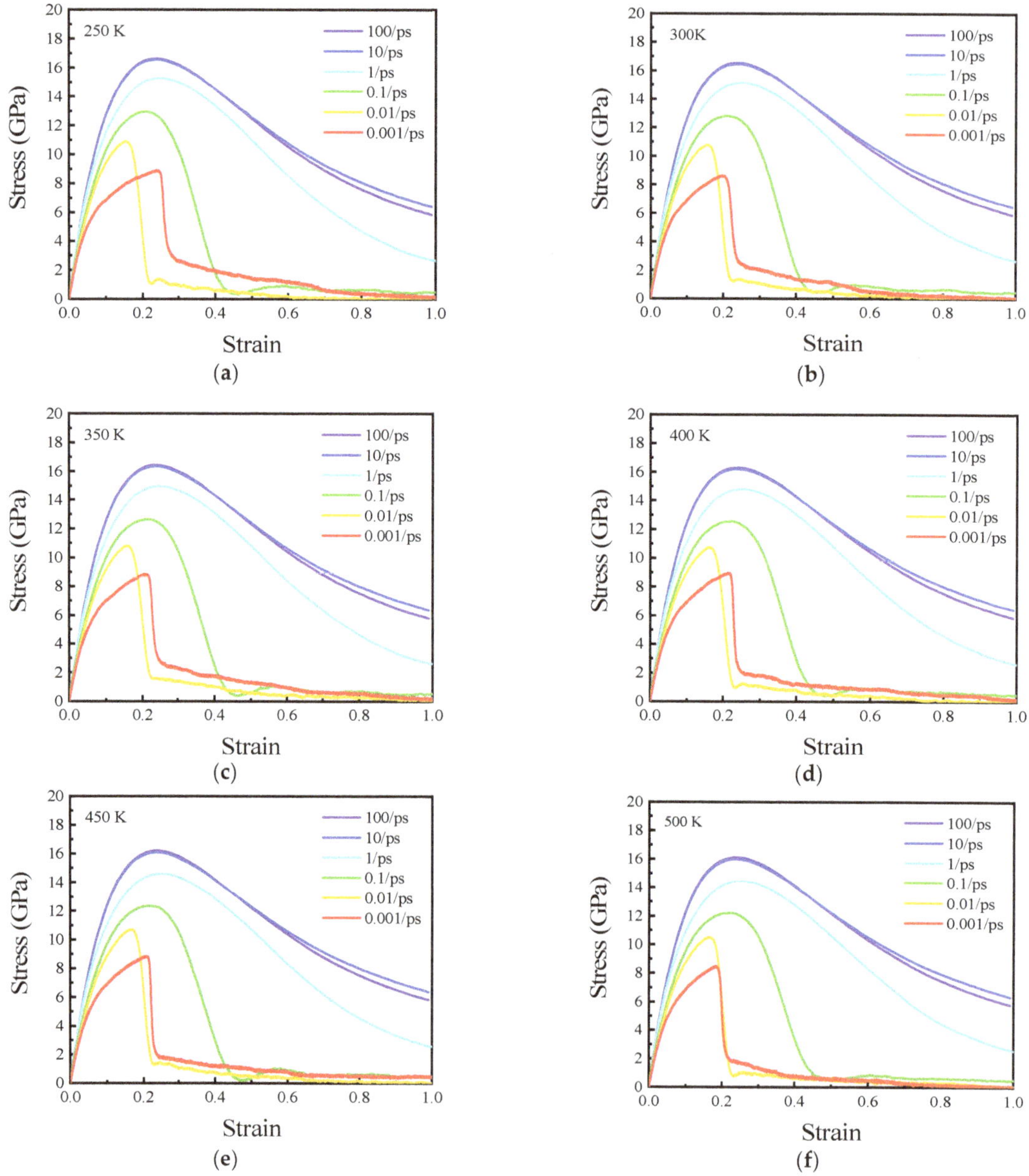

Figure 10. Stress–strain curves with strain rate changing from 0.001 to 100 ps^{-1} at (**a**) 250 K, (**b**) 300 K, (**c**) 350 K, (**d**) 400 K, (**e**) 450 K, and (**f**) 500 K.

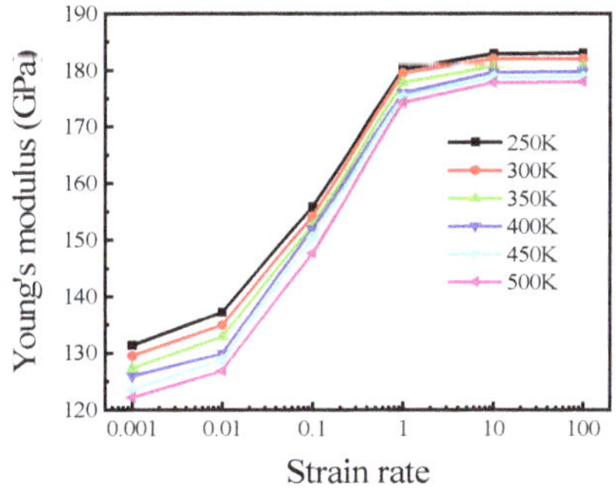

Figure 11. Young's modulus at different strain rates.

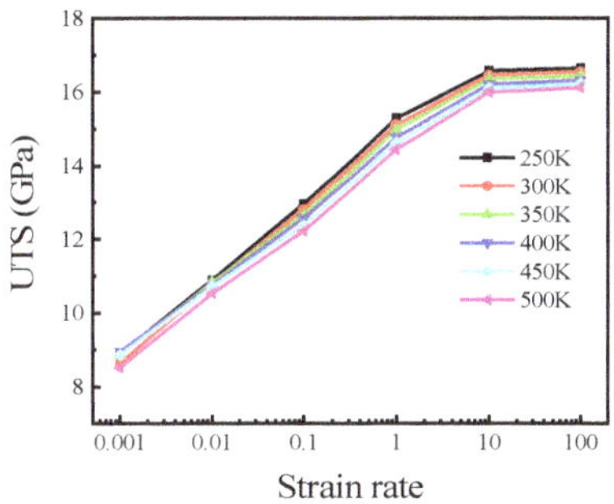

Figure 12. UTS at different strain rates.

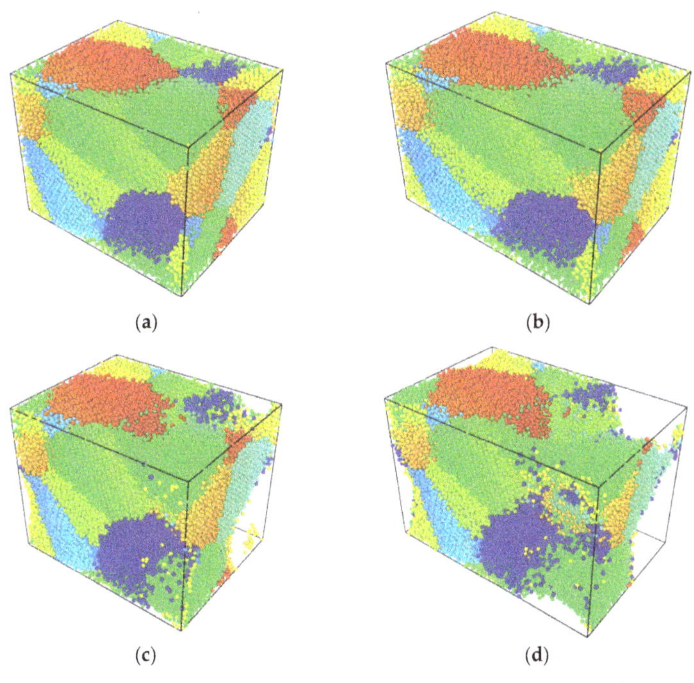

Figure 13. Polycrystals at different strain rates of (**a**) $\varepsilon = 0.25$, $\dot{\varepsilon} = 1\ \text{ps}^{-1}$; (**b**) $\varepsilon = 0.5$, $\dot{\varepsilon} = 1\ \text{ps}^{-1}$; (**c**) $\varepsilon = 0.25$, $\dot{\varepsilon} = 0.01\ \text{ps}^{-1}$; (**d**) $\varepsilon = 0.5$, $\dot{\varepsilon} = 0.01\ \text{ps}^{-1}$.

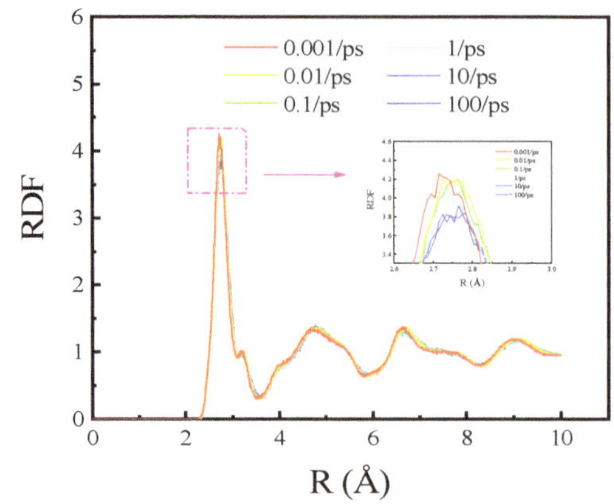

Figure 14. RDF of 10% strain at different strain rates.

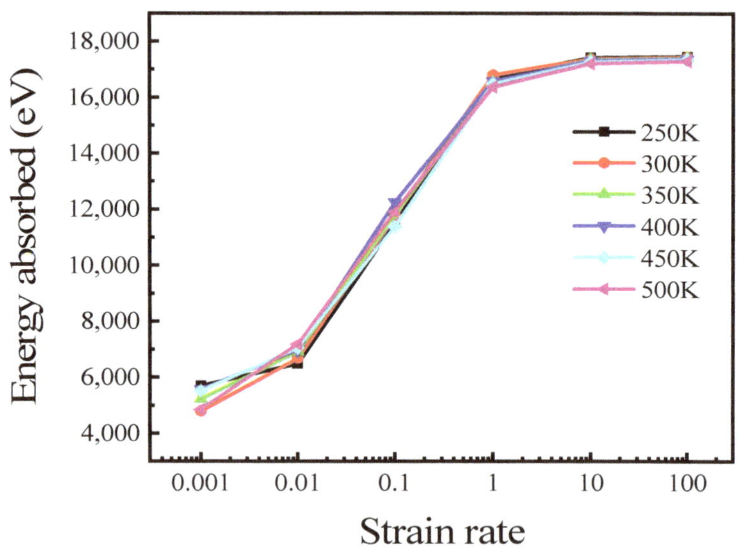

Figure 15. Energy absorbed at UTS.

3.5. Low Strain Rate Effects on Tensile Properties of Polycrystalline Cu_6Sn_5

The stress-strain curves at 300 K with the strain rate from 0.00001 to 0.0005 ps^{-1} are shown in Figure 16. When the strain rate is lower than 0.001 ps^{-1}, the stress does not decrease rapidly after the UTS but shows an obvious stage of plastic deformation. Within this strain rate range, the UTS still decreases with a decrease in the strain rate. However, at high strain rates, the relationship between UTS and strain rate is approximately logarithmic, while at low strain rates, the relationship between UTS and strain rate is approximately quadratic, as shown in Figure 17. As the strain rate decreases, the UTS gradually approaches a constant. The polycrystals when the strain rate is 0.00001 and the strains are 0.2 and 0.4 are shown in Figure 18. It can be found that there are no obvious dislocations when stretching at this strain rate. This is because when the tensile velocity is too low, there is sufficient time for stress relaxation. As seen from the local magnification of Figure 14, with a decrease in the strain rate, the plastic deformation of the polycrystalline Cu_6Sn_5 becomes more significant.

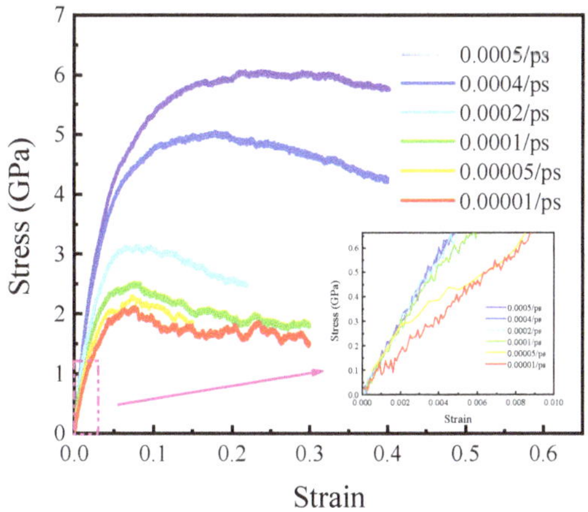

Figure 16. Stress–strain curve at low strain rates.

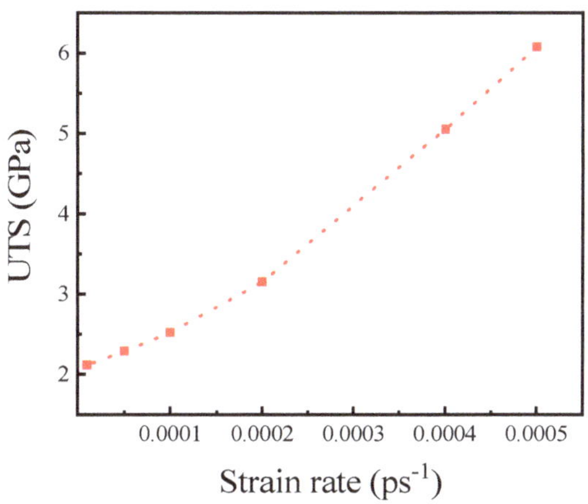

Figure 17. UTS at different strain rates.

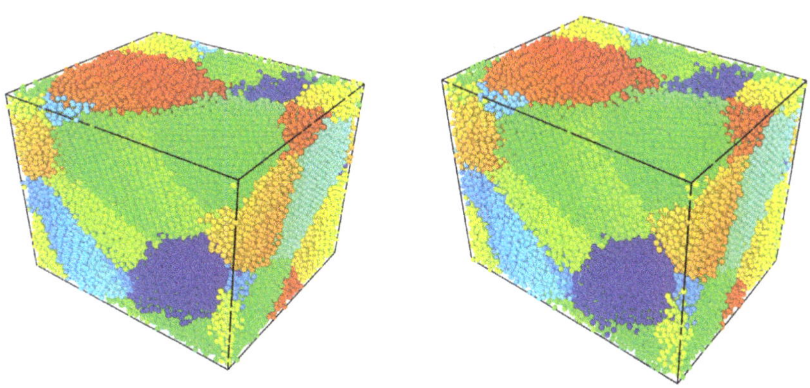

(**a**) (**b**)

Figure 18. Polycrystals at strain of (**a**) 0.2; (**b**) 0.4.

4. Conclusions

The tensile properties of the polycrystalline Cu_6Sn_5 were investigated at different temperatures and strain rates via the MD simulation. Based on the polycrystalline Cu_6Sn_5 created through Voronoi tessellation by Atomsk, the stress–strain curves at different temperatures and strain rates were obtained, from which Young's modulus and UTS were extracted. The stress–strain response could be affected by both the temperature and strain rate. The conclusions are as follows:

(1) Young's modulus and UTS decreased approximately linearly when the temperature increased from 250 to 500 K at high strain rates. The effect of temperature was relatively small compared with that of strain rate. The effect of temperature on mechanical properties was approximately negative linear, while the effect of strain rate on mechanical properties was approximately exponential with high strain rates.
(2) The strain rate affected the deformation characteristics of the IMC. When it ranged from 0.001 to 100 ps^{-1}, the elastic deformation was the dominant deformation in the stretching process. On the other hand, with a decrease in the strain rate, the plastic deformation was gradually more significant. When the strain rate was between 0.00001 and 0.0005 ps^{-1}, the plastic deformation characteristics gradually appeared.
(3) The strain rate affected the tensile strength of the IMC. At high strain rates (from 0.001 to 1 ps^{-1}), Young's modulus and UTS decreased with decreasing strain rate, and they exhibited exponential relationships with the strain rate. At low strain rates (from 0.00001 to 0.0005 ps^{-1}), the relation between the UTS and strain rate was quadratic.

Author Contributions: Conceptualization, W.H. and K.P.; funding acquisition, K.P. and Y.G.; investigation, W.H. and J.Z.; methodology, W.H. and K.P.; software, W.H.; validation, J.Z.; writing—original draft preparation, W.H.; writing—review and editing, K.P. and J.Z. All authors have read and agreed to the published version of the manuscript.

Funding: This study was sponsored by the National Natural Science Foundation of China (NSFC 61474032), the National Defense Basic Scientific Research Program of China under Grant JSZL2018204 B003, and Self-Topic Fund of Engineering Research Center of Electronic Information Materials and Devices Nos. EIMD-AA202007.

Data Availability Statement: All the data supporting reported results can be found in this manuscript.

Conflicts of Interest: The authors declare that they have no conflict of interest.

References

1. Qiu, H.; Hu, X.; Li, S.; Wan, Y.; Li, Q. Shear strength and fracture surface analysis of lead-free solder joints with high fraction of IMCs. *Vacuum* **2020**, *180*, 109611. [CrossRef]
2. Zhao, M.; Zhang, L.; Liu, Z.Q.; Xiong, M.Y.; Sun, L. Structure and properties of Sn-Cu lead-free solders in electronics packaging. *Sci. Technol. Adv. Mater.* **2019**, *20*, 421–444. [CrossRef] [PubMed]
3. Xiong, M.Y.; Zhang, L. Interface reaction and intermetallic compound growth behavior of Sn-Ag-Cu lead-free solder joints on different substrates in electronic packaging. *J. Mater. Sci.* **2018**, *54*, 1741–1768. [CrossRef]
4. Hu, X.; Xu, T.; Keer, L.M.; Li, Y.; Jiang, X. Shear strength and fracture behavior of reflowed Sn3.0Ag0.5Cu/Cu solder joints under various strain rates. *J. Alloys Compd.* **2017**, *690*, 720–729. [CrossRef]
5. Yin, Z.; Sun, F.; Guo, M. Investigation of Elevated Temperature Mechanical Properties of Intermetallic Compounds in the Cu-Sn System Using Nanoindentation. *J. Electron. Packag.* **2020**, *142*, 021004. [CrossRef]
6. Yu, C.F.; Cheng, H.C.; Chen, W.H. Molecular dynamics calculations and nanoindentation testing of the strain-rate and size dependent material properties of Cu_3Sn IMC. In Proceedings of the 2010 5th International Microsystems Packaging Assembly and Circuits Technology Conference, Taipei, Taiwan, 20–22 October 2010. [CrossRef]
7. Xiao, G.S.; Yang, X.X.; Yuan, G.Z.; Li, Z.G.; Shu, X.F. Mechanical properties of intermetallic compounds at the Sn-3.0Ag-0.5Cu/Cu joint interface using nanoindentation. *Mater. Des.* **2015**, *88*, 520–527. [CrossRef]
8. Bi, X.; Hu, X.; Jiang, X.; Li, Q. Effect of Cu additions on mechanical properties of Ni3Sn4-based intermetallic compounds: First-principles calculations and nano-indentation measurements. *Vacuum* **2019**, *164*, 7–14. [CrossRef]
9. Haseeb, A.; Rahman, A.; Chia, P.Y. Nanoindentation creep on Cu_3Sn, Cu_6Sn_5 and $(Cu, Ni)_6Sn_5$ intermetallic compounds grown in electrodeposited multilayered thin film. *J. Mater. Sci. Mater. Electron.* **2018**, *29*, 1258–1263. [CrossRef]
10. Marques, V.M.F.; Johnston, C.; Grant, P.S. Nanomechanical characterization of Sn–Ag–Cu/Cu joints—Part 1: Young's modulus, hardness and deformation mechanisms as a function of temperature. *Acta Mater.* **2013**, *61*, 2460–2470. [CrossRef]

11. Rahman, A.Z.M.S.; Chia, P.Y.; Haseeb, A.S.M.A. Mechanical properties of intermetallic compounds in electrodeposited multilayered thin film at small scale by nanoindentation. *Mater. Lett.* **2015**, *147*, 50–53. [CrossRef]
12. Huang, W.; Pan, K.L.; Zhang, J.; Gong, Y.B. Effect of In-Doping on Mechanical Properties of Cu_6Sn_5-Based Intermetallic Compounds: A First-Principles Study. *J. Electron. Mater.* **2021**, *50*, 4164–4171. [CrossRef]
13. Zhang, W.W.; Ma, Y.; Zhou, W.; Wu, P. The Structural, Elastic and Electronic Properties of Ni3−xCuxSn4 (x = 0, 0.5, 1 and 1.5) Intermetallic Compounds via Ab Initio Calculations. *J. Electron. Mater.* **2019**, *48*, 4533–4543. [CrossRef]
14. Liang, L.; Zhang, J.; Xu, Y.; Zhang, Y.; Wang, W.; Yang, J. The effect of pressure and orientation on Cu-Cu_3Sn interface reliability under isothermal ageing and monotonic traction via molecular dynamics investigation. *Mater. Des.* **2018**, *149*, 194–204. [CrossRef]
15. Liu, B.H.; Chen, Y.L.; Hsu, Q.C. Study on Bonding and Shear Flow Phenomena of Shear Probe Test for BGA Solder Joint in Nano-Scale Analysis. In Proceedings of the ASME 2016 International Mechanical Engineering Congress and Exposition, Phoenix, AZ, USA, 11–17 November 2016. [CrossRef]
16. Cheng, H.C.; Yu, C.F.; Chen, W.H. Size, Temperature, and Strain-Rate Dependence on Tensile Mechanical Behaviors of Ni3Sn4Intermetallic Compound Using Molecular Dynamics Simulation. *J. Nanomater.* **2014**, *2014*, 214510. [CrossRef]
17. Gao, F.; Qu, J. Calculating the diffusivity of Cu and Sn in Cu_3Sn intermetallic by molecular dynamics simulations. *Mater. Lett.* **2012**, *73*, 92–94. [CrossRef]
18. Chen, W.H.; Yu, C.F.; Cheng, H.C.; Lu, S.T. Crystal size and direction dependence of the elastic properties of Cu_3Sn through molecular dynamics simulation and nanoindentation testing. *Microelectron. Reliab.* **2012**, *52*, 1699–1710. [CrossRef]
19. Li, L.H.; Wang, W.L.; Wei, B. First-principle and molecular dynamics calculations for physical properties of Ni-Sn alloy system. *Comput. Mater. Sci.* **2015**, *99*, 274–284. [CrossRef]
20. Choudhury, S.F.; Ladani, L. Local shear stress-strain response of Sn-3.5Ag/Cu solder joint with high fraction of intermetallic compounds: Experimental analysis. *J. Alloys Compd.* **2016**, *680*, 665–676. [CrossRef]
21. Qin, F.; An, T.; Chen, N. Strain Rate Effects and Rate-Dependent Constitutive Models of Lead-Based and Lead-Free Solders. *J. Appl Mech* **2009**, *77*, 011008. [CrossRef]
22. Cheng, H.C.; Yu, C.F.; Chen, W.H. Strain- and strain-rate-dependent mechanical properties and behaviors of Cu_3Sn compound using molecular dynamics simulation. *J. Mater. Sci.* **2012**, *47*, 3103–3114. [CrossRef]
23. Fan, J.T. High-rate squeezing process of bulk metallic glasses. *Sci. Rep.* **2017**, *7*, 45051. [CrossRef]
24. Fan, J.T.; Yang, L.Q. Damage mechanisms of bulk metallic glasses under high-rate compression. *Int. J. Impact. Eng.* **2017**, *106*, 217–222. [CrossRef]
25. Yang, L.; Fan, J.; Nam, V.B.; Rabczuk, T. A nanoscale study of the negative strain rate dependency of the strength of metallic glasses by molecular dynamics simulations. *Phys. Chem. Chem. Phys.* **2018**, *20*, 26552–26557. [CrossRef] [PubMed]
26. Plimpton, S. Fast Parallel Algorithms for Short-range Molecular-Dynamics. *J. Comput. Phys.* **1995**, *117*, 1–19. [CrossRef]
27. Baskes, M.I. Modified Embedded-Atom Potentials for Cubic Materials and Impurities. *Phys. Rev. B* **1992**, *46*, 2727–2742. [CrossRef]
28. Larsson, A.K.; Stenberg, L.; Lidin, S. The superstructure of domain-twinned η'-Cu_6Sn_5. *Acta Crystallogr. Sect. B* **1994**, *50*, 636–643. [CrossRef]
29. Aguilar, J.F.; Ravelo, R.; Baskes, M.I. Morphology and dynamics of 2D Sn-Cu alloys on (100) and (111) Cu surfaces. *Modell. Simul. Mater. Sci. Eng.* **2000**, *8*, 335–344. [CrossRef]
30. Hirel, P. Atomsk: A tool for manipulating and converting atomic data files. *Comput. Phys. Commun.* **2015**, *197*, 212–219. [CrossRef]
31. Lee, N.T.S.; Tan, V.B.C.; Lim, K.M. First-principles calculations of structural and mechanical properties of Cu_6Sn_5. *Appl. Phys. Lett.* **2006**, 88. [CrossRef]
32. Ghosh, G.; Asta, M. Phase stability, phase transformations, and elastic properties of Cu_6Sn_5: Ab initio calculations and experimental results. *J. Mater. Res.* **2005**, *20*, 3102–3117. [CrossRef]
33. Thomas, R.; Yooseob, S.; Wenbin, W.; Shruti, R.; George, Z.V.; Jeffrey, W.K. Order in polycrystalline plasticity deformation fields: Short-range intermittency and long-range persistency. *Int. J. Plast.* **2020**, *128*, 102674. [CrossRef]
34. Ikeda, H.; Qi, Y.; Cagin, T.; Samwer, K.; Johnson, W.L.; Goddard, W.A. Strain rate induced amorphization in metallic nanowires. *Phys. Rev. Lett.* **1999**, *82*, 2900–2903. [CrossRef]
35. Zhou, M. A new look at the atomic level virial stress: On continuum-molecular system equivalence. *Proc. Math. Phys. Eng. Sci.* **2003**, *459*, 2347–2392. [CrossRef]

crystals

MDPI

Article

Rolling Texture of Cu–30%Zn Alloy Using Taylor Model Based on Twinning and Coplanar Slip

Shih-Chieh Hsiao, Sin-Ying Lin, Huang-Jun Chen, Ping-Yin Hsieh and Jui-Chao Kuo *

Department of Materials Science and Engineering, National Cheng Kung University, Tainan 70101, Taiwan; sshawnnccatt@hotmail.com (S.-C.H.); annie30207@gmail.com (S.-Y.L.); ironcobra@livemail.tw (H.-J.C.); freeonmaxs@yahoo.com.tw (P.-Y.H.)

* Correspondence: jckuo@mail.ncku.edu.tw

Abstract: A modified Taylor model, hereafter referred to as the MTCS (Mechanical-Twinning-with-Coplanar-Slip)-model, is proposed in the present work to predict weak texture components in the shear bands of brass-type fcc metals with a twin–matrix lamellar (TML) structure. The MTCS-model considers two boundary conditions (i.e., twinning does not occur in previously twinned areas and coplanar slip occurs in the TML region) to simulate the rolling texture of Cu–30%Zn. In the first approximation, texture simulation using the MTCS-model revealed brass-type textures, including Y{1 1 1} <1 1 2> and Z{1 1 1} <1 1 0> components, which correspond to the observed experimental textures. Single orientations of C(1 1 2)$[\bar{1}\,\bar{1}\,1]$ and S′(1 2 3)$[4\,\bar{1}\,2]$ were applied to the MTCS-model to understand the evolution of Y and Z components. For the Y orientation, the C orientation rotates toward T(5 5 2)[1 1 5] by twinning after 30% reduction and then toward Y(1 1 1)[1 1 2] by coplanar slip after over 30% reduction. For the Z orientation, the S′ orientation rotates toward T′(3 2 1)$[2\,\bar{1}\,\bar{4}]$ by twinning after 30% reduction and then toward Z(1 1 1)$[1\,0\,\bar{1}]$ by coplanar slip after over 30% reduction.

Keywords: brass-type shear band; twin–matrix lamellae; coplanar slip; Taylor model; Cu–Zn alloy; cold-rolling texture; X-ray diffraction

Citation: Hsiao, S.-C.; Lin, S.-Y.; Chen, H.-J.; Hsieh, P.-Y.; Kuo, J.-C. Rolling Texture of Cu–30%Zn Alloy Using Taylor Model Based on Twinning and Coplanar Slip. *Crystals* **2021**, *11*, 1351. https://doi.org/10.3390/cryst11111351

Academic Editor: Wojciech Polkowski

Received: 27 October 2021
Accepted: 5 November 2021
Published: 7 November 2021

1. Introduction

In the industry applications, the formability of metals plays a significant role, which mainly depends on the crystallographic texture and microstructure [1]. Moderate-strain metals and alloys including copper exhibit laminar microstructures called shear bands (SBs), a form of plastic instability. Duggan et al. [2] and Fargette et al. [3] investigated the formation mechanism of SBs in Cu–Zn alloys. In rolled metals, the SBs form as thin planar sheets that are parallel to the transverse direction (TD) and inclined at ~35° to the rolling direction (RD). Brass-type SBs and copper-type SBs have been found in low stacking fault energy (SFE) and intermediate to high SFE materials [2,4], respectively. Hatherly and Malin [5] defined low SFE as <20 mJ/m^2, intermediate SFE between 20 and 40 mJ/m^2 and high SFE as >40 mJ/m^2. Considering the low stacking fault energy (SFE) of fcc metals, the SBs exhibit the structure of twin–matrix lamellae (TML) composed of twin and matrix lamella layers. The TML structure was reported in the observation of deformed Cu–30%Zn by Duggan et al. [2] and Fargette et al. [3]. Malin and Hatherly [6] reported the TML structure in pure copper. An abnormal slip system parallel to twinning planes also occurs in the lamellar layer structure [2].

After large thickness reductions for α-brass, the SBs substantially change the texture from copper-type to brass-type because of the formation of the fine TML structure [2]. Wassermann et al. [7] reported that cold-rolled α-brass with C{1 1 2} <1 1 1>-oriented grains exhibit twin orientation T{5 5 2} <1 1 5> after twinning; consequently, the T-oriented grains rotate toward the B{1 1 0} <1 1 2> orientation by dislocation slip. Hirsch et al. [4] found that the T-oriented grains rotate to the transition orientation of {1 1 1} <1 1 2> (brass^R) due to mechanical twinning (MT) after small reduction, and the brass[R] grains rotate to the G {1 1 0} <0 0 1> orientation through shear banding after large reductions. Paul et al. [8,9] studied low-SFE fcc single crystals initially oriented with copper and revealed that the two coplanar slip systems adjusted by the initiation of shear banding play an important role in the formation of brass-type textures.

Sevillano et al. [10] and van Houtte et al. [11,12] predicted copper-type texture using full constraint (FC)- and relaxed constraint (RC)-Taylor models, respectively. Leffers [13,14], Hirsch and Lücke [15] and van Houtte [16] estimated the brass-type texture using various Taylor models that consider MT. According to Chin's study [17], Kallend [18] and van Houtte introduced MT into FC- [16] and RC-Taylor models [11,12]. Hirsch et al. [15] quantitatively compared the rolling texture between FC and RC Taylor models. Leffers [19,20] used a modified Sachs model to predict brass-type texture. Kalidindi [21] proposed a crystal plasticity model considering deformation twinning and observed that twinning is difficult to occur in the twinned regions. In addition, Kalidindi [22] modelled with shear banding to predict the texture transition from Cu-type to brass-type. Lebensohn and Tomé [23,24] utilized the VPSC model to simulate brass-type texture including comprehensive relative activity of slip and twin systems. Toth et al. [25] used a Taylor version of the VPSC model considering dislocation slip and twinning to simulate the deformed texture of TWIP Steel with fcc structure. Chalapathi et al. [26] proposed a modified LAMEL model to simulate the rolling texture of an fcc steel. Among all models, the present work aimed to incorporate the experimental observations of TML [27–31] in the Taylor model while considering MT to predict the rolling texture of Cu–30%Zn. The modified Taylor model was compared with conventional Taylor models such as FC and RC Taylor models in terms of rolling texture.

2. Materials and Methods

The dimensions of the as-received Cu–30%Zn alloy were reduced to 60.0 × 20.0 × 20.0 mm^3 by using an abrasive cutting machine and annealed at 600 °C for 1 h to homogenize, and then cold-rolled up to 90% thickness reduction. A two-high non-reversing mill with roll diameter of 590 mm was set with rolling speed of 9 rpm to conduct the cold rolling experiment. Drops of lubricant oil were applied on the rolls to reduce to friction and heat during rolling. The alloy was rolled 10 times to reduce the thickness from 20.0 to 14.0 mm (30% reduction). Then, 1/3 length of the alloy was cut off and continued to roll 12 times to reduce the thickness from 14.0 to 8.0 mm (60% reduction). Finally, 1/2 length of the alloy was cut off and continued to roll 15 times to reduce the thickness from 8.0 to 2.0 mm (90% reduction). The middle regions of the rolled materials were selected and cut out in dimensions of 20.0 mm width and 20.0 mm length for further texture analysis. The specimens were prepared by grinding using #400, #800, #1500, #2500, and #4000 SiC papers.

The texture of Cu–30%Zn alloy was examined on the surface parallel to the out-of-plane direction called ND direction by using Bruker (Germany) D8 ADVANCE in NCKU with CuK$_\alpha$ radiation of λ = 1.5406 Å at 40 kV and 40 mA. Three incomplete pole figures of {1 1 1}, {2 0 0}, and {2 2 0} were recorded by varying the tilting angle of 0°–70° and the rotation angle of 0°–360° with a scanning step of 5°. Defocusing correction was then employed on the measurement of random powder Cu–30%Zn alloy. Orientation distribution function (ODF) and complete {1 1 1} pole figure were calculated using LaboSoft (Poland) LaboTex ver.3.0 software in NCKU.

3. Modelling

The rolling texture of Cu–30%Zn alloy was modelled using modified Taylor models in the Matlab software. FC-Taylor model, RC-Taylor model, RC-Taylor model considering MT called MT-model, and RC-Taylor model considering MT and coplanar slip called MTCS-model were constructed as follows.

3.1. FC-Model

According to the method of van Houtte et al. [12,32] for establishing the FC-Taylor model, a given displacement gradient e_{ij} in a grain is composed of a symmetric matrix ε_{ij} (called the strain tensor) and an antisymmetric matrix ω_{ij} (called the rotation tensor). Here, the given strain tensor ε_{ij} of 90% reduction (corresponding true strain, $\left|In\frac{2.0}{20.0}\right| = 2.3$) is expressed in the macroscopic frame for plane strain condition as follows:

$$\varepsilon_{ij} = \begin{pmatrix} 2.3 & 0 & 0 \\ 0 & 0 & 0 \\ 0 & 0 & -2.3 \end{pmatrix} \quad (1)$$

An incremental strain of 0.01 was set for each step in the simulation. In the crystal frame, the symmetric matrix ε^{S}_{ij}, which relates the shear on the slip system s, is described for the slip systems of {1 1 1} <1 1 0> by:

$$\varepsilon^{s}_{ij} = \sum_{s=1}^{n^s} \frac{\gamma^s}{2}(b^s \otimes m^s + m^s \otimes b^s) \text{ and} \quad (2)$$

$$\varepsilon_{ij} = \varepsilon^{s}_{ij} \quad (3)$$

where, b^s, m^s and γ^s denote the direction of the Burgers vector, the normal to the slip plane, and the shear on the slip system s, respectively, in the crystal frame. The symbol $\otimes$ denotes the dyadic product of two vectors. The antisymmetric matrix ω^{S}_{ij} is expressed by:

$$\omega^{s}_{ij} = \sum_{s=1}^{n^s} \frac{\gamma^s}{2}(b^s \otimes m^s - m^s \otimes b^s) \quad (4)$$

and

$$\omega_{ij} = \omega^{s}_{ij} + \Omega^s \quad (5)$$

where Ω^s is the lattice rotation in the macroscopic frame.

Five linear equations are needed to solve Equation (2) with 792 combinations for a given strain. According to the Taylor assumption [33], the minimum work corresponds to the minimum sum of the activated five absolute shears and is expressed as:

$$w = \sum_{s=1}^{n^s} \tau^{s}_{c}|\gamma^s| \quad (6)$$

The critical resolved shear stress is denoted as τ^{s}_{c} for all 12 slip systems, and the number of the activated slip systems is $n^s = 5$. The lattice rotation Ω^s in the macroscopic frame can be obtained using Equation (5). As a result, the new orientation matrix g^* is expressed by:

$$g^* = (I - \Omega^s)g \quad (7)$$

where g is the initial orientation matrix before deformation.

3.2. RC-Model

After high reduction, the deformation texture shows discrepancies from that predicted by FC-Taylor model. Van Houtte [11] proposed the concept of partly constrained deformation of crystallites and labelled it as RC-model because of the observation that grains become flattened and elongated after high rolling reduction. In the RC-Taylor model, only four slip systems are activated. Hence, the relaxed constraint of ε_{XZ} shear strain can be calculated in Equation (1), where the X direction is the rolling direction and Z the normal direction, and the number of the activated slip system is $n^s = 4$ in Equations (2) and (4).

The given strain tensor ε_{ij} is expressed in the macroscopic frame for plane strain condition as follows:

$$\varepsilon_{ij} = \begin{pmatrix} 2.3 & 0 & \varepsilon_{XZ} \\ 0 & 0 & 0 \\ \varepsilon_{XZ} & 0 & -2.3 \end{pmatrix} \tag{8}$$

where the ε_{XZ} shear strain is unconstrained. Equations (2)–(7) are the same in the case of the FC-Taylor model.

3.3. MT-Model

Following their observation of MT in low-SFE metals, van Houtte [16] and Chin et al. [17] proposed a modified Taylor model that assumes that the shear in the crystal frame is due to the MT in {1 1 1} <1 1 2> twin systems for FC- and RC-models and not {1 1 1} <1 1 0> slip systems. Following the concept of van Houtte [16], the present study employed an RC-model combining three slip systems and one twin system. This model is referred to as the MT-model and differs from the RC-model in that the latter considers four slip systems. The flowchart of the subroutine is shown in Figure 1.

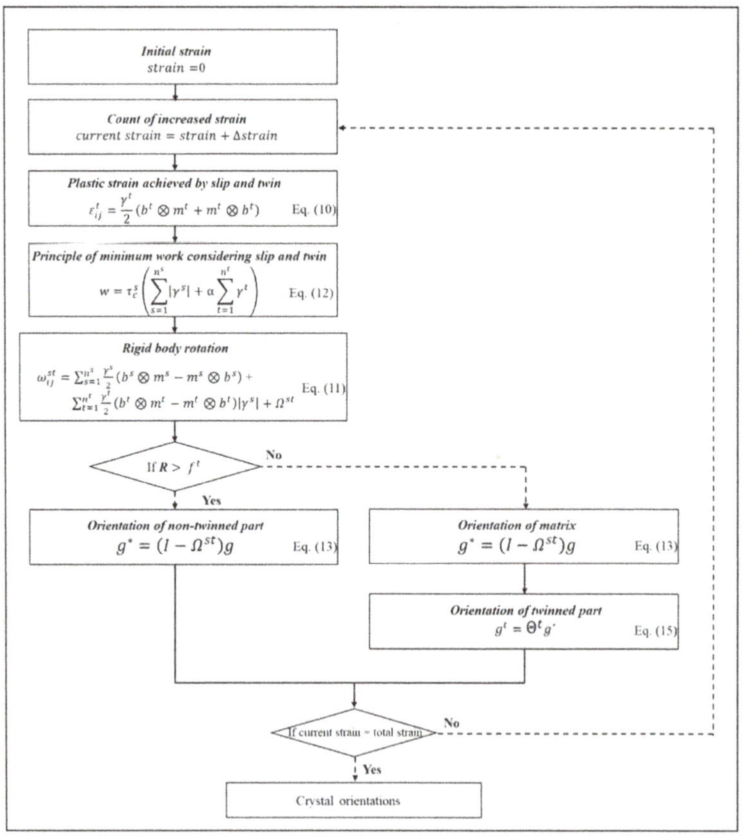

Figure 1. Flowchart of subroutine in MT-model.

In the MT-model, the strain tensor resulting from three slip systems and one twinning system is expressed in the crystal frame as follows:

$$\varepsilon_{ij}^{st} = \sum_{s=1}^{n^s} \frac{\gamma^s}{2}\left(b^s \otimes m^s + m^s \otimes b^s\right) + \sum_{t=1}^{n^t} \frac{\gamma^t}{2}\left(b^t \otimes m^t + m^t \otimes b^t\right) \tag{9}$$

where the numbers of the activated slip and twin systems are $n^s = 3$ and $n^t = 1$, respectively. The former and latter parts of the symmetric matrix corresponding to the shear on the slip systems of {1 1 1} <1 1 0> in Equation (2) and the twin systems of {1 1 1} <1 1 2> are expressed as:

$$\varepsilon_{ij}^{t} = \frac{\gamma^t}{2}\left(b^t \otimes m^t + m^t \otimes b^t\right) \tag{10}$$

where b^t, m^t, and γ^t denote the direction of the Burgers vector, the normal to the twin plane, and the shear on the twin system t, respectively. The slip and twin systems used in the models are listed in Table 1. The rotational antisymmetric matrix in the crystal frame is expressed as:

$$\omega_{ij}^{st} = \sum_{s=1}^{n^s} \frac{\gamma^s}{2}\left(b^s \otimes m^s - m^s \otimes b^s\right) + \sum_{t=1}^{n^t} \frac{\gamma^t}{2}\left(b^t \otimes m^t - m^t \otimes b^t\right)|\gamma^s| + \Omega^{st} \tag{11}$$

where Ω^{st} is the lattice rotation in the macroscopic frame.

Table 1. Twelve slip and twin systems of fcc metals used in the models.

Slip System	Slip Plane	Slip Direction	Twin System	Twin Plane	Twin Direction
SS1	$(1\,1\,1)$	$[0\,1\,\bar{1}]$	TS1	$(1\,1\,1)$	$[1\,1\,\bar{2}]$
SS2	$(1\,1\,1)$	$[\bar{1}\,0\,1]$	TS2	$(1\,1\,1)$	$[\bar{2}\,1\,1]$
SS3	$(1\,1\,1)$	$[1\,\bar{1}\,0]$	TS3	$(1\,1\,1)$	$[1\,\bar{2}\,1]$
SS4	$(\bar{1}\,\bar{1}\,1)$	$[0\,1\,1]$	TS4	$(\bar{1}\,\bar{1}\,1)$	$[1\,1\,2]$
SS5	$(\bar{1}\,\bar{1}\,1)$	$[1\,0\,1]$	TS5	$(\bar{1}\,\bar{1}\,1)$	$[2\,\bar{1}\,1]$
SS6	$(\bar{1}\,\bar{1}\,1)$	$[\bar{1}\,1\,0]$	TS6	$(\bar{1}\,\bar{1}\,1)$	$[\bar{1}\,2\,1]$
SS7	$(\bar{1}\,1\,1)$	$[0\,1\,\bar{1}]$	TS7	$(\bar{1}\,1\,1)$	$[\bar{1}\,1\,\bar{2}]$
SS8	$(\bar{1}\,1\,1)$	$[1\,0\,1]$	TS8	$(\bar{1}\,1\,1)$	$[2\,1\,1]$
SS9	$(\bar{1}\,1\,1)$	$[1\,1\,0]$	TS9	$(\bar{1}\,1\,1)$	$[\bar{1}\,\bar{2}\,1]$
SS10	$(1\,\bar{1}\,1)$	$[0\,1\,1]$	TS10	$(1\,\bar{1}\,1)$	$[1\,\bar{1}\,\bar{2}]$
SS11	$(1\,\bar{1}\,1)$	$[\bar{1}\,0\,1]$	TS11	$(1\,\bar{1}\,1)$	$[\bar{2}\,\bar{1}\,1]$
SS12	$(1\,\bar{1}\,1)$	$[1\,1\,0]$	TS12	$(1\,\bar{1}\,1)$	$[1\,2\,1]$

Considering the contributions of slip and twinning, the minimum work of the MT-model is expressed in terms of $\alpha = \frac{\tau_c^t}{\tau_c^s}$ as follows:

$$w = \tau_c^s\left(\sum_{s=1}^{n^s}|\gamma^s| + \alpha\sum_{t=1}^{n^t}\gamma^t\right) \tag{12}$$

where the numbers of the activated slip and twinning systems are $n^s = 3$ and $n^t = 1$, respectively, and the critical resolved shear stresses for twinning and slip in all 12 twin and slip systems are denoted τ_c^t and τ_c^s, respectively. The CRSS values of the slip and twin systems are assumed to be identical, that is, $\alpha = 1$. Thus, the new orientation of matrix g^* is expressed by:

$$g^* = \left(I - \Omega^{st}\right)g \tag{13}$$

where g is the initial orientation matrix prior to deformation.

After twinning deformation, the orientation number of the fine TML structure increases twofold after each simulation step, which leads to increases in computation time.

To address the problem of time-consuming computations, van Houtte assumed a simplified method with only one orientation; here, either the matrix orientation or the twin orientation is selected as the new orientation of the matrix and twin area in the TML region. This orientation depends on the relation between a random number R ranging from 0 to 1 and the volume fraction of the twin region f^t, which is expressed as:

$$f^t = \frac{\gamma^t}{\gamma_0} \quad (14)$$

where the constant of the twinning shear of fcc metals is denoted $\gamma_0 = \frac{1}{\sqrt{2}}$. If R is greater than f^t, then the new orientation is determined by Equation (11). If $R < f^t$, then the new orientation g^* is given by Equation (11). The twinned orientation g^t obtained after twinning is given by Equation (13):

$$g^t = \Theta^t g^* \quad (15)$$

where g^* is the new matrix orientation and Θ^t is a matrix that transforms the matrix orientation into a twin orientation, which is expressed as follows:

$$\Theta^t = 2m^t \otimes m^{t^T} - I \quad (16)$$

where T denotes the transpose of the matrix m^t.

3.4. *MTCS-Model*

Considering the coplanar slip in the TML region reported by Hirsch et al. [4,34], the current work presents another modified Taylor model, hereafter referred to as the MTCS-model, that combines the MT-model with the concept of coplanar slip in the TML region. A major difference between the MTCS- and MT-models is the addition of two assumptions resulting in different textures in the former. The first assumption in the MTCS-model is that further twinning is forbidden in a priori twinned regions. The second assumption in this model is that the deformation of twinned grains contributes to two coplanar slip systems. The former assumption is based on the perspective that twinning cannot easily occur in previously twinned areas, as reported by Kalidindi [21]; in other words, secondary or further twinning is forbidden in previous twinning areas. The latter assumption is based on the coplanar slip observed in the TML region by Hirsch et al. [4]. The coplanar slip forms on the plane of activated twin systems. The two other slip systems were selected from non-coplanar slip systems. Therefore, the key difference between the MT- and MTCS-models lies in changes in the twinned orientation.

Following the procedures of the MT-model, we calculate rigid body rotation for the MTCS-model by taking the plastic strain and minimum work into account using Equations (10)–(12), as shown in Figure 2. After each deformation step, a new orientation is determined at random by selecting a number R between 0 and 1. Here the possibility of twinning is equal to the fraction of the twinning area f^t. When $R \leq f^t$, the new orientation is determined by twinning by using Equations (7) and (15); when $R > f^t$, the new orientation is calculated by applying Equation (7) because of the deformation of the slip, as shown in Figure 2. This procedure for orientation determination leads to indicating the twinned and non-twinned orientations.

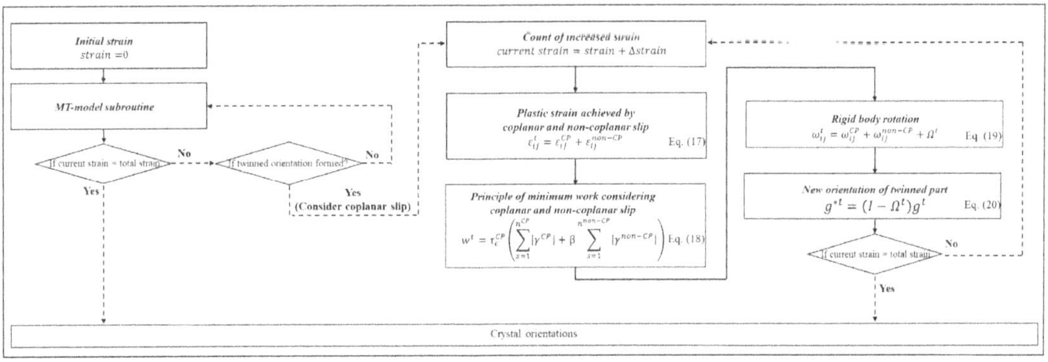

Figure 2. Flowchart of subroutine in MTCS-model.

The two assumptions are then implemented in the MTSC-model. The first assumption is that secondary twinning, that is, further twinning in a previously twinned orientation, is excluded. For the non-twinned orientation indicated, rigid body rotation is calculated by considering the plastic strain and minimum work by using Equations (8)–(10). However, the twinned orientation indicated does not change according to Equation (13) but follows the right route in Figure 2 to avoid the formation of secondary twinning. This phenomenon corresponds to the assumption that the twinning orientation occurs only once. The second assumption is that the coplanar slip occurring in the TML region is implemented in the reorientation calculation of the twinned orientation. The activated systems in twinned orientations occur on twinning planes called coplanar slip systems at the first twinning. Thus, the 12 slip systems used in the models could be classified into coplanar and non-coplanar slip systems on the basis of the twinning planes. Therefore, two activated slip systems are selected from the coplanar slip systems, and another two slip systems are selected from the non-coplanar slip systems by using the RC-model. Thus, in addition to non-coplanar slip, coplanar slip can contribute to the plastic strain of twinned orientations via the relation:

$$\varepsilon_{ij}^{t} = \varepsilon_{ij}^{CP} + \varepsilon_{ij}^{non-CP} \tag{17}$$

where ε_{ij}^{CP} and $\varepsilon_{ij}^{non-CP}$ are the plastic strains resulting from the coplanar slip and non-coplanar slip, respectively.

Considering the contribution of coplanar and non-coplanar slips, the optimization of the minimum work in the MTCS-model is expressed in terms of $\beta = \frac{\tau_c^{non-CP}}{\tau_c^{CP}}$ as:

$$w^{t} = \tau_c^{CP}\left(\sum^{n^{CP}}_{s=1}\left|\gamma^{CP}\right| + \beta\sum^{n^{non-CP}}_{s=1}\left|\gamma^{non-CP}\right|\right) \tag{18}$$

where the numbers of the activated slip and twinning systems are $n^{CP} = 2$ and $n^{non-CP} = 2$, respectively, and the critical resolved shear stresses for the coplanar and non-coplanar slip systems are denoted τ_c^{CP} and τ_c^{non-CP}, respectively. The CRSS values of the slip and twin systems are assumed to be identical, that is, $\beta = 1$. The antisymmetric matrix ω_{ij}^{t} is expressed as:

$$\omega_{ij}^{t} = \omega_{ij}^{CP} + \omega_{ij}^{non-CP} + \Omega^{t} \tag{19}$$

where Ω^{t} is the lattice rotation in the macroscopic frame.

Thus, the new orientation of matrix g^{*t} is expressed as:

$$g^{*t} = (I - \Omega^{t})g^{t} \tag{20}$$

where g^{t} is the twinned orientation matrix before deformation resulting from the coplanar slip.

For the four simulation models, the orientation number of grains is 5000, each strain step is 0.01, the total strain is 90% reduction, and the 5000 grains initially show random orientations. The simulated textures were analyzed, including ODF, complete {1 1 1} pole figures, and volume fraction by using LaboTex software. The Euler angles follow the definition of Bunge.

Single crystal C$(1\,1\,2)[\bar{1}\,\bar{1}\,1]$ and S′$(1\,2\,3)[\bar{4}\,\bar{1}\,2]$ orientations were also employed for the initial orientations in MTCS-model to predict the components of Y and Z, respectively.

4. Results and Discussion

This section is divided by subheadings. It provides a concise and precise description of the experimental results, their interpretation, as well as the experimental conclusions that can be drawn.

4.1. Effect of Twinning on the Rolling Texture

FC-, RC-, MT- and MTCS-models were applied to understand the influence of twinning on the rolling texture. The shear mechanism of FC- and RC-models is related to pure slip, and that of MT- and MTCS-models is related to a combination of partial slip and twin. The {1 1 1} pole figures of FC-, RC-, MT-, and MTCS-models are shown in Figure 3a–e, presenting the {1 1 1} pole figure of Cu–30%Zn obtained from XRD measurements after 90% cold rolling. The {1 1 1} pole figures of FC- and RC-models belong to a typical copper texture, and those of MT and experiment show a typical brass texture. In addition, the {1 1 1} pole figure of MTCS-model indicates a combination of copper and brass textures.

ODFs were obtained from simulation and experiment to quantify the orientation distributions of four models. As shown in Figure 4, the blue symbols indicate the orientations of C{1 1 2} <1 1 1>, S{1 2 3} <6 3 4>, B{1 1 0} <1 1 2>, and G{1 1 0} <0 0 1>, and the red symbols indicate the orientations of T{2 5 5} <5 1 1>, Y{1 1 1} <1 1 2>, and Z{1 1 1} <1 1 0>. Table 2 shows the preferred orientations simulated by the four models. The FC- and RC-models predict a β-fiber after cold rolling in Figure 4a,b, the MT-model shows an α-fiber, and the MTCS-model exhibits a combination of α- and β-fibers in Figure 4c,d. Comparison of pole figures and ODFs revealed that the FC- and RC-models show β-fiber, the MT-model reveals α-fiber, and the MTCS-model indicates α-and β-fibers.

The relative activities of the slip and twin systems were analyzed. Figure 5 shows the relative activity of the systems as a function of deformation strain for the slip or twin systems considered in the MT and MTCS models. The relative activities of slip and twin systems for each step of simulation are determined using Equations (21) and (22), respectively [28].

In the case of the RC-model, only slip systems were activated, that is, the activity of slip has a constant value of 1. In addition, the activity of the 12 slip systems generally ranges from 8% to 9% throughout the simulation.

In the case of the MT-model, the activity of the twin system calculated from the simulation results is larger than that of the slip system at $\varepsilon < 0.2$; by contrast, the activity of the slip system is greater than that of the twin system at $\varepsilon > 0.2$. These results indicate that the deformation mechanism of the MT-model is dominated by twinning at $\varepsilon < 0.2$. Under the condition of $\varepsilon < 0.2$, the deformation mechanism of the MT-model is dominated by the twinning of slip system TS1$(1\,1\,1)[1\,1\,\bar{2}]$. The deformation mechanism of the MT-model is dominated by the slipping of slip systems SS4$(\bar{1}\,\bar{1}\,1)[0\,1\,1]$ and SS5$(\bar{1}\,\bar{1}\,1)[1\,0\,1]$ at $\varepsilon > 0.2$ and by the slipping of slip systems SS1$(1\,1\,1)[0\,1\,\bar{1}]$ and SS2$(1\,1\,1)[\bar{1}\,0\,1]$ at $\varepsilon > 0.7$. At $\varepsilon > 0.7$, the activities of the slip and twin systems are stabilized in the MT-model. Overall, in the case of the MT-model, the respective activated shear fractions of the dominant slip systems of SS4$(\bar{1}\,\bar{1}\,1)[0\,1\,1]$ and SS5$(\bar{1}\,\bar{1}\,1)[1\,0\,1]$ are 24.5% and 24.5% of the slip and twin contribution, corresponding to 27.9% and 27.9% of the slip contribution, respectively. The activated shear fraction of the dominant twin system of TS1$(1\,1\,1)[1\,1\,\bar{2}]$ is 12.3% of the slip and twin contribution, which corresponds to 99.9% of the twin contribution.

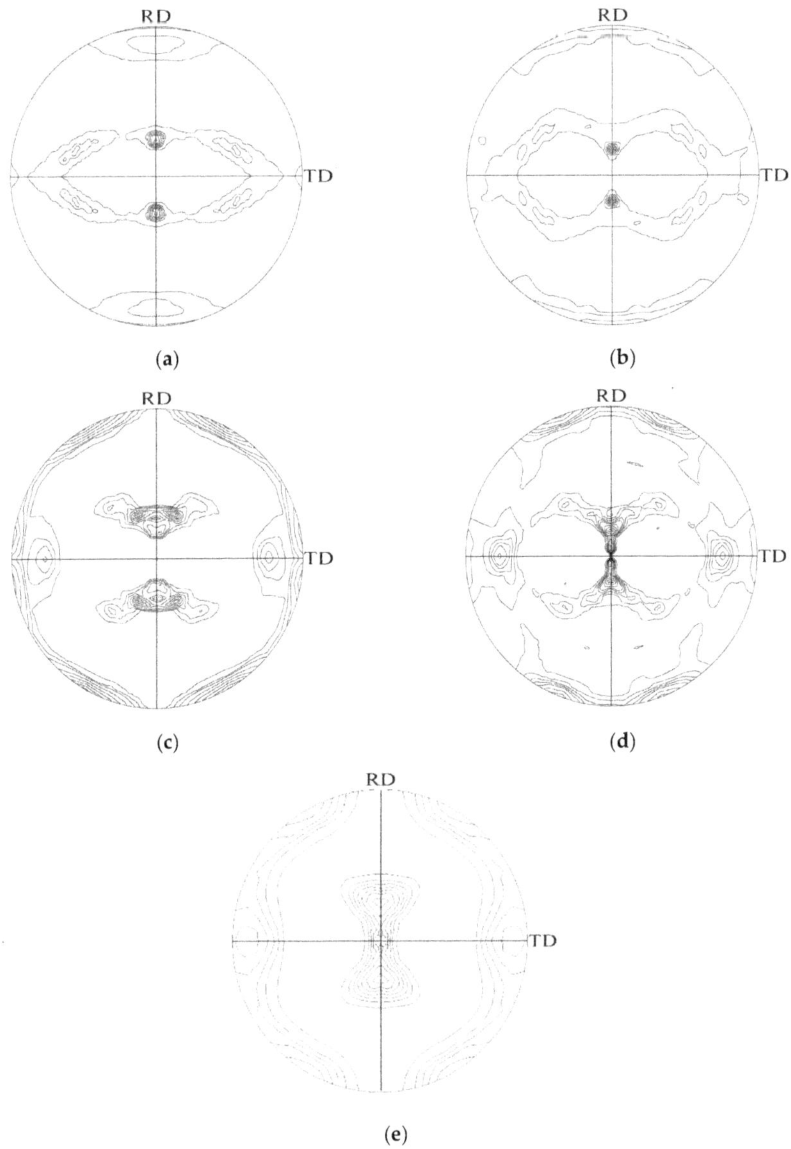

Figure 3. The {1 1 1} pole figures of fcc metals simulated using (**a**) FC-, (**b**) RC-, (**c**) MT-, and (**d**) MTCS-models with random orientations up to 90% reduction and (**e**) after 90% cold-rolled in Cu–30%Zn alloy obtained from XRD measurements.

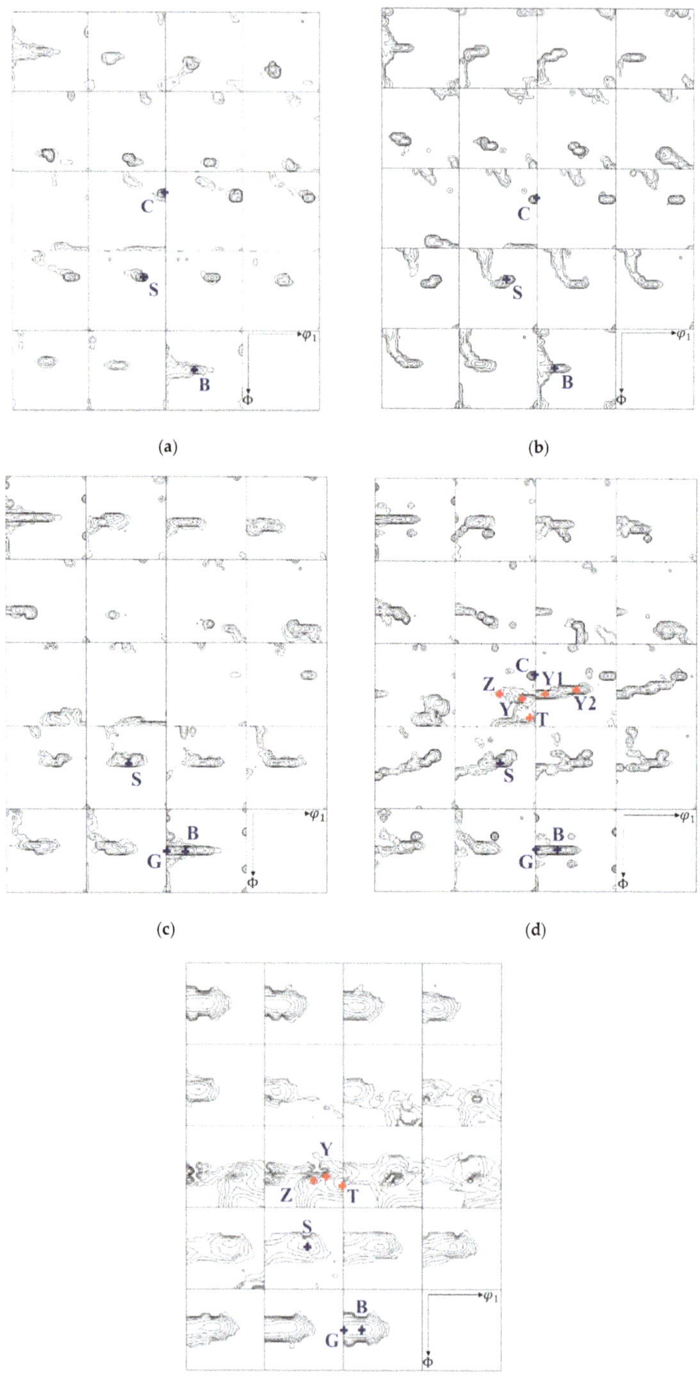

Figure 4. ODFs of fcc metals simulated using (**a**) FC-, (**b**) RC-, (**c**) MT-, and (**d**) MTCS-models with random orientations to 90% reduction. (**e**) ODF of 90% cold-rolled Cu–30%Zn from XRD measurements. The letters indicate specific orientations.

Table 2. Major orientation components predicted with random orientations using FC-, RC-, MT-, and MTCS-models after 90% reduction.

Reduction	Model	Preferred Orientation					
		C	S	B	T	Y	Z
90%	FC	(90.0, 24.9, 45.0) ≈ $(1\,1\,3)[\bar{3}\,\bar{3}\,2]$	(64.9, 30.0, 65.0) ≈ $(2\,1\,4)[\bar{6}\,\bar{8}\,5]$	(34.6, 45.2, 0.0) ≈ $(1\,0\,1)[\bar{1}\,\bar{2}\,1]$	-	-	-
	RC	(90.0, 35.1, 45.0) ≈ $(1\,1\,2)[\bar{1}\,\bar{1}\,1]$	(54.9, 35.1, 65.0) ≈ $(6\,3\,10)[\bar{7}\,\overline{16}\,9]$	(19.8, 45.1, 90.0) ≈ $(1\,0\,1)[\bar{1}\,\bar{4}\,1]$	-	-	-
	MT	-	(50.1, 40.0, 65.0) ≈ $(6\,3\,8)[\bar{3}\,\overline{10}\,6]$	(24.8, 45.0, 90.0) ≈ $(1\,0\,1)[\bar{1}\,\bar{3}\,1]$	-	-	-
	MTCS	(90.0, 35.2, 45.0) ≈ $(1\,1\,2)[\bar{1}\,\bar{1}\,1]$	(50.8, 40.0, 65.0) ≈ $(11\,5\,14)[\bar{9}\,\overline{25}\,16]$	(25.2, 45.1, 90.0) ≈ $(1\,0\,1)[\bar{1}\,\bar{3}\,1]$	(85.1, 80.1, 45.0) ≈ $(3\,5\,1)[0\,\bar{1}\,5]$	(75.4, 60.2, 45.0) ≈ $(5\,5\,4)[\bar{1}\,\bar{3}\,5]$	(49.8, 55.1, 45.0) ≈ $(1\,1\,1)[2\,\overline{11}\,9]$

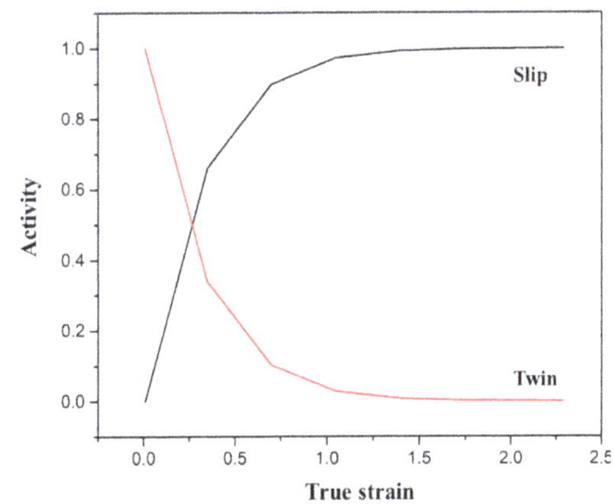

(**a**)

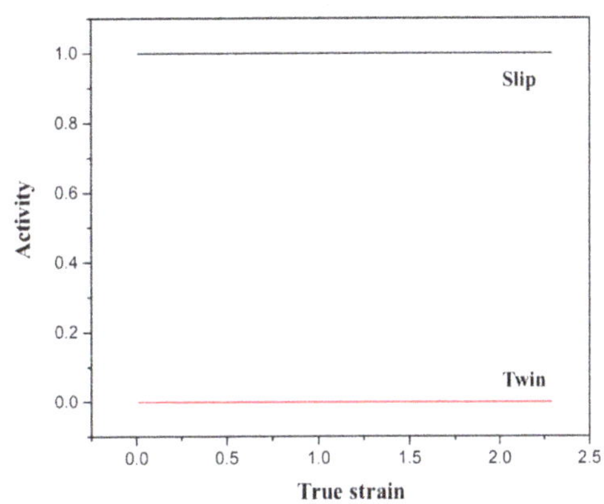

(**b**)

Figure 5. The activity plot of the slip and twin systems for (**a**) MT-model and (**b**) MTCS-model.

$$Relative\ activity\ of\ slip = \frac{\sum_{k=1}^{5000}\left(\sum_{i=1}^{12}\left|\gamma_{ik}^{s}\right|\right)}{\sum_{k=1}^{5000}\left(\sum_{i=1}^{12}\left|\gamma_{ik}^{s}\right| + \sum_{j=1}^{12}\left|\gamma_{jk}^{t}\right|\right)} \quad (21)$$

$$Relative\ activity\ of\ twin = \frac{\sum_{k=1}^{5000}\left(\sum_{j=1}^{12}\left|\gamma_{jk}^{t}\right|\right)}{\sum_{k=1}^{5000}\left(\sum_{i=1}^{12}\left|\gamma_{ik}^{s}\right| + \sum_{j=1}^{12}\left|\gamma_{jk}^{t}\right|\right)} \quad (22)$$

The deformation mechanism of the MTCS-model is dominated by the slipping of slip systems SS1$(1\ 1\ 1)\left[0\ 1\ \overline{1}\right]$ and SS2$(1\ 1\ 1)\left[\overline{1}\ 0\ 1\right]$, which show similarly stable activities of approximately 50% throughout the simulation. The activity of twin systems is close to zero throughout the simulation (i.e., ~10^{-18}) because of the model's assumption of restricted secondary twinning (i.e., twinning may occur only once). Madhavan et al. [35] reported that the evolution of Cu-type rolling textures after up to 95% reduction may be completely attributed to slip. Overall, in the case of the MTCS-model, the activated shear fractions of the dominant slip systems of SS1$(1\ 1\ 1)\left[0\ 1\ \overline{1}\right]$ and SS2$(1\ 1\ 1)\left[\overline{1}\ 0\ 1\right]$ are approximately 50% and 50%, respectively, of the slip and twin contributions; these values correspond to approximately 50% of the slip contribution. The activated shear fraction of the dominant twin system of TS11$(1\ \overline{1}\ 1)\left[\overline{2}\ \overline{1}\ 1\right]$ is 0% of the slip and twin contribution, which corresponds to 33.2% of the twin contribution.

The volume fraction of major components was calculated to quantify the orientation components of the four models. Given their pure slip mechanism, the volume fractions of C and S orientations are 15.52% and 11.81% for FC-model and 13.07% and 12.56% for RC-model, as shown in Figure 6a,b. In both cases of FC- and RC- models, the volume fraction of B and G components are relatively low because twinning mechanism was not considered. When considering partial slip and twinning for the MT-model, the volume fraction of C and S are reduced to 0.75% and 8.04%, while the volume fraction of brass-type components, B and G components, increase. Furthermore, taking the coplanar slip mechanism into account in the MTCS-model, the C and S orientations are stabilized due to the restriction of secondary twinning, which leads to higher volume fractions than those in MT-model. The major volume fraction of 10.95% calculated from MTCS-model in Figure 6d reveals the S orientation, while in the case of MT-model in Figure 6c that of 11.36% is the B orientation. This observation suggests that the deformation mechanism of coplanar slip may lead to the orientation change from B to S orientation because the deformation mechanism of coplanar slip is considered in the MTCS-model. Furthermore, the volume fraction of T orientation is 0.37% for MT-model (in Figure 6c), and those of T, Y, and Z orientations are 1.64%, 2.34%, and 2.36%, respectively, for the MTCS-model (in Figure 6d). The results reveal that the difference between MT- and MTCS-models lies on the orientation prediction of Y and Z, where the volume fractions of both orientations are small.

Wassermann et al. [7] observed that on cold-rolled α-brass, the twin orientation T{5 5 2} <1 1 5> is formed after twinning and consequently rotates to B{1 1 0} <1 1 2>. Hirsch et al. [4] observed that in the low SFE Cu–Zn alloys, texture transition occurs at the intermediate of high strains; a decrease in C orientation leads to an increase in T orientation. The experimental result indicates the onset of twinning by the decrease of C orientation and the increase of G. After 70% reduction, the G orientation is stable and the rest of T orientation shifts toward Y. Madhavan et al. [35] observed the texture evolution of cold-rolled Ni–40%Co. At the early stage, the deformation is achieved by slip and MT. At higher reductions, high fraction of Cu-type shear bands was observed, which leads to final G orientation with high volume fraction.

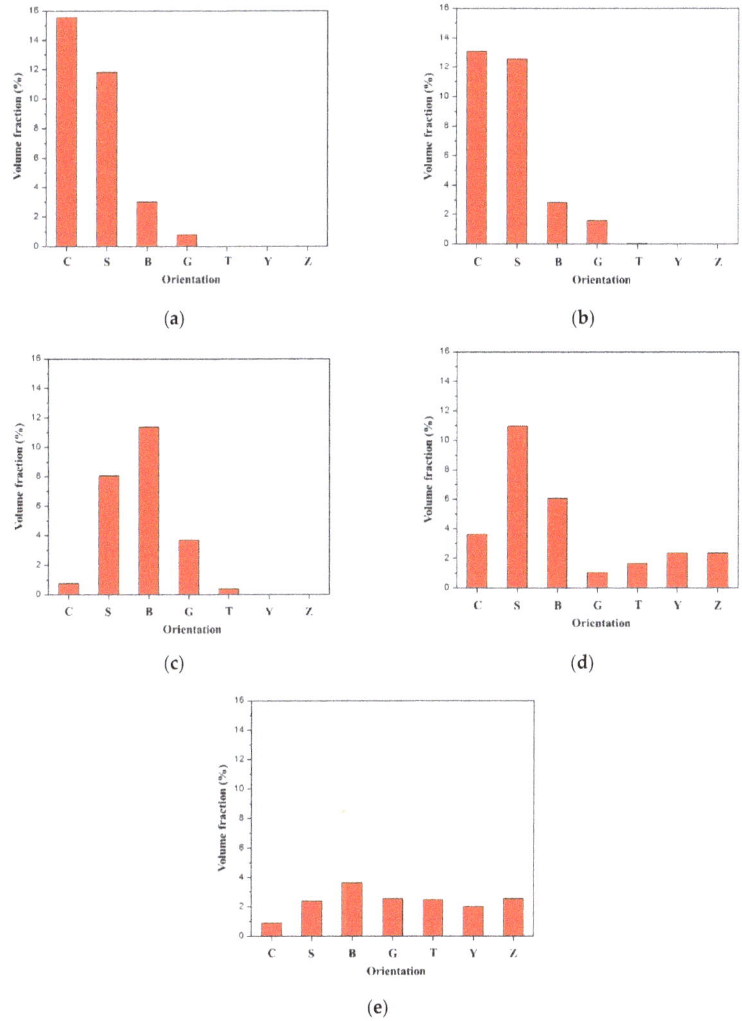

Figure 6. Volume fractions of specific orientations of rolling texture simulated using (**a**) FC-, (**b**) RC-, (**c**) MT-, and (**d**) MTCS-models and (**e**) from XRD measurements in Cu–30%Zn alloy after 90% reduction.

For FC- and RC-models, we only assume that the plastic deformation results from slipping on the slip systems. For MT- and MTCS-models, we consider that deformation occurs due to MT. This finding suggests that the deformation mode, either due to slip or twinning, changes the texture. The MTCS-model can predict the orientation components of Y and Z with the initial random orientations.

With the use of a single crystal, the development and formation of Y and Z orientations were successfully estimated. Thus, the MTCS-model in combination with the initial orientations of C and S′ was employed in the following sections.

4.2. Formation of Y Orientation

Hirsch et al. [4] reported that the T orientation is formed at low strain of <60% reduction due to the twin mechanism in Cu–30%Zn. At 70% reduction, this T orientation rotates oppositely to Y orientation due to coplanar slip instead of rotating toward G.

Duggan et al. [2] observed that the T orientation and the matrix orientation of C rotate toward Y orientation. On this basis, the present work used the MTCS-model with the C orientation of $(1\,1\,2)[\bar{1}\,\bar{1}\,1]$ to understand the development of Y orientation. Figures 7 and 8 show the simulated {1 1 1} and {2 0 0} pole figures of 30%, 60%, and 90% reductions with initial single C orientation using the MTCS-model.

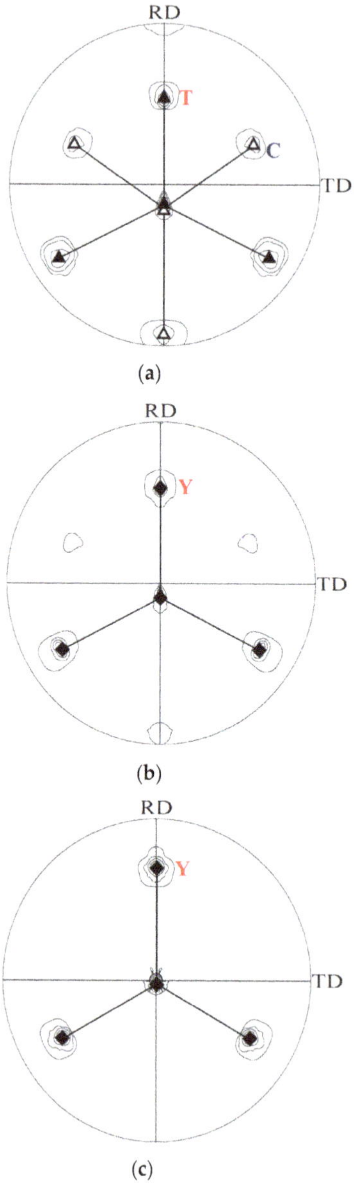

Figure 7. The {1 1 1} pole figures of fcc metals simulated using MTCS-model with single C orientation after the reduction ratios of (**a**) 30%, (**b**) 60%, and (**c**) 90%. △, ▲, and ◆ are the pole positions on {1 1 1} pole figure of the orientation close to $C(1\,1\,2)[\bar{1}\,\bar{1}\,1]$, $T(5\,5\,2)[1\,1\,\bar{5}]$, and $Y(1\,1\,1)[1\,1\,\bar{2}]$, respectively.

At 30% reduction, some of the initial C orientations either stay close to the C orientation of (90°, 37.7°, 45°) with 2.7° [1 0 1] misorientation or rotate to the orientation (270°, 69°, 45°) near T with 5.2° [1 0 1] misorientation due to twinning. The rotation angle and axis between simulated C and T is 60.0° [9 8 9] after 30% reduction.

At 60% reduction, most of the orientations rotate close to Y(270°, 63.6°, 45°) with a misorientation of 8.9° from the Y orientation because of slip. According to Asbeck et al. [36] and Hirsch et al. [4], the orientation of (270°, 74.2°, 45°)rotates toward the Y orientation of (270°, 54.7°, 45°) instead of moving toward the G orientation of (270°, 90°, 45°). The simulation results obtained from the MTCS-model are in agreement with those reported by Asbeck et al. and Hirsch et al. After 90% reduction, the major orientations rotate toward the Y orientation of (270°, 57.1°, 45°) with a misorientation of 2.4°. This path of orientation change is in agreement with the study of Hirsch et al. [4]. In summary, the C orientation of (1 1 2)[$\bar{1}$ $\bar{1}$ 1] rotates to the T orientation on (1 1 1) plane due to twinning at 30% reduction. After 60% reduction, the T orientation rotates toward Y orientation, which requires the coplanar slip systems of SS1(1 1 1)[0 1 $\bar{1}$], SS2(1 1 1)[$\bar{1}$ 0 1], and SS3(1 1 1)[1 $\bar{1}$ 0] on the (1 1 1) plane. This change in orientation is shown in Figure 9.

4.3. Formation of Z Orientation

The formation of Z orientation is attributed to the twinning of S′(133.1°, 36.7°, 26.6°), which is close to S(121.0, 36.7, 26.6). Hirsch et al. [4] reported that after twinning, the S′ orientation leads to T′(313.1°, 36.7°, 26.6°), one of the symmetrically equivalent variants of S′. As a result, the TS′ and S′ orientations rotate toward Z orientation by coplanar slip. Thus, the MTCS-model with the S′ orientation of (1 2 3) [4 $\bar{1}$ 2] was used to understand the development of Z orientation. Figures 10 and 11 show the simulated {1 1 1} and {2 0 0} pole figures of 30%, 60%, and 90% reductions with an initial single S′ orientation using the MTCS-model. After 30% reduction, some of the initial S′ orientations remain near S′(131.3°, 38.8°, 26.7°) with 2.9° [4 3 5] from the initial orientation. Owing to twinning, the other orientations rotate near T′(294°, 70.9°, 55.4°) with 4.0° [9 6 11] from the S′ orientation. The rotation angle and axis between simulated S′ and T′ is 60.0° [9 8 9] after 30% reduction. At 60% reduction, the S′ orientation of (127.5°, 38.7°, 26.8°) is still observed, and the other orientation is close to T′(294.3°, 68.7°, 53.2°). The former has a misorientation of 5.3° away from the S′ orientation, and the latter has a misorientation of 6.9° away from the T′ orientation. With increasing reduction from 30% to 60%, the T′ orientation rotates toward Z and decreases the misorientations from 18.7° to 16.3°.

At 90% reduction, most of the orientations rotate close to Z(292.8°, 63.1°, 48.9°) with a misorientation of 10.4°, and the orientation of S′ is still found. The increase in reduction from 60% to 90% decreases the misorientations between T′ and Z. Hirsch et al. [4] observed that the peak shift of S orientation leads to a large φ_2 angle and a small φ_1 angle. In the present simulation of the MTCS-model, the φ_2 angle increases from the initial 26.7° to 31.5°, and the φ_1 angle decreases from the initial 133.1° to 116.9°. This trend is in agreement with the observation of Hirsch et al.

The S′ orientation of (1 2 3) [$\bar{4}$ $\bar{1}$ 2] rotates to the T′ orientation because of twinning on the twin plane of (1 1 1) at 30% reduction. At above 30% reduction, the T′ orientation rotates toward the final Z orientation as explained by the coplanar slip of SS1(1 1 1)[0 1 $\bar{1}$], SS2(1 1 1)[$\bar{1}$ 0 1], and SS3(1 1 1)[1 $\bar{1}$ 0] further gliding on the (1 1 1) plane. This change in orientation is shown in Figure 12. Therefore, the combination of MT and coplanar slip in the TML region can be successfully simulated by the Taylor model to reveal the formation of Y and Z orientations observed in the experiments.

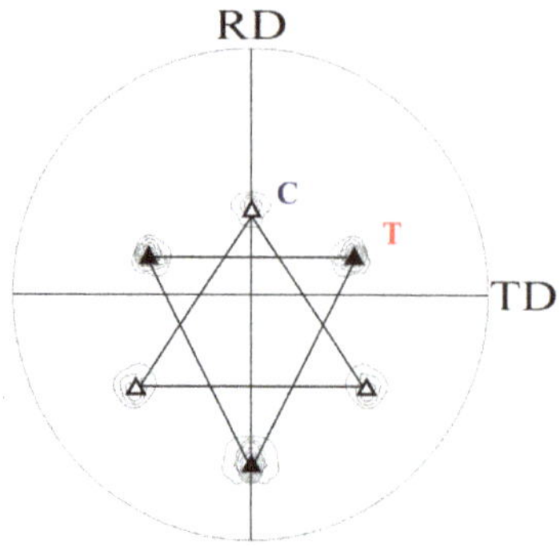

(**a**)

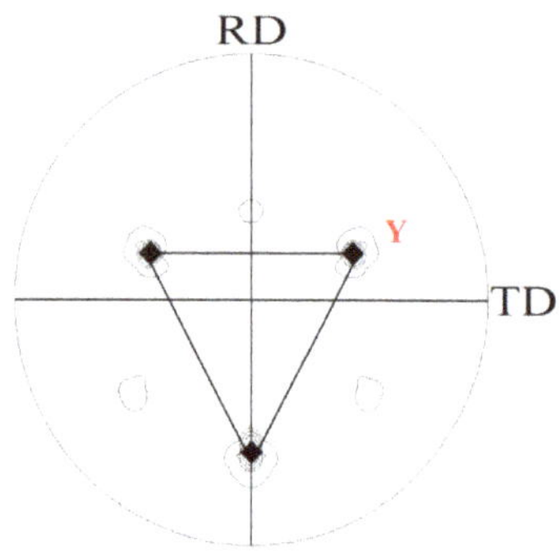

(**b**)

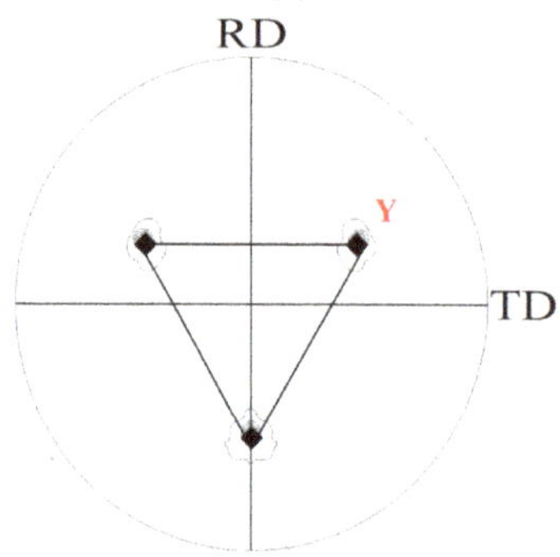

(**c**)

Figure 8. The {2 0 0} pole figures of fcc metals simulated using MTCS-model with single C orientation after the reduction ratios of (**a**) 30%, (**b**) 60%, and (**c**) 90%. △, ▲, and ◆ are the pole positions on {2 0 0} pole figure of the orientation close to C(1 1 2) $[\bar{1}\,\bar{1}\,1]$, T(5 5 2) $[1\,1\,\bar{5}]$, and Y(1 1 1) $[1\,1\,\bar{2}]$, respectively.

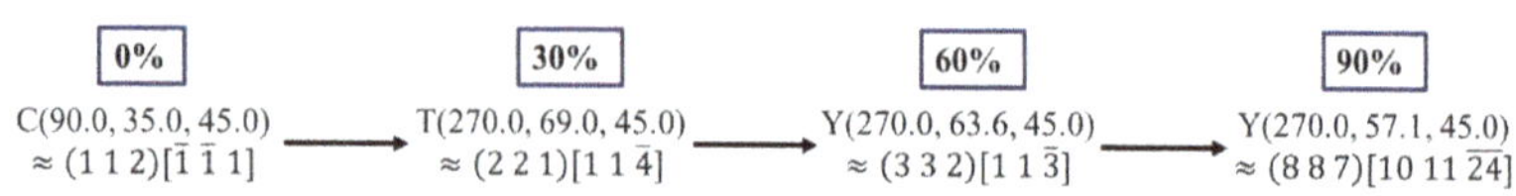

Figure 9. Orientation evolution of Y orientation simulated with MCTS-model.

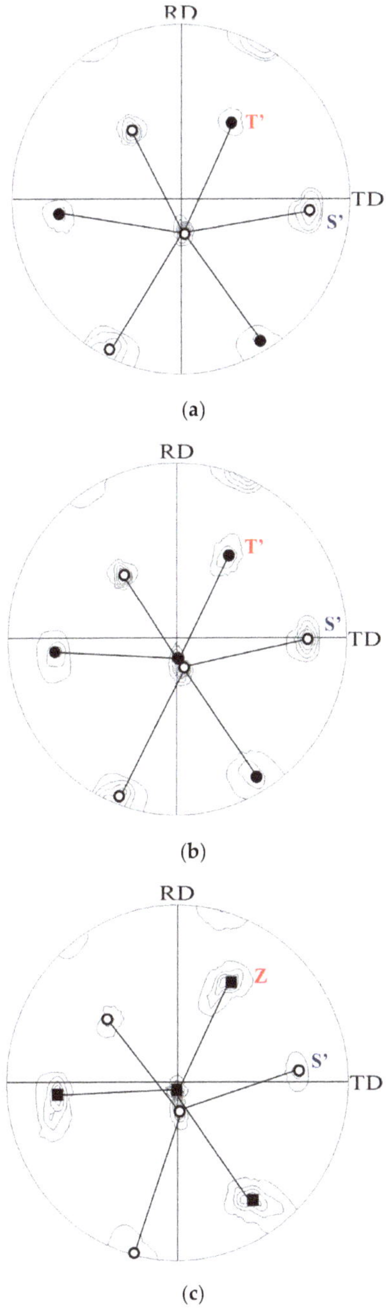

Figure 10. The {1 1 1} pole figures of fcc metal simulated using MTCS-model with single S′ orientation after the reduction ratios of (**a**) 30%, (**b**) 60%, and (**c**) 90%. ○, •, and ■ are the pole positions on {1 1 1} pole figure of the orientation close to S′$(1\,2\,3)[\bar{4}\,\bar{1}\,2]$, T′$(3\,2\,1)[2\,\bar{1}\,\bar{4}]$, and Z$(1\,1\,1)[1\,0\,\bar{1}]$, respectively.

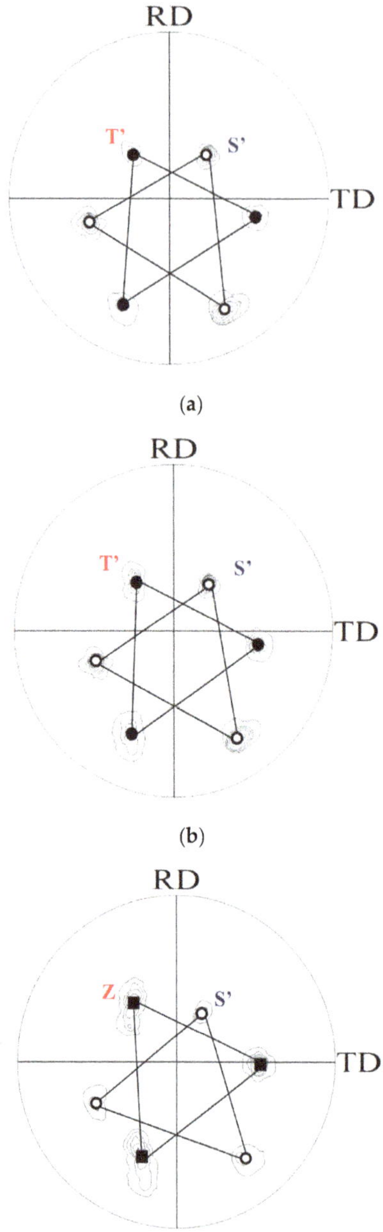

(c)

Figure 11. The {2 0 0} pole figures of fcc metal simulated using MTCS-model with single S′ orientation after the reduction ratios of (**a**) 30%, (**b**) 60%, and (**c**) 90%. ○, •, and ■ are the pole positions on {2 0 0} pole figure of the orientation close to S′$(1\,2\,3)[\bar{4}\,\bar{1}\,2]$, T′$(3\,2\,1)[2\,\bar{1}\,\bar{4}]$, and Z$(1\,1\,1)[1\,0\,\bar{1}]$, respectively.

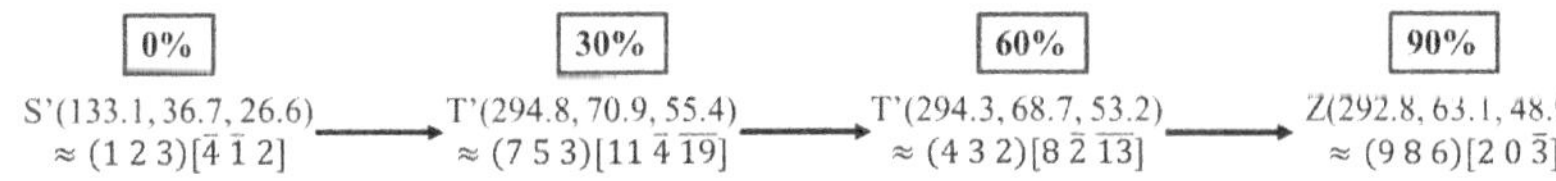

Figure 12. Orientation evolution of Z orientation simulated with MCTS-model.

5. Conclusions

Conventional Taylor models including FC- and RC-models considering pure slip mechanism simulate strong copper-type textures. Both FC- and RC-models display preferred orientations close to C, S and B. Among the components, the volume fraction of C orientations is 2.45% higher in FC-model. With consideration of MT mechanism, the condition of partial slip and twinning leads to partial brass-type textures. Significantly decreased volume fraction of C and S were determined with 12.32% and 4.52%. In the meantime, the increased volume fraction of B and G with 8.53% and 2.11% indicates the formation of brass-type texture.

Considering MT and coplanar slip in the TML region in this study, a Taylor-based MTCS-model is proposed to simulate the rolling texture of Cu–30%Zn. Comparing with the results of MT-model, the volume fraction of C and S orientations was determined with 2.87% and 2.91% increase, respectively. In the meantime, the decreased volume fraction of B with 5.31% indicates the instability of B orientation. In addition to the β-fibers, the simulated results of the MTCS-model display the experimentally observed texture components including T, Y, and Z orientations, with corresponding volume fractions of 1.62%, 2.34%, and 2.36%, respectively.

Furthermore, we can successfully predict the reorientations of C-Y-T and S′-T′-Z by additionally considering twinning and then coplanar slip in the proposed MTCS-model. The Y and Z orientations, however, were not observed in the FC-, RC- and MT-models, but were found in the MTCS-model. Evolution of single C and S′ orientations further suggests the texture transition from copper-type to brass-type texture. The texture transitions from C to Y and from S′ to Z reveal the following. Considering the texture transition from C to Y, the C orientation of $(1\,1\,2)[\bar{1}\,\bar{1}\,1]$ rotates toward $T(2\,2\,1)[1\,1\,\bar{4}]$ because of twinning after 30% reduction, after which the T orientation rotates toward $Y(3\,3\,2)[1\,1\,\bar{3}]$ and $Y(8\,8\,7)[10\,11\,24]$ because of continued coplanar slip after reductions of 30% and 60%, respectively. In the case of the texture transition from S′ to Z, the S′ orientation of $(1\,2\,3)[\bar{4}\,\bar{1}\,2]$ rotates toward $T'(7\,5\,3)[11\,\bar{4}\,\overline{19}]$ by twinning after 30% reduction, after which the T′ orientation rotates toward $Z(4\,3\,2)[8\,\bar{2}\,\overline{13}]$ and $Z(9\,8\,6)[2\,0\,\bar{3}]$ because of continued coplanar slip after reductions of 30% and 60%, respectively.

Author Contributions: Conceptualization, S.-C.H. and J.-C.K.; methodology, S.-C.H., S.-Y.L. and J.-C.K.; software, S.-C.H., S.-Y.L. and P.-Y.H.; validation, S.-C.H. and J.-C.K.; formal analysis, S.-C.H.; investigation, S.-C.H., S.-Y.L. and J.-C.K.; resources, S.-C.H., S.-Y.L. and H.-J.C.; data curation, S.-C.H. and J.-C.K.; writing—original draft preparation, S.-C.H.; writing—review and editing, S.-C.H.; visualization, S.-C.H.; supervision, J.-C.K.; project administration, J.-C.K.; funding acquisition, J.-C.K. All authors have read and agreed to the published version of the manuscript.

Funding: This research was funded by Ministry of Science and Technology (Taiwan), grant number MOST 107-2221-E-006-018 and MOST 109-2221-E-006-131-MY2.

Data Availability Statement: Data available on request due to restrictions. The data presented in this study may be available on request from the corresponding author. The data are not publicly available due to the large amount of the data.

Acknowledgments: The authors would gratefully like to thank the Ministry of Science and Technology for supporting the fund of the project under MOST 107-2221-E-006-018 and MOST 109-2221-E-006-131-MY2.

Conflicts of Interest: The funders had no role in the design of the study; in the collection, analyses, or interpretation of data; in the writing of the manuscript, or in the decision to publish the results.

References

1. Engler, O.; Randle, V. Introduction. In *Introduction to Texture Analysis: Macrotexture, Microtexture, and Orientation Mapping*, 2nd ed.; CRC Press: Boca Raton, FL, USA, 2010; pp. 5–7.
2. Duggan, B.J.; Hatherly, M.; Hutchinson, W.B.; Wakefield, P.T. Deformation structures and textures in cold-rolled 70: 30 brass. *Met. Sci.* **1978**, *12*, 343–351. [CrossRef]
3. Fargette, B.; Whitwham, D. Plastic Deformation of the Brass CuZn 30 by Heavy Rolling Reductions. *Mem. Sci. Rev. Metall.* **1976**, *73*, 197–206.
4. Hirsch, J.; Lücke, K.; Hatherly, M. Overview no. 76: Mechanism of deformation and development of rolling textures in polycrystalline fcc metals—III. The influence of slip inhomogeneities and twinning. *Acta Metall.* **1988**, *36*, 2905–2927. [CrossRef]
5. Hatherly, M.; Malin, A. Shear bands in deformed metals. *Scr. Metall.* **1984**, *18*, 449–454. [CrossRef]
6. Malin, A.; Hatherly, M. Microstructure of cold-rolled copper. *Met. Sci.* **1979**, *13*, 463–472. [CrossRef]
7. Wassermann, G. Der einfluss mechanischer zwillingsbildung auf die entstehung der walztexturen kubisch flachenzentrierter metalle. *Z. Metallkunde* **1963**, *54*, 61–65.
8. Paul, H.; Morawiec, A.; Piątkowski, A.; Bouzy, E.; Fundenberger, J.J. Brass-type shear bands and their influence on texture formation. *Metall. Mater. Trans. A* **2004**, *35*, 3775–3786. [CrossRef]
9. Paul, H.; Driver, J.H.; Maurice, C.; Jasieński, Z. Shear band microtexture formation in twinned face centred cubic single crystals. *Mater. Sci. Eng. A* **2003**, *359*, 178–191. [CrossRef]
10. Sevillano, J.G.; van Houtte, P.; Aernoudt, E. Large strain work hardening and textures. *Prog. Mater. Sci.* **1980**, *25*, 69–134. [CrossRef]
11. Van Houtte, P. A comprehensive mathematical formulation of an extended Taylor–Bishop–Hill model featuring relaxed constraints, the Renouard–Wintenberger theory and a strain rate sensitivity model. *Texture Stress Microstruct.* **1988**, *8*, 313–350. [CrossRef]
12. van Houtte, P. On the equivalence of the relaxed Taylor theory and the Bishop-Hill theory for partially constrained plastic deformation of crystals. *Mater. Sci. Eng.* **1982**, *55*, 69–77. [CrossRef]
13. Leffers, T.; Ray, R. The brass-type texture and its deviation from the copper-type texture. *Prog. Mater. Sci.* **2009**, *54*, 351–396. [CrossRef]
14. Hirsch, J.; Lücke, K. Overview no. 76: Mechanism of deformation and development of rolling textures in polycrystalline fcc metals—II. Simulation and interpretation of experiments on the basis of Taylor-type theories. *Acta Metall.* **1988**, *36*, 2883–2904. [CrossRef]
15. Leffers, T. The brass-type texture–how close are we to understand it. *Mater. Sci. Forum* **2012**, *702*, 216–223. [CrossRef]
16. van Houtte, P. Simulation of the rolling and shear texture of brass by the Taylor theory adapted for mechanical twinning. *Acta Metall.* **1978**, *26*, 591–604. [CrossRef]
17. Chin, G.; Hosford, W.; Mendorf, D. Accommodation of constrained deformation in fcc metals by slip and twinning. *Proc. R. Soc. Lond. Ser. A* **1969**, *309*, 433–456.
18. Kallend, J.S. *Quantitative Analysis of Texture Data*; Cambridge University Press: Cambridge, UK, 1971.
19. Leffers, T. The brass-type texture once again. In Proceedings of the 11th International Conference on Textures of Materials, Xian, China, 16–20 September 1996; International Academic Publishers: Beijing, China, 1996; pp. 299–306.
20. Leffers, T. On the development of the brass-type texture. *Textures Microstruct.* **1970**, *22*, 53–58. [CrossRef]
21. Kalidindi, S.R. Incorporation of deformation twinning in crystal plasticity models. *J. Mech. Phys. Solids* **1998**, *46*, 267–290. [CrossRef]
22. Kalidindi, S.R. Modeling anisotropic strain hardening and deformation textures in low stacking fault energy fcc metals. *Int. J. Plast.* **2001**, *17*, 837–860. [CrossRef]
23. Lebensohn, R.A.; Tomé, C. A self-consistent viscoplastic model: Prediction of rolling textures of anisotropic polycrystals. *Mater. Sci. Eng. A* **1994**, *175*, 71–82. [CrossRef]
24. Lebensohn, R.A.; Tomé, C.; Castaneda, P.P. Self-consistent modelling of the mechanical behaviour of viscoplastic polycrystals incorporating intragranular field fluctuations. *Philos. Mag.* **2007**, *87*, 4287–4322. [CrossRef]
25. Toth, L.S.; Haase, C.; Allen, R.; Lapovok, R.; Molodov, D.A.; Cherkaoui, M.; el Kadiri, H. Modeling the effect of primary and secondary twinning on texture evolution during severe plastic deformation of a twinning-induced plasticity steel. *Materials* **2018**, *11*, 863. [CrossRef]
26. Chalapathi, D.; Sivaprasad, P.; Kanjarla, A.K. Role of deformation twinning and second phase on the texture evolution in a duplex stainless steel during cold rolling: Experimental and modelling study. *Mater. Sci. Eng. A* **2020**, *780*, 139–155. [CrossRef]
27. Leffers, T.; Bilde-Sørensen, J. Intra-and intergranular heterogeneities in the plastic deformation of brass during rolling. *Acta Metall. Mater.* **1990**, *38*, 1917–1926. [CrossRef]
28. Duggan, B.; Hutchinson, W.; Hatherly, M. Recrystallization studies in 70: 30 brass using a high voltage electron microscope. *Scr. Metall.* **1978**, *12*, 293–295. [CrossRef]
29. Hutchinson, W.; Duggan, B.; Hatherly, M. Development of deformation texture and microstructure in cold-rolled Cu–30Zn. *Met. Technol.* **1979**, *6*, 398–403. [CrossRef]
30. Donadille, C.; Valle, R.; Dervin, P.; Penellel, R. Development of texture and microstructure during cold-rolling and annealing of FCC alloys: Example of an austenitic stainless steel. *Acta Metall.* **1989**, *37*, 1547–1571. [CrossRef]

31. Paul, H.; Morawiec, A.; Driver, J.H.; Bouzy, E. On twinning and shear banding in a Cu–8 at.% Al alloy plane strain compressed at 77 K. *Int. J. Plast.* **2009**, *25*, 1588–1608. [CrossRef]
32. van Houtte, P.; Aernoudt, E. Considerations on crystal and strain symmetry in calculation of deformation textures with taylor theory. *Mater. Sci. Eng.* **1976**, *23*, 11–22. [CrossRef]
33. Taylor, G.I. Plastic strain in metals. *J. Inst. Metals* **1938**, *62*, 307–324.
34. Hirsch, J.; Lucke, K. Mechanism of Deformation and Development of Rolling Textures in Polycrystalline, F.C.C. Metals. I. Description of Rolling Texture Development in Homogeneous CuZn Alloys. *Acta Metall.* **1988**, *36*, 2863–2882. [CrossRef]
35. Madhavan, R.; Ray, R.K.; Suwas, S. Texture transition in cold-rolled nickel–40 wt.% cobalt alloy. *Acta Mater.* **2014**, *74*, 151–164. [CrossRef]
36. Asbeck, H.; Mecking, H. Influence of friction and geometry of deformation on texture inhomogeneities during rolling of Cu single crystals as an example. *Mater. Sci. Eng.* **1978**, *34*, 111–119. [CrossRef]

crystals

MDPI

Article

Effects of the Rare Earth Y on the Structural and Tensile Properties of Mg-based Alloy: A First-Principles Study

Yan Gao [1,*], Chuang Wu [1], Wenjiang Feng [2], Yan He [2], Haisheng He [1], Jingyu Yang [1] and Xiuyan Chen [2]

[1] Experimental Teaching Center, Shenyang Normal University, Shenyang 110034, China; meto085@163.com (C.W.); hehaisheng1002@163.com (H.H.); jingyuyang525@163.com (J.Y.)

[2] College of Physics Science and Technology, Shenyang Normal University, Shenyang 110034, China; fengwj@synu.edu.cn (W.F.); lisaheyan@163.com (Y.H.); chenxy@synu.edu.cn (X.C.)

* Correspondence: gaoy@synu.edu.cn

Abstract: In order to investigate the effect of the rare earth element Y on the strengthening potency of magnesium alloys and its strengthening mechanism under tension. In this paper, the solid solution structures with Y atom content of 1.8 at.% and 3.7 at.% were built, respectively, and their cohesive energies and stress-strain curve were calculated in the strain range of 0–20%. The calculation results of the cohesive energies showed that the structure of element Y is more stable with the increase of strains. The calculation results of stress and strain showed that Y element can improve the yield strength and tensile strength of the Mg-based alloy, and the strengthening effect is better when the Y content is 3.7 at.%.

Keywords: Mg-based alloy; first principles; structural properties; tensile properties

Citation: Gao, Y.; Wu, C.; Feng, W.; He, Y.; He, H.; Yang, J.; Chen, X. Effects of the Rare Earth Y on the Structural and Tensile Properties of Mg-based Alloy: A First-Principles Study. *Crystals* **2021**, *11*, 1003. https://doi.org/10.3390/cryst11081003

Academic Editors: Andreas Hermann and Wojciech Polkowski

Received: 8 July 2021
Accepted: 19 August 2021
Published: 22 August 2021

1. Introduction

In recent years, magnesium alloys have been widely used in aerospace, the automotive industry, computers, the chemical industry and national defense and military industries due to their excellent properties, such as low density, light weight and high specific strength [1–3]. However, the defects of magnesium alloys are also very obvious [4]. The mechanical properties of ordinary magnesium alloys in high temperature environments are not good, which seriously restricts their further development and application [5].

Alloying is the most commonly used among a variety of magnesium alloy strengthening methods [6]. A new type of composite magnesium alloy is formed by adding alloying elements to make it a high-strength, high-toughness, high-performance magnesium alloy [7,8]. Among the many alloying elements of magnesium alloys, rare earth elements perform best [9]. Rare earth elements have the functions of deoxidizing by removing hydrogen and improving casting performance, and also have the ability to enhance alloy strength and high temperature creep resistance [10,11]. In addition, the large size of the rare earth atoms can prevent the α-Mg crystal grains from becoming larger, and help to refine the crystal grains, reduce the tendency of hot cracking due to the looseness of the microstructure, and improve the casting and welding performance of magnesium alloys [12]. Therefore, in recent years, rare-earth magnesium alloys have gradually become a hot spot for development.

The element Y is a widely used rare earth element in heat-resistant magnesium alloys, and it has the same hexagonal close-packed crystal structure as Mg atoms [13]. Magnesium alloy containing Y has the characteristics of high temperature resistance, high plastic toughness and high strength [14]. After the high melting point Y element is added to the magnesium alloy, on the one hand, it can increase the nucleation rate of the alloy and play the role of grain refinement [15], on the other hand, dispersed second-phase particles can be precipitated in the magnesium alloy, which can effectively hinder the movement of dislocations and grain boundary slip. It can improve the creep resistance of magnesium alloy at high temperature and achieve the purpose of enhancing the strength of magnesium alloy [16,17].

The first-principles tensile test method can calculate the stress value of a crystal structure under different stresses (the stress-strain relationship). The theoretical tensile strength of the crystal structure can be predicted by analyzing the stress value at the yield or fracture of the crystal structure. Wang et al. [18] calculated the stress-strain curves of solid solution structures $Mg_{53}Al$ and $Mg_{51}Al_3$. It was found that the strong covalent bond between Al and Mg and the rearrangement of the electron charge density could improve the tensile strength of the Mg-based alloy, and the $Mg_{51}Al_3$ unit cell could increase the tensile strength of the Mg_{54} unit cell by 9.4%. Zhang et al. [19] calculated and studied the tensile strength of the Al unit cell. The calculation results showed that the theoretical tensile strength of the Al grain boundary was 9.5 GPa at the strain of 16%. Wang et al. [20] calculated the influence of the distribution of Al and Zn atoms on the strength of Mg alloys. It was found that the structure with uniform distribution of alloying elements has greater ideal tensile strength than the structure with separate distribution of alloying elements. Luo et al. [21] calculated the stress-strain curve of Mg-based alloy solid solution in which Al, Zn and Y atoms were dissolved. It was found that the Al, Zn and Y atoms all have a solid solution strengthening effect on the Mg-based alloy, and the Y atom has the best solid solution strengthening effect, which conforms to the experimental rules.

The ideal tensile strength of the crystal structure is an important index parameter to measure material properties and evaluate material quality [22,23]. Therefore, the first-principles tensile calculation method can be used to study the solid solution strengthening effect of alloying elements on Mg-based alloy, and which has an important guiding value for the development and application of magnesium alloys. Considering that the solid solubility of Y atoms in the magnesium alloy does not exceed 3.75 at.%. In this research, $Mg_{53}Y_1$ and $Mg_{52}Y_2$ with Y atom solid solubility of 1.8% at.% and 3.7 at.% were used as the research object. The stress-strain curves and electronic structure changes of the Mg_{54}, $Mg_{53}Y_1$ and $Mg_{52}Y_2$ structures under 0–20% strains were calculated, and the improvement of the yield strength and tensile strength of the Mg-based alloy by the element Y and the strengthening mechanism were analyzed.

2. Computational Methods

The research content in this paper was assembled using the CASTEP software, which is based on the first-principles density functional theory [24,25]. Tensile tests were performed by applying stress to the c-axis direction of the crystal structures of Mg_{54}, $Mg_{53}Y_1$ and $Mg_{52}Y_2$, with a 2% strain increment. In order to obtain accurate stress and strain conditions, the strain interval is 1% between 6% and 10% strain, and the upper limit of the applied strain is 20%. The crystal structure must be geometrically optimized after each strain is applied. The optimize cell option was not checked during the relaxation process, therefore, the lattice constant was not optimized, and only the atomic coordinates in the supercell were optimized. After geometric optimization, the energy, stress and strain values and electronic structure of the supercell structure were calculated.

The CASTEP software parameter setting includes the following contents: Considering that the number of atoms in the unit cell is relatively large, in order not to affect the calculation speed exchange correlation, the function option selects the PW91 functional in the approximate form of GGA. In the convergence setting of the optimized crystal structure, the convergence value of the total energy is 1.0×10^{-5} eV/atom. The convergence value of the force between atoms is 0.03 eV/nm, the maximum internal stress is 0.05 GPa, and the tolerance offset value is set 0.001 Å. In the electronic setting, the cut-off energy is 340 eV, and the number of K points is $3 \times 3 \times 1$. The correlation between particles is set to Ultrasoft super soft pseudopotential. The SCF self-consistent iteration tolerance value is 1.0×10^{-6} eV/atom, the number of convergence steps for geometric mechanism optimization and electronic properties calculation is 150, and the electronic minimizer is set to the default density mixing method and Pulay correction.

3. Results and Discussion

3.1. Structure Properties

As shown in Figure 1a the unit cell of pure magnesium with two Mg atoms has a close-packed hexagonal structure, which space group is P63/MMC, and the lattice constants are a = b = 0.3209 nm, and c = 0.5211 nm. Based on the pure magnesium unit cell and considering factors such as calculation time and accuracy, a 3 × 3 × 3 supercell is built as shown in Figure 1b. The cell contains 54 Mg atoms, and is therefore hereinafter referred to as Mg_{54}, and the corresponding supercell lattice constants are a = b = 0.9628 nm, c = 1.5632 nm. Replacing the magnesium atom at coordinates x = 0.5556, y = 0.4444, z = 0.5833 with an Y atom we obtain the $Mg_{53}Y_1$ crystal structure, that is, a Mg-based alloy solid solution structure with a Y atom content of 1.8 at.%. This structure is shown in Figure 1c. By replacing the magnesium atoms at coordinates x = 0.5556, y = 0.4444, z = 0.5833 and x = 0.5556, y = 0.4444, z = 0.2500 with Y atoms, as shown in Figure 1d, the crystal structure of $Mg_{52}Y_2$ can be obtained, that is, a Mg-based alloy solid solution structure with a Y atoms content of 3.7 at.%.

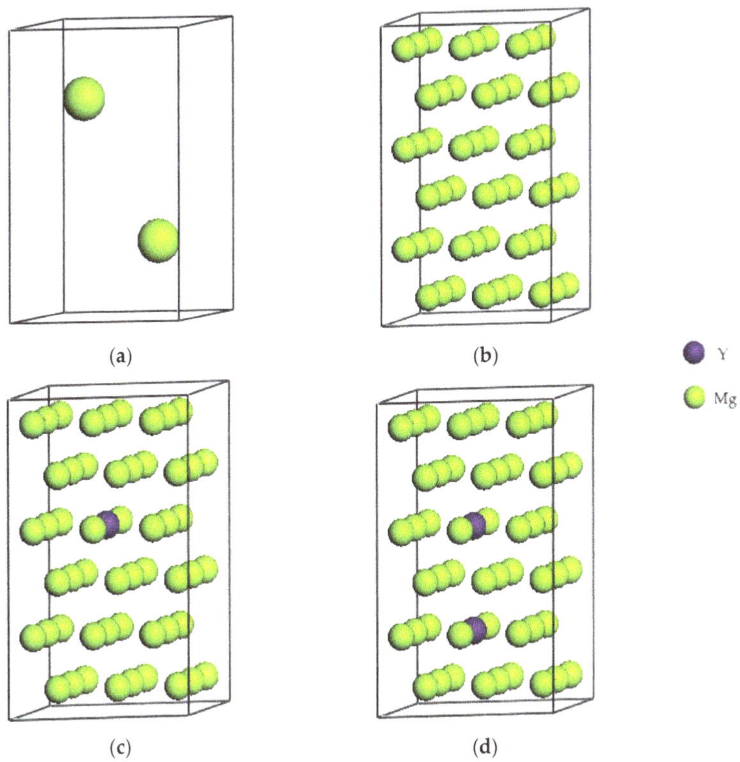

Figure 1. Crystal structures of Mg (**a**), Mg_{54} (**b**), $Mg_{53}Y_1$ (**c**) and $Mg_{52}Y_2$ (**d**).

Based on the Nielsen-Martin calculation method, the first-principles stretching calculation is carried out, and the stress acting on the supercell is the average stress, which can be expressed as [26]:

$$\sigma_{\alpha\beta} = \frac{1}{\Omega}\frac{\partial E_{tot}}{\partial \varepsilon_{\alpha\beta}} \quad (1)$$

In this formula, $\sigma_{\alpha\beta}$ is the average stress acting on the unit cell, Ω is the unit cell volume, E_{tot} is the total energy of the unit cell, and $\varepsilon_{\alpha\beta}$ is the strain tensor. A tensile strain is applied in the direction of the C axis of the unit cell:

$$\varepsilon = \frac{(l_\varepsilon - l_0)}{l_0} \times 100\% \tag{2}$$

In Equation (2), l_0 is the initial cell c-axis length when no strain is applied, and l_ε is the c-axis length of the cell corresponding to the applied strain. It is worth noting that in order to save calculation time, the simulation calculation in this section does not consider the influence of Poisson effect, that is, ignores the influence of tensile strain on the other two axial lattice constants, and considers its value to be fixed.

The stability of the crystal structure depends on its cohesive energy. The definition of binding energy is: if the crystal is split into single free atoms, the work done by the outside world is the cohesive energy of the compound. The cohesive energy of a stably existing compound is negative, and the lower the cohesive energy value, the more stable the crystal structure of the compound. The calculation formula of cohesive energy is as follows [27]:

$$E_{coh} = \frac{E_{tot}^{AB} - N_A E_{atom}^{A} - N_B E_{atom}^{B}}{N_A + N_B} \tag{3}$$

In the above formula, E_{coh} is the cohesive energy of the intermetallic compound, E_{tot} is the total energy of the compound, E_{atom}^{A} and E_{atom}^{B} are the energies of the A and B atoms in the free state, respectively. The free state atomic energyies of Mg and Y are −972.5822 eV/atom and −188.5729 eV/atom, respectively: N_A and N_B are the corresponding numbers of atoms in the unit cell.

The cohesive energy of Mg_{54}, $Mg_{53}Y_1$, and $Mg_{52}Y_2$ at 0–20% strains were calculated, as shown in Table 1. First, the cohesive energy of the three structures are all negative at zero strain, indicating that the three structures can exist stably. Further analysis found that in the range of 0–20% strains, the cohesive energy of the three structures all increase with the increase of strain. Since the larger the absolute value of the cohesive energy, the more stable the structure, so it can be determined that the stability of the three structures decreases with the increase of strain. The reason why the structures become unstable may be the weakening of the chemical bonds between atoms due to stretching. It is worth noting that although the stability of the three structures has decreased, but the cohesive energy values are always negative, indicating that the three structures can still remain stable within the range of 0–20% strains. In addition, it can also be found that the stability of the $Mg_{53}Y_1$, and $Mg_{52}Y_2$ structures are stronger than that of Mg_{54}, and the $Mg_{52}Y_2$ structure is the most stable.

Table 1. The cohesive energy of Mg_{54}, $Mg_{53}Y_1$ and $Mg_{52}Y_2$ under different strains.

Strain (%)	**E_{coh}, kJ·mol^{-1}**		
	Mg_{54}	**$Mg_{53}Y_1$**	**$Mg_{52}Y_2$**
0	−193.46	−199.66	−205.63
2	−193.01	−199.40	−205.52
4	−192.44	−198.95	−205.15
6	−191.70	−198.28	−204.49
7	−191.41	−197.97	−204.33
8	−191.37	−197.74	−204.08
9	−190.90	−197.43	−203.77
10	−190.47	−196.97	−203.32
12	−189.77	−196.15	−202.43
14	−188.75	−195.08	−201.29
16	−187.46	−193.77	−199.92
18	−186.00	−192.32	−198.84
20	−184.60	−190.84	−197.64

3.2. Tensile Properties

In production applications, structural parts often fail due to excessive plastic deformation. For components with strict requirements, plastic deformation is generally not allowed. For components with less stringent requirements, materials are often selected based on the yield strength value σ_s. Under 0–20% strains, the stress values of Mg_{54}, $Mg_{53}Y_1$, and $Mg_{52}Y_2$ solid solution structure are listed in Table 2, and the stress-strain curve is drawn at the same time, as shown in Figure 2. The abscissa represents the applied strains, and the ordinate represents the stress values corresponding to the strains.

Table 2. The Stress values of Mg_{54}, $Mg_{53}Y_1$, $Mg_{53}Y_2$ under different strains.

Strains (%)	Stress, GPa		
	Mg_{54}	**$Mg_{53}Y_1$**	**$Mg_{52}Y_2$**
0	0.00	0.00	0.00
2	1.74	1.23	0.75
4	2.35	1.94	1.83
6	2.69	2.67	2.63
7	1.21	1.49	1.57
8	1.85	1.99	2.04
9	2.22	2.36	2.40
10	2.47	2.76	2.92
12	3.11	3.39	3.65
14	4.15	4.25	4.47
16	4.97	4.93	5.15
18	5.15	5.24	5.63
20	5.02	5.37	4.22

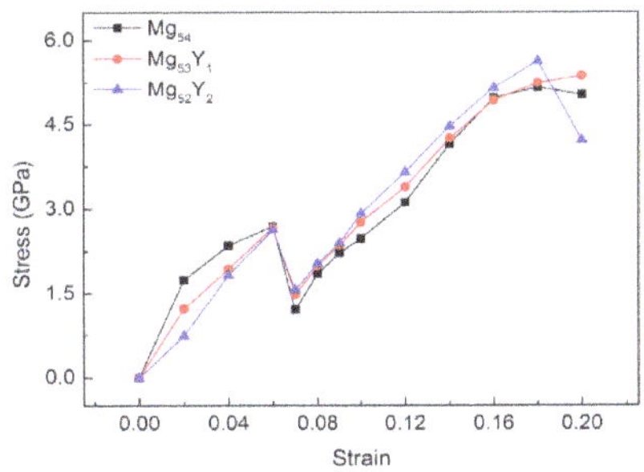

Figure 2. Stress-strain curve of Y Mg_{54}, $Mg_{53}Y_1$ and $Mg_{52}Y_2$.

It can be found from Figure 2 that the three types of structures are all elastic-uneven plastic-uniform plastic deformation types. When the structure is subjected to elastic deformation, obvious upper and lower yield points appear. In the initial stage, when the stress is small, the elongation of the structure changes in proportion to the stress. At this time, the material undergoes elastic deformation after stress, and the material can return to its original length when there is no stress. It can be seen from Figure 2 that the linear variation range of Mg_{54}, $Mg_{53}Y_1$, and $Mg_{52}Y_2$ is very small.

As the stress increases, the tensile strains experienced by the material continues to increase. At this time, both elastic deformation and plastic deformation occur, and it is difficult for the material to fully recover to its original length after the stress is unloaded. When the strain reaches 6%, the upper yield points of the three structures appear at the same time. The upper yield strengths of Mg_{54}, $Mg_{53}Y_1$, and $Mg_{52}Y_2$ are 2.69 GPa, 2.67 GPa and 2.63 GPa, respectively. At the same time, with the application of stress, the lower yield point appears at 7% strain. The lower yield strengths of Mg_{54}, $Mg_{53}Y_1$, and $Mg_{52}Y_2$ are 1.21 GPa, 1.49 GPa, and 1.57 GPa, respectively. For structures with upper and lower yield points, the lower yield strength is usually selected as the yield strength of the material structure, which is generally expressed by σ_s [28]. The yield strength is a unique strength index for materials with yield phenomena. The yield strengths of $Mg_{53}Y_1$, and $Mg_{52}Y_2$ are higher than that of Mg_{54}, and are increased by 23.14% and 29.75%, respectively, compared to Mg_{54}. It shows that the rare earth element Y can increase the yield strength of Mg-based alloy alloys, and the strengthening effect on Mg-based alloy alloys is stronger when the Y content is 3.7 at.%.

After the material is stretched to the yield stage, there will be a plastic deformation interval, and the resistance of the material against external force stretching will increase with the growth of the plastic deformation, until the stress reaches the tensile strength σ_b [29]. Tensile strength is the ability of a material to resist damage under the action of external force. After reaching the tensile strength, the stress values will decrease as the strains increases. At this time, the deformation strengthening effect of the material structure can no longer compensate for the reduced load-bearing capacity due to the reduction of the cross-section. The tensile strength represents the maximum stress value that a material can withstand under tensile deformation, and which is an important indicator of the material's resistance to tensile deformation. It can be seen from Figure 2 that when the yield strength is reached, several structures begin to undergo plastic deformation. With the gradual increase of strains, the stress values of several structures increase rapidly until the tensile strength is reached. The tensile strength values of Mg_{54}, $Mg_{53}Y_1$, and $Mg_{52}Y_2$ are 5.15 GPa, 5.37 GPa and 5.63 GPa, respectively. It shows that the rare earth Y element can improve the tensile strength of Mg-based alloy alloys, and the enhancement effect is best when the Y content is 3.7 at.%.

Based on the above analysis, it can be found that the Y element can enhance the strength of the Mg-based alloys. When the Y atom content is 3.7 at.%, the strengthening effect is greater than 1.8 at.%. This conclusion is consistent with the experimental results [30]. It was found that the Y element can increases the tensile strength of magnesium alloys, but the plasticity will be decreased. Therefore, although the addition of Y reduces the elastic deformability of the Mg-based alloy. However, both the yield strength and the tensile strength are improved, and the strengthening effect becomes stronger as the solid solubility increases.

4. Conclusions

The first-principles method is used to investigate the effect of rare earth Y element on the tensile properties of Mg-based alloys. Under the strains of 0–20%, the crystal structures of Mg_{54}, $Mg_{53}Y_1$, and $Mg_{52}Y_2$ can all remain stable. The structure of $Mg_{52}Y_2$ is more stable than that of Mg_{54} and $Mg_{53}Y_1$, indicating that rare earth element Y can enhance the stability of Mg-based alloya. When the Y atoms are dissolved in the Mg-based alloy at a content of 1.8 at.% and 3.7 at.%, the yield strength and tensile strength of the Mg-based alloy can be promoted. The theoretical tensile strength values of Mg_{54}, $Mg_{53}Y_1$, and $Mg_{52}Y_2$ are 5.15 GPa, 5.37 GPa and 5.63 GPa, respectively. When the Y atom content is 3.7 at.%, the enhancement effect on the Mg-based alloy is better than 1.8 at.%.

Author Contributions: Writing—original draft preparation, Y.G.; software, C.W., W.F. and Y.H.; writing—review and editing, X.C., H.H. and J.Y.; project administration Y.G. and Y.H.; All authors have read and agreed to the published version of the manuscript.

Funding: This research was funded by the Scientific Research Project of Liaoning Provincial Department of Education, grant number LQN202001; the Liaoning Provincial Natural Science Foundation Program Guidance Plan; grant number 2019-ZD-0484.

Institutional Review Board Statement: Not applicable.

Informed Consent Statement: Not applicable.

Data Availability Statement: Not applicable.

Conflicts of Interest: The authors declare that they have no conflict of interest.

References

1. Patel, H.A.; Chen, D.L.; Bhole, S.D.; Sadayappan, K. Microstructure and tensile properties of thixomolded magnesium alloys. *J. Alloys Compd.* **2010**, *496*, 140–148. [CrossRef]
2. Bakhsheshi-Rad, H.R.; Abdellahi, M.; Hamzah, E.; Ismail, A.F.; Bahmanpour, M. Modelling corrosion rate of biodegradable magnesium-based alloys: The case study of Mg-Zn-RE-xCa (x=0, 0.5, 1.5, 3 and 6wt%) alloys. *J. Alloys Compd.* **2016**, *687*, 630–642. [CrossRef]
3. Kurzynowski, T.; Pawlak, A.; Smolina, I. The potential of SLM technology for processing magnesium alloys in aerospace industry. *Arch. Civ. Mech. Eng.* **2020**, *20*, 1–13. [CrossRef]
4. Wu, M.M.; Jiang, Y.; Wang, J.W.; Wu, J.; Tang, B.Y.; Peng, L.M.; Ding, W.J. Structural, elastic and electronic properties of mg(cu 1 x zn x) 2 alloys calculated by first-principles. *J. Alloys Compd.* **2011**, *509*, 2885–2890. [CrossRef]
5. Wang, M.; Xu, X.Y.; Wang, H.Y.; He, L.; Huang, M. Evolution of dislocation and twin densities in a Mg alloy at quasi-static and high strain rates. *Acta Mater.* **2020**, *201*, 102–113. [CrossRef]
6. Luo, Z.; Chen, X.H.; Song, K.; Liu, C.Q.; Dai, Y.; Zhao, D.; Pan, F.S. Effect of Alloying Element on Electromagnetic Interference Shielding Effectiveness of Binary Magnesium Alloys. *Acta Metall. Sin.* **2019**, *32*, 817–824. [CrossRef]
7. Yang, C.; Pan, F.; Chen, X.; Luo, N. Effects of Sm addition on electromagnetic interference shielding property of Mg-Zn-Zr alloys. *Appl. Phys. A* **2017**, *123*, 400.1–400.7. [CrossRef]
8. Guan, K.; Li. B.; Yang, Q.; Qiu, X.; Tian, Z.; Meng, J. Effects of 1.5 wt% samarium (Sm) addition on microstructures and tensile properties of a Mg-6.0Zn-0.5Zr alloy. *J. Alloys Compd.* **2018**, *735*, 1737–1749. [CrossRef]
9. Zucchi, F.; Grassi, V.; Frignani, A.; Monticelli, C.; Trabanelli, G. Electrochemical behaviour of a magnesium alloy containing rare earth elements. *J. Appl. Electrochem.* **2006**, *36*, 195–204. [CrossRef]
10. Rao, J.; Li, H. Oxidation and ignition behavior of a magnesium alloy containing rare earth elements. *Int. J. Adv. Manuf. Tech.* **2010**, *51*, 225–231. [CrossRef]
11. Tsai, Y.; Chou, C.; Jeng, R.; Lee, S.; Lin, C. Effect of rare earth elements addition on microstructures and mechanical properties of A356 alloy. *Int. J. Cast Metal. Res.* **2011**, *24*, 83–87. [CrossRef]
12. Coy, A.E.; Viejo, F.; Skeldon, P.; Thompson, G.E. Susceptibility of rare-earth-magnesium alloys to micro-galvanic corrosion. *Corros. Sci.* **2010**, *52*, 3896–3906. [CrossRef]
13. Yang, M.B.; Hou, M.D.; Zhang, J.; Pan, F.S. Effects of Ce, Y and Gd additions on as-cast microstructure and mechanical properties of Mg-3Sn-2Sr magnesium alloy. *Trans. Nonferr. Metal. Soc.* **2014**, *24*, 2497–2506. [CrossRef]
14. Gorny, A.; Bamberger, M.; Katsman, A. High temperature phase stabilized microstructure in Mg-Zn-Sn alloys with Y and Sb additions. *J. Mater. Sci.* **2007**, *42*, 10014–10022. [CrossRef]
15. Yang, M.; Pan, F. Effects of Y addition on as-cast microstructure and mechanical properties of Mg-3Sn-2Ca (wt.%) magnesium alloy. *Mat. Sci. Eng. A* **2009**, *525*, 112–120. [CrossRef]
16. Huang, X.; Suzuki, K.; Saito, N. Textures and stretch formability of Mg-6Al-1Zn magnesium alloy sheets rolled at high temperatures up to 793 K. *Scr. Mater.* **2009**, *60*, 651–654. [CrossRef]
17. Chen, H.L.; Lin, L.; Mao, P.L.; Liu, Z. Phase stability, electronic, elastic and thermodynamic properties of Al-RE intermetallics in Mg-Al-RE alloy: A first principles study. *J. Magnes. Alloys* **2015**, *3*, 197–202. [CrossRef]
18. Wang, C.; Han, P.; Zhang, L.; Zhang, C.; Yan, X.; Xu, B. The strengthening effect of Al atoms into Mg-Al alloy: A first-principles study. *J. Alloys Compd.* **2009**, *482*, 540–543. [CrossRef]
19. Zhang, Y.; Lü, G.H.; Deng, S.H.; Wang, T.M. First-principles computational tensile test on an Al grain boundary. *Acta Phys. Sin.* **2006**, *55*, 2901–2907. (In Chinese)
20. Wang, C.; Huang, T.L.; Wang, H.Y.; Xue, X.N.; Jiang, Q.C. Effects of distributions of Al, Zn and Al+Zn atoms on the strengthening potency of Mg alloys: A first-principles calculations. *Comp. Mater. Sci.* **2015**, *104*, 23–28. [CrossRef]
21. Luo, S.Q.; Tang, A.T.; Pan, F.S.; Song, K.; Wang, W.Q. Effect of mole ratio of Y to Zn on phase constituent of Mg-Zn-Zr-Y alloys. *Trans. Nonferr. Metal. Soc.* **2011**, *21*, 795–800. [CrossRef]
22. Tong, L.B.; Li, X.H.; Zhang, H.J. Effect of long period stacking ordered phase on the microstructure, texture and mechanical properties of extruded Mg-Y-Zn alloy. *Mater. Sci. Eng. A* **2013**, *563*, 177–183. [CrossRef]
23. Giusepponi, S.; Celino, M. The ideal tensile strength of tungsten and tungsten alloys by first-principles calculations. *J. Nucl. Mater.* **2013**, *435*, 52–55. [CrossRef]
24. Mao, P.; Yu, B.; Liu, Z.; Wang, F.; Ju, Y. First-principles calculations of structural, elastic and electronic properties of AB_2 type intermetallics in Mg-Zn-Ca-Cu alloy. *J. Magnes Alloys* **2013**, *1*, 256–262. [CrossRef]
25. Mao, P.L.; Yu, B.; Liu, Z.; Wang, F.; Ju, Y. Mechanical, electronic and thermodynamic properties of Mg_2Ca Laves phase under high pressure: A first-principles calculation. *Comp. Mater. Sci.* **2014**, *88*, 61–70. [CrossRef]
26. Gironcoli, S.D.; Baroni, S.; Resta, R. Piezoelectric properties of III-V semiconductors from first-principles linear-response theory. *Phys. Rev. Lett.* **1989**, *62*, 2853–2856. [CrossRef] [PubMed]
27. Wang, F.; Sun, S.J.; Wang, Z.; Yu, B.; Mao, P.L.; Liu, Z. Microstructure, mechanical properties and first-principle analysis of vacuum die-cast Mg-7Al alloy with Sn addition. *Rare Metals* **2015**, *15*, 1–7. [CrossRef]
28. Jing, Z.; Zhang, X.; Li, W.; Pan, F.; Guo, Z. Partition of Er among the constituent phases and the yield phenomenon in a semi-continuously cast Mg-Zn-Zr alloy. *Scr. Mater.* **2010**, *63*, 367–370.

29. Lin, L.; Chen, L.; Liu, Z. Tensile strength improvement of an Mg-12Gd-3Y (wt%) alloy processed by hot extrusion and free forging. *J. Mater. Sci.* **2008**, *43*, 4493. [CrossRef]
30. Zheng, A. Effect of Rare Earth Y Content on Microstructure and Mechanical Property of Mg Alloy. *Hot Working Technol.* **2016**, *4*, 40–42. (In Chinese)

crystals

Review

Thermodynamic Modeling and Mechanical Properties of Mg-Zn-{Y, Ce} Alloys: Review

Mohammad Aljarrah [1,2,*], Jasim Alnahas [2] and Mohammed Alhartomi [3]

1 Industrial Engineering Department, Faculty of Engineering, The Hashemite University, P.O. Box 330127, Zarqa 13133, Jordan
2 Industrial Engineering Department, Faculty of Engineering, University of Tabuk, Tabuk 71491, Saudi Arabia; jalnahas@ut.edu.sa
3 Electrical Engineering Department, Faculty of Engineering, University of Tabuk, Tabuk 71491, Saudi Arabia; amalhartomi@ut.edu.sa
* Correspondence: maljarrah@hu.edu.jo

Abstract: Magnesium alloys are a strong candidate for various applications in automobile and aerospace industries due to their low density and specific strength. Micro-alloying magnesium with zinc, yttrium, and cerium enhances mechanical properties of magnesium through grain refinement and precipitation hardening. In this work, a critical review of magnesium-based binary systems including Mg-Zn, Mg-Y, Mg-Ce, Zn-Y, and Zn-Ce is presented. Based on the CALPHAD approach and first-principles calculations, thermodynamic modeling of Mg-Zn-Y and Mg-Zn-Ce ternary phase diagrams have been summarized. The influence of micro-alloying (yttrium and cerium) on the mechanical properties of magnesium is discussed. A comparison between mechanical properties of magnesium commercial alloys and magnesium–zinc–{yttrium and cerium} have been summarized in tables.

Keywords: thermodynamic modeling; magnesium; phase diagram; liquidus projection

Citation: Aljarrah, M.; Alnahas, J.; Alhartomi, M. Thermodynamic Modeling and Mechanical Properties of Mg-Zn-{Y, Ce} Alloys: Review. *Crystals* **2021**, *11*, 1592. https://doi.org/10.3390/cryst11121592

Academic Editor: Wojciech Polkowski

Received: 8 December 2021
Accepted: 15 December 2021
Published: 20 December 2021

1. Introduction

The need for weight reduction in automobile and aerospace industries makes magnesium alloys attractive due to their low density and high strength-to-weight ratio [1–3]. However, the use of magnesium alloys in structural parts is limited because of their poor mechanical properties at elevated temperatures [4–16]. Many researchers investigated the effect of micro-alloying on magnesium to enhance its mechanical performance [17–76]. The addition of rare-earth (RE) elements are attractive and receive increasing attention because of their excellent properties such as better creep resistance, grain refinement, improved ductility, enhanced formability, and strength [40,43–45,54,55,60,61,67–72,76].

Micro-alloying magnesium with RE such as zinc and yttrium resulted in promising mechanical properties [27,31,35,40,44,46,55,58,60]. RE elements enhance mechanical properties due to precipitation hardening through precipitation of nanoparticles of ternary phases [27–71]. These phases have an ability to inhibit the growth of deformation twins [18–23]. Furthermore, the addition of RE elements to Mg-Zn promote activation of prismatic slip and increase the stacking fault energy, therefore weakening the texture of magnesium alloys [35–40,44,51,61,66,72]. Micro-alloying magnesium with zinc increases its fluidity in casting [77], whereas yttrium addition has a remarkable effect on aging precipitation and high solid solution strengthening [78–80]. Moreover, cerium tends to precipitate a thermally high stable compound (Mg_2Ce) in magnesium rich region, which improve microstructure stability at elevated temperatures. Diluting zinc in Mg-Ce alloy significantly improves stretch formability by modifying the basal plane texture through solid solution hardening mechanism [81–85]. Moreover, the highest zinc in Mg-Ce alloy improves yield strength and ultimate tensile strength through precipitation of intermetallic compounds. Whereas the ratio of Ce/Zn increases, grain refinements and loss of formability occurs [71,81–88]. Mg-Zn-Y alloys display promising mechanical properties because of precipitates of thermally stable ternary compounds (W-Mg_3Y solid solution, I-Mg_3YZn_6, and LPSO-phase $Mg_{12}ZnY$) as well as high solubility of yttrium in magnesium.

To better understand phase stability, phase relation, and the effect of precipitation on age hardening, knowledge of binary and ternary phase diagrams is essential. Additionally, accurate prediction of phase diagram plays an important role in materials development and alloy design. Phase diagram is a tool used to predict the equilibrium phase(s) and phase(s) percentage at certain temperatures for specified alloys and simulate the phase consistency and solidification process of individual alloys. Moreover, the percentage of the predicted phase(s) that exist in the microstructure can be calculated. This will enable us to track particular alloys during solidification and subsequent heat treatment by predicting phase composition and distribution. Therefore, binary sub-systems of Mg-Zn-{Y, Ce} including Mg-Zn, Mg-Y, Mg-Ce, Zn-Y, and Zn-Ce phase diagram have been critically reviewed. In addition, ternary phase diagrams of Mg-Zn-Ce and Mg-Zn-Y have been assessed. A comparison between mechanical properties of commercial Mg-based alloys and Mg-Zn-{Ce,Y} alloys has been reported.

The CALPHAD approach is a well-known method to predict phase equilibria in a multi-component system based on Gibbs free energy of the phases [89–91]. Solid solutions were modeled using compound energy formalism with sublattice [92]. The modified quasi-chemical (MQC) solution model precisely describes short-range ordering in the liquid phase; therefore, liquid phase was optimized using MQC to treat configurational entropy [93]. The main novelty of the current work is to critically review phase equilibria of Mg-Zn-{Y, Zn} systems and mechanical properties based on the experimental investigations reported in the literature.

2. Zinc-Yttrium Phase Diagram

Chiotti et al. [94] largely examined phase diagram and thermodynamic data of Zn-Y phase diagram using DTA, metallographic, and XRD. Mason and Chiotti [95] subsequently reviewed the work of [94] and measured phase relation and thermodynamic properties of the intermetallic compounds using eight samples. In the work of [94,95], tantalum containers were unsuccessful because of the penetration of Y-Zn liquid at high zinc contents. Mason and Chiotti [95] reported three intermetallic compounds that melt congruently: YZn, YZn_2, and Y_2Zn_{17} ($YZn_{8.5}$) at 1105, 1080, and 890 °C, respectively. Thermodynamic modeling of Y-Zn binary phase diagram in the work of [96–98] presented a polymorphic transformation in the YZn_2 at 750 °C, which is in accord with [95,99]. Mason and Chiotti [95] found five intermetallic compounds that decompose peritectically: YZn_3, Y_3Zn_{11} ($YZn_{3.67}$), $Y_{13}Zn_{58}$ ($YZn_{4.46}$), YZn_6, and YZn_{12} at 905, 896, 882, 872, and 685 °C, respectively. Mason and Chiotti [95] determined the thermodynamic properties of the intermetallic compounds using dewpoint method. The large number of intermetallic compounds found in the RE-Zn system was similar and related to RE-coordination number [100]. Crystal structure data of Y-Zn compounds were determined by [100–103]. Gibbs energy of formation of the intermediate compounds in the Y-Zn system was investigated by [104–108]. The most accurate description of Y-Zn binary phase diagram was established by Zhu and Pelton [109] based on experimental data [94,95] as shown in Figures 1 and 2. The optimized Y-Zn phase diagram presented by Zhu and Pelton [109] presented some amendment to the work of Spencer et al. [98]. The calculated enthalpy and Gibbs energies of formation of the intermetallic compound presented in the work of [98] are in good agreement with the experimental data of [95,104,105,108].

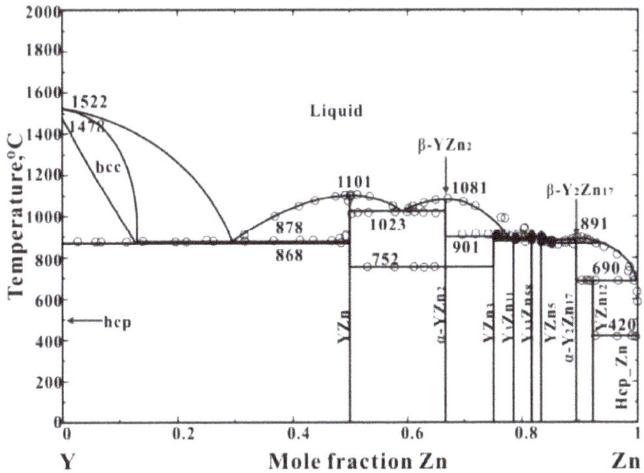

Figure 1. Yttrium–zinc phase diagram [109].

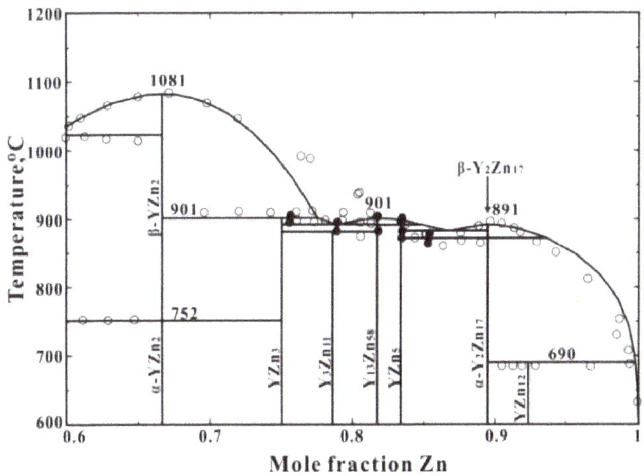

Figure 2. Yttrium–zinc phase diagram in Zn-rich region [109].

3. Zinc–Cerium Phase Diagram

The first Zn-Ce phase diagram was published by Hansen and Anderko [110]. Subsequently, Veleckis et al. [111] reported eight intermediate phases; $CeZn_{11}$, Ce_2Zn_{17}, CeZn, $CeZn_{8.8\text{-}6.2}$, $CeZn_2$, $CeZn_7$, Ce_2Zn, and Ce_4Zn. Okamoto and Hiroaki [112] suggested the existence of nine intermediate phases, namely CeZn, $CeZn_2$, $CeZn_3$, $CeZn_{3.67}$, $CeZn_{4.5}$, $CeZn_{5.25}$, $CeZn_7$, Ce_2Zn_{17}, and $CeZn_{11}$. The discrepancies in the stoichiometry and phase boundary reported by [110–112] were because of the delayed nucleation of these phases. Investigating the phase boundary and similarity of the Zn-Ce system to another Zn-RE phase diagram (such as Zn-Pr, Zn-Nd, Zn-Y, and Zn-Pm), nine intermetallic compounds were suggested [101,112–114]: CeZn, $CeZn_2$, $CeZn_3$, Ce_3Zn_{11}, $Ce_{13}Zn_{58}$, $CeZn_5$, Ce_3Zn_{22}, Ce_2Zn_{17}, and $CeZn_{11}$. A detailed investigation on the crystallographic data of intermetallic phases was presented in [114]. These intermediate compounds were included in the thermodynamic modeling of Zn-Ce phase diagram in the work of Wang et al. [115], Spencer et al. [98], and Zhu and Pelton [109]. The work of Chiotti and Mason [116] was the only experimental phase diagram data that could be found in the literature. Chiotti and Mason [116] investigated Zn-Ce phase diagram using metallography, differential thermal analysis (DTA), X-ray diffraction, and vapor pressure measurements. Johnson and Yonco [117] reported the standard Gibbs free energy of formation of the $CeZn_{11}$ phase, which was in accord with [116]. Chiotti and Mason [116] used dewpoint method to derive standard Gibbs free energy of formation for the intermetallic compounds. Johnson and Yonco [118] used the equation of standard Gibbs free energy to derive enthalpy of formation of the intermediate compounds.

Spencer et al. [98] and Zhu and Pelton [109] used modified quasi-chemical model to optimize liquid phase. Zn-Ce phase diagram published by [109] was an improvement to the work of Zhu and Pelton [109]. Zn-Ce phase diagram presented by Zhu and Pelton [109] is shown in Figure 3.

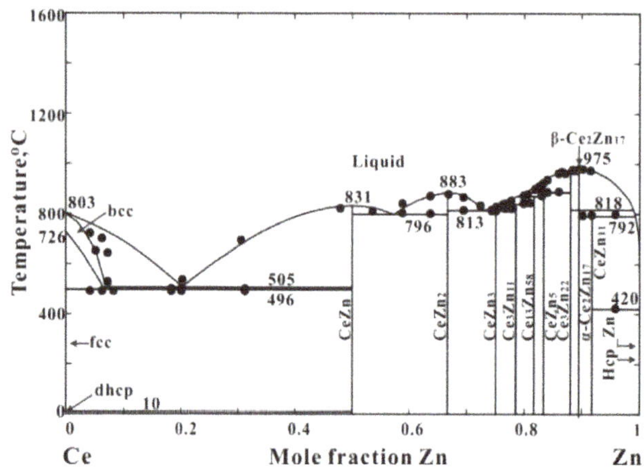

Figure 3. Zinc–cerium phase diagram calculated by [109].

4. Mg-Zn, Mg-Y, and Mg-Ce Phase Diagrams

Based on the literature, many researchers modeled liquid phase using a random solution model. This model is only anticipated at a very high temperature when the entropy term overwhelms any tendency for ordering or clustering of atoms. Therefore, the configurational entropy of mixing should vary with temperature. The modified quasi-chemical solution model has a better treatment of configurational entropy that accounts for a non-random distribution of atoms. Therefore, no model based on the random mixing can properly describe the influence of short-range ordering, because they do not solve the problem of the configurational entropy. The description of short-range ordering can be taken into account with bond energy models by considering the interactions between atoms that extend beyond the nearest neighbor's approximation. This problem has been treated using the modified quasi-chemical model. Liquid phase in the work of [77] was optimized using the modified quasi-chemical model (MQM). This model has been used to describe the liquid phase as this is the only scientific model that accounts for the presence of short-range ordering. Therefore, the reported phase diagrams in the work of [77] adequately describe thermodynamic properties of these systems. Islam et al. [77] critically reviewed and assessed thermodynamic data and phase diagrams of Mg-Zn, Mg-Y, and Mg-Ce systems. Figures 4–6 presented the most accurate calculated binary phase diagrams for these systems [77]. It is worth mentioning that the liquid phase was optimized using a modified quasi-chemical model to accurately describe short range ordering in the liquid.

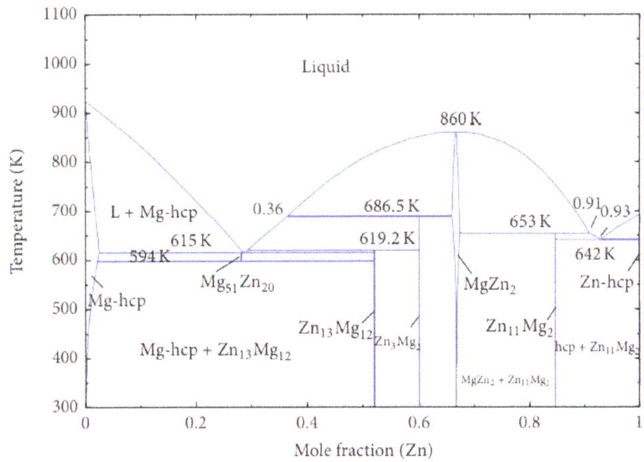

Figure 4. Mg-Zn phase diagram [119].

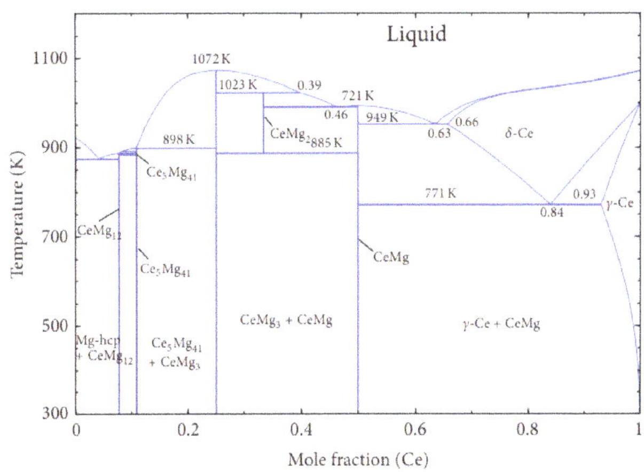

Figure 5. Mg-Ce phase diagram [77].

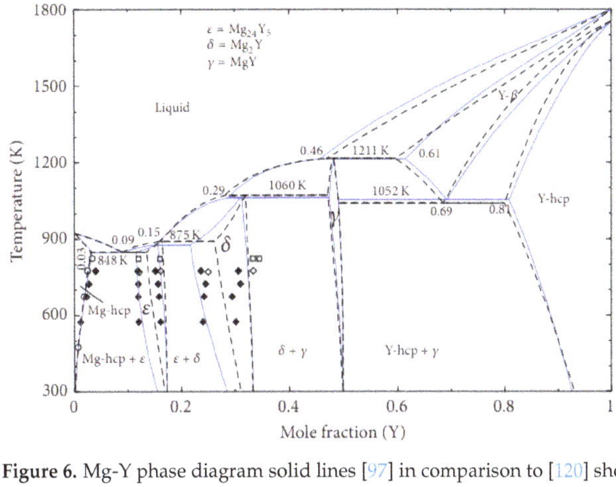

Figure 6. Mg-Y phase diagram solid lines [97] in comparison to [120] showed in dotted line [77].

5. Magnesium-Zinc-Yttrium Ternary Phase Diagram

Gröbner et al. [121] investigated the Mg-Zn-Y ternary system using ten ternary alloys by DSC, SEM/EDXS, and TEM. Based on their experimental results and assessment to the stoichiometric of ternary phases reported in the literature [96,122–140], Gröbner et al. [121] calculated liquidus projections and isothermal sections at 400, 500, and 600 °C. In 2015, Zhu and Pelton [140] calculated liquidus projection and isothermal sections at 400, 500, and 600 °C. Zhu and Pelton [140] defined ternary phase diagrams of Mg-Zn-RE systems using the Kohler model to estimate ternary properties of Mg-Zn-RE systems. It is worth mentioning that liquidus projections of Zhu and Pelton [140] and Gröbner et al. [121] are the only works that could be found in the literature. Gröbner et al. [121] modelled five ternary compounds: 18R, 14H, W, I, and Z, and one ternary solid solution (H). However, Zhu and Pelton [140] reported four ternary compounds (τ_5, H, X, and I phases) and three ternary solid solutions (Y(Mg,Zn), $Y_2(Mg,Zn)_{17}$, and τ_3 ($YMg(Mg,Zn)_2$).

Chemical compositions and notations of the ternary phases were confusing as described in the literature [96,121–140]. Many of the ternary phases reported in the literature were considered as metastable phases according to the work of Zhu and Pelton [140]. The slow kinetics of transformation of ternary phase, long-period stacking ordered (LPSO), has been described in the literature with different notations and chemical compositions [32–36,39,40,52,67,69,70,112,126,127,138]. This ternary phase exists in many Mg-Zn-RE systems which corresponds to $Mg_{12}ZnY_2$ [40,140] and was designated in the literature as X-phase with simplified composition $Mg_{12}YZn$ [96,127,140]. Ternary phase with notation of I-phase was reported by Tsai et al. [124] as $Mg_{30}Zn_{60}Y_{10}$ and later simplified as Mg_3Zn_6Y [138] and adopted in thermodynamic modeling in the work of [96,121,140]. Moreover, W-phase was reported in the work of [96] with composition of $Mg_3Zn_3Y_2$ and $Mg_{25}Zn_{60}Y_{14}$ [128], while Zhu and Pelton [140] and Gröbner et al. [121] described this phase as a ternary solid solution of yttrium in (MgZn) binary phase where yttrium may substitute magnesium and zinc element in the sublattice. Ternary phase designated as H-phase and composition of $Mg_{15}Zn_{70}Y_{15}$ [124] was accepted in the work of Zhu and Pelton [140]. Similarly to other Mg-Zn-RE ternary systems, this phase has been modeled as stoichiometric ternary compound. However, Gröbner et al. [121] describe this phase as ternary solid solubility of Mg in (YZn_5): $Y(Mg, Zn)_{1.5}Zn_{3.5}$ using the experimental data of [138]. Zhu and Pelton [140] treated H-phase differently because the crystallographic data (lattice constants) significantly differ from those of YZn_5 phase. Ternary solid solubility of Mg in $Zn_{17}Y_2$ binary phase, reported in the work of Zhu and Pelton- [140], was not observed in the liquidus projections of Gröbner et al. [121]. Based on the above confusion of the chemistry of ternary compounds, as well as ternary solid solutions in the Mg-Y-Zn system, further experimental investigation is required to resolve the discrepancies in the literature. Liquidus projections of the ternary Mg-Zn-Y phase diagram reported by Gröbner et al. [121] and Zhu and Pelton [140] are shown in Figure 7a,b, respectively.

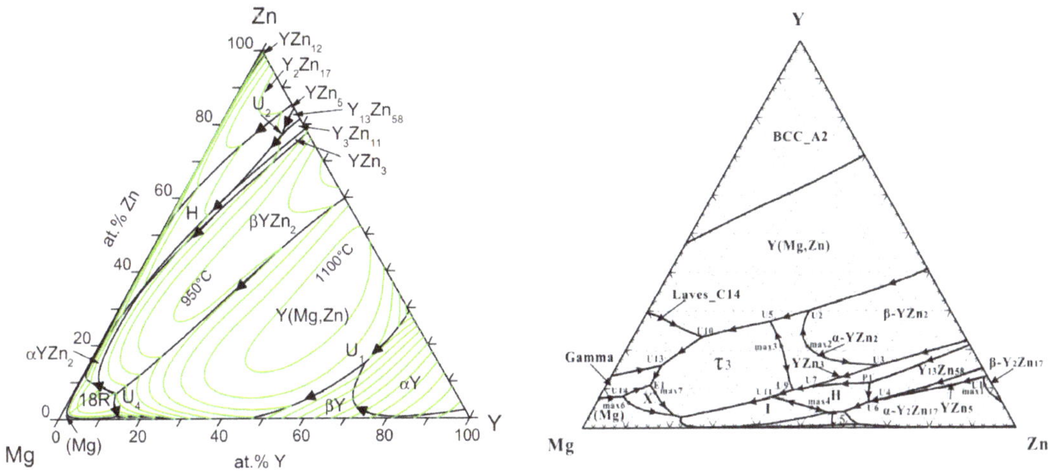

(**a**) (**b**)

Figure 7. Liquidus projections of the ternary Mg-Zn-Y phase diagram; (**a**) Gröbner et al. [121] and (**b**) Zhu and Pelton [140].

6. Magnesium–Zinc–Cerium Ternary Phase Diagram

Experimental investigation and thermodynamic modeling of the Mg-Zn-Ce received considerable attention by many researchers [141–153]. However, the reported ternary phases and ternary solid solutions were confusing. Table 1 summarizes the reported ternary phases in the Mg-Zn-Ce system [151].

Table 1. Reported ternary phases in the Mg-Zn-Ce system in comparison with literature.

Phase	[149]	[151]	[150]	[146]	[145]	[144]
$Mg_{11}Zn_{83}Ce_6$	$Ce(Mg_{1-x}Zn_x)_{11}$	$Ce(Mg_{1-y}Zn_y)_{11}$	$Ce(Mg_{0.14}Zn_{0.86})_{11}$		$Ce(Mg_xZn_{1-x})_{10.1}$	
(Mg,Zn)Ce	(Mg,Zn)Ce	$MgZn_4Ce$	$Mg_{19}Zn_{81}Ce_{20}$			
$(Mg,Zn150)_3Ce$	$(Mg,Zn)_3Ce$	$Mg_{2.3-x}Zn_{12.8+x}Ce$	Mg_xZn_yCe 1.2 x 2.3 12.8 y 13.9			$(Mg,Zn)_3Ce$
$Mg_{13}Zn_{30}Ce_3$	$Mg_7Zn_{12}Ce$	$Mg_7Zn_{12}Ce$			$Mg_7Zn_{12}Ce$	$Mg_7Zn_{12}Ce$
$Mg_5Zn_9Ce_2$	Mg_3Zn_5Ce	Mg_3Zn_5Ce	$Mg_{2.5}Zn_{4.5}Ce$			Mg_3Zn_5Ce
$Mg_{12}Ce$	$(Mg,Zn)_{12}Ce$	$Mg_3Zn_3Ce_2$	$Mg_{1+x}Zn_{2-x}Ce$	$Ce_{6.21}Mg_xZn_y$ 7.52 x 14.56 79.23 y 86.27		$Ce(Mg_xZn_{1-x})_9$
$Mg_{53}Zn_{45}Ce_2$	$Mg_{53}Zn_{45}Ce_2$	$Mg_{29}Zn_{25}Ce$	$Mg_{29.2}Zn_{24.8}Ce$	$Mg_{53.14}Zn_{45.04}Ce_{1.82}$		
$(Mg,Zn)_2Zn_9Ce_3$	$MgZn_2Ce$	$Mg_3Zn_{19}Ce_6$	$Mg_{13}Zn_{30}Ce_3$			
$Mg_{19}Zn_{81}Ce_{20}$						
Mg_3Ce	$(Mg,Zn)_2Ce$					

Ternary solid solution designated by Mg_3Ce was reported by [149,152] where zinc atom substitutes magnesium atom in the binary Mg_3Ce phase. The solubility of zinc in Mg_3Ce and crystallographic data are shown in Table 2. Recently, this phase was verified in the work of Shi et al. [153].

Table 2. Crystallographic data and Zn solubility in Mg_3Ce.

Phase Name	Pearson Symbol-Space Group	Zinc Solubility	Reference
Mg_3Ce	$cF16 - Fm\bar{3}m$	28 at.% at 300 °C	[149]
		28.4 at.% at 300 °C	[151]
		30 at.% at 300 °C	[144]
		36 at.% at 350 °C	[146]
		40 at.% at 197 °C	[150]
		6.4 at.% at 350 °C	[153]

Ternary solid solution, denoted as τ_3 in the work of [152,153], was reported by [144,150] with a prototype of $AlMnCu_2$. The percentage of zinc in this phase was 38 to 50 at.% [150], whereas Kevorkov and Pekguleryuz [148] reported this percentage as 45 to 50 at.%. Moreover, Mel'nik et al. [144] and Chiu et al. [149] reported that this phase contained 35 to 45, and 0 to 48 at.% Zn, respectively. It is worth noting that Chiu et al. [149] reported Mg_3Ce solid solution with two different prototypes, namely BiF_3 and $AlMnCu_2$. Whereas Shi et al. [153] verified the existence of τ_3 and indicated the difficulty to distinguish between τ_3 and Mg_3Ce because of the structural similarity. The authors of [153] declared the difficulty of detecting the difference using XRD. According to the BSE image in the work of [153], Mg_3Ce showed diamond shape whereas τ_3 exhibited irregular shape. Meanwhile, τ_3 exists in equilibrium with Mg_3Ce in accord with [152,153]. Reported comparisons between the crystallographic data and solid solubility for other ternary phases were summarized in the work of Chiu et al. [149] and Zhu et al. [152].

Liquidus projections of the Mg-Zn-Ce ternary phase diagram were calculated by [148–153]. The calculated ternary phase diagram by Chiu et al. [149] proposed that τ_3 and Mg_3Ce were similar phase, which contradicts the findings in [152,153]. Moreover, ternary phase diagram presented by Zhu et al. [152] was an amendment to the work of Mostafa and Medraj [151]. Primary crystallization regions of ternary phases could not be found in the liquidus projection of Shi et al. [153]. Liquidus projections of Mg-Zn-Ce calculated by Zhu et al. [152] and Shi et al. [153] are shown in Figure 8a,b, respectively.

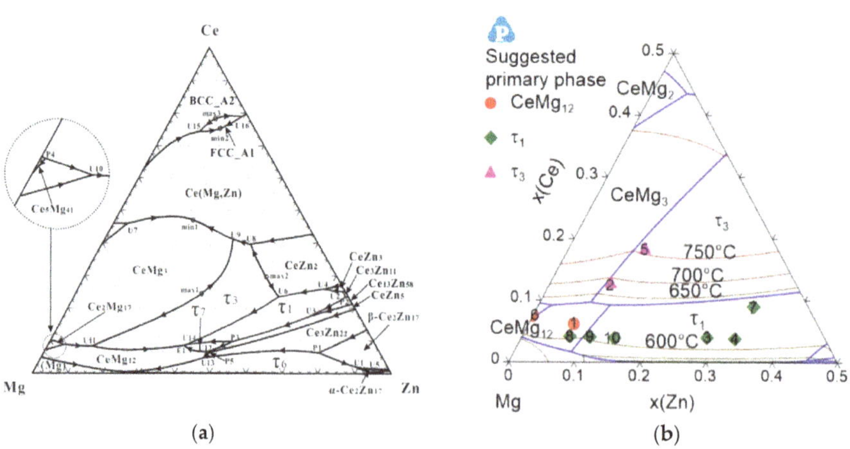

Figure 8. Calculated liquidus projections of Mg-Zn-Ce presented in (**a**) Zhu et al. [152] and (**b**) Shi et al. [153].

7. Mechanical Properties of Mg-Zn-{Ce, Y} Alloys

Alloying Mg-Zn with rare earth elements is promising in modifying magnesium texture. Among rare-earth element, many researchers reported that micro-alloying Mg-Zn with yttrium or cerium exhibited a comparable ductility and formability with commercial magnesium alloys. Tables 3 and 4 summarize the mechanical properties of the published alloys in Mg-Zn-Y [154–157] and Mg-Zn-Ce [48,71,81,85,156–158], respectively.

Table 3. Mechanical properties of Mg-Zn-Y alloys.

Nominal Composition (wt.%)	Yield Strength (MPa)	Ultimate Tensile Strength (MPa)	Ductility	Process Conditions
Mg-2Zn+0.4Y	160	240	30%	Samples were cast at 690 °C and then extruded at 310 °C [154]
Mg-14.4Zn-3.3Y	365 ± 3.5	380	8%	Samples were cast and solutionized at 480 °C for 24 h followed by extrusion at 430 °C, then aged at 150 °C [155]
Mg-14.4Zn-3.3Y	171	320	12	Samples were cast and solutionized at 480 °C for 24 h followed by extrusion at 430 °C [155]
Mg-1.5Zn-0.2Y	135	238	17%	The ingots were homogenized at 450 °C for 12 h, then rolled at 400 °C, and after that sheet annealed at 350 °C for 1 h [156]
Mg-6.0Zn-1.0Y	268.3		12.9%	Alloys were solutionized at 480 °C and then extruded at 390 °C [157]
Mg-6.0Zn-1.0Y	288.7		17.3%	Alloys were solutionized at 480 °C and then extruded at 390 °C and aged at 150 °C for 48h [157]
Mg-3.0Zn-0.5Y	262	18.3%		Alloys were solutionized at 480 °C and then extruded at 350 °C [157]

Table 4. Mechanical properties of Mg-Zn-Ce alloys.

Nominal Composition (wt.%)	Yield Strength (MPa)	Ultimate Tensile Strength (MPa)	Ductility	Process Conditions
Mg-2Zn+0.4Ce	190	255	18%	Samples were cast at 690 °C and then extruded at 310 °C [48]
Mg-2Zn-0.2Ce (ZE20)	69	170	31%	
Mg-5Zn-0.2Ce (ZE50)	135	247	15%	Samples were cast at 700 °C and then extruded at 400 °C [48]
Mg-8Zn-0.2Ce (ZE80)	136	289	16%	

Table 4. *Cont.*

Nominal Composition (wt.%)	Yield Strength (MPa)	Ultimate Tensile Strength (MPa)	Ductility	Process Conditions
Mg-1.5Zn-0.2Ce	140	240	19%	The ingots were homogenized at 450 °C for 12h, then rolled at 400 °C and after that sheet annealed at 350° C for 1 h [156]
Mg-6Zn-0.2Ce	225	270	30%	Alloys were cast at 750 °C and homogenized at 350 °C for 12 h. After extrusion, alloys aged at 175 °C from 0.5 to 80 h [157]
Mg-2%Zn-0.5%Ce (ZE20)	199.2	~245	6%	Samples were prepared by continuous casting, then homogenized at 823 K for 8 h. Sheets were rolled by conventional rolling at 673 K [158]
Mg-2%Zn-0.5%Ce (ZE20)	125	~235	13.8%	Samples were prepared by continuous casting, then homogenized at 823 K for 8 h. Sheets were rolled by conventional rolling at 673 K. Sheets were annealed at 673 K [158]
Mg-2%Zn-0.5%Ce (ZE20)	~170	~240	28.23	Samples were prepared by continuous casting, then homogenized at 823 K for 8 h. Sheets were rolled by packed rolling at 673 K. Sheets were annealed at 673 K [158]
Mg-2%Zn-0.5%Ce (ZE20)	~165	~236	33.4%	Samples were prepared by continuous casting, then homogenized at 823 K for 8 h. Sheets were rolled by packed rolling at 723 K. Sheets were annealed at 723 K [158]
Mg-0.5Zn-0.2Ce	133	213	25%	Samples were heated at 723 K for 20 min and sheets were rolled by unidirectional rolling at 353 K. Then, sheets were annealed at 623 K for 90 min [81]
Mg-1.0Zn-0.2Ce	110	202	23%	
Mg-1.5Zn-0.2Ce	116	206	29%	
Mg-2.0Zn-0.2Ce	118	222	25%	
Mg-2.5Zn-0.2Ce	131	228	16%	
Mg-1.5Zn-0.2Ce	153	231	26%	Samples were extruded at 703 K and then annealed at 623K for 90 min [85]

Table 4. *Cont.*

Nominal Composition (wt.%)	Yield Strength (MPa)	Ultimate Tensile Strength (MPa)	Ductility	Process Conditions
Mg-1.5Zn-0.2Ce	194	248	20%	Samples were extruded at 573K and then annealed at 623K for 90 min [85]
Mg-1.0Zn-1.0Ce	95	191	22%	Samples were annealed at 350 °C and then annealed at 450 °C for 1 h [71]
Mg-2.0Zn-1.0Ce	101	197	26.2%	
Mg-4Zn-1.0Ce	109	220	18%	
Mg-1.0Zn-0.5Ce	95	191	30%	

The addition of yttrium to Mg-Zn alloys enhances the formation of magnesium solid solution in ternary systems due to high solid solubility of yttrium in magnesium. Meanwhile, precipitation of nano-scale ternary phases as a result enhances mechanical properties. It is worth noting that the ratio of Zn/Y and heat treatment conditions play a significant role in mechanical properties as can be seen in Table 3. Unlike yttrium, the micro-addition of cerium to Mg-Zn reduces magnesium solid solution and precipitates a binary Mg12Ce phase as well as ternary nano-scale phases. Maximum solid solubility of cerium in magnesium is 0.5 wt.% at 590 °C and Mg12Ce precipitate up to 32.4 wt.% Ce. The existence of cerium and zinc in magnesium resulted in weakening structure and therefore enhanced ductility of magnesium alloys. Meanwhile, the percentage of cerium magnesium alloys must be in small amounts to hinder precipitation of high intensity of Mg12Ce phase; besides, heat treatment conditions and weight fraction of cerium play a remarkable role in mechanical properties of Mg-Zn alloys as shown in Table 4.

8. Conclusions

To reduce oil consumption in the automobile industry, designers are interested in lightweight alternative materials. Among lightweight materials is magnesium, and products of Mg-Zn alloys used in the automobile industry include transmission housings, heads, and engine blocks. In the current work, thermodynamic modeling of yttrium-zinc and yttrium-cerium phase diagrams were critically assessed, and the most appropriate phase diagrams were presented. Crystallographic data and solid solubilities of ternary phases in Mg-Zn-Y and Mg-Zn-Ce systems were evaluated. Lack of experimental data on ternary Mg-Zn-Y required further experimental investigations. Based on the recent findings, liquidus projections of the Mg-Zn-Y and Mg-Zn-Ce ternary phase diagrams were given. Ternary intermetallic phases and ternary solid solution reported in the literature were confusing, and additional key experiments are needed to resolve the discrepancies on the existence and chemical compositions of these phases. Mechanical properties reported in the literature of the two ternary systems were summarized.

Author Contributions: Conceptualization, M.A. (Mohammad Aljarrah); methodology, M.A. (Mohammad Aljarrah), J.A. and M.A. (Mohammed Alhartomi).; software, M.A. (Mohammad Aljarrah); validation, M.A. (Mohammad Aljarrah), J.A. and M.A. (Mohammed Alhartomi); formal analysis, M.A. (Mohammad Aljarrah), J.A. and M.A. (Mohammed Alhartomi); investigation, M.A. (Mohammad Aljarrah); resources, M.A. (Mohammad Aljarrah), J.A. and M.A. (Mohammed Alhartomi); data curation, M.A. (Mohammad Aljarrah); writing—original draft preparation, M.A. (Mohammad Aljarrah), J.A. and M.A. (Mohammed Alhartomi); writing—review and editing, M.A. (Mohammad Aljarrah); visualization, J.A.; supervision, M.A. (Mohammad Aljarrah); project administration, M.A. (Mohammad Aljarrah). All authors have read and agreed to the published version of the manuscript.

Funding: This research received no external funding.

Data Availability Statement: Not applicable.

Conflicts of Interest: The authors declare no conflict of interest.

References

1. Mordike, B.L.; Ebert, T. Magnesium: Properties-applications-potential. *Mater. Sci. Eng. A* **2001**, *302*, 37–45. [CrossRef]
2. Luo, A.A.; Sachdev, A.K.; Powell, B.R. Advance casting technology for lightweight automotive applications. *China Foundry* **2010**, *7*, 463–469.
3. Pan, F.S.; Zhang, J.; Wang, J.F.; Yang, M.B.; Han, E.H.; Chen, R.S. Key R & D activities for development of new types of wrought magnesium alloys in China. *Trans. Nonferr. Met. Soc.* **2010**, *20*, 1249–1258.
4. Hong, Z.; Wei, L.; Xing, W.M.; Yu, Z. Study on ignition proof AZ91D magnesium alloy chips with cerium addition. *J. Rare Earths* **2005**, *23*, 466–469.
5. Wang, R.; Eliezer, A.; Gutman, E. An investigation on the microstructure of an AM50 magnesium alloy. *Mater. Sci. Eng. A* **2003**, *355*, 201–207. [CrossRef]
6. Kiani, M.; Gandikota, I.; Rohano, M.R.; Motoyama, K. Design of lightweight magnesium car body structure under crash and vibration constraints. *J. Magnes. Alloy* **2014**, *2*, 99–108. [CrossRef]
7. Wen, K.; Liu, K.; Wang, Z.; Li, S.; Du, W. Effect of microstructure evaluation on mechanical property of extruded Mg-$_{12}$Gd-$_{2}$Er-$_{1}$Xn-$_{0.6}$ Zr alloys. *J. Magnes. Alloy* **2015**, *3*, 23–28. [CrossRef]
8. Barnett, M.R.; Nave, M.D.; Bettles, C.J. Deformation microstructures and textures of some cold rolled mg alloys. *Mater. Sci. Eng. A* **2004**, *386*, 205–211. [CrossRef]
9. Stanford, N.; Atwell, D.; Beer, A.; Davies, C.; Barnett, M.R. Effect of microalloying with rare-earth elements on the texture of extruded magnesium-based alloys. *Scr. Mater.* **2008**, *59*, 772–775. [CrossRef]
10. Zhang, H.; Huang, G.; Wang, L.; Roven, H.J.; Xu, Z.; Pan, F. Improved ductility of magnesium alloys by a simple shear process followed by annealing. *Scr. Mater.* **2013**, *69*, 49–52. [CrossRef]
11. Sakamoto, M.; Akiyama, S.; Ogi, K. Suppression of Ignition and Burning of Molten Mg Alloys by Ca Bearing Stable Oxide Film. *J. Mater. Sci. Lett.* **1997**, *16*, 1048–1050. [CrossRef]
12. Stanford, N. The Effect of Calcium on the Texture, and Mechanical Properties of Extruded Mg–Mn–Ca Alloys. *Mater. Sci. Eng. A* **2010**, *528*, 314–322. [CrossRef]
13. Jing, B.; Yangshan, S.; Shan, X.; Feng, X.; Tianbai, Z. Microstructure and Tensile Creep Behavior of Mg–4Al Based Magnesium Alloys with Alkaline-Earth Elements Sr and Ca Additions. *Mater. Sci. Eng. A* **2006**, *419*, 181–188. [CrossRef]
14. Yoo, M.H. Slip, twinning, and fracture in hexagonal close-packed metals. *Metall. Trans. A* **1981**, *12*, 409–418. [CrossRef]
15. Tu, T.; Chen, X.; Zhao, C.; Yuan, Y.; Pan, F. A simultaneous increase of elastic modulus and ductility by Al and Li additions in Mg-Gd-Zn-Zr-Ag alloy. *Mater. Sci. Eng. A* **2020**, *771*, 138576. [CrossRef]
16. Zhang, A.; Kang, R.; Wu, L.; Pan, H.; Xie, H.; Huang, Q.; Liu, Y.; Ai, Z.; Ma, L.; Ren, Y.; et al. A new rare-earth-free Mg-Sn-Ca-Mn wrought alloy with ultra-high strength and good ductility. *Mater. Sci. Eng. A* **2019**, *754*, 269–274. [CrossRef]
17. Shin, D.H.; Kim, I.; Kim, J.; Kim, Y.S.; Semiatin, S.L. Microstructure development during equal-channel angular pressing of titanium. *Acta Mater.* **2003**, *51*, 983–996. [CrossRef]
18. Sun, H.Q.; Shi, Y.N.; Zhang, M.X.; Lu, K. Plastic strain-induced grain refinement in the nanometer scale in a Mg alloy. *Acta Mater.* **2007**, *55*, 975–982. [CrossRef]
19. Robson, J.D.; Henry, D.T.; Davis, B. Particle effects on recrystallization in magnesium–manganese alloys: Particle-stimulated nucleation. *Acta Mater.* **2009**, *2009*, 2739–2747. [CrossRef]
20. Karakulak, E. A review: Past, present, and future of grain refining of magnesium castings. *J. Magnes. Alloy.* **2019**, *7*, 355–369. [CrossRef]
21. Zhao, D.; Chen, X.; Ye, J.; Chen, T.; Dai, Y.; Liu, C.; Luo, Z.; Gao, S.; Zhang, J.; Yao, J.; et al. Simultaneously improving elastic modulus and damping capacity of extruded Mg-Gd-Y-Zn-Mn alloy via alloying with Si. *J. Alloys Compd.* **2019**, *810*, 425–435. [CrossRef]
22. Shen, Y.F.; Guan, R.G.; Zhao, Z.Y.; Misra, R.D.K. Ultrafine-grained Al–$_{0.2}$Sc–$_{0.1}$Zr alloy: The mechanistic contribution of nano-sized precipitates on grain refinement during the novel process of accumulative continuous extrusion. *Acta Mater.* **2015**, *100*, 247–255. [CrossRef]
23. Chen, Y.; Yang, Y.; Feng, Z.; Zhao, G.; Huang, B.; Luo, X.; Zhang, Y.; Zhang, W. Microstructure, microtexture and precipitation in the ultrafine-grained surface layer of an Al-Zn-Mg-Cu alloy processed by sliding friction treatment. *Mater. Charact.* **2017**, *123*, 1890197. [CrossRef]
24. Fu, L.; Li, Y.; Jiang, F.; Huang, J.; Xu, G.; Yin, Z. On the role of Sc or Er micro-alloying in the microstructure evolution of Al-Mg alloy sheets during annealing. *Mater. Charact.* **2019**, *157*, 109918. [CrossRef]
25. Lim, T.S.; Ryu, H.S.; Hong, S.-H. Plasma electrolytic oxidation/cerium conversion composite coatings for the improved corrosion protection of AZ31 Mg alloys. *J. Electrochem. Soc.* **2012**, *160*, C77.
26. Castellanos, A.; Altube, A.; Vega, J.M.; García-Lecina, E.; Díez, J.A.; Grande, H.J. Effect of different post-treatments on the corrosion resistance and tribological properties of AZ91D magnesium alloy coated PEO. *Surf. Coat. Tech.* **2015**, *278*, 99–107. [CrossRef]
27. Pérez, P.; Cabeza, S.; Garcés, G.; Adeva, P. Influence of long period stacking ordered phase arrangements on the corrosion behaviour of extruded $Mg_{97}Y_2Zn_1$ alloy. *Corros. Sci.* **2016**, *107*, 107–112. [CrossRef]
28. Lapovok, R.; Gao, X.; Nie, J.-F.; Estrin, Y.; Mathaudhu, S.N. Enhancement of properties in cast Mg–Y–Zn rod processed by severe plastic deformation. *Mater. Sci. Eng. A* **2014**, *615*, 198–207. [CrossRef]

29. Liu, C.; Chen, X.; Zhang, W.; Zhang, Y.; Pan, F. Microstructure, creep behavior and corrosion resistance in the ultrafine-grained surface layer of Mg-6Zn-0.2Y-0.4Ce-0.5Zr alloy processed by surfacing friction treatment. *Mater. Sci. Eng. A* **2020**, *776*, 138995 [CrossRef]
30. Laleh, M.; Kargar, F.; Rouhaghdam, A.S. Investigation of rare earth sealing of porous micro-arc oxidation coating formed on AZ91D magnesium alloy. *J. Rare Earths* **2012**, *30*, 1293–1297. [CrossRef]
31. Chen, B.; Lin, D.; Zeng, X.; Lu, C. Effects of yttrium and zinc addition on the microstructure and mechanical properties of Mg–Y–Zn alloys. *J. Mater. Sci.* **2010**, *45*, 2510–2517. [CrossRef]
32. Yamasaki, M.; Matsushita, M.; Hagihara, K.; Izuno, H.; Abe, E.; Kawamura, Y. Highly ordered 10H-type long-period stacking order phase in a Mg–Zn–Y ternary alloy. *Scr. Mater.* **2014**, *78–79*, 13–16. [CrossRef]
33. Lyu, J.; Kim, J.; Liao, H.; She, J.; Song, J.; Peng, J.; Pan, F.; Jiang, B. Effect of substitution of Zn with Ni on microstructure evolution and mechanical properties of LPSO dominant Mg–Y–Zn alloys. *Mater. Sci. Eng. A* **2020**, *773*, 138735. [CrossRef]
34. Hagihara, K.; Kinoshita, A.; Sugino, Y.; Yamasaki, M.; Kawamura, Y.; Yasuda, H.Y.; Umakoshi, Y. Effect of long-period stacking ordered phase on mechanical properties of $Mg_{97}Zn_1Y_2$ extruded alloy. *Acta Mater.* **2010**, *58*, 6282–6293. [CrossRef]
35. Tong, L.B.; Li, X.H.; Zhang, H.J. Effect of long period stacking ordered phase on the microstructure, texture, and mechanical properties of extruded Mg–Y–Zn alloy. *Mater. Sci. Eng. A* **2013**, *563*, 177–183. [CrossRef]
36. Zhu, S.M.; Lapovok, R.; Nie, J.F.; Estrin, Y.; Mathaudhu, S.N. Microstructure and mechanical properties of LPSO phase dominant $Mg_{85.8}Y_{7.1}Zn_{7.1}$ and $Mg_{85.8}Y_{7.1}Ni_{7.1}$ alloys. *Mater. Sci. Eng. A* **2017**, *692*, 35–42. [CrossRef]
37. Victoria-Hernández, J.; Yi, S.; Klaumünzer, D.; Letzig, D. Recrystallization behavior and its relationship with deformation mechanisms of a hot rolled Mg-Zn-Ca-Zr alloy. *Mater. Sci. Eng. A* **2019**, *761*, 138054. [CrossRef]
38. Zhang, W.; Wei, Q.; Huo, W.T.; Lu, J.W.; Hu, J.J.; Zhang, Y.S. Dynamic recrystallization in nanocrystalline AZ31 Mg-alloy. *Vacuum* **2017**, *143*, 236–240. [CrossRef]
39. Xu, D.; Han, E.; Xu, Y. Effect of long-period stacking ordered phase on microstructure, mechanical property, and corrosion resistance of Mg alloys: A review. *Proc. Natl. Sci.-Mater.* **2016**, *26*, 117–128. [CrossRef]
40. Kawamura, Y.; Yamasaki, M. Formation and mechanical properties of $Mg_{97}Zn_1RE_2$ alloys with long-period stacking ordered structure. *Mater. Trans.* **2007**, *48*, 2986–2992. [CrossRef]
41. Patel, V.; Li, W.; Liu, X.; Wen, Q.; Su, Y.; Shen, J.; Fu, B. Tailoring grain refinement through thickness in magnesium alloy via stationary shoulder friction stir processing and copper backing plate. *Mater. Sci. Eng. A* **2020**, *784*, 139322. [CrossRef]
42. Shahnam, A.; Karimzadeh, F.; Golozar, M.A.; Hosseini, S.N. Microstructure evolution of ultra-fine-grained AZ31 B magnesium alloy produced by submerged friction stir processing. *J. Mater. Eng. Perform.* **2019**, *28*, 4593–4601. [CrossRef]
43. Sun, W.T.; Qiao, X.G.; Zheng, M.Y.; Xu, C.; Kamado, S.; Zhao, X.J.; Chen, H.W.; Gao, N.; Starink, M.J. Altered ageing behaviour of a nanostructured Mg-$_{8.2}$Gd-$_{3.8}$Y-$_{1.0}$Zn$_{0.4}$Zr alloy processed by high pressure torsion. *Acta Mater.* **2018**, *151*, 260–270. [CrossRef]
44. Yamasaki, M.; Hashimoto, K.; Hagihara, K.; Kawamura, Y. Effect of multimodal microstructure evolution on mechanical properties of Mg–Zn–Y extruded alloy. *Acta Mater.* **2011**, *59*, 3646–3658. [CrossRef]
45. Liu, H.; Ju, J.; Lu, F.; Yan, J.; Bai, J.; Jiang, J.; Ma, A. Dynamic precipitation behavior and mechanical property of an Mg94Y4Zn2 alloy prepared by multi-pass successive equal channel angular pressing. *Mater. Sci. Eng. A* **2017**, *682*, 255–259. [CrossRef]
46. Li, B.; Hou, X.; Teng, B. Effects of friction stir process and subsequent aging treatment on the microstructure evolution and mechanical properties of Mg-Gd-Y-Zn-Zr alloy. *Mater. Charact.* **2019**, *155*, 109832. [CrossRef]
47. Essadiqi, E.; Shehata, M.T.; Javaid, A.; Shen, G.; Aljarrah, M.; Verma, R.; Mishra, R. *Alloying and Process Design of Mg Sheet*; CANMET Materials: Ottawa, ON, Canada, 2011.
48. Jian, W.W.; Cheng, G.M.; Xu, W.Z.; Koch, C.C.; Wang, Q.D.; Zhu, Y.T.; Mathaudhu, S.N. Physics and model of strengthening by parallel stacking faults. *Appl. Phys. Lett.* **2013**, *103*, 133108. [CrossRef]
49. Shang, W.; Chen, B.; Shi, X.; Chen, Y.; Xiao, X. Electrochemical corrosion behavior of composite MAO/sol–gel coatings on magnesium alloy AZ91D using combined micro-arc oxidation and sol–gel technique. *J. Alloys Compd.* **2009**, *474*, 541–545. [CrossRef]
50. Zhang, Y.S.; Zhang, P.X.; Niu, H.Z.; Chen, C.; Wang, G.; Xiao, D.H.; Chen, X.H.; Yu, Z.T.; Yuan, S.B.; Bai, X.F. Surface nanocrystallization of Cu and Ta by sliding friction. *Mater. Sci. Eng. A* **2014**, *607*, 351–355. [CrossRef]
51. Zhang, W.; Lu, J.; Huo, W.; Zhang, Y.; Wei, Q. Microstructural evolution of AZ31 magnesium alloy subjected to sliding friction treatment. *Philos. Mag.* **2018**, *98*, 1576–1593. [CrossRef]
52. Jiang, M.G.; Xu, C.; Yan, H.; Fan, G.H.; Nakata, T.; Lao, C.S.; Chen, R.S.; Kamado, S.; Han, E.H.; Lu, B.H. Unveiling the formation of basal texture variations based on twinning and dynamic recrystallization in AZ31 magnesium alloy during extrusion. *Acta Mater.* **2018**, *157*, 53–71. [CrossRef]
53. González, S.; Pérez, P.; Garcés, G.; Adeva, P. Influence of the processing route on the mechanical properties at high temperatures of Mg-Ni-Y-RE alloys containing LPSO phases. *Mater. Sci. Eng. A* **2016**, *673*, 266–279. [CrossRef]
54. Li, X.; Liu, X.; Luan, B.L. Corrosion and wear properties of PEO coatings formed on AM60B alloy in $NaAlO_2$ electrolytes. *Appl. Surf. Sci.* **2011**, *257*, 9135–9141. [CrossRef]
55. Garces, G.; Muñoz-Morris, M.A.; Morris, D.G.; Perez, P.; Adeva, P. Optimization of strength by microstructural refinement of MgY_2Zn_1 alloy during extrusion and ECAP processing. *Mater. Sci. Eng. A* **2014**, *614*, 96–105. [CrossRef]
56. Aljarrah, M.; Essadiqi, E.; Kang, D.H.; Jung, I.H. Solidification microstructure and mechanical properties of hot rolled and annealed Mg sheet produced through twin roll casting route. *Mater. Sci. Forum* **2011**, *690*, 331–334. [CrossRef]

57. Huo, W.T.; Zhang, W.; Lu, J.W.; Zhang, Y.S. Simultaneously enhanced strength and corrosion resistance of Mg–$_3$Al–$_1$Zn alloy sheets with nano-grained surface layer produced by sliding friction treatment. *J. Alloys Compd.* **2017**, *720*, 324–331. [CrossRef]
58. Zhang, W.; Huo, W.T.; Lu, J.W.; Hu, J.J.; Wei, Q.; Zhang, Y.S. Gradient shear banding in a magnesium alloy induced by sliding friction treatment. *Vacuum* **2017**, *143*, 95–97. [CrossRef]
59. Zhao, C.; Li, Z.; Shi, J.; Chen, X.; Tu, T.; Luo, Z.; Cheng, R.; Atrens, A.; Pan, F. Strain hardening behavior of Mg–Y alloys after extrusion process. *J. Magnes. Alloy.* **2019**, *7*, 672–680. [CrossRef]
60. Zhao, D.; Chen, X.; Wang, X.; Pan, F. Effect of impurity reduction on dynamic recrystallization, texture evolution and mechanical anisotropy of rolled AZ31 alloy. *Mater. Sci. Eng. A* **2020**, *773*, 138741. [CrossRef]
61. Aljarrah, M.; Essadiqi, E. On the precipitates and mechanical properties of magnesium-yttrium sheets. *Alex. Eng. J.* **2013**, *52*, 221–225. [CrossRef]
62. Xu, S.W.; Zheng, M.Y.; Kamado, S.; Wu, K.; Wang, G.J.; Lv, X.Y. Dynamic microstructural changes during hot extrusion and mechanical properties of a Mg–$_{5.0}$ Zn–$_{0.9}$ Y–$_{0.16}$ Zr (wt.%) alloy. *Mater. Sci. Eng. A* **2011**, *528*, 4055–4067. [CrossRef]
63. Wang, C.; Liu, Y.; Lin, T.; Luo, T.; Zhao, Y.; Hou, H.; Yang, Y. Hot compression deformation behavior of Mg-$_5$Zn-$_{3.5}$Sn-$_1$Mn-$_{0.5}$Ca-$_{0.5}$Cu alloy. *Mater. Charact.* **2019**, *157*, 109896. [CrossRef]
64. Sitdikov, R.K.O. Dynamic recrystallization in pure magnesium. *Mater. Trans.* **2001**, *42*, 1928–1937. [CrossRef]
65. Chen, Y.; Yang, Y.; Feng, Z.; Huang, B.; Luo, X.; Zhang, W. The depth-dependent gradient deformation bands in a sliding friction treated Al-Zn-Mg-Cu alloy. *Mater. Charact.* **2017**, *132*, 269–279. [CrossRef]
66. Wei, Y.H.; Liu, B.S.; Hou, L.F.; Xu, B.S.; Liu, G. Characterization and properties of nanocrystalline surface layer in Mg alloy induced by surface mechanical attrition treatment. *J. Alloys Compd.* **2008**, *452*, 336–342. [CrossRef]
67. Woo, W.; Choo, H.; Brown, D.W.; Liaw, P.K.; Feng, Z. Texture variation and its influence on the tensile behavior of a friction-stir processed magnesium alloy. *Scr. Mater.* **2006**, *54*, 1859–1864. [CrossRef]
68. Nishioka, T.; Yamamoto, Y.; Kimura, K.; Hagihara, K.; Izuno, H.; Happo, N.; Hosokawa, S.; Abe, E.; Suzuki, M.; Matsushita, T.; et al. In-plane positional correlations among dopants in 10H type long period stacking ordered $Mg_{75}Zn_{10}Y_{15}$ alloy studied by X-ray fluorescence holography. *Materialia* **2018**, *3*, 256–259. [CrossRef]
69. Kawamura, Y.; Hayashi, K.; Inoue, A.; Masumoto, T. Rapidly solidified powder metallurgy $Mg_{97}Zn_1Y_2$ alloys with excellent tensile yield strength above 600 MPa. *Mater. Trans.* **2001**, *42*, 1172–1176. [CrossRef]
70. Zhang, J.; Xu, J.; Cheng, W.; Chen, C.; Kang, J. Corrosion behavior of Mg–Zn–Y alloy with long-period stacking ordered structures. *J. Mater. Sci. Tech.* **2012**, *28*, 1157–1162. [CrossRef]
71. Li, C.Q.; Xu, D.K.; Zeng, Z.R.; Wang, B.J.; Sheng, L.Y.; Chen, X.B.; Han, E.H. Effect of volume fraction of LPSO phases on corrosion and mechanical properties of Mg-Zn-Y alloys. *Mater. Des.* **2017**, *121*, 430–441. [CrossRef]
72. Aljarrah, M.; Essadiqi, E.; Fouad, R.H.; Rababah, M.; Almagableh, A. The effect of annealing conditions and alloying elements on the microstructure stability and mechanical properties of Mg-Zn-Ce sheets. *Appl. Mech. Mater.* **2014**, *472*, 937–947. [CrossRef]
73. Liu, L.; Pan, F.; Chen, X.; Huang, Y.; Song, B.; Yang, H.; Hort, N. The effect of Y addition on recrystallization and mechanical properties of Mg–$_6$Zn–xY–$_{0.5}$Ce–$_{0.4}$Zr alloys. *Vacuum* **2018**, *155*, 445–455. [CrossRef]
74. Mingo, B.; Arrabal, R.; Mohedano, M.; Llamazares, Y.; Matykina, E.; Yerokhin, A.; Pardo, A. Influence of sealing post-treatments on the corrosion resistance of PEO coated AZ91 magnesium alloy. *Appl. Surf. Sci.* **2018**, *433*, 653–667. [CrossRef]
75. Jain, V.; Su, J.Q.; Mishra, R.S.; Verma, R.; Javaid, A.; Aljarrah, M.; Essadiqi, E. *Microstructure and Mechanical Properties of Mg-$_{1.7}$ Y-$_{1.2}$ Zn Sheet Processed by Hot Rolling and Friction Stir Processing*; Springer: Cham, Switzerland, 2011; pp. 565–570.
76. Liu, M.; Uggowitzer, P.J.; Nagasekhar, A.V.; Schmutz, P.; Easton, M.; Song, G.-L.; Atrens, A. Calculated phase diagrams and the corrosion of die-cast Mg–Al alloys. *Corros. Sci.* **2009**, *51*, 602–619. [CrossRef]
77. Wang, G.; Huang, G.; Chen, X.; Deng, Q.; Tang, A.; Jiang, B.; Pan, F. Effects of Zn addition on the mechanical properties and texture of extruded Mg-Zn-Ca-Ce magnesium alloy sheets. *Mater. Sci. Eng. A* **2017**, *705*, 46–54. [CrossRef]
78. Islam, M.M.; Mostafa, A.O.; Medraj, M. Essential magnesium alloys binary phase diagrams and their thermodynamic data. *J. Mater.* **2014**, *2014*, 704283.
79. Stanford, N.; Cottam, R.; Davis, B.; Robson, J. Evaluating the Effect of Yttrium as a Solute Strengthener in Magnesium Using in Situ Neutron Diffraction. *Acta Mater.* **2014**, *78*, 1–13. [CrossRef]
80. Gao, L.; Chen, R.S.; Han, E.H. Solid Solution Strengthening Behaviors in Binary Mg–Y Single Phase Alloys. *J. Alloys Compd.* **2009**, *472*, 234–240. [CrossRef]
81. Gao, L.; Chen, R.S.; Han, E.H. Effects of Rare-Earth Elements Gd and Y on the Solid Solution Strengthening of Mg Alloys. *J. Alloys Compd.* **2009**, *481*, 379–384. [CrossRef]
82. Chino, Y.; Huang, Z.; Suzuki, K.; Sassa, K.; Mabuchi, M. Influence of Zn concentration on stretch formability at room temperature of Mg-Zn-Ce alloy. *Mater. Sci. Eng. A* **2010**, *528*, 566–572. [CrossRef]
83. Zhou, T.; Xia, H.; Chen, Z.H. Effect of Ce on microstructures and mechanical properties of rapidly solidified Mg–Zn alloy. *Mater. Sci. Tech.* **2011**, *27*, 1198–1205. [CrossRef]
84. Langelier, B.; Esmaeili, S. Effects of Ce additions on the age hardening response of Mg–Zn alloys. *Mater. Charact.* **2015**, *101*, 1–8. [CrossRef]
85. Luo, A.A.; Mishra, R.K.; Sachdev, A.K. High-ductility magnesium-zinc-cerium extrusion alloys. *Scr. Mater.* **2011**, *64*, 410–413. [CrossRef]

86. Chino, Y.; Huang, X.; Suzuki1, K.; Sassa, K.; Mabuchi, M. Microstructure, Texture and Mechanical Properties of Mg-Zn-Ce Alloy Extruded at Different Temperatures. *Mater. Trans.* **2011**, *52*, 1104–1107. [CrossRef]
87. Mackenzie, L.W.F.; Pekguleryuz, M.O. The recrystallization and texture of magnesium-zinc-cerium alloys. *Scr. Mater.* **2008**, *59*, 665–668. [CrossRef]
88. Guo, X.; Remennik, S.; Xu, C.; Shechtman, D. Development of Mg-6.0%Zn-1.0%Y-0.6%Ce-0.6% Zr magnesium alloy and its microstructural evolution during processing. *Mater. Sci. Eng. A* **2008**, *473*, 266–273. [CrossRef]
89. Guo, X.; Kinstler, J.; Glazman, L.; Shechtman, D. High strength Mg-Zn-Y-Ce-Zr alloy bars prepared by RD and extrusion technology. *Mater. Sci. Forum* **2005**, *488–489*, 495–498. [CrossRef]
90. Kaufmann, L.; Bernstein, H. *Computer Calculation of Phase Diagrams with Special Reference to Refractory Metals*; Academic Press: Cambridge, MA, USA, 1970.
91. Lukas, H. *Computational Thermodynamics: The Calphad Method*; Cambridge University Press: Cambridge, UK, 2007.
92. Hillert, M. The compound energy formalism. *J. Alloys Compd.* **2001**, *320*, 161–176. [CrossRef]
93. Pelton, A.D.; Degterov, S.A.; Eriksson, G.; Robelin, C.; Dessureault, Y. The modified quasichemical model I—binary solutions. *Metall. Mater. Trans. B* **2000**, *31*, 651–659. [CrossRef]
94. Chiotti, P.; Mason, J.T.; Gill, K.J. Phase Diagram and Thermodynamic Properties of the Yttrium—Zinc System. *Trans. TMS-AIME* **1963**, *227*, 910–916.
95. Mason, J.T.; Chiotti, P. Phase diagram and thermodynamic properties of the yttrium-zinc system. *Metall. Trans. A* **1976**, *7*, 287–291. [CrossRef]
96. Shao, G.; Varsani, V.; Fan, Z. Thermodynamic modelling of the Y-Zn and Mg-Zn-Y systems. *Calphad* **2006**, *30*, 286–295, Erratum in *Calphad* **2007**, *31*, 313. [CrossRef]
97. Liu, X.J.; Wen, M.Z.; Wang, C.P.; Pan, F.S. Thermodynamic assessment of the Zn-Y and Al-Zn-Y systems. *J. Alloys Compd.* **2008**, *452*, 283–290. [CrossRef]
98. Spencer, P.J.; Pelton, A.D.; Kang, Y.B.; Chartrand, P.; Fuerst, C.D. Thermodynamic assessment of the Ca-Zn, Sr-Zn, Y-Zn and Ce-Zn systems. *Calphad* **2008**, *32*, 423–431. [CrossRef]
99. Harsha, K.S. The Crystal Structures of Intermetallic Compounds in the Yttrium-Zinc System. Ph.D. Thesis, Pennsylvania State University, State College, PA, USA, 1964.
100. Veleckis, E.; Schablaske, R.V.; Johnson, I.; Feder, H.M. Intermetallic phases in the systems of zinc with lanthanum, cerium, praseodymium, neodymium, and yttrium. *Trans. TMS-AIME* **1967**, *239*, 58.
101. Bruzzone, G.; Fornasini, M.L.; Merlo, F. Rare-earth intermediate phases with zinc. *J. Less-Common Met.* **1970**, *22*, 253–264. [CrossRef]
102. Fornasini, M.L. Crystal structnre of (Ho-, Er-, Tm-, Lu-, Y-) Zn, and $ThCd_5$ intermetallic compounds. *J. Less-Common Met.* **1971**, *25*, 329–333. [CrossRef]
103. Ryba, E. Transformation in AB2 Intermetallic Compounds, US Atomic Energy Comm. Report COO-3415-3. 1963. Available online: https://www.osti.gov/servlets/purl/4272069 (accessed on 7 December 2021).
104. Butorov, V.P.; Nichkov, I.F.; Novikov, E.A.; Raspopin, S.P. Izv. Vyssh. Ucheb. Zaved. *Tsvetn. Met.* **1973**, *15*, 96.
105. Hoshino, Y.; Plambeck, J.A. Electrochemical studies of yttrium and ytterium-zinc alloys in fused LiCl-KCl eutectic. *Can. J. Chem.* **1970**, *48*, 685–687. [CrossRef]
106. Yamschikov, L.F.; Lebedev, V.A.; Nichkov, N.F. Metally. *Izv. Akad. Nauk SSSR* **1979**, *83*.
107. Marquina, C.; Kim-Ngan, N.H.; Bakker, K.; Radwanski, R.J.; Jacobs, T.H.; Buschow, K.H.J.; Franse, J.J.M.; Ibarra, M.R. Specific heats of R_2Zn_{17} intermetallic compounds. *J. Phys. Condens. Matter.* **1993**, *5*, 2009. [CrossRef]
108. Morishita, M.; Yamamoto, H.; Tsuboki, K.; Horike, T. Standard Gibbs energy of formation of $Zn_{17}Y2$ and $Zn_{12}Y$ determined by solution calorimetry and measurement of heat capacity near zero kelvin. *Int. J. Mater. Res.* **2007**, *98*, 10–15. [CrossRef]
109. Zhu, Z.; Pelton, A.D. Critical assessment and optimization of phase diagrams and thermodynamic properties of RE–Zn systems—Part II—Y–Zn, Eu–Zn, Gd–Zn, Tb–Zn, Dy–Zn, Ho–Zn, Er–Zn, Tm–Zn, Yb–Zn and Lu–Zn. *J. Alloys Compd.* **2015**, *641*, 261–271. [CrossRef]
110. Hansen, M.; Anderko, K. *Constitution of Binary Alloys*, 2nd ed.; McGraw-Hill Book Go., Inc.: New York, NY, USA, 1958.
111. Veleckis, E.; Rosen, C.L.; Feder, H.M. A recording effusion balance for phase diagram investigation: U-Cd, U-Zn and Ce-Zn systems. *Phys. Chem.* **1961**, *65*, 2127–2131. [CrossRef]
112. Okamoto, H.; Hiroaki, T.B. *Binary Alloy Phase Diagrams*; ASM International, Materials Park: Geauga County, OH, USA, 1990.
113. Kripyakevich, P.I.; Kuz'ma, Y.B.; Ugrin, N.S. Crystal Structures of the compounds Ce3Zn22, La3Zn22, and Pr3Zn22. *J. Struct. Chem.* **1967**, *8*, 632–633. [CrossRef]
114. Malik, Z.P. On the Quaternary Systems Ce-Ni-Zn-{B, Si}. Ph.D. Thesis, University of Vienna, Vienna, Austria, 2012.
115. Wang, C.P.; Chen, X.; Liu, X.J.; Pan, F.S.; Ishida, K. Thermodynamic modeling of the Ce-Zn and Pr-Zn systems. *J. Alloys Compd.* **2008**, *458*, 166–173. [CrossRef]
116. Chiotti, P.; Mason, J.T. Phase relations and thermodynamic properties for the cerium-zinc systems. *Trans. AIME* **1965**, *233*, 786–795.
117. Johnson, I.; Yonco, R.M. *Atomic Energy Commission Research and Development Report ANL 6231*; Office of Technical Services, U.S. Department of Commerce: Washington, DC, USA, 1960; p. 78.

118. Johnson, I.; Yonco, R.M. Thermodynamics of Cadmium- and Zinc-Rich Alloys in the Cd-La, Cd-Ce, Cd-Pr, Zn-La, Zn-Ce and Zn-Pr Systems. *Met. Trans.* **1970**, *1*, 905–910.
119. Ghosh, P.; Mezbahul-Islam, M.; Medraj, M. Critical assessment and thermodynamic modeling of Mg-Zn, Mg-Sn, Sn-Zn and Mg-Sn-Zn systems. *Calphad* **2012**, *36*, 28–43. [CrossRef]
120. Islam, M.M.; Kevorkov, D.; Medraj, M. The equilibrium phase diagram of the magnesium-copper-yttrium system. *J. Chem. Thermodyn.* **2008**, *40*, 1064–1076. [CrossRef]
121. Gröbner, J.; Kozlov, A.; Fang, X.Y.; Geng, J.; Nie, J.F.; Schmid-Fetzer, R. Phase equilibria and transformations in ternary Mg-rich Mg-Y-Zn alloys. *Acta Mater.* **2012**, *60*, 5948–5962. [CrossRef]
122. Ping, D.H.; Hono, K.; Kawamura, Y.; Inoue, A. Local chemistry of a nanocrystalline high-strength $Mg_{97}Y_2Zn_1$ alloy. *Philos. Mag. Lett.* **2002**, *82*, 543–551. [CrossRef]
123. Villars, P.; Calvert, L.D. *Pearson's Handbook of Crystallographic Data for Intermetallic Phases*; ASM International, Materials Park: Geauga County, OH, USA, 1985.
124. Tsai, A.P.; Murakami, Y.; Niikura, A. The Zn-Mg-Y phase diagram involving quasicrystals. *Philos. Mag. A* **2000**, *80*, 1043–1054. [CrossRef]
125. Singh, A.; Watanabe, M.; Kato, A.; Tsai, A.P. Formation of icosahedral-hexagonal H phase nano-composites in Mg-Zn-Y alloys. *Scr. Mater.* **2004**, *51*, 955–960. [CrossRef]
126. Zhu, Y.M.; Morton, A.J.; Nie, J.F. The 18R and 14H long-period stacking ordered structures in Mg-Y-Zn alloys. *Acta Mater.* **2010**, *58*, 2936–2947. [CrossRef]
127. Pedezhnova, E.M.; Mel'nik, E.V.; Miliyevskiy, R.A.; Dobatkina, T.V.; Kinzhibalo, V.V. Investigation of the Mg-Zn-Y System. *Russ. Metall.* **1982**, *4*, 185–188.
128. Matsuda, M.; Ii, S.; Kawamura, Y.; Ikuhara, Y.; Nishida, M. Variation of long-period stacking order structures in rapidly solidified $Mg_{97}Zn_1Y_2$ alloy. *Mater. Sci. Eng. A* **2005**, *393*, 269–274. [CrossRef]
129. Deng, D.W.; Kuo, K.H.; Luo, Z.P.; Miller, D.J.; Kramer, M.J.; Dennis, K.W. Crystal structure of the hexagonal Zn_3MgY. *J. Alloys Compd.* **2004**, *373*, 156–160. [CrossRef]
130. Takakura, H.; Sato, A.; Yamamoto, A.; Tsai, A.P. Crystal structure of a hexagonal phase and its relation to a quasicrystalline phase in Zn-Mg-Y alloy. *Philos. Mag. Lett.* **1998**, *78*, 263–270. [CrossRef]
131. Luo, Z.P.; Zhang, S.Q. High-resolution electron microscopy on the X-$Mg_{12}ZnY$ phase in ahigh strength Mg-Zn-Zr-Y magnesium alloy. *J. Mater. Sci. Lett.* **2000**, *19*, 813–815. [CrossRef]
132. Luo, Z.P.; Zhang, S.; Tang, Y.; Zhao, D. Quasicrystals in as-cast Mg-Zn-RE alloys. *Scr. Metall. Mater.* **1993**, *28*, 1513–1518. [CrossRef]
133. Luo, Z.P.; Zhang, S.Q.; Tang, Y.L.; Zhao, D.S. On the stable quasicrystals in slowly cooled Mg-Zn-Y alloys. *Scr. Metall. Mater.* **1995**, *32*, 1411–1416. [CrossRef]
134. Luo, Z.P.; Sui, H.X.; Zhang, S.Q. On the stable Mg-Zn-Y quasicrystals. *Metall. Mater. Trans. A* **1996**, *27*, 1779–1784. [CrossRef]
135. Tsai, A.P.; Nijkura, A.; Inoue, A.; Masumoto, T. Stoichiometric icosahedral phase in the Zn-Mg-Y system. *J. Mater. Res.* **1997**, *12*, 1468–1471. [CrossRef]
136. Abe, E.; Takakura, H.; Singh, H.; Tsai, A.P. Hexagonal superstructures in the Zn-Mg-rare-earth alloys. *J. Alloys Compd.* **1999**, *283*, 169–172. [CrossRef]
137. Li, M.R.; Kuo, K.H. Intermetallic phases and phase reactions in Zn-Mg (<40 at.%)-Y (<20 at.%) region. *J. Alloys Compd.* **2007**, *432*, 81–89. [CrossRef]
138. Shao, G.; Varsani, V.; Wang, Y.; Qian, M.; Fan, Z. On the solidification microstructure of $Mg_{-30}Zn_{-2.5}Y$ metal-intermetallic alloy. *Intermetallics* **2006**, *14*, 596–602. [CrossRef]
139. Li, M.R.; Zou, X.D.; Kuo, K.H. Anew hexagonal phase displaying pseudo-icosahedral symmetry in Zn-Mg-Y alloy. *Scr. Mater.* **2009**, *60*, 683–686. [CrossRef]
140. Zhu, Z.; Pelton, A.D. Thermodynamic modeling of the Y-Mg-Zn, Gd-Mg-Zn, Tb-Mg-Zn, Dy-Mg-Zn, Ho-Mg-Zn, Er-Mg-Zn, Tm-Mg-Zn and Lu-Mg-Zn systems. *J. Alloys Compd.* **2015**, *652*, 426–443. [CrossRef]
141. Huang, M.; Li, H.; Ding, H.; Ren, Y.; Qin, G.; Hao, S. Partial phase relationships of Mg-Zn-Ce system at 350 °C. *Trans. Nonferr. Met. Soc.* **2009**, *19*, 681–685. [CrossRef]
142. Huang, M.; Li, H.; Ding, H.; Bao, L.; Ma, X.; Hao, S. Intermetallics and phase relations of Mg-Zn-Ce alloys at 400 °C. *Trans. Nonferr. Met. Soc.* **2012**, *22*, 539–545. [CrossRef]
143. Korolkov, A.M.; Saldau, Y.P. Solubility of Zn and Ce in Mg in the solid state. *Izv Sekt. Fiz-Khim Anal.* **1946**, *16*, 295–306.
144. Mel'nik, E.V.; Kostina, M.F.; Yarmlyuk, Y.P.; Zmii, O.F. Study of the magnesium-zinc-cerium and magnesium-zinc-calcium ternary systems. *Magnievye Splavy. Mater. Vses. Soveshch. Issled. Razrab. Primen. Magnievyhk Splavov* **1978**, *9599*.
145. Drits, M.E.; Drozdova, E.I.; Korolkova, I.G.; Kinzhibalo, V.V.; Tyvanchuk, A.T. Investigation of polythermal sections of the Mg-Zn-Ce system in the Mg-rich region. *Russ. Metall.* **1989**, *2*, 195–197.
146. Kolitsch, U.; Bellen, P.; Kaesche, S.; Maccio, D.; Bochvar, N.; Liberov, Y.; Rogl, P. Cerium-magnesium-zinc. In *Ternary Alloys—A Comprehensive Compendium of Evaluated Constitutional Data and Phase Diagrams*; Effenberg, G., Petzow, G., Eds.; VCH Verlagsgesellscahft, MSI GmbH: Stuttgart, Germany, 2000; pp. 168–176.
147. Agarwal, R.; Fries, S.G.; Lukas, H.L.; Petzow, G.; Sommer, F.; Chart, T.G.; Effenberg, G. Assessment of the Mg-Zn system. *Z. Für Metall.* **1992**, *83*, 216–223. [CrossRef]

148. Kevorkov, D.; Pekguleryuz, M. Experimental study of the Ce-Mg-Zn phase diagram at 350 °C via diffusion couple techniques. *J. Alloys Compnd.* **2009**, *478*, 427–436. [CrossRef]
149. Chiu, C.; Gröbner, J.; Kozlov, A.; Schmid-Fetzer, R. Experimental study and thermodynamic assessment of ternary Mg-Zn-Ce phase relations focused on Mg-rich region. *Intermetallics* **2010**, *18*, 399–405. [CrossRef]
150. Pavlyuk, V.; Marciniak, B.; Różycka, E.S. The isothermal section of the phase diagram of Ce–Mg–Zn ternary system at 470 K. *Intermetallics* **2012**, *20*, 8–15. [CrossRef]
151. Mostafa, A.; Medraj, M. Phase equilibria of the Ce-Mg-Zn ternary system at 300 °C. *Metals* **2014**, *4*, 168–195. [CrossRef]
152. Zhu, Z.; Gharghouri, M.A.; Medraj, M.; Lee, S.Y.; Pelton, A.D. Thermodynamic modelling and in-situ neutron diffraction investigation of the (Ce + Mg + Zn) system. *J. Chem. Thermodyn.* **2016**, *93*, 242–254. [CrossRef]
153. Shi, H.; Li, Q.; Zhang, J.; Luo, Q.; Chou, K. Re-assessment of the Mg-Zn-Ce system focusing on the phase equilibria in Mg-rich corner. *Calphad* **2020**, *68*, 101742. [CrossRef]
154. Le, Q.C.; Zhang, Z.Q.; Shao, Z.W.; Cui, J.Z.; Xie, Y. Microstructures and mechanical properties of Mg-2%Zn-0.4%Re alloys. *Trans. Nonferr. Met. Soc. China* **2010**, *20*, 352–356. [CrossRef]
155. Singh, A.; Somekawa, H.; Mukai, T. High temperature processing of Mg–Zn–Y alloys containing quasicrystal phase for high strength. *Mater. Sci. Eng. A* **2011**, *528*, 6647–6651. [CrossRef]
156. Liu, P.; Jiang, H.; Cai, Z.; Kang, Q.; Zhang, Y. The effect of Y, Ce and Gd on texture, recrystallization and mechanical property of Mg-Zn alloys. *J. Magnes.* **2016**, *4*, 188–196. [CrossRef]
157. Singh, A. Tailoring microstructure of Mg-Zn-Y alloys with quasicrystal and related phases for high mechanical strength. *Sci. Technol. Adv. Mater.* **2014**, *15*, 44803. [CrossRef]
158. Du, Y.; Zheng, M.; Qiao, X.; Jiang, B. Enhancing the strength and ductility in Mg-Zn-Ce alloy through achieving high density precipitates and texture weakening. *Adv. Eng. Mater.* **2017**, *19*, 1700487. [CrossRef]

crystals

MDPI

Article

Study on Phase Transformation Orientation Relationship of HCP-FCC during Rolling of High Purity Titanium

Fengmei Bai [1,2], Qingliang Zhu [1,2], Jiaming Shen [3], Zhihan Lu [1,2], Liqiang Zhang [1,2,*], Naqash Ali [1,2], Hongwei Zhou [3,4,*] and Xianghua Liu [5]

1 Anhui Province Key Laboratory of Metallurgical Engineering & Resources Recycling, Anhui University of Technology, Ma'anshan 243002, China; baifengmei@ahut.edu.cn (F.B.); sergey.kurochkin1991@gmail.com (Q.Z.); Lzh_h@outlook.com (Z.L.); naqash.sanaullah322@gmail.com (N.A.)
2 School of Metallurgical Engineering, Anhui University of Technology, Ma'anshan 243032, China
3 School of Materials Science and Engineering, Anhui University of Technology, Ma'anshan 243002, China; sjm15068158690@outlook.com
4 State Key Laboratory of Metastable Materials Science and Technology, Yanshan University, Qinhuangdao 066004, China
5 State Key Laboratory of Rolling and Automation, Northeastern University, Shenyang 110189, China; liuxh@mail.neu.edu.cn
* Correspondence: zhangsir@ahut.edu.cn (L.Z.); hwzhou@ahut.edu.cn (H.Z.)

Abstract: High purity titanium (Ti) thin strip was prepared by rolling with large deformation and was characterized by the means of Transmission Electron Microscopy (TEM), selected area diffraction (SAED) pattern, high-resolution (HRTEM) analysis, as well as Transmission Kikuchi Diffraction (TKD). It is found that there are face-centered cubic (FCC) Ti laths formed within the matrix of hexagonal close packing (HCP) Ti. This shows that the HCP-FCC phase transition occurred during the rolling, and a specific orientation relationship (OR) between HCP phase and FCC phase obeys $\langle 0001\rangle_{\alpha}//\langle 001\rangle_{FCC}$ and $\{10\bar{1}0\}_{\alpha}//\{110\}_{FCC}$. The ORs of HCP-FCC phase transition are deeply studied by TKD pole figure and phase transformation matrix. It is found that the derived results via pole figure and transformation matrix are equivalent, and are consistent with TEM-SAED analysis results, which proves that these two methods can effectively characterize the ORs of HCP-FCC phase transition and predict possible FCC phase variants.

Keywords: high purity titanium; FCC phase; TKD; pole figure; transformation matrix

Citation: Bai, F.; Zhu, Q.; Shen, J.; Lu, Z.; Zhang, L.; Ali, N.; Zhou, H.; Liu, X. Study on Phase Transformation Orientation Relationship of HCP-FCC during Rolling of High Purity Titanium. *Crystals* **2021**, *11*, 1164. https://doi.org/10.3390/cryst11101164

Academic Editor: Wojciech Polkowski

Received: 20 August 2021
Accepted: 20 September 2021
Published: 24 September 2021

1. Introduction

Pure titanium (Ti) is a rare metal and widely used in aerospace, automobile, Micro-Electro-Mechanical System (MEMS) and other engineering fields. It is well known that pure Ti has HCP crystal structure and body-centered cubic (BCC) structure at ambient and high temperature conditions, respectively [1]. Pure Ti is easily oxidized during rolling process at high temperature, while room temperature is required to ensure its purity. However, the stable Ti phase at ambient conditions is HCP phase, which shows a poor ductility. In recent years, some new phase transitions of pure Ti have been discovered, such as an allotropic phase transformation from the HCP α-Ti to the hexagonal ω-Ti during high-pressure torsion (HPT) [2] and the HCP-Ti to FCC-Ti transition [3]. However, the latter is of great interest for the researchers. Hong et al. [3] refined pure titanium grains with equal-channel angular pressing (ECAP), and found the transition from HCP-Ti to FCC-Ti. The FCC-Ti is nucleated in the deformation cross-slip zone with the implementation of tensile deformation of pure titanium foil, which belongs to a stress-induced phase transformation. Other authors also have reported the HCP-Ti to FCC-Ti transformation during the deposition of pure Ti thin films [4], plastic deformation [1,5–9], and heat treatment process [1,10–13]. The stability of FCC-Ti in pure titanium has been proved by molecular dynamics theory [14] and first principles calculation [15,16]. The HCP-Ti to FCC-Ti transition also appears in titanium alloys [17,18] and in Nb-Ti-Si based alloys [19].

Niu et al. [20] found that the deformation mechanism was composed of dislocation slip, twins and HCP→FCC phase transition. The phase transition from HCP to FCC expands the volume of unit cell ~10.5%, and it is expected that the phase transition contributes about 3.6% to the applied strain. Bai et al. [21] considered that in the rolling process of high purity titanium, apart from twinning [22–24] and dislocation mechanism [25], the HCP-FCC phase transition has an important contribution to its elongation process. These show that the HCP-FCC phase transition can occur easily when pure titanium undergoes large plastic deformation at room temperature, which coordinates the deformation.

There will be a certain orientation relationship (OR) between the new phase and the matrix after phase transformation. Two kinds of ORs between HCP (α) and FCC have been discovered [6,9], known as B type: $\langle 1\bar{2}10\rangle_{\alpha}//\langle 1\bar{1}0\rangle_{FCC}$, $\{0001\}_{\alpha}//\{111\}_{FCC}$ and P type: $\langle 0001\rangle_{\alpha}//\langle 001\rangle_{FCC}$, $\{10\bar{1}0\}_{\alpha}//\{110\}_{FCC}$. Zheng [6] and Zu [26] found that the P type is accomplished by shuffle and shear shuffle mechanism, while B type depends on the mechanism of Shockley partial dislocation. Ren et al. [14] performed the molecular dynamics simulation of uniaxial tension of titanium single crystal nanocolumns along the $\{0001\}_{\alpha}$ plane, which revealed that the phase transition is mainly controlled by the local dislocation slip and stacking fault. In order to study the formation mechanism and characteristics of FCC new phase in matrix, the deep dicussion of OR between two phases is necessary.

Most researchers [3,6,20,26] study the phase transition and orientation relationship from HCP-Ti to FCC-Ti via TEM. In this paper, we utilize both TEM and TKD to observe the phase transition ORs. TKD clarified the ORs, which was consistent with that via TEM. Meanwhile, the pole figure and phase transition matrix of the ORs was obtained, based on the results of TEM and TKD analysis. When the phase transition between HCP and FCC obeys the specific OR, the parallel crystal directions and planes of the two phases can be accurately expressed by pole figures and phase transition matrix, and the variants meeting the OR can be predicted.

2. Experimental Material and Procedure

The raw material of high purity Ti (HP-Ti) was supplied by Ningbo Chuangrun New Materials CO., LTD, Ningbo, China. Sheet samples of 11 mm × 30 mm × 60 mm size were cut from the raw material, and multi-pass symmetrical rolling was performed to prepare a thin strip with a thickness of 0.86 mm with about 92% total reduction in thickness, as shown in Figure 1. Details of the equipment and processing for cold-rolling can be found in the previous literature reports [27,28]. Talos F200X TEM (Thermo Fisher, Waltham, MA, USA) examinations were used to investigate the FCC-Ti in HCP matrix. A 0.3 mm slice was cut from the sheets by wire cutting, then mechanically polished, and a sample with a diameter of 3 mm and a thickness of 50 μm was treated by ion milling in order to prepare the sample for TEM analysis.

The FCC-Ti is nanoscale in size and TKD was used to characterize the ORs between HCP-Ti and FCC-Ti. For efficient TKD testing, the cold-rolled strips were annealed at 500 °C for 1 h under Argon protection. A Zeiss AURIGA FIB-SEM workstation equipped with TKD capability was carried out to characterize the TEM foils. TKD testing was performed at 30 kV and a step size of 50 nm.

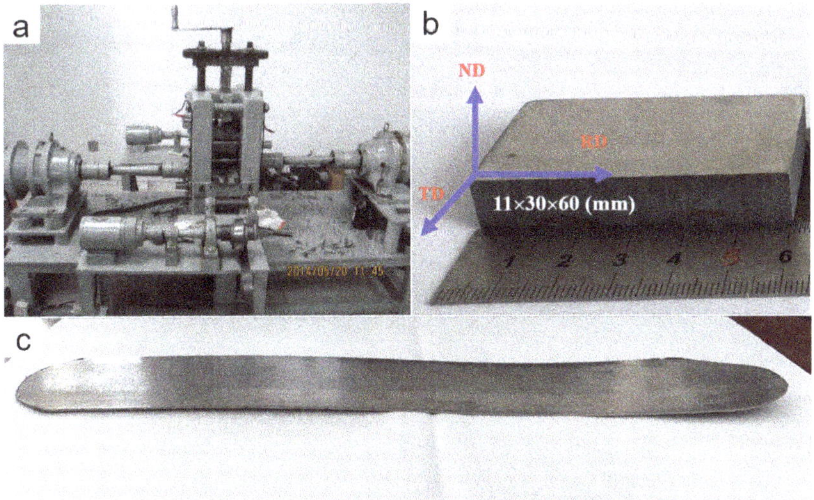

Figure 1. Micro-rolling mill and thin strip samples. (**a**) Experimental setup; (**b**) raw material of high purity; (**c**) a thin strip with a thickness of 0.86 mm.

3. Results and Discussion

3.1. TEM Observations

Figure 2 shows the TEM images of the samples after cold-rolling. Some deformation lath structure with the width ranging from 30 nm to 50 nm was found in Figure 2a. It observed that there is a deformation lath in a sub-grain with a size of about 1 μm, which divides the subgrain into two parts. The deformation zone almost runs through the whole grain and sub-grain, which plays a significant role for the grain refinement, and is similar to the twinning. Figure 1b shows the SAED pattern of deformation lath and its adjacent matrix. The analysis shows that the zone axis of hexagonal phase (α-HCP) is $[\bar{1}2\bar{1}3]_{\alpha}$, while the deformation band has FCC crystal structure, and the zone axis is $[112]_{FCC}$. The lattice parameters of HCP phase are calculated as $a = 0.298$ nm and $c = 0.471$ nm and the lattice parameter of FCC phase is $a_{FCC} = 0.425$ nm. These lattice parameters are consistent with the other studies [3,29]. This shows that the deformation laths in Figure 1a are FCC phase structure, instead of deformed twins at this deformation range [9,30]. In recent years, many researchers have reported that the pure titanium changes from HCP phase to FCC phase under large deformation along with the formation of FCC phase deformation lath [3,7,20,31]. The OR of the two phases obtained from Figure 2b is as follow:

$$[\bar{1}2\bar{1}3]_{\alpha}//[112]_{FCC} \quad (10\bar{1}0)_{\alpha}//(\bar{2}20)_{FCC} \quad (1\bar{2}12)_{\alpha}//(11\bar{1})_{FCC} \tag{1}$$

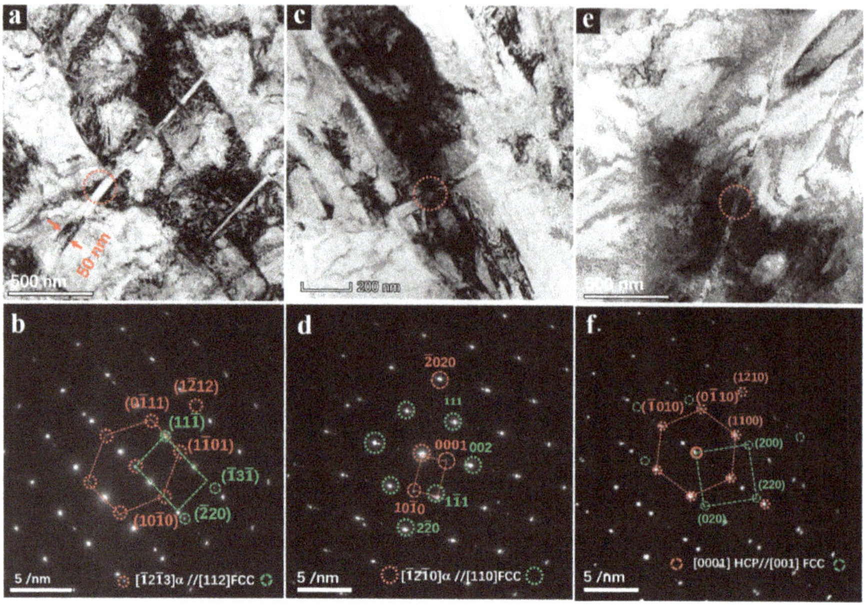

Figure 2. (**a**,**c**,**e**) TEM pictures and (**b**,**d**,**f**) SAED patterns of the FCC-Ti lath structure and matrix, three determination positions of SAED were remarked by circles in TEM pictures, respectively.

Figure 2d shows the SAED pattern analysis in which the deformation lath is marked by a circle in Figure 1c is also FCC-Ti phase, and its lattice constant is a = 0.425 nm. The OR between the two phases in Figure 2c,d is calibrated as follows:

$$[1\bar{2}10]_{\alpha}//[110]_{FCC} \quad (0002)_{\alpha}//(002)_{FCC} \quad (10\bar{1}0)_{\alpha}//(2\bar{2}\bar{0})_{FCC} \tag{2}$$

Figure 2e is another typical TEM picture of the FCC-Ti lath. The lattice parameters of HCP and FCC phases are consistent with the results in Figures 2b and 1d. The OR between HCP-Ti and FCC-Ti is:

$$[0001]_{\alpha}//[001]_{FCC} \quad (10\bar{1}0)_{\alpha}//(220)_{FCC} \quad (1\bar{2}10)_{\alpha}//(2\bar{2}0)_{FCC} \tag{3}$$

The β-Ti alloys mainly exhibit ω phase transformation, which is claimed to be caused of heterogeneous nucleation of the α phase [32]. Recently, the phase transformation from the HCP α-Ti to the hexagonal ω-Ti was appeared during the high-pressure torsion (HPT) [2]. In this work, hexagonal ω-Ti phase transition is not found by TEM analysis, which is examined by the difference of SAED patterns between hexagonal ω-Ti and HCP-Ti.

The previous literature reports show that the FCC phase in pure-Ti is Ti-H hydride, not FCC-Ti [33–35]. In our previous work [21], it is found that FCC phase has better plasticity than HCP-Ti, and FCC phase can reverse to HCP phase during deformation. If the FCC phase is Ti-H hydride, its plasticity will be poor. Meanwhile, the transition from FCC Ti-H to HCP phase will not exist during deformation. Therefore, it can be confirmed that the FCC phase in this work is FCC-Ti instead of FCC hydride.

Figure 3 shows the HRTEM images of the FCC phase lath. It can be seen in Figure 3a that the HCP/FCC interface is straight. Filter inverse fast Fourier transform (FFT) image of FCC and HCP phases are illustrated in Figure 3b,c. The FFT patterns in Figure 3e,f also proves the existence of two phases, and satisfies the OR in Expression (3). This result is consistent with other studies [1,3,4,11,12,14,18,31,36,37]. In Figure 3d, the inverse FFT analysis of HCP phase near to the FCC phase shows that the HCP matrix phase contains a quite few edge dislocations, and the extra half planes are parallel to the $\{10\bar{1}0\}$ planes. This means that the mechanism of phase transition from HCP to FCC is related with the dislocation slipping.

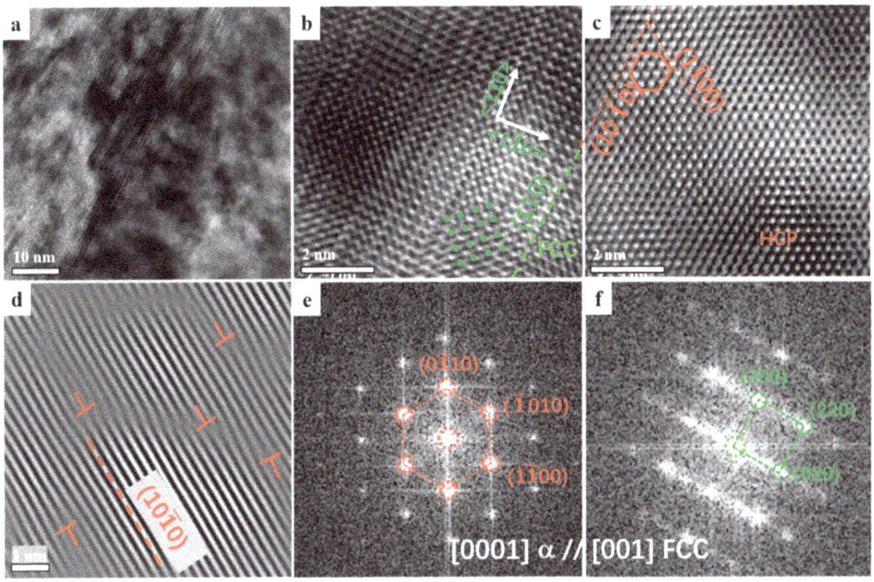

Figure 3. HRTEM and FFT of FCC-Ti laths and HCP matrix. (**a**) HRTEM; (**b**,**c**) Filter inverse FFT image; (**d**) dislocation in HCP-Ti lath; (**e**,**f**) FFT patterns of two phases in (**a**).

The microscopic mechanism of the phase transition from HCP to FCC is shown by the schematic diagram in Figure 4. The phase transition is caused by the Shockley partial dislocation sliding along the direction $\frac{1}{6}\langle\bar{1}2\bar{1}0\rangle$ on the prism surface $(10\bar{1}0)_\alpha$, which is consistent with the results of molecular dynamics theory calculation [14], first principles calculation and TEM analysis [3,8,15]. Therefore, the transition can improve the plastic deformation capability of HP-Ti.

The results of two-phase OR obtained by the SAED patterns analysis are counted in Table 1. There are three kinds of relationship formulas with different parallel orientation and planes in Figure 2, which are named as P_1, P_2, and P_3, respectively.

Table 1. The ORs deduced from the SAED patterns.

Type	ORs	Induced From
P_1	$[\bar{1}2\bar{1}3]_\alpha//[112]_{FCC}$ $(10\bar{1}0)_\alpha//(2\bar{2}0)_{FCC}$ $(1\bar{2}12)_\alpha//(11\bar{1})_{FCC}$	Figure 2b
P_2	$[1\bar{2}10]_\alpha//[110]_{FCC}$ $(0002)_\alpha//(\bar{2}00)_{FCC}$ $(10\bar{1}0)_\alpha//(2\bar{2}0)_{FCC}$	Figure 2d
P_3	$[0001]_\alpha//[001]_{FCC}$ $(10\bar{1}0)_\alpha//(220)_{FCC}$ $(1\bar{2}10)_\alpha//(2\bar{2}0)_{FCC}$	Figure 2f

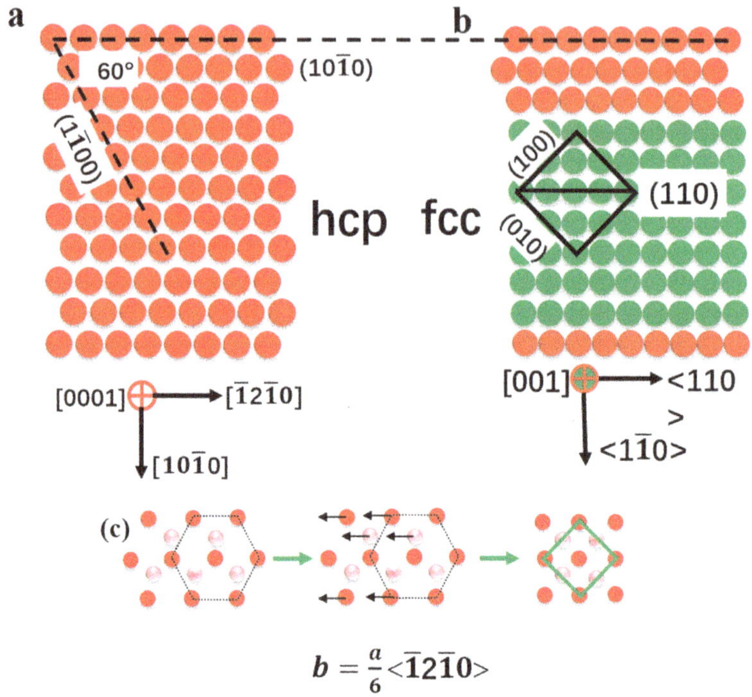

Figure 4. Schematic of HCP→FCC phase transient. (**a**) HCP-Ti phase structure; (**b**) FCC-Ti phase structure; (**c**) HCP to FCC phase transient by Shockley partial dislocation.

3.2. Pole Figures

It is well known that the conventional Electron backscattered diffraction (EBSD) can be used to determine the ORs between the matrix and the second phase with micron or even millimeter scale, for example, ferrite precipitates with its neighboring grain holding K-S OR via EBSD in medium carbon steel [38]. TEM results show that the formation of the FCC-Ti has been proposed to preferred the ORs with the HCP-Ti matrix. At the same time, the width of FCC-Ti laths was ranging from 30 nm to 50 nm. TKD has a significantly higher spatial resolution than the conventional EBSD [39]. TKD was widely applied in crystal ORs determination and phase mapping on a nanoscale in recent years [40–42].

Therefore, it is quite feasible for identification of FCC-Ti phase. Figure 5 illustrates the FCC-Ti grains, characterized by the TKD testing. As shown in Figure 5a–c, the FCC-Ti laths were located both within the α grains and at the grain boundaries. The contoured $\{0001\}$, $\{10\bar{1}0\}$, $\{1\bar{2}10\}$, $\{1\bar{2}11\}$, $\{1\bar{2}12\}$, $\{20\bar{2}3\}$, $\{10\bar{1}3\}$ poles of the HCP-Ti grain and $\{100\}$, $\{110\}$, $\{3\bar{1}0\}$, $\{2\bar{2}1\}$, $\{1\bar{1}1\}$, $\{112\}$ poles of the FCC-Ti grain from the marked regions in Figure 5c are plotted in Figure 5d. The parallel alignments of low-index directions and planes are marked by the red arrows, which are the same as those listed in Table 1. Pole figures (PFs) in Figure 5d obey the P-type OR between HCP-Ti and FCC-Ti phases: $\langle 0001\rangle_{\alpha}//\langle 001\rangle_{FCC}$, $\{10\bar{1}0\}_{\alpha}//\{110\}_{FCC}$, which was in agreement with SAED results in Table 1 and PFs result [10].

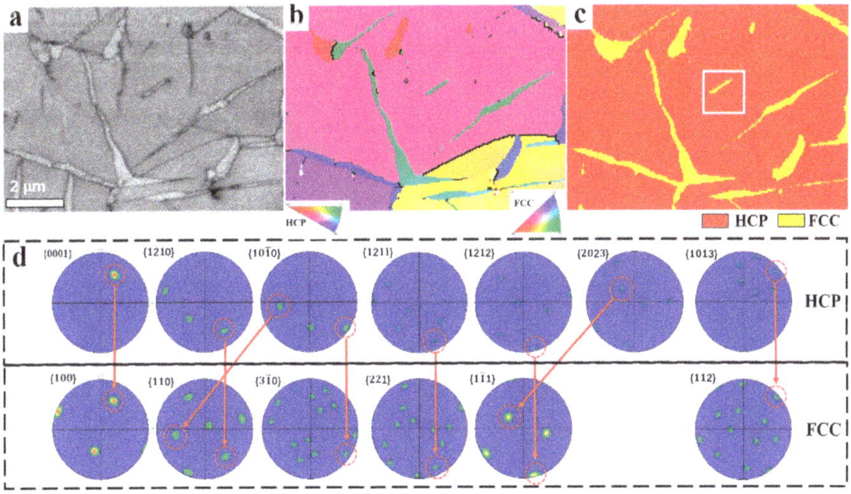

Figure 5. TKD of the HCP-Ti and FCC-Ti phases. (**a**) Band contrast (BC); (**b**) inverse pole figure (IPF), (**c**) Phase mapping, the red part is HCP-Ti phase and the yellow part is the FCC-Ti phase; (**d**) PFs of the two phases.

According to the TKD results in Figure 5, the composite PFs representation of HCP-Ti and FCC-Ti are satisfying the P-type OR is illustrated in Figure 6. Due to symmetry, only a quarter of the whole PF is shown, which is consistent with other study [9]. FCC-Ti phase may have several variants due to its high symmetry, for example, P_1 ($[\bar{1}2\bar{1}3]_{\alpha}//[112]_{FCC}$) and P_1' ($[1\bar{2}13]_{\alpha}//[1\bar{1}2]_{FCC}$) in Figure 6a, P_2 ($(10\bar{1}0)_{\alpha}//(2\bar{2}0)_{FCC}$) and P_2' ($(10\bar{1}0)_{\alpha}//(110)_{FCC}$) in Figure 6b are all equivalent. It can be concluded from the PF that P_1, P_2, and P_3, the ORs induced from the SAED patterns in Figure 2 are equivalent, but the zone axis is different from each other during TEM analysis. Therefore, the PF is an important method to study the OR and variations of phase transition in HCP and FCC phases.

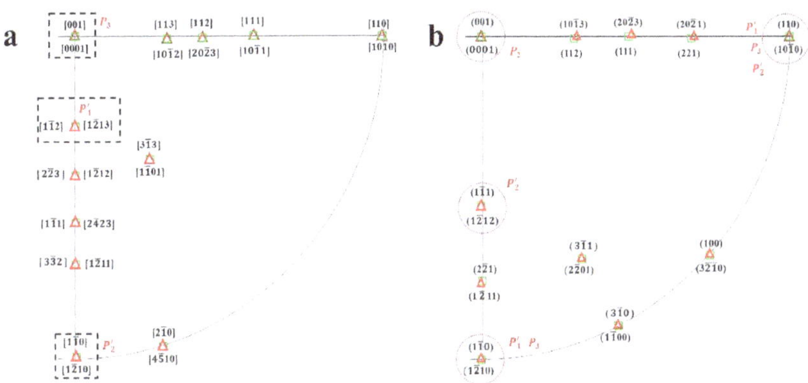

Figure 6. Composite PFs representation of HCP and FCC satisfying the ORs. (**a**) Parallel crystal orientation indexes of two phase, (**b**) parallel plane indexes.

3.3. Matrix of Orientation Relationships

ORs mean that the second phase (**S**) and matrix (**M**) obey the conditions of parallel crystal orientation [uvw] and parallel crystal plane [hkl]. The transformation between crystal orientation and crystal plane, and the relationship between real space and reciprocal space of crystal can be characterized by the transformation matrix [29,43]. For the sake of simplicity, uppercase (HKL) and lowercase (hkl) represent the crystal plane index of **M** and **S**, respectively, and uppercase [UVW] and lowercase [uvw] represent the crystal orientation index of matrix and second phase, respectively.

It is known that the transformation matrix [**B**] of the parallel crystal orientation index of **M** and **S** is [44]:

$$[\mathbf{B}] = \begin{pmatrix} H_1 & K_1 & L_1 \\ H_2 & K_2 & L_2 \\ H_3 & K_3 & L_3 \end{pmatrix}^{-1} \begin{pmatrix} d_1/D_1 & 0 & 0 \\ 0 & d_2/D_2 & 0 \\ 0 & 0 & d_3/D_3 \end{pmatrix} \begin{pmatrix} h_1 & k_1 & l_1 \\ h_2 & k_2 & l_2 \\ h_3 & k_3 & l_3 \end{pmatrix} \quad (4)$$

In which, $(H_iK_iL_i)$ and $(h_ik_il_i)$ (i = 1,2,3) represent three groups of parallel crystal planes of **M** and **S**, respectively, d_i is the interplanar spacing of the **S** phase $(h_1k_1l_1)$ crystal planes, D_i is the plane spacing of **M** phase $(H_iK_iL_i)$ crystal planes.

The parallel crystal plane index [UVW] and [uvw] of **M** and **S** satisfying the special ORs is converted as follows:

$$\begin{bmatrix} U \\ V \\ W \end{bmatrix} = [\mathbf{B}] \begin{bmatrix} u \\ v \\ w \end{bmatrix} \quad (5)$$

It can be deduced that the transformation matrix of the parallel crystal plane indices of M and S phases is $[\mathbf{R}] = [\mathbf{B}]^T$, and the transformation relationship between matrix and the parallel crystal plane of the second phase (HKL) and (hkl) is as follows [44]:

$$\begin{bmatrix} h \\ k \\ l \end{bmatrix} = [\mathbf{R}] \begin{bmatrix} H \\ K \\ L \end{bmatrix} = [\mathbf{B}]^T \begin{bmatrix} H \\ K \\ L \end{bmatrix} \quad (6)$$

Formulas (5) and (6) show that if the M phase crystal plane (HKL) and crystal orientation [UVW] are known, the **S** phase crystal plane (hkl) and crystal orientation [uvw] can be obtained and vice versa.

As long as the matrix [**B**] is obtained, the crystal plane index or orientation index of **M** and **S** phases with specific ORs can be obtained. According to the ORs obtained by SEAD in Figure 1, three groups of parallel crystal planes $(H_1K_1L_1)//(h_1k_1l_1)$, $(H_2K_2L_2)//(h_2k_2l_2)$, $(H_3K_3L_3)//(h_3k_3l_3)$ of α-HCP and the second FCC phase are, respectively, taken as $(10\bar{1}0)_\alpha//(220)_{FCC}$, $(1\bar{2}10)_\alpha//(2\bar{2}0)_{FCC}$, $(0002)_\alpha//(002)_{FCC}$.

The lattice parameters of HCP phase a_α = 0.298 nm, c = 0.471 nm; The lattice parameter of FCC phase is a_{FCC} = 0.425 nm. The above parameters are substituted into formula (4) to obtain matrix [**B**] and transpose matrix $[\mathbf{B}]^T$.

$$[\mathbf{B}] = \begin{bmatrix} 1.208305 & 1.208305 & 0 \\ -0.442270 & 1.650576 & 0 \\ 0 & 0 & 0.940299 \end{bmatrix} \quad (7)$$

$$[\mathbf{R}] = [\mathbf{B}]^T = \begin{bmatrix} 1.208305 & -0.442270 & 0 \\ 1.208305 & 1.650576 & 0 \\ 0 & 0 & 0.940299 \end{bmatrix} \quad (8)$$

Tables 2 and 3 show the possible parallel relationship between crystal plane and orientation predicted by [**B**] and [**R**] transformation matrix. Only some low orientation indexes and plane indexes are listed in the Tables 2 and 3, which correspond to the points in the pole maps in Figure 5. It should be pointed out that the ORs obtained by transformation matrix is different from the result of edge-to-edge matching model prediction [45]. This model has been used to predict the OR between HCP Mg alloys and the FCC-precipitates via diffusion-controlled phase transformations.

Table 2. Parallel crystal orientation indexes predicted by transformation matrix and PFs.

$(HKIL)_{HCP}$	$(HKL)_{HCP}$	$(hkl)_{FCC}$ via Matrix Calculation	$(hkl)_{FCC}$ via TKD
(0001)	(001)	$[0\ 0\ 0.95]$	(001)
$(1\bar{2}10)$	$(1\bar{2}0)$	$[2.10\ \overline{2.10}\ 0]$	(110)
$(1\bar{2}11)$	$(1\bar{2}1)$	$[2.09\ \overline{2.09}\ 0.94]$	$(2\bar{2}1)$
$(1\bar{2}12)$	$(1\bar{2}2)$	$[2.09\ \overline{2.09}\ 1.88]$	$(1\bar{1}1)$
$(10\bar{1}0)$	(100)	$[1.21\ 1.21\ 0]$	(110)
$(20\bar{2}3)$	(203)	$[2.42\ 2.42\ 2.85]$	(111)
$(10\bar{1}3)$	$[103]$	$[1.21\ 1.21\ 2.82]$	(112)
$(20\bar{2}1)$	$[201]$	$[2.42\ 2.42\ 0.94]$	(221)
$(2\bar{2}01)$	$(2\bar{2}1)$	$[3.30\ -0.88\ 0.94]$	$(3\bar{1}1)$
$(3\bar{2}\bar{1}0)$	$(3\bar{2}0)$	$[4.51\ 0.32\ 0]$	(100)
$(1\bar{1}00)$	$(1\bar{1}0)$	$[1.65\ -0.44\ 0]$	$(3\bar{1}0)$

Table 3. Parallel crystal plane indexes predicted by transformation matrix and PFs.

$(uvw)_{FCC}$	$(UVW)_{HCP}$ by Matrix Calculation	$(UVtW)_{HCP}$	$(UVtW)_{HCP}$ via TKD
$[1\bar{1}0]$	$[0\ \overline{-2.10}\ 0]$	$[0.70\ \overline{1.40}\ 0.70\ 0]$	$[1\bar{2}10]$
$[\bar{3}32]$	$[0\ 6.29\ 1.90]$	$[\overline{2.10}\ 4.20\ \overline{2.10}\ 1.9]$	$[1\bar{2}11]$
$[1\bar{1}1]$	$[0\ \overline{-2.10}\ 0.95]$	$[0.70\ \overline{1.40}\ 0.70\ 0.95]$	$[2\bar{4}23]$
$[22\bar{3}]$	$[0\ \overline{4.20}\ \overline{2.85}]$	$[1.40\ \overline{2.80}\ 1.40\ \overline{2.85}]$	$[1\bar{2}1\bar{2}]$
$[\bar{1}12]$	$[0\ 2.10\ 1.90]$	$[\overline{0.70}\ 1.40\ \overline{0.70}\ 1.9]$	$[\bar{1}2\bar{1}3]$
$[110]$	$[2.42\ 1.21\ 0]$	$[1.21\ 0\ \overline{1.21}\ 0]$	$[10\bar{1}0]$
$[3\bar{1}3]$	$[2.42\ \overline{2.98}\ 2.85]$	$[2.61\ \overline{2.79}\ 0.18\ 2.85]$	$[1\bar{1}01]$
$[111]$	$[2.42\ 1.21\ 0.95]$	$[1.21\ 0\ \overline{1.21}\ 0.95]$	$[10\bar{1}1]$
$[112]$	$[2.42\ 1.21\ 1.9]$	$[1.21\ 0\ \overline{1.21}\ 1.90]$	$[20\bar{2}3]$
$[113]$	$[2.42\ 1.21\ 2.84]$	$[1.21\ 0\ \overline{1.21}\ 2.85]$	$[10\bar{1}2]$
$[1\bar{2}0]$	$[\overline{1.21}\ \overline{3.75}\ 0]$	$[0.44\ \overline{2.10}\ 1.65\ 0]$	$[1\bar{5}40]$
$[001]$	$[0\ 0\ 0.95]$	$[0\ 0\ 0\ 0.95]$	$[0001]$

Even though **[B]** and **[R]** transformation matrix is obtained by P3 OR, the calculation results in Tables 2 and 3 are consistent with P1 and P2 ORs. These results are also consistent with the P-type OR reported in literatures. The P-type OR has the same mechanism of phase transition as the Shockley partial dislocation sliding along the direction $\frac{1}{6}\langle\bar{1}2\bar{1}0\rangle$ on the prism surface $(10\bar{1}0)_{\alpha}$ shown in Figures 3 and 4.

Therefore, when the phase transition between HCP and FCC satisfies the specific crystallographic OR, the parallel crystal orientation and crystal plane indexes of the two phases can be accurately expressed by PFs and transformation matrix, and the second phase variants satisfying the OR can be predicted.

4. Conclusions

(1) A large number of FCC-Ti laths are found in the rolling structure of high purity titanium, and the OR between HCP and FCC phases is P type: $\langle 0001\rangle_{\alpha}//\langle 001\rangle_{FCC}$, $\{10\bar{1}0\}_{\alpha}//\{110\}_{FCC}$. The width of FCC-Ti laths is from 30 nm to 50 nm.

(2) According to the OR, the deformation mechanism of the phase transformation process is mainly realized by dislocation slip mechanism, and the phase transformation is caused by the Shockley partial dislocation sliding along the direction $\frac{1}{6}\langle\bar{1}2\bar{1}0\rangle$ on the prism surface $\{10\bar{1}0\}_{\alpha}$.

(3) Phase mapping and PF functions of TKD were quite feasible for the OR identification of a nanoscale FCC-Ti phase. When the phase transition between HCP and FCC obeys the specific crystallographic OR, the parallel crystal directions and planes of the two phases can be accurately expressed by pole figures and phase transition matrix, and the variants meeting the OR can be predicted.

Author Contributions: Conceptualization, F.B., L.Z. and H.Z.; data curation, F.B. and Q.Z.; writing—original draft preparation, F.B. and N.A.; writing—review and editing, N.A., L.Z. and H.Z.; formal analysis, F.B. and Q.Z.; supervision, L.Z.; funding acquisition, L.Z. and H.Z.; investigation, F.B. and J.S.; methodology, F.B. and Z.L.; project administration, L.Z., H.Z. and X.L.; resources, X.L. All authors have read and agreed to the published version of the manuscript.

Funding: This research was supported by the National Natural Science Foundation of China (No. 51874001), the Natural Science Foundation of Anhui Province (KJ2020A0252), and the Provincial college students' innovative entrepreneurial training plan (Nos. S201910360162 and S202110360149).

Acknowledgments: The authors are thankful for providing material by Ningbo Chuangrun New Materials Co., Ltd., Ningbo, China.

Conflicts of Interest: The authors declare no conflict of interest.

References

1. Yu, Q.; Kacher, J.; Gammer, C.; Traylor, R.; Samanta, A.; Yang, Z.Z.; Minor, A.M. In situ TEM observation of FCC Ti formation at elevated temperatures. *Scr. Mater.* **2017**, *140*, 9–12. [CrossRef]
2. Shirooyeh, M.; Xu, J.; Langdon, T.G. Microhardness evolution and mechanical characteristics of commercial purity titanium processed by high-pressure torsion. *Mater. Sci. Eng.* **2014**, *614*, 223–231. [CrossRef]
3. Hong, D.H.; Lee, T.W.; Lim, S.H.; Kim, W.Y.; Hwang, S.K. Stress-induced hexagonal close-packed to face-centered cubic phase transformation in commercial-purity titanium under cryogenic plane-strain compression. *Scr. Mater.* **2013**, *69*, 405–408. [CrossRef]
4. Chakraborty, J.; Kumar, K.; Ranjan, R.; Chowdhury, S.G.; Singh, S.R. Thickness-dependent fcc–hcp phase transformation in polycrystalline titanium thin films. *Acta Mater.* **2011**, *59*, 2615–2623. [CrossRef]
5. Hao, P.D.; Chen, P.; Deng, L.; Li, F.X.; Yi, J.H.; Şopu, D.; Eckert, J.; Tao, J.M.; Liu, Y.C.; Bao, R. Anisotropic elastic and thermodynamic properties of the HCP-Titanium and the FCC-Titanium structure under different pressures. *J. Mater. Res. Technol.* **2020**, *9*, 3488–3501. [CrossRef]
6. Zheng, X.D.; Gong, M.Y.; Xiong, T.; Ge, H.L.; Yang, L.X.; Zhou, Y.T.; Zheng, S.J.; Wang, J.; Ma, X.L. Deformation induced FCC lamellae and their interaction in commercial pure Ti. *Scr. Mater.* **2019**, *162*, 326–330. [CrossRef]
7. Zhao, H.; Ding, N.J.; Ren, Y.P.; Xie, H.B.; Yang, B.; Qin, G.W. Shear-induced hexagonal close-packed to face-centered cubic phase transition in pure titanium processed by equal channel angular drawing. *J. Mater. Sci.* **2019**, *54*, 7953–7960. [CrossRef]
8. Kou, W.J.; Sun, Q.Y.; Xiao, L.; Sun, J. Plastic deformation-induced HCP-to-FCC phase transformation in submicron-scale pure titanium pillars. *J. Mater. Sci.* **2020**, *55*, 2193–2201. [CrossRef]
9. Wei, B.Q.; Ni, S.; Liu, Y.; Song, M. Three dimensional crystallographic orientation relationships for hexagonal close packed structure to face centered cubic structure transformation in pure titanium. *Scr. Mater.* **2019**, *169*, 46–51. [CrossRef]
10. Li, Z.; Cheng, X.; Li, J.; Wang, H.M. Formation of face-centered cubic titanium in laser surface re-melted commercially pure titanium plate. *J. Mater. Sci. Technol.* **2018**, *34*, 767–773. [CrossRef]
11. Jing, R.; Liu, C.Y.; Ma, M.Z.; Liu, R.P. Microstructural evolution and formation mechanism of FCC titanium during heat treatment processing. *J. Alloys Compd.* **2013**, *552*, 202–207. [CrossRef]
12. Jing, R.; Liang, S.X.; Liu, C.Y.; Ma, M.Z.; Liu, R.P. Aging effects on the microstructures and mechanical properties of the Ti-20Zr-6.5Al-4V alloy. *Mater. Sci. Eng.* **2013**, *559*, 474–479. [CrossRef]
13. Bolokang, A.S.; Phasha, M.J.; Motaung, D.E.; Cummings, F.R.; Muller, T.F.G.; Arendse, C.J. Microstructure and phase transformation on milled and unmilled Ti induced by water quenching. *Mater. Lett.* **2014**, *132*, 157–161. [CrossRef]
14. Ren, J.Q.; Sun, Q.Y.; Xiao, L.; Ding, X.D.; Sun, J. Phase transformation behavior in titanium single-crystal nanopillars under [0001] orientation tension: A molecular dynamics simulation. *Comp. Mater. Sci.* **2014**, *92*, 8–12. [CrossRef]
15. Yang, J.X.; Zhao, H.L.; Gong, H.R.; Song, M.; Ren, Q.Q. Proposed mechanism of HCP → FCC phase transition in titianium through first principles calculation and experiments. *Sci. Rep.* **2018**, *8*, 1992. [CrossRef] [PubMed]
16. Chen, P.; Wang, F.; Li, B. Transitory phase transformations during {101¯2} twinning in titanium. *Acta Mater.* **2019**, *171*, 65–78. [CrossRef]
17. Wei, B.; Ni, S.; Liu, Y.; Liao, X.; Song, M. Phase transformation and structural evolution in a Ti-5at.% Al alloy induced by cold-rolling. *J. Mater. Sci. Technol.* **2020**, *49*, 211–223. [CrossRef]
18. Liu, Y.G.; Li, M.Q.; Liu, H.J. Deformation induced face-centered cubic titanium and its twinning behavior in Ti–6Al–4V. *Scr. Mater.* **2016**, *119*, 5–8. [CrossRef]
19. Ma, X.; Guo, X.; Fu, M.; Qiao, Y. In-situ TEM observation of hcp-Ti to fcc-Ti phase transformation in Nb-Ti-Si based alloys. *Mater. Charact.* **2018**, *142*, 332–339. [CrossRef]
20. Niu, L.; Wang, S.; Chen, C.; Qian, S.F.; Liu, R.; Li, H.; Liao, B.; Zhong, Z.H.; Lu, P.; Wang, M.P. Mechanical behavior and deformation mechanism of commercial pure titanium foils. *Mater. Sci. Eng.* **2017**, *707*, 435–442. [CrossRef]
21. Bai, F.; Yin, L.; Zhao, W.; Zhou, H.; Song, M.; Liu, Y.; Liu, X. Deformational behavior of face-centered cubic (FCC) phase in high-pure titanium. *Mater. Sci. Eng.* **2021**, *800*, 140287. [CrossRef]

22. Ren, L.; Xiao, W.; Kent, D.; Wan, M.; Ma, C.; Zhou, L. Simultaneously enhanced strength and ductility in a metastable β-Ti alloy by stress-induced hierarchical twin structure. *Scr. Mater.* **2020**, *184*, 6–11. [CrossRef]
23. Fu, Y.; Xiao, W.L.; Kent, D.; Dargusch, M.S.; Wang, J.S.; Zhao, X.Q.; Ma, C.L. Ultrahigh strain hardening in a transformation-induced plasticity and twinning-induced plasticity titanium alloy. *Scr. Mater.* **2020**, *187*, 285–290. [CrossRef]
24. Zhou, P.; Zhu, G.-Z. Strain Accommodations among Twin Variants in Ti and Mg. *Crystals* **2021**, *11*, 453. [CrossRef]
25. Dyakonov, G.S.; Mironov, S.; Semenova, I.P.; Valiev, R.Z.; Semiatin, S.L. EBSD analysis of grain-refinement mechanisms operating during equal-channel angular pressing of commercial-purity titanium. *Acta Mater.* **2019**, *173*, 174–183. [CrossRef]
26. Zu, Q.; Guo, Y.-F.; Yao, X. Surface and orientation effects on stress-induced hcp-fcc phase transformation in Ti nanopillars. *Appl. Surf. Sci.* **2020**, *509*, 145234. [CrossRef]
27. Yu, Q.B.; Liu, X.H.; Tang, D.L. Extreme Extensibility of Copper Foil under Compound Forming Conditions. *Sci. Rep.* **2013**, *3*, 3556. [CrossRef] [PubMed]
28. Chen, S.D.; Liu, X.H.; Liu, L.Z.; Song, M. Crystal plasticity finite element simulation of slip and deformation in ultrathin copper strip rolling. *Acta Metall. Sin.* **2016**, *52*, 120–128. [CrossRef]
29. Wang, S.; Niu, L.; Chen, C.; Pang, Y.; Liao, B.; Zhong, Z.H.; Lu, P.; Li, P.; Wu, X.D.; Coenen, J.W. Size effects on the tensile properties and deformation mechanism of commercial pure titanium foils. *Mater. Sci. Eng.* **2018**, *730*, 244–261. [CrossRef]
30. Bai, F.M.; Ye, X.; Zhang, H.Y.; Zhou, H.W.; Song, M.; Sun, Y.X.; He, Y.Z. A significant increase in the hardness of nanotwinned titanium alloys prepared via the martensitic phase transformation. *Mater. Lett.* **2019**, *255*, 126507. [CrossRef]
31. Wu, H.C.; Kumar, A.; Wang, J.; Bi, X.F.; Tomé, C.N.; Zhang, Z.; Mao, S.X. Rolling-induced Face Centered Cubic Titanium in Hexagonal Close Packed Titanium at Room Temperature. *Sci. Rep.* **2016**, *6*, 24370. [CrossRef] [PubMed]
32. Zheng, Y.; Williams, R.E.A.; Wang, D.; Shi, R.; Nag, S.; Kami, P.; Sosa, J.M.; Banerjee, R.; Wang, Y.; Fraser, H.L. Role of ω phase in the formation of extremely refined intragranular α precipitates in metastable β-titanium alloys. *Acta Mater.* **2016**, *103*, 850–858. [CrossRef]
33. Wen, J.; Allain, N.; Fleury, E. Determination of orientation relationships between FCC-hydride and HCP-titanium and their correlation with hydrides distribution. *J. Alloys Compd.* **2020**, *817*, 153297. [CrossRef]
34. Conforto, E.; Caillard, D. A fast method for determining favourable orientation relationships and interface planes: Application to titanium–titanium hydrides transformations. *Acta Mater.* **2007**, *55*, 785–798. [CrossRef]
35. Chang, Y.; Zhang, S.; Liebscher, C.H.; Dye, D.; Ponge, D.; Scheu, C.; Dehm, G.; Raabe, D.; Gault, B.; Lu, W. Could face-centered cubic titanium in cold-rolled commercially-pure titanium only be a Ti-hydride? *Scr. Mater.* **2020**, *178*, 39–43. [CrossRef]
36. Sarkar, R.; Ghosal, P.; Prasad, K.S.; Nandy, T.K.; Ray, K.K. An FCC phase in a metastable β-titanium alloy. *Phil. Mag. Lett.* **2014**, *94*, 311–318. [CrossRef]
37. Manna, I.; Chattopadhyay, P.P.; Nandi, P.; Banhart, F.; Fecht, H.-J. Formation of face-centered-cubic titanium by mechanical attrition. *J. Appl. Phys.* **2003**, *93*, 1520–1524. [CrossRef]
38. Ali, N.; Zhang, L.; Zhou, H.; Zhao, A.; Zhang, C.; Fu, K.; Cheng, J. Effect of soft reduction technique on microstructure and toughness of medium carbon steel. *Mater. Today Commun.* **2021**, *26*, 102130. [CrossRef]
39. Zhao, P.; Chen, B.; Kelleher, J.; Yuan, G.; Guan, B.; Zhang, X.; Tu, S. High-cycle-fatigue induced continuous grain growth in ultrafine-grained titanium. *Acta Mater.* **2019**, *174*, 29–42. [CrossRef]
40. Liu, C.; Li, G.; Gu, H.; Yuan, F.; Han, F.; Ali, M.; Zhang, Y.; Guo, W. Observation of FCC-Zr phase in as-cast Zircaloy-4 alloy. *Mater. Lett.* **2020**, *267*, 127551. [CrossRef]
41. Liu, J.; Yu, H.; Karamched, P.; Hu, J.; He, G.; Goran, D.; Hughes, G.M.; Wilkinson, A.J.; Lozano-Perez, S.; Grovenor, C.R.M. Mechanism of the α-Zr to hexagonal-ZrO transformation and its impact on the corrosion performance of nuclear Zr alloys. *Acta Mater.* **2019**, *179*, 328–341. [CrossRef]
42. Xiao, B.; Xu, L.; Cayron, C.; Xue, J.; Sha, G.; Logé, R. Solute-dislocation interactions and creep-enhanced Cu precipitation in a novel ferritic-martensitic steel. *Acta Mater.* **2020**, *195*, 199–208. [CrossRef]
43. Li, X.L.; Deng, X.T.; Lei, C.S.; Wang, Z.D. New orientation relationship with low interfacial energy in MC/ferrite system observed in Nb-Ti bearing steel during isothermal quenching process. *Scr. Mater.* **2019**, *163*, 101–106. [CrossRef]
44. Rong, Y. *Introduction to Analytical Electron Microscopy*, 2nd ed.; Higher Education Press: Beijing, China, 2015.
45. Zhang, M.X.; Kelly, P.M. Edge-to-edge matching and its applications: Part II. Application to Mg–Al, Mg–Y and Mg–Mn alloys. *Acta Mater.* **2005**, *53*, 1085–1096. [CrossRef]

crystals

Article

A Constitutive Relation Based on the Johnson–Cook Model for Ti-22Al-23Nb-2(Mo, Zr) Alloy at Elevated Temperature

Yanju Wang [1], Duo Zhou [2], Yi Zhou [1], Aixue Sha [1], Huaxing Cheng [2] and Yabin Yan [2,*]

[1] Materials Evaluation Center for Aeronautical and Aeroengine Application, AECC Beijing Institute of Aeronautical Materials, Beijing 100095, China; yanjuwang@biam.ac.cn (Y.W.); yizhou@biam.ac.cn (Y.Z.); aixue.sha@biam.ac.cn (A.S.)

[2] Key Laboratory of Pressure Systems and Safety, Ministry of Education, School of Mechanical Power and Engineering, East China University of Science and Technology, Shanghai 200237, China; y45190047@mail.ecust.edu.cn (D.Z.); y81200043@mail.ecust.edu.cn (H.C.)

* Correspondence: yanyabin@ecust.edu.cn

Abstract: Although several schemes have been proposed to modify the classical Johnson–Cook (J-C) model, the effect of temperature on the flow stress of materials at different temperatures has not been clarified. In the current study, to investigate the deformation behavior of Ti-22Al-23Nb-2(Mo, Zr) alloy at different temperatures, uniaxial tension experiments were performed at both room (RT, 28 °C) and elevated temperatures, and a modified J-C model was developed to describe the temperature-dependent plastic flow. In tensile experiments, Ti_2AlNb-based alloy showed a continuous work hardening until reaching the ultimate strength at RT, while an apparent drop appeared in the flow stress after the peak stress at elevated temperature. Moreover, the experimental peak stress significantly depends on the testing temperature. To correctly describe the different variations of flow stresses at different temperatures, a parameter, S, which represents the softening behavior of flow stress, is integrated into the classical J-C model. In addition, the applicability and validity of the proposed J-C model were verified by calibration with experimental curves of different temperatures. On the other hand, the fractography of post-test specimens was examined to interrupt the increased fracture brittleness of Ti_2AlNb-based alloy at elevated temperatures. The proposed constitutive relation based on the J-C model is applicable to predict the deformation behavior of other Ti_2AlNb-based alloys at different temperatures.

Keywords: Ti_2AlNb alloy; mechanical properties; constitutive relation; Johnson–Cook model

Citation: Wang, Y.; Zhou, D.; Zhou, Y.; Sha, A.; Cheng, H.; Yan, Y. A Constitutive Relation Based on the Johnson–Cook Model for Ti-22Al-23Nb-2(Mo, Zr) Alloy at Elevated Temperature. *Crystals* **2021**, *11*, 754. https://doi.org/10.3390/cryst11070754

Academic Editors: Wojciech Polkowski and Pavel Lukáč

Received: 4 June 2021
Accepted: 24 June 2021
Published: 28 June 2021

1. Introduction

The ordered orthorhombic Ti_2AlNb, which was discovered in a Ti_3Al-xNb alloy in the late 1980s by Banerjee et al. [1], has attracted considerable attention as potential high-temperature structural materials for aero-engines owing to its superior performances [2–4], such as low density, high Young's modulus, specific strength, good high-temperature fatigue performance, and creep resistance. Because the actual application in aero-engines, Ti_2AlNb alloy is expected to be subject to various mechanical loads at high temperatures from 400 °C to even 1000 °C. It is critically important to precisely predict the mechanical behaviors of Ti_2AlNb alloy, including deformation and fracture, at elevated temperature for assuring the structural integrity and stability of aero-engines.

Up to now, many efforts have been devoted to the development of material systems, including composition design [5–8] and phase equilibrium and transformation [9,10], and thermal-mechanical processing techniques [11,12]. Recently, the mechanical behaviors of Ti_2AlNb alloy, such as tension [13,14], low cycle fatigue (LCF) [15], and creep [16,17], have also been discussed, and the effect of the evolution of the microstructure [18] and modification of the alloying element [19,20] was also investigated, where Mo addition was found to improve the creep resistance of Ti_2AlNb alloy [19], and Mo and Fe additions led to the increment of the room temperature ductility of Ti_2AlNb alloys [21]. Furthermore, the stress–strain curves of Ti_2AlNb alloys with different alloy compositions and microstructures were experimentally achieved using the uniaxial tension experiments at both room and elevated temperatures [13]. The stress–strain curve at elevated temperature is usually characterized by a linear increment to a peak followed by a significant drop in flow stress, i.e., the softening of flow stress [13,22], which indicates that the representation of stress–strain curves of Ti_2AlNb alloy strongly depends on the temperature during the mechanical experiments. Therefore, the proposal of an appropriate theoretical model, which can correctly describe the stress–strain curves of Ti_2AlNb alloys at different temperatures, is a critical issue for further application in aero-engines. However, to the best knowledge of the authors, a rare investigation has been performed to develop the theoretical model to predict the deformation of Ti_2AlNb alloys at different temperatures.

In 1983, Johnson and Cook [23] proposed a general empirical constitutive model for materials subjected to a high temperature and loading rate. Until now, the J-C model has been widely applied to evaluate various metallic materials with quasi-static and high-speed impact loads at various temperatures [24–26]. However, it should be noted that the work hardening, strain rate hardening, and thermal softening are separately considered in J-C models, i.e., an uncoupled model [27–30]. In fact, during the deformation of metals at elevated temperature, the work hardening or/and strain rate hardening and thermal softening exist simultaneously and affect each other. Although the classical J-C model is applicable to the special case where the flow stress possesses a linear relation with the temperature and strain rate [23], unavoidable deviation will exist when the J-C model describes the deformation behavior under a relatively high temperature or strain rate, causing a significant nonlinear relationship with the flow stress.

In the past decade, different types of constitutive relations based on damage mechanics have been proposed to investigate the deformation behaviors of metals at elevated temperatures. Considering the effect of dislocation density, grain size, phase volume fraction, and the evolution of damage and plastic evolution, Liu et al. [31] proposed a uniform viscous plastic constitutive equation to analyze the plastic flow of Ti_2AlNb alloys at elevated temperatures of 910–970 °C. In addition, the evolution behaviors of the dislocation density, recrystallization, and grain size during thermoplastic deformation were clarified using the uniform viscous plastic relation [32]. In addition, the nucleation and growth of the intergranular cavity induced by the increment of dislocation density and grain size were discussed based on the superplasticity theory considering the evolution mechanism of damage [33]. On the other hand, besides the aforementioned constitutive relations regarding the deformation mechanisms at elevated temperatures, several modification schemes have been proposed for the classical J-C model to analyze the deformation of metals under a high temperature and strain rate [24–26,34–37]. Lin et al. [34] proposed a modified J-C model, where there was a coupling effect between yield stress and the working hardening. In addition, the influence of temperature and strain rate on the flow stress was also considered. Huh et al. [35] modified the strain rate term into a quadratic polynomial form to represent the nonlinear relation between the flow stress and the strain rate. Ulacia et al. [36] further modified the one proposed by Hub et al. [35] and defined the strain hardening exponent as a function of the strain rate, which was adopted to investigate the sensitivity of work hardening to the strain rate of magnesium alloys. By studying the stress–strain curves of magnesium alloys with different strain rates, Tan et al. [37] found that the strain rate hardening coefficient in the classical J-C model was a function of both the plastic strain and strain rate. Compared with the strain rate, the influence of temperature on the work hardening of flow stress during deformation has not been clarified. In addition, the sensitivity of the strain rate varies with the external temperature, which is induced by the change of the deformation mechanism at different temperatures. Therefore, considering the effect of temperature on the work hardening of flow stress in the J-C model is another key issue for its application in describing the deformation behaviors of materials at elevated temperatures.

In the current study, the deformation behaviors of Ti-22Al-23Nb-2(Mo, Zr) alloys (abbreviated as "Ti_2AlNb-based alloys") were experimentally investigated using uniaxial tension at both room and elevated temperatures, and corresponding stress–strain curves were obtained as well. In addition, by calibrating with the experiments, a modified J-C model considering the significant effect on the flow stress at different temperatures is developed, and its validity was examined by predicting the deformation of other specimens. Finally, the fractography of post-test specimens was carefully observed using a high-resolution field emission scanning electron microscopy (FE-SEM), and the increment of the brittle nature of fracture for Ti_2AlNb-based alloys with an increasing temperature was interrupted from the characteristic of microstructures of the fracture trace.

2. Experiments

The as-received material was a hot-rolled alloy plate with a nominal composition of Ti-22Al-23Nb-2(Mo, Zr) (at.%), and the final rolling temperature was 1020~1075 °C. After hot-rolling, the alloy was subjected to air cooling with 900~980 °C and a subsequent 2 h of annealing at 650~830 °C, respectively. Figure 1 shows the microstructure of the current Ti_2AlNb-based alloy in the SEM-BSE mode. The fine needle-like O phases are uniformly distributed on the B2 matrix in an interwoven network, where the O phase possesses a typical length of 2~5 μm and a width less than 300 nm.

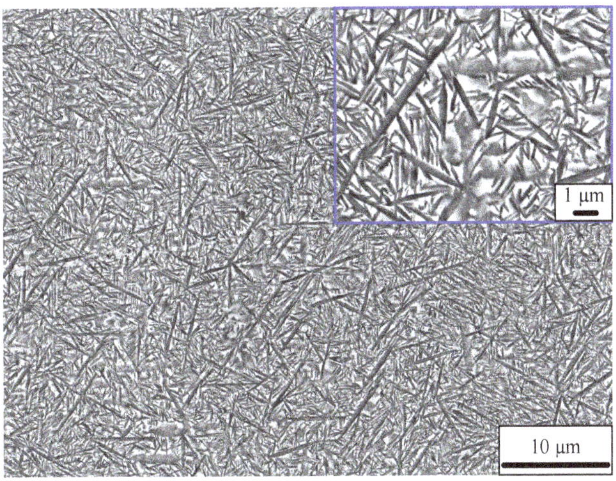

Figure 1. SEM micrograph of present Ti_2AlNb-based alloy.

Figure 2 shows the geometric dimensions of the tension specimen. The deformation behavior of the alloy was investigated by uniaxial tensile tests performed on a tension machine (Instron 5985) with a high-temperature furnace at room temperature (RT, 28 °C), 500, 550 and 650 °C, respectively. The uniaxial tension experiments were performed with a manner of displacement-control according to China National Standard of GB/T228-2010 and GB/T 228.2-2015, and the loading rate was set to 0.006 mm/mm/min for all specimens. In addition, the strain gauge was removed when the strain exceeded 5% and the loading rate was subsequently adjusted to 0.02 mm/mm/min. On the other hand, the fracture morphology of all tested specimens at different temperatures was carefully examined using a high-resolution SEM (Zeiss, Crossbeam FIB-SEM, Jena, Germany).

Figure 3 shows the true stress–strain curves of all tested specimens at RT and elevated temperature. It can be found that the deformation behavior of Ti_2AlNb-based alloy obviously changes with the increment of the temperature. At RT, the flow stress continuously increases with the applied strain until the ultimate tension strain is larger than 4%, which indicates an apparent working hardening at RT. The work hardening is also clearly identified in the flow stress of tension specimens at 500 °C, although the ultimate fracture strain is reduced to around 2%. On the other hand, with a higher temperature of 550 °C, the flow gradually increases to a peak and subsequently drops down, indicating the appearance of softening due to the high temperature. In addition, the ultimate fracture strain is further reduced to 1.5%, which is significantly smaller than the ones at RT and 500 °C. For the flow stress of the tension experiment at 650 °C, it quickly reaches a peak followed by a sharp and linear drop, showing abnormal softening. Furthermore, the ultimate fracture strain is almost the same as the one at 500 °C. Therefore, the temperature possesses a dominant effect on the deformation behavior, i.e., the flow stress, of Ti_2AlNb-based alloys, and the softening replaces the work hardening in the flow stress when the temperature is higher than 500 °C for the present study. This means that a critical transition temperature exists for the deformation of Ti_2AlNb-based alloys, and the microscopic mechanism for the deformation behavior across such a critical temperature is expected to be different as well.

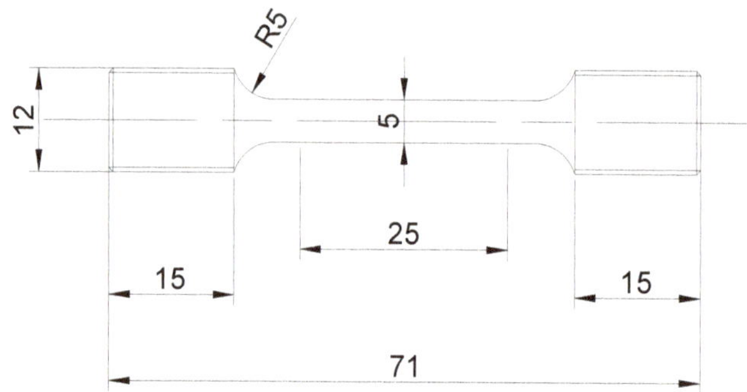

Figure 2. Geometric dimensions of the tension specimen (unit in mm).

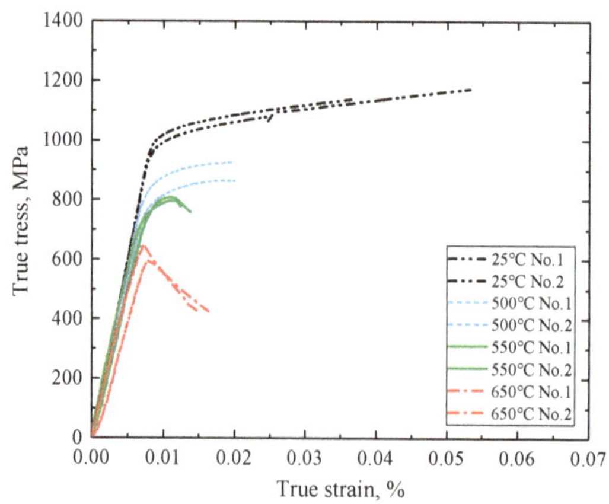

Figure 3. The true stress–strain curves of all specimens at different temperatures.

To further clarify the deformation properties of Ti_2AlNb-based alloys at different temperatures, the variation of Young's modulus and yield stress with the temperature is discussed and shown in Figure 4. As shown in Figure 4a, Young's modulus, which is determined from the slope of the experimental stress–strain curves, gradually decreases with the increment of the temperature. It is worth noting that a significant drop of Young's modulus emerges after 550 °C, which indicates the softening of Ti_2AlNb-based alloys at such elevated temperatures. On the other hand, Figure 4b illustrates the variation trend of the yield stress with the temperature. As no obvious yield point can be precisely identified from the experimental curve, the yield point is taken as the stress at which 0.2% plastic deformation occurs, named $\sigma_{0.2}$. The yield stress also decreases with the elevation of the temperature, and a dramatic decrement appears after 550 °C as well. The similar temperature dependence of both Young's modulus and yield stress is found to be induced by the partial disordering of the O phase at the elevated temperature [38–40]. In addition, the significant drop in Young's modulus and yield stress after 550 °C indicates an intrinsic change of the deformation mechanism from the RT to the relatively high temperature.

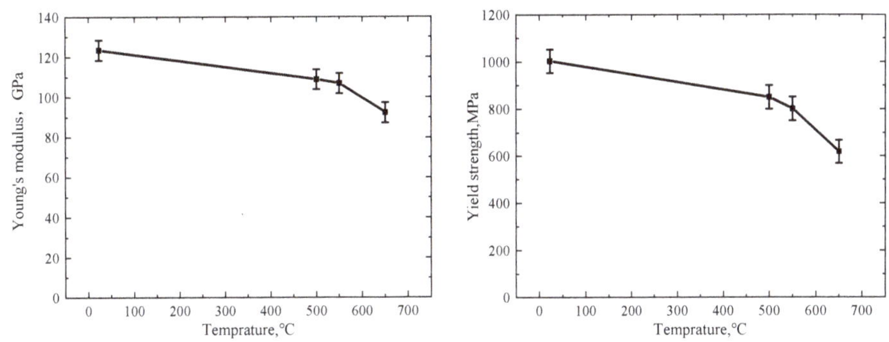

(a) (b)

Figure 4. Variation of (**a**) Young's modulus and (**b**) yield stress of Ti_2AlNb-based alloys with the temperature.

3. Constitutive Relation Based on Modified J-C Model

3.1. Classical Johnson–Cook Model

Because the tension experiments of all specimens were performed with the same loading rate, the effect of the strain rate for the J-C model is ignored in the present study. By setting the experimental strain rate to be the reference one, the classical J-C model can be simplified as:

$$\sigma = \left(A + B\varepsilon_p^n\right)\left(1 - T^{*m}\right) \tag{1}$$

Here, σ is the stress, ε_p is the plastic strain. In addition, $T^* = \frac{T-T_r}{T_m-T_r}$, where T is the experimental temperature, T_r is the room temperature, T_m is the melting temperature of materials, and A, B, m, and n are the parameters that need to be determined.

The undetermined parameters can be obtained as follows. The experimental condition of $T = T_r$ Equation (1) can be rewritten as:

$$\sigma = \left(A + B\varepsilon_p^n\right) \tag{2}$$

When the plastic strain $\varepsilon_p = 0$, $A = \sigma_{0.2}$, Equation (2) can be expressed as:

$$Ln(\sigma - A) = LnB + nLn\varepsilon_p \tag{3}$$

For the current study, based on the experimental stress–strain curve at RT, the parameters of B and n can be determined from the linear fitting of $Ln(\sigma - A)$ and $Ln\varepsilon_p$. On the other hand, for $\varepsilon_p = 0$, Equation (2) can be written as:

$$\sigma_{0.2} = A\left(1 - T^{*m}\right) \tag{4}$$

According to the yield strength in the experimental stress–strain curve at 550 °C, the temperature-sensitive parameter m can be obtained.

Figure 5 shows the comparison between the predicted stress–strain curves using the classical J-C model and experimental ones. The J-C model gives a correct prediction of the deformation behaviors of Ti_2AlNb-based alloys at both RT and 500 °C. However, for the deformation behaviors at 550 and 650 °C, where an obvious linear drop exists in the flow stress, a significant deviation appears between the theoretical and experimental curves, indicating the inapplicability of the classical J-C model for the deformation of Ti_2AlNb-based alloys at relatively high temperatures.

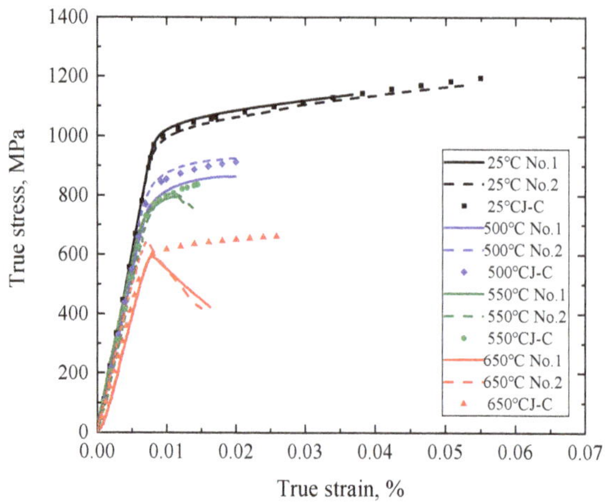

Figure 5. Comparison between the experimental and predicted stress–strain curves using the classical J-C model.

Figure 6 shows the relative difference between the experimental and theoretical flow stresses obtained from the classical J-C model. For specimens tested at the same temperature, the difference almost linearly increases with the increment of the plastic strain. On the other hand, with the increment of the temperature, the enlargement of the relative difference becomes much quicker, i.e., the slope of the relative difference–plastic strain curve becomes larger. Therefore, the relative difference shows a linear relationship with the plastic strain, and the slope of the linear relationship significantly increases with the increment of the temperature.

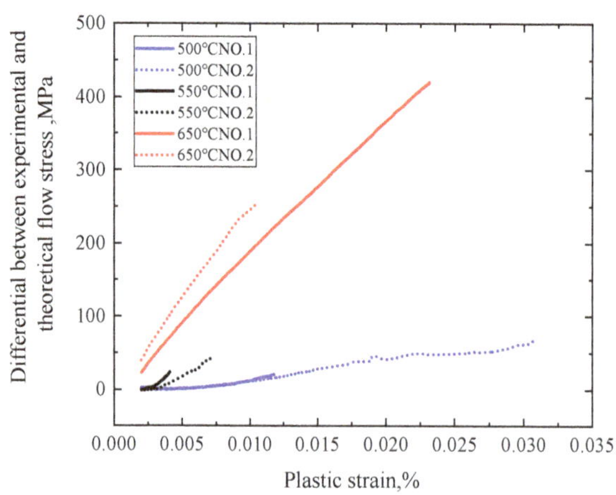

Figure 6. Variation of the difference between experimental and theoretical flow stresses at different temperatures.

3.2. Modified Johnson–Cook Model

In the classic J-C model, by regarding the yield strength at RT and the referencing strain rate to be a benchmark, the variation of stress is a function of the strain, strain rate, and temperature. Although the classic J-C model can describe the deformation behavior of materials in the temperature range from RT to the melting point, the working hardening coefficient B and n, and the strain rate hardening coefficient C are obtained from the tension tests at RT. Therefore, the deviation between the theoretical prediction from the J-C model and the experimental results enlarges with the increment of the temperature. To improve the precision of theoretical prediction, we proposed a modified J-C model considering the effect of temperature on flow stress.

Firstly, Equation (1) is written in the following form:

$$1-\frac{\sigma}{1-\left(A+B\varepsilon_p^n\right)\left(1+CLn\dot{\varepsilon}^*\right)}=T^{*m} \tag{5}$$

where $\dot{\varepsilon}^*=\frac{\dot{\varepsilon}}{\dot{\varepsilon}_0}$. $\dot{\varepsilon}$ is the experimental strain rate and $\dot{\varepsilon}_0$ is the reference strain rate. By setting $\dot{\varepsilon}^*=1$, the flow stress at RT is defined as $\sigma_r=\left(A+B\varepsilon_p^n\right)$. In addition, when the temperature reaches the melting point, the flow stress σ_m should be zero. Then, we have the following relation:

$$1-\frac{\sigma}{\sigma_r}=T^{*m} \tag{6}$$

By re-writing the form of Equation (6), we have:

$$\frac{\sigma_r-\sigma}{\sigma_r-\sigma_m}=\left(\frac{T-T_r}{T_m-T_r}\right)^m \tag{7}$$

$$\sigma=\sigma_r-(\sigma_r-\sigma_m)T^{*m} \tag{8}$$

If the effect of strain rate is considered, the flow stress is obtained as:

$$\sigma=\left[\sigma_r-(\sigma_r-\sigma_m)T^{*m}\right]\left(1+CLn\dot{\varepsilon}^*\right) \tag{9}$$

where $T\in[T_r,T_m]$. As aforementioned, the classical J-C model can only offer a correct prediction of the deformation behavior of Ti_2AlNb based at the temperature lower than 500 °C, which is much lower than its melting point (T_m = 1690 °C). However, a large error exists in the theoretical prediction of the J-C model for relatively high temperatures, i.e., $T > 550$ °C. Therefore, we define the expression of stress within a relatively high-temperature range of $T\in[T_l,T_h]$ as:

$$\sigma=\left[\sigma_l-(\sigma_l-\sigma_h)T^{*m}\right]\left(1+CLn\dot{\varepsilon}^*\right) \tag{10}$$

where T_l and T_h are the lower and upper limits of the range of the elevated temperature for applying the J-C model, and σ_l, σ_h are the corresponding flow stresses with $\dot{\varepsilon}^*=1$, which are the functions of ε_p. According to the classical J-C model, the flow stresses σ_l and σ_h can be obtained as:

$$\sigma_l=\mathrm{F}(\varepsilon_p)=\left(A_l+B_l\varepsilon_p^{n_1}\right) \tag{11}$$

$$\sigma_h=\mathrm{F}(\varepsilon_p)=\left(A_h+B_h\varepsilon_p^{n_2}\right) \tag{12}$$

To correctly predict the observed linear softening of flow stress of Ti_2AlNb-based alloys at an elevated temperature, a parameter S, representing the softening of flow stress, is introduced into the model, and Equations (11) and (12) are derived as:

$$\sigma_l=\mathrm{F}(\varepsilon_p)=\left(A_l+B_l\varepsilon_p^{n_1}\right)-S_l\varepsilon_p \tag{13}$$

$$\sigma_h = F(\varepsilon_p) = (A_h + B_h \varepsilon_p^{n_2}) - S_h \varepsilon_p \tag{14}$$

Therefore, a modified J-C model considering the effect of elevated temperature on the flow stress is achieved as:

$$\sigma = \{(A_l + B_l \varepsilon_p^{n_1} - S_l \varepsilon_p) - [(A_l + B_l \varepsilon_p^{n_1} - S_l \varepsilon_p) - (A_h + B_h \varepsilon_p^{n_2} - S_h \varepsilon_p)] T^{*m}\} \left(1 + CLn\dot{\varepsilon}^*\right) \tag{15}$$

The parameters in Equations (13) and (14) can be determined using nonlinear curve fitting based on the experimental stress–strain curves at the lower and upper temperature limits. In addition, the temperature-sensitive parameter m can be derived from Equation (9) using the experiment results at the temperature limits as well. Until now, all the parameters in the modified J-C model are determined, and the flow stress at any temperature in the temperature range can be directly obtained using Equation (15). Table 1 lists all the parameters in the classical and modified J-C models.

Table 1. The definition of all parameters adopted in the current study.

Term		Term	
σ	True stress (MPa)	R	Correlation coefficient
ε_p	Plastic strain (%)	T_l	Lower limits of the range of elevated temperature
T	Experimental temperature (°C)	T_h	Upper limits of the range of elevated temperature
T_r	Room temperature (°C)	σ_l, σ_h	Corresponding flow stresses with T_l and T_h
T_m	Melting temperature (°C)	S_l, S_h	Softening constants with T_l and T_h
A	Yield stress in J-C model (MPa)	$\dot{\varepsilon}$	Experimental strain rate
B, m, n	Material constants of J-C model	$\dot{\varepsilon}_0$	Reference strain rate
$\sigma_{0.2}$	Yield stress (MPa)		

In the current study, the lower and upper temperature limits were set to 500 and 650 °C. Therefore, by using Equations (9), (13), and (14), the predicted stress–strain curves using the newly proposed modified J-C model were obtained and are shown in Figure 7. The proposed J-C model precisely predicts the total stress–strain curves, including the variation of flow stress of Ti_2AlNb-based alloys at both lower (500 °C) and upper (650 °C) temperature limits, which indicates the validity of the proposed J-C model. On the other hand, the critical transition temperature for the deformation behavior is included in the specific temperature range of 500~650 °C. However, as the detailed value of the transition temperature was not determined in the current study, the prediction of the deformation behavior at the transition temperature remains for further study.

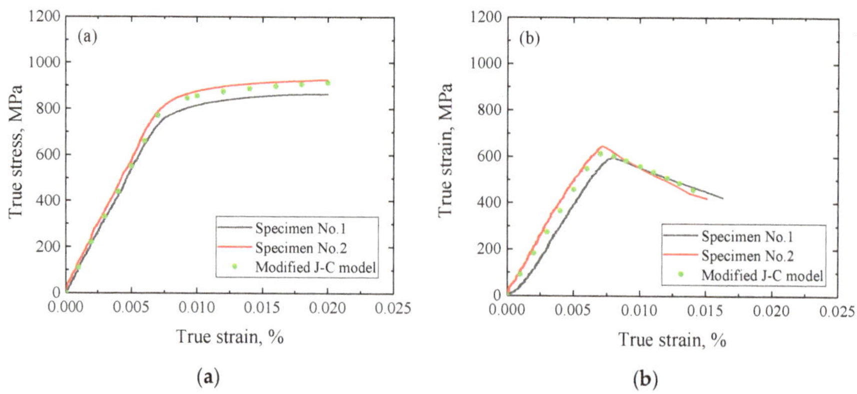

Figure 7. Comparison between the experimental and the predicted stress–strain curves using the modified J-C model at (**a**) 500 °C and (**b**) 650 °C.

3.3. Validity of the Modified J-C Model

To further verify the validity and applicability of the modified J-C model, the deformation behavior of Ti_2AlNb based at 550 °C was predicted by substituting T = 550 °C into Equation (15). Figure 8 shows the comparison between the theoretically predicted and experimental stress–strain curves at 550 °C. The modified J-C model gives a very close prediction of the stress–strain curve at 550 °C as well. Therefore, it can be concluded that the modified J-C model can precisely describe the deformation behaviors of Ti_2AlNb-based alloys at any temperature within the specified range.

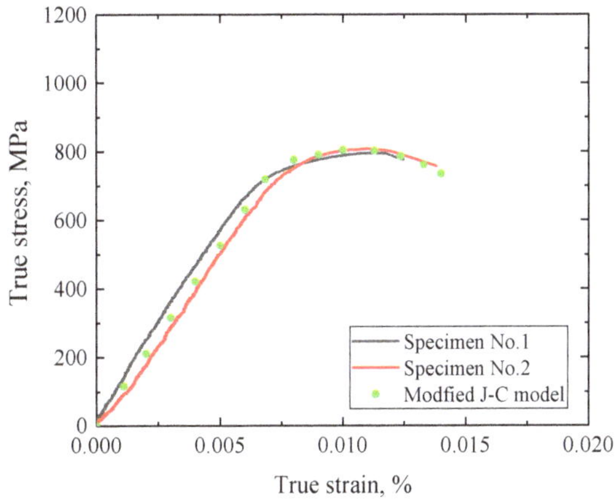

Figure 8. Predicted stress–strain curve from the modified J-C model and its comparison with the experimental result at 550 °C.

Figure 9 shows the comparison between the experimental and theoretical prediction with the classical and modified J-C models. The quality of the developed J-C model was examined using the standard statistical quantities determined by Equations (16)–(18) [41,42]. The magnitudes of R, AARE, and RMSE are 0.999, 1.3%, and 11.0 MPa determined from the modified J-C model, while 0.870, 14.5%, and 113.7 MPa were obtained from the classical model, respectively. Therefore, the proposed modified J-C model possesses an excellent precision on the prediction of deformation of Ti_2AlNb-based alloys at different temperatures.

$$\text{correlation coeffecient}\,(R) = \frac{\sum_{i=1}^{N}(E_i - E')(P_i - P')}{\sqrt{\sum_{i=1}^{N}(E_i - E')^2 \sum_{i=1}^{N}(P_i - P')^2}} \tag{16}$$

$$\text{average absolute relative error}\,(AARE) = \frac{1}{N}\sum_{i=1}^{N}\left|\frac{E_i - P_i}{E_i}\right| \tag{17}$$

$$\text{root mean square error}\,(RMSE) = \sqrt{\frac{1}{N}\sum_{i=1}^{N}(E_i - P_i)^2} \tag{18}$$

On the other hand, to further examine the robustness of the modified J-C model, the deformation behaviors of another titanium alloy of VT14 at different elevated temperatures [42] were predicted using the proposed model, and the comparison between the experimental [42] and theoretical results is shown in Figure 10. After the parameters of the modified J-C model were determined using the experimental flow stresses at the upper and lower limits of temperature range of 800–875 °C, the variation of flow stress of VT14 at 850 °C was precisely predicted by the proposed model. Therefore, the modified J-C model in the current study can predict different materials at elevated temperatures.

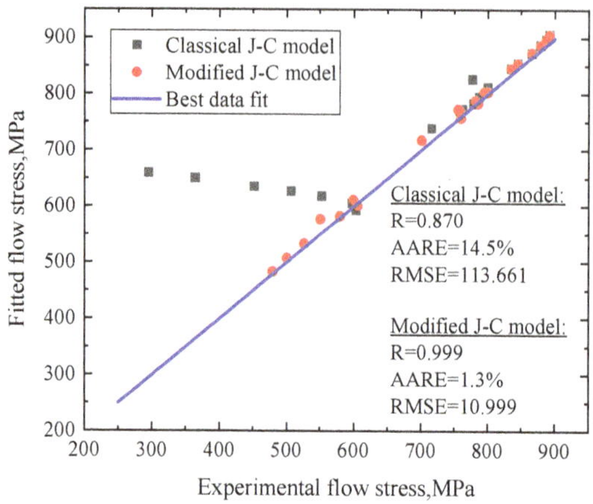

Figure 9. Correlation between experimental and fitted flow stress with the classical and modified J-C models, respectively.

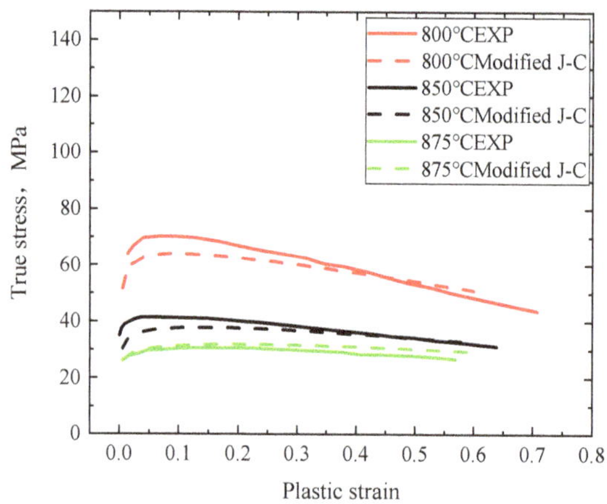

Figure 10. Comparison between experimental [42] and theoretical flow stresses of titanium alloy of VT14.

4. Temperature Dependence of Fractography with SEM Characterization

As shown in Figure 3, significant softening exists at the flow stress of Ti_2AlNb-based alloys at elevated temperatures of 550 and 650 °C in the current study. It has been well known that the dislocation motion dominates the plastic deformation of metals. With an increase of the temperature, the motion resistance due to Peierls–Nabarro stress is reduced and thus the dislocation motion becomes easier than that at RT [43–45]. On the other hand, as the yield strength is sensitive to the Peierls–Nabarro stress [44], and thus the yield strength of materials becomes lower as well. Therefore, the weakened resistance of dislocation motion induced by the elevated temperature causes softening in the flow stress of Ti_2AlNb-based alloys.

On the other hand, to further clarify the significant drop of flow stress at an elevated temperature, we examined the fractography of post-test specimens at different temperatures using FE-SEM. Figure 11 shows the overall views of fracture surfaces and the high magnification observation of corresponding microscopic characteristics of the tested specimens at RT, 550 °C, and 650 °C, respectively. For the fracture surface at RT, a typical mixture of intergranular and transgranular fractures is identified from the overall view (Figure 11a). The appearance of cleavage steps further confirms the existence of intergranular fracture

(Figure 11b). Therefore, the co-existence of intergranular and transgranular fractures results in relatively good plasticity and a large ultimate fracture strain of Ti_2AlNb-based alloys at RT. On the other hand, at the elevated temperature, an apparent drop emerges in the flow stress and the ultimate fracture strain is reduced as well. As shown in Figure 11c, the overall view of fracture surface at 550 °C indicates that the proportion of transgranular fracture is largely increased, and secondary cracks also appear. A close observation of the typical zone of the fracture surface, as shown in Figure 11d, shows that dimples appear, and the cleavage fracture still exists. This means that, although the plastic deformation indicated by dimples is strengthened, the transgranular fracture and appearance of secondary cracks cause the brittle nature of Ti_2AlNb-based alloys at 550 °C. Finally, the fracture surface at 650 °C in Figure 11e,f clearly illustrates the predominance of transgranular fracture and the increased number of secondary cracks, which causes the significant softening of flow stress at 650 °C. Therefore, the SEM characterization of the fractography indicates the transition of the deformation behavior of Ti_2AlNb-based alloy at different temperatures is mainly represented by the variation of the fracture mechanism from the mixture of intergranular and transgranular fractures to the dominance of the trangranular fracture and secondary crack. In addition, the linearity of the relative difference between the experimental and theoretical results with the temperature suggests that the proposed parameter can be adopted to approximately evaluate the transition of the fracture mechanism during the increment of the testing temperature.

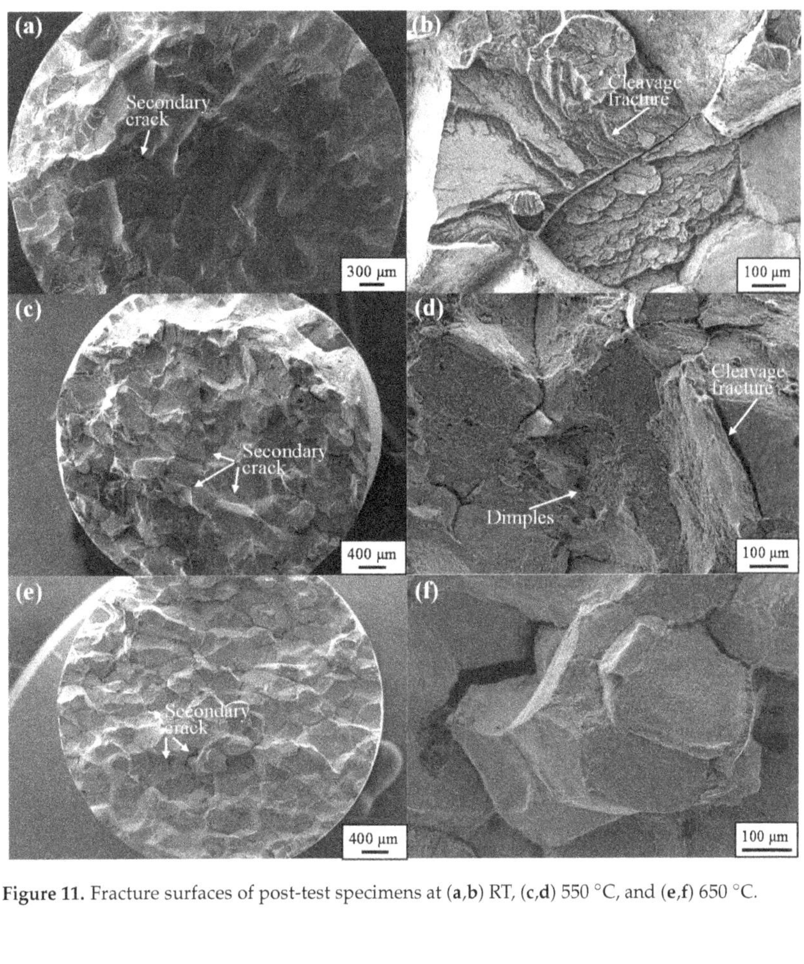

Figure 11. Fracture surfaces of post-test specimens at (**a**,**b**) RT, (**c**,**d**) 550 °C, and (**e**,**f**) 650 °C.

On the other hand, it was found that Ti_2AlNb-based alloys generally show high hardness and significant brittleness due to the characteristic complex of long-range-ordered crystal structures [46]. In addition, cleavage or intergranular fracture dominates the failure of Ti_2AlNb-based alloys at low homologous temperatures, i.e., the brittle-to-ductile transition temperature (DBTT). In the current study, although the tension experiments in the current study were performed at elevated temperatures of up to 650 °C, the testing temperature in the current study was still lower than the DBTT of Ti_2AlNb-based alloys [46]. Therefore, although the elevated temperature improved the motion of dislocation and induced the decrement of yield stress, the fracture of all tested specimens were brittle ones. In addition, the analysis of the fractography of all tested specimens further confirms the brittleness of fracture of all specimens in the current study.

5. Conclusions

The deformation behavior of Ti_2AlNb-based alloys was investigated using uniaxial tension experiments at RT and elevated temperature, and a modified J-C model was proposed to describe the observed variation of flow stress at different temperatures. The main results are summarized as follows.

1. Uniaxial tension experiments at different temperatures revealed the dominant effect of temperature on the deformation of Ti_2AlNb-based alloys. Both Young's modulus and yield strength decreased with the increment of the temperature. Furthermore, different from RT, obvious softening of the flow stress was observed in experiments at 550 and 650 °C;
2. The classical J-C model was found to fail to describe the softening of flow stress at the relatively high temperature, although it gave a correct prediction of the deformation of Ti_2AlNb-based alloys at RT;
3. A modified J-C model was developed by introducing an extra parameter to represent the linear drop of flow stress at elevated temperatures. After determining the corresponding parameters of the J-C model with the experimental results of the lower and upper temperature limits, the deformation behavior at any temperature belonging to the specific range can be correctly described.
4. The fractography of post-test specimens at different temperatures was characterized by FE-SEM. Besides the decrement of resistance to dislocation motion due to the elevated temperature, the softening of flow stress was strongly dependent on the increased proportion of transgranular fracture and the existence of a secondary crack in Ti_2AlNb-based alloys.

This work provides important insight into the deformation and fracture behaviors of Ti_2AlNb alloys at both RT and elevated temperature, and a powerful and precise theoretical model to predict its deformation at any temperature within a required range, which is the critical issue for the application Ti_2AlNb alloys at elevated temperatures.

Author Contributions: Writing—original draft preparation Conceptualization, Y.W.; methodology, resources, Y.Z. and A.S.; formal analysis, visualization, D.Z. and H.C.; Conceptualization, validation, writing—review and editing, supervision, Y.Y.; All authors have read and agreed to the published version of the manuscript.

Funding: This research was funded by Natural Science Foundation of Shanghai (Grant No. 19ZR1413200), Program for Professor of Special Appointment (Eastern Scholar) at Shanghai Institutions of Higher Learning, Major Project for Aero Engine-Gas Turbine of Ministry of Industry and Information Technology, China (KH1B191503), Stably supporting key project of State Administration of Science, Technology and Industry for National Defense, China (KZ0C191708).

Informed Consent Statement: Not applicable.

Data Availability Statement: The raw/processed data required to reproduce these findings cannot be shared at this time as the data also forms part of an ongoing study.

Acknowledgments: This work was supported by Natural Science Foundation of Shanghai (Grant No. 19ZR1413200), Program for Professor of Special Appointment (Eastern Scholar) at Shanghai Institutions of Higher Learning, Major Project for Aero Engine-Gas Turbine of Ministry of Industry and Information Technology, China (KH1B191503), Stably supporting key project of State Administration of Science, Technology and Industry for National Defence, China (KZ0C191708).

Conflicts of Interest: The authors declare no conflict of interest. The funders had no role in the design of the study; in the collection, analyses, or interpretation of data; in the writing of the manuscript, or in the decision to publish the results.

References

1. Banerjee, D.; Gogia, A.K.; Nandi, T.K. A new ordered orthorhombic phase in a Ti_3AlNb alloy. *Acta Metall.* **1988**, *36*, 871–882. [CrossRef]
2. Wu, X. Review of alloy and process development of TiAl alloys. *Intermetallics* **2006**, *14*, 1114–1122. [CrossRef]
3. Bo, W.; Zinkevich, M.; Aldinger, F.; Chu, M.; Shen, J. Prediction of the ordering behaviours of the orthorhombic phase based on Ti_2AlNb alloys by combining thermodynamic model with ab initio calculation. *Intermetallics* **2008**, *16*, 42–51.
4. Dey, S.R.; Roy, S.; Suwas, S. Annealing response of the intermetallic alloy Ti-22Al-25Nb. *Intermetallics* **2010**, *18*, 1122–1131. [CrossRef]
5. Mao, Y.; Hagiwara, M.; Emura, S. Creep behavior and tensile properties of Mo- and Fe-added orthorhombic Ti-22Al-11Nb-2Mo-1Fe alloy. *Scr. Mater.* **2007**, *57*, 261–264. [CrossRef]
6. Chen, X.; Weidong, Z.; Wei, W.; Xiaobo, L.; Jianwei, Z. The enhanced tensile property by introducing bimodal size distribution of lamellar O for O+B2 Ti_2AlNb based alloy. *Mater. Sci. Eng. A* **2013**, *587*, 54–60. [CrossRef]
7. Germann, L.; Banerjee, D.; Guedou, J.Y. Effect of composition on the mechanical properties of newly developed Ti_2AlNb-based titanium aluminide. *Intermetallics* **2005**, *13*, 920–924. [CrossRef]
8. Wen, Y.F.; Wang, L.; Liu, H.L.; Song, L. Ab initio study of the elastic and mechanical properties of B19 TiAl. *Crystals* **2017**, *7*, 39. [CrossRef]
9. Shao, B.; Zong, Y.; Wen, D.; Tian, Y.; Shan, D. Investigation of the phase transformations in Ti-22Al-25Nb alloy. *Mater. Charact.* **2016**, *114*, 75–78. [CrossRef]
10. Shao, B.; Wan, S.; Xu, W.; Shan, D.; Guo, B.; Zong, Y. Formation mechanism of an $\alpha 2$ phase-rich layer on the surface of Ti-22Al-25Nb alloy. *Mater. Charact.* **2018**, *145*, 205–209. [CrossRef]
11. Wu, Y.; Wang, D.; Liu, Z.; Liu, G. A unified internal state variable material model for Ti_2AlNb-alloy and its applications in hot gas forming. *Int. J. Mech. Sci.* **2019**, *164*, 105126. [CrossRef]
12. Xue, C.; Zeng, W.; Wang, W. Quantitative analysis on microstructure evolution and tensile property for the isothermally forged Ti_2AlNb based alloy during heat treatment. *Mater. Sci. Eng. A* **2013**, *573*, 183–189. [CrossRef]
13. Lin, P.; He, Z.; Yuan, S. Tensile deformation behavior of Ti-22Al-25Nb alloy at elevated temperatures. *Mater. Sci. Eng. A* **2012**, *556*, 617–624. [CrossRef]
14. Cai, J.; Wang, K.; Zhai, P. A Modified Johnson-Cook Constitutive Equation to Predict Hot Deformation Behavior of Ti-6Al-4V Alloy. *J. Mater. Eng. Perform.* **2015**, *24*, 32–44. [CrossRef]
15. Yinling, Z.; Aihan, F.; Shoujiang, Q. Microstructure and low cycle fatigue of a Ti_2AlNb-based lightweight alloy. *J. Mater. Sci. Technol.* **2020**, *44*, 140–147.
16. Nandy, T.K.; Banerjee, D. Creep of the orthorhombic phase based on the intermetallic Ti_2AlNb. *Intermetallics* **2000**, *8*, 915–928. [CrossRef]
17. Jiao, X.; Liu, G.; Wang, D. Creep behavior and effects of heat treatment on creep resistance of Ti-22Al-24Nb-0.5Mo alloy. *Mater. Sci. Eng. A* **2017**, *680*, 182–189. [CrossRef]
18. Zhang, N.; Han, X.; Sun, D. Microstructure evolution and mechanical properties of LaB_6-modified Ti_2AlNb alloy fabricated by blended elemental powder metallurgy. *Powder Technol.* **2020**, *369*, 334–344. [CrossRef]
19. Wei, W.; Zeng, W.; Dong, L.; Zhu, B.; Zheng, Y.; Liang, X. Microstructural evolution and tensile behavior of Ti_2AlNb alloys based $\alpha 2$-phase decomposition. *Mater. Sci. Eng. A* **2016**, *662*, 120–128.
20. Zhang, Y.; Liu, Y.; Yu, L. Microstructures and tensile properties of Ti_2AlNb and Mo-modified Ti_2AlNb alloys fabricated by hot isostatic pressing. *Mater. Sci. Eng. A* **2020**, *776*, 139043. [CrossRef]
21. Emura, S.; Tsuzaki, K.; Tsuchiya, K. Improvement of room temperature ductility for Mo and Fe modified Ti_2AlNb alloy. *Mater. Sci. Eng. A* **2010**, *528*, 355–362. [CrossRef]
22. Wang, H.P.; Liu, D.; Wang, J.G.; Shi, Y.Z.; Zheng, Y.; Hu, Y. Investigation on the thermal deformation behavior of the nickel-based superalloy strengthened by γ' phase. *Crystals* **2019**, *9*, 125. [CrossRef]
23. Johnson, G.R.; Cook, W.H. A constitutive model and data for metals subjected to large strains high strain rates and high temperatures. In Proceedings of the Seventh International Symposium on Ballistics, The Hague, The Netherlands, 19–21 April 1983; Volume 21, pp. 541–548.
24. Zhang, D.; Shangguan, Q.; Xie, C. A modified Johnson-Cook model of dynamic tensile behaviors for 7075-T6 aluminum alloy. *J. Alloys Compd.* **2015**, *619*, 186–194. [CrossRef]

25. Li, H.Y.; Wang, X.F.; Duan, J.Y. A modified Johnson Cook model for elevated temperature flow behavior of T24 steel. *Mater. Sci. Eng. A* **2013**, *577*, 138–146. [CrossRef]
26. Mirkoohi, E.; Tran, H.C.; Lo, Y.L.; Chang, Y.C.; Lin, H.Y.; Liang, S.Y. Analytical modeling of residual stress in laser powder bed fusion considering part's boundary condition. *Crystals* **2020**, *10*, 337. [CrossRef]
27. Shrot, A.; Bäker, M. Determination of Johnson–Cook parameters from machining simulations. *Comput. Mater. Sci.* **2012**, *52*, 298–304. [CrossRef]
28. Lin, Y.C.; Liu, G. A new mathematical model for predicting flow stress of typical high-strength alloy steel at elevated high temperature. *Comput. Mater. Sci.* **2010**, *48*, 54–58. [CrossRef]
29. Samantaray, D.; Mandal, S.; Bhaduri, A.K. A comparative study on Johnson-Cook, modified Zerilli-Armstrong and Arrhenius type constitutive model to predict elevated temperature flow behaviour in modified 9Cr-1Mo steel. *Comput. Mater. Sci.* **2019**, *47*, 568–576. [CrossRef]
30. Samantaray, D.; Mandal, S.; Borah, U. A thermo-viscoplastic constitutive model to predict elevated-temperature flow behaviour in a titanium-modified austenitic stainless steel. *Mater. Sci. Eng. A* **2009**, *526*, 1–6. [CrossRef]
31. Liu, Z.; Wang, X.; Jiao, X.; Wu, Y.; Liu, G. Prediction of microstructure evolution during hot gas forming of Ti_2AlNb-based alloy tubular component with square cross-section. *Procedia Manuf.* **2018**, *15*, 1156–1163. [CrossRef]
32. Lin, J.; Liu, Y. A set of unified constitutive equations for modelling microstructure evolution in hot deformation. *J. Mater. Process. Technol.* **2003**, *143*, 281–285. [CrossRef]
33. Lin, J.; Cheong, B.H.; Yao, X. Universal multi-objective function for optimising superplastic-damage constitutive equations. *J. Mater. Process. Technol.* **2002**, *125*, 199–205. [CrossRef]
34. Lin, Y.C.; Chen, X.M.; Liu, G. A modified Johnson-Cook model for tensile behaviors of typical high-strength alloy steel. *Mater. Sci. Eng. A* **2010**, *527*, 6980–6986. [CrossRef]
35. Huh, H.; Kang, W.J.; Han, S.S. A tension split Hopkinson bar for investigating the dynamic behavior of sheet metals. *Exp. Mech.* **2002**, *42*, 8–17. [CrossRef]
36. Ulacia, I.; Salisbury, C.P.; Hurtado, I. Tensile characterization and constitutive modeling of AZ31B magnesium alloy sheet over wide range of strain rates and temperatures. *J. Mater. Process. Tech.* **2011**, *211*, 830–839. [CrossRef]
37. Tan, J.Q.; Zhan, M.; Liu, S. A modified Johnson-Cook model for tensile flow behaviors of 7050-T7451 aluminum alloy at high strain rates. *Mater. Sci. Eng. A* **2015**, *631*, 214–219. [CrossRef]
38. Chu, F.; Mitchell, T.E.; Majumdar, B.; Miracle, D.; Nandy, T.K.; Banerjee, D. Elastic properties of the O phase in Ti-Al-Nb alloys. *Intermetallics* **1997**, *5*, 147–156. [CrossRef]
39. Banerjee, D. The intermetallic Ti_2AlNb. *Prog. Mater. Sci.* **1997**, *42*, 135–158. [CrossRef]
40. Kumpfert, J. Intermetallic alloys based on orthorhombic titanium aluminide. *Adv. Eng. Mater.* **2001**, *3*, 851–864. [CrossRef]
41. Mosleh, A.O.; Mikhaylovskaya, A.V.; Kotov, A.D.; Kwame, J.S. Experimental, modelling and simulation of an approach for optimizing the superplastic forming of Ti-6%Al-4%V titanium alloy. *J. Manuf. Process.* **2019**, *45*, 262–272. [CrossRef]
42. Mosleh, A.O.; Mestre-Rinn, P.; Khalil, A.M.; Kotov, A.D.; Mikhaylovskaya, A.V. Modelling approach for predicting the superplastic deformation behaviour of titanium alloys with strain hardening/softening characterizations. *Mater. Res. Express* **2020**, *7*, 016504. [CrossRef]
43. Wu, Z.; Hu, R.; Zhang, T. Microstructure determined fracture behavior of a high Nb containing TiAl alloy. *Mater. Sci. Eng. A* **2016**, *666*, 297–304. [CrossRef]
44. Hertzberg, R.W.; Hauser, F.E. Deformation and Fracture Mechanics of Engineering Materials. *J. Mater. Sci. Technol.* **1997**, *19*, 283–286. [CrossRef]
45. Zhou, M.; Clode, M.P. Constitutive equations for modelling flow softening due to dynamic recovery and heat generation during plastic deformation. *Mech. Mater.* **1998**, *27*, 63–76. [CrossRef]
46. Ritchie, R.O. Mechanisms of fatigue-crack propagation in ductile and brittle solids. *Int. J. Fract.* **1999**, *100*, 55–83. [CrossRef]

crystals

MDPI

Article

Strain-Gradient Crystal Plasticity Finite Element Modeling of Slip Band Formation in α-Zirconium

Omid Sedaghat and Hamidreza Abdolvand *

Department of Mechanical & Materials Engineering, Western University, Spencer Engineering Building, London, ON N6A 5B9, Canada; osedagha@uwo.ca
* Correspondence: hamid.abdolvand@uwo.ca

Abstract: Two methods for the determination of geometrically necessary dislocation (GND) densities are implemented in a lower-order strain-gradient crystal plasticity finite element model. The equations are implemented in user material (UMAT) subroutines. Method I has a direct and unique solution for the density of GNDs, while Method II has unlimited solutions, where an optimization technique is used to determine GND densities. The performance of each method for capturing the formation of slip bands based on the calculated GND maps is critically analyzed. First, the model parameters are identified using single crystal simulations. This is followed by importing the as-measured microstructure for a deformed α-zirconium specimen into the finite element solver to compare the numerical results obtained from the models to those measured experimentally using the high angular resolution electron backscatter diffraction technique. It is shown that both methods are capable of modeling the formation of slip bands that are parallel to those observed experimentally. Formation of such bands is observed in both GND maps and plastic shear strain maps without pre-determining the slip band domain. Further, there is a negligible difference between the calculated grain-scale stresses and elastic lattice rotations from the two methods, where the modeling results are close to the measured ones. However, the magnitudes and distributions of calculated GND densities from the two methods are very different.

Keywords: zirconium; strain-gradient crystal plasticity; slip band; geometrically necessary dislocation; statistically stored dislocation; high angular resolution electron back scatter diffraction

Citation: Sedaghat, O.; Abdolvand, H. Strain-Gradient Crystal Plasticity Finite Element Modeling of Slip Band Formation in α-Zirconium. *Crystals* **2021**, *11*, 1382. https://doi.org/10.3390/cryst11111382

Academic Editor: Wojciech Polkowski

Received: 28 October 2021
Accepted: 10 November 2021
Published: 12 November 2021

1. Introduction

Crystal plasticity is a class of constitutive equations that can be used to describe deformation mechanisms of polycrystalline materials at the grain scale [1–3]. It has been used in many studies to understand the evolution of heterogeneous deformation fields within individual grains in "real" time scales [4–7]. A significant number of crystal plasticity studies have focused on modeling deformation via slip, which is controlled by the movement of dislocations on a particular plane known as the slip plane and in a particular direction known as the slip direction. The strength of the obstacles interacting with dislocations determines the resistance of a slip system to the movement of the dislocations and the consequent material hardening. These obstacles can be categorized into short-range and long-range barriers [8]. The primary short-range obstacles are generally assumed to be the other dislocations that intersect the slip plane. The long-range obstacles may include the elastic stress field induced by far field dislocations or grain boundaries. It is generally assumed that the long-range obstacles are affected by the density of the geometrically necessary dislocations (GNDs), while the evolution of both statistically stored dislocations (SSDs) and GNDs during crystallographic slip equally increase the short-range interactions [9,10]. When conventional crystal plasticity models are used, the difference between these two obstacles is generally ignored and the critical resolved shear stress (CRSS) for the movement of the dislocations does not depend on the deformation state in the neighboring points. However, the plastic response of a polycrystal depends on both the local state of deformation and the gradients of the plastic strain. Incorporation of the plastic strain gradient in crystal plasticity constitutive equations makes the model response dependent on the neighboring elements—this is generally investigated in the non-local crystal plasticity approach [11–15]. The strain-gradient plasticity theories are formulated in two different forms, i.e., the lower-order and higher-order approach [16]. In the lower-order approach [17–19], the plastic strain gradient terms are only incorporated in the hardening laws. However, in the higher-order approach, the equilibrium equations are also adjusted [20–23]. This paper focuses on examining the methods available in the literature for calculating GND densities in the lower-order crystal plasticity framework by providing a direct comparison between the results of each method to those measured experimentally.

GND densities are inherently linked to the lattice curvature and can be determined using the gradients of plastic strain [24]. The derivation of the dislocation densities from the "curl" of the deformation gradient in the Nye equation has been accompanied with inconsistencies in the literature, some of which have been reported by Das et al. [25]. Generally, two methods are used in the literature for extracting GNDs using Nye equations. In the first method, the contribution of each slip system, e.g., α, to GND density is assumed to be proportional to the plastic shear accommodated on the same slip system α. Hence, the number of linear equations that should be solved is equal to the number of unknowns, i.e., GNDs on each slip system α. This is called the direct method, where the density of GNDs on each slip system is determined unambiguously. The direct method was initially proposed by Dai [26] and was subsequently implemented in crystal plasticity finite element (CPFE) models [10,18,27–29]. In the second method, the cumulative contribution of plastic shears on all slip systems to the total GND densities on all slip systems is assumed to be proportional to the Nye tensor. This usually results in an under-determined system of equations where the number of unknowns is more than the number of equations. Two main approaches are suggested for determining the unknowns. In the first approach, known as L^1, a solution is found by minimizing the dislocation line energy. In the second approach, known as L^2, a solution is found by minimizing the sum of the squares of the dislocation densities [30]. Das et al. showed that the use of the L^2 method leads to a solution where dislocation densities are evenly distributed on all slip systems, while the use of the L^1 method leads to capturing some variations of GNDs between the slip systems [25]. Negligible difference was found between the magnitudes of the total dislocation densities obtained from the two minimization approaches. The L^2 minimization approach has been widely used in the CPFE models [31–34]. In contrast to the direct method, which provides a unique solution for the GND density on each slip system, both minimization approaches lead to non-unique solutions. Therefore, an investigation is required to compare the magnitudes and distributions of the obtained GND densities from these two methods.

The measurement of internal stress and dislocation densities helps validate dislocation-based crystal plasticity models. Diffraction-based experimental methods are mainly used for measuring internal lattice strains [35–38] and localized stress fields [38,39] as well as the density of dislocations [40]. For example, the dislocation densities in the vicinity of slip bands in HCP polycrystals were measured using micro X-ray diffraction [41]. Although a three-dimensional view of the dislocation densities is provided, only a few grains can be studied in this technique. A high angular resolution electron back scatter diffraction (HR-EBSD) technique can be used to measure "relative" elastic strains and rotations and "absolute" GND densities for many grains, yet close to the sample surface. Diffraction patterns are collected and then cross correlated using a reference pattern. The deformation gradient tensor is then calculated and used to determine the relative elastic strain tensor [42–45] and GND densities using the Nye tensor [46]. Many studies have used HR-EBSD to compare the numerically obtained GND densities from CPFE models to the measured ones [33,47–49].

The objectives of this paper are to compare the GND densities calculated using the direct method to those obtained from the minimization-based method and investigate the capability of the GND-based CPFE models in capturing the formation of slip bands in an α-zirconium polycrystal. Hence, the direct and the minimization-based methods for calculating GND densities are formulated and implemented in a lower-order strain-gradient CPFE model. After calibrating single crystal parameters, the evolution of GND and SSD densities predicted using each method is studied in detail using single crystal FE models. The capability of the developed non-local models in capturing the experimentally observed slip bands of a deformed polycrystalline α-zirconium is subsequently investigated. This is followed by comparing the results of the models for GND densities, stresses, and elastic rotations to those measured via HR-EBSD.

2. Sample Preparation and Experimental Set Up

Blocks of α-zirconium material were firstly annealed at 700 °C and then air cooled to relieve residual stresses as much as possible. A dog bone sample was subsequently made with a gauge length of 20 mm. The sample was mechanically polished down to 4000 grit, followed by polishing in a 50 nm colloidal silica suspension and electropolishing in a −30 °C solution of 90% methanol and 10% perchloric acid for 60 s at 25 V. The cross section of the sample after polishing was measured to be 0.5×0.5 mm^2.

The sample was uniaxially deformed at room temperature under strain control to a tensile strain of 1.2% using a strain rate of 2.6×10^{-5} s^{-1}. The EBSD measurement was performed in a Zeiss MERLIN field emission gun scanning electron microscope (FEG-SEM) with 20 keV beam energy, 15 nA probe current, and at the working distance of 18 mm. Diffraction patterns were collected every 0.5 μm using a Bruker high resolution EBSD detector. Following the method developed by Wilkinson et al. [42], elastic strain and lattice rotations were calculated by cross correlating the collected 800 × 800 pixels Kikuchi patterns [50]. An EBSD map of the deformed sample with the coordinate system used is shown in Figure 1a. As shown, the X-axis coincides with the loading direction, the Z-axis is along the electron beam direction and sample thickness, and the Y-axis is defined by the cross product of the other two axes. In addition, as shown in Figure 1b, the c-axis of the HCP crystals is oriented towards the z-axis with a little spread towards the x-axis. Since the specimen was only deformed to 1.2% and the c-axis of most of the crystals is perpendicular to the loading direction, deformation twinning was not active in this experiment. No twin was observed in the EBSD and the ARGUS images taken from sample after deformation. As such, the effects of twinning are ignored in all mathematical formulation and CPFE modeling.

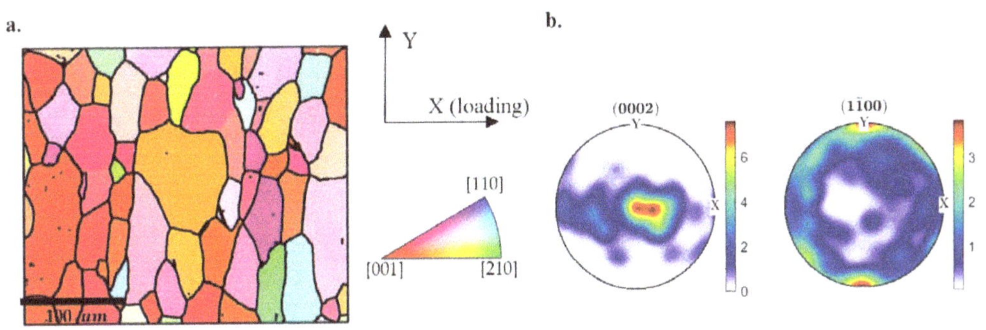

Figure 1. (**a**) An inverse pole figure-Z of the sample with the corresponding pole figure shown in (**b**).

3. Crystal Plasticity Formulation and Input Model

In this section the constitutive equations implemented for calculating GND densities using both the direct and minimization-based methods are described. Here, the direct and minimization-based methods are called "Method I" and "Method II", respectively. These equations are implemented in a User MATerial (UMAT) subroutine originally developed by Abdolvand et al. [51] and recently updated by Sedaghat and Abdolvand [47]. After describing the implemented constitutive equations, the steps followed to prepare the FE input model are presented.

3.1. Crystal Plasticity Constitutive Equations

The total deformation gradient (F) can be decomposed to the elastic (F^e) and plastic (F^p) parts:

$$F = F^e F^p \tag{1}$$

The total velocity gradient tensor (L) in the current configuration can be divided into the elastic (L^e) and plastic (L^p) parts as:

$$L = \dot{F}F^{-1} = \dot{F}^e F^{e-1} + F^e \dot{F}^p F^{p-1} F^{e-1} = L^e + L^p \tag{2}$$

The total velocity gradient tensor can be divided into its symmetric part, i.e., deformation rate tensors (D^e, D^p), and asymmetric part, i.e., spin tensors (Ω^e, Ω^p). The plastic part of the velocity gradient tensor is calculated using the following equation [52]:

$$L^p = D^p + \Omega^p = \sum_{\alpha=1}^{N} \dot{\gamma}^\alpha \vec{m}^\alpha \otimes \vec{n}^\alpha \tag{3}$$

where $\vec{m}^\alpha$, $\vec{n}^\alpha$, and $\dot{\gamma}^\alpha$ respectively represent the slip direction, the slip plane normal, and the shear rate on the αth slip system for the N number of active slip systems. $\vec{m}^\alpha \otimes \vec{n}^\alpha$ is known as the Schmid tensor of the slip system α. The shear rate on the slip system α is calculated based on the resolved shear stress (τ^α) acting on the same slip system [52]:

$$\dot{\gamma}^\alpha = \dot{\gamma}_0^\alpha \,\mathrm{sign}\left(\frac{\tau^\alpha}{g^\alpha}\right)\left|\frac{\tau^\alpha}{g^\alpha}\right|^n \tag{4}$$

where $\dot{\gamma}_0$ is a reference shear strain rate, n represents the sensitivity of the material to a strain rate, and g^α is the CRSS of the slip system α. The resolved shear stress acting on the slip system α is proportional to Kirchhoff stress (ψ) through the following equation [53]:

$$\tau^\alpha = P^\alpha : \psi \tag{5}$$

where P^α is the symmetric part of the Schmid tensor. The Jaumann rate of the Kirchhoff stress tensor ($\check{\psi}$) is related to the elastic part of the deformation rate tensor (D^e) and the elastic stiffness tensor ($\mathbb{C}$) of the HCP crystal after rotation to the deformed configuration as:

$$\check{\psi} = \mathbb{C} : D^e \tag{6}$$

The elastic modulus of zirconium HCP crystals used in this study is the one determined by Fisher and Renken [54]: C_{11} = 143.5 GPa, C_{33} = 164.9 GPa, C_{12} = 72.5 GPa, C_{13} = 65.4 GPa, and C_{44} = 32.1 GPa. The objective rate of Kirchhoff stress, in Equation (6), is defined as:

$$\check{\psi} = \dot{\psi} - \Omega^e \psi + \psi \Omega^e \tag{7}$$

The strength of each slip system (g^α) in Equation (4) follows a dislocation-based hardening law [10–12,27]:

$$g^\alpha = g^\alpha_{*,0} + \frac{H^\alpha}{\sqrt{D}} + \xi G b^\alpha \sqrt{\sum_{\beta=1}^{N} q^{\alpha\beta}\left(\rho^\beta_{GND} + \rho^\beta_{SSD}\right)} = g^\alpha_{*,0} + \frac{H^\alpha}{\sqrt{D}} + \xi G b^\alpha \sqrt{\rho_T} \tag{8}$$

where g^α is the updated CRSS, $g^\alpha_{*,0}$ is the initial CRSS, D is the equivalent grain diameter, H^α is the Hall-Petch parameter for slip system α, ξ is a material constant, G is the shear modulus, b^α is the size of the Burgers vector for the slip system α, and ρ^β_{GND} and ρ^β_{SSD} are the CND and SSD densities on the slip system β. A hardening matrix $q^{\alpha\beta}$ is introduced to consider the effects of self- and latent-hardening. For the sake of simplicity in deriving a solution, the term under the radical is called ρ_T. It is assumed that the GND and SSD densities have the same hardening matrix $q^{\alpha\beta}$. The shear modulus, G in Equation (8), is assumed to be the average value of C_{44}, C_{55}, and C_{66} [55]. The material constant ξ is assumed to be 0.5.

It is assumed that the evolution of SSD densities follows the equation below [10,56]:

$$\dot{\rho}^{\alpha}_{SSD} = \frac{\left|\dot{\gamma}^{\alpha}\right|}{b^{\alpha}}\left(K^{\alpha}\sqrt{\rho^{\alpha}_{SSD}+\rho^{\alpha}_{GND}} - 2y^{\alpha}_{c}\,\rho^{\alpha}_{SSD}\right) \quad (9)$$

where K^{α} is the dislocation accumulation constant and y^{α}_{c} is the dislocation annihilation length. The first term in the right-hand side of the Equation (9) is the dominant term at the early stages of plasticity, while with further loading, the effects of the second term will be non-negligible. In the following sub-sections, the two methods used for calculating GND densities are presented.

3.1.1. Method I

In the first method, the GND density of the slip system α is uniquely correlated with the resolved shear strain accumulated on the same system [26]:

$$\dot{\rho}^{\alpha}_{GND} = \frac{\dot{\gamma}^{\alpha}}{b^{\alpha}}\left|Curl\left[\vec{n}^{\alpha}_{j}\,\boldsymbol{F}^{p}_{ji}\right]\right| = \frac{\dot{\gamma}^{\alpha}}{b^{\alpha}}\left|\epsilon_{ikq}\frac{\partial\left(\vec{n}^{\alpha}_{j}\,\boldsymbol{F}^{p}_{jq}\right)}{\partial x_k}\right| \quad (10)$$

where ϵ_{ikq} is the permutation tensor. Hence, ρ^{α}_{GND} has a unique solution. For implementing Equations (8)–(10) into the UMAT, the time derivative of Equation (8) is firstly calculated and coupled with Equations (9) and (10) to solve for the increments of shear strains using the Newton-Raphson iterative algorithm. The evolution of CRSS can be reformulated by taking the time derivative of Equation (8):

$$\dot{g}^{\alpha} = \xi G b^{\alpha}\frac{\dot{\rho}_T}{2\sqrt{\rho_T}}, \text{ where } \dot{\rho}_T = \sum_{\beta=1}^{N} q^{\alpha\beta}\left(\dot{\rho}^{\beta}_{GND}+\dot{\rho}^{\beta}_{SSD}\right) \quad (11)$$

By substituting Equations (9) and (10) into Equation (11) we have:

$$\dot{g}^{\alpha} = \sum_{\beta=1}^{N}\left[q^{\alpha\beta}\frac{\xi G b^{\alpha}}{2b^{\beta}\sqrt{\rho_T}}\left(\left|\epsilon_{ikq}\frac{\partial\left(\vec{n}^{\alpha}_{j}\,\boldsymbol{F}^{p}_{jq}\right)}{\partial x_k}\right| + K^{\beta}\sqrt{\rho^{\beta}_{SSD}+\rho^{\beta}_{GND}} - 2y^{\beta}_{c}\,\rho^{\beta}_{SSD}\right)\right]\dot{\gamma}^{\beta} \quad (12)$$

The complexity of implementation of this method in the UMAT subroutine is calculation of the gradient term, which was originally developed by Abdolvand [57].

3.1.2. Method II

In the second method, the density of GNDs on the slip system α is defined as [30]:

$$\left(curl\left(\boldsymbol{F}^{p^T}\right)\right)^T = \sum_{\alpha=1}^{N}\left(\rho^{\alpha}_{GND,s}\,\vec{b}^{\alpha}\otimes\vec{m}^{\alpha} + \rho^{\alpha}_{GND,et}\,\vec{b}^{\alpha}\otimes\vec{t}^{\alpha} + \rho^{\alpha}_{GND,en}\,\vec{b}^{\alpha}\otimes\vec{n}^{\alpha}\right) \quad (13)$$

ρ^{α}_{GND} for each slip system can be decomposed into three components, one for screw type dislocation $\rho^{\alpha}_{GND,s}$ and two edge type dislocations, $\rho^{\alpha}_{GND,en}$ and $\rho^{\alpha}_{GND,et}$ with $t^{\alpha} = m^{\alpha}\times n^{\alpha}$, respectively. In contrast to Method I, Method II has infinite solutions where a minimization scheme is usually used to find a solution. Equation (13) can be solved by minimizing the sum of the squares of GND densities [19]:

$$\{\rho^{\alpha}_{GND}\} = A^T(AA^T)^{-1}B \quad (14)$$

where $\{\rho^{\alpha}_{GND}\}$ is 3N × 1 column vector including the components of GND for slip system α, B is a 9 × 1 vector, containing the curl function components in Equation (13), and A is a 9 × 3N matrix with the basis tensors of Equation (13). The time derivative of Equation (13) is determined as:

$$\left(curl(\dot{F}^{p^T})\right)^T = \sum_{\alpha=1}^{N} (\dot{\rho}^{\alpha}_{GND,\,s}\, \vec{b}^{\alpha} \otimes \vec{m}^{\alpha} + \dot{\rho}^{\alpha}_{GND,et}\, \vec{b}^{\alpha} \otimes \vec{t}^{\alpha} + \dot{\rho}^{\alpha}_{GND,en}\, \vec{b}^{\alpha} \otimes \vec{n}^{\alpha}) \quad (15)$$

By substituting Equations (15) and (9) into Equation (11), we have:

$$\dot{g}^{\alpha} = \sum_{\beta=1}^{N} \frac{q^{\alpha\beta}\xi G b^{\alpha}(K^{\beta}\sqrt{\rho^{\beta}_{SSD} + \rho^{\beta}_{GND}} - 2y^{\beta}_{c}\rho^{\beta}_{SSD})}{2b^{\beta}\sqrt{\rho_T}} . \left|\dot{\gamma}^{\beta}\right| + \sum_{\beta=1}^{N} \frac{q^{\alpha\beta}\xi G b^{\alpha}(\dot{\rho}^{\beta}_{GND})}{2b^{\beta}\sqrt{\rho_T}} \quad (16)$$

Similar to Method I, the first term in the right-hand side of Equation (16) shows an explicit correlation between the shear strain rate $\dot{\gamma}^{\beta}$ and $\dot{g}^{\alpha}$, but the second term represents the effects of GND evolution rate on $\dot{g}^{\alpha}$. Therefore, both GND density and the rate of GND density evolution should be determined in Method II. Further, a minimization approach must be followed to determine the two tensors. Accordingly, the implementation of Method II is relatively more complicated and computationally costly compared to Method I. Here, the method originally developed by Abdolvand [57] was adopted to calculate the curl of $\boldsymbol{F^p}$, which was subsequently implemented in the UMAT as described in [47].

The single crystal parameters used for HCP zirconium for each method are provided in Table 1. The single crystal parameters for the Method II model were calibrated using a comprehensive data set measured in a series of in-situ neutron diffraction experiments conducted on HCP zirconium polycrystals [58]. Measured macroscopic stress-strain curves, lattice strains, and texture development were used to calibrate the model parameters as explained in [47]. The model parameters for Method I are calibrated such that they reproduce identical stress-strain curves to those from the model with Method II. The initial SSD density is set at 10^{10} m^{-2} for both methods. The effect of twinning is ignored in the simulations since no twin was observed in the experiment. Here, only the single crystal parameters for Method I require calibration. The process of calibration is discussed in Section 4.1.

Table 1. Single crystal parameters used in the hardening law for HCP zirconium from Sedaghat and Abdolvand [47].

		Prism	Basal	Pyramidal
n		20	20	20
$\dot{\gamma}^{\alpha}_{0}$ (S^{-1})		3.5×10^{-4}	3.5×10^{-4}	1.0×10^{-4}
q^{ss} (self)		1	1	1
q^{st} (t = prism)		1	1	0
q^{st} (t = basal)		1	1	0
q^{st} (t = pyramidal)		0	0	1
Burgers vector (nm)		0.323	0.323	0.608
H^{α} $(MPa\sqrt{m})$		0.109	0.146	0.292
$g^{\alpha}_{*,0}$ (MPa)		95	135	266
MI	K^{α}	0.07	0.07	0.30
	y^{α}_{c} (nm)	5	5	10
MII	K^{α}	0.05	0.05	0.30
	y^{α}_{c} (nm)	5	5	10

3.2. *Input Models*

3.2.1. Single Crystal

The evolution of GND and SSD densities as well as the CRSS of each slip system is studied using the single crystal model. This is to provide a comparison between the two methods used. A cube-shaped single crystal with a side length of 20 μm is used for the FE input model. The cube is deformed uniaxially to 5% applied strain at the strain rate of 10^{-5} s^{-1}. Uniaxial tensile strain is applied on the surface DCGH along the X-direction (Figure 2a). Fixed displacement boundary conditions are applied on the ABFE, AEHD, and EFGH surfaces along X, Y, and Z directions, respectively. The model is discretized with 1000 quadratic brick elements with reduced integration points (C3D20R), with 10 elements in each direction. The c-axis of the HCP crystal was rotated with respect to the loading direction from 0° to 90° at the step size of 22.5°. As shown in Figure 2a, the misorientation between the loading direction and crystal c-axis is presented by θ.

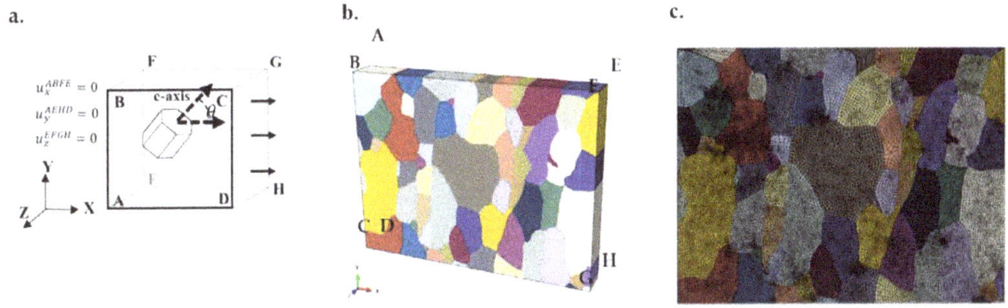

Figure 2. (**a**) Applied boundary conditions for the single crystal model. (**b**) The EBSD-measured microstructure imported into the finite element solver with the meshed map shown in (**c**).

3.2.2. Polycrystal

The measured EBSD map shown in Figure 1a is imported into the ABAQUS FE solver. The imported polycrystal model is shown in Figure 2. The as-measured strain and strain rates are applied to the FE model. The boundary conditions used are $u_x = 0$ and $u_y = 0$ on the ABCD and CGHD surfaces, respectively, and $u_z = 0$ on the AD and DH edges. The model is extruded in the z direction to the thickness of 50 μm to be consistent with the measured average grain size of the specimen. After conducting a mesh sensitivity study to ensure that FE results are converged, the model was discretized to 65,157 C3D20R brick elements. Three elements along the model thickness are used. The EBSD-measured grain orientations are assigned to each modeled grain. The model was deformed in two steps; in step 1, 1.2% uniaxial strain was applied to the model in the global x-direction, and in step 2, the model is unloaded and allowed to relax. In all simulations presented, the maximum time increment allowed is 5 s to ensure that the calculated GND densities are independent from the size of the selected time increment and are converged.

4. Results

A single crystal model is firstly used to calibrate the single crystal parameters for Method I of the CPFE model. This is followed by comparing the results of the CPFE models to those measured using HR-EBSD for stress, lattice rotations, and GND densities within the polycrystal. The capability of each method in determining slip activity and capturing slip bands is also discussed.

4.1. Single Crystal

The single crystal model shown in Figure 2a is used to calibrate the single crystal parameters for Method I of the CPFE model. The calculated stress of this method is compared to the one from Method II of the CPFE model. In addition, the calculated slip activities and GND densities are compared between the two methods. This is to understand the differences between each method at the single crystal level and to establish a foundation for the differences observed in the polycrystalline simulations.

In Figure 3a, the stress-strain curves calculated with the CPFE model using both Method I and II at five different crystal orientations are shown. In addition, the calculated relative activities of each slip system from both methods are provided in Figure 3b. Relative activities are the resolved shear strain (γ^{α}) for each slip set divided by the total shear strain at the applied strain of 5%. It is shown that identical stress-strain curves and slip activities are obtained for different crystal orientations.

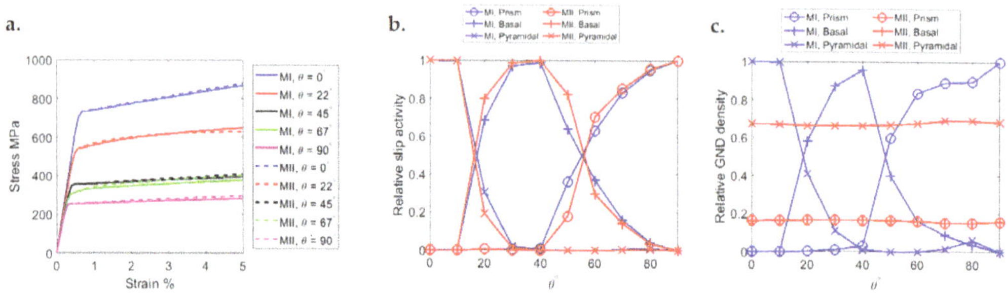

Figure 3. (**a**) The average stress-strain curve, (**b**) relative slip activity, and (**c**) relative GND density calculated for a single crystal with different c-axis misorientation with the loading direction (θ). Comparisons are made for Method I (M I) as well as Method II (M II) of the CPFE model.

Figure 3c compares the calculated relative GND density of each slip system from both methods of the CPFE model. Relative GND densities are determined by dividing the total GND density calculated for each slip set by the total GND density calculated for all slip systems. The calculated relative GND densities from Method I follow the same trends captured for the slip activities. However, the relative GND densities of Method II follow a nearly horizontal line, which is independent from the misorientation angle θ and the calculated slip system activity. Moreover, the relative GND density of the pyramidal slip system is almost four times higher than those from prism and basal slip systems. The relative GND densities calculated for prism and basal slip systems from Method II are identical. These two curves overlapp in Figure 3c. The relative GND densities calculated using Method II for each slip set is proportional to the number of variants for each slip set. That is, GND densities are evenly distributed on all variants of a slip system when Method II is used.

The evolution of the average GND and SSD densities of each slip set from the two methods implemented in the CPFE framework is compared in Figure 4. For each slip set, the total calculated dislocation density, either GND or SSD, is divided by the number of variants in the same slip set, i.e., M in $\frac{\sum_{\alpha=1}^{M}\rho_{GND}^{\alpha}}{M}$ is 3, 3, and 12 for prism, basal, and pyramidal slip systems, respectively. Results are for three different misorientation angles (θ). In addition, the calculated total dislocation densities, GNDs plus SSDs, from the two methods are compared. Although both methods predict identical average stress-strain curves (Figure 3a), a significant difference in the predicted GND densities is observed. The SSD and GND densities from Method II have the same order of magnitude. However, in Method I, SSD densities are at least two orders of magnitude higher than GND densities. This means that the hardening of slip systems from Method I is mainly driven by SSD densities as the contribution of GND densities is almost negligible. The cumulative dislocation densities calculated from both SSDs and GNDs are almost equal to those calculated from Method II. This results in calculating the same evolution for CRSS using both methods.

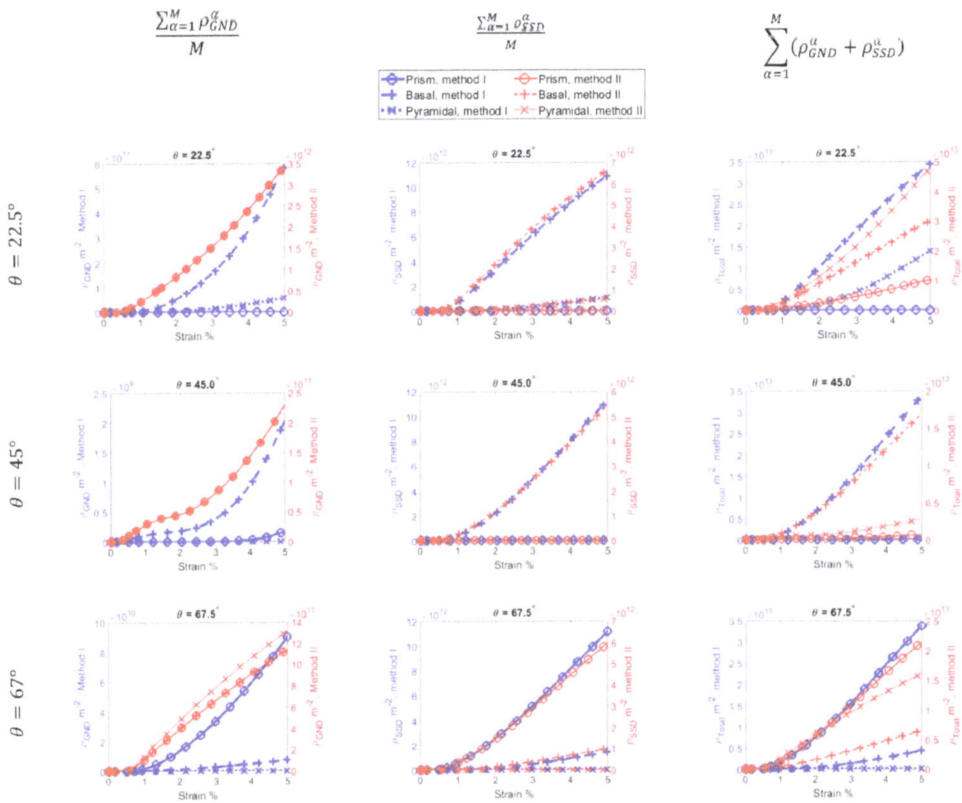

Figure 4. The evolution of the average GND density, average SSD density, and total dislocation density of prism, basal and pyramidal slip systems at different crystal c-axis misorientations with the loading direction (θ). Results are for CPFE model using Method I and II. M is 3, 3, and 12 for prism, basal, and pyramidal slip systems.

4.2. Polycrystal

Results presented in this section are from the CPFE model using both Method I and II. The CPFE results for GND densities are presented and compared to those measured via HR-EBSD. In addition, the capability of both methods in simulating formation of slip bands is evaluated by comparing the numerical results to those from SEM images of the deformed specimen. Finally, the distribution of calculated stresses and elastic lattice rotations are compared to those measured using HR-EBSD.

4.2.1. GND Density and Slip Activity

The distributions of total GND densities after the unload step for Method I and II are shown in Figure 5a,b, respectively. The total GND density from Method I is about an order of magnitude smaller than that calculated using Method II. For example, in grains G2, G3, and G5, marked in Figure 5a, a significant portion of the grains have GND densities lower than 10^9 m^{-2} when Method I is used. Moreover, the maximum GND density from Method I is about 4×10^{12} m^{-2}, but it is ~7×10^{14} m^{-2} when Method II is used. Although there is a noticeable difference in the magnitudes of calculated GNDs, the predicted trends are quite similar between the two methods. For example, most of the grain boundaries have high GND densities. In addition, both methods predict highly localized GND concentration sites at the top-right sections of grains G1 and G2, upper side of grain G3, and right side of grain G6. These areas are shown by the white circles in Figure 5a. HR-EBSD measurement also shows high GND density at these regions, as shown in Figure 5c. However, the magnitudes of GND densities from HR-EBSD are higher than those from both CPFE methods, but much closer to those from Method II. In comparison to Method I, the magnitude of GND densities from Method II is closer to those measured with HR-EBSD. This is not surprising because in the calculation of GND densities from HR-EBSD measurements, Method II is used along with the L^1 optimization scheme. As a result, it is expected to have a better agreement between HR-EBSD measurements for GNDs and those from Method II of the CPFE model. The effects of using L^1 and L^2 optimization schemes in determining GND densities from HR-EBSD are discussed elsewhere [59,60]. In addition, Method I cannot be used for HR-EBSD measurement as the plastic resolved shear strain on each slip system (γ^α) cannot be directly extracted with this technique (see Equation (10)).

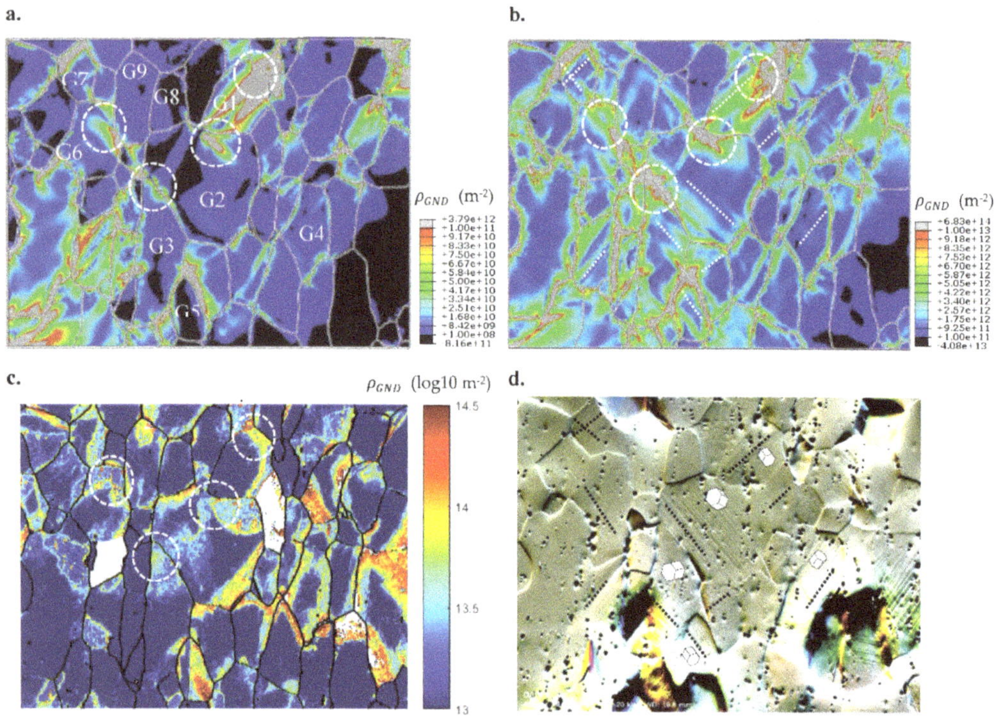

Figure 5. The distribution of total GND density from (**a**) CPFE model using Method I, (**b**) CPFE model using Method II. and (**c**) HR-EBSD. (**d**) SEM-ARGUS image of the deformed sample. In (**d**) slip bands are shown with the black dashed lines and crystal orientations are shown using HCP crystals.

Figure 5d shows an ARGUS image of the deformed polycrystal, in which slip bands can be observed in most of the grains. For better visualization, the orientations of slip bands are shown with the black dotted lines. Interestingly, CPFE results from both Method I and II also show GND density localization in the form of slip bands. For example, in grains G1, G2, G3, and G4, a distinct slip band is formed that spans from one side of the grain to another, as shown with the white dotted lines in Figure 5b. The orientations of these bands are in agreement with those observed in the SEM image in Figure 5d. In addition, both CPFE methods have captured the crossing of the two separate slip bands observed in grain G7. The two slip bands can also be observed in the ARGUS image. The CPFE results reveal that these slip bands are associated with the localization of plastic shear on the prism planes. This is also consistent with the measured grain orientations as shown in Figure 5d. For example, in grains G2, G3, and G4, the observed slip bands of the ARGUS image are parallel to ones of the variants of the prism planes of the illustrated HCP crystals. Therefore, both Method I and II are equally effective when predicting the direction of slip bands.

In order to determine the relationship between the observed slip bands and active slip systems, the average resolved shear strain and GND density of each slip set are extracted from the CPFE results and are shown in Table 2. These values are calculated by taking the volume average of the resolved shear strains, or GNDs, calculated at all integration points assigned to each grain. The calculated average value for each slip set is the summation over all variants, i.e., 3 variants for basal or prism, and 12 variants for pyramidal <c + a>. Table 2 summarizes the amount of slip and GND density accumulated on the prism, basal, and pyramidal slip systems of grains G1 to G5. The predominant slip system with the highest value is presented in bold. Both methods confirm that the predominant slip system for all five selected grains is slip on the prism planes. For example, for grain G4 almost all plastic deformation is accommodated by prism slip systems whereas for grain G3 both basal and prism slip systems are active with prism being the most active one. Although these trends are also observed for the GND densities obtained from Method I, the pyramidal slip system accommodates most of the GNDs calculated in Method II. In fact, the GND density from method II is almost evenly distributed on all eighteen variants, but since the pyramidal slip system has 12 variants, it always has the highest GND density once the summation is done over all 12 variants. These results are consistent with the trends observed for the single crystal model. The heterogeneous distribution of the GND densities on the different slip variants in Method I is more reasonable compared to Method II, where the GND densities on individual slip variants are always equal, regardless of the applied loading conditions.

Table 2. Average resolved shear strains and GND densities from the CPFE models. Results are for grains G1 to G5 of the polycrystalline model at 1.2% applied strain. The predominant slip system is shown in bold.

Grain ID	GND Method	Cumulative Resolved Shear Strain			GND Density m^{-2}			
		Prism	Basal	Pyr	Prism	Basal	Pyr	Total
1	I	**4.8×10^{-2}**	3.6×10^{-3}	$<10^{-4}$	**1.6×10^{11}**	1.5×10^{9}	1.3×10^{6}	1.6×10^{11}
	II	**4.4×10^{-2}**	2.4×10^{-3}	$<10^{-4}$	3.8×10^{11}	3.8×10^{11}	**2.1×10^{12}**	2.9×10^{12}
2	I	**2.4×10^{-2}**	1.2×10^{-3}	$<10^{-4}$	**3.7×10^{9}**	1.4×10^{8}	4.1×10^{5}	3.9×10^{9}
	II	**2.3×10^{-2}**	1.1×10^{-3}	$<10^{-4}$	1.8×10^{11}	1.8×10^{11}	**9.1×10^{11}**	9.5×10^{11}
3	I	**7.0×10^{-2}**	3.2×10^{-2}	$<10^{-4}$	**1.2×10^{10}**	**1.4×10^{10}**	4.5×10^{6}	2.6×10^{10}
	II	**6.9×10^{-2}**	3.0×10^{-2}	$<10^{-4}$	1.9×10^{12}	1.9×10^{12}	**8.8×10^{12}**	1.3×10^{13}
4	I	**1.1×10^{-1}**	5.5×10^{-3}	$<10^{-4}$	**7.7×10^{10}**	5.1×10^{9}	9.2×10^{7}	8.2×10^{10}
	II	**1.2×10^{-1}**	5.3×10^{-3}	$<10^{-4}$	2.1×10^{12}	2.1×10^{12}	**1.2×10^{13}**	1.6×10^{13}
5	I	**2.4×10^{-2}**	1.2×10^{-3}	1.2×10^{-4}	**2.6×10^{9}**	1.2×10^{9}	4.5×10^{8}	4.3×10^{9}
	II	**2.3×10^{-2}**	0.9×10^{-3}	1.1×10^{-4}	4.2×10^{11}	4.2×10^{11}	**2.5×10^{12}**	3.3×10^{12}

Figure 6 shows the distribution of resolved shear strains on the first and second most active slip variants. Results are for grains G2, G3, and G4. All variants are related to the slip on prism, except for grain G3 where its second dominant variant belongs to the basal system (see Table 2). Shear strain localization in the form of slip bands is observed in all three grains. These slip bands are shown with a red dotted lines in Figure 6. These bands are parallel to the ones observed in the GND maps in Figure 5a,b as well as those observed in the ARGUS image of Figure 5d.

4.2.2. Stress

The normal stress component along the loading direction (σ_{11}) and the in-plane shear stress component (σ_{12}) of grains G3, G4, and G5 are shown in Figures 7 and 8, respectively. The distribution of σ_{11} is almost identical between Method I and II. After unloading, the normal stress drops significantly. The distributions of σ_{11} from HR-EBSD are also shown in the last row of Figure 7. In an HR-EBSD measurement, a reference point is selected within each grain to measure "relative" stresses and elastic lattice rotations with respect to this point. These reference points are shown with the red dots in Figure 7. Since stresses from HR-EBSD are relative, the calculated stresses at all integration points assigned to a grain are reduced from those calculated at the integration point that coincides with the HR-EBSD reference point. This is to provide a like-to-like comparison between CPFE results and the HR-EBSD measurement.

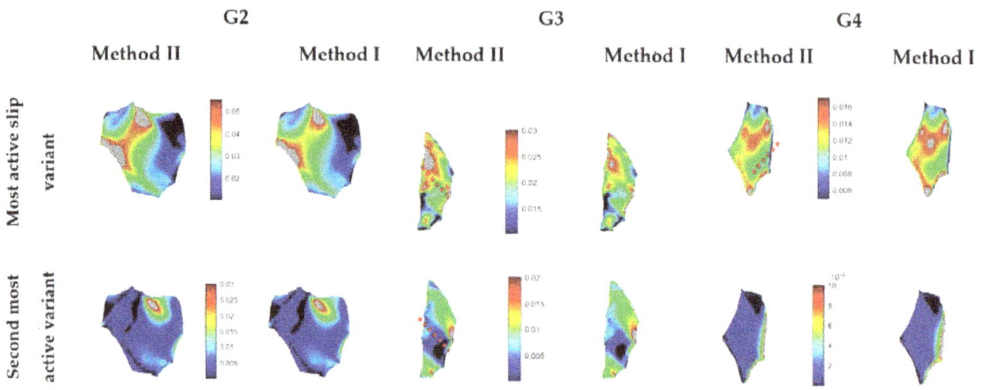

Figure 6. The distribution of resolved shear strain on the most active (**first row**) and the second most active (**second row**) slip systems. Results are for the CPFE model with Method I shown in the right column and Method II in left column. Grain IDs are shown on the top of each grain.

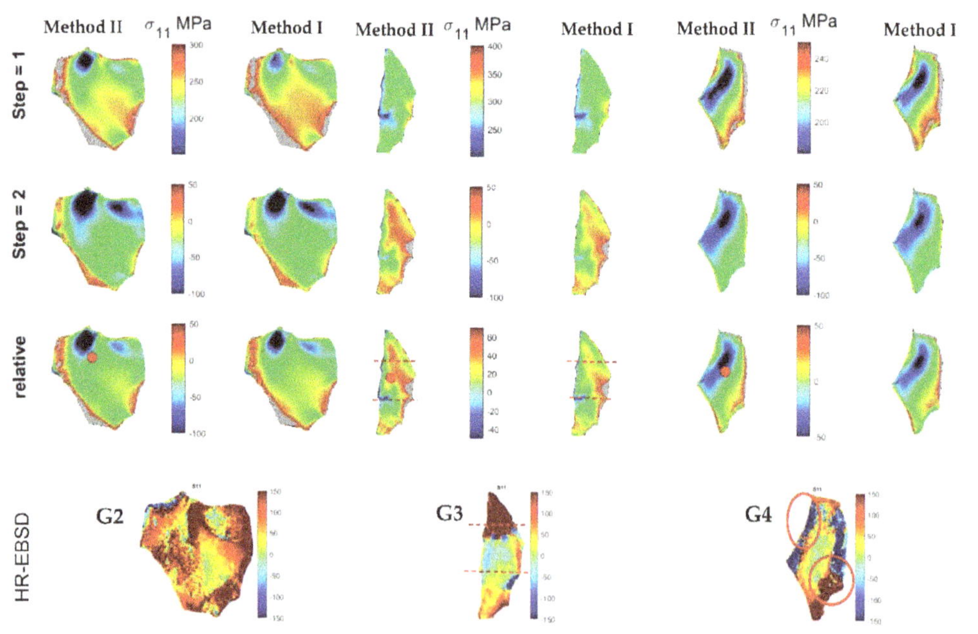

Figure 7. The normal stress component along the loading direction for grain G2, G3, and G4. A comparison between Method I (**right column**) and Method II (**left column**) of the CPFE model and HR-EBSD (**last row**).

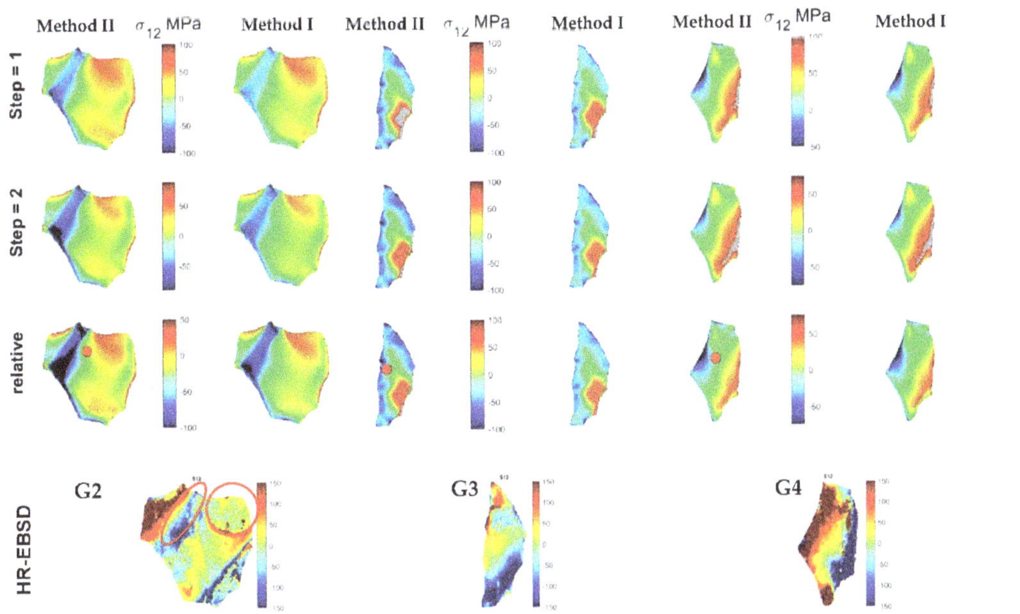

Figure 8. The in-plane shear stress component of grains G2, G3, and G4. A comparison between Method I (**right column**) and Method II (**left column**) of the CPFE model and HR-EBSD (**last row**).

For grain G2 of Figure 7, a tensile stress field is predicted in the lower half of the grain where the stress magnitude increases toward the grain boundary. This is in agreement with HR-EBSD measurement. In addition, two compressive stress fields are observed in the upper half of the grain. In comparison to the measurement, the size of these two fields is overestimated in CPFE results, although the trends are the same. Grain G3 can be divided into three regions, i.e., top, bottom, and middle regions. Calculated stress σ_{11} is tensile in the top and bottom regions, while it is compressive in the middle region. This is in agreement with the HR-EBSD measurement. For grain G4, the left side of the grain has a compressive stress field, and the bottom side of the grain has a tensile stress field in both the model and experiment.

In Figure 8, it is shown that the distribution of shear stress σ_{12} is nearly identical between Method I and II. For grain G2, a negative shear stress band is observed in the CPFE results on the top-left side of the grain. This band is also observed in the HR-EBSD measurement. In addition, a positive σ_{12} region, at the top-right side of the grain, is calculated in both CPFE models, which is also observed in the HR-EBSD measurement. However, there are regions in grains G3 and G4 where the calculated shear stress σ_{12} does not match with those from the HR-EBSD measurement. The magnitude and distribution of calculated normal (Figure 7) and shear stresses (Figure 8) are quite similar between the Method I and II.

4.2.3. Lattice Rotations

The in-plane elastic rotation component of grains G2, G3, and G4 are shown in Figure 9 and the results of Method I and II are compared with those from HR-EBSD. Since measurements were done at the surface, only the in-plane component of the lattice rotation (Ω^{e}_{12}) is shown. The relative values with respect to the reference point are presented in the last row of the simulation results. No difference is observed between the two methods in all the three grains after the loading and unloading steps. This means that although different GND densities are calculated in these two methods, the parameters are calibrated well, such that the resulting deformation tensors are almost identical from the two methods. The relative values with respect to the reference point of the EBSD follow the same trend as the HR-EBSD measurement. For example, both CPFE and HR-EBSD results show positive elastic rotation Ω_{12} in the left half of grain G2. In addition, in both the model and experiment, the top side of grain G3 has negative elastic rotation Ω_{12}. Furthermore, the CPFE results show negative Ω_{12} on the top-right side and positive Ω_{12} on the lower left side of grain G4, which agrees with the HR-EBSD measurement.

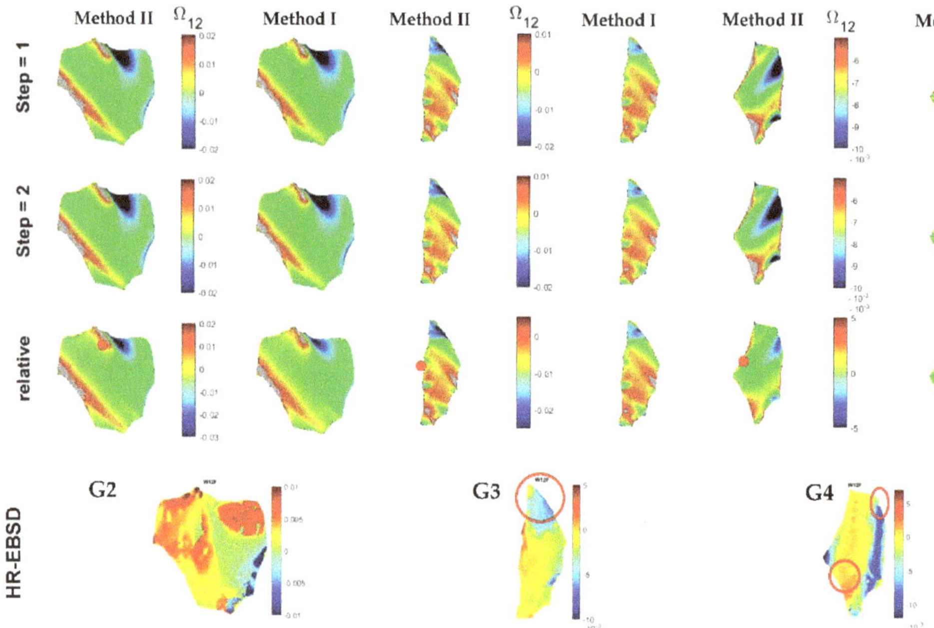

Figure 9. The elastic rotation component Ω_{12} of grains G2, G3, and G4. A comparison between Method II (**left column**) and Method 1 (**right column**) of the CPFE model and HR-EBSD (**last row**).

5. Discussion

A lower-order strain-gradient crystal plasticity finite element model was developed in which two different methods were used to describe the evolution of GND densities. In Method I, a definite value was calculated for the magnitude of GNDs on each slip system, whereas in Method II, a minimization approach was used to solve for an under-determined system of equations. The performance of CPFE models was examined both for single crystal models and for polycrystalline models. In addition, the as-measured microstructure of a deformed α-zirconium sample was imported to the FE solver to compare the calculated grain-scale stresses, elastic lattice rotations, and GNDs to those measured using HR-EBSD.

Results shown in all previous sections were for single or polycrystalline models with an average grain size of more than 20 μm. The results of the strain-gradient CPFE model, however, are length-scale-dependent and are highly sensitive to the size of the studied grains [47]. Such a size dependency originates from the accumulated GND density, which is a function of plastic strain gradients. Since the magnitudes of GND densities from Method I and II are very different, it is important to check the effects of grain size on the results obtained. To do so, the single crystal model used in Section 3.1 is also used here, but the model is scaled so that different grain sizes can be studied. The model is scaled to ten different sizes from 0.125 μm to 64 μm and the calculated stresses, GND, and SSD densities from Method I and II are compared. The misorientation between the c-axis of the grain and the loading direction is set at 22.5°, 45°, and 67.5°. The macroscopic strain applied to the polycrystal is 5%. The Hall-Petch effects are switched off to only study the geometrical effects on the material hardening.

Figure 10a–c show that the average stress for the single crystal with a misorientation angle of 22.5°, 45°, 67.5°, respectively. It is shown that the stress magnitude is highly affected by the model size when Method II is used, particularly when the grain size is smaller than 10 µm. Although Method I also shows some dependency on the grain size, the magnitude of the stress is only significantly affected by the grain size for the grains smaller than 1 µm. This is the reason why the stresses obtained from the two methods for the polycrystal, presented in Section 4.2.2 with an average grain size of 50 µm, are not that different. Figure 10d–f show that the average GND and SSD density for the single crystal with a misorientation angle of 22.5°, 45°, 67.5°, respectively. It is shown that the calculated GND density has a linear relation with the grain size in the log-log scale. Interestingly, the slopes of these two lines are the same for the two methods used. In contrast to GNDs, SSDs are almost unaffected by the grain size. Therefore, the hardening of bigger grains is SSD driven, while that of smaller grains is GND driven. This means that SSDs are as important in determining stress development. Hence, when the formulation provided here is used to study fracture of polycrystals at grain scale [31,32], it is very important to fully characterize the contribution of both GNDs and SSDs to material plastic deformation.

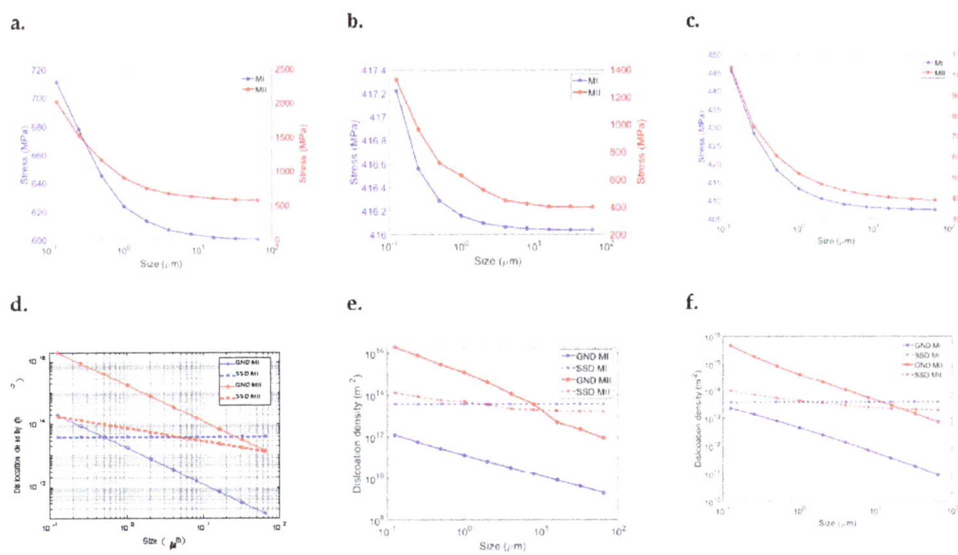

Figure 10. A comparison between the results of the CPFE model, using Method I and II, for the single crystal model in Figure 2a at ten different grain sizes. Stress along the loading direction for the misorientation angles of (**a**) 22.5°, (**b**) 45°, and (**c**) 67.5°. GND and SSD densities for the misorientation angles of (**d**) 22.5°, (**e**) 45°, and (**f**) 67.5°. The values on the x-axis represent the size of the single crystal. Hall-Petch effects are switched off.

A lower-order strain-gradient crystal plasticity model is used here, where the material hardening law is modified by including the strain gradient terms. One of the disadvantages of this approach is the formation of unrealistic, localized deformation fields as described in [16]. Further, when lower-order formulation is used, CPFE results are mesh-size dependent; on the other hand, when higher-order formulation is used, CPFE results are almost mesh independent [61].

The simulation results obtained from CPFE models showed several discrepancies with respect to those from HR-EBSD. The observed discrepancies could be due to ignoring the sub-surface grain structures in the CPFE results. As shown in Figure 2, a columnar grain structure was assumed in the CPFE model by extruding the measured EBSD map along the z-axis. This is because the EBSD measurement provides the information about the sub-surface grains. Although ignoring the sub-surface grain structures may affect the stress values at the free surface in the CPFE results [62–64], Zhang et al. showed that the calculated trends in the GND densities and strains are not much affected by the sub-surface microstructure [48].

Table 3 summarizes the results of the comparisons made between Method I and II. The effectiveness of the two methods can be assessed using two different perspectives: implementation and performance. From the implementation point of view, Method I is more straightforward and can be used for developing higher-order non-local CPFE models. This method is also computationally efficient since no optimization is required to determine the density of GNDs. In terms of performance, the use of Method II leads to closer numbers to those measured with HR-EBSD for GND density but results in uniform distribution of GNDs on all slip variants. Finally, when larger grain sizes are used, typically more than 10 μm, there is no difference between the two methods.

Table 3. A comparison between Method I and II for determining GND densities.

Implementation	Lower-order non-local CPFE	Both methods can be implemented, but the implementation of Method I is more straightforward than Method II
	Higher-order non-local CPFE	Only Method I
Performance	Total GND density	Results from Method II are closer to those measured with HR-EBSD.
	GND density on each slip system	Method I: the calculated values are proportional to the cumulative slip for each slip variant Method II: almost equal for all slip systems

6. Conclusions

A user material (UMAT) subroutine is updated to include the strain-gradient effects. Two different methods are used for determining GND densities. The performance of the two methods for simulating the formation of slip bands in a deformed α-zirconium specimen is investigated. Numerical results are compared with those measured with HR-EBSD. It is shown that:

1. The GND maps calculated from the strain-gradient CPFE models using both methods show formation of localized GND lines within the grains of the polycrystal. These GND lines are parallel to the slip bands observed in the deformed specimen using electron microscopy. The slip bands can also be seen in the CPFE-calculated shear strain maps.
2. The use of the minimization-based approach for the determination of GND densities (Method II) leads to a uniform distribution of GNDs on all slip systems, whereas when the direct approach (Method I) is used, the magnitude of the calculated GNDs on each slip system is proportional to the plastic shear strain accumulated on the same slip system.
3. Although the magnitudes of GND densities are different from the two methods, the trends observed for the calculated grain-scale stresses and lattice rotations are the same. This is because for the studied microstructure, where the average grain size is 50 μm, the calculated total dislocation density, GND plus SSD, from the two methods is almost the same.
4. When a smaller grain size is used, the calculated average stresses from the two methods are different. The critical grain size below which the geometrical effects become significant is higher in Method II, compared to Method I.
5. The dislocation-based hardening law used here is SSD driven for larger grains and GND driven for smaller grains, and accurate implementation of both mechanisms is important when different grain sizes exist in the microstructure.

The formation of the slip bands can also be simulated using higher-order strain-gradient crystal plasticity models. The incorporation of higher order terms to the CPFE model is in the scope of our future studies.

Author Contributions: Conceptualization, O.S. and H.A.; methodology, O.S. and H.A; formal analysis and investigation, O.S and H.A.; writing—original draft preparation, O.S.; writing—review and editing, H.A.; supervision, H.A.; funding acquisition, H.A. All authors have read and agreed to the published version of the manuscript.

Funding: This work was funded by a Discovery Grant (#RGPIN/04969-2017) from the Natural Sciences and Engineering Research Council of Canada (NSERC).

Data Availability Statement: Not applicable.

Conflicts of Interest: The authors declare no conflict of interest.

References

1. Rovinelli, A.; Proudhon, H.; Lebensohn, R.A.; Sangid, M.D. Assessing the reliability of fast Fourier transform-based crystal plasticity simulations of a polycrystalline material near a crack tip. *Int. J. Solids Struct.* **2020**, *184*, 153–166. [CrossRef]
2. Petkov, M.P.; Hu, J.; Tarleton, E.; Cocks, A.C.F. Comparison of self-consistent and crystal plasticity FE approaches for modelling the high-temperature deformation of 316H austenitic stainless steel. *Int. J. Solids Struct.* **2019**, *171*, 54–80. [CrossRef]
3. Ma, R.; Truster, T.J.; Puplampu, S.B.; Penumadu, D. Investigating mechanical degradation due to fire exposure of aluminum alloy 5083 using crystal plasticity finite element method. *Int. J. Solids Struct.* **2018**, *134*, 151–160. [CrossRef]
4. Lu, X.; Zhang, X.; Shi, M.; Roters, F.; Kang, G. Dislocation mechanism based size-dependent crystal plasticity modeling and simulation of gradient nano-grained copper. *Int. J. Plast.* **2019**, *113*, 52–73. [CrossRef]
5. Han, F.; Roters, F.; Raabe, D. Microstructure-based multiscale modeling of large strain plastic deformation by coupling a full-field crystal plasticity-spectral solver with an implicit finite element solver. *Int. J. Plast.* **2020**, *125*, 97–117. [CrossRef]
6. Kasemer, M.; Dawson, P. A finite element methodology to incorporate kinematic activation of discrete deformation twins in a crystal plasticity framework. *Comput. Methods Appl. Mech. Eng.* **2020**, *358*, 1–29. [CrossRef]
7. Wijnen, J.; Peerlings, R.H.J.; Hoefnagels, J.P.M.; Geers, M.G.D. A discrete slip plane model for simulating heterogeneous plastic deformation in single crystals. *Int. J. Solids Struct.* **2021**, *228*, 111094. [CrossRef]
8. Nemat-nasser, S.; Ni, L.; Okinaka, T. A constitutive model for fcc crystals with application to polycrystalline OFHC copper. *Mech. Mater.* **1998**, *30*, 325–341. [CrossRef]
9. Kocks, U.F.; Argon, A.S.; Ashby, M.F. *Thermodynamics and Kinetics of Slip*; Pergamon Press: Cambridge, UK, 1975.
10. Evers, L.P.; Brekelmans, W.A.M.; Geers, M.G.D. Non-local crystal plasticity model with intrinsic SSD and GND effects. *J. Mech. Phys. Solids* **2004**, *52*, 2379–2401. [CrossRef]
11. Ashby, M.F. The deformation of plastically non-homogeneous materials. *Philos. Mag.* **1970**, *21*, 399–424. [CrossRef]
12. Taylor, G.I. The Mechanism of Plastic Deformation of Crystals. Part 1. Theoretical. *Proc. R. Soc.* **1934**, *538*, 362–387.
13. Arsenlis, A.; Parks, D.M. Modeling the evolution of crystallographic dislocation density in crystal plasticity. *J. Mech. Phys. Solids* **2002**, *50*, 1979–2009. [CrossRef]
14. Arsenlis, A.; Parks, D.M.; Becker, R.; Bulatov, V.V. On the evolution of crystallographic dislocation density in non-homogeneously deforming crystals. *J. Mech. Phys. Solids* **2004**, *52*, 1213–1246. [CrossRef]
15. Gurtin, M.E. A gradient theory of single-crystal viscoplasticity that accounts for geometrically necessary dislocations. *J. Mech. Phys. Solids* **2002**, *50*, 5–32. [CrossRef]
16. Niordson, C.F.; Hutchinson, J.W. On lower order strain gradient plasticity theories. *Eur. J. Mech. A/Solids* **2003**, *22*, 771–778. [CrossRef]
17. Bassani, J.L. Incompatibility and a simple gradient theory of plasticity. *J. Mech. Phys. Solids* **2001**, *49*, 1983–1996. [CrossRef]
18. Busso, E.P.; Meissonnier, F.T.; Dowd, N.P.O. Gradient-dependent deformation of two-phase single crystals. *J. Mech. Phys. Solids* **2000**, *48*, 2333–2361. [CrossRef]
19. Arsenlis, A.; Parks, D.M. Crystallographic aspects of geometrically nesesary and statistically stored dislocation density. *Acta Mater.* **1999**, *47*, 1597–1611. [CrossRef]
20. De Borst, R.; Mühlhaus, H.-B. Gradient-dependent plasticity: Formulation and algorithmic aspects. *Int. J. Numer. Methods Eng.* **1992**, *35*, 521–539. [CrossRef]
21. Gao, H.; Huang, Y.; Nix, W.D.; Hutchinson, J.W. Mechanism-based strain gradient plasticity—I. Theory. *J. Mech. Phys. Solids* **1999**, *47*, 1239–1263. [CrossRef]
22. Yun, G.; Hwang, K.C.; Huang, Y.; Wu, P.D. A reformulation of mechanism-based strain gradient plasticity. *Philos. Mag.* **2005**, *85*, 4011–4029. [CrossRef]
23. Gurtin, M.E. On the plasticity of single crystals: Free energy, microforces, plastic-strain gradients. *J. Mech. Phys. Solids* **2000**, *48*, 989–1036. [CrossRef]
24. Nye, J. Some geometrical relations in dislocated crystals. *Acta Metall.* **1953**, *1*, 153–162. [CrossRef]

25. Das, S.; Hofmann, F.; Tarleton, E. Consistent determination of geometrically necessary dislocation density from simulations and experiments. *Int. J. Plast.* **2018**, *109*, 18–42. [CrossRef]
26. Dai, H. Geometrically-Necessary Dislocation Density in Continuum Plasticity Theory. Ph.D. Thesis, Massachusetts Institute of Technology, Cambridge, MA, USA, 1997.
27. Ma, A.; Roters, F.; Raabe, D. A dislocation density based constitutive model for crystal plasticity FEM including geometrically necessary dislocations. *Acta Mater.* **2006**, *54*, 2169–2179. [CrossRef]
28. Dunne, F.P.E.; Rugg, D.; Walker, A. Lengthscale-dependent, elastically anisotropic, physically-based hcp crystal plasticity: Application to cold-dwell fatigue in Ti alloys. *Int. J. Plast.* **2007**, *23*, 1061–1083. [CrossRef]
29. Waheed, S.; Zheng, Z.; Balint, D.S.; Dunne, F.P.E. Microstructural effects on strain rate and dwell sensitivity in dual-phase titanium alloys. *Acta Mater.* **2019**, *162*, 136–148. [CrossRef]
30. Arsenlis, A. Modeling Dislocation Density Evolution in Continuum Crystal Plasticity. Available online: http://dspace.mit.edu/handle/1721.1/36679%5Cnhttps://dspace.mit.edu/handle/1721.1/36679 (accessed on 16 July 2001).
31. Chen, B.; Jansenss, K.; Dunne, F.P.E. Role of geometrically necessary dislocation density in multiaxial and non-proportional fatigue crack nucleation. *Int. J. Fatigue* **2020**, *135*, 105517. [CrossRef]
32. Wilson, D.; Dunne, F.P.E. A mechanistic modelling methodology for microstructure-sensitive fatigue crack growth. *J. Mech. Phys. Solids* **2019**, *124*, 827–848. [CrossRef]
33. Dunne, F.P.E.; Kiwanuka, R.; Wilkinson, A.J. Crystal plasticity analysis of micro-deformation, lattice rotation and geometrically necessary dislocation density. *Proc. R. Soc. A Math. Phys. Eng. Sci.* **2012**, *468*, 2509–2531. [CrossRef]
34. Cheng, J.; Ghosh, S. A crystal plasticity FE model for deformation with twin nucleation in magnesium alloys. *Int. J. Plast.* **2015**, *67*, 148–170. [CrossRef]
35. Wu, W.; An, K.; Huang, L.; Lee, S.Y.; Liaw, P.K. Deformation dynamics study of a wrought magnesium alloy by real-time in situ neutron diffraction. *Scr. Mater.* **2013**, *69*, 358–361. [CrossRef]
36. Xu, F.; Holt, R.A.; Daymond, M.R.; Rogge, R.B.; Oliver, E.C. Development of internal strains in textured Zircaloy-2 during uni-axial deformation. *Mater. Sci. Eng. A* **2008**, *488*, 172–185. [CrossRef]
37. Louca, K.; Abdolvand, H. Accurate determination of grain properties using three-dimensional synchrotron X-ray diffraction: A comparison with EBSD. *Mater. Charact.* **2021**, *171*, 110753. [CrossRef]
38. Abdolvand, H.; Louca, K.; Mareau, C.; Majkut, M.; Wright, J. On the nucleation of deformation twins at the early stages of plasticity. *Acta Mater.* **2020**, *196*, 733–746. [CrossRef]
39. Abdolvand, H.; Wright, J.P.; Wilkinson, A.J. On the state of deformation in a polycrystalline material in three-dimension: Elastic strains, lattice rotations, and deformation mechanisms. *Int. J. Plast.* **2018**, *106*, 145–163. [CrossRef]
40. Wallis, D.; Parsons, A.J.; Hansen, L.N. Quantifying geometrically necessary dislocations in quartz using HR-EBSD: Application to chessboard subgrain boundaries. *J. Struct. Geol.* **2019**, *125*, 235–247. [CrossRef]
41. Ungar, T.; Drakopoulos, M.; Be, J.L.; Chauveau, T.; Castelnau, O.; Riba, G.; Snigirev, A.; Snigireva, I.; Schroer, C.; Bacroix, B. Grain to grain slip activity in plastically deformed Zr determined by X-ray micro-diffraction line profile analysis. *Acta Mater.* **2007**, *55*, 1117–1127. [CrossRef]
42. Wilkinson, A.J.; Meaden, G.; Dingley, D.J. High-resolution elastic strain measurement from electron backscatter diffraction patterns: New levels of sensitivity. *Ultramicroscopy* **2006**, *106*, 307–313. [CrossRef] [PubMed]
43. Wilkinson, A.J. Measurement of elastic strains and small lattice rotations using electron back scatter diffraction. *Ultramicroscopy* **1996**, *62*, 237–247. [CrossRef]
44. Troost, K.Z.; Van Der Sluis, P.; Gravesteijn, D.J. Microscale elastic-strain determination by backscatter Kikuchi diffraction in the scanning electron microscope. *Appl. Phys. Lett.* **1993**, *62*, 1110–1112. [CrossRef]
45. Guo, Y.; Abdolvand, H.; Britton, T.B.; Wilkinson, A.J. Growth of {112¯2} twins in titanium: A combined experimental and modelling investigation of the local state of deformation. *Acta Mater.* **2017**, *126*, 221–235. [CrossRef]
46. Britton, T.B.; Wilkinson, A.J. High resolution electron backscatter diffraction measurements of elastic strain variations in the presence of larger lattice rotations. *Ultramicroscopy* **2012**, *114*, 82–95. [CrossRef] [PubMed]
47. Sedaghat, O.; Abdolvand, H. A non-local crystal plasticity constitutive model for hexagonal close-packed polycrystals. *Int. J. Plast.* **2021**, *136*, 102883. [CrossRef]
48. Zhang, Z.; Lunt, D.; Abdolvand, H.; Wilkinson, A.J.; Preuss, M.; Dunne, F.P.E. Quantitative investigation of micro slip and localization in polycrystalline materials under uniaxial tension. *Int. J. Plast.* **2018**, *108*, 88–106. [CrossRef]
49. Abdolvand, H.; Sedaghat, O.; Guo, Y. Nucleation and growth of {112¯2} twins in titanium: Elastic energy and stress fields at the vicinity of twins. *Materialia* **2018**, *2*, 58–62. [CrossRef]
50. Abdolvand, H.; Wright, J.; Wilkinson, A.J. Strong grain neighbour effects in polycrystals. *Nat. Commun.* **2018**, *9*, 171. [CrossRef]
51. Abdolvand, H.; Wilkinson, A.J. On the effects of reorientation and shear transfer during twin formation: Comparison between high resolution electron backscatter diffraction experiments and a crystal plasticity finite element model. *Int. J. Plast.* **2016**, *84*, 160–182. [CrossRef]
52. Asaro, R.J.; Needleman, A. Overview No. 42 Texture development and strain hardening in rate dependent polycrystals. *Acta Metall.* **1985**, *33*, 923–953. [CrossRef]
53. Asaro, R.J. Crystal Plasticity. *J. Appl. Mech.* **1983**, *50*, 921–934. [CrossRef]

54. Fisher, E.S.; Renken, C.J. Single crystal elastic moduli and the hcp → bcc transformation in Ti, Zr, and Hf. *Phys. Rev.* **1964**, *135*, A482–A494. [CrossRef]
55. Beyerlein, I.J.; Tomé, C.N. A dislocation-based constitutive law for pure Zr including temperature effects. *Int. J. Plast.* **2008**, *24*, 867–895. [CrossRef]
56. Cheong, K.S.; Busso, E.P.; Arsenlis, A. A study of microstructural length scale effects on the behaviour of FCC polycrystals using strain gradient concepts. *Int. J. Plast.* **2005**, *21*, 1797–1814. [CrossRef]
57. Abdolvand, H. Progressive modelling and experimentation of hydrogen diffusion and precipitation in anisotropic polycrystals. *Int. J. Plast.* **2019**, *116*, 39–61. [CrossRef]
58. Abdolvand, H.; Daymond, M.R.; Mareau, C. Incorporation of twinning into a crystal plasticity finite element model: Evolution of lattice strains and texture in Zircaloy-2. *Int. J. Plast.* **2011**, *27*, 1721–1738. [CrossRef]
59. Wallis, D.; Hansen, L.N.; Britton, T.B.; Wilkinson, A.J. High-Angular Resolution Electron Backscatter Diffraction as a New Tool for Mapping Lattice Distortion in Geological Minerals. *J. Geophys. Res. Solid Earth* **2019**, *124*, 6337–6358. [CrossRef]
60. Wallis, D.; Hansen, L.N.; Ben Britton, T.; Wilkinson, A.J. Geometrically necessary dislocation densities in olivine obtained using high-angular resolution electron backscatter diffraction. *Ultramicroscopy* **2016**, *168*, 34–45. [CrossRef]
61. Kuroda, M.; Tvergaard, V. Studies of scale dependent crystal viscoplasticity models. *J. Mech. Phys. Solids* **2006**, *54*, 1789–1810. [CrossRef]
62. Baudoin, P.; Hama, T.; Takuda, H. Influence of critical resolved shear stress ratios on the response of a commercially pure titanium oligocrystal: Crystal plasticity simulations and experiment. *Int. J. Plast.* **2019**, *115*, 111–131. [CrossRef]
63. Zeghadi, A.; Forest, S.; Gourgues, A.F.; Bouaziz, O. Ensemble averaging stress-strain fields in polycrystalline aggregates with a constrained surface microstructure-part 2: Crystal plasticity. *Philos. Mag.* **2007**, *87*, 1425–1446. [CrossRef]
64. St-Pierre, L.; Héripré, E.; Dexet, M.; Crépin, J.; Bertolino, G.; Bilger, N. 3D simulations of microstructure and comparison with experimental microstructure coming from O.I.M analysis. *Int. J. Plast.* **2008**, *24*, 1516–1532. [CrossRef]

crystals

MDPI

Article

Microstructural Evolution and Tensile Testing of a Bi–Sn (57/43) Alloy Processed by Tube High-Pressure Shearing

Chuan-Ting Wang [1], Zheng Li [2], Yong He [1,*], Jing-Tao Wang [2] and Terence G. Langdon [3,4,*]

1 School of Mechanical Engineering, Nanjing University of Science & Technology, Nanjing 210094, China; ctwang@njust.edu.cn
2 School of Materials Science and Engineering, Nanjing University of Science & Technology, Nanjing 210094, China; lizheng@njust.edu.cn (Z.L.); jtwang@njust.edu.cn (J.-T.W.)
3 Departments of Aerospace & Mechanical Engineering and Materials Science, University of Southern California, Los Angeles, CA 90089-1453, USA
4 Materials Research Group, Department of Mechanical Engineering, University of Southampton, Southampton SO17 1BJ, UK
* Correspondence: yonghe1964@163.com (Y.H.); langdon@usc.edu (T.G.L.)

Abstract: Tube high-pressure shearing (t-HPS) processing was performed on a eutectic Bi–Sn (57/43) alloy for 0.25, 1, 5 and 20 turns. The selected samples were stored at room temperature for up to 56 days to examine the strain weakening and self-annealing behavior of the alloy. The results showed that t-HPS processing gradually refined the microstructure and led to decreasing of microhardness, but microhardness increased slowly during the subsequent storage at room temperature. Shear localization of the eutectic structure during t-HPS processing was observed as large amounts of narrow dense lamellar zones were visible in the deformed microstructures. The Bi–Sn (57/43) alloy processed by t-HPS exhibited significantly enhanced superplastic properties with elongations up to >1800% in a sample after t-HPS processing for 20 turns. This high elongation is attributed to the breaking of the lamellar structure and the very small grain size.

Keywords: Bi–Sn alloy; microhardness; self-annealing; superplasticity; tube high-pressure shearing

Citation: Wang, C.-T.; Li, Z.; He, Y.; Wang, J.-T.; Langdon, T.G. Microstructural Evolution and Tensile Testing of a Bi–Sn (57/43) Alloy Processed by Tube High-Pressure Shearing. *Crystals* **2021**, *11*, 1229. https://doi.org/10.3390/cryst11101229

Academic Editor: Wojciech Polkowski

Received: 30 September 2021
Accepted: 9 October 2021
Published: 12 October 2021

1. Introduction

Considerable interest has developed over the last two decades to producing and measuring the mechanical properties of materials with exceptionally small grain sizes. Ultrafine-grained (UFG) metals are defined specifically as metals where the average grain size is less than ~1 μm [1] and these small grains divide into the two categories of submicrometer grains with sizes of 100–1000 nm and nanometer grains where the average size is <100 nm. An advantage in producing these materials is that they exhibit high strength and there is a potential for achieving superplastic elongations at exceptionally rapid strain rates which would be advantageous for manufacturing parts in the superplastic forming industry [2]. Materials with UFG structures are generally produced using procedures based on the application of severe plastic deformation (SPD) where the material is subjected to a very high strain but without any significant changes in the overall dimensions of the workpiece. As discussed in a comprehensive review [3], the two main techniques of SPD processing are equal channel angular pressing (ECAP), where a rod or a bar is pressed through a die constrained within a channel bent through a sharp angle [4], and high-pressure torsion (HPT), where the sample is generally in the form of a thin disk subjected to high pressure and concurrent torsional straining [5].

Both ECAP and HPT are effective for processing to produce metals with very small grains, but the procedures are different because ECAP is a discontinuous and labor-intensive process in which the sample is removed and then reinserted into the die between each pass whereas HPT is a continuous process where the sample is torsionally strained up to the required maximum numbers of rotations. In practice, the experiments show that processing by HPT has two advantages over ECAP because it produces smaller grain sizes [6,7] and a larger fraction of grain boundaries with large angles of misorientation [8–12]. Nevertheless, HPT processing has a significant disadvantage because the processed samples are generally very small and therefore they can be used in small-scale applications such as electronic devices but cannot be utilized in large-scale industrial applications.

To overcome this deficiency while at the same time using a processing method based on the application of shearing, an alternative approach was developed which is known as tube high-pressure shearing (t-HPS). This is a new and novel SPD technique that was used effectively to process commercial purity aluminum [13] and also to produce a multilayered structure using pure Pb and pure Sn as prototype materials [14,15].

To date, very few results are available on the processing of materials by t-HPS although the procedure appears promising for fabricating UFG metals. Accordingly, the objective of this study was to examine the microstructural evolution associated with t-HPS, consider the effect of the occurrence of self-annealing after the processing operation and conduct tensile testing to evaluate the potential for achieving good superplastic properties. The experiments were conducted using the Bi–Sn eutectic alloy where this material was chosen for two reasons. First, a very recent study using HPT showed that this alloy exhibited extensive self-annealing but with good superplastic behavior [16,17]. Second, a very early investigation representing the first report of true superplasticity in any metal showed that the Bi–Sn eutectic alloy in an extruded condition exhibited exceptionally high elongations to failure in tensile testing [18].

2. Experimental Material and Procedures

Tubular samples of the Bi–Sn (57/43 wt%) eutectic alloy were cast from a mixture of Bi and Sn beads with purities of 99.99%. The principle of the t-HPS process was introduced earlier [13] and basically the procedure in t-HPS is that a tubular sample with an outer radius of around 47.7 mm and a wall thickness of around 2.7 mm is placed in the gap between the central mandrel and the outer cylinder of the t-HPS facility and high hydrostatic pressure is introduced to the sample by compressing the pressure rings placed at the upper and lower ends of the tubular sample. As the outer cylinder rotates, the large frictional force at the interfaces between the central mandrel/outer cylinder and the sample drags the sample to shear. The rotation speed in these experiments was set to about 0.2 turns per minute. The samples were processed by t-HPS through a total number (n) of 1/4, 1, 5 and 20 turns. Following t-HPS, self-annealing was conducted by storing the samples at room temperature (RT) for various times up to a maximum of 56 days.

The samples were etched using a solution of $FeCl_3$ after polishing. X-ray diffraction (XRD) (Cu Ka radiation) was employed to examine the Bi–Sn samples after processing with the diffractometer set to around 0.015° per step. The microstructures of the selected samples were observed with a scanning electron microscope (SEM) JEOL JSM-7001 operating at a voltage of 15 kV. Microhardness values were recorded using a Vickers hardness tester with a load of 50 gf and dwell times of 10 s. Focused ion beam (FIB) was employed to investigate the nature of the Sn particles within the Bi phase after HPT processing and storage at RT. The ion beam voltage of the FIB was 30 kV and the final milling current was 50 pA. The cross-section of the disk sample was milled so that the Sn particles became visible.

After t-HPS processing, the tube samples were firstly flattened slowly to form a long sample. Then, tensile samples with gauge lengths of 3 mm and widths of 2 mm were machined from the long samples. The thickness of these samples was around 2 mm. Tensile tests were performed on these samples using an Instron testing facility under initial strain rates of 1.0×10^{-2}, 1.0×10^{-3} and 1.0×10^{-4} s^{-1} at room temperature.

3. Experimental Results

Figure 1 shows the microstructures of the Bi–Sn alloy (a) in the as-cast condition and after t-HPS processing for (b) 0.25, (c) 1, (d) 5 and (e) 20 turns, where the SEM observations were conducted after storage at RT for 8 h following t-HPS processing. Inspection showed the as-cast Bi–Sn alloy has a typical eutectic structure where the Bi phase and the Sn phase are complementary and packed together. As the etchant preferentially attacked the Sn phase, large amounts of very small Bi precipitate particles became visible within the Sn-rich areas after etching. After t-HPS processing for 0.25 and 1 turn, no significant microstructural change was observed despite some slight rotation of the lamellar structure, and the overall eutectic structure was not destroyed. As the number of turns increased to 5 and 20, it was observed that each phase started to fracture. The edges of both phases became sharper and grains with equiaxed shapes became visible.

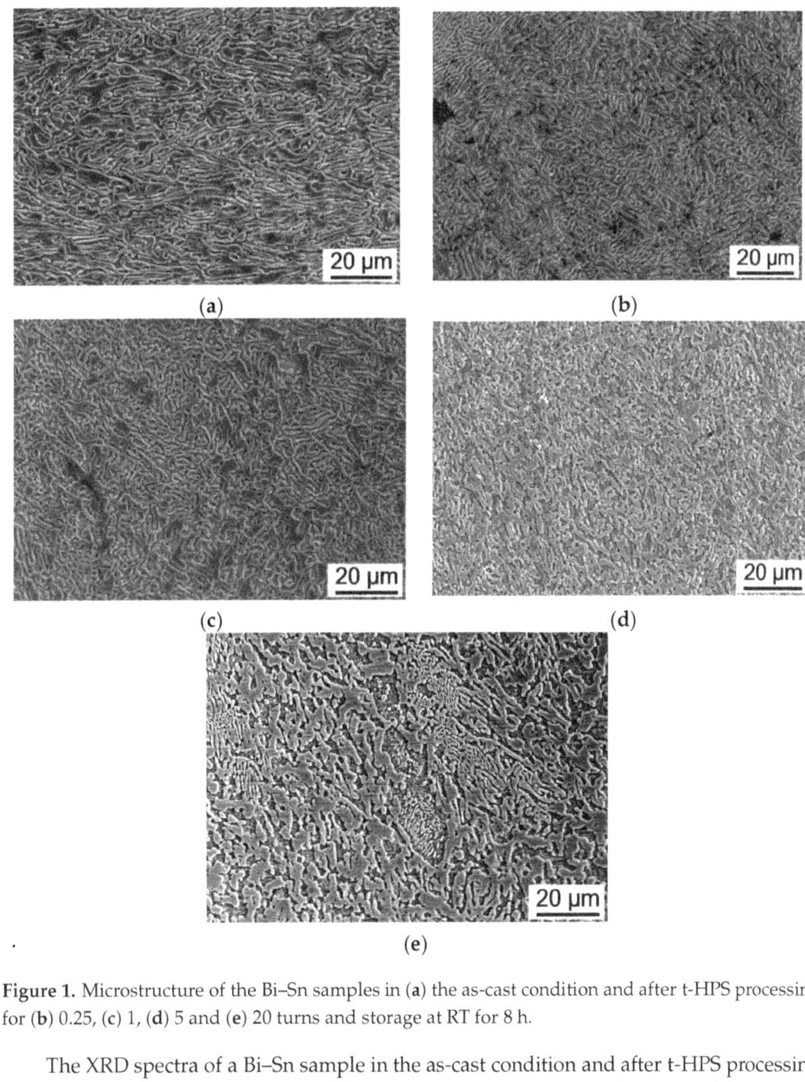

Figure 1. Microstructure of the Bi–Sn samples in (**a**) the as-cast condition and after t-HPS processing for (**b**) 0.25, (**c**) 1, (**d**) 5 and (**e**) 20 turns and storage at RT for 8 h.

The XRD spectra of a Bi–Sn sample in the as-cast condition and after t-HPS processing for five turns are shown in Figure 2 where the peaks of pure Bi and Sn are indicated. The XRD pattern reveals that the Bi–Sn (57/43) alloy is composed of pure Bi and pure Sn elements only and no binary compound is detected.

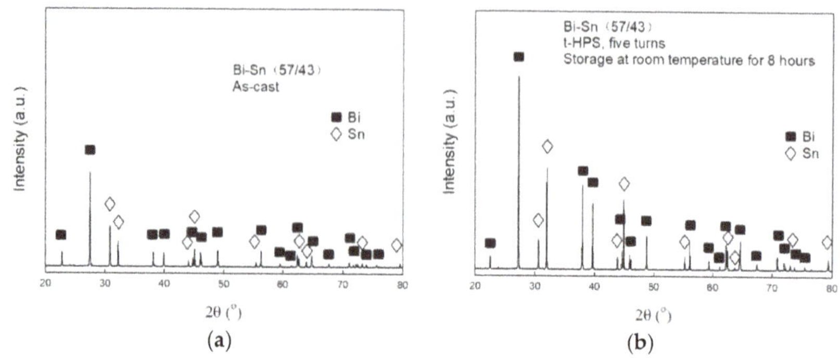

Figure 2. XRD spectrum of the Bi–Sn samples in (**a**) the as-cast condition and (**b**) after t-HPS processing for five turns and storage at RT for 8 h.

When the microstructure was observed under a lower magnification, it was noticed that the shear deformation was not homogeneous and there were some narrow dense lamellar bands visible in the matrix. As shown in Figure 3a, the width of these bands was around 10–20 μm after five turns of t-HPS processing. The phase inside such bands preserved the original lamellar structure of the as-cast sample while the phase structure outside these bands consisted of equiaxed grains. After 20 turns of deformation, the length of dense lamellar bands was much shorter.

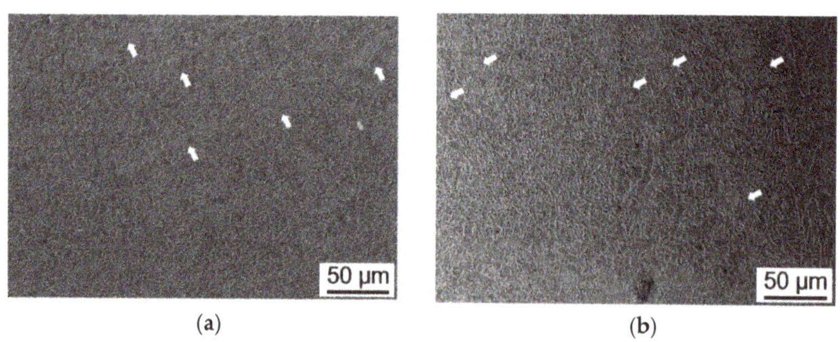

Figure 3. Microstructure of the Bi–Sn samples after t-HPS processing for (**a**) 5 and (**b**) 20 turns and storage at RT for 8 h, with arrows indicating the lamellar bands.

A Bi–Sn sample processed by t-HPS for five turns was used to evaluate the evolution of microhardness during storage at RT. Microhardness indentations were recorded for the sample and these measured values were then plotted against the number of days of storage as shown in Figure 4. The as-cast Bi–Sn (57/43) alloy had microhardness of ~22.5 ± 0.5 Hv. The microhardness values were recorded as ~11.1 ± 0.9 Hv after t-HPS processing and storage for 8 h at RT. Microhardness increased gradually during storage, reaching ~13.0 ± 0.5 Hv after 7 days of storage, and further increased to ~14.4 ± 0.3 Hv and ~14.9 ± 0.7 Hv after 42 days and 56 days of storage, respectively. Similar behavior was widely reported in low melting temperature materials processed by severe plastic deformation, for example, in the Zn–Al, Sn–Pb and Bi–Sn alloys [16,17,19–22]. However, it is worth noting that microhardness of the Bi–Sn (57/43) alloy after t-HPS processing is more stable compared to the condition after processing by HPT. As shown in an earlier study, microhardness of the Bi–Sn alloy dropped from ~25.2 Hv to around 8 Hv after processing by HPT for five turns, and thereafter microhardness increased to around 17 Hv after 7 days of storage at room temperature [16,17].

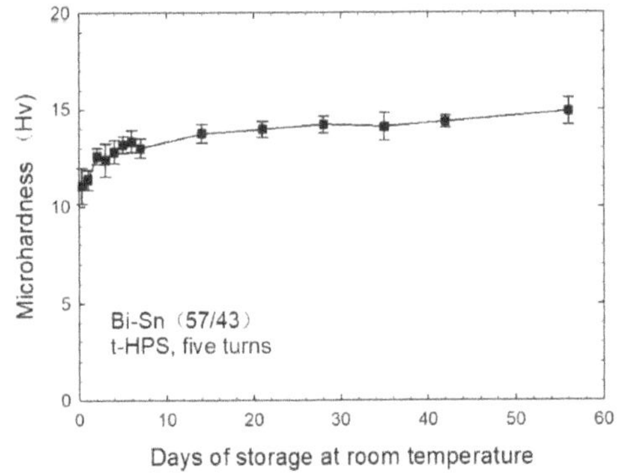

Figure 4. Microhardness of the Bi–Sn samples versus the number of days of storage at RT after t-HPS processing for five turns.

Figure 5 shows the microstructures of the Bi–Sn samples after storage at RT for various periods. The Sn phase was etched away by the etchant so that the Bi phase is clearly visible in the images. As shown in Figure 5a, the phase structure was broken by the shear stain and there were large numbers of Bi phase islands with essentially equiaxed shape. As the number of storage days increased, the Bi phase grew and connected together again so that the total number of small Bi islands decreased. Moreover, it is observed in all of these images that pitting holes existed inside the Bi phase where these pitting holes had sizes of around several hundred nanometers. It is reasonable to expect that these pitting holes were pre-resolved Sn particles within the Bi phase which were etched away by the etchant.

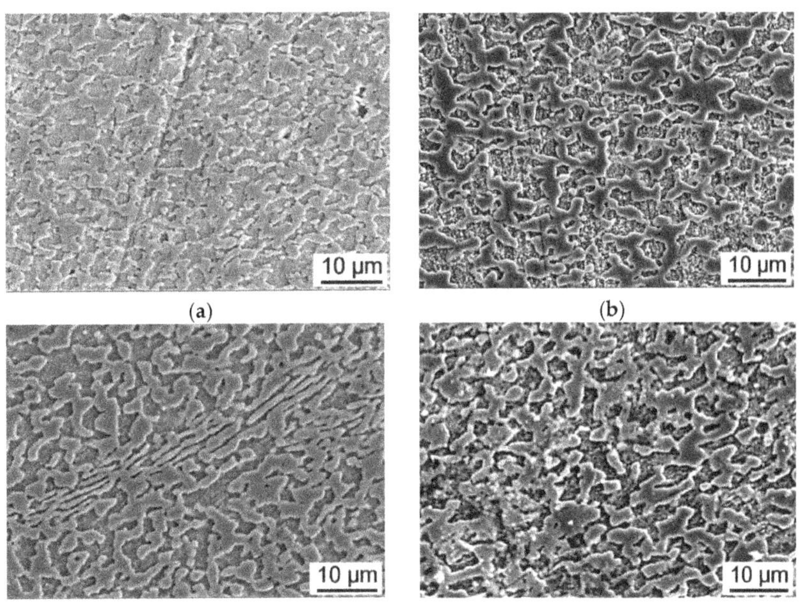

Figure 5. Microstructures of the Bi–Sn samples after t-HPS processing for five turns and storage at RT for (**a**) 8 h, (**b**) 2 days, (**c**) 7 days and (**d**) 21 days.

FIB was employed to mill the samples and observe these Sn particles within the Bi phase. The solubility of the second phase was higher in the melted state compared to the solid state, therefore these Sn second-phase particles may participate in solidification and become trapped within the Bi phase. As the material is subjected to t-HPS processing, tips of the large Sn phase may be broken and mixed into the Bi phase through the flow of the materials under shear deformation. This is shown in Figure 6b where the number of visible Sn particles inside the Bi phase is larger compared to Figure 6a. Unfortunately, it was not feasible to compare the mean size of these particles as the number of visible particles was not sufficient. It is also observed that the grain boundaries inside each phase were visible after fine milling.

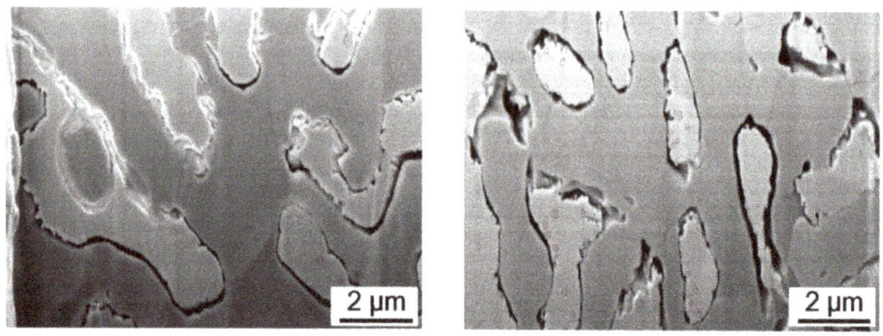

Figure 6. Microstructures (**a**) of the as-cast Bi–Sn sample and (**b**) after t-HPS processing for five turns and storage at RT for 8 h.

Tensile tests were conducted on the Bi–Sn samples and the main results of each sample under various initial strain rates are shown in Table 1. The results demonstrate that the superplasticity of the Bi–Sn alloy was significantly improved by t-HPS processing and elongations to failure of more than 1500% and 1800% were achieved in the sample processed by t-HPS for 5 and 20 turns, respectively. For all the samples, smaller strain rates were favorable for having higher elongations to failure.

Table 1. Tensile properties of the Bi–Sn samples under various initial strain rates at RT.

Strain Rate	**1.0×10^{-2} s^{-1}**		**1.0×10^{-3} s^{-1}**		**1.0×10^{-4} s^{-1}**	
	UTS, MPa	**Elongation**	**UTS, MPa**	**Elongation**	**UTS, MPa**	**Elongation**
As-cast	70	~40%	60	~80%	53	~130%
0.25 turns	69	~80%	48	~280%	33	~430%
1 turn	69	~100%	49	~270%	31	~490%
5 turns	/	/	42	~1170%	23	~1530%
20 turns	/	/	26	~1060%	12	~1820%

In this study, an elongation of >1800% was achieved in the Bi–Sn (57/43) sample after t-HPS processing for 20 turns. In an earlier study, a Bi–Sn (58/42) sample processed by HPT for 10 turns featured a superplastic elongation of around 1220% under a strain rate of 1×10^{-4} s^{-1} at RT. Thus, the tensile samples of this study were generally more superplastic and exhibited larger elongations to failure than those processed by HPT [16]. Nevertheless, it should be noted that an elongation of 1900% was achieved in a very early study using larger rod samples of the same alloy after extrusion [18].

4. Discussion

During t-HPS processing, the imposed shear strains are around 16.5, 66, 330 and 1300 for 0.25, 1, 5 and 20 turns of processing, respectively [13]. The tube sample used in this investigation had a large diameter of around 47 mm and the wall thickness of the tube sample was around 2.7 mm. It is reasonable to expect that during t-HPS processing the tube sample deformed very close to simple shear. By contrast, during HPT processing, the sample is in the form of a thin disk which normally has a diameter of 10 mm and thickness of around 0.8 mm so that the deformation along the radius of the disk is inhomogenously distributed. This geometry and the overall inhomogeneous nature would lead to a significant disorder in the manner of deformation since the material located at a larger radial position on the disk must drag the material which is located closer to the center. Such a combined deformation of shear and rotation during HPT is more efficient in refining the microstructure of materials. Thus, during t-HPS processing of the Bi–Sn alloy, the original lamellar structure is randomly packed together. These lamellae lie perpendicular to the shear direction and tend to fracture first to accumulate the shear strain [23,24]. Moreover, the Bi–Sn alloy exhibits a significant strain-induced softening behavior and this behavior leads to a lower hardness value within the shear-localized zones which may in turn aggravate the shear inhomogeneity. As shown in Figure 3, the preserved lamellar bands are mostly parallel to each other. It is apparent that the Bi–Sn sample processed by t-HPS for 20 turns had not yet reached a saturated state as shown in Figure 3b, although the imposed shear strain was as large as 1300. More strain is therefore needed until all of these preserved lamellar structures disappear.

As shown in the SEM images, lamellar structures of the Bi phase and the Sn phase were mutually contained within each other, as shown in Figure 1a. Such a lamella-dominating structure exists after t-HPS for low numbers of turns as shown in Figure 1b,c. The absolute melting temperature, T_m, of the Bi–Sn eutectic alloy is around 412 K and therefore the ambient room temperature is around 0.7 T_m for the alloy [25–29]. It is widely recognized that the ductility of materials in tensile testing at elevated temperatures is related to the dislocation movement and/or grain boundary sliding behavior [30]. For this alloy, it is rather difficult for dislocations to pass through the interfaces at phase boundaries. Moreover, boundary sliding at the phase boundaries is also very difficult in the as-cast Bi–Sn alloy as the Bi phase and the Sn phase lamellar structures are closely interlocked within each other [31–34]. It is reasonable to expect, therefore, that the ductility of the as-cast Bi–Sn alloy will be poor due to its lamellar structure, and this is confirmed in these experiments where the as-cast alloy had an elongation of only around 130% under a strain rate of 1.0×10^{-4} s^{-1}. Nevertheless, after t-HPS processing for large numbers of turns, the two phases experience diastrophism under the action of a large shear force. This large shear force gradually destroys the lamellar locking between the two phases and large numbers of grain boundaries are formed inside each phase. In practice, these new grain boundaries probably act as favorable sites for grain boundary sliding so that, as a result, the ductility of the Bi–Sn sample becomes significantly improved after t-HPS processing for large numbers of turns. In this study, the recorded elongations of >1000% easily fulfilled the requirement of an elongation of >400% in order to achieve true superplastic flow [35].

Grain growth and phase growth also occur in the Bi–Sn alloy during storage, giving the self-annealing effect. At the same time, the second-phase particles continuously precipitate out from the matrix. These grain growth and phase growth are the dominating factors controlling the increasing hardness during storage at RT.

5. Summary and Conclusions

1. Experiments on a Bi–Sn (57/43) eutectic alloy showed that t-HPS processing gradually refined the microstructure and led to decreasing microhardness.
2. Shear localization of the eutectic structure during t-HPS processing was observed, but some preserved dense lamellar bands were visible even after t-HPS processing for 20 turns.
3. The Bi–Sn (57/43) alloy processed by t-HPS showed significantly enhanced superplasticity, with elongations up to >1000%. This is attributed to the breaking of the lamellar structure and the presence of a refined grain size.

Author Contributions: Methodology, Y.H., J.-T.W. and T.G.L.; validation, C.-T.W. and Z.L.; investigation, C.-T.W.; resources, Y.H., J.-T.W. and T.G.L.; data curation, C.-T.W. and Z.L.; writing—original draft preparation, C.-T.W.; writing—review and editing, Y.H., J.-T.W. and T.G.L.; supervision, T.G.L.; project administration, T.G.L.; funding acquisition, Y.H., J.-T.W. and T.G.L. All authors have read and agreed to the published version of the manuscript.

Funding: This work was supported by the National Science Foundation of the United States (grant No. DMR-1160966), the European Research Council (ERC Grant Agreement No. 267464-SPDMETALS) and the Natural Science Foundation of China (grant No. 52074160).

Data Availability Statement: The raw and processed data generated during this study will be made available upon reasonable request.

Acknowledgments: The authors thank L.Y. Li and Y. Yang of NJUST for assistance with t-HPS processing.

Conflicts of Interest: The authors declare no conflict of interest.

References

1. Valiev, R.Z.; Estrin, Y.; Horita, Z.; Langdon, T.G.; Zehetbauer, M.J.; Zhu, Y.T. Producing bulk ultrafine-grained materials by severe plastic deformation. *JOM* **2006**, *58*, 33–39. [CrossRef]
2. Barnes, A.J. Superplastic forming 40 years and still growing. *J. Mater. Eng. Perform.* **2007**, *16*, 440–454. [CrossRef]
3. Langdon, T.G. Twenty-five years of ultrafine-grained materials: Achieving exceptional properties through grain refinement. *Acta Mater.* **2013**, *61*, 7035–7059. [CrossRef]
4. Valiev, R.Z.; Langdon, T.G. Principles of equal-channel angular pressing as a processing tool for gain refinement. *Prog. Mater. Sci.* **2006**, *51*, 881–981. [CrossRef]
5. Zhilyaev, A.P.; Langdon, T.G. Using high-pressure tosrion for metal processing: Fundamentals and applications. *Prog. Mater. Sci.* **2008**, *53*, 893–979. [CrossRef]
6. Zhilyaev, A.P.; Kim, B.K.; Nurislamova, G.V.; Baró, M.D.; Szpunar, J.A.; Langdon, T.G. Orientation imaging microscopy of ultrafine-grained nickel. *Scripta Mater.* **2002**, *46*, 575–580. [CrossRef]
7. Zhilyaev, A.P.; Nurislamova, G.V.; Kim, B.K.; Baró, M.D.; Szpunar, J.A.; Langdon, T.G. Experimental parameters influencing grain refinement and microstructural evolution during high-pressure torsion. *Acta Mater.* **2003**, *51*, 753–765. [CrossRef]
8. Wongsa-Ngam, J.; Kawasaki, M.; Langdon, T.G. A comparison of microstructures and mechanical properties in a Cu-Zr alloy processed using different SPD techniques. *J. Mater. Sci.* **2013**, *48*, 4653–4660. [CrossRef]
9. Brodova, I.; Rasposienko, D.; Shirinkina, I.; Petrova, A.; Akopyan, T.; Bobruk, E. Effect of severe plastic deformation on structure refinement and mechanical properties of the Al-Zn-Mg-Fe-Ni Alloy. *Metals* **2021**, *11*, 296. [CrossRef]
10. Edalati, K.; Li, H.-W.; Kilmametov, A.; Floriano, R.; Borchers, C. High-Pressure Torsion for Synthesis of High-Entropy Alloys. *Metals* **2021**, *11*, 1263. [CrossRef]
11. Nocivin, A.; Raducanu, D.; Vasile, B.; Trisca-Rusu, C.; Cojocaru, E.; Dan, A.; Irimescu, R.; Cojocaru, V. Tailoring a Low Young Modulus for a Beta Titanium Alloy by Combining Severe Plastic Deformation with Solution Treatment. *Materials* **2021**, *14*, 3467. [CrossRef]
12. Svirid, A.; Pushin, V.; Kuranova, N.; Makarov, V.; Ustyugov, Y. Structural and Phase Transformations and Physical and Mechanical Properties of Cu-Al-Ni Shape Memory Alloys Subjected to Severe Plastic Deformation and Annealing. *Materials* **2021**, *14*, 4394. [CrossRef]
13. Wang, J.T.; Li, Z.; Wang, J.; Langdon, T.G. Principles of severe plastic deformation using tube high-pressure shearing. *Scr. Mater.* **2012**, *67*, 810–813. [CrossRef]
14. Li, Z.; Zhang, P.F.; Yuan, H.; Lin, K.; Liu, Y.; Yin, D.L.; Wang, J.T.; Langdon, T.G. Principle of one-step synthesis for multilayered structures using tube high-pressure shearing. *Mater. Sci. Eng. A* **2016**, *658*, 367–375. [CrossRef]
15. Meng, J.J.; Li, Z.; Liu, Y.; Bin Zhu, Y.; Wang, S.; Lin, K.; Tao, J.Q.; Wang, J.T. Investigation on the Strain Distribution in Tube High-Pressure Shearing. *Metals* **2019**, *9*, 1117. [CrossRef]
16. Wang, C.T.; He, Y.; Langdon, T.G. The significance of strain weakening and self-annealing in a superplastic Bi–Sn eutectic alloy processed by high-pressure torsion. *Acta Mater.* **2019**, *185*, 245–256. [CrossRef]
17. Wang, C.T.; Langdon, T.G. An examination of strain weakening and self-annealing in a Bi-Sn alloy processed by high-pressure torsion. *Mater. Lett.* **2021**, *301*, 130321. [CrossRef]
18. Pearson, C.E. The viscous properties of extruded eutectic alloys of lead-tin and bismuth-tin. *J. Inst. Met.* **1934**, *54*, 111–124.
19. Kawasaki, M.; Ahn, B.; Langdon, T.G. Microstructural evolution in a two-phase alloy processed by high-pressure torsion. *Acta Mater.* **2009**, *58*, 919–930. [CrossRef]
20. Zhang, N.X.; Kawasaki, M.; Huang, Y.; Langdon, T.G. Microstructural evolution in two-phase alloys processed by high-pressure torsion. *J. Mater. Sci.* **2012**, *48*, 4582–4591. [CrossRef]
21. Zhang, N.X.; Chinh, N.Q.; Kawasaki, M.; Huang, Y.; Langdon, T.G. Self-annealing in a two-phase Pb-Sn alloy after processing by high-pressure torsion. *Mater. Sci. Eng. A* **2016**, *666*, 350–359. [CrossRef]

22. Zhang, N.X.; Kawasaki, M.; Huang, Y.; Langdon, T.G. An examination of microstructural evolution in a Pb–Sn eutectic alloy processed by high-pressure torsion and subsequent self-annealing. *Mater. Sci. Eng. A* **2020**, *802*, 140653. [CrossRef]
23. Wang, J.; Kang, S.-B.; Kim, H. Microstructure transformation from lamellar to equiaxed microduplex through equal-channel angular pressing in an Al-33 pct Cu eutectic alloy. *Met. Mater. Trans. A* **2004**, *35*, 279–286. [CrossRef]
24. Wang, J.; Kang, S.-B.; Kim, H. Shear features during equal channel angular pressing of a lamellae eutectic alloy. *Mater. Sci. Eng. A* **2004**, *383*, 356–361. [CrossRef]
25. Massalski, T.B. *Binary Alloy. Phase Diagrams*, 2nd ed.; ASM International: Materials Park, OH, USA, 1990; pp. 794–796.
26. Ren, G.; Collins, M.N. Improved Reliability and Mechanical Performance of Ag Microalloyed Sn58Bi Solder Alloys. *Metals* **2019**, *9*, 462. [CrossRef]
27. Wang, K.; Wang, F.; Huang, Y.; Qi, K. Comprehensive Properties of a Novel Quaternary Sn-Bi-Sb-Ag Solder: Wettability, Interfacial Structure and Mechanical Properties. *Metals* **2019**, *9*, 791. [CrossRef]
28. Liu, Y.; Tu, K.N. Low melting point solders based on Sn, Bi, and In elements. *Mater. Today Adv.* **2020**, *8*, 100115. [CrossRef]
29. Kang, H.; Rajendran, S.H.; Jung, J.P. Low melting temperature Sn-Bi solder: Effect of alloying and nanoparticle addition on the microstructural, thermal, interfacial bonding, and mechanical characteristics. *Metals* **2021**, *11*, 364. [CrossRef]
30. Langdon, T.G. A unified approach to grain boundary sliding in creep and superplasticity. *Acta Metall. Mater.* **1994**, *42*, 2437–2443. [CrossRef]
31. Hu, X.; Li, K.; Ai, F. Research on lamellar structure and micro-hardness of directionally solidified Sn-58Bi eutectic alloy. *China Foundry* **2012**, *9*, 360–365.
32. Lee, H.B.; Kim, Y.W.; Kim, S.H.; Park, S.H.; Choi, J.-P.; Aranas, C., Jr. A Modular Solder System with Hierarchical Morphology and Backward Compatibility. *Small* **2018**, *14*, e1801349. [CrossRef]
33. Kim, S.H.; Yeon, S.-M.; Kim, J.H.; Park, S.J.; Lee, J.E.; Park, S.-H.; Choi, J.-P.; Aranas, C.J.; Son, Y. Fine Microstructured In–Sn–Bi Solder for Adhesion on a Flexible PET Substrate: Its Effect on Superplasticity and Toughness. *ACS Appl. Mater. Interfaces* **2019**, *11*, 17090–17099. [CrossRef] [PubMed]
34. Xu, K.-K.; Zhang, L.; Gao, L.-L.; Jiang, N.; Zhang, L.; Zhong, S.-J. Review of microstructure and properties of low temperature lead-free solder in electronic packaging. *Sci. Technol. Adv. Mater.* **2020**, *21*, 689–711. [CrossRef] [PubMed]
35. Langdon, T.G. Seventy-five years of superplasticity: Historic developments and new opportunities. *J. Mater. Sci.* **2009**, *44*, 5998–6010. [CrossRef]

crystals

MDPI

Article

An Integrative Simulation for Mixing Different Polycarbonate Grades with the Same Color: Experimental Analysis and Evaluations

Jamal Alsadi *, Rabah Ismail and Issam Trrad

Engineering College, University of Jadara, Irbid 21110, Jordan; r.ismail@jadara.edu.jo (R.I.); i.trrad@jadara.edu.jo (I.T.)

* Correspondence: jamal.alsadi@ontariotechu.net

Abstract: The processing parameters' impact such as temperature (Temp.), feed rate (F.R.), and speed (S.) at three distinct grades of the same color was explored in this study. To investigate the effect of the characteristics on color formulations, they were each adjusted to five different levels. For these grades, which were all associated with the same color, an intermeshing twin-screw extruder (TSE) was used. The compounded materials were molded into flat coupons then evaluated with a spectrophotometer for their CIE (L*, a*, b*, and dE*) values. A spectrophotometer was used to determine the color of a compounded plastic batch, which measured three numbers indicating the tristimulus values (CIE L*a*b*). The lightness axis, which ranged from 0 (black) to 100 (white), is known as the L*-axis (white). Redness-greenness and yellowness-blueness were represented by the other two coordinates, a* and b*, respectively. The color difference deviation (Delta E*) from a target was dimensionless, when dE* approached zero. However, the most excellent favorable color difference value occurred and different processing impact factors on polycarbonate grade were investigated. Using the response service design (RSD) software of Stat-Ease Design-Expert® (Minneapolis, MN, USA), historical data were gathered and evaluated. To reduce the value of dE*, the impacts of these processing factors were investigated with the three processing parameters. The whole tristimulus color value could be simulated. Parameters were adjusted on 45 different treatments, using a five-level controlled response method to investigate their impact on color and detect non-optimal responses. The ANOVA for each grade was used to build the predicted regression models. The significant processing parameters were subjected to experimental running to simulate the regression models and achieve the best color, reducing waste.

Keywords: different grades; RSD; simulate r egression models; processing and parameters; analysis of variance; resin pigment blends

Citation: Alsadi, J.; Ismail, R.; Trrad, I. An Integrative Simulation for Mixing Different Polycarbonate Grades with the Same Color: Experimental Analysis and Evaluations. *Crystals* **2022**, *12*, 423. https://doi.org/10.3390/cryst12030423

Academic Editor: Wojciech Polkowski

Received: 6 February 2022
Accepted: 14 March 2022
Published: 18 March 2022

1. Introduction

Plastics are relatively new materials for producing colored materials. As a result, there are few scientific data on plastic color mismatching and its long-term consequences. Plastic fabrication allows for the creation of robust, lightweight plastics in various shapes. In many situations, plastic shapes are favored over metal shapes. Polycarbonate is a rigid, transparent polymer used in a variety of applications. Some factors may significantly impact the color of plastics intended for outdoor use. As a result, it is critical to comprehend how numerous elements can influence material compounding. This research aims to see how processing settings affect color matching for a few grade-color. To generate the proper color with minimal waste, the plastic industry has spent the last few decades seeking to understand the significant challenges involved in plastic color matching procedures. Lambert's law claims that the amount of light absorbed is proportional to the concentration of the absorbing substance. Still, Beers law states that the amount of light absorbed is proportional to the thickness of the absorbing material [1]. Manufacturing technology makes colored plastic for plastic process prototyping on a small and medium scale. As a result, the plant receives orders that must be completed in a matter of days.

To summarize, an object's color appearance is determined by the total amount and type of scattering and absorption that occurs. As a result, the item will seem white if there is no absorption and nearly equal levels of scattering at all visible wavelengths; and if the visible light is absorbed by the pigment [2].

Several hundred ingredients are divided into three categories: resins, additives, and pigments. Ingredients and additives combine to create a specific grade of plastic. The pigments give the plastic its color. Because of their surfaces and orientations, the pigments absorb certain hues, while reflecting others randomly.

White light is created by mixing all visible spectrum wavelengths in roughly equal quantities [3]. The light source and observer are replaced with color measurement tools such as a colorimeter or spectrophotometer to standardize color evaluation in the plastic compounding industry [4].

As color is represented in codes or values, this allows for more uniform color recognition. There were two data mining approaches used. One was a decision tree classifier, and the other was online analytical processing (OLAP) (DTC). OLAP assisted in identifying a relationship between factors that resulted in failed batches and parameters with a high rate of alteration. The DTC was proposed as a decision assistance tool for detecting combinations of characteristics that could cause a color mismatch. The DTC investigates characteristics that could lead to color mismatch issues in compounded polymers. To find such factors in the past, OLAP and data mining methodologies were applied [5–7].

Overall, DTC was utilized in this study to investigate possible correlations between the components of color, grade, kind, product, and line. To the best of our knowledge, no research has used DTC for color mismatch analysis, according to our literature review. DTC is used in some relevant manufacturing articles (semi-conductors [8]). Other researchers have used neuro-networks to forecast output colors based on past data [9].

In previous studies, the artificial neural network (ANN) was utilized to eliminate mistakes in polycarbonate color values [10]. The neural network in this paper was used to reduce the errors in color tristimulus values (L*, a*, b*), which directly affect the D.E. calculated [11].

The problem cannot be solved by concentrating on a few situations because the colors' nature is constantly changing. Therefore, this research proposal focuses on determining the fundamental causes of color mismatches in compounded plastics. As a result, plastics firms will reduce waste and boost production. More importantly, it will improve knowledge of the technical challenges of color matching in plastic production. Focusing on resolving challenges for a single product is complex and potentially fruitless because that product may not be duplicated in the future. Compared to the paint industry, color mismatch issues have not been investigated deeply through the plastic compounding business. The parameter(s) creating first-pass color opportunities must be discovered to limit material rejects. Researchers mixed three different titanium dioxide pigments into heavily loaded polyethylene masterbatches, each with a different surface treatment.

They discovered that the three grades' best screw design and operating conditions were considerably different. Processing circumstances or certain combinations of modifiers and additives in the resin system were shown to have a negative impact on the final desired hue [12,13]. Paints and coatings have had a lot of research done on pigment dispersion, but plastics have not gotten nearly as much attention [14,15].

The high shear rates, processing temperatures, and processing pressures used in plastics manufacturing operations [16,17] significantly contrast the two dispersion mechanisms. Various scholars have conducted several investigations, during compounding, on the effect of processing parameters on color [18,19]. The minimum processing time is advised to achieve excellent gloss, brilliance, and blend uniformity. Furthermore, for each item, an optimal loading should be utilized; too much pigment is not only expensive but also hazardous because it diminishes impact resistance [20,21]. Increase the duration of the mixing and decrease the viscosity of resin to solve problems with dispersion or achieving a homogeneous mixture [22,23]. Various researchers have conducted a few investigations on the influence of processing factors on dynamic mixing in a screw extrusion during polymer compounding [18,24].

Many scientists have reviewed polymer blending as an essential field of polymer science. As Sanchez et al. [25] demonstrated, the PC/PBT blends are transparent in the melt stage and somewhat miscible blends in the solid state.

Liang and Gupta (2000) investigated the rheological qualities of a recycled P.C. blended with virgin P.C., concluding that separated P.C. could be added to pure P.C. up to 15% without significantly affecting its properties [26]. Lee S. et al. investigated the rheological and phase behavior of P.C./Polyester blends. They discovered, however, that the combinations do not obey the mixing rule, which is standard in all investigations. They discovered, however, that the combinations do not obey the mixing rule, which is standard in all investigations [27]. Other researchers' experiments on extruders showed that single screw extruders could reach dispersive mixing capabilities comparable to twin-screw extruders [28].

A 45-mm diameter single-screw extruder with eight glass panes was used in another investigation to investigate the color mixing process [29]. The researchers determined where color mixing began and finished by using such an extruder. The quality of mixing was shown to be directly proportional to the maximum processing pressure in the extruder. Furthermore, earlier research has examined how the screw shape and operating conditions affect dispersion performance and torque loading during twin-screw compounding [30].

One of the most significant color matching components taken from a remote location is spectrophotometric measures to create a suitable color standard. Spectrophotometers are valuable quality control equipment for measuring color and defining color variations numerically. However, their function as a device is to reduce a color target to a collection of numbers, which are subsequently sent to a color formulator as a matching target [31,32]. CIELAB is the name of the color measurement method. The values utilized by CIE are named L*, a*, and b*. L* denotes the difference between light (L* = 100) and dark (L* = 0), a* denotes the green (−a*) and red (+a*) difference, and b* indicates the yellow (+b*) and blue (−b*) difference [33,34]. dE* is used to express deviations in L*, a*, and b*, where:

$$dE^* = \sqrt{(\Delta L^*)^2 + (\Delta a^*)^2 + (\Delta b^*)^2} \quad (1)$$

The color difference's amplitude, not its direction, is represented by dE*. As a quality control measure, colored materials are compared to a standard when being manufactured. Color discrepancies are employed instead of absolute color values. The total color change, dE*, shows the color difference in the CIELAB color space [17,35].

The findings of designed experiments were analyzed and discussed in this study, which highlights individual and combined influences on output color of three process parameters.

The experimental data confirm the statistical model's fitness [36] by systematically examining resins, additives, and pigments, and how processing conditions and diverse interactions impact them. More precisely, the scientific concerns surrounding the twin co-rotating screw process processing parameters on different grades of the same color were explored. The study's main aim was to develop an equation that might be used to determine differences between the two samples and could be used to any color at any time. To explore the impact of parameters on color and detect non-optimal responses, a five-level controlled response method was used on 45 different treatments. The anticipated regression models were built using the ANOVA for three different grades. Speed, temperature, and F.R. were among the processing characteristics studied. To provide a foundation for process improvement recommendations, experimental data were collected, and statistical analysis was undertaken.

2. Materials and Methods

The three classes with the highest adjustment when dealing with red pigments were discovered based on preliminary data mining results from the first few months of 2009. In this study, these grades were denoted by the numbers 1, 2, and 3. For the dispersion of color in parts per 100 among these grades, a mixture of two polycarbonate resins and four distinct pigments were utilized (PPH). As indicated in Table 1, all three grades utilized the same color, as were shown in Figure 1.

Table 1. Compounding formulation used for three grades.

Resin/Color	Grade–Color (1)		Grade–Color (2)		Grade–Color (3)	
Type	pph	gms	pph	gms	pph	gms
Resin 1	30	1800	30	1800	–	–
Resin 2	70	4200	70	4200	100	6000
White Pigment	1.925	115.5	1.76	105.6	1.76	105.6
Black Pigment	0.11	6.60	0.00968	0.5808	0.00968	0.58
Red Pigment	0.1875	11.25	0.01602	0.9612	0.01602	0.96
Yellow Pigment	0.1075	6.45	0.1084	6.504	0.1084	6.50

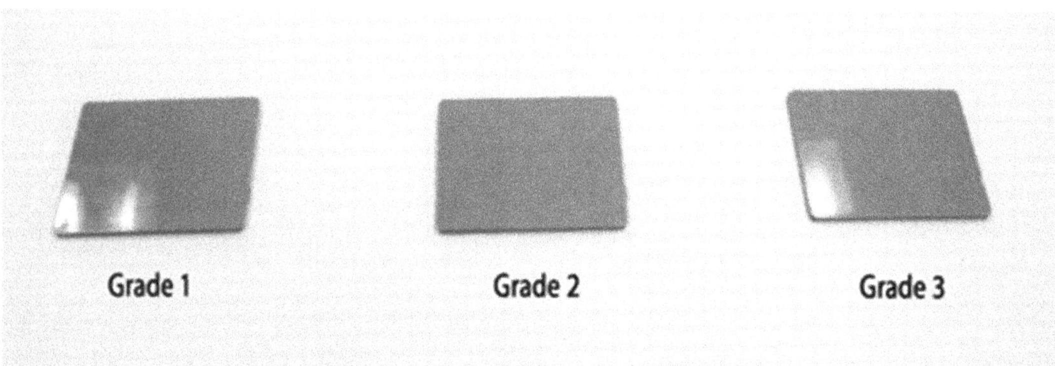

Figure 1. Three grades have different formulations but the same color.

Grades 1 and 2 used a mixture of two polycarbonates resins with various weights of the same pigments, while grade 3 used one poly car resin with the same weight of pigments as grade 2. As a result, resin 1 had a melt flow index (MFI) of 25 g/min, while resin 2 had an MFI of 6.5 g/10 min, where the weights were heavier than water, and the temperature for autoignition was 630 °C for all grades. At the industrial plant, three grades were subjected to testing. The materials were extruded at L/D ratios of 37 and Do/Di ratios of 1.55, respectively, utilizing a twin-screw extruder (25.5 mm, 27 kW). There were ten heating zones on the extruder, nine designated on the barrel, and one at the die.

The extruded melt was then pelletized after being quenched in cold water. These pellets were subsequently formed into rectangular chips ($3 \times 2 \times 0.1$ inches), which were measured against a target value via injection molding. Three coupons were created for each experiment at each of the five-parameter values to assure accuracy. Then each voucher was given three readings. The total simulating design data for the tristimulus color value with the three processing parameters were 45 runs, as recorded in Table 2.

Table 2. Response surface design 45 runs for 3 grades.

Nos.	Temp	RPM	kg/h	Grade	L*	a*	b*	dE*
1	230	750	25	Grade 1	67.26	1.52	4.545	0.435
2	240	750	25	Grade 1	67.1767	1.5	4.5	0.456
3	255	750	25	Grade 1	67.285	1.43	4.453	0.533
4	270	750	25	Grade 1	67.185	1.511	4.56167	0.4
5	280	750	25	Grade 1	66.735	1.547	4.63	0.5
6	255	700	25	Grade 1	67.055	1.48167	4.41167	0.55
7	255	725	25	Grade 1	67.0333	1.46667	4.34667	0.62
8	255	750	25	Grade 1	67.286	1.49	4.45	0.54
9	255	775	25	Grade 1	66.995	1.44167	4.30167	0.66
10	255	800	25	Grade 1	67.1033	1.45167	4.30833	0.65
11	255	750	20	Grade 1	67.0183	1.55	4.78	0.22
12	255	750	23	Grade 1	66.81	1.423	4.41	0.63
13	255	750	25	Grade 1	67.285	1.5	4.45	0.54
14	255	750	27	Grade 1	66.7583	1.43	4.41	0.65
15	255	750	30	Grade 1	66.915	1.43	4.47	0.53
16	230	750	25	Grade 2	66.44	1.57	4.71	1.29
17	240	750	25	Grade 2	66.33	1.54	4.63	1.28
18	255	750	25	Grade 2	66.37	1.56	4.77	1.25
19	270	750	25	Grade 2	66.47	1.54	4.65	1.24
20	280	750	25	Grade 2	66.21	1.55	4.68	1.23
21	255	700	25	Grade 2	66.3533	1.55167	4.74167	1.228
22	255	725	25	Grade 2	66.4183	1.54833	4.73	1.16
23	255	750	25	Grade 2	66.3733	1.56	4.77667	1.21
24	255	775	25	Grade 2	66.3017	1.57167	4.80833	1.278
25	255	800	25	Grade 2	66.5067	1.55667	4.76	1.21
26	255	750	20	Grade 2	66.575	1.56667	4.67167	1.018
27	255	750	23	Grade 2	66.465	1.582	4.676	1.128
28	255	750	25	Grade 2	66.3733	1.56	4.77667	1.21
29	255	750	27	Grade 2	66.345	1.585	4.71	1.24
30	255	750	30	Grade 2	66.4783	1.58667	4.69333	1.11
31	230	750	25	Grade 3	67.715	1.63	5.115	0.4
32	240	750	25	Grade 3	67.515	1.686667	5.236667	0.51
33	255	750	25	Grade 3	67.515	1.686667	5.236667	0.46
34	270	750	25	Grade 3	67.605	1.641667	5.113333	0.38
35	280	750	25	Grade 3	67.525	1.68	5.235	0.51
36	255	700	25	Grade 3	67.48	1.66	5.18	0.438
37	255	725	25	Grade 3	67.5583	1.615	5.10167	0.41
38	255	750	25	Grade 3	67.515	1.68667	5.23667	0.46
39	255	775	25	Grade 3	67.8467	1.60833	5.03833	0.42
40	255	800	25	Grade 3	67.525	1.68	5.235	0.39
41	255	750	20	Grade 3	67.6283	1.63833	5.065	0.346
42	255	750	23	Grade 3	67.5733	1.63833	5.10667	0.378
43	255	750	25	Grade 3	67.515	1.68667	5.23667	0.463
44	255	750	27	Grade 3	67.635	1.64167	5.13	0.413
45	255	750	30	Grade 3	67.54	1.63	5.12	0.38

In CIE L*, a*, and b* values, L* = 67.57, a* = 1.43, and b* = 4.8 were chosen as the required color output, while the permitted dE* was 0.85. Using the Software of Design-Expert, Version 8, Stat-Ease Inc. (Minneapolis, MN, USA), the statistical data were established. Then the data were used to compare and analyze the factors' effect on grades. The ANOVA determined which parameters were significant and whether there was any interaction between them. As previously stated, the study's goal: develop an equation that could help in expecting the L*, a*, and b* tristimulus values.

3. Results

The design of the experiment was used to do statistical analysis and ANOVA. Using Stat-Ease Design Expert® Version 8 software, the influence of parameters on L*, a*, b*, and dE* was investigated, as seen in Table 3.

Table 3. Summary of the design data.

Type	Response Surface	Run	45	Response Surface Design
Design Type	Historical Data	Blocks	No Blocks	
Design Model	Quadratic	Build time	59.3	
Factor	Name	Units	Type	Sub-Type
A	Temp	°C	Numeric	continuous
B	Speed	RPM	Numeric	continuous
C	Feed rate	kg/h	Numeric	continuous
D	grade		Categoric	Nominal
Factor	Min	max	Coded	Values
A	230	280	−1	1
B	700	800	−1	1
C	20	30	−1	1
D	B	A		
RESPONSE	Name	Obs	Analysis	Model
Y1	L*	37	Polynomial	R Linear
Y2	a*	37	Polynomial	Quadratic
Y3	b*	37	Polynomial	R^2 Fi
Y4	dE*	37	Polynomial	Quadratic
RESPONSE	Min	Max	Mean	Std.Dev
Y1	66.21	68	67.02	0.5
Y2	1.43	1.7	1.56	0.07
Y3	4.3	5.2	4.7	0.299
Y4	0.22	1.3	0.7	0.36

Note: A Temp, B Speed, C Feed Rate, D grade, Y1 L*, Y2 a*, Y3 b* and Y4 dE*.

3.1. Analysis of Variance (ANOVA)

Sequential F-tests were run using a linear model as a starting point and adding terms (quadratic and linear if appropriate). The F-statistic was assessed for each model type, and the highest degree and critical elements model was picked. The same procedure was used for all tristimulus values, and only the significant terms were included. The ANOVA table for the sum of squares of a sequential model for dE* characterization is shown in Table 4. The quadratic model with the Prob > F was < 0.05, the most considerable condition. Furthermore, it was statistically significant (Prob > F was less than 0.0001) because it had a high F value (184.4). As a result, this model was suitable for the dE* response. The model's adjusted R-square value (97%) also corroborated this, as seen in Table 4.

The adjusted R-square measure was the same as the R-square measure, except that it was scaled down to account for the number of variables in the model. Both measures represent the model's capacity to explain variation in the answer. For example, the observed adjusted R-square value of 97% showed that the model explained roughly 97% of the variability in dE*. In contrast, about 3% of the variability in dE* was unknown.

The adjusted R-square value of 0.96 was reasonably close to the predicted R-square value of 0.95. A signal-to-noise ratio was used in the Adeq Precision measurement. It is ideal to have a ratio of more than four. The observed percentage of 35.5 specified that the observed variance is significant compared to the fitted model's underlying uncertainty. In other words, the observed variance was significant in proportion to the fitted model's underlying uncertainty. The design space systems can also be generated by using the exact modeling.

Table 4. ANOVA for the color of three grades.

Tristimulus Values	Processing Factors	F-Statistic Value	Probability Values	R^2	Adjacent R^2	Predicted R^2	Adequate Precision
L*	Model	193.82	0.0001	0.9463	0.9414	0.9318	34.082
	A	5.21	0.029				
	D	288.12	0.0001				
a*	Model	37.71	0.0001	0.901	0.8771	0.8385	20.869
	A	0.24	0.626				
	C	3.46	0.073				
	D	122.62	0.0001				
	CD	5.2	0.0117				
	A^2	3.32	0.079				
b*	Model	75.96	0.0001	0.9245	0.9124	0.8553	23.831
	C	1.05	0.3126				
	D	185.69	0.0001				
	CD	2.98	0.0654				
dE*	Model	184.47	0.0001	0.9736	0.9683	0.9532	35.528
	C	10.63	0.0028				
	D	538.19	0.0001				
	CD	3.13	0.0583				
	C^2	21.44	0.0001				

Feed rate (C) and grade (D) had significant effects on dE*, as shown in Table 4. Their p-values (Prob > F) were equal or less than 0.05 (typically ≤ 0.05), indicating that they were statistically significant models. On the other hand, temperature (A) and speed (B) had large p-values, indicating that they were not statistically significant for the dE* response. The interaction between feed rate (C) and the investigated grades (D) would be statistically significant if a confidence level of 90% was assumed; the p-value for this interaction (CD) in the model fitted to dE* was 0.0583 (see Table 4).

3.2. Simulate Regression Models

The expected response for each response was determined using multiple linear regression analysis. Equations (2)–(13) depict the response functions for grade 1, 2, and 3 for L*, a*, b*, and dE*, respectively, as shown in Table 5.

Table 5. Simulate regression models.

Response	Regression Model		
	Grade 1	Grade 2	Grade 3
L*	$68.04585 - 3.931478 \times 10^3 \times$ Temp ... ... (2)	$67.41030 - 3.93147 \times 10^3 \times$ Temp ... ... (6)	$68.59188 - 3.93147 \times 10^3 \times$ Temp ... ... (10)
a*	$3.97525 - 0.017160 \times$ Temp $- 0.013080 \times$ Feed Rate $+ 3.399968 \times 10^5 \times$ Temp2 ... ... (3)	$3.67696 - 0.017160 \times$ Temp $+ 1.82759 \times 10^3 \times$ Feed Rate $+ 3.39996 \times 10^5 \times$ Temp2 ... ... (7)	$3.82467 - 0.017160 \times$ Temp $- 6.02931 \times 10^4 \times$ Feed rate $+ 3.39996 \times 10^3 \times$ Temp2 ... (11)
b*	$5.26525 - 0.031351 \times$ Feed Rate ... ... (4)	$4.62909 + 3.03966 \times 10^3 \times$ Feed Rate. ... ... (8)	$5.00853 + 5.54586 \times 10^3 \times$ Feed Rate ... ... (12)
dE*	$-3.68725 + 0.30417 \times$ Feed Rate $- 5.409068 \times 10^3 \times$ Feed Rate2 ... ... (5)	$-2.44834 + 0.28225 \times$ Feed Rate $- 5.409068 \times 10^3 \times$ Feed Rate2 ... ... (9)	$-3.04190 + 0.27459 \times$ Feed Rate $- 5.409068 \times 10^3 \times$ Feed Rate2 ... ... (13)

3.3. Point Prediction

The response surface method was optimized using a "numerical optimizer" for the lowest color value (dE*) in the feasible region. The Design-Expert response®'s (Minneapolis, MN, USA) optimizer calculated numerous local (feasible area) variables. For each grade, Table 6 provides the predicted tristimulus color values of CIE (L*, a*, b*, and dE*). Minor deviations were detected in the color values acquired by the optimization process. These discrepancies could be due to a lack of precise temperature control during the extrusion process, which affects the viscosity of the polymer, as well as the pigment dispersion and ability to obtain the required color.

Table 6. Simulate tristimulus color solutions.

Grade	Process. Parameters			Tristimulus Values			
	Temp	Screw Speed	Feed Rate	L*	a*	b*	dE*
	°C	rpm	kg/h	Black/White	Red/Green	Yellow/Blue	Color O.P.
1	250.9	750	25.16	67.15	1.48	4.47	0.54
2	243.56	750	21.21	66.9	1.55	4.69	1.1
3	257.34	750	24.38	67.58	1.64	5.14	0.43

3.4. Effect of Processing Parameters through 3 Grades

Figures 2–11 show the impact of process conditions on color output over all three grades in terms of CIE dE* values, as created by Design-Expert® L*, a*, and b* represents CIE tristimulus data, related graphics were also created. However, because dE* takes precedence in this paper, the figures for these tristimulus values are limited. A common occurrence for the grades responsible for the reddest pigment modifications, a design of the experiment was carried out. The goal of the tests was to figure out what processing and material characteristics were producing color discrepancies. The color difference and the processing factors were explored for correlations. With Design-Expert®, general trends were charted, and trials were carried out and statistically analyzed. The experiment was designed with temperature, speed, and flow rate are three processing conditions for a total of 45 runs; these runs were done in three grades (1, 2, and 3) for one parameter while keeping the other two constants, as shown in Table 2.

Five levels were employed to conduct experiments on these three parameters. For example, temperatures reached 230, 240, 255, 270, and 280 °C. For each grade and level, values for the four responses (L*, a*, b*, and dE*) were taken from three different coupons and three different positions on each of these coupons. Figures 1–9 show the variation in color output for grades 1, 2, and 3 as a function of the three processing settings.

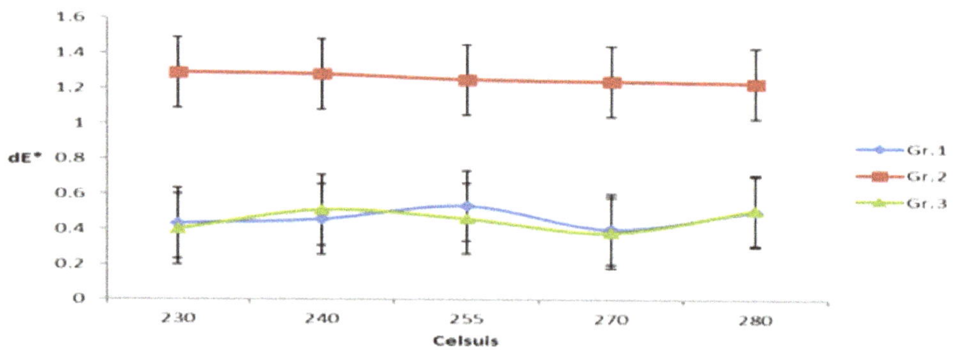

Figure 2. Temperature effects on dE* for Grade 1, Grade 2 and Grade 3.

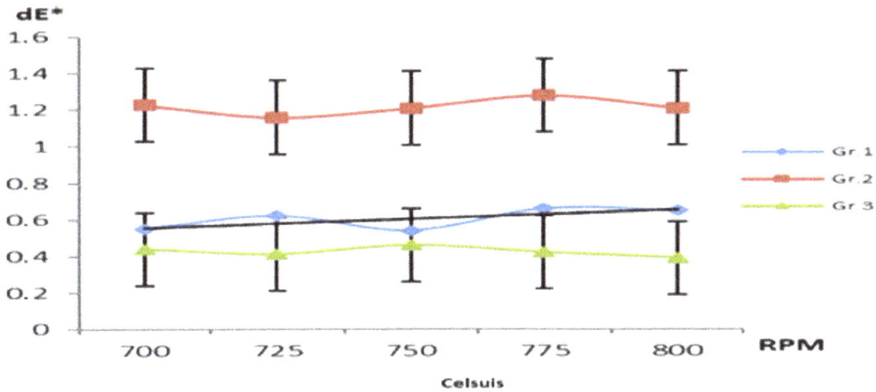

Figure 3. Effect of screw speed on dE* for Grade 1, Grade 2 and Grade 3.

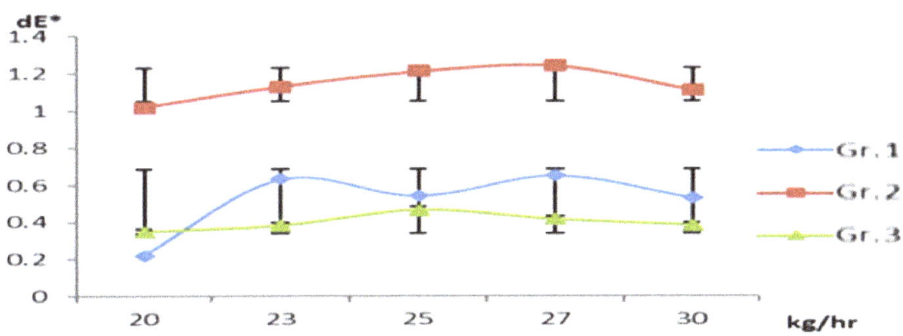

Figure 4. The impact of feeding rate on color.

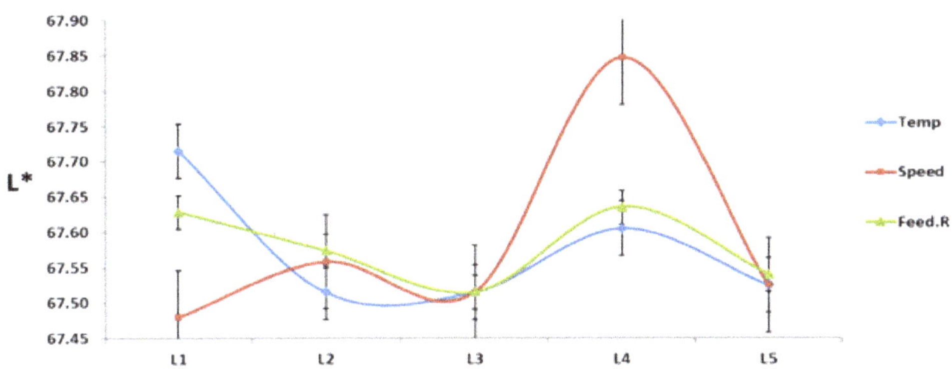

Figure 5. Effect of processing parameters on grade 3, L*.

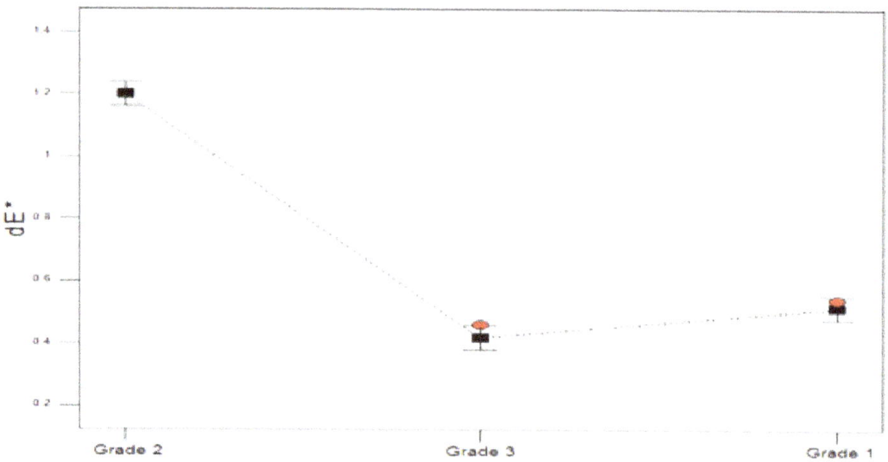

Figure 6. Interaction effect of Grade 1, Grade 2 and Grade 3.

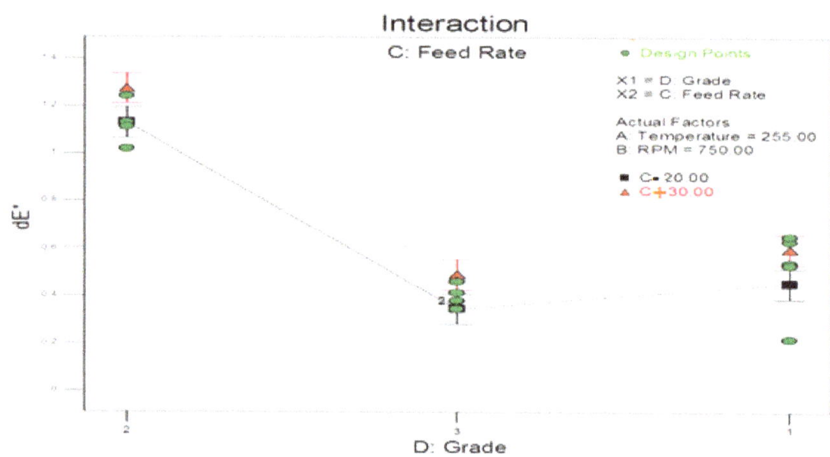

Figure 7. Effect of feed rate on dE* for Grade 1, Grade 2 and Grade 3.

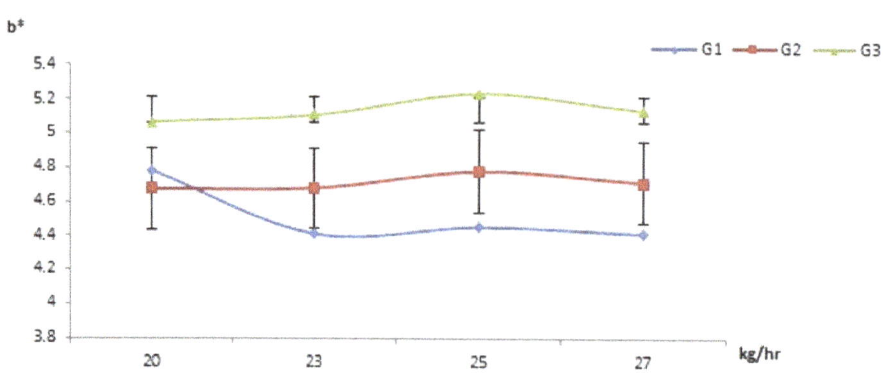

Figure 8. Effect of feed rate on b*.

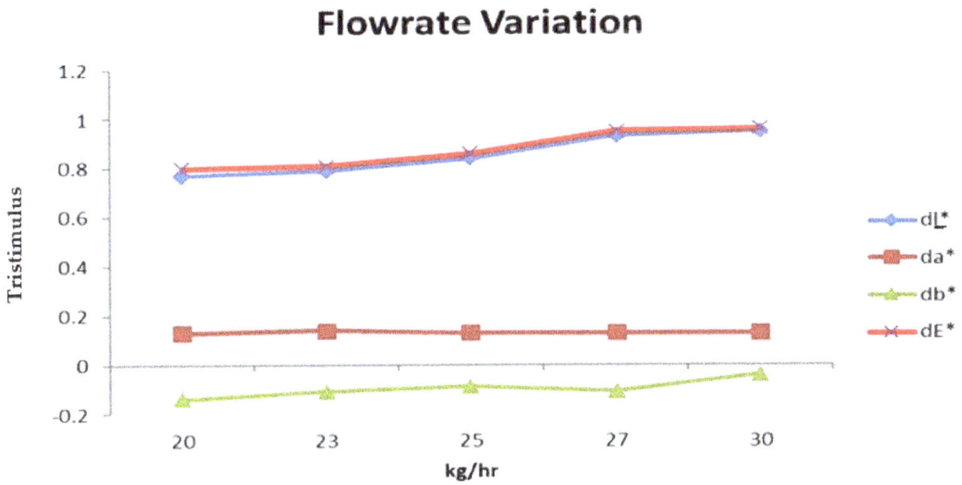

Figure 9. Tristimulus versus feed rate on grade 3.

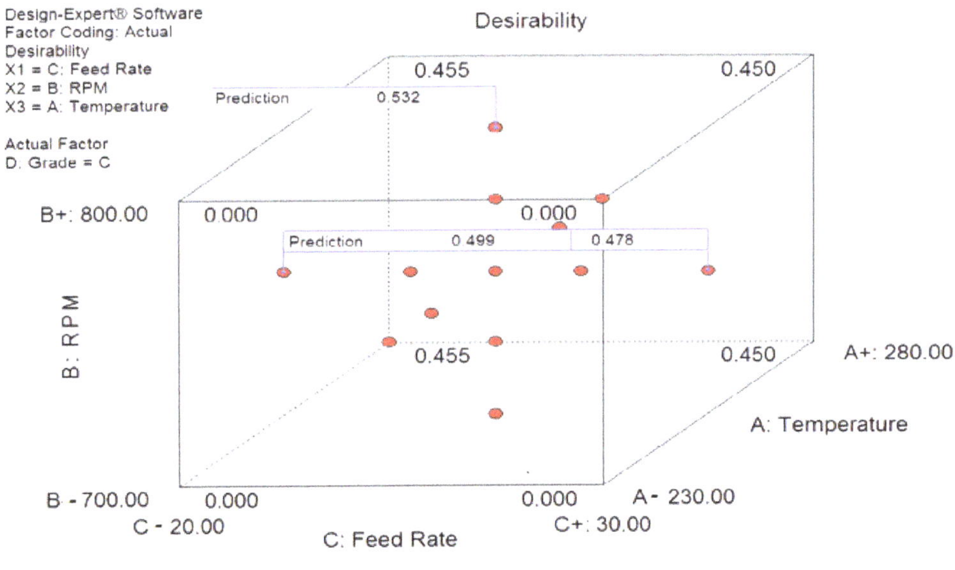

Figure 10. Cube desirability between grade 3 and dE*.

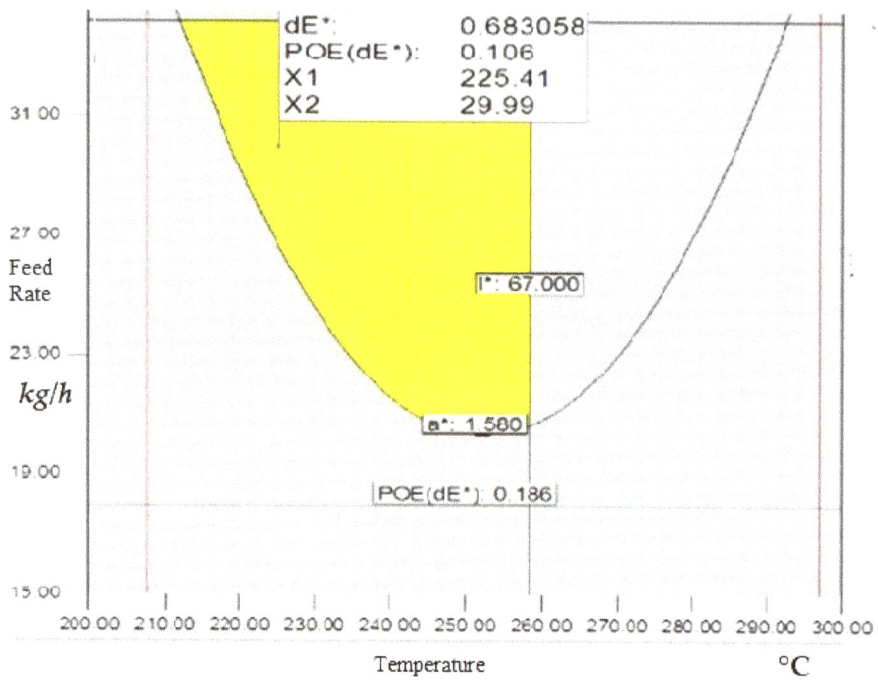

Figure 11. Interaction and overlay plot at 750 rpm.

*3.5. Effect of Temperature on dE**

Figure 2 depicts the fluctuation in dE* about temperature, demonstrating that color output deviation increases somewhat for grades 1 and 3. However, at 280 °C (level five), it became more prominent for both classes—because of the high temperature causing some breakdown in the formulation's resin or additives. When comparing grade 2 to the previous grades, the color output divergence increased dramatically.

*3.6. Screw Speed Effect on dE**

Figure 3 shows that at 775 and 800 rpm, the responses of grades 1, 2, and 3 to screw speed change are very comparable. This appears to be responsive to screw speed changes due to a higher shear rate, which increases pigment particle dispersion in the extruded material. It is also worth noting that for grade 1, level three had the lowest dE* value. For the three grades, color output begins to improve above 775 rpm. Furthermore, the color output divergence increases dramatically when comparing grade 2 to grades 1 and 3. The shear rate has the same effect on grade 2 as it does on grades 1 and 3.

3.7. Feed Rate Effect on Color

Similarly, Figure 4 illustrates the fluctuation in dE* as a function of feed rate. Grades 1, 2, and 3 showed a similar trend to temperature and speed fluctuation (Figures 1 and 2). The feed rate varied between 20 and 30 kg/h. for levels one and five, all three grades demonstrated slight variances from the optimum color output. This could be attributed to a rise in the shear rate at both the maximum and lowest feed rates. In the compounding mixer, this improves the flowability and dispersion of pigments. Furthermore, it shows how they may have a lowering of dE* values differ depending on the feed rate. This could be related to enhanced pigment dispersion, as higher feed rates cause stronger shear, which leads to better pigment dispersion [37,38].

3.8. Interactions Effect of L* Values

The interactions between L* and the processing parameters are depicted in Figure 5. All other parameters were fixed, while each was adjusted to five different levels. Only the interactions for L* are shown for the sake of brevity. Figure 5 shows that variations in processing circumstances impact L* values almost like they affect dE* values. A similar trend variation was seen (L1, L2, L3, L4, and L5).

However, the L* values for L3 showed the slightest variance and the same value for the three processing parameters for the target color output, especially for grade 3.

Furthermore, Figure 5 depicts the relationship between processing conditions and (L*) in distinct ways.

Increasing the temperature and weight percentage of PC1 with (higher melt flow index) showed a significant effect in lower viscosity value and decreased color matching values (dE^*). The formulation and processes effectively controlled the viscosity, and microtomed plastic sections performed characterizing to different thicknesses and temperatures. The optimal number of particles was increased at higher temperatures and thickness [39]. Characterization of polycarbonate formulation at different temperatures was also analyzed. They were rheologically characterized using the rotational rheometer [40].

3.9. Effect of Grades on dE*

Figure 6 shows that dE* values for grades 1 and 2 were greater than those for grade 3. This could be because grade 3 used fewer pigments and had better pigment dispersion than grades 1 and 2.

3.10. Grades and Feed Rate Interactions

Figure 7 shows the effect of feed rate on color output across the three grades (when screw speed is 750 rpm and temperature is 255 °C). Figure 7 and Table 7 show that feed rate appears to have the most significant impact on dE* for all three grades. This can be seen at both low and high feed rates, and it could be due to better dispersion.

Table 7. Optimum processing values for color output.

Optimum Processing Values for the Three Grades								
Feed Rate Parameter at Fixed Temp and Screw Speed (RPM)					Temp at Fixed RPM and kg/h		Screw Speed at Fixed Temp and Feed Rate	
GRADES	**Feed Rate**	**dE***	**Feed Rate**	**dE***	**Temp**	**dE***	**Speed**	**dE***
GRADE 1	20	0.22	30	0.53	270	0.4	750	0.54
GRADE 2	20	1.08	30	1.11	280	1.23	725	1.16
GRADE 3	20	0.34	30	0.38	270	0.38	800	0.39

In addition, Table 7 confirms the optimum processing readings for the three grades. The optimum color values are for grade 3 and grade 1.

3.11. Effect on b* Values

Figure 8 shows that grades 2 and 3 to b* change responses were comparable. In their formulas, both grades had the same amount of pigment. As a result, this emphasizes the need for having the same pigment composition and the precision of minute pigment loading. It is provided to demonstrate how minor modifications in a formulation can result in major color variations, leading to lot rejection. More precise measurements should be adopted when weighing any pigment amount, especially when working with sensitive formulations. An in-depth study and understanding of pigment interactions can improve first-pass color production [41,42].

This paper aimed to evaluate the influence of various processing parameters on the dispersion quality of polycarbonate compounds. In addition, the influences of parameters, pigment size distribution, and morphology on the pigment dispersion were also studied [43].

This study reviews the impact of scanning microscopic methods to evaluate the influence of various processing parameters on the dispersion quality of the polycarbonate compound. Experimental data were compared with historical data records [44]. Figure 9 indicates that the feed rate increased the tristimulus color values of dL*, da*, db*, and dE* for grade 3. It indicates the federate has a significant response on color output.

3.12. Desirability and Overlay Plots

The following figures show the desirability and overlay plot between processing parameters and grades. Figure 10 is a 3-D view of the predicted desirability for the interaction of processing parameters in terms of dE* = 0.478. It can be created for each optimal discovery.

Figure 11 illustrates the overlay plot between temperature and feed rate, while screw speed was constant at 750 rpm. In the factor space, the graphical optimization showed the area of possible response values. The yellow region represents the area that meets the required target value, while the gray area represents the area that does not. The points represent the optimum L*, a*, b*, and dE* values for the grades under consideration. The optima occur at 225 °C, 750 rpm, and 29 kg/h, achieving the optimum average value of dE* (0.68) for all grades.

4. Conclusions

The current study reveals that different grades respond differently to the desired color output under operating conditions. It is also clear that grade 3 had the best color output value. This could be due to the decreased number of pigments utilized in grade 3's material formulation and, hence, better dispersion of these pigments than in grades 1 or 2. As a result, grade 3 had the lowest MFR readings. A high MFR causes a decrease in viscosity, ultimately breaking the bonds and increasing the flowability of the mixing material.

The impacts of processing factors on color outputs of various grades were investigated, and statistical analysis was used to find correlations between the inputs and outputs. Experiments using general trends (G.T.) and response surface methods (RSM) based on the design of experiments were used to determine the optimum of extrusion settings and color values (DOE). Color outputs and optimal processing conditions were predicted with predictive models.

Different grades produced different color outputs under the same or similar operating conditions.

Finally, it is evident that the three grades had various formulae, but they all had the same color. The optimal color output values and grades were chosen based on our simulation.

From the ANOVA, the *F*-value implied the model was significant for dE*. Feed rate seems to have the most significant effect on dE* of the output color for grade 3. A lowering of dE* values was observed at higher and lower feed rates. This may be due to better dispersion; at increased feed rates, the higher flow generates higher shear, which was associated with better dispersion of pigments. This, in turn, improves the mixture's homogeneity and effectively improves the dispersion of pigments and the quality of the output color. In many cases, mixing resins is required to produce the desired outcomes. However, different resins have different flow rates. The addition of one resin may increase the viscosity of the masterbatch and must go through a melting stage for testing and controlling the quality of the incoming material, which has a substantial impact on color matching. Further research will identify the best processing parameters for various grades and color formulas, resulting in significant waste reduction and faster delivery times for small numbers and prototypes.

Author Contributions: J.A. and R.I. were in charge of the study's design. J.A., R.I. and I.T. performed the statistical analyses, and all authors contributed to the interpretation of the results. The final document, which was drafted by J.A., R.I. and I.T. All authors have read and agreed to the published version of the manuscript.

Funding: This research received no external funding.

Institutional Review Board Statement: Not applicable.

Informed Consent Statement: Not applicable.

Data Availability Statement: Not applicable.

Conflicts of Interest: The authors declare no conflict of interest.

References

1. Charvat, R.A. (Ed.) Color as a Science. In *Coloring of Plastics*; John Wiley & Sons: Hoboken, NJ, USA, 2004.
2. Sabic. Scattering, Absorption/Opacity, Transparency, and Translucency. Internal Scattering. Available online: https://www.sabicip.com/staticcxp/user/images/learnaboutcolor/transparency_and_translucency_1.jpg (accessed on 19 March 2014).
3. Callister, W.D.; Rethwisch, D.G. The spectrum of electromagnetic radiation. In *Materials Science, and Engineering: An Introduction*, 7th ed.; John Wiley & Sons: Hoboken, NJ, USA, 2007.
4. Judd, D.B.; Wyszeki, G. *Color in Business, Science and Industry*, 2nd ed.; John Wiley & Sons: Hoboken, NJ, USA, 1975.
5. Bourennani, F.; Alsadi, J.; Rizvi, G.M.; Ross, D. Manufacturing Processing Improvements Using Business Intelligence. *J. Inf. Technol. Rev.* **2011**, *2*, 125–131.
6. Bourennani, F.; Alsadi, J.; Rizvi, G.M.; Ross, D. Decision Tree Classifier for Analysis of Parameters Association Causing Polymer Color Mismatch. In Proceedings of the Annual Technical Conference of the Society of Plastics Engineers (ANTEC), Orlando, FL, USA, 2–4 April 2012.
7. Bourennani, F.; Rizvi, G.M.; Ross, D. Plastic color mismatch causes identification using OLAP and data mining. In Proceedings of the ICDIM 2010, Thunday-Bay, ON, Canada, 5–8 July 2010.
8. Braha, D.; Shmilovici, A. Data mining for improving a cleaning process in the semi-conductor industry. *IEEE Trans. Semicond. Manuf.* **2012**, *15*, 91–101. [CrossRef]
9. Kuo, C.F.J.; Huang, Y.J.; Su, T.L.; Shih, C.Y. Computerized Color Distinguishing System for Color Printed Fabric by Using the Approach of Probabilistic Neural Network. *Polym-Plast. Technol. Eng.* **2008**, *47*, 264–272. [CrossRef]
10. Saeed UAlsadi, J.; Ahmed, S.; Rizvi, G.; Ross, D. Neural Network: A potential approach for error reduction in color values of polycarbonate. *J. Adv. Polym. Technol.* **2013**, *33*, 21402. [CrossRef]
11. Saeed UAlsadi, J.; Ahmed, S.; Rizvi, G.; Ross, D. Polymer Color Properties: Neural Network Modelling. *Adv. Polym. Technol.* **2014**, *33*, 21462. [CrossRef]
12. Mulholland, B.M. Effect of Additives on the Color & Appearance of Plastics. In *SPE ANTEC*; Ticona Engineering Polymers: Irving, TX, USA, 2007.
13. Markarian, J. North American compounders seek growth in innovation. *Plast. Addit. Compd.* **2006**, *8*, 42–44. [CrossRef]
14. Abrams, R.; Ali, M.; Denton, P.; Igualada, J.; Groen, M.; Gschwind, E. Colouring plastics: Fundamentals and trends. *Plast. Addit. Compd.* **2001**, *3*, 18–25. [CrossRef]
15. Markarian, J. US compounding industry faces challenging times. *Plast. Addit. Compd.* **2008**, *10*, 38–41. [CrossRef]
16. Charvat, R.A. (Ed.) The Color Compounding Process. In *Coloring of Plastics: Fundamentals*, 2nd ed.; John Wiley & Sons: Hoboken, NJ, USA, 2004.
17. Hunterlab. Observer. The Basics of Color Perception and Measurement 2001. Available online: https://www.hunterlab.com/media/documents/basics-of-color-theory.pdf (accessed on 17 February 2015).
18. Rwei, S.P. Distributive Mixing in a Single-Screw Extruder—Evaluation in the Flow Direction. *Polym. Eng. Sci.* **2001**, *41*, 1665–1673. [CrossRef]
19. Kimura, K.; Nakayama, Y.; Kajiwara, T. Distributive Mixing Characteristics Of A Dulmage-Type Screw For A Single-Screw Extruder: Experimental And Numerical Evaluation. *Chem. Eng. J. Adv.* **2021**, *7*, 100137. [CrossRef]
20. Wong, A.Y.; Liu, T. Screw Configuration Effects on the Color Mixing Characteristics of Polymer in Single-Screw Extrusion. In *Technical Papers of the Annual Technical Conference-Society of Plastics Engineers Incorporated*; Society of Plastics Engineers Inc.: Danbury, CT, USA, 1988.
21. Wong, A.Y.; Lam, Y.; Wong, A.C.M. Quantification of dynamic mixing performance of single screws of different configurations by visualization and image analysis. *Adv. Polym. Technol.* **2009**, *28*, 1–15. [CrossRef]
22. Charvat, R.A. (Ed.) Metallic pigments. In *Coloring of Plastics: Fundamentals*, 2nd ed.; John Wiley & Sons: Hoboken, NJ, USA, 2004.
23. Duce, C.; Bernazzani, L.; Bramanti, E.; Spepi, A.; Colombini, M.P.; Tine, M.R. Alkyd artists' paints: Do pigments affect the stability of the resin? A TG and DSC study on fast-drying oil colors. *Polym. Degrad. Stab.* **2014**, *105*, 48–58. [CrossRef]
24. Wong, A.Y.; Lam, Y. Visualization study on the dynamic mixing quality during single-screw extrusion. *J. Polym. Res.* **2008**, *15*, 11–19. [CrossRef]

25. Sanchez, P.; Remiro, P.M.; Nazabal, J. Physical properties and structure of unreacted PC/PBT blends. *J. Appl. Polym. Sci.* **1993**, *50*, 995–1005. [CrossRef]
26. Liang, R.F.; Gupta, R.K. Society of Plastic Engineering. In Proceedings of the ANTEC, Orlando, FL, USA, 7–11 May 2000.
27. Lee, S.; Mather, P.T.; Pearson, D.S. Phase behavior and rheology of blends containing polycarbonate and a thermotropic polyester. *J. Appl. Polym. Sci.* **1996**, *59*, 243–250. [CrossRef]
28. Rauwendaal, C. *Polymer Mixing, A Self-Study Guide*; Hanser Publishers: Munich, Germany, 1998.
29. Alemaskin, K.; Manas-Zloczower, I.; Kaufman, M. Color Mixing in Single Screw Extruder: Simulation & Experimental Validation. In Proceedings of the Annual Technical Conference-ANTEC, Conference Proceedings, Boston, MA, USA, January 2005.
30. Mudalamane, R.; Niedenzu, P.; Sedar, W. Impact of Titanium Dioxide Surface Characteristics on Extrusion Processing. In Proceedings of the SPE ANTEC, Wilmington, DE, USA, 5–6 October 2004.
31. Meade, D.I. Introduction to colorant selection and Application Technology. In *Coloring of Plastics: Fundamentals*, 2nd ed.; Charvat, R.A., Ed.; John Wiley & Sons: Hoboken, NJ, USA, 2004.
32. Müller, A. *Coloring of Plastics*; Carl Hanser Verlag: Munich, Germany, 2003.
33. Gardner, B.K. Instruments Catalogue: Gloss Color. In *Physical Testing*; Gardner: Columbia, MD, USA, 1999.
34. Elliot, A.J.; Maier, M.A.; Moller, A.C.; Friedman, R.; Meinhardt, J. Color and psychological functioning: The effect of red on performance attainment. *J. Exp. Psychol. Gen.* **2007**, *136*, 154–168. [CrossRef]
35. *ASTM D 2244-93*; Standard Test Method for Calculation of Color Differences from Instrumentally Measured Color Coordinates. Aluminum Sun Shade: Reston, VA, USA, 2021.
36. Ahmad SAlsadi, J.; Saeed, U.; Rizvi, G.; Ross, D.; Clarke, R.; Price, J. Process Optimization through Designed Experiments to Achieve Consistency in Output Color of a Compounded Plastic Grade. *Qual. Eng.* **2015**, *27*, 144–160.
37. Alsadi, J.A.; Saeed, U.; Ahmad, S.; Rizvi, G.; Ross, D. Processing issues of color mismatch: Rheological characterization of polycarbonate blends. *Polym. Eng. Sci.* **2015**, *55*, 1994–2001. [CrossRef]
38. Alsadi, J.A. Designing Experiments: Three Level Full Factorial Design and Variation of Processing Parameters Methods for Polymer Colors. *Adv. Sci. Technol. Eng. Syst. J.* **2018**, *3*, 109–115.
39. Alsadi, J.A. Analysis of Material Viscosity Variations: Mix Processing PC1/PC2 Composites. *J. Qual. Inquiry.* **2021**, *12*, 9790–9803.
40. Al Sadi, J.; Hawary, A. The Ideal Temperature Setting of Polymer Blends: Investigational Characterization Effects of Color Matching. *Mater. Sci. Eng.* **2021**, *1194*, 012003. [CrossRef]
41. Alsadi, J.A. Investigation of the effects of Formulation, process parameters, Dispersions, and Rheology on using combined Modelling and experimental Simulations. *Mater. Today* **2019**, *13*, 530–540. [CrossRef]
42. Alsadi, J. Experimental Assessment of Pigment Dispersion in Compounding of Plastics: Rheological Characterization at the Crossover Points. *Mater. Today Proc.* **2021**, *45*, 7344–7351. [CrossRef]
43. Alsadi, J. Study on the effect of dispersion and processing parameters in the microscopically evaluated color of plastic grade. *J. Am. Inst. Phys.* **2019**, *2139*, 110007.
44. Alsadi, J. Systematic review: Impact of Processing Parameters on Dispersion of Polycarbonate Composites, and Pigment Characterized by Different techniques. *Mater. Today Proc.* **2020**, *27*, 3254–3264. [CrossRef]

crystals

MDPI

Communication

Synthesis and Characterization of Mechanically Alloyed Nanostructured (Ti,Cr)C Carbide for Cutting Tools Application

Mohsen Mhadhbi [1] and Wojciech Polkowski [2,*]

[1] Laboratory of Useful Materials, National Institute of Research and Physicochemical Analysis, Technopole Sidi Thabet, Ariana 2020, Tunisia
[2] Łukasiewicz Research Network—Krakow Institute of Technology, 30-418 Kraków, Poland
* Correspondence: wojciech.polkowski@kit.lukasiewicz.gov.pl; Tel.: +48-12-2618-324

Abstract: (Ti,Cr)C is a novel additive for high-performance cermets. In this work, a $(Ti_{0.8}Cr_{0.2})C$ nanostructured solid solution was synthesized via Mechanical Alloying (MA) from the mixture of of Ti, Cr, and C powders. The MA process was carried out at room temperature under argon atmosphere with a duration limited to 20 h. Phase changes and microstructure evolution of the powders during the MA process were characterized by X-ray diffraction (XRD), scanning electron microscopy (SEM), and transmission electron microscopy (TEM) techniques. The results of XRD analysis demonstrated the synthesis of (Ti,Cr)C solid solution with a crystallite size of about 10 nm that were micro-strained to about 1.34%. The crystallite size displays a decreasing trend with increasing milling time. The results of direct observations of structural features by TEM method in 20 h MAed samples shows a good agreement with the results from the XRD analyses.

Keywords: Nano (Ti,Cr)C powder; mechanical alloying (MA); nanostructure; X-ray diffraction

Citation: Mhadhbi, M.; Polkowski, W. Synthesis and Characterization of Mechanically Alloyed Nanostructured (Ti,Cr)C Carbide for Cutting Tools Application. *Crystals* **2022**, *12*, 1280. https://doi.org/10.3390/cryst12091280

Academic Editor: Umberto Prisco

Received: 16 August 2022
Accepted: 7 September 2022
Published: 9 September 2022

Publisher's Note: MDPI stays neutral with regard to jurisdictional claims in published maps and institutional affiliations.

1. Introduction

Titanium carbide (TiC) is widely used in industrial applications as a hard coating to protect the surface of cutting tools from wear and erosion, resulting in an extended tool life [1]. It exhibits high strength, high hardness, good wear resistance, high melting point, high chemical stability, and low friction coefficient [2–5]. In particular, nanosized TiC particles are considered as promising microstructural modifiers and mechanical strengtheners for particle dispersed composite alloys, since fine TiC dispersoids in the metallic matrix improve the overall properties of the materials without an adverse effect on their ductility or toughness [6,7].

So far, several methods have been used to prepare nanostructured carbides including a carbothermal reduction [8], mechanical alloying (MA) [9], spark plasma sintering (SPS) [10], chemical vapor deposition (CVD) [11], etc. Among the aforementioned techniques, the MA is an easy and cost-efficient route usually used to prepare nanostructured materials, and especially for manufacturing composite powders [12].

As described in literature reports, the main idea behind using a (Ti,*M*)C (*M* = transition refractory metal) solid solution instead of TiC is to improve the toughness of cermets. Park and Kang [13] have synthesized nanocrystalline $(Ti_{1-x}W_x)C$ solid solutions, with a homogenous microstructure and improved properties by the SPS process. Kim et al. [14] fabricated homogeneous (Ti,W)C nanocomposite powders by a high-energy ball milling of a mixture of Ti, W, and C powders. The obtained nanopowders were then SPSed to receive fully densified sinters having a uniform microstructure with a mean grain size of 500 nm. Kwon et al. [15] prepared (Ti,V)C solid solution powders by the MA of Ti-V alloy and graphite powder mixture. The MA process was carried out in a high-energy planetary ball mill for up to 20 h under an argon atmosphere. Additionally, Bandyopadhyay et al. [16] investigated the effect of Ti substitution by W on the microstructure of the $Ti_{0.9}W_{0.1}C$ carbide. They reported that TiWC alloy was formed after 50 min of milling and a fully nanocrystalline single phase cubic $Ti_{0.9}W_{0.1}C$ compound with a particle size of 11 nm was formed after 8 h of milling. Hence, the effect of ball milling on microstructural change of (Ti,W)C solid solution was experimentally studied by Yang et al. [17]. They found that with increasing milling time up to 108 h, the initial crystallite size decreased from 38.6 to 19.2 nm. Analogously, Dutta et al. [18] reported the formation of $Ti_{0.9}Al_{0.1}C$ nano-carbide after 3 h of milling. They found that that the results of crystal structure examinations obtained by TEM technique are in a good agreement with those derived from the XRD measurements. Wang et al. [19] also produced a (Ti, Mo)C carbide reinforced Fe-based surface composite coating by the laser cladding technique. It has been concluded that (Ti, Mo)C particles with the FCC structure and various shapes are obtained after the solidification. Recently, Yildiz et al. [20] prepared via the SPS a novel multi-component (Ti,Zr,Hf,W)C ceramic with a nano-hardness of 32.7 GPa and a fracture toughness of 5 MPa $m^{1/2}$. Recently, Vorotilo et al. [21] proposed that a solid solution (Ti,Cr)C, while retaining the main advantages of TiC, possesses higher oxidation resistance owing to the formation of Cr_2O_3. The (Ti,Cr)C cermets have been produced by a combustion synthesis driven by the Self-propagated High Temperature Synthesis (SHS) [21] or by High-Velocity Air Plasma Spraying [22]. However, to our best knowledge, there are no available reports on the MA synthesis of this type of solid solution-based cermet. Therefore, by taking into account a documented feasibility of the SPS process in a fabrication of high performance cermets, as well as a high impact of the batch powders on the resulted properties of final sinters; the main goal of our this work is to obtain a $(Ti_{0.8}C_{0.2})C$ composite powder with a suitable particle size that can be employed in the SPS process. Specifically, we are showing for the first time the results of systematic studies on the effect of milling time on the structural evolution of a novel mechanically alloyed $(Ti_{0.8}Cr_{0.2})C$ powders.

2. Materials and Methods

A mixture of elemental Ti (<40 μm, 99.9%, Prolabo, Bankgkok, Thailand), carbon (5 μm, 99.9%, Fischer Scientific, Waltham, MA, USA) and chromium (100–300 μm, 99.9%, ACROS, Boston, MA, USA) powders was sealed into a stainless-steel vial (having a volume of 45 mL) with 5 stainless steel balls (15 mm in diameter and 14 g in mass each) in a glove box filled with purified argon to prevent oxidation. The ball to powder weight ratio was 70:1. The mechanical alloying (MA) procedure of up to 20 h was performed at room temperature using a high-energy ball mill (Fritsch Pulverisette 7 planetary mill, Weimar, Germany), as shown in Figure 1.

Figure 1. P7 planetary ball mill and milling media used in the experiments.

The crystalline properties of the powders were characterized by the XRD method, by using a Panalytical XPERT PRO MPD (Cambridge, UK) diffractometer with CuK_α radiation. The existing phases were determined by the High Score Plus program based on the ICDD PDF2 data base, while a crystallite size was determined from acquired diffraction data by the FullProf program (Saclay, France) [23] employing the Rietveld method [24]. The morphology of powders was investigated by scanning electron microscopy (SEM, JEOL JSM-5400 model, Tokyo, Japan) coupled with energy dispersive X-ray analysis (EDX). The transmission electron microscopy (TEM) was used for the investigation of internal nanostructure of powders. The structural analysis was complemented by using FEI Tecnai G2 high-resolution TEM (Massachusetts, USA) operated at 200 kV with a spatial resolution of 1.9 Angstrom.

3. Results and Discussion

3.1. XRD Characterization

Figure 2a shows the XRD patterns of samples prepared by the MA after various milling times. The acquired XRD pattern of the un-milled powder mixture consists of Bragg diffraction peaks of Ti, Cr, and C. After 1 h 30 min of MA, all the carbon peaks disappeared totally, while the appearance of $TiCr_2$ peak and the reflections of Ti and Cr became broad without any shift in peak positions. Consequently, it is considered that the high-energy impact leads to the transformation of crystalline graphite into amorphous carbon (which is also supported by the literature reports [25,26]). After 3 h of MA, a disappearance of Cr peaks and appearance of (110) peak related to austenite iron (α-Fe) contamination resulted from the abrasive wear between vial and balls, was observed. When the MA time was extended to 5 h, the Ti peaks disappeared completely and the diffraction peaks were broadened and shifted into lower angles as a result of the formation of (Ti,Cr)C solid solution due to refinement of the grain size and the introduction of internal stresses. An analogous behavior was also reported in [27–29]. After continuing the MA for 10 h, the (Ti,Cr)C peaks broadened further due to the decrease in the crystallite sizes and an accumulation of lattice strains caused by fracturing and cold welding of the powder particles during the milling process. Between 10 and 20 h of MA, the structure remains almost unchanged: the (Ti,Cr)C solid solution maintains the NaCl type structure with a space group of Fm-3m. These results are similar to those reported by Shon et al. [30]. They have revealed that the average crystallite size of (Ti,Cr)C powders, obtained from the mixture of Ti-Cr alloy and graphite, MA for 20 h was about 34 nm. Figure 2b presents the Rietveld refinement plot of TiCrC powder after 20 h of the MA process. The calculated profile is represented by the black line, the observed profile by the red dots, and the difference pattern by the blue continuous line. The vertical lines (green) represent the positions of all possible Bragg reflections. The figure indicates that Cr substitutes Ti in the TiC matrix and consequently lead to a decrease in lattice parameter.

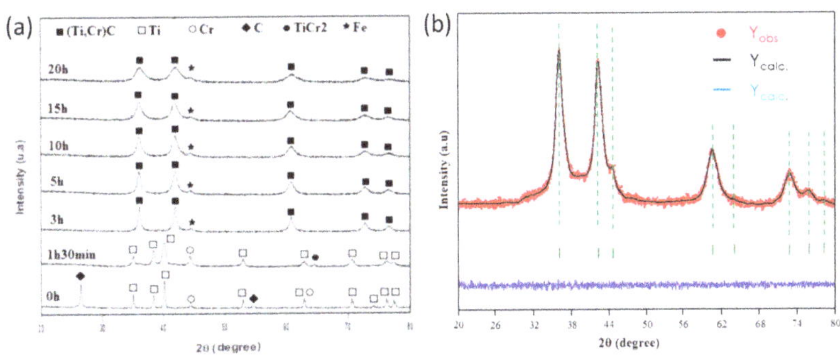

Figure 2. XRD patterns of samples prepared by MA for different milling times (**a**). Rietveld refinement plot of the sample milled for 20 h (**b**).

The quantitative analysis of microstructural parameters derived from Rietveld refinement of XRD patterns taken from samples MAed for different times, are illustrated in Table 1. It is found that the average crystallite size of (Ti, Cr)C decreases with increasing milling time, from 18 nm after 3 h of the MA and remain nearly constant (about 10 nm) for 15 and 20 h of MA. Thus, all identified phases, namely Ti, Cr, $TiCr_2$, (Ti,Cr)C and Fe, present the same behavior. This decrease can be explained by the refinement of structure and the creation of a large number of crystal defects, such as dislocations and stacking faults [31]. These findings are in line with reports by Zhu et al. [32] who have also connected a decrease in crystallite size with increased dislocation density; reflected as a broadening of XRD peaks. From these observations, it is evident that the reduced particle size plays significant role in inclusion of Cr atoms in TiC matrix. As it is listed in Table 1, the values of mean microstrains are higher for the (Ti, Cr)C phase as compared to other phases. Moreover, the average microstrain of (Ti, Cr)C increases with the MA time, from 0.25 ± 0.01% after 3 h of MA to reach 1.34 ± 0.04% after 20 h of MA. The increase of the microstrain with milling time is generated by the increase of dislocation density and grain boundary fraction, which due to severe plastic deformation generated from the high-energy induced during milling [33]. Furthermore, the calculated lattice parameter of (Ti, Cr)C, obtained from XRD data, decreased from 0.44169 to 0.43036 nm for the powder samples milled for 3 and 20 h, respectively. The value is consistent with that reported by Kwon et al. (a = 0.43075 nm) [34]. This suggest that the lattice parameter was decreased by the substitution of Cr because the atomic radius of Cr (166 pm) is smaller than that of Ti (176 pm) [35]. The decrease in lattice parameter can be explained by the substitution of Ti by Cr and, therefore, by the formation of the $Ti_{0.8}Cr_{0.2}C$ carbide. This finding is consistent with the reported literature for other similar compounds, like TiC_x, Cu-TiC_x, and $Cr_{2-x}M_xS_3$ (M = Ti, V, Sn) [36,37].

Table 1. The results of quantitative analysis of microstructural parameters derived from Rietveld refinement for XRD patterns of samples MA for different times.

MA Time (h)	Phase	Space Group	Lattice Parameter (nm)	Crystallite Size (nm)	Microstrain (%)
1.5	Ti Cr $TiCr_2$	Fm-3m Im-3m $P6_3/mmc$	a = 0.43281(4) a = 0.29236(1) a = 0.69011(3)	55 40 31	0.25 ± 0.01 0.94 ± 0.01 0.15 ± 0.01
3	(Ti,Cr)C Fe	Fm-3m Im-3m	a = 0.44169(5) a = 0.2873(5)	18 17	0.12 ± 0.02 0.21± 0.01
5	(Ti,Cr)C Fe	Fm-3m Im-3m	a = 0.44033(4) a = 0.2874(3)	15 15	0.25 ± 0.03 0.22 ± 0.01
10	(Ti,Cr)C Fe	Fm-3m Im-3m	a = 0.44949(4) a = 0.2876(2)	13 13	1.28 ± 0.04 0.23 ± 0.01
15	(Ti,Cr)C Fe	Fm-3m Im-3m	a = 0.43033(2) a = 0.2877(5)	11 11	1.33 ± 0.04 0.25 ± 0.01
20	(Ti,Cr)C Fe	Fm-3m Im-3m	a = 0.43036(3) a = 0.2879(4)	10 11	1.34 ± 0.04 0.26 ± 0.01

3.2. SEM Characterization

Figure 3 shows SEM micrographs of the $Ti_{0.8}Cr_{0.2}C$ powders after the MA for 5, 10, and 20 h. After 5 h of the MA, the grains are formed by an assembly of large particles exhibiting an irregular shape and having an average particle size of about 1–2 μm (Figure 3a). When the MA time was increased to 10 h, we observed significant particle refinement. The initial powders were transformed into fine particles with an average particle size of about 0.4 μm (Figure 3b). The MA up to 20 h results in a further particle size reduction and a formation of homogenous powders with fine and agglomerated particles having the average particle size less than 1 μm (Figure 3c). It is well accepted that the MA process leads to continuous cold welding, fracturing, and rewelding of powder particles [35]. Figure 3d shows an exemplary EDX spectrum acquired from the powder milled for 20 h. The results indicate a presence of major elements (i.e., Ti, Cr, and C). In addition, the results of EDX analyzes revealed trace amounts of Fe, probably coming from the applied milling materials of balls and vial. Furthermore, the results of EDX analyses indicate that the carbide contains 47.59 C, 40.49 Ti, 9.95 Cr, and 1.97 Fe (at.%). Thus, the alloying ratio is Ti:Cr:C = 0.82:0.201:1, which indicates that carbides may be $Ti_{0.8}Cr_{0.2}C$.

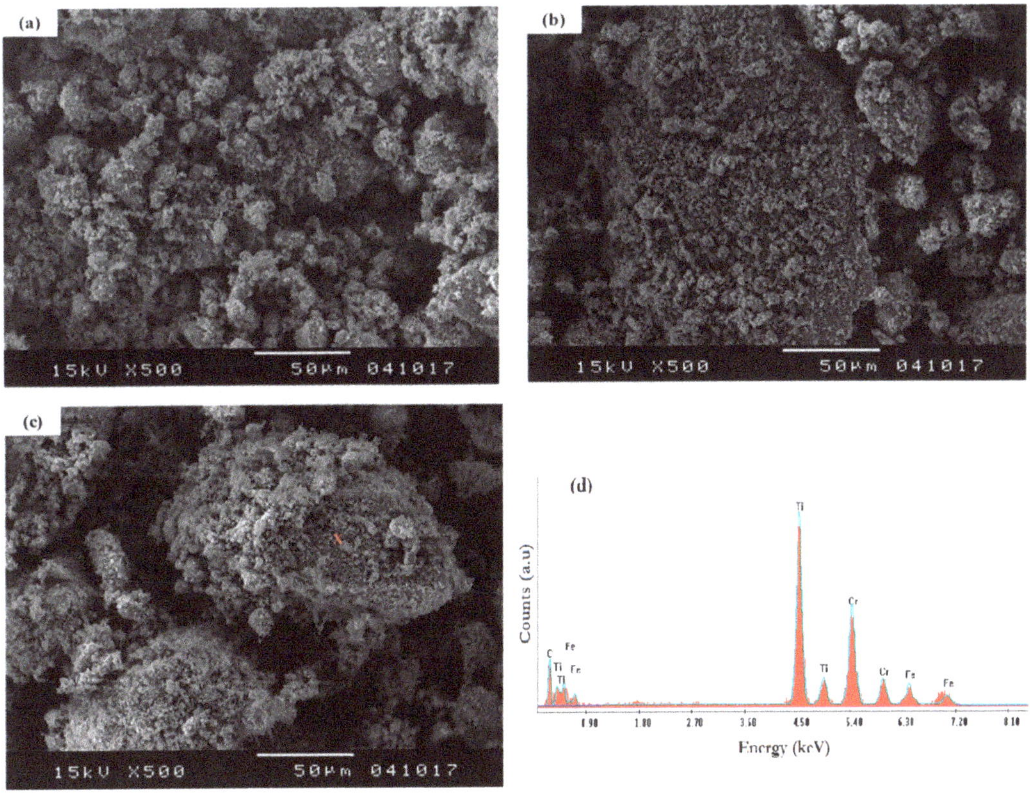

Figure 3. SEM micrographs of samples after MA for 5 h (**a**); 10 h (**b**); and 20 h (**c**). (**d**) An exemplary EDX spectrum taken form the sample milled for 20 h (zone marked with X).

3.3. TEM Characterization

Figure 4 shows a TEM micrograph and selected area electron diffraction (SAED) pattern of the sample after the MA for 20 h. Figure 4a reveals that the particle powders were irregular in shape, while the average size was refined to about 15 nm and the agglomeration was high. Therefore, the figure reveals the existence of many grains (i.e., the polycrystalline structure) in nanoscale powder particles. This result was in a good agreement with those obtained by the XRD method. Figure 4b shows the indexed SAED showing only the presence of cubic (Ti,Cr)C solid solution phase in the 20 h milled sample.

To summarize: based on our original experimental findings coming from the pioneering systematic studies, we confirm that a nanostructured (Ti,Cr)C solid solution having an adequate microstructural features and morphology to be applied in the SPS process, can be successfully obtained after 20 h of MA of Ti, Cr, and C elemental powders. The as-obtained nanopowders can be further consolidated through SPS in order to obtain a dense bulk (Ti,Cr)C nanocarbide with enhanced properties compared to TiC carbide. Thus, three parameters in terms of crystallite size, microstrain, and lattice parameter are obtained from XRD line profiles. The values are typical to these of batch powders for the SPS process [36]. Thus, the milling time should be carefully selected taking to account the particle size and shape of the prepared powders. It can be observed that both crystallite size and lattice parameter decrease with increasing milling time, whereas the microstrain increases. These results are in a good agreement with the reference work [38]. As shown in the SEM images, the morphological evolution during the MA process includes three typical steps for a mechanical synthesis of multicomponent intermetallic/interstitial compounds, i.e., (i) a cold welding, (ii) a fracturing, and (iii) a rewelding between particles for longer milling times (between 15 and 20 h). Both applied diffraction-based techniques (TEM/SAED and XRD) confirmed the morphology and structure of MAed powders.

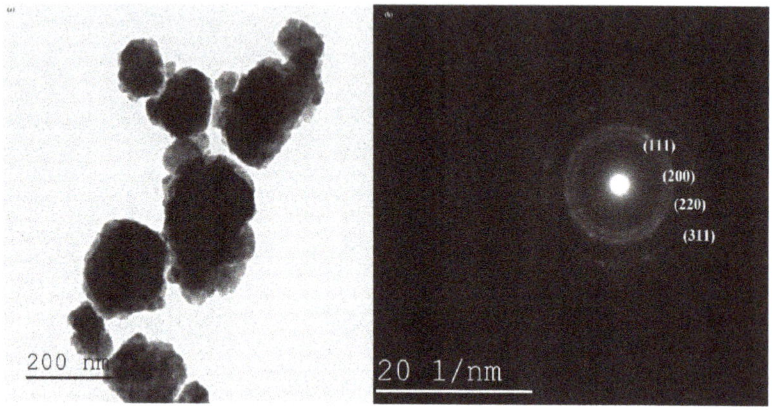

Figure 4. (**a**) HRTEM image and (**b**) SAED patterns of the powder after MA for 20 h.

4. Conclusions

Based on the results of the present study supported by a proper literature data, the following conclusions can be drawn:

1. Nanocrystalline solid solution (Ti,Cr)C with nano-sized crystallites was fabricated by mechanical alloying of Ti, Cr, and graphite elemental powders. The average grain size of the (Ti,Cr)C was 10 nm, and the average micro-strain was 1.34% for sample MA during 20 h.
2. The (Ti,Cr)C solid solution having a NaCl type structure and a space group of Fm-3m, was formed after 3 h of the MA process.
3. The crystallite size and the lattice parameter decrease by increasing milling time down to 10 nm and 0.43036 nm after 20 h of MA, respectively. This decrease can be attributed to the formation of crystal defects induced by high-energy ball milling. The microstrain increases with increasing milling time, which can be related to the strain-enhanced solubility of Cr in TiC matrix.
4. The crystallite size obtained from XRD is well correlated with the results of TEM analyses.

Author Contributions: M.M. have designed and performed the experiments. M.M. and W.P. have discussed and analyzed the data. M.M has written the draft version of manuscript. W.P. has edited the draft and prepared the final version of manuscript. All authors have read and agreed to the published version of the manuscript.

Funding: This research received no external funding.

Data Availability Statement: Raw experimental data are available on demand from M.M.

Acknowledgments: The authors would like to thank to all who have contributed to this work.

Conflicts of Interest: The authors declare no conflict of interest.

References

1. He, X.M.; Zhi, W.; Li, H.D. Investigation of TiC Films Synthesized by Low Energy Ion Bombardment. *J. Mater. Res.* **1994**, *9*, 2355–2361. [CrossRef]
2. Dong, Z.J.; Li, X.K.; Yuan, G.M.; Cui, Z.W.; Cong, Y.; Westwood, A. Tensile Strength, Oxidation Resistance and Wettability of Carbon Fibers Coated with a TiC Layer Using a Molten Salt Method. *Mater. Des.* **2013**, *50*, 156–164. [CrossRef]
3. Boonruang, C.; Thongtem, S. Fast Processing Technique for TiC Coatings on Titanium. *Chiang Mai J. Sci.* **2010**, *37*, 206–212.
4. Brama, M.; Rhodes, N.; Hunt, J.; Ricci, A.; Teghil, R.; Migliaccio, S.; Rocca, C.D.; Leccisotti, S.; Lioi, A.; Scandurra, M.; et al. Effect of Titanium Carbide Coating on the Osseointegration Response in Vitro and in Vivo. *Biomaterials* **2007**, *28*, 595–608. [CrossRef]
5. Saba, F.; Raygan, S.; Abdizadeh, H.; Dolatmoradi, A. Preparing TiC Coating on AISI D2 Steel Using Mechanical Milling Technique. *Powder Technol.* **2013**, *246*, 229–234. [CrossRef]
6. Lazarova, R.; Petrov, R.H.; Gaydarova, V.; Alexeev, A.; Manchev, M.; Manolov, V. Microstructure and Mechanical Properties of P265GH Cast Steel after Modification with TiCN Particles. *Mater. Des.* **2011**, *32*, 2734–2741. [CrossRef]
7. Han, Y.; Shi, J.; Xu, L.; Cao, W.Q.; Dong, H. TiC Precipitation Induced Effect on Microstructure and Mechanical Properties in Low Carbon Medium Manganese Steel. *Mater. Sci. Eng. A* **2011**, *530*, 643–651. [CrossRef]
8. Jung, J.; Kang, S. Sintered (Ti,W)C Carbides. *Scr. Mater.* **2007**, *56*, 561–564. [CrossRef]
9. Mhadhbi, M. Structural and Morphological Studies of Nanostructured TiC Powders Prepared by Mechanical Alloying. *Mod. App. Mater. Sci.* **2021**, *3*, 445–450. [CrossRef]
10. Liu, S.; Hu, W.; Xiang, J.; Wen, F.; Xu, B.; Yu, D.; He, J.; Tian, Y.; Liu, Z. Mechanical Properties of Nanocrystalline TiC-ZrC Solid Solutions Fabricated by Spark Plasma Sintering. *Ceram. Int.* **2014**, *40*, 10517–10522. [CrossRef]
11. Hayashi, M.; Mamyouda, T.; Habuka, H.; Ishiguro, A.; Ishii, S.; Daigo, Y.; Ito, H.; Mizushima, I.; Takahashi, Y. Design of a Silicon Carbide Chemical Vapor Deposition Reactor Cleaning Process Using Chlorine Trifluoride Gas Accounting for Exothermic Reaction Heat. *ECS J. Solid State Sci. Technol.* **2020**, *9*, 104008. [CrossRef]
12. Suryanarayana, C. Mechanical Alloying and Milling. *Prog. Mater. Sci.* **2001**, *46*, 1–184. [CrossRef]
13. Park, S.; Kang, S. Toughened Ultra-fine (Ti,W)(CN)-Ni Cermets. *Scr. Mater.* **2005**, *52*, 129–133. [CrossRef]
14. Kim, Y.K.; Shim, J.H.; Yang, H.S.; Park, J.K. Mechanochemical Synthesis of Nanocomposite Powder for Ultrafine (Ti, Mo)C-Ni Cermet Without Core-rim Structure. *Int. Ref. Met. Hard. Mater.* **2004**, *22*, 193–196. [CrossRef]
15. Kwon, H.; Jung, S.A.; Suh, C.Y.; Roh, K.M.; Kim, W. Mechanical Properties of (Ti,V)C-Ni Composite Prepared Using Ultrafine Solid-Solution (Ti,V)C Phase. *Ceram. Int.* **2014**, *40*, 12579–12583. [CrossRef]
16. Bandyopadhyay, S.; Dutta, H.; Pradhan, S.K. XRD and HRTEM Characterization of Mechanosynthesized $Ti_{0.9}W_{0.1}C$ Cermet. *J. Alloys Compd.* **2013**, *581*, 710–716. [CrossRef]
17. Yang, M.; Guo, Z.; Xiong, J.; Liu, F.; Qi, K. Microstructural Changes of (Ti,W)C Solid Solution Induced by Ball Milling. *Int. Refract. Met. Hard Mater.* **2017**, *66*, 83–87. [CrossRef]
18. Dutta, H.; Sen, A.; Pradhan, S.K. Microstructure Characterization of Ball-mill Prepared Ternary $Ti_{0.9}Al_{0.1}C$ by X-ray Diffraction and Electron Microscopy. *J. Alloys Compd.* **2010**, *501*, 198–203. [CrossRef]
19. Wang, X.; Zhang, M.; Qu, S. Development and Characterization of (Ti, Mo)C Carbides Reinforced Fe-based Surface Composite Coating Produced by Laser Cladding. *Opt. Lasers Eng.* **2010**, *48*, 893–898. [CrossRef]
20. Yildiz, A.B.; Yixuan, H.; Babu, R.P.; Hansen, T.C.; Eriksson, M.; Reddy, K.M.; Hedström, P. Design, Synthesis, Structure, and Stability of Novel Multi-Principal Element (Ti,Zr,Hf,W)C Ceramic with a Miscibility Gap. *J. Eur. Ceram. Soc.* **2022**, *42*, 4429–4435. [CrossRef]
21. Vorotilo, S.; Kiryukhantsev-Korneev, P.V.; Seplyarskii, B.S.; Kochetkov, R.A.; Abzalov, N.I.; Kovalev, I.D.; Lisina, T.G.; Zaitsev, A.A. (Ti,Cr)C-Based Cermets with Varied Nicr Binder Content via Elemental SHS for Perspective Cutting Tools. *Crystals* **2020**, *10*, 412. [CrossRef]
22. Borisov, Y.S.; Borisova, A.L.; Kolomytsev, M.V.; Masyuchok, O.P.; Timofeeva, I.I.; Vasilkovskaya, M.A. High-Velocity Air Plasma Spraying of (Ti, Cr)C–32 wt.% Ni Clad Powder. *Powder Metall. Met. Ceram.* **2005**, *56*, 305–315. [CrossRef]
23. Carvajal, R. Recent Advances in Magnetic Structure Determination by Neutron Powder Diffraction. *J. Phys. B* **1993**, *192*, 55–69. [CrossRef]
24. Rietveld, H.M. A Profile Refinement Method for Nuclear and Magnetic Structures. *J. Appl. Crystallogr.* **1969**, *2*, 65–71. [CrossRef]
25. Loshe, B.H.; Calka, A.; Wexler, D. Synthesis of TiC by Controlled Ball Milling of Titanium and Carbon. *J. Mater. Sci.* **2007**, *42*, 669–675. [CrossRef]
26. Arao, Y.; Tanks, J.D.; Aida, K.; Kubouchi, M. Exfoliation Behavior of Large Anionic Graphite Flakes in Liquid Produced by Salt-Assisted Ball Milling. *Processes* **2020**, *8*, 28. [CrossRef]
27. Hübler, D.; Gradt, T. Effect of Different Binders and Secondary Carbides on NbC Cermets. *Forsch. Ingenieurwes.* **2022**, *86*, 197–211. [CrossRef]
28. Cheng, Y.N.; Nie, W.Y.; Guan, R.; Jia, W.K.; Yan, F.G. Study on Damage Behavior of Carbide Tool for Milling Difficult-to-Machine Material Show Less. *Proc. Inst. Mech. Eng. Part C J. Mech. Eng. Sci.* **2018**, *233*, 735–747. [CrossRef]
29. Shon, I.J.; Oh, H.S.; Lim, J.W.; Kwon, H. Mechanical Properties and Consolidation of Binderless Nanostructured (Ti,Cr)C from Mechanochemically-Synthesized Powder by High-Frequency Induction Heating Sintering. *Ceram. Int.* **2013**, *39*, 9721–9726. [CrossRef]

30. Mohammed, S.M.A.K.; Chen, D.L. Carbon Nanotube-Reinforced Aluminum Matrix Composites. *Adv. Eng. Mater.* **2020**, *22*, 12–27. [CrossRef]
31. Zhu, Y.T.; Huang, J.Y.; Gubicza, J.; Ungar, T.; Wang, Y.M.; Ma, E.; Valiev, R.Z. Nanostructures in Ti Processed by Severe Plastic Deformation. *J. Mater. Res.* **2003**, *18*, 1908–1917. [CrossRef]
32. Mhadhbi, M.; Khitouni, M.; Escoda, L.; Suñol, J.J.; Dammak, M. Characterization of Mechanically Alloyed Nanocrystalline Fe(Al): Crystallite Size and Dislocation Density. *J. Nanometer.* **2010**, *2010*, 712407. [CrossRef]
33. Slater, J.C. Atomic Radii in Crystals. *J. Chem. Phys.* **1964**, *41*, 3199–3204. [CrossRef]
34. Zarrinfar, N.; Shipway, P.H.; Kennedy, A.R.; Saidi, A. Carbide Stoichiometry in TiC_x and Cu-TiC_x Produced by Self-Propagating High-Temperature Synthesis. *Scr. Mater.* **2002**, *46*, 121–126. [CrossRef]
35. Groß, H.; Ekici, Y.; Poschmann, M.; Croeneveld, D.; Dankwort, T.; Koenig, J.D.; Bensch, W.; Kienle, L. Does a Low Amount of Substituents Improve the Thermoelectric Properties of $Cr_{2-x}M_xS_3$ (M = Ti, V, Sn)? *J. Electron. Mater.* **2022**, *51*, 3510–3520. [CrossRef]
36. Suryanarayana, C. Mechanical Alloying: A Novel Technique to Synthesize Advanced Materials. *Research* **2019**, *2019*, 4219812. [CrossRef]
37. Kwon, H.; Jung, S.A.; Kim, W. (Ti,Cr)C Synthesized In Situ by Spark Plasma Sintering of TiC/Cr_3C_2 Powder Mixtures. *Mater. Trans.* **2015**, *56*, 264–268. [CrossRef]
38. Pradhan, B.; Ghosh, S.; Roy, D.; Samanta, L.K. Preparation of Nanocrystalline $CuAlFeS_2$-Mixed Chalcopyrite by High-energy Ball Milling. *Physica* **2006**, *33*, 66–68. [CrossRef]

crystals

MDPI

Communication

Microstructure and Superelastic Properties of FeNiCoAlTi Single Crystals with the <100> Orientation under Tension

Li-Wei Tseng [1,*], Chih-Hsuan Chen [2], Yu-Chih Tzeng [3], Po-Yu Lee [1], Nian-Hu Lu [2] and Yury Chumlyakov [4]

1 Department of Mechatronics Engineering, National Changhua University of Education, Changhua 50007, Taiwan; d0951003@gm.ncue.edu.tw
2 Department of Mechanical Engineering, National Taiwan University, Taipei 10617, Taiwan; chchen23@ntu.edu.tw (C.-H.C.); f06522712@ntu.edu.tw (N.-H.L.)
3 Department of Power Vehicle and Systems Engineering, Chung Cheng Institute of Technology, National Defense University, Taoyuan City 33550, Taiwan; s933003@ccit.ndu.edu.tw
4 Department of Physics, Siberian Physical Technical Institute, Tomsk State University, 634050 Tomsk, Russia; y_chymlyakov@hotmail.com
* Correspondence: lwtseng@cc.ncue.edu.tw

Abstract: The microstructure and superelastic response of an $Fe_{41}Ni_{28}Co_{17}Al_{11.5}Ti_{2.5}$ (at.%) single crystal along the <100> orientation was investigated under tension at room temperature after aging at 600 °C for 24 h. From the superelastic results, the samples aged at 600 °C for 24 h exhibited 4.5% recoverable strain at room temperature. The digital image correlation (DIC) method was used to observe the strain distribution during the 6.5% applied strain loading. The DIC results showed that the strain was uniformly distributed during the loading and unloading cycles. Only one martensite variant was observed from the DIC results. This was related to the aging heat treatment times. The martensite morphology became a single variant with a longer aging time. The thermo-magnetization results indicated that the phase transformation and temperature hysteresis was around 36 °C. Increasing the magnetic field from 0.05 to 7 Tesla, the transformation temperatures increased. The maximum magnetization was 160 emu/g under the magnetic field of 7 Tesla. From the transmission electron microscopy results, the $L1_2$ precipitates were around 10 nm in size, and they were high in Ni content and low in Fe content.

Keywords: superelasticity; FeNiCoAlTi; shape memory alloys

Citation: Tseng, L.-W.; Chen, C.-H.; Tzeng, Y.-C.; Lee, P.-Y.; Lu, N.-H.; Chumlyakov, Y. Microstructure and Superelastic Properties of FeNiCoAlTi Single Crystals with the <100> Orientation under Tension. *Crystals* **2022**, *12*, 548. https://doi.org/10.3390/cryst12040548

Academic Editor: Wojciech Polkowski

Received: 12 March 2022
Accepted: 11 April 2022
Published: 14 April 2022

1. Introduction

Shape memory alloys (SMAs) possess two unique properties: the shape memory effect and superelasticity. Commercial Nickel–Titanium (NiTi) SMAs have a high cost in terms of material and difficulty to manufacture, which limits the wide application of these alloys. In contrast to NiTi alloys, iron-based SMAs have a low material cost and good workability, and they have drawn attention and interest in both industry and academia.

Recently, FeNiCoAlXB (X: Ta, Nb, Ti) alloys have been reported to have greater than 4% superelastic strain [1–3] and 1.6% shape memory strain [4], and single crystals have exhibited at least a 5% recoverable strain [5–8]. The martensitic transformation (MT) of the FeNiCoAlXB (X: Ta, Nb, Ti) alloy system is face-centered cubic (fcc) (austenite) to body-centered tetragonal (bct) (martensite). Iron-based SMAs are required to undergo aging heat treatment to introduce the $L1_2$ precipitates into the austenite matrix. The purposes of $L1_2$ precipitates are to strengthen the austenite matrix, change iron-based SMAs from non-thermo-elastic to thermo-elastic transformation, and assist martensitic transformation during the superelastic tests [1].

Previous aging heat-treatment studies of FeNiCoAlXB (X: Ta, Nb, Ti) alloys have reported that the optical aging temperatures of these alloys, to obtain superelastic properties, ranged from 600 to 700 °C [1–8]. Tanaka et al. [1] found that $Fe_{40.95}Ni_{28}Co_{17}Al_{11.5}Ta_{2.5}B_{0.05}$

(at.%) alloys aged at 600 °C for 72 h show a 13.5% superelastic strain at room temperature, and the diameter of the precipitate was 3 nm. Ma et al. [5] reported that 3.5% superelastic strain at 0 °C under tension was achieved in $Fe_{41}Ni_{28}Co_{17}Al_{11.5}Ta_{2.5}$ (at.%) single crystals with the <100> orientation aged at 600 °C for 90 h. The precipitate size was around 3–5 nm. In addition, FeNiCoAlTa single crystals aged, for the first stage of aging, at 700 °C for 0.5 h and then at 700 °C for 3 h, exhibiting 15% superelastic strain at −196 °C [9]. The size of precipitates was 2–6 nm.

Tseng et al. [6] found that replacing Ta with Ti could reduce the aging heat treatment times. FeNiCoAlTi single crystals with <100> orientation aged at 600 °C for 4 h possessed 6% superelastic strain at −80 °C. The precipitate size was around 5 nm. Chumlyakov et al. [7] used a two-stage aging heat treatment (600 °C for 4 h and 600 °C for 2 h) and achieved a 3% superelastic strain at −100 °C. In previous studies, good superelastic properties could only be obtained at low temperatures. When the test temperature was at room temperature, the superelastic strain was deteriorated due to very low transformation temperatures [5,6,9]. As a result, the motivation of this study was to obtain the aging condition to show good superelastic properties at room temperature. In this study, $Fe_{41}Ni_{28}Co_{17}Al_{11.5}Ti_{2.5}$ (at.%) single crystals, with the <100> orientation after aging heat treatment, were investigated to determine their microstructure, thermo-magnetization, and superelastic properties under tension.

2. Materials and Methods

$Fe_{41}Ni_{28}Co_{17}Al_{11.5}Ti_{2.5}$ (at.%) single crystals were grown using the Bridgman technique. Tensile samples with gauge dimensions of 1.5 mm × 3 mm × 8 mm were cut by wire electro-discharge machining (EDM) from single crystal alloys with the tension axis parallel to the <100> crystal orientation. The single crystals were first sealed in a quartz tube under an argon atmosphere. The single crystals were solution heat treated at 1300 °C for 24 h to homogenize the sample and then quenched in water. Subsequently, the homogenized samples were aged at 600 °C for 24 h to allow precipitation.

The crystal structure and size of the precipitates were analyzed by transmission electron microscopy (TEM) at room temperature. TEM samples were prepared using a focused ion beam (FIB). TEM observations were conducted with a JEOL JEM-F200 electron microscope. The composition of the precipitates was analyzed by energy dispersive spectroscopy (EDS). The martensitic transformation temperatures were measured with a Superconducting Quantum Interference Device (SQUID) magnetometer under magnetic fields of 0.05 and 7 Tesla (T). The heating and cooling rates for this experiment were 5 °C/min. The magnetic results were used to determine the thermo-martensitic transformation temperatures for calculation of the temperature hysteresis. A sample was first heated to 120 °C under zero magnetic field. Then, it was cooled to 260 °C and subsequently heated to around 120 °C under a constant magnetic field of 0.05 T. After the 0.05 T, the magnetic field was increased to 7 T. The sample was then cooled from 120 °C to −260 °C and heated to 120 °C again to complete another cycle.

The superelastic responses of the aged single crystals were characterized by a superelastic experiment under tension. A tensile test was performed with a universal tensile testing machine (AG-IS 50KN, Shimadzu, Kyoto, Japan). The tensile test was conducted in strain-controlled mode with a strain rate of $2 \times 10^{-4}\ s^{-1}$. The sample was cyclic deformed to reach different successive target strains. A virtual optical extensometer (Vic-Gauge 2D, Correlated Solutions, Irmo, SC, USA) tracked the in situ strain during the tensile cycles. A single crystal was tested at room temperature in an incremental superelastic experiment. In this experiment, the sample was loaded to a 0.5% strain and then unloaded. This process was then repeated at increasingly higher strain levels until the sample fractured. A speckle pattern was applied on the sample surface for digital image correlation (DIC) analyses. During the superelastic experiments, the deformation of the sample was recorded with a Complementary Metal-Oxide-Semiconductor (CMOS) at a frame rate of 5 Hz. The strain distribution of the sample during the superelastic cycle was analyzed in commercial VIC-2D software. The optical microscope of the fractured tensile sample was observed by using an Olympus digital optical microscope. The sample was etched with 93% C_2H_5OH + 7% HNO_3 solution for 1 min.

3. Results and Discussion

3.1. TEM Results

Figure 1a presents a room temperature bright field TEM image of a $Fe_{41}Ni_{28}Co_{17}Al_{11.5}Ti_{2.5}$ single crystal with <100> orientation aged at 600 °C for 24 h. The high density of precipitates in this aging condition can clearly be observed. The TEM result shows the high density of precipitates. Figure 1b shows a high-resolution TEM image and corresponding selected-area electron diffraction (SAED) pattern of a sample aged at 600 °C for 24 h. In this sample, the diameters of the precipitates were around 8–10 nm. Insert Figure 1b shows the diffraction pattern of austenite and precipitates. The major diffraction spots are from the austenite matrix and are identified in the fcc phase. The spot with the red circle is the diffraction pattern of a precipitate. The diffraction pattern shows that the precipitates had the $L1_2$ structure. The diameters of the precipitates were 5 nm for a sample aged at 600 °C for 4 h [6]. The results indicate that the size of the precipitates increased from 5 to 10 nm when the aging times were increased from 4 to 24 h. The results of the high-magnification STEM examination combined with EDS line scanning for the spherical precipitates are shown in Figure 1c. Its composition via EDS analysis is also tabulated in Table 1. It is found that the main element of the composition for the spherical precipitates were enriched in Ni and Al and had lower Fe content. Therefore, it can be inferred that the spherical precipitation is in the Ni-rich precipitates phase. Such Ni-rich precipitates have been observed in several alloys systems such as FeNiCoAlTa and FeMnAlNi [5,10].

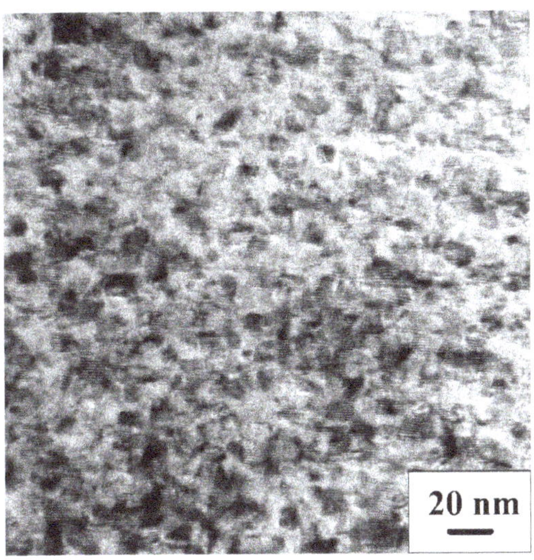

(a)

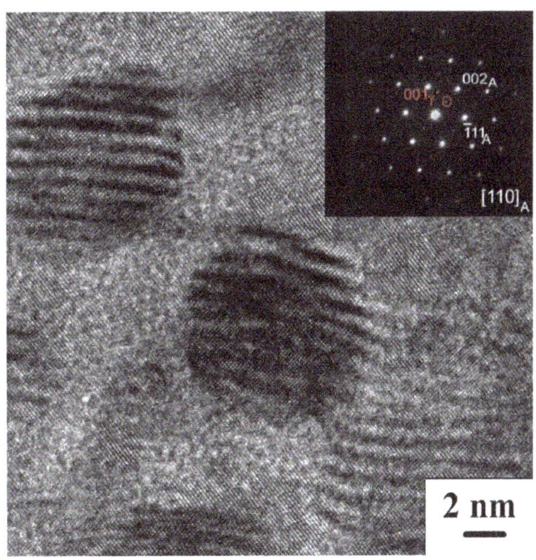

(b)

Figure 1. *Cont.*

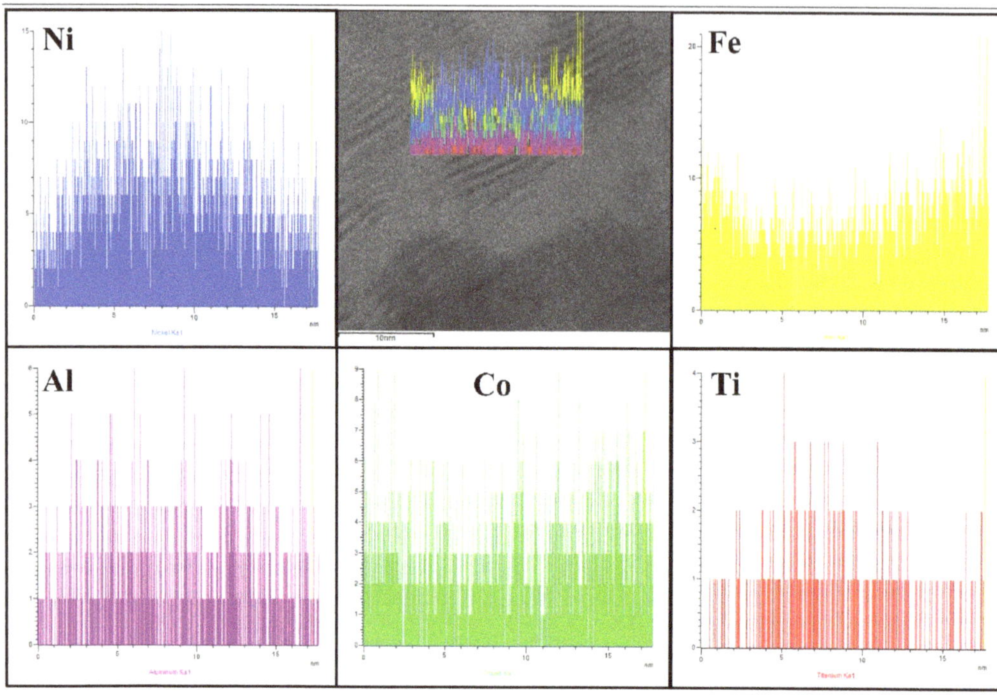

(c)

Figure 1. TEM images of the precipitates in a FeNiCoAlTi single crystal aged at 600 °C for 24 h: (**a**) bright field TEM image, (**b**) high resolution TEM image of the precipitates and the corresponding selected-area electron diffraction pattern, and (**c**) TEM mapping results.

Table 1. Composition of the precipitates found by EDS.

	Al (at.%)	Ti (at.%)	Fe (at.%)	Co (at.%)	Ni (at.%)
Measured	12.5	2.67	37.1	16.57	31.16
Nominal	11.5	2.5	41	17	28

3.2. Thermo-Magnetization Results

Figure 2a,b display the thermo-magnetization results of an $Fe_{41}Ni_{28}Co_{17}Al_{11.5}Ti_{2.5}$ single crystal aged at 600 °C for 24 h under magnetic fields of 0.05 and 7 T. The thermo-elastic martensitic transformation temperatures of the samples were determined from the magnetic field results of the 0.05 T test. The tangent line method was used to determine the transformation temperatures as indicated in Figure 2a. From the results, the martensitic transformation temperatures were determined to be as follows: austenite finish temperature (A_f) = −122 °C, austenite start temperature (A_s) = −163 °C, martensite start temperature (M_s) = −158 °C, and martensite finish temperature (M_f) = −207 °C. The temperature hysteresis was defined as $|A_f - M_s|$ and calculated to be 36 °C. The thermo-martensitic transformation was not observed in the previous single crystal aged at 600 °C for 4 h because the transformation temperatures of this aging condition were too low, and the phase transformations were not observed during the heating and cooling cycles by SQUID measurement [6]. From the magnetic results of the 7 T magnetic field test, the magnetization in this aging condition was 160 emu/g. The M_s was around −128 °C. The results showed that the martensitic transformation temperatures increase with the increase in the magnetic field from 0.05 T to 7 T. In Figure 2a, it can be seen that As was close to Ms. Based on the theory of thermo-elastic martensitic transformation [8], we can conclude that the elastic energy generated because the transformation was equal to twice the dissipation energy. Therefore, the heat treatment created the conditions for martensitic transformation.

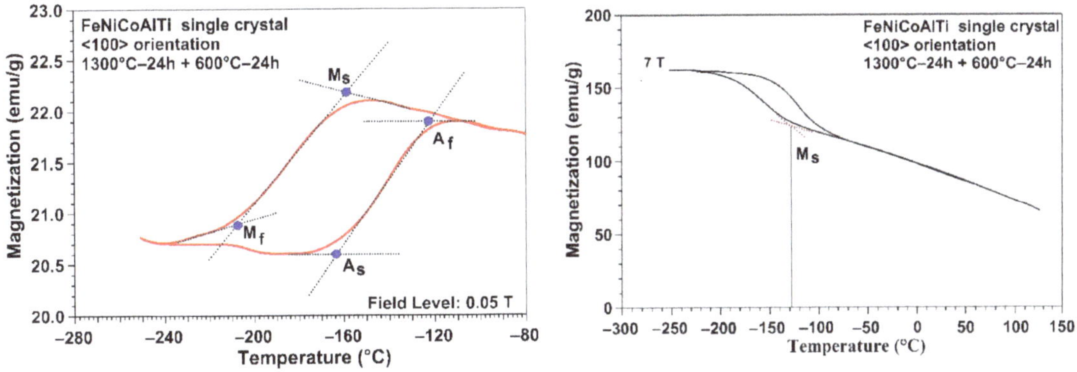

(a) (b)

Figure 2. Magnetization responses of the FeNiCoAlTi single crystals as a function of temperature under magnetic fields of (**a**) 0.05 T and (**b**) 7 T for single crystals aged at 600 °C for 24 h.

3.3. Superelastic Results

Figure 3a shows the stress–strain curves of $Fe_{41}Ni_{28}Co_{17}Al_{11.5}Ti_{2.5}$ single crystals. The critical stress was determined, by the tangent line method, to be around 323 MPa. The tensile sample fractured during the 7% applied strain test. The fracture stress was around 635 MPa. The recoverable and irrecoverable strains extracted from incremental strain tests were plotted as a function of applied strain in Figure 3b. The maximum values of superelastic strain were about 4.8%. The recoverable strain was limited due to the next stress level being close to the fracture stress.

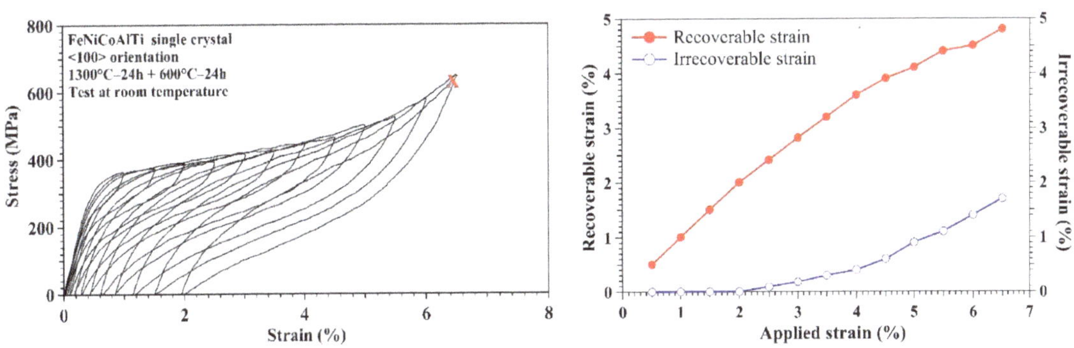

(a) (b)

Figure 3. *Cont.*

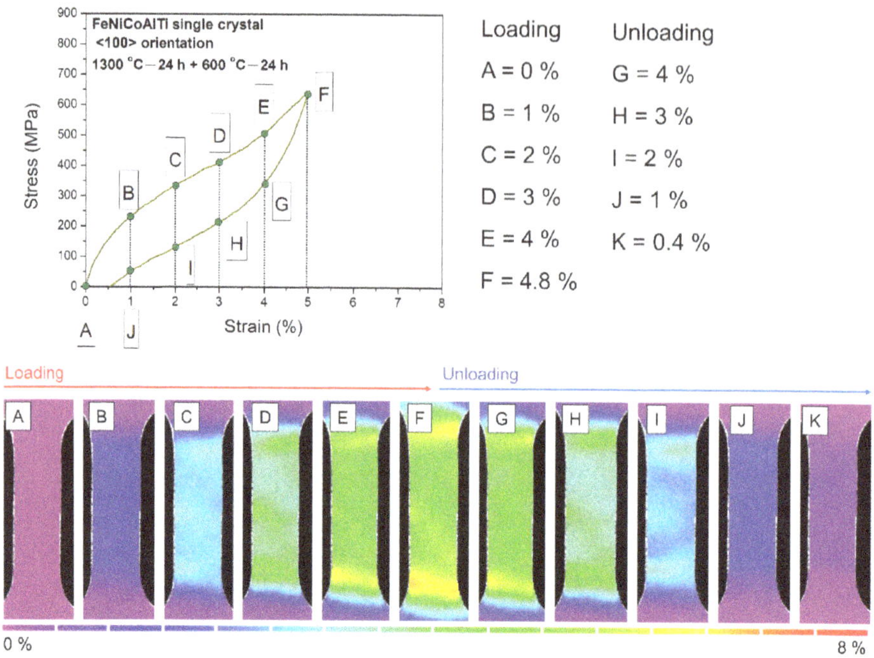

(c)

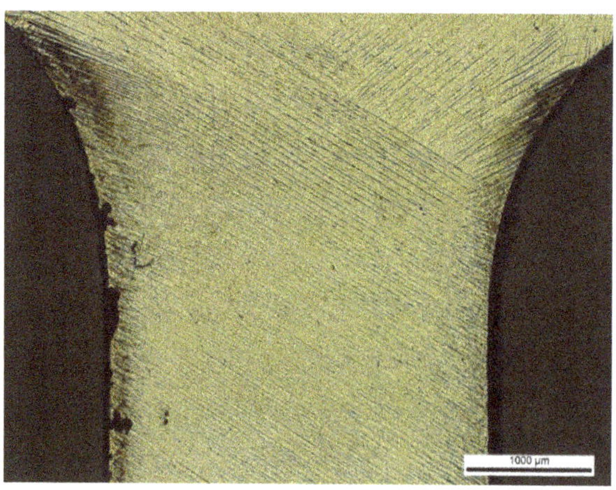

(d)

Figure 3. (**a**) Incremental strain superelastic experiment at room temperature in <100> oriented FeNiCoAlTi single crystal aged at 600 °C for 24 h and (**b**) variation of recoverable strain and irrecoverable strain as a function of the applied strain. "X" marks the point of fracture. (**c**) Contour plot of the strain field during 6.5% applied strain loading/unloading determined by DIC method. (**d**) Optical image of a fractured FeNiCoAlTi tensile sample.

The digital image correlation (DIC) method was used to observe the strain distribution during the 6.5% applied strain loading. In the preceding 6% applied strain test, the residual strain was around 1.5%. To clearly show the loading/unloading process, the residual strain was subtracted from the applied strain of 6.5%, leaving 5% strain, as shown in

Figure 3c. The DIC results indicate that the strain was uniformly distributed, and no severe localized strain concentrations were found in the gauge section. These results indicated that the deformation was uniform during the loading and unloading cycles. These DIC results were different from those reported by Abuzai and Sehitoglu [11]. Their DIC results indicated the presence of three martensite variants during the superelastic test. In the current study, only one martensite variant was observed from the DIC results. This difference is attributed to the different precipitate morphologies resulting from the different aging conditions. In their study, the sample was aged at 600 °C for 3 h. Their result suggests that the martensite adopts a twinned morphology in a short aging time during tensile deformation. In contrast, the sample in the current study was aged at 600 °C for 24 h, and only one variant was observed during the tensile deformation. Figure 3d shows the optical micrograph of the fractured $Fe_{41}Ni_{28}Co_{17}Al_{11.5}Ti_{2.5}$ tensile sample aged at 600 °C for 24 h. From the optical microscope results, traces of martensitic bands were declined about 60° to the tensile direction. This result suggests that, with a longer aging time, the martensite morphology becomes a single variant. Moreover, the critical stress of the sample aged at 600 °C for 3 h was around 800 MPa [11], whereas the critical stress of our sample aged at 600 °C for 24 h was 323 MPa. These results imply that the transformation temperature increases with an increase in aging time from 3 to 24 h, so less stress is required to induce martensitic transformation. The undirected role of variant–variant interactions on mechanical hysteresis is demonstrated by the comparison of the mechanical hysteresis in crystals aged for 3 h [11] and the crystals aged for 24 h in the current study. In the sample aged for 24 h, one variant of martensite created conditions for very low mechanical hysteresis of about 150 MPa at 2% transformation deformation. In those aged for 3 h, when three variants were created under stress, the mechanical hysteresis was three times higher than that of the crystals aged for 24 h.

4. Conclusions

In summary, the microstructure and superelastic properties of FeNiCoAlTi single crystals with the <100> orientation aged at 600 °C for 24 h under tension were investigated in this study. The main conclusions are listed as follows:

1. The precipitates were around 10 nm in size. The precipitate structure was the $L1_2$ structure, and the precipitates had high Ni and Al contents.
2. The thermo-magnetization results indicated phase transformations, and the temperature hysteresis was 36 °C.
3. The superelastic test results showed the recoverable strain to be around 4.5% at room temperature. The DIC results showed that the sample was uniformly deformed, and one martensite variant was observed during the loading/unloading.

Author Contributions: Conceptualization, L.-W.T.; methodology, L.-W.T., C.-H.C., Y.-C.T. and Y.C.; validation, L.-W.T., C.-H.C. and Y.-C.T.; investigation, L.-W.T., C.-H.C., Y.-C.T., P.-Y.L. and N.-H.L.; resources, L.-W.T. and C.-H.C.; data curation, L.-W.T., C.-H.C., Y.-C.T., Y.C., P.-Y.L. and N.-H.L.; writing—original draft preparation, L.-W.T.; writing—review and editing, L.-W.T.; visualization, L.-W.T. and C.-H.C.; supervision, L.-W.T.; project administration, L.-W.T. All authors have read and agreed to the published version of the manuscript.

Funding: This research was funded by the Ministry of Science and Technology (MOST) grant number MOST 109-2221-E-018-010-MY2, and the Young Scholar Fellowship Program of MOST under grant number MOST 110-2636-E-002-005. This study was also supported by the Tomsk State University Development Program.

Institutional Review Board Statement: Not applicable.

Informed Consent Statement: Not applicable.

Data Availability Statement: Not applicable.

Acknowledgments: The authors gratefully acknowledge the Siberian Physical Technical Institute, Tomsk State University, Yury Chumlyakov for the fabrication of single crystal samples. The authors would like to thank Yung-Sheng Chen at the Instrumentation Center, National Tsing Hua University for SQUID measurements.

Conflicts of Interest: The authors declare no conflict of interest.

References

1. Tanaka, Y.; Himuro, Y.; Kainuma, R.; Sutou, Y.; Omori, T.; Ishida, K. Ferrous polycrystalline shape-memory alloy showing huge superelasticity. *Science* **2010**, *27*, 1488–1490. [CrossRef] [PubMed]
2. Omori, T.; Abe, S.; Tanaka, Y.; Lee, D.; Ishida, K.; Kainuma, R. Thermoelastic martensitic transformation and superelasticity in Fe–Ni–Co–Al–Nb–B polycrystalline alloy. *Scr. Mater.* **2013**, *69*, 812–815. [CrossRef]
3. Lee, D.; Omori, T.; Kainuma, K. Ductility enhancement and superelasticity in Fe–Ni–Co–Al–Ti–B polycrystalline alloy. *J. Alloys Compd.* **2014**, *617*, 120–123. [CrossRef]
4. Tseng, L.W.; Chen, C.H.; Chen, W.C.; Cheng, Y.; Lu, N.H. Shape memory properties and microstructure of new iron-based FeNiCoAlTiNb shape memory alloys. *Crystals* **2021**, *11*, 1253. [CrossRef]
5. Ma, J.; Hornbuckle, B.; Karaman, I.; Thompson, G.B.; Luo, Z.; Chumlyakov, Y. The effect of nanoprecipitates on the superelastic properties of FeNiCoAlTa shape memory alloy single crystals. *Acta Mater.* **2013**, *61*, 3445–3455. [CrossRef]
6. Tseng, L.W.; Ma, J.; Karaman, I.; Wang, S.J.; Chumlyakov, Y. Superelastic response of the FeNiCoAlTi single crystals under tension and compression. *Scr. Mater.* **2015**, *101*, 1–4. [CrossRef]
7. Chumlyakov, Y.I.; Kireeva, I.V.; Pobedennaya, Z.V.; Krooß, P.; Niendorf, T. Shape memory effect and superelasticity of [001]-Oriented FeNiCoAlNb single crystals aged under and without stress. *Metals* **2021**, *11*, 943. [CrossRef]
8. Chumlyakov, Y.I.; Kireeva, I.V.; Pobedennaya, Z.V.; Krooβ, P.; Niendorf, T. Rubber-like behaviour and superelasticity of [001]-oriented FeNiCoAlNb single crystals containing γ'-and β-phase particles. *J. Alloys Compd.* **2021**, *856*, 15815. [CrossRef]
9. Czerny, M.; Maziarz, W.; Cios, G.; Wojcik, A.; Chumlyakov, Y.I.; Schell, N.; Fitta, M.; Chulist, R. The effect of heat treatment on the precipitation hardening in FeNiCoAlTa single crystals. *Mater. Sci. Eng. A* **2020**, *784*, 139327. [CrossRef]
10. Tseng, L.W.; Ma, J.; Hornbuckle, B.; Karaman, I.; Thompson, G.B.; Luo, Z.; Chumlyakov, Y. The effect of precipitates on the superelastic response of [100] oriented FeMnAlNi single crystals under compression. *Acta Mater.* **2015**, *97*, 234–244. [CrossRef]
11. Abuzaid, W.; Sehitoglu, H. Superelasticity and functional fatigue of single crystalline FeNiCoAlTi iron-based shape memory alloy. *Mater. Des.* **2018**, *160*, 642. [CrossRef]

crystals

MDPI

Article

Shape Memory Properties and Microstructure of New Iron-Based FeNiCoAlTiNb Shape Memory Alloys

Li-Wei Tseng [1,*], Chih-Hsuan Chen [2], Wei-Cheng Chen [1], Yu Cheng [2] and Nian-Hu Lu [2]

1 Department of Mechatronics Engineering, National Changhua University of Education, Changhua 50007, Taiwan; earth520ya@gmail.com
2 Department of Mechanical Engineering, National Taiwan University, Taipei 10617, Taiwan; chchen23@ntu.edu.tw (C.-H.C.); f06543027@ntu.edu.tw (Y.C.); f06522712@ntu.edu.tw (N.-H.L.)
* Correspondence: lwtseng@cc.ncue.edu.tw

Abstract: The shape memory properties and microstructure of $Fe_{41}Ni_{28}Co_{17}Al_{11.5}(Ti+Nb)_{2.5}$ (at.%) cold-rolled alloys were studied at the first time using the values reported in constant stress thermal cycling experiments in a three-point bending test. Thermo-magnetization curves of 97% cold-rolled and solution-treated sample aged at 600 °C for 24, 48 and 72 h showed evidence of the martensitic transformation, and the transformation temperatures increased their values from 24 to 72 h. The alloy cold-rolled to 97% and then solution-treated at 1277 °C for 1 h showed that most grains were aligned near <100> in the rolling direction in the recrystallization texture. The intensity of texture was 13.54, and an average grain size was around 400 μm. The sample aged at 600 °C for 48 h showed fully recoverable strain up to 1.6% at 200 MPa stress level in the three-point bending test. However, the experimental recoverable strain values were lower than the theoretical values, possibly due to the small volume fraction of low angle grain boundary, the formation of brittle grain boundary precipitates, and a grain boundary constraint lower than the expected intensity of texture in the samples.

Keywords: shape memory alloys; FeNiCoAlTiNb; microstructure; shape memory effect; martensitic transformation

Citation: Tseng, L.-W.; Chen, C.-H.; Chen, W.-C.; Cheng, Y.; Lu, N.-H. Shape Memory Properties and Microstructure of New Iron-Based FeNiCoAlTiNb Shape Memory Alloys. *Crystals* **2021**, *11*, 1253. https://doi.org/10.3390/cryst11101253

Academic Editor: Wojciech Polkowski

Received: 14 September 2021
Accepted: 13 October 2021
Published: 15 October 2021

1. Introduction

Shape memory alloys (SMAs) are small materials with functional and intelligent properties. SMAs exhibit two unique properties: (i) the shape memory effect, in which the material can recover the deformation upon heating or magnetization, and (ii) superelasticity, which is stress-induced during martensitic transformation, and the material can recover the deformation after removing the load. Currently, nickel–titanium (NiTi) SMAs are the most well-developed SMAs and are commonly used in industries including aerospace, automotive, biomedical, robotics, actuators and sensors due to their high recoverable strain in both tension and compression [1–3]. However, the materials used in NiTi SMAs have a high cost and they have difficulty in cold rolling. These factors limit the applications of large quantities of NiTi SMAs due to the cost of materials and processing issues [4]. Compared to NiTi alloys, iron-based SMAs are commercially attractive because they have a low material cost and good machinability and workability. As a result, iron-based SMAs have attracted attention and interest in the industry and academic fields.

The martensitic transformation (MT) of Fe-based SMAs is classified into three groups based on their martensite and austenite structures: (1) face-centered cubic (FCC) (austenite) to body-centered cubic (BCC) or body-centered tetragonal (BCT) (martensite). This transformation system can be observed in an FeNi-based alloy system [5–10]. (2) The martensitic transformation in FeMnSi SMAs are FCC-HCP [11,12]: FCC (austenite) to hexagonal close-packed (HCP) (martensite). (3) The transformation system reported in FeMnAl [13], FeMnGa [14,15] and FeMnAlNi [16] SMAs is BCC (austenite) to FCC (martensite).

In general, four distinct morphologies of martensite in iron-based SMAs can be observed: lenticular, lath, butterfly and thin plate [1,17]. Lenticular martensite is formed in FePt SMAs. In FeNi alloys, thin plate martensite forms at the lowest temperature, lenticular martensite forms at intermediate temperatures and lath martensite forms at the highest temperature [17]. The formation of lath martensite typically contains a high density of internal dislocations due to accommodation distortion. Unlike the lath martensite, thin plate types of martensite contain a high density of transformation twins. For butterfly type of martensite, the substructure comprises a mixture of dislocations and twins [17]. Among four types of martensite morphology, only the thin plate martensite shows the shape memory effect [1].

Most Fe-based SMAs, such as FeMnAl, FeNiCoTi, FeMnSi and FeMnGa SMAs, show less than 1% recoverable strain of superelasticity [1,3,5,11,13]. Recently, two new Fe-based SMAs systems were presented, with superelastic strains above 4% at room temperature. The first system is $Fe_{40.95}Ni_{28}Co_{17}Al_{11.5}Ta_{2.5}B_{0.05}$ (at.%), which shows a 13.5% superelastic strain [9], and in a similar system, FeNiCoAlXB (X: Ta, Nb, Ti) presents a 5% superelastic strain in polycrystalline alloys [9,10,18,19], single crystals [20–24] and wire [25]. The second system is $Fe_{43.5}Mn_{34}Al_{15}Ni_{7.5}$ (at.%), and this system shows a 5% superelastic strain in polycrystalline alloys [16,26,27], polycrystalline wires [28–30] and single crystals [31,32]. In FeNiCoAlXB (X: Ta, Nb, Ti) alloy systems, three conditions are required to achieve this excellent superelastic property. (1) A strong texture is required in these alloys. If the texture is random, it shows brittle behavior. In FeNiCoAlTaB and FeNiCoAlTiB alloys, the strong texture is in the <100> orientation [8,10]. In FeNiCoAlNbB alloys, the strong texture is in the <110> orientation [9]. (2) A large grain size is required to decrease the grain boundary and triple junction constraints during the superelastic tests. In FeNiCoAlTaB alloys, the average grain size is around 400 μm [8,33]. (3) The alloy is required to undergo aging heat treatment to obtain $L1_2$ precipitates. The nano size precipitates not only strengthen the matrix but also change the material from non-thermoelastic to thermoelastic transformation [8]. The precipitate size is around 3 nm for the 600 °C–72 h sample [8].

Studies regarding the shape memory behavior of <100>-oriented $Fe_{41}Ni_{28}Co_{17}Al_{11.5}Ta_{2.5}$ (at.%) single crystals have been carried out. Ma et al. [20] reported a 3.75% recoverable strain in tension at the stress level of 50 MPa. The precipitate size was around 5 nm for a 600 °C–90 h sample. Czerny et al. [34,35] found that two stages of aging heat treatment could improve the compressive properties. A FeNiCoAlTa single crystal underwent the first stage of aging at 700 °C for 0.5 h and the second stage of aging at 700 °C for 3 h, showing a 15% superelastic strain at −196 °C [35].

Tseng et al. [23] found that using Ti to replace Ta could reduce the aging heat treatment times. A FeNiCoAlTi single crystal with <100> orientation aged at 600 °C for 4 h possessed 6% superelastic strain when the test temperature was at −80 °C. Chumlyakov et al. [36–38] showed that adding Nb to this alloy could increase the hardness of the austenite matrix and improve their superelastic properties. FeNiCoAlNb single crystals with <100> orientation in two-step aging (700 °C–0.5 h + 700 °C–3 h) demonstrated 7% superelastic strain in tension and 13.5% in compression. The precipitates size ranged from 5 to 8 nm [38].

Poklonov et al. [39] reported <100>-oriented FeNiCoAlTiNb single crystals that underwent a thermal cycling tensile test and obtained 2.2% recoverable strain at 50 MPa when they were heat treated at 1277 °C for 10 h and subsequently aged at 600 °C for 4 h. However, a large irrecoverable strain was observed when the applied stress was 100 MPa. The sample fractured at a stress level of 150 MPa. Tseng et al. [40] investigated the microstructure of FeNiCoAlTiNb alloy and found the homogeneous solution heat treatment condition was 1277 °C for 24 h. Second phases were formed in the sample when the heat treatment temperature was above 1300 °C.

In this study, $Fe_{41}Ni_{28}Co_{17}Al_{11.5}(Ti{+}Nb)_{2.5}$ (at.%) polycrystalline alloys that underwent cold rolling were first investigated for their microstructure, texture, thermo- magnetization and shape memory properties using the three-point bending test. In this paper, we discuss the aging effect on microstructure, transformation temperatures and shape memory properties in the three-point bending test.

2. Materials and Methods

Iron, cobalt, nickel, aluminum, niobium and titanium (99.9 wt%) were used as raw materials. Ingots with the nominal composition of $Fe_{41}Ni_{28}Co_{17}Al_{11.5}(Ti+Nb)_{2.5}$ (at.%) were fabricated by vacuum induction melting (VIM). Wire electrical discharge machining (EDM) was used to cut the bar to the desired block, the dimensions of which were 100 × 25 × 25 mm (length, width and thickness). The block was used for the cold-rolling experiment.

As-received $Fe_{41}Ni_{28}Co_{17}Al_{11.5}(Ti+Nb)_{2.5}$ (at.%) ingots were prepared using several conditions of thermo-mechanical processing to obtain polycrystalline sheets. The processing conditions are listed below:

(1) Solution heat treatment (SHT 1) at 1277 °C for 24 h (h), followed by water quench (WQ). The first stage of solution heat treatment was used to homogenize the alloys;
(2) Cold rolling (CR) at room temperature with a reduction ratio of 97% to a thickness of 0.7 mm, and three-point bending samples were cut from the cold-rolled sheet;
(3) Solution heat treatment (SHT 2) at 1277 °C for 1h and water quenching;
(4) Aging or precipitation heat treatment (PHT) at 600 °C for 24, 48 and 72h to introduce $L1_2$ precipitates and water quenching. For the sake of brevity, we refer to the alloys aged at 600 °C for x hours as "600 °C–h".

The thermo-mechanical processing setup used in this study is summarized in Figure 1.

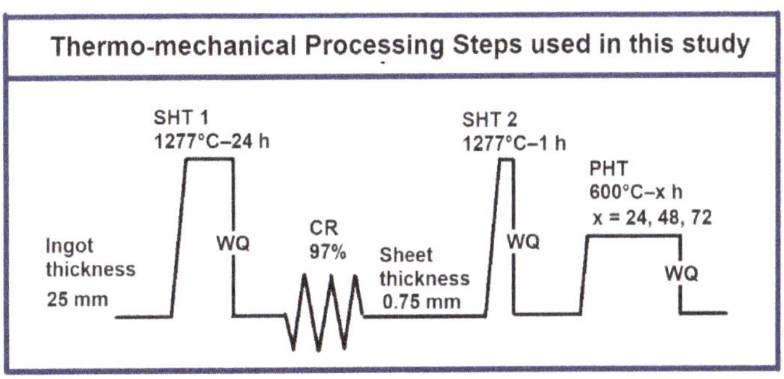

Figure 1. Illustration of the thermo-mechanical processes to obtain $Fe_{41}Ni_{28}Co_{17}Al_{11.5}(Ti+Nb)_{2.5}$ (at.%) alloy sheets. SHT 1, WQ, CR, SHT 2 and AHT indicate first stage of solution heat treatment, water quenching, cold rolling, second stage of solution heat treatment and aging heat treatment, respectively.

The FeNiCoAlTiNb alloys were ground to 0.05 μm and the surfaces were etched with the solution, which was composed of 7% nitric acid and 93% ethanol. The observation of microstructure was made using the Olympus digital optical microscope. Vickers microhardness measurements were made using a microhardness tester FM-310. The samples for the OM tests were (1) 97%CR + 1277 °C–1 h, (2) 97%CR + 1277 °C–1 h + 600 °C–24 h, (3) 97%CR + 1277 °C–1 h + 600 °C–48 h and (4) 97%CR + 1277 °C–1 h + 600 °C–72 h. Hardness values of measurement were made at room temperature.

A high-resolution X-ray diffractometer with Cu-Kα radiation (λ = 0.1542 nm) was used to obtain the crystal structures of 1277 °C–24 h, 97%CR and 97%CR after heat treatment at 1277 °C for 1 h samples. The as-received, 1277 °C–24 h, 97%CR and 97%CR samples after heat treatment at 1277 °C for 1 h were cut out for texture measurement using a D5000 X-ray diffraction device (Siemens, Aubrey, TX, USA). The FCC{111}, FCC{200}, and FCC{220} pole figure data were selected to plot the orientation distribution functions (ODFs) [19]. The microstructure and texture of the cold-rolled specimens after being treated with solution at 1277 °C for 1h were determined using electron backscatter diffraction (EBSD). The etching solution used in preparing the sample surface was 10% $HClO_4$ + 90% C_2H_5OH.

The SEM images of fractured surfaces for the 600 °C–48 h sample after three-point bending test were obtained using a JSM-7800F device (JEOL, Musashino, Akishima, Tokyo).

In order to find the martensitic transformation temperatures and the thermo- magnetization properties, the cold-rolled samples that were aged at 600 °C for 24, 48 and 72 h were measured by using a superconducting quantum interference device (SQUID) under fields of 0.05 and 7 Tesla. The cooling and heating rate was determined to be 5 K/min. The magnetic results were used to determine the martensitic transformation temperatures used to calculate the temperature hysteresis. The sample was first heated up to 110 °C under a zero magnetic field, and then cooled down to −260 °C and heated up around to 110 °C under a constant magnetic field of 0.05 Tesla. After the 0.05 Tesla test, the magnetic field was increased to 7 Tesla. The sample was cooled from 120 °C to −260 °C and heated up to 120 °C again to complete one cycle. In thermo-magnetization curve, magnetization was used to describe how a material responds to an applied magnetic field and one magnetization (emu/g) corresponded to 1 Am^2/kg.

The shape memory effects of FeNiCoAlTiNb samples were determined using the three-point bending method with an ElectroForce 3230 (TA Instruments, New Castle, DE, USA) equipped with a hot/cold chamber. The support span was 20 mm. Three-point bending shape memory samples were cut from the cold-rolled sheet. The samples were solution- treated at 1277 °C for 1 h and underwent aging heat treatment at 600 °C for 24, 48 and 72 h. The samples were straight before the bending test. Figure 2 presents the setup of the bending test. For this test, the sample was first loaded with a stress level of 50 MPa at room temperature and was then cooled and heated between −150 °C and room temperature to complete one thermal cycle. This process was repeated for increasing stress levels. A heating/cooling rate was selected as 5 K/min. When the sample completed one thermal cycle, the applied stress level increased to 100 MPa and the next heating/cooling cycle was carried out. The same procedure was applied using the stress levels of 150 and 200 MPa. The 600 °C–48 h sample failed when the applied stress level was 250 MPa. The 600 °C−72 h sample fractured during the 150 MPa applied stress level.

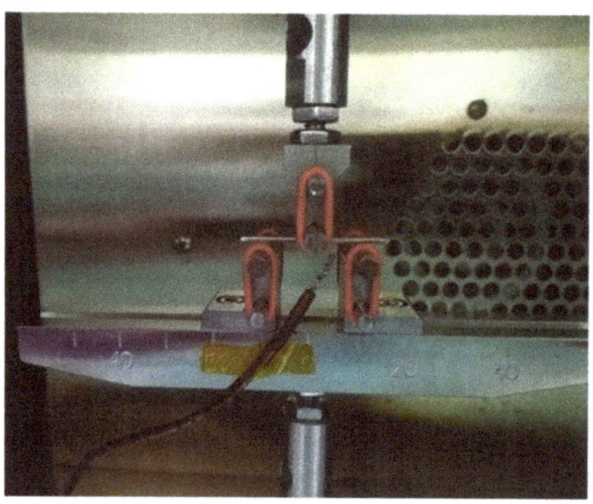

Figure 2. Setup for the bending test.

3. Results

3.1. Microstructure and Microhardness Results

Figure 3a–d show the microstructure of the FeNiCoAlTiNb polycrystalline in different thermo-mechanical processing conditions. Figure 3a shows the 97%CR sample after being heat treated at 1277 °C for 1 h. The grain boundaries were clearly seen in this condition. Figure 3b–d display the 97%CR sample after being treated with solution at 1277 °C for 1 h and aging heat treatment at 600 °C for 24, 48 and 72 h. The microstructure results show that leaf-like β phases accumulate in the grain boundary, especially in triple junctions.

Increasing the aging times from 24 to 72 h, large number of β phases form in the aging sample. The generation of β phases can also be observed in FeNiCoAlXB (X: Ta, Ti, Nb) systems [8–10,12].

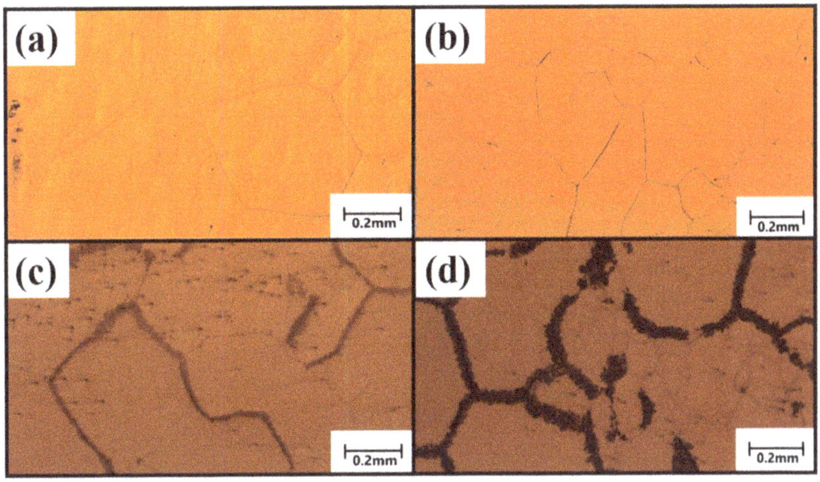

Figure 3. Microstructure of FeNiCoAlTiNb alloy in different thermo-mechanical processing conditions: (**a**) 97%CR+1277 °C–1 h, (**b**) 97%CR+1277 °C–1 h + 600 °C–24 h, (**c**) 97%CR + 1277 °C–1 h + 600 °C–48 h and (**d**) 97%CR + 1277 °C–1 h + 600 °C–72 h.

Table 1 shows the Vickers microhardness of the FeNiCoAlTiNb alloy in different conditions of thermomechanical processing. The hardness value of the 97%CR sample is around 430HV. After the solution heat treatment, the hardness is 260HV. For the aging heat treatment condition, the microhardness values are 462 HV, 510 HV and 552 HV for 600 °C–24 h, 600 °C–48 h and 600 °C–72 h, respectively. The microhardness results of three aging conditions are summarized in Figure 4. From the results, as the aging times increase, the hardness values increase their values due to the formation of precipitation hardening. In FeNiCoAlTa single crystal studies, the hardness values increased when the aging times increased from 24 to 90 h at an aging temperature of 600 °C, which suggests precipitates strengthen the austenite matrix [21].

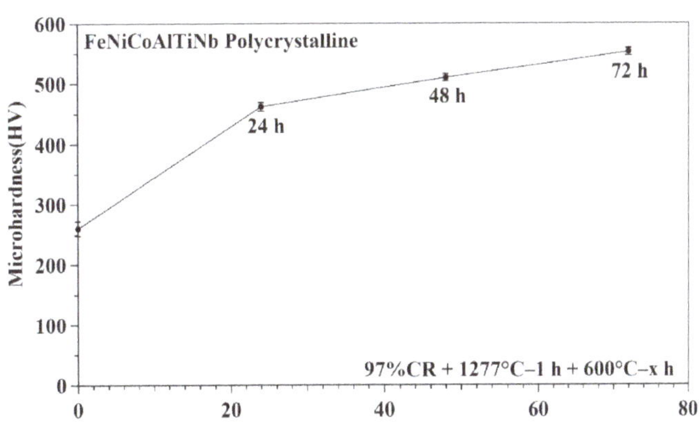

Figure 4. Hardness results of FeNiCoAlTiNb samples under different aging conditions.

Table 1. Microhardness results of FeNiCoAlTiNb alloy in different thermo-mechanical processing conditions.

Thermo-Mechanical Processing	Vickers Hardness (HV)
97%CR	430 ± 12
97%CR + 1277 °C–1 h	260 ± 6.5
97%CR + 1277 °C–1 h + 600 °C–24 h	462 ± 5.5
97%CR + 1277 °C–1 h + 600 °C–48 h	510 ± 5
97%CR + 1277 °C–1 h + 600 °C–72 h	552 ± 6

3.2. XRD Results

Figure 5 presents the XRD results for samples' conditions: 1277 °C–24 h, 1277 °C–24 h + 97%CR and 1277 °C–24 h + 97%CR + 1277 °C–1 h. The XRD profile results show that face-centered cubic (FCC) phases are found in three conditions. In the cold-rolled samples after 1277 °C–1 h treatment, a strong peak intensity is observed in the (111) plane.

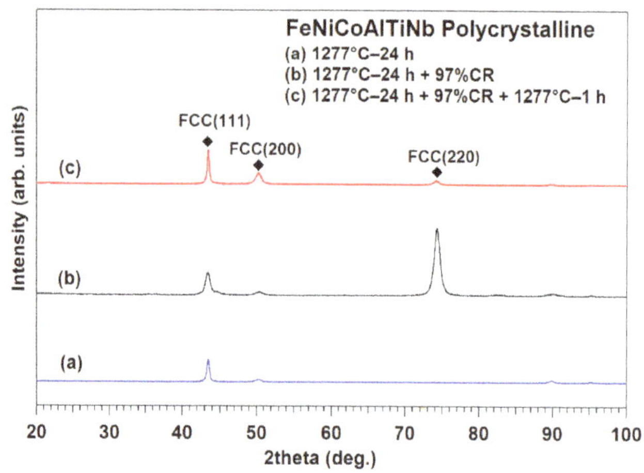

Figure 5. XRD results of FeNiCoAlTiNb samples under different processing conditions: 1277 °C–24 h, 1277 °C–24 h + 97%CR and 1277 °C–24 h + 97%CR + 1277 °C–1 h.

3.3. Magnetization Results

The transformation temperatures of the cold-rolled samples after being aged at 600 °C for 24, 48 and 72 h were determined using the magnetic field test. After solution heat treatment, the 97% cold-rolled sample does not show here, since martensitic transformation is not observed in this condition. Figure 6a–f display the thermo-magnetization results of FeNiCoAlTiNb that underwent aging heat treatment at 600 °C for 24, 48 and 72 h under the magnetic fields of 0.05 and 7 Tesla. The martensitic transformation temperatures of the sample aged at 600 °C for 24, 48 and 72 h were determined using the magnetic field results of 0.05 Tesla, as shown in Figure 6a,c,e. The tangent line method was used to determine the transformation temperatures. From the result, the martensitic transformation temperatures of the 600 °C–24 h aging sample were: austenite finish temperature (Af) = −103 °C, austenite start temperature (As) = −147 °C, martensite start temperature (Ms) = −135 °C and martensite finish temperature (Mf) = −189 °C. The temperature hysteresis was defined as $\Delta T = |Af - Ms|$, and was calculated to be 36 °C. For the 600 °C–48 h aging sample, Af = −68 °C, As = −105 °C, Ms = −106 °C and martensite finish temperature (Mf) = −154 °C. The temperature hysteresis was 30 °C. For the 600 °C–72 h aging sample, Af = −53 °C, As = −100 °C, Ms = −90 °C, and martensite finish temperature (Mf) = −134 °C. The temperature hysteresis was 37 °C. The transformation temperatures and temperature hysteresis of the FeNiCoAlXB (X: Ta, Nb, Ti) alloys and the FeNiCoAlTiNb alloy reported by the present authors [8–10] are summarized in Table 2.

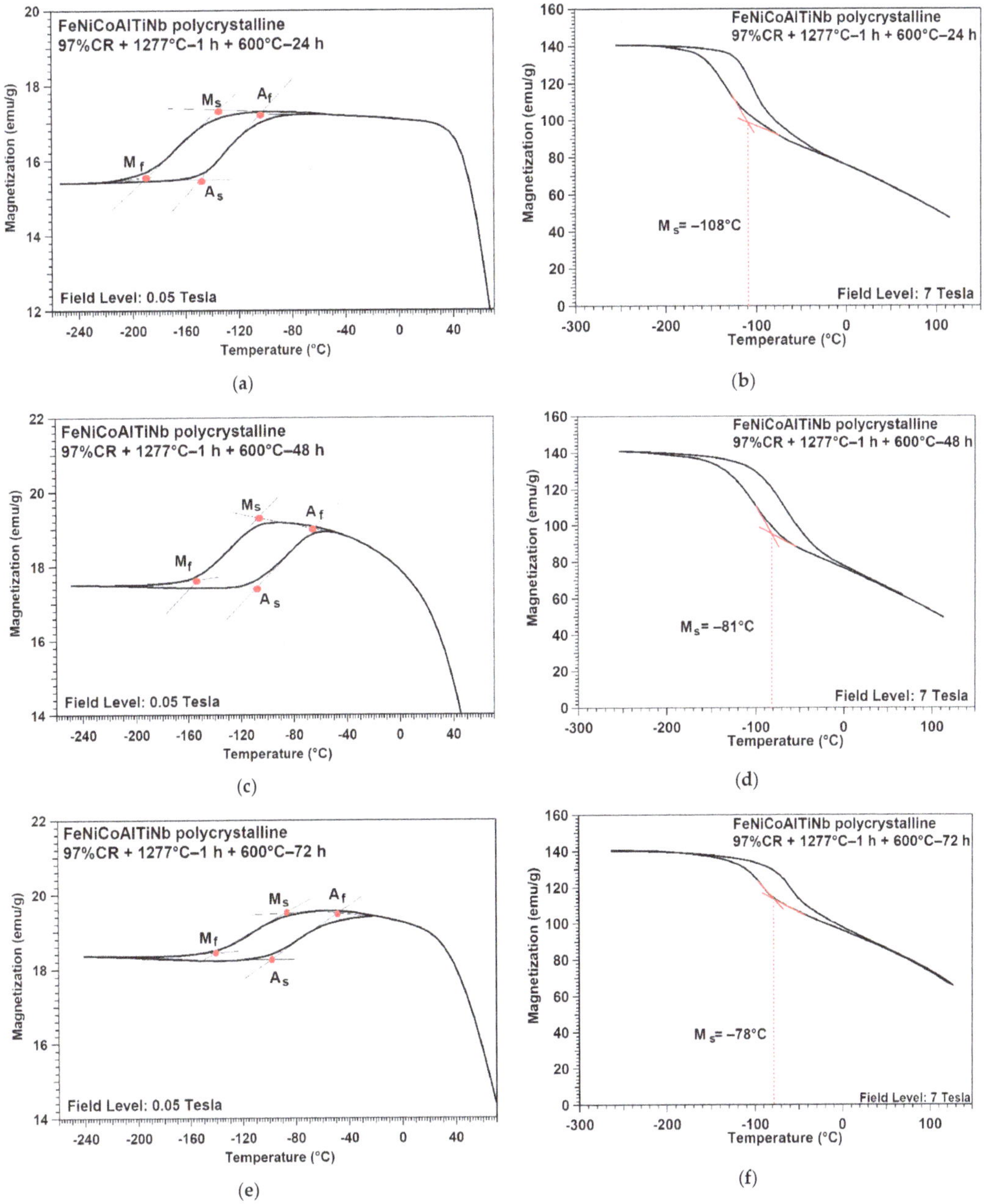

Figure 6. Thermo-magnetization curves of FeNiCoAlTiNb aging sample. (**a**,**b**) 0.05 and 7 Tesla for 600 °C−24 h, (**c**,**d**) 7 Tesla for 600 °C−48 h, (**e**,**f**) 0.05 and 7 Tesla for 600 °C−72 h.

Figure 6b,d,f present the magnetic results of the 7 Tesla experiments; the magnetization of both aging conditions is 140 emu/g. The Ms can also be obtained using tangent line methods, and Ms was −108 °C, −81 °C and −78 °C in 600 °C–24 h, 600 °C–48 h and 600 °C–72 h conditions, respectively. The results show the transformation temperatures move to higher temperatures as the aging times increase. A similar trend was reported in FeNiCoAlTa single crystals [21].

Table 2. Transformation temperatures and temperature hysteresis of the present FeNiCoAlTiNb alloy in comparison with data for FeNiCoAlXB (X: Ta, Nb, Ti) alloys [8–10].

Alloys	Transformation Temperature (°C)	Temperature Hysteresis (°C)
FeNiCoAlTaB (600 °C–72 h)	Af = −62 and Ms = −86	24
FeNiCoAlTa (600 °C–90 h)	Af = −63 and Ms = −98	35
FeNiCoAlNbB (600 °C–80 h)	Af = −30 and Ms = −70	40
FeNiCoAlTiB (550 °C–24 h)	Af = −4 and Ms = −35	31
FeNiCoAlTiNb (600 °C–24 h)	Af = −103 and Ms = −135	36
FeNiCoAlTiNb (600 °C–48 h)	Af = −68 and Ms = −106	32
FeNiCoAlTiNb (600 °C–72 h)	Af = −53 and Ms = −90	37

3.4. EBSD and ODFs Results

The recrystallization texture of the 97% cold-rolled samples after undergoing solution treatments at 1277 °C for 1 h was obtained using EBSD. Figure 7 shows a quasi-colored orientation map of the 97% cold-rolled samples after solution treatment, where the color indicates the crystal orientation presented in the stereographic triangle. In the figure, the RD, TD and ND represent the rolling direction, transverse direction and normal direction, respectively. From the results of orientation maps, the average grain size is around 402 μm. In the FeNiCoAlTaB alloy, the average grain size is around 400 μm [8]. From the inverse pole figure result, strong <001> orientation is observed in the rolling direction. The intensity of texture is 13.64. The recrystallization texture of the present alloy is similar to the <100> texture of the FeNiCoAlTaB and FeNiCoAlTiB alloy [8,10] but different from the <110> texture of the Fe–Ni–Co–Al–Nb–B alloy [9].

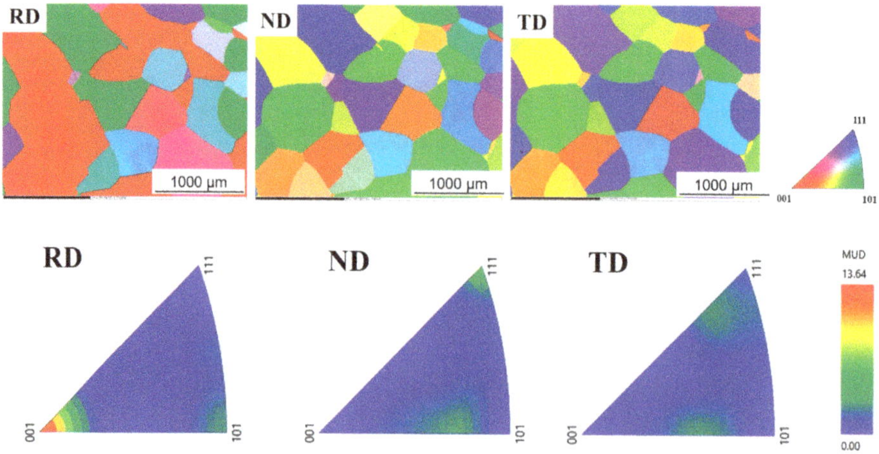

Figure 7. Quasi-colored orientation maps of microstructure and inverse pole figures of the cold-rolled alloy after being solution treated at 1277 °C for 1 h (RD: rolling direction, TD: transverse direction, ND: normal direction).

Figure 8 shows ODFs (phi 2= 0°, 45°, and 60° sections) for the alloy with different thermo-mechanical processing conditions. The texture of the as-received sample is cubic {001}⟨100⟩, Goss/brass (G/B) {110}⟨115⟩ and brass ({110}⟨112⟩. The sample after solution treatment shows cubic {001}⟨100⟩, rotated cube (Rt-C) {001}⟨110⟩ and rotated Cu (Rt-Cu) {112}⟨011⟩ texture. The 97% cold-rolled sample shows Goss (G) {110}⟨001⟩) and Goss/brass (G/B) {110}⟨115⟩ textures. The cold-rolled sample that underwent 1277C–1h heat treatment shows a Goss (G) {110}⟨001⟩ texture. Therefore, cold-rolling deformation and recrystallization at 1277 for 1h give rise to a preferred ⟨100⟩ recrystallized orientation.

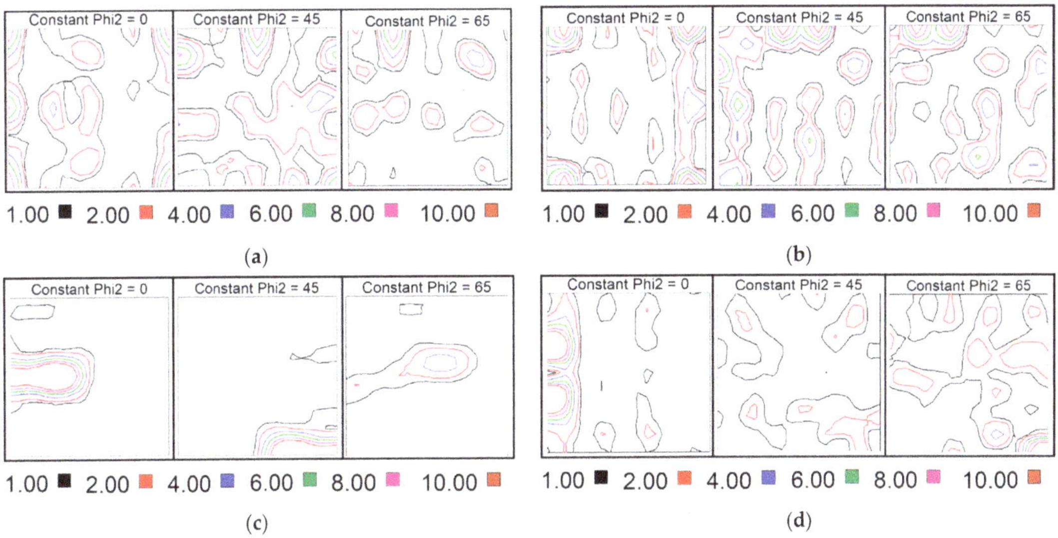

Figure 8. ODFs results of FeNiCoAlTiNb polycrystalline under different thermo-mechanical processing conditions: (**a**) as-received, (**b**) 1277C–24 h, (**c**) 97%CR and (**d**) 97%CR + 1277 °C for 1 h.

3.5. Three-Point Bending Test Results

The results from the thermal cycling under the three-point bending test at different stress levels are displayed in Figure 9. Figure 9a shows the thermal cycling results of sample aged at 600 °C for 24 h. When the stress level is 100 MPa, the martensitic transformation occurs at around −150 °C. The complete loop of phase transformation does not observe because the transformation temperature is too low, and the device can only cool down to −150 °C, and the stress level is not high enough to induce martensitic transformation to higher temperature. The sample fails during the 200 MPa thermal cycle test. The early failure of thermal cycling specimens was probably caused by the variant–variant interactions during transformation process. During the cooling process, the interaction between the growing martensitic phase increases variant interaction and leads to fracture [20].

Figure 9b presents the thermal cycling results of the sample aged at 600 °C for 48 h. The phase transformation was observed when the stress level was 50 MPa. The sample failed when the stress level increased to 250 MPa. Transformation temperatures, recoverable strain and temperature hysteresis values can be obtained through this test. The maximum recoverable strain was around 1.6% at a stress level of 200 MPa. The transformation temperatures (Af and Ms) are extracted in Figure 9b and plotted as a function of applied stress, as shown in Figure 9c. From the results, when the applied stress levels increased, the Af and Ms slightly increased their temperatures. The Ms was −110 °C at 50 MPa and increased to −100 °C at 200 MPa. The Af was −70 °C at 50 MPa and increased to −56 °C at 200 MPa. In addition, the lines for each transformation temperature could be extrapolated down to zero stress to determine the stress-free transformation temperatures. The Ms was

−112 °C and Af was −74 °C, which were values close to the transformation temperatures determined by the SQUID results.

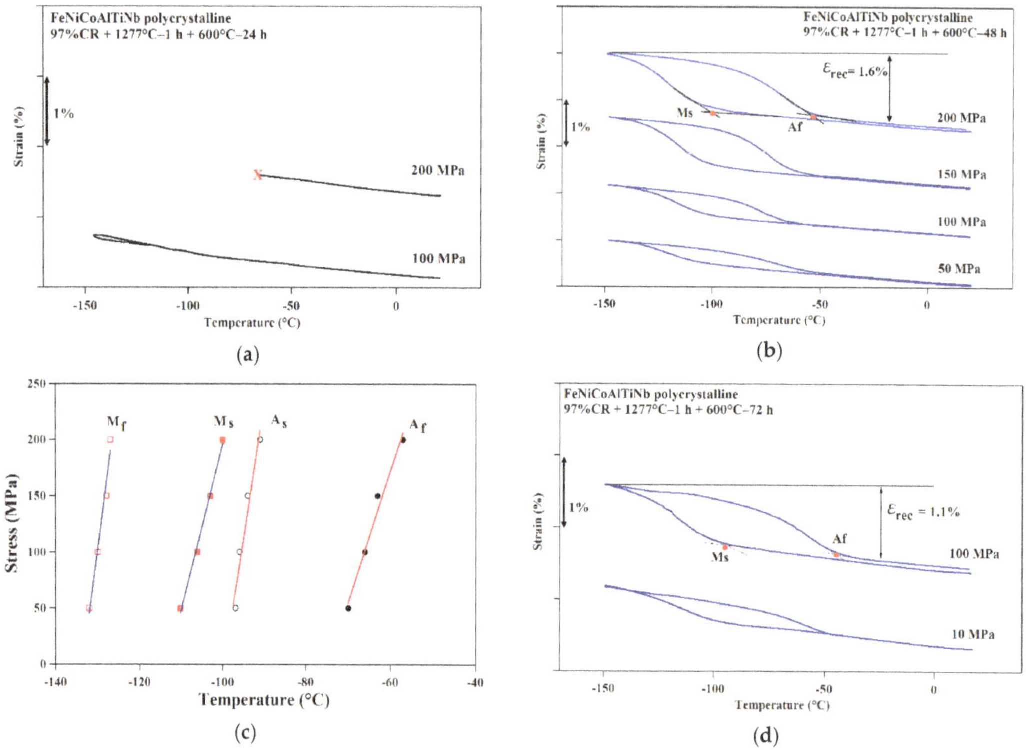

Figure 9. Shape memory effect of FeNiCoAlTiNb 97% cold-rolled sample after being aged at 600 °C for x hours using three-point bending test. (**a**) 600 °C–24 h aging sample, (**b**) 600 °C–48 h aging sample, (**c**) stress versus temperature phase diagram for a 600 °C–48 h aging sample. The values for each transformation temperature are obtained from Figure 7b results. (**d**) 600 °C–72 h aging sample.

Figure 9d displays the thermal cycling results of the sample aged at 600 °C for 72 h. The maximum recoverable strain was 1.1%. The sample failed during the 150 MPa thermal cycle test. The sample was brittle due to the formation of β phases at the grain boundaries.

4. Discussion

From the three-point bending test results, it is clear that the recoverable strain of the 600 °C–48 h aged sample is smaller than the theoretical calculated values of FeNiCoAlTaB [8]. One possible reason for this is the smaller fraction of low angle grain boundary (LAB). Figure 10 shows the grain boundary misorientation of the 97% cold-rolled alloy after solution treatment at 1277 °C for 1 h. Based on the calculation, the fraction of LABs is about 11%. The FeNiCoAXB (X: Ta, Nb, Ti) polycrystalline alloy systems are required to undergo aging heat treatment to increase the strength of the austenite matrix and obtain the nano size $L1_2$ precipitates [8–10,25]. Meanwhile, β phases (grain boundary precipitates) are generated along the grain boundary, especially at the triple junction. β phases prefer to form at high angle grain boundaries (HABs) due to higher energy [9,10]. From Figure 3c results, it is seen that the β phases form at the grain boundaries. As the fraction of LABs is 11% in the 97% cold-rolled sample after solution treatment at 1277 °C for 1 h, β phases can be seen in the grain boundaries for the 600 °C–48 h sample. β phases deteriorate ductility and shape memory properties. As a consequence, the recoverable strain of the shape memory effect is smaller than the theoretical calculated values. Although a small number of β phases is formed in the 600 °C–24 sample, the transformation curve of the shape memory effect is not observed even under the 2000 MPa stress level due to its lowest transformation temperatures. For the 600 °C–72 sample, more β phases were generated in this aging condition, as shown in Figure 3d. As a consequence, the sample failed during the 150 MPa test.

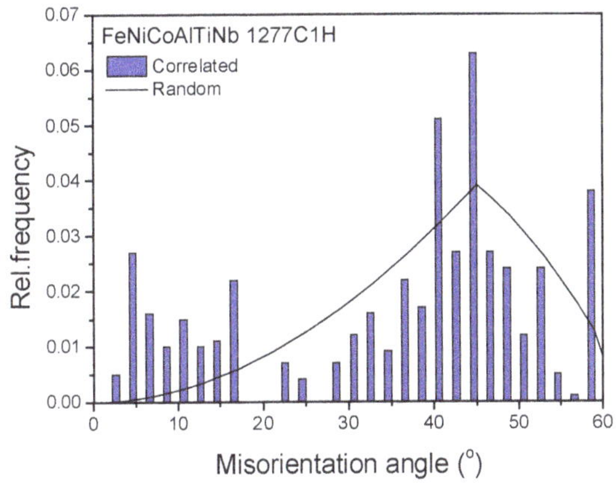

Figure 10. Grain boundary misorientation of the cold-rolled alloy after solution treatment.

Another possible reason for the significantly lower shape memory recoverable strain observed in this study is related to the texture intensity of the <100> orientation in the rolling direction. In this study, the texture intensity of the <100> component in the RD direction is 13.64, as shown in Figure 7. The texture intensity of same component and orientation in FeNiCoAlTaB is 25.4 [8]. The value of texture intensity in this sample is lower than that in the FeNiCoAlTaB specimen. Texture development can improve the recoverable strain of shape memory and superelasticity [8–10]. In iron-based SMAs, reducing the grain constraint is crucial due to martensite variant selection. Based on the Bain distortion theory, only three variants can assist the deformation, which is different from the NiTi SMAs, in which twenty-four variants are able to assist deformation [1].

In iron-based SMAs, the theoretical transformation strains are 8.7%, 4.1% and 2.1% in <100>, <110> and <111> orientations, respectively [8]. Based on the results in Figure 7, Quasi-colored orientation maps in the rolling direction show that some grains are oriented in the <110> and <111> direction. The large difference in the crystal orientation between surrounding grains increases the grain constraint. When the applied stress level is increased, the incompatibility between adjacent grains becomes intense and results in the deterioration of ductility. Figure 11a presents SEM images of the fracture surface for the 600 °C–48 h sample after three-point bending test. The fracture is along the grain boundary. Figure 11b shows an SEM image of a cross section. The fracture surfaces of the sample are smooth. The results suggest that the sample has a brittle intergranular fracture, and the fracture is along a grain boundary.

The final possible reason for the smaller strain of this polycrystalline material is the addition of boron. The volume fraction of LABs is 11% in the present alloy and 11% in FeNiCoAlNbB [9]. Since both alloys have a similar fraction of LABs, the aging condition for the FeNiCoAlNbB alloy is 600 °C for 96 h. A small amount of grain boundary precipitates is observed in this aging condition. Since B element tends to lower the grain boundary energy, this retards β phases' growth in the FeNiCoAlNbB alloy. The recrystallization texture of the FeNiCoAlNbB alloy near <110> in the rolling direction is different from the FeNiCoAlTaB [8,9] and present alloy.

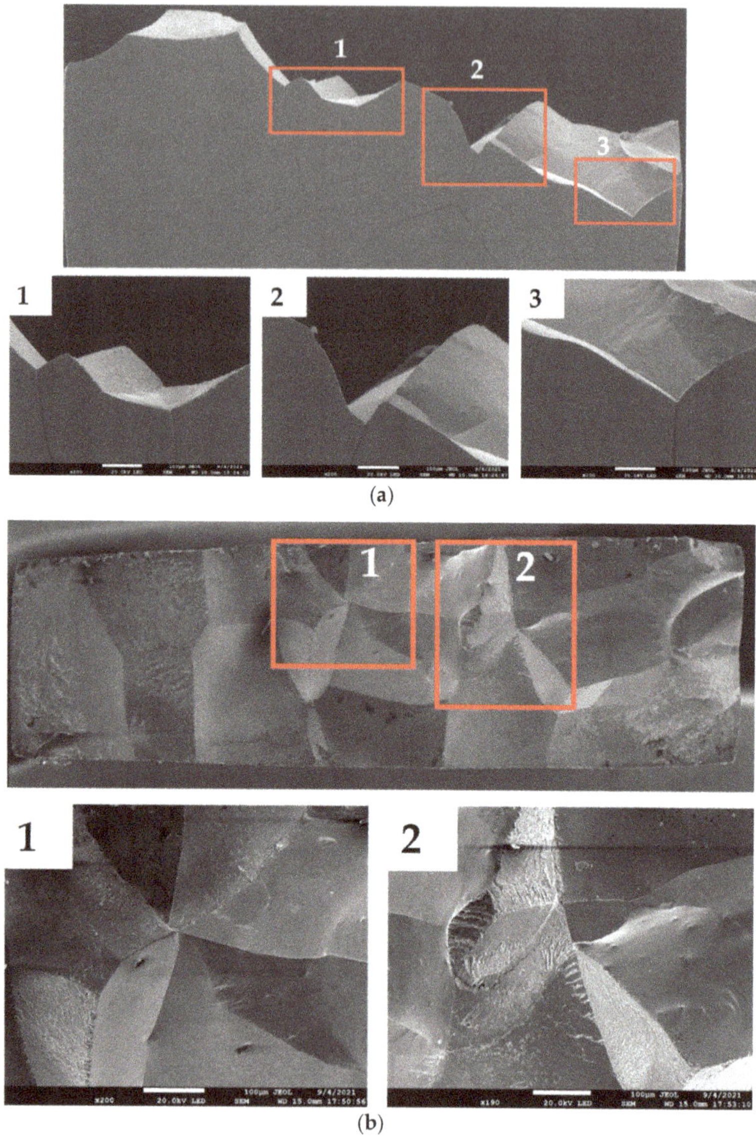

Figure 11. SEM images of 600 °C–48 h sample after three-point bending test. (**a**) Fracture surface of sample and (**b**) cross section of fracture surface.

5. Conclusions

In summary, the microstructure, thermo-magnetization, and shape memory properties of new cold-rolled FeNiCoAlTiNb SMA samples were investigated for the first time. The main conclusions can be listed as follows:

(1) The β phases were generated along the grain boundaries when the aging times were 48 and 72 h at 600 °C. The XRD results of cold-rolled samples after 1277 °C–1 h showed a strong peak intensity in the (111) plane.

(2) The cold-rolled sample aged at 600 °C for 24, 48 and 72 h showed phase transformation under the magnetic field of 0.05 and 7 Tesla. As the aging durations were

increased from 24 to 72 h, the martensitic transformation temperature shifted to a higher temperature, which indicates the transformation temperatures increase with an increase in the aging times.

(3) Recrystallization textures of near <100> orientation were developed using a combination of cold rolling with a reduction ratio of 97% and subsequent annealing at 1277 °C for 1 h. The average grain size was around 400 μm, and the intensity of texture was 13.64.

(4) Shape memory behavior was not observed in the 600 °C–24 h sample at a 2000 MPa stress level due to the fact that it possessed the lowest transformation temperature. The shape memory properties in the 600 °C–48 h sample with a recoverable strain of 1.6% at 200 MPa stress level were obtained using a three-point bending test. The observed recoverable strain values were lower than the theoretical values, possibly owing to the smaller fraction of low angle grain boundary, the formation of grain boundary precipitates and grain boundary constraint due to a slightly lower than expected intensity of texture in the samples. The thermal hysteresis was around 40 °C. The recoverable strain in the 600 °C–72 h sample was 1.1% at 100 MPa. The sample fractured at a stress level of 150 MPa due to the formation of a large amount of brittle β phases.

Author Contributions: Conceptualization, L.-W.T.; methodology, L.-W.T. and C.-H.C.; validation, L.-W.T. and C.-H.C.; investigation, L.-W.T., C.-H.C., W.-C.C., Y.C. and N.-H.L.; resources, L.-W.T. and C.-H.C.; data curation, W.-C.C., Y.C. and N.-H.L.; writing—original draft preparation, L.-W.T.; writing—review and editing, L.-W.T.; visualization, L.-W.T. and C.-H.C.; supervision, L.-W.T.; project administration, L.-W.T. All authors have read and agreed to the published version of the manuscript.

Funding: This research was funded by the Ministry of Science and Technology (MOST), grant numbers MOST 109-2221-E-018-010-MY2. This work was financially supported by the Young Scholar Fellowship Program of the Ministry of Science and Technology (MOST) in Taiwan, grant number MOST 110-2636-E-002-005.

Acknowledgments: The FeNiCoAlTiNb alloys were fabricated by the National Chung-Shan Institute of Science and Technology (NCSIST), which is gratefully acknowledged. The authors would like to thank Yung-Sheng Chen at the Instrumentation Center, National Tsing Hua University for the SQUID measurements. Thanks Ko-Kai Tseng at the High Entropy Materials Center, National Tsing Hua University for the cold rolling experiment.

Conflicts of Interest: The authors declare no conflict of interest.

References

1. Otsuka, K.; Wayman, C.M. *Shape Memory Materials*; Cambridge University Press: Cambridge, UK, 1998; pp. 117–132.
2. Jani, J.M.; Leary, M.; Subic, A.; Gibson, M.A. A review of shape memory alloy research, applications and opportunities. *Mater. Des.* **2014**, *56*, 1078–1113. [CrossRef]
3. Omori, T.; Kainuma, R. Martensitic transformation and superelasticity in Fe-Mn-Al-Based shape memory alloys. *Shape Mem. Superelasticity* **2017**, *3*, 334–355. [CrossRef]
4. Khalil, W.; Mikolajczak, A.; Bouby, C.; Zineb, T.B. A constitutive model for Fe-based shape memory alloy considering martensitic transformation and plastic sliding coupling: Application to a finite element structural analysis. *J. Intell. Mater. Syst. Struct.* **2012**, *23*, 1143–1160. [CrossRef]
5. Maki, T.; Kobayashi, K.; Minato, M.; Tamura, I. Thermoelastic martensite in an ausaged Fe-Ni-Ti-Co alloy. *Scr. Metall.* **1984**, *18*, 1105–1109. [CrossRef]
6. Sehitoglu, H.; Karaman, I.; Zhang, X.; Chumlyakov, Y.; Maier, H.J. Deformation of FeNiCoTi shape memory single crystals. *Scr. Mater.* **2001**, *44*, 779–784. [CrossRef]
7. Sehitoglu, H.; Zhang, X.Y.; Kotil, T.; Canadinc, D.; Chumlyakov, Y.; Maier, H.J. Shape memory behavior of FeNiCoTi single and polycrystals. *Metall. Mater. Trans. A* **2002**, *33*, 3661–3672. [CrossRef]
8. Tanaka, Y.; Himuro, Y.; Kainuma, R.; Sutou, Y.; Omori, T.; Ishida, K. Ferrous polycrystalline shape-memory alloy showing huge superelasticity. *Science* **2010**, *27*, 1488–1490. [CrossRef]
9. Omori, T.; Abe, S.; Tanaka, Y.; Lee, D.; Ishida, K.; Kainuma, R. Thermoelastic martensitic transformation and superelasticity in Fe-Ni-Co-Al-Nb-B polycrystalline alloy. *Scr. Mater.* **2013**, *69*, 812–815. [CrossRef]

10. Lee, D.; Omori, T.; Kainuma, K. Ductility enhancement and superelasticity in Fe-Ni-Co-Al-T-B polycrystalline alloy. *J. Alloy. Compd.* **2014**, *617*, 120–123. [CrossRef]
11. Otsuka, H.; Yamada, H.; Maruyama, T.; Tanahashi, H.; Matsuda, S.; Murakami, H. Effects of alloying additions on Fe-Mn-Si shape memory alloys. *ISIJ Int.* **1990**, *30*, 674–679. [CrossRef]
12. Sato, A.; Chishima, E.; Yamaji, Y.; Mori, T. Orientation and composition dependencies of shape memory effect in Fe-Mn-Si Alloys. *Acta Metall.* **1984**, *32*, 539–547. [CrossRef]
13. Ando, K.; Omori, T.; Ohnuma, T.; Kainuma, R.; Ishida, K. Ferromagnetic to weak-magnetic transition accompanied by bcc to fcc transformation in Fe-Mn-Al alloy. *Appl. Phys. Lett.* **2009**, *95*, 212504. [CrossRef]
14. Omori, T.; Watanabe, K.; Umetsu, R.Y.; Kainuma, R.; Ishida, K. Martensitic transformation and magnetic field-induced strain in Fe-Mn-Ga shape memory alloy. *Appl. Phys. Lett.* **2009**, *95*, 082508. [CrossRef]
15. Zhu, W.; Liu, E.K.; Feng, L.; Tang, X.D.; Chen, J.L.; Wu, G.H.; Liu, H.Y.; Meng, F.B.; Luo, H.Z. Magnetic-field-induced transformation in FeMnGa alloys. *Appl. Phys. Lett.* **2009**, *95*, 222512. [CrossRef]
16. Omori, T.; Ando, K.; Okano, M.; Xu, X.; Tanaka, Y.; Ohnuma, I.; Kainuma, R.; Ishida, K. Superelastic effect in polycrystalline ferrous alloys. *Science* **2011**, *333*, 68–71. [CrossRef] [PubMed]
17. Sawaguchi, T.; Maruyama, T.; Otsuka, H.; Kushibe, A.; Inoue, Y.; Tsuzaki, K. Design concept and applications of Fe-Mn-Si-based alloys -from shape-memory to seismic response control. *Mater. Trans.* **2016**, *57*, 283–293. [CrossRef]
18. Geng, Y.; Lee, D.; Xu, X.; Nagasako, M.; Jin, X.; Omori, T.; Kainuma, R. Coherency of ordered γ' precipitates and thermoelastic martensitic transformation in FeNiCoAlTaB alloys. *J. Alloy. Compd.* **2015**, *628*, 287–292. [CrossRef]
19. Fu, H.; Zhao, H.; Shilei, S.; Zhang, Z.; Xie, J. Evolution of the cold-rolling and recrystallization textures in FeNiCoAlNbB shape memory alloy. *J. Alloy. Compd.* **2016**, *628*, 287–292. [CrossRef]
20. Ma, J.; Kockar, B.; Evirgen, A.; Karaman, I.; Luo, Z.; Chumlyakov, Y. Shape memory behavior and tension–compression asymmetry of a FeNiCoAlTa single-crystalline shape memory alloy. *Acta Mater.* **2012**, *60*, 2186–2195. [CrossRef]
21. Ma, J.; Hornbuckle, B.; Karaman, I.; Thompson, G.B.; Luo, Z.; Chumlyakov, Y. The effect of nanoprecipitates on the superelastic properties of FeNiCoAlTa shape memory alloy single crystals. *Acta Mater.* **2013**, *61*, 3445–3455. [CrossRef]
22. Krooß, P.; Holzweissig, M.J.; Niendorf, T.; Somsen, C.; Schaper, M.; Chumlyakov, Y.I.; Maier, H.J. Thermal cycling behavior of an aged FeNiCoAlTa single-crystal shape memory alloy. *Scr. Mater.* **2014**, *81*, 28–31. [CrossRef]
23. Tseng, L.W.; Ma, J.; Karaman, I.; Wang, S.J.; Chumlyakov, Y. Superelastic response of the FeNiCoAlTi single crystals under tension and compression. *Scr. Mater.* **2015**, *101*, 1–4. [CrossRef]
24. Chumlyakov, Y.I.; Kireeva, I.V.; Kutz, O.A.; Turabi, A.S.; Karaca, H.E.; Karaman, I. Unusual reversible twinning modes and giant superelastic strains in FeNiCoAlNb single crystals. *Scr. Mater.* **2016**, *119*, 43–46. [CrossRef]
25. Choi, W.S.; Pang, E.L.; Choi, P.P.; Schuh, C.A. FeNiCoAlTaB superelastic and shape-memory wires with oligocrystalline grain structure. *Scr. Mater.* **2020**, *188*, 1–5. [CrossRef]
26. Abuzaida, W.; Sehitoglu, H. Shape memory effect in FeMnNiAl iron-based shape memory alloy. *Scr. Mater.* **2019**, *169*, 57–60. [CrossRef]
27. Vollmer, M.; Arold, T.; Kriegel, M.J.; Klemm, V.; Degener, S.; Freudenberger, J.; Niendorf, T. Promoting abnormal grain growth in Fe-based shape memory alloys through compositional adjustments. *Nat. Commun.* **2019**, *10*, 2337. [CrossRef]
28. Omori, T.; Okano, M.; Kainuma, R. Effect of grain size on superelasticity in Fe-Mn-Al-Ni shape memory alloy wire. *APL Mater.* **2013**, *1*, 032103. [CrossRef]
29. Ozcan, H.; Ma, J.; Wang, S.J.; Karaman, I.; Chumlyakov, Y.; Brown, J.; Noebe, R.D. Effects of cyclic heat treatment and aging on superelasticity in oligocrystalline Fe-Mn-Al-Ni shape memory alloy wires. *Scr. Mater.* **2017**, *134*, 66–70. [CrossRef]
30. Ozcan, H.; Ma, J.; Karaman, I.; Chumlyakov, Y.I.; Santamarta, R.; Brown, J.; Noebe, R.D. Microstructural design considerations in Fe-Mn-Al-Ni shape memory alloy wires: Effects of natural aging. *Scr. Mater.* **2018**, *142*, 153–157. [CrossRef]
31. Tseng, L.W.; Ma, J.; Wang, S.J.; Karaman, I.; Kaya, M.; Luo, Z.P.; Chumlyakov, Y.I. Superelastic response of a single crystalline FeMnAlNi shape memory alloy under tension and compression. *Acta Mater.* **2015**, *89*, 374–383. [CrossRef]
32. Tseng, L.W.; Ma, J.; Hornbuckle, B.; Karaman, I.; Thompson, G.B.; Luo, Z.; Chumlyakov, Y. The effect of precipitates on the superelastic response of [100] oriented FeMnAlNi single crystals under compression. *Acta Mater.* **2015**, *97*, 234–244. [CrossRef]
33. Zhang, C.; Zhu, C.; Shin, S.; Casalena, L.; Vecchio, K. Multifunctional non-equiatomic high entropy alloys with superelastic, high damping, and excellent cryogenic properties. *Adv. Eng. Mater.* **2019**, *21*, 1800941. [CrossRef]
34. Czerny, M.; Cios, G.; Maziarz, W.; Chumlyakov, Y.; Chulist, R. Studies on the two-step aging process of Fe-based shape memory single crystals. *Materials* **2020**, *13*, 1724. [CrossRef]
35. Czerny, M.; Maziarz, W.; Cios, G.; Wojcik, A.; Chumlyakov, Y.I.; Schell, N.; Fitta, M.; Chulist, R. The effect of heat treatment on the precipitation hardening in FeNiCoAlTa single crystals. *Mater. Sci. Eng. A* **2020**, *784*, 139327. [CrossRef]
36. Chumlyakov, Y.I.; Kireeva, I.V.; Pobedennaya, P.; Krooβ, P.; Niendorf, T. Rubber-like behaviour and superelasticity of [1]-oriented FeNiCoAlNb single crystals containing γ- and β-phase particles. *J. Alloy. Compd.* **2021**, *856*, 158158. [CrossRef]
37. Chumlyakov, Y.I.; Kireeva, I.V.; Kuksgauzen, D.A.; Niendorf, T.; Krooβ, P. Tension-compression asymmetry of the superelastic behavior of high-strength [1]-oriented FeNiCoAlNb crystals. *Mater. Lett.* **2021**, *289*, 129395. [CrossRef]
38. Chumlyakov, Y.I.; Kireeva, I.V.; Pobedennaya, Z.V.; Krooβ, P.; Niendorf, T. Shape memory effect and superelasticity of [1]-oriented FeNiCoAlNb single crystals aged under and without stress. *Metals* **2021**, *11*, 943. [CrossRef]

39. Poklonov, V.; Chumlyakov, Y.; Kireeva, I.; Lyamkind, S. Thermoelastic martensitic transformation in single crystals of FeNiCoAlTiNb alloy. *AIP Conf. Proc.* **2017**, *1909*, 020174.
40. Tseng, L.W.; Tzeng, Y.C.; Tsai, Y.L.; Chumlyakov, Y.I. Microstructure investigation of new iron-based FeNiCoAlTiNb shape memory alloys. *Results Mater.* **2021**, *10*, 1001881–1001887.

crystals

MDPI

Communication

Investigation on the Effects of Grain Boundary on Deformation Behavior of Bicrystalline Pillar by Crystal Plasticity Finite Element Method

Hui Zhou, Pei Wang * and Shanping Lu *

Institute of Metal Research, Chinese Academy of Sciences, Shenyang 110016, China; hzhou15b@alum.imr.ac.cn
* Correspondence: pwang@imr.ac.cn (P.W.); shplu@imr.ac.cn (S.L.);
Tel.: +86-24-23970106 (P.W.); +86-24-23971429 (S.L.)

Abstract: A dislocation density–grain boundary interaction scheme coupled with the dislocation density-based crystalline plasticity finite element method has been established and used to investigate the deformation behavior of bicrystalline pillars with the same grain boundary misorientation angle but different crystal orientations. It is found that the angle between the activated slip systems, which is determined by the crystal orientations, rather than the grain boundary misorientation angle, influences the interactions between the plastic slip and the grain boundary, which further influence the heterogeneous deformation of bicrystalline specimens.

Keywords: grain boundary; crystallographic orientation; crystallographic plasticity; deformation behavior

Citation: Zhou, H.; Wang, P.; Lu, S. Investigation on the Effects of Grain Boundary on Deformation Behavior of Bicrystalline Pillar by Crystal Plasticity Finite Element Method. *Crystals* **2021**, *11*, 923. https://doi.org/10.3390/cryst11080923

Academic Editor: Wojciech Polkowski

Received: 11 July 2021
Accepted: 5 August 2021
Published: 9 August 2021

1. Introduction

Grain boundaries (GBs) have significant influence on the mechanical properties of polycrystalline materials by interacting with dislocations and other defects [1–3]. Therefore, the interactions between GBs and dislocations have attracted much attention for many decades [4–9]. Due to the complexity, the detailed interaction between GB and dislocation has not been understood thoroughly to date [3,10,11]. However, it is well accepted that the interactions between dislocations and GBs have a close relationship with the GB misorientation angle [2,3]. Dislocations trend to pile up at high-angle GBs (HAGBs) and pass through low-angle GBs (LAGBs) in popular opinion [12–14]. Therefore, the angle of GBs is employed to estimate the effects of GBs on deformation behavior in many studies.

In recent years, bicrystalline micro-pillar experiments have been widely used to investigate the effects of GBs on the deformation behavior of polycrystals [15–17]. Kheradmand et al. [18,19] have observed slip transfer LAGBs in a bicrystalline micro-pillar, which is consistent with the previous understanding. However, there are some phenomena that contradict the common view of dislocation transmission in the bicrystalline micro-pillar with HAGBs. For example, although the work of [19] has shown no slip transfer in the HAGB in a bicrystalline pillar, as anticipated, the works of [10,11,20] have successfully revealed slip transfer HAGBs in some bicrystalline pillars. Although all the GBs are HAGBs, the orientations of component crystals and the directions of load are randomly distributed, which may induce the different activation behavior of slip systems and further influence the slip transfer [17,21]. This may be the reason for the contradictory effects of HAGBs on slip transfer. It implies that the consideration of GB misorientation angle alone would not lead to fully understanding the slip transfer mode of bicrystalline specimens.

Due to the complicated development of stress and strain formed in the deformation process of bicrystals, simulations have been widely used to investigate the stress and strain evolution [22]. Raabe and Lu et al. [14,23] have employed phenomenological crystal plasticity models to qualitatively analyze the effects of GB misorientation angle on the local plastic deformation of bicrystals and found the blocking effect of GBs on plastic strain. However, the interactions between dislocations and GBs were not considered in the models. Therefore, those investigations cannot reveal the microscopic mechanism behind the effect of GBs on the plastic deformation behavior [22]. Generally, plastic deformation is the result of dislocation motion [9]. Therefore, it is necessary to establish a model that based on the evolution of dislocations to reveal the underlying mechanism. Zikry et al. [24,25] have established a dislocation density-based crystal plasticity finite element method, and coupled it with dislocation density–grain boundary interaction (DDGBI) schemes [5,26,27]. Using this model, Zikry et al. [24,28] have revealed that the different plastic slips between the component crystals influence the pile-up of dislocations at GBs in bicrystalline specimens [24]. This implies that to understand the effects of GBs on the deformation behavior of specimens, it is necessary to gain an insight into the interactions between the GB and activated slip systems. The activity of slip systems is directly influenced by crystal orientations. However, the effects of crystal orientations on the interactions between slip and GBs in bicrystalline specimens have been overlooked. Therefore, the effects of GBs on the deformation behavior of bicrystalline specimens with HAGBs are still unclear. In fact, this unclear point is also found in polycrystal material.

For a detailed understanding of the effects of GBs on the deformation of bicrystalline specimens, a DDGBI scheme coupled with the dislocation density-based crystalline plasticity finite element method has been established and used to investigate the deformation behavior of three high-purity nickel (99.99%) bicrystalline pillars with different crystal orientations but an identical GB misorientation angle (38.7°). The global coordinate systems and schematic of the bicrystalline pillar are shown in Figure 1a. The detailed orientations of the component crystals of each bicrystalline pillar are listed in Table 1. The loading direction is along [001] of global coordinate systems. The diameter and height of samples are 1.4 μm and 2.1 μm, respectively.

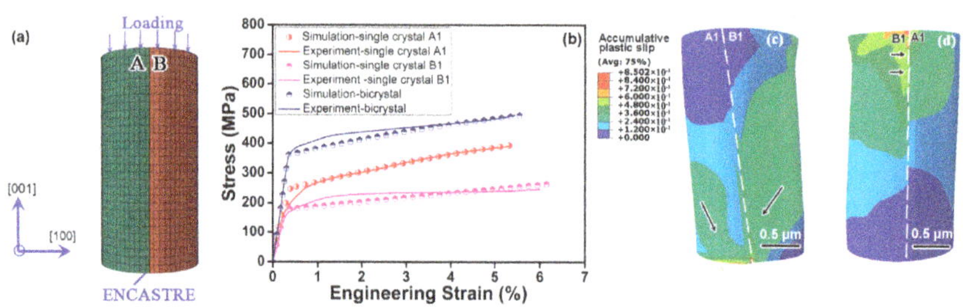

Figure 1. (**a**) Schematic of bicrystalline pillar, (**b**) engineering stress–strain curves of specimens that are used to determine the material parameters. The contours of the accumulative plastic slip on the surface of specimen BC1, (**c**) on the front side and (**d**) on the reverse side.

Table 1. Crystal orientations of the three bicrystalline specimens.

Specimen	Crystal	Direction	Crystal Orientation	Crystal	Direction	Crystal Orientation
BC1	*A*1	[001] [100]	[1 5 3] [1 −2 3]	*B*1	[001] [100]	[6 5 5] [5 −17 11]
BC2	*A*2	[010] [100]	[1 5 3] [1 −2 3]	*B*2	[010] [100]	[6 5 5] [5 −17 11]
BC3	*A*3	[100] [001]	[1 5 3] [1 −2 3]	*B*3	[100] [001]	[6 5 5] [5 −17 11]

2. Crystal Plasticity Simulation Model

The constitutive model used in this study is based on the theories proposed by Asaro [29] and Zikry [30]. The equations of the shear strain rate, reference stress of slip systems and dislocation density have been introduced in detail in [24,31–33]. Here, only the DDGBI scheme is described briefly. To obtain a detailed understanding of DDGBI, a transmission scalar for each slip system, $\lambda_{AB}^{\alpha\beta}$, was proposed [9] as

$$\lambda_{AB}^{\alpha\beta} = \left(n_A^{\alpha} \cdot n_B^{\beta}\right)\left(s_A^{\alpha} \cdot s_B^{\beta}\right) \tag{1}$$

where s is the current slip vector and n is the normal vector. The superscript α and β correspond to the slip system α and β in grain A and B, respectively. Grain A and B are separated by GBs. Due to the orthogonality of the unit vector, $\lambda_{AB}^{\alpha\beta}$ varies from zero to one, where complete dislocation blockage and complete dislocation transmission correspond to a value of zero and one, respectively. Therefore, $\lambda_{AB}^{\alpha\beta}$ can be employed as an index to indicate whether the slip on the α slip system in grain A transmits across the GB to the β slip system in grain B or not [24].

Additionally, slip transmission only occurs between the activated slip systems [21]. Therefore, the second criterion of slip transmission between slip systems in the adjacent grains can be expressed as

$$\left|\tau_A^{(\alpha)}\right| \geq \tau_c^{(\alpha)},\ \left|\tau_B^{(\beta)}\right| \geq \tau_c^{(\beta)} \tag{2}$$

where τ is the resolved shear stress, and the subscript c corresponds to the critical stress to drive the motion of dislocations. Moreover, definition of the slip systems used in the crystal plasticity model and the Schmid factors of the slip system for the three specimens are listed in Table 2.

Table 2. Definition of the slip systems used in the crystal plasticity model and the Schmid factors of the slip system for the three specimens.

	Slip Plane	Slip Direction	Schmid Factor					
			*A*1	*A*2	*A*3	*B*1	*B*2	*B*3
γ_1	1 1 1	0 −1 1	0.197	0.082	0.292	0.060	0.026	0.026
γ_2	1 1 1	1 0 −1	0.208	0.326	0.117	0.065	0.080	0.006
γ_3	1 1 1	−1 1 0	0.405	0.245	0.175	0.005	0.055	0.021
γ_4	−1 1 1	1 0 1	0.337	0.326	0	0.209	0.046	0.165
γ_5	−1 1 1	1 1 0	0.486	0.489	0	0.209	0.323	0.124
γ_6	−1 1 1	0 −1 1	0.149	0.163	0	0.022	0.277	0.289
γ_7	1 −1 1	0 1 1	0.069	0.082	0.175	0.285	0.101	0.186
γ_8	1 −1 1	1 1 0	0.052	0.245	0.175	0.313	0.059	0.371
γ_9	1 −1 1	1 0 −1	0.017	0.326	0.350	0.028	0.160	0.186
γ_{10}	1 1 −1	0 1 1	0.278	0.163	0.117	0.285	0.412	0.129
γ_{11}	1 1 −1	1 0 1	0.146	0.326	0.467	0.313	0.037	0.251
γ_{12}	1 1 −1	−1 1 0	0.132	0.489	0.350	0.029	0.437	0.475

Figure 1a shows the finite element mesh of the pillar for uniaxial compress deformation and the boundary condition in commercial ABAQUS software. A 3D solid element, C3D8, with eight nodes and one integration point, was employed. The mesh density in the regions adjacent to GBs was slightly higher than other regions. In order to verify the mesh convergence, mesh with different sizes in the regions adjacent to GBs was considered. The mesh size of the model was about 50 nm. The displacement boundary conditions were applied to specimens as shown in Figure 1a. For uniaxial compression, a prescribed displacement was imposed on the top surface of specimens, and the fixed condition (ENCASTRE) was applied on the bottom surface of specimens. Additionally, the strain rate was 0.001 s^{-1} during the loading.

To obtain the material parameters to be used in simulation, deformation behavior of the single crystalline pillars, crystal *A*1 and *B*1, was simulated first. Based on the relationships of the dislocation density evolution coefficients [24] and engineering stress–strain curves of the specimens [19], the material parameters were obtained. Thereafter, the accuracy of material parameters and DDGBI model were verified by simulating the deformation of specimen BC1 first. The engineering stress–strain curves of these samples are shown in Figure 1b. It can be observed that the simulations are in good agreement with the experiments. Figure 1c shows the accumulative plastic slip on the surface of specimen BC1 from the front side of the specimen. It is clear that the plastically deformed area in grain *B*1 is larger than that in grain *A*1. Moreover, the area close to the GB in grain *B*1 has higher plastic slip than other areas. The similar phenomenon is also observed from the reverse side (Figure 1d). Those phenomena are consistent with the distribution of slip bands in experiments reported in [19]. Those results suggest that the DDGBI scheme and the material coefficients can be used to study the heterogeneous deformation of bicrystals. All the parameters that were used in the simulation are listed in Table 3.

Table 3. Material parameters used in the crystal plasticity finite element simulations.

Parameter	Symbol	Value	Unit	Reference
Elastic moduli	C_{11}	246,500	MPa	
	C_{12}	147,300	MPa	
	C_{44}	124,700	MPa	
Reference strain rate	$\dot{\gamma}_{ref}^{(\alpha)}$	0.001	1/s	[33]
Rate sensitivity parameter	m	20	1	[33]
Static yield stress	τ_y	120	MPa	
Interaction coefficient	$\alpha_{\alpha\beta}$	0.2	1	[32]
	$\alpha_{\alpha\alpha}$	0.6	1	[32]
Initial dislocation density	ρ_{im0}	10,000	mm^{-2}	[24]
	ρ_{m0}	10	mm^{-2}	[24]
Dislocation density evolution coefficient	g_{sour}	4.52×10^{-5}	1	
	g_{minter}	5.53	1	
	g_{immob}	1.626×10^{-3}	1	
	g_{recor}	6.67	1	[24]
Burgers vector	b	3.0×10^{-7}	mm	[24]

3. Results

Figure 2 shows the contours of elastic stress on the longitudinal section of the three specimens after 0.05% engineering strain. For specimen BC1, elastic stress on crystal *B*1 is higher than that on crystal *A*1 (Figure 2a), while the difference in the elastic stress between crystal *A*2 and *B*2 is slightly smaller in specimen BC2 (Figure 2b). For specimen BC3, the contours of elastic stress on crystal *A*3 and *B*3 are almost same (Figure 2c). Those phenomena reveal that crystal orientations have great influence on the distribution of elastic stress in bicrystalline pillars.

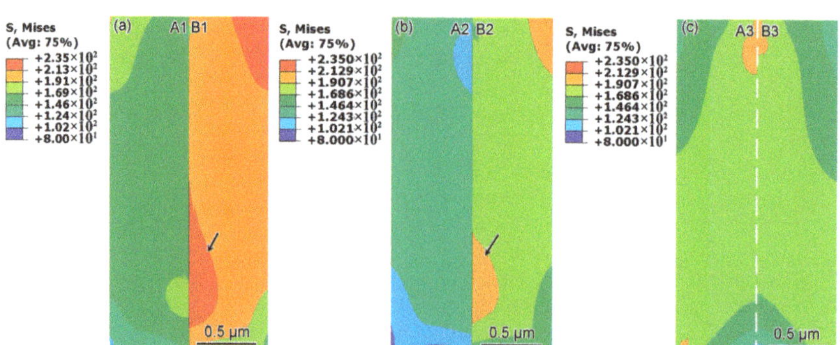

Figure 2. Contours of stress of the three specimens when the engineering strain is 0.05%. (**a**) BC1, (**b**) BC2 and (**c**) BC3.

In the elastic regime, the stresses distributed on the component crystals are influenced by their elastic moduli in the loading direction, which can be given [34] by Equation (3):

$$\sigma_A = \frac{\sigma_T A_T}{A_A + \frac{E_B A_B}{E_A}}, \ \sigma_B = \frac{\sigma_T A_T}{A_B + \frac{E_A A_A}{E_B}} \tag{3}$$

where σ_T is the total stress applied to the bicrystalline pillar, A_T, A_A and A_B are the cross-section areas of the bicrystalline pillar and the component crystal A and B, respectively. E_A and E_B are the elastic moduli for crystal A and B in the loading direction, respectively. In this simulation, $A_A = A_B$. Therefore, Equation (3) can be expressed as

$$\sigma_A = \frac{2E_A\sigma_T}{E_B + E_A}, \ \sigma_B = \frac{2E_B\sigma_T}{E_B + E_A} \tag{4}$$

The orientation dependence of the elastic moduli E in a uniaxial stress state $\vec{\chi}$ direction is given [34] by

$$\frac{1}{E_{\vec{x}}} = \frac{C_{11} + C_{12}}{(C_{11} - C_{12})(C_{11} + 2C_{12})} + \left(\frac{1}{C_{44}} - 2\frac{1}{C_{11} - C_{12}}\right)\frac{\left(x_1^2x_2^2 + x_2^2x_3^2 + x_3^2x_1^2\right)}{\|\chi\|^4} \tag{5}$$

Based on Equations (4) and (5), the elastic moduli of the component crystals in a uniaxial loading direction, and the ratios of the stresses on the component crystals to the total stresses of specimens, were calculated and are listed in Table 4. The difference in the distributed stresses between crystal A and B in specimen BC1, BC2 and BC3 is $0.346\sigma_T$, $0.272\sigma_T$ and $0.022\sigma_T$, respectively. Therefore, the difference in the stress distribution between the component crystals in these three specimens decreases gradually (Figure 2). The above analysis illustrates that crystal orientation determines the elastic moduli and further influences stress proportions of the components in the elastic deformation stage.

Table 4. Elastic moduli and the ratio of the applied stresses on the component crystals to the total stress.

Sample	**BC1**		**BC2**		**BC3**	
Crystal	*A*1	*B*1	*A*2	*B*2	*A*3	*B*3
E(MPa)	182,000	258,000	156,000	205,000	208,300	203,600
σ/σ_T	0.827	1.173	0.864	1.136	1.011	0.989

The different stress proportions of the components in the elastic deformation stage will induce different levels of initiation and development of plastic deformation. To reveal the initiation of plastic deformation, the evolutions of plastic slip at two special nodes close to the GB were analyzed. The locations of the nodes on the three specimens are shown in Figure 3a–c. For specimen BC1, node L1 starts plastic deformation when the engineering strain is 0.14% (Figure 3d). For node R1, plastic deformation begins when the engineering strain is 0.167% (Figure 3e), which is later than node L1. This is because the largest Schmid factor of crystal $A1$ is 0.486 on γ_5, and the largest Schmid factor of grain $B1$ is 0.313 on γ_8 and γ_{11}, as listed in Table 2. Combining the different stress proportions with the Schmid factors of the component crystals, the resolved shear stress on γ_5 of crystal $A1$ is higher than those on γ_8 and γ_{11} of crystal $B1$. After the initiation of plastic deformation, plastic slips interact with the GB and influence the evolution of the plastic slip. For specimen BC1, the transmission scalars $\lambda^{5,8}_{A1,B1}$ and $\lambda^{5,11}_{A1,B1}$ are 0.126, and the other transmission scalars are equal to zero. This means the plastic slip on γ_5 of crystal $A1$ almost cannot transmit to γ_8 and γ_{11} of crystal $B1$. Therefore, the distribution of plastic slip on γ_5 of crystal $A1$ is very different from the plastic slip on γ_8 and γ_{11} of crystal $B1$ on the two sides of the GB, i.e., the plastic deformation is not continuous in the vicinity of the GB. This is confirmed by the contours of the plastic slip on the GB of specimen BC1, as shown in Figure 4a–c. It implies that the GB is an obstacle to the development of plastic slip when the transmission scalars of the activated slip systems are low. Therefore, the accumulative plastic slip on specimen BC1 is heterogeneous and there is a sharp change in the distribution of plastic slip between *B*1 and *A*1 (Figure 3a).

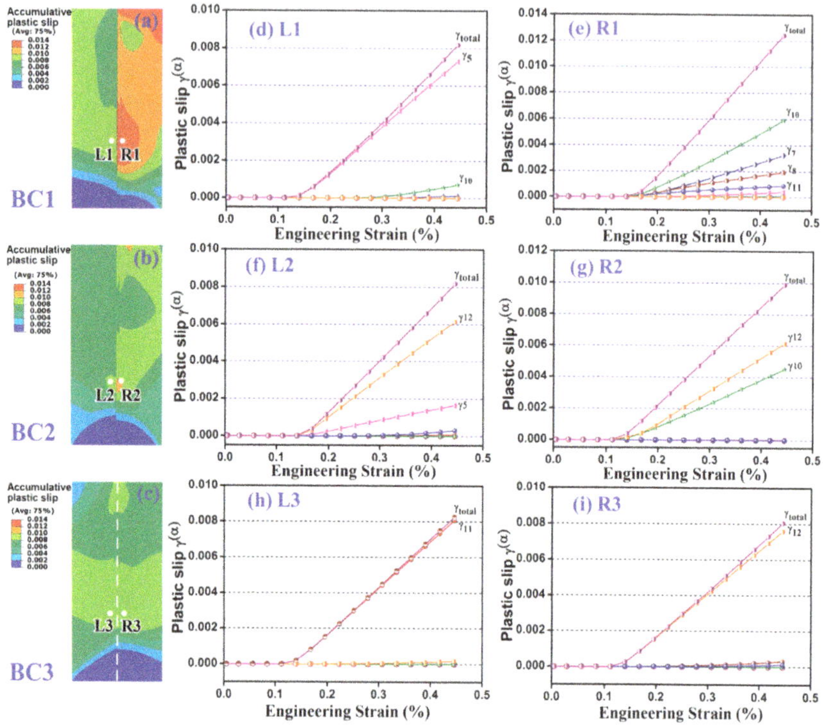

Figure 3. (**a**–**c**) Field of accumulative plastic slip on the longitudinal section of BC1, BC2 and BC3, respectively, after 0.4% strain. L1 (**d**), R1 (**e**), L2 (**f**), R2 (**g**), L3 (**h**) and R3 (**i**).

Similarly, the combined effects of Schmid factors and the stress proportion lead to the plastic deformation at node R2 beginning earlier than that at L2, as shown in Figure 3f,g. Additionally, the plastic deformation at R2 is mainly contributed to by the plastic slip on γ_{12} and γ_{10}, while it is γ_{12} and γ_5 at L2. For specimen BC2, the transmission scalars $\lambda_{A2,B2}^{12,10}$ and $\lambda_{A2,B2}^{12,12}$ are 0.887 and 0.826, respectively. Due to the higher transmission scalar, the plastic slip on γ_{12} of crystal *A*2 can transmit to γ_{10} and γ_{12} of crystal *B*2, as marked by arrows in Figure 4d–f, which promotes continuous plastic deformation in the vicinity of the GB. Therefore, the accumulative plastic deformation on the component crystals is nearly homogeneous (Figure 3b).

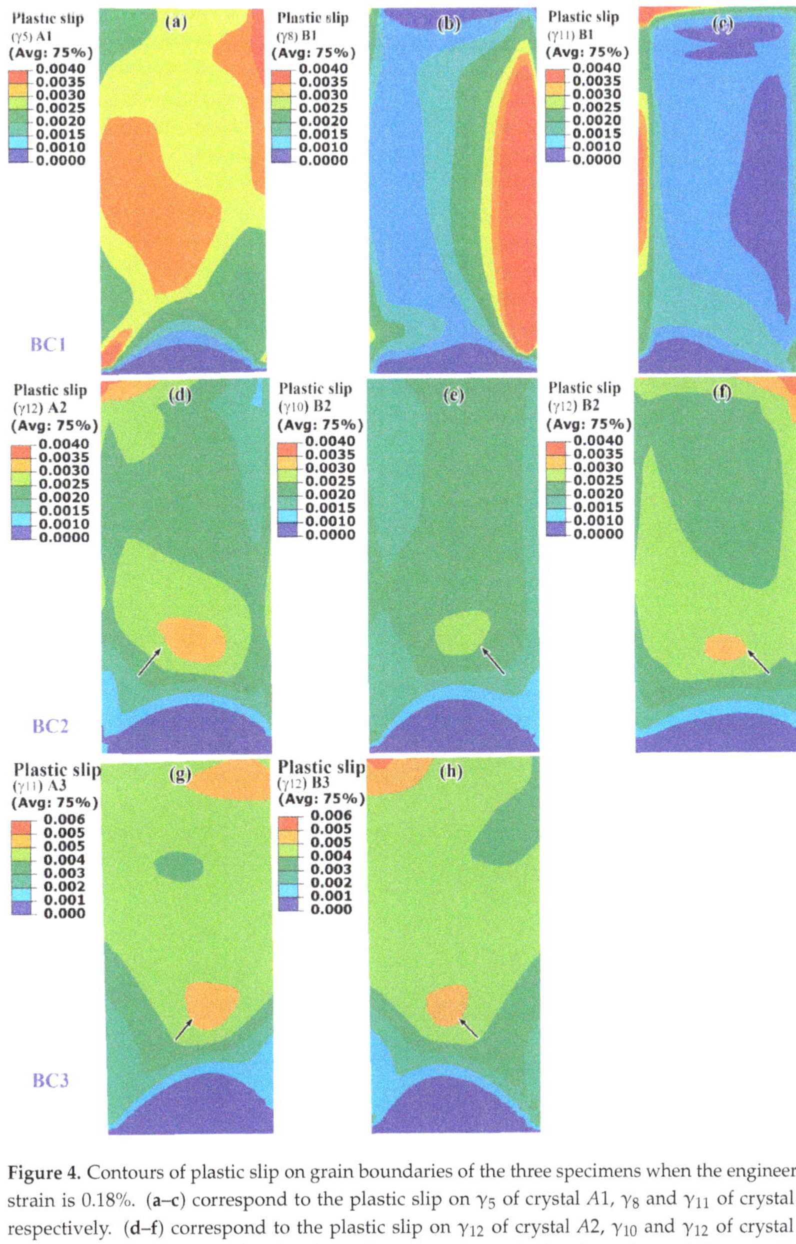

Figure 4. Contours of plastic slip on grain boundaries of the three specimens when the engineering strain is 0.18%. (**a–c**) correspond to the plastic slip on γ_5 of crystal $A1$, γ_8 and γ_{11} of crystal $B1$, respectively. (**d–f**) correspond to the plastic slip on γ_{12} of crystal $A2$, γ_{10} and γ_{12} of crystal $B2$, respectively. (**g**,**h**) correspond to the plastic slip on γ_{11} of crystal $A3$ and γ_{12} of crystal $B3$, respectively.

For specimen BC3, the plastic deformation begins almost simultaneously on the two component crystals, as shown in Figure 3h,i. The plastic deformation of crystal $A3$ and $B3$ is mainly contributed to by the plastic slip on γ_{11} and on γ_{12}, respectively. As the transmission scalar $\lambda_{A3,B3}^{11,12}$ is 0.891, the plastic slip on γ_{11} of crystal $A3$ and on γ_{12} of crystal $B3$ can transmit the GB easily. Therefore, the accumulative plastic slips on γ_{11} of crystal $A3$ and γ_{12} of crystal $B3$ are nearly the same, as marked by the arrow in Figure 4g,h, which induces continuous accumulative plastic slip on the GB in the BC3 specimen (Figure 3c). The differences in the accumulative plastic slip among the three bicrystalline pillars reveal that the crystal orientations determine the activation of slip systems and further influence the interaction between plastic slip and GBs, and the angle between the activated slip systems of component crystals are the critical factor influencing the homogeneity of deformation.

Besides these above three specimens, we performed the simulations on other bicrystalline micro-pillar specimens with different degrees of HAGBs, such as 45°, 55°, 75°, etc. Additionally, the same conclusion has been reached that the angle between the activated slip systems of component crystals are the critical factor influencing the homogeneity of deformation. Motivated by the discussion, the underlying mechanism of the effects of GBs on the heterogeneous plastic deformation of specimens can be understood from the evolution of the activated slip systems. For the specimens with LAGBs, the orientations of crystals are close to each other, thus the component crystals are stressed almost identically [19] in the elastic stage. Moreover, the Schmid factors of the same slip systems in the component crystals are nearly the same. The combined effects of the Schmid factors and applied stress induce same activated slip systems of the component crystals. Meanwhile, the angles between the slip systems of the component crystals are low. This ensures that the slip on the activated slip systems of the component crystals can transmit the GBs easily [35]. In contrast, for the specimens with HAGBs, the stress on the component crystals is always unequal and the Schmid factors of the same slip systems of the component crystals are different. The combined effects of applied stresses and Schmid factors may induce different activated slip systems in the two component crystals, as shown in Figure 3. The angles between these activated slip systems of the two components are varied, thus the corresponding transmission scalars are random, just as for the transmission scalars of specimen BC1–BC3. That is why the plastic deformation of the specimens with the same HAGBs are different (Figure 3a–c). This analysis suggests that besides the angle of GBs, the angle between the activated slip systems should also be considered to understand the plastic deformation of specimens.

4. Conclusions

In summary, the effects of GBs on the heterogeneous deformation of bicrystalline pillars have been investigated by the crystalline plasticity finite element method coupled with a dislocation density grain boundary interaction scheme. It is found that the orientations of the crystals determine the elastic moduli and stress proportions of the component crystals at the elastic deformation stage. The combined effects of stress proportions and Schmid factors further influence the initiation of plastic deformation. Additionally, then, the angle between the activated slip systems of the component crystals determines the interaction of slip with GBs, which finally influences the heterogeneous deformation of specimens.

Author Contributions: Conceptualization, H.Z., P.W. and S.L.; methodology, H.Z. and P.W.; software, P.W. and H.Z.; validation, H.Z., P.W. and S.L.; formal analysis, H.Z., P.W. and S.L.; investigation, P.W. and H.Z; resources, H.Z., P.W. and S.L.; data curation, H.Z., P.W. and S.L.; writing—original draft preparation, H.Z. and P.W.; writing—review and editing, H.Z. and P.W.; visualization, H.Z., P.W. and S.L.; supervision, P.W. and S.L.; project administration, P.W. and S.L.; funding acquisition, S.L. All authors have read and agreed to the published version of the manuscript.

Funding: This work was funded by the National Natural Science Foundation of China, Grant No. 51474203 and the Youth Innovation Promotion Association, Chinese Academy of Sciences, Grant No. 2013126.

Acknowledgments: The authors thank the financial support received from the National Natural Science Foundation of China (grant No. 51474203) and the Youth Innovation Promotion Association, Chinese Academy of Sciences (grant No. 2013126). The authors are grateful to Sheng Gang for helping with the subroutine.

Conflicts of Interest: The authors declare no conflict of interest.

References

1. Hamid, M.; Lyu, H.; Schuessler, B.J.; Wo, P.C.; Zbib, H.M. Modeling and characterization of grain boundaries and slip transmission in dislocation density-based crystal plasticity. *Crystals* **2017**, *7*, 152. [CrossRef]
2. Bieler, T.; Eisenlohr, P.; Zhang, C.; Phukan, H.; Crimp, M. Grain boundaries and interfaces in slip transfer. *Curr. Opin. Solid State Mater. Sci.* **2014**, *18*, 212–226. [CrossRef]
3. Hu, J.; Zhuang, Z.; Liu, F.; Liu, X.; Liu, Z. Investigation of grain boundary and orientation effects in polycrystalline metals by a dislocation-based crystal plasticity model. *Comput. Mater. Sci.* **2018**, *159*, 86–94. [CrossRef]
4. Shen, Z.; Wagoner, R.H.; Clark, W.A.T. Dislocation and grain boundary interactions in metals. *Acta Mater.* **1988**, *36*, 3231–3241. [CrossRef]
5. Bayerschen, E.; McBride, A.; Reddy, B.D.; Bohlke, T. Review on slip transmission criteria in experiments and crystal plasticity models. *J. Mater. Sci.* **2015**, *51*, 2243–2258. [CrossRef]
6. Adams, D.W.; Fullwood, D.T.; Wagoner, R.H.; Homer, E.R. Atomistic survey of grain boundary-dislocation interactions in FCC nickel. *Comput. Mater. Sci.* **2019**, *164*, 171–185. [CrossRef]
7. Patala, S. Understanding grain boundaries—The role of crystallography, structural descriptors and machine learning. *Comput. Mater. Sci.* **2019**, *162*, 281–294. [CrossRef]
8. Singh, D.; Parashar, A. Effect of symmetric and asymmetric tilt grain boundaries on the tensile behavior of bcc-Niobium. *Comput. Mater. Sci.* **2018**, *143*, 126–132. [CrossRef]
9. Malyar, N.; Micha, J.; Dehm, G.; Kirchlechner, C. Size effect in bi-crystalline micropillars with a penetrable high angle grain boundary. *Acta Mater.* **2017**, *129*, 312–320. [CrossRef]
10. Malyar, N.; Dehm, G.; Kirchlechner, C. Strain rate dependence of the slip transfer through a penetrable high angle grain boundary in copper. *Scr. Mater.* **2017**, *138*, 88–91. [CrossRef]
11. Kacher, J.; Eftink, B.; Cui, B.; Robertson, I.M. Dislocation interactions with grain boundaries. *Curr. Opin. Solid State Mater. Sci.* **2014**, *18*, 227–243. [CrossRef]
12. Guo, Y.Z.; Li, F.D.; Suo, T.; Tang, Z.B.; Li, Y.L. A close observation on the deformation behavior of bicrystal copper under tensile loading. *Mech. Mater.* **2013**, *62*, 80–89. [CrossRef]
13. Field, D.P.; Alankar, A. Observation of deformation and lattice rotation in a Cu bicrystal. *Met. Mater. Trans. A* **2010**, *42*, 676–683. [CrossRef]
14. Zaefferer, S.; Kuo, J.-C.; Zhao, Z.; Winning, M.; Raabe, D. On the influence of the grain boundary misorientation on the plastic deformation of aluminum bicrystals. *Acta Mater.* **2003**, *51*, 4719–4735. [CrossRef]
15. Chen, C.R.; Li, S.X.; Wang, Z.G. Characteristic of strain and resolved shear stress in a bicrystal with the grain boundary oeroendicular to the tensile axis. *Mater. Sci. Eng. A* **1998**, *247*, 15–22. [CrossRef]
16. Kiener, D.; Motz, C.; Dehm, G. Micro-compression testing: A critical discussion of experimental constraints. *Mater. Sci. Eng. A* **2009**, *505*, 79–87. [CrossRef]
17. Tiba, I.; Richeton, T.; Motz, C.; Vehoff, H.; Berbenni, S. Incompatibility stresses at grain boundaries in Ni bicrystalline micropillars analyzed by an anisotropic model and slip activity. *Acta Mater.* **2015**, *83*, 227–238. [CrossRef]
18. Kheradmand, N.; Vehoff, H.; Barnoush, A. An insight into the role of the grain boundary in plastic deformation by means of a bicrystalline pillar compression test and atomistic simulation. *Acta Mater.* **2013**, *61*, 7454–7465. [CrossRef]
19. Kheradmand, N.; Knorr, A.F.; Marx, M.; Deng, Y. Microscopic incompatibility controlling plastic deformation of bicrystals. *Acta Mater.* **2016**, *106*, 219–228. [CrossRef]
20. Weaver, J.S.; Li, N.; Mara, N.A.; Jones, D.R.; Cho, H.; Bronkhorst, C.A.; Fensin, S.J.; Gray, G.T. Slip transmission of high angle grain boundaries in body-centered cubic metals: Micropillar compression of pure Ta single and bi-crystals. *Acta Mater.* **2018**, *156*, 356–368. [CrossRef]
21. Mayeur, J.R.; Beyerlein, I.J.; Bronkhorst, C.A.; Mourad, H.M. Incorporating interface affected zones into crystal plasticity. *Int. J. Plast.* **2015**, *65*, 206–225. [CrossRef]
22. Su, Y.; Zambaldi, C.; Mercier, D.; Eisenlohr, P.; Bieler, T.R.; Crimp, M.A. Quantifying deformation processes near grain boundaries in α titanium using nanoindentation and crystal plasticity modeling. *Int. J. Plast.* **2016**, *86*, 170–186. [CrossRef]
23. Lu, F.; Guang, Z.; Ke-Shi, Z. Grain boundary effects on the inelastic deformation behavior of bicrystals. *Mater. Sci. Eng. A* **2003**, *361*, 83–92. [CrossRef]
24. Zikry, M.; Kao, M. Inelastic microstructural failure mechanisms in crystalline materials with high angle grain boundaries. *J. Mech. Phys. Solids* **1996**, *44*, 1765–1798. [CrossRef]
25. Wu, Q.; Zikry, M.A. Microstructural modeling of transgranular and intergranular fracture in crystalline materials with coincident site lattice grain-boundaries: $\Sigma 3$ and $\Sigma 17b$ bicrystals. *Mater. Sci. Eng. A* **2016**, *661*, 32–39. [CrossRef]
26. Livingston, J.D.; Chalmers, B. Multiple slip in bicrystal deformation. *Acta Met.* **1957**, *5*, 322–327. [CrossRef]
27. Koning, M.D.; Miller, R.; Bulatov, V.V.; Abraham, F.F. Modelling grain boundary resistance in intergranular dislocation slip transmission. *Philos. Mag. A* **2009**, *82*, 2511–2527. [CrossRef]
28. Ashmawi, W.M.; Zikry, M.A. Prediction of grain-boundary interfacial mechanisms in polycrystalline materials. *J. Eng. Mater. Technol.* **2002**, *124*, 88–96. [CrossRef]
29. Asaro, R.J.; Rice, J.R. Strain localization in ductile single crystals. *J. Mech. Phys. Solids* **1977**, *25*, 309–338. [CrossRef]

30. Zikry, M.A. An accurate and stable algorithm for high strain-rate finite strain plasticity. *Comput. Struct.* **1994**, *50*, 337–350. [CrossRef]
31. Ashmawi, W.M.; Zikry, M.A. Effects of grain boundaries and dislocation density evolution on large strain deformation modes in FCC crystalline materials. *J. Comput. Mater. Des.* **2000**, *7*, 55–62. [CrossRef]
32. Shanthraj, P.; Zikry, M.A. Dislocation density evolution and interactions in crystalline materials. *Acta Mater.* **2011**, *59*, 7695–7702. [CrossRef]
33. Zhou, H.; Zhang, X.; Wang, P.; Lu, S. Crystal plasticity analysis of cylindrical holes and their effects on the deformation behavior of Ni-based single-crystal superalloys with different secondary orientations. *Int. J. Plast.* **2019**, *119*, 249–272. [CrossRef]
34. Hook, R.E.; Hirth, J.P. The deformation behavior of non-isoaxial bicrystals of Fe-3% Si. *Acta Met.* **1967**, *15*, 1099–1110. [CrossRef]
35. Bieler, T.R.; Alizadeh, R.; Peña-Ortega, M.; Llorca, J. An analysis of (the lack of) slip transfer between near-cube oriented grains in pure Al. *Int. J. Plast.* **2019**, *118*, 269–290. [CrossRef]

crystals

MDPI

Article

Some Issues on Crystal Plasticity Models Formulation: Motion Decomposition and Constitutive Law Variants

Peter Trusov, Alexey Shveykin * and Nikita Kondratev

Laboratory of Multilevel Structural and Functional Materials Modeling, Perm National Research Polytechnic University, 614990 Perm, Russia; tpv@matmod.pstu.ac.ru (P.T.); kondratevns@gmail.com (N.K.)
* Correspondence: alexey.shveykin@gmail.com

Abstract: In this paper, kinematic relations and constitutive laws in crystal plasticity are analyzed in the context of geometric nonlinearity description and fulfillment of thermodynamic requirements in the case of elastic deformation. We consider the most popular relations: in finite form, written in terms of the unloaded configuration, and in rate form, written in terms of the current configuration. The presence of a corotational derivative in the relations formulated in terms of the current configuration testifies to the fact that the model is based on the decomposition of motion into the deformation motion and the rigid motion of a moving coordinate system, and precisely the stress rate with respect to this coordinate system is associated with the strain rate. We also examine the relations of the mesolevel model with an explicit separation of a moving coordinate system and the elastic distortion of crystallites relative to it in the deformation gradient. These relations are compared with the above formulations, which makes it possible to determine how close they are. The results of the performed analytical calculations show the equivalence or similarity (in the sense of the response determined under the same influences) of the formulation and are supported by the results of numerical calculation. It is shown that the formulation based on the decomposition of motion with an explicit separation of the moving coordinate system motion provides a theoretical framework for the transition to a similar formulation in rate form written in terms of the current configuration. The formulation of this kind is preferable for the numerical solution of boundary value problems (in a case when the current configuration and, consequently, contact boundaries, are not known a priori) used to model the technological treatment processes.

Keywords: crystal plasticity; multilevel models; large strain; motion decomposition; constitutive law; anisotropic materials; corotational derivative

Citation: Trusov, P.; Shveykin, A.; Kondratev, N. Some Issues on Crystal Plasticity Models Formulation: Motion Decomposition and Constitutive Law Variants. *Crystals* **2021**, *11*, 1392. https://doi.org/10.3390/cryst11111392

Academic Editor: Wojciech Polkowski

Received: 13 October 2021
Accepted: 11 November 2021
Published: 15 November 2021

1. Introduction

In recent decades, extensive research efforts have been made to develop a multilevel approach to constructing constitutive models for metals and alloys via the introduction of internal variables [1–4]. The crystal plasticity models obtained within the framework of this approach provide an opportunity to explicitly describe changes in the structure of the material (its anisotropic physical and mechanical macro-properties) and deformation mechanisms [5–9]. This validates the application of these models for studying and improvement of materials processing technologies, as well as for solving the problems associated with the development of new functional materials.

Crystal plasticity models can be divided into three classes. In statistical models aimed at describing the behavior of representative macrovolumes, consideration is given to a sample of uniformly deformed grains (reviews of the models of this class are provided in [10–12]. The application of modified statistical models makes it possible to take into account the interaction of neighboring grains, e.g., during the explicit description of the motion of dislocations across the boundary [13]. Self-consistent models are used to investigate the behavior of uniformly deformed grains in a matrix representing the effective characteristics of polycrystals. These models provide the fulfillment of equilibrium conditions; the effective characteristics are determined using statistical averaging methods. Reviews of the self-consistent models are given in [7,10,14]. Direct models based on the finite element method are recognized to be the most effective models for analyzing the non-uniform stress-strain state of grains (reviews of these models can be found in [7–9,15,16].

The crystal plasticity models mentioned above differ in the way of aggregation (combining the elements of a lower scale level into one element of a higher scale level), but all these models are based on the mesolevel relations used to describe crystallites, which led to the name "crystal plasticity".

In the last decade, numerous attempts have been made to develop crystal plasticity models based on generalized, mainly gradient, continua [17–20]. However, most authors assumed that the material under study is a simple (first-order) material [21]. In this case, the basic structure of the mesolevel constitutive model aimed at taking into account intragranular dislocation slip and lattice rotation is composed of the following equations: constitutive law for the relationship between stresses and elastic component of deformation, kinematic relations (including equations for determining the lattice spin), relation for the determination of the inelastic component of deformation and equations for the description of hardening.

The description of hardening (usually through the evolutionary relations for the introduced internal variables "critical shear stresses") is the important element of the model; there are many works devoted to the development of crystal plasticity in this direction (we mention here only some of the well-known works [22–24]). To date, other ratios of the basic structure of the model can be considered as the typical ratios, and therefore they are often dropped from the consideration in papers. However, researchers use several different types of basic formulations that differ in kinematic and constitutive relations. In our opinion, it is important to undertake a thorough analysis of these relations, which will enable their comparison in the context of geometric nonlinearity description and fulfillment of thermodynamic requirements. This is especially important given the fact that an intensively developing direction in crystal plasticity is related to the application of constitutive models for describing other, side by side with intragranular dislocation sliding, significant mechanisms and processes, such as twinning [8,25–27], phase transitions [8,28,29], grain boundary sliding [13,30–32], recrystallization [33–35]. In order to create advanced models, it is necessary to make a reasonable choice of a platform (the basic structure of the model).

In Section 2, popular formulations of mesolevel models are presented and analyzed. In the finite form, the models are formulated in the unloaded configuration, and in the rate form, in the actual configuration. The incorporation of a corotational derivative into the models formulated in the actual configuration means that the motion is decomposed into a rigid motion of the moving coordinate system and a deformation motion of the medium relative to it; the stress rate relative to the moving coordinate system is associated with the rate of this deformation. Section 3 considers the structure of kinematic and constitutive relations in the case when the component responsible for the rigid rotation of a moving coordinate system, relative to which the elastic distortions of the lattice are determined, is explicitly separated in the deformation gradient. An analytical comparison of different formulations and the results of numerical calculations are given in Section 4.

2. Some Well-Known Variants of Base Mesolevel Relations

In a number of works, linear constitutive relations with isotropic elastic law are used to describe small deformations [36]. However, since crystal plasticity models are usually focused on describing the behavior of materials at large displacement gradients, including changes in anisotropic properties at the macrolevel (texture formation), such approximation is rarely used. In addition, the application of anisotropic elastic law causes significant errors in the calculation of residual stress [37].

Let us consider the well-known variants of geometrically nonlinear anisotropic relations of the basic structure of crystal plasticity models, including the description of intragranular dislocation slip and crystallite lattice rotation. Single and double contractions are denoted by "·" and ":", respectively, and the outer tensor product is denoted by "⊗".

2.1. Relations in Terms of the Unloaded Configuration (U-Model)

In many studies, the classical decomposition of the Kröner–Lee deformation gradient f is explicitly utilized [38–40]:

$$f = f^e \cdot f^p, \tag{1}$$

where f^e, f^p denote the elastic and plastic components of the deformation gradient.

The crystal lattice rotation is associated with the orthogonal vector r^e from the polar decomposition $f^e = r^e \cdot u^e = v^e \cdot r^e$, so the lattice spin is described by the relation:

$$\overline{\omega} = \dot{r}^e \cdot r^{eT}, \tag{2}$$

where the dot indicates the time derivative.

Constitutive relations are formulated in the "classical" unloaded configuration found from the actual configuration by applying an affine transformation with $(f^e)^{-1}$:

$$k = \pi_0 : c^e, \tag{3}$$

where $k = J(f^e)^{-1} \cdot \sigma \cdot (f^e)^{-T}$ is the second Piola–Kirchhoff tensor defined in the unloaded configuration (σ is the Cauchy stress tensor, $J = \det f^e$), $c^e = 1/2\left((f^e)^T \cdot f^e - I\right)$ is the elastic Green-Lagrange strain defined in the unloaded configuration, $\pi_0 = \pi^{ijmn} k_{0i} \otimes k_{0j} \otimes k_{0m} \otimes k_{0n}$ is the elastic tensor, in which k_{0i} denotes the crystallographic basis in the reference configuration.

The plastic component of the deformation gradient f^p is obtained from the relation:

$$\dot{f}^p \cdot (f^p)^{-1} = \sum_{k=1}^{K} \dot{\gamma}^{(k)} \overset{o}{b}^{(k)} \otimes \overset{o}{n}^{(k)}, \tag{4}$$

where $\overset{o}{b}^{(k)}, \overset{o}{n}^{(k)}$ are the unit vectors of the slip direction and the normal to the slip plane in the reference configuration, and K is the number of the slip systems in a crystallite. Here, positive and negative dislocations (doubling of slip systems) are considered separately, which makes the specification of hardening laws more variable, and therefore we make use of this technique in our study.

In most models, the shear rates $\dot{\gamma}^{(k)}$ on the slip systems are determined by viscoplastic relations [41–43]:

$$\dot{\gamma}^{(k)} = \dot{\gamma}_0 \left(\frac{\tau^{(k)}}{\tau_c^{(k)}} \right)^m H(\tau^{(k)} - \tau_c^{(k)}), \tag{5}$$

where $\tau^{(k)}$, $\tau_c^{(k)}$ are the resolved and critical shear stresses on the k-th slip system, $\dot{\gamma}_0$ is the shear rate in the slip system in the case when the tangential stress approaches the critical shear stress, and m is the strain rate sensitivity exponent of the material, and $H(\cdot)$ is the Heaviside function. There are some papers, the authors of which propose different ways to determine the resolved shear stress; this issue deserves a separate discussion because it lies, strictly speaking, beyond the scope of this paper. For definiteness, we suppose that $\tau^{(k)} = k : n^{(k)} \otimes b^{(k)}$, where k is the weighted Kirchhoff stress tensor ($k = J\sigma$), $b^{(k)}, n^{(k)}$ are the unit vectors of the slip direction and the normal to the slip plane in the actual configuration.

In many works, Relation (5) are often supplemented with a term $\exp\left(-\frac{U}{\kappa\theta}\right)$, where κ is the Boltzman constant, and U is the model parameter (activation energy). This allows one to consider the dislocation motion as a function of temperature. There is also an approach that involves determination of the temperature-dependent parameters $\dot{\gamma}_0$, m [44] or formulations of corresponding relations for the critical stresses $\tau_c^{(k)}$ [24]. In any case, the initial critical stress values $\tau_c^{(k)}(0)$ should appear to be temperature-dependent.

Based on the literature review, we can conclude that the strain rate dependence of stresses in crystal plasticity can be taken into account in different ways. In most of the works $\dot{\gamma}_0 = const$, and therefore the strain rate sensitivity is defined by the parameter m. The value of $\dot{\gamma}_0$ should match the strain rate in order to avoid the sawtooth stress-strain response (curve) of the material during the numerical implementation of the model. It should also be noted that in the general case, there is an interaction between the parameters $\dot{\gamma}_0$ and m. At high m, the elastoviscoplastic model is almost similar to the elastoplastic model. Relation (5) provides a point in the stress space in a small neighborhood of the yield surface defined by Schmid's criterion: when the point is trying to move at great distance away from the neighborhood, significant relaxation of stresses is realized, and the point gets back to it. One can also use the relation $\dot{\gamma}_0 = \eta d_u, d_u = \sqrt{\frac{2}{3}\boldsymbol{d}' : \boldsymbol{d}'}$ is the strain rate intensity, $\boldsymbol{d}'$ is the deviator of the stretching tensor $\boldsymbol{d}$, η is the model parameter (usually $\eta = 1$) [24,45]. For the active slip systems, the shear stresses tend towards critical stresses [45]. In the framework of this approach, the effect of the strain rate on the material response is embedded into the evolutionary relations for critical stresses $\tau_c^{(k)}$.

The aim of this study is to analyze the kinematic relations and constitutive laws. To this end, we use, for definiteness, Relation (5) with constant $\dot{\gamma}_0$ in all cases under study and describe a hardening behavior according to a simple and well-known law [46–48] as:

$$\begin{gathered}\dot{\tau}_c^{(k)} = \sum_{l=1}^{K} h^{(kl)}\dot{\gamma}^{(l)}, \\ h^{(kl)} = \left[q_{l\,at} + (1 - q_{l\,at})\delta^{(kl)}\right]h^{(l)}, h^{(l)} = h_0\left|1 - \tau_c^{(l)}/\tau_{sat}\right|^a,\end{gathered} \tag{6}$$

where the latent hardening parameter $q_{l\,at}$ takes the values 1 and 1.4 for coplanar and noncoplanar slip systems (with numbers k and l), respectively, and $\delta^{(kl)}$ is the Kronecker delta.

Thus, for definiteness, all the base models considered in our study involve Relations (5) and (6) so that the inelastic strain rate can be determined.

We will call the model, which uses the kinematic relations and constitutive laws (1)–(4), the "U-model" (the unloading configuration type of the model). The given base structure of the relations is employed in numerous works, for instance, in [8,47–53]. In [54], this model is termed the Kalidindi–Bronkhorst–Anand formulation [55].

2.2. Relations in Terms of the Actual Configuration with a Taylor Spin (T-Model)

The models are often formulated using the Jaumann type rate of the Cauchy stress. The law of elasticity is written in this case as [56,57]:

$$\sigma^J = \dot{\sigma} - \overline{\omega}_T \cdot \sigma + \sigma \cdot \overline{\omega}_T = \boldsymbol{\pi} : (\boldsymbol{d} - \boldsymbol{d}^p)\,, \tag{7}$$

where $\boldsymbol{\pi}$ is the elastic tensor in the actual configuration (defined for the current crystallographic orientation), $\boldsymbol{d}$ is the stretching tensor (presented on the basis of the additive decomposition $\boldsymbol{d} = \boldsymbol{d}^p + \boldsymbol{d}^e$), the plastic part of which is determined as:

$$\boldsymbol{d}^p = \sum_{k=1}^{K} \frac{1}{2}\dot{\gamma}^{(k)}\left(\boldsymbol{b}^{(k)} \otimes \boldsymbol{n}^{(k)} + \boldsymbol{n}^{(k)} \otimes \boldsymbol{b}^{(k)}\right). \tag{8}$$

The spin $\overline{\omega}_T$ is calculated in the framework of the Taylor's rotation model [58]:

$$\overline{\omega}_T = \frac{1}{2}(\boldsymbol{l} - \boldsymbol{l}^T) + \sum_{\kappa=1}^{K} \frac{1}{2}\dot{\gamma}^{(k)}(\boldsymbol{n}^{(k)} \otimes \boldsymbol{b}^{(k)} - \boldsymbol{b}^{(k)} \otimes \boldsymbol{n}^{(k)}), \tag{9}$$

where $\boldsymbol{l}$ is the velocity gradient.

Some recent studies (e.g., [59,60]) have used the law of elasticity written as:

$$\sigma^J = \dot{\sigma} - \overline{\omega}_T \cdot \sigma + \sigma \cdot \overline{\omega}_T = \pi : \left(\dot{\varepsilon} - \dot{\varepsilon}^p\right) - \sigma tr(\dot{\varepsilon}) \tag{10}$$

Apparently, these studies suggest that any strain rate can be defined as $\dot{\varepsilon}$, yet $\dot{\varepsilon} = \boldsymbol{d}$ remains the most common option. The $\dot{\varepsilon}^p$ is determined by making use of (8).

It is easy to see that, for $\dot{\varepsilon} = \boldsymbol{d}$, Relation (10) can be written in the form:

$$\boldsymbol{k}^J = \dot{\boldsymbol{k}} - \overline{\omega}_T \cdot \boldsymbol{k} + \boldsymbol{k} \cdot \overline{\omega}_T = \tilde{\boldsymbol{\pi}} : (\boldsymbol{d} - \boldsymbol{d}^p), \tag{11}$$

where $\overline{\boldsymbol{\pi}}$ is the elastic tensor for establishing the relation between the stress rate and elastic strain rate shown above, $\tilde{\boldsymbol{\pi}} = J\boldsymbol{\pi}$.

Relation (11) and equivalent Relation (10) are supposed to be used instead of (7) since the internal energy density (per unit mass) is defined as:

$$u = \int_0^\tau \left(\frac{1}{\hat{\rho}}\sigma : \boldsymbol{d}\right) dt = \frac{1}{\overset{\circ}{\rho}} \int_0^\tau (J\sigma : \boldsymbol{d}) dt = \frac{1}{\overset{\circ}{\rho}} \int_0^\tau (\kappa : \boldsymbol{d}) dt, \tag{12}$$

According to (12), at small deformations (at $\overline{\omega}_T = \boldsymbol{0}$ and for in case of change in volume) when (11) is fulfilled, there will be no energy dissipation during the purely elastic deformation, which does not provide (7). A detailed analysis of the fulfillment of the thermodynamic constraints is given in Section 2.3. Note that, due to the smallness of volume changes in metals and alloys, (11) and (7) give similar results.

The model that involves the kinematic relations and constitutive laws (8), (9), and (11) will be further called the "T-model" (Taylor spin model). The relations of such conceptual structure have been used in many papers, in particular, in [59–61]. Some variants of self-consistent models [7] are based on the mesolevel equations of the same structure but with correction of spin (9) via subtraction of the antisymmetric part of the Eshelby tensor for considered grain.

2.3. Analysis of the Formulations with an Emphasis on Describing Geometric Nonlinearity and Fulfillment of Thermodynamic Constraints

In many works cited above, it was shown that, due to the structure of the model relations, the dissipation of energy under inelastic deformation is positive within the framework of both U- and T-models.

The U-model elastic law (3) involves the energetic conjugate stress-strain measures, and this guarantees the absence of energy dissipation during purely elastic deformation. Besides, the finite form of the relations ensures that, after the purely elastic deformation of the material (being in the unloaded configuration) along any closed deformation trajectories, the stresses will be zero. Thus, the conservation conditions for purely elastic deformation [62] are fulfilled within the framework of the U-model.

At the same time, the U-model implies the necessity to perform numerical integration of (4) so that $\boldsymbol{f}^p$ can be determined. This can violate the condition of isohoricity of plastic deformation and require corrective techniques. Therefore, ref. [54] proposes an alternative formulation for crystal plasticity that is based on the Kröner–Lee decomposition (also) and logarithmic strain measure, whose rate was decomposed additionally by using elastic corrector rates (defined on the basis of the plastic flow characteristics—shear rates on the slip systems). A major advantage is the lack of a necessity to calculate the exponent when determining $\boldsymbol{f}^p$ at the end of each step. Meanwhile, we need to point out here that the interpretation of the physical meaning of the elastic law when considering it as a relation between the introduced complex measures of the stress-strain state (not only in the context of thermodynamics) is rather complicated. In addition, the fulfillment of the additivity condition for the components of the strain rate measure is guaranteed only if one slip system is active; in the general case, this condition is fulfilled approximately [54].

On the other hand, for the study of technological processes of thermomechanical processing, it is more convenient to formulate the boundary value problem in terms of the actual configuration in the rate form. This enables one to use numerical methods: a step-by-step solution with a redefinition of the configuration of the computational domain (including the contacting surfaces) can be realized. Another significant advantage of this formulation is an additive decomposition of the rate of inelastic deformation into contributions from various mechanisms. So, advanced crystal plasticity models taking into account grain boundary sliding were constructed on the basis of this structure [13,30–32]. The T-model, referring to this type of formulation, uses a transparent measure of the strain rate—stretching tensor $\boldsymbol{d}$.

It should be noted that the use of the corotational derivative in Relation (11) suggests that the linear (verified) constitutive relation is assumed to be valid for the observer related to a moving coordinate system rotating with the Taylor spin [63]. This suggests that the motion is decomposed into the rigid motion of the moving coordinate system and the deformation motion with respect to it [64]. Note that, to be fully consistent with the issue, we should write the measure of the rate of elastic deformation on the right-hand side of (11) as $(\boldsymbol{l} - \boldsymbol{l}^p - \overline{\boldsymbol{\omega}}_T)$, $\boldsymbol{l}^p = \sum_{k=1}^{K} \dot{\gamma}^{(k)} \boldsymbol{b}^{(k)} \otimes \boldsymbol{n}^{(k)}$ [65]. However, due to the tensor symmetry, the results will be similar to those obtained when using $(\boldsymbol{d} - \boldsymbol{d}^p)$. In some studies, the authors explicitly say that they use linear constitutive relations written in the local coordinate systems of crystallites [66,67].

It is evident that, for the relations in terms of the actual configuration in the rate form, the key question is how to determine the spin of the moving coordinate system.

The determination of a hypoelastic law of the form (11), for which the above-mentioned conservation conditions are fulfilled under purely elastic deformation, is a well-known problem in the mechanics of solids [62]. The popular Zaremba–Jaumann [68,69] and Green–Naghdi [70] corotational derivatives do not guarantee the fulfillment of the conservation conditions. Therefore, a logarithmic corotational derivative without this drawback was developed [62,71,72]. However, in [63,73], one can find examples that illustrate the difficulties which can be encountered in modeling the multistage deformation processes (including unloading): if the purely elastic cyclic loading is realized upon completion of elastoplastic deformation and unloading, then the stress hysteresis will be observed. In [73,74], the logarithmic spin, called the kinetic logarithmic spin (k-logspin), was improved. The modernized spin makes it possible to meet the conservation conditions in the simulation of multi-stage deformation processes, including unloading, but the consideration is limited to anisotropic material. Thus, the problem of formulating a hypoelastic constitutive relation of the Form (11) for an anisotropic material that ensures the fulfillment of the conservative conditions has not yet been completely solved. Obviously, this is also the case for the crystal plasticity T-model.

So, both the U-model and the T-model have "a certain extent roughness" formulations, which still arouses the interest of researchers in finding a more elegant solution, with the leveling of these "roughness" formulations (see, e.g., [54]).

Note that, when formulating the T-model, the moving coordinate system (MCS) must be related to the crystallographic coordinate system since the components of the elastic tensor are assumed to be constant in the MCS. In other words, the MCS must be associated with the symmetry elements of the crystal.

According to the Taylor constrained rotation model, the displacement gradient at the mesoscale (grain level) is represented in the form [58]:

$$\boldsymbol{l} = \overline{\boldsymbol{\omega}}_T + \sum_{\kappa=1}^{K} \dot{\gamma}^{(k)} \boldsymbol{b}^{(k)} \otimes \boldsymbol{n}^{(k)}, \tag{13}$$

whence, by taking the antisymmetric part, we obtain the relation for the lattice spin (9).

Traditionally, the Taylor model is used for rigid-plastic models (this follows from the symmetrized Relation (13)—all deformation is associated with the shears on the slip systems). The closeness of the spin (9) to the "material rotation" spin (2) was demonstrated [9,75]. In both cases, one cannot speak of a direct relation between the MCS and the symmetry elements of the lattice. We also note that during deformation, the crystallographic coordinate system (CSC) can be distorted, while the MCS remains rigid, and therefore, strictly speaking, their identification is an approximation of the mathematical model.

In [63,65,76], an approach to the formulation of geometrically nonlinear relations is proposed. The approach is based on a modified multiplicative decomposition of the deformation gradient with an explicit selection of rigid rotation of the MCS, which is associated with the symmetry elements of crystallites. In the next section, the equations of the corresponding mesoscale formulations are given in a brief form (in the finite form in the lattice unloaded configuration and in the rate form in the current configuration). In Section 4, these ratios are compared with the U-model and T-model and are utilized to compare these formulations to each other.

3. Modification of Mesolevel Relations with an Explicit Separation of the Rigid Moving Coordinate System with Respect to Which Elastic Distortion Is Defined

3.1. Relations in Terms of Lattice Unloaded Configuration (LU-Model)

In [63,76], it was proposed to modify the decomposition of motion at a mesolevel, i.e., to realize a multiplicative representation of the deformation gradient f with an explicit separation of the rigid moving coordinate system (MCS) $\mathrm{Op^1p^2p^3}$:

$$f = f^e \cdot f^p = \bar{f}^e \cdot r \cdot f^p. \tag{14}$$

Here, $r = \hat{k}_i \otimes \left(\hat{k}^i \Big|_{t=0} \right) \equiv \hat{k}_i \otimes k^i$ is the proper orthogonal tensor which converts the reference basis MCS) k_i into the current basis $\hat{k}_i$; in other words, r is the tensor of rotation of the MCS from reference to actual configuration; f^p is the component of the deformation gradient that transforms the initial configuration into the plastically deformed configuration, $\bar{f}^e$ is the component of the deformation gradient component that transforms the plastically deformed configuration into the actual configuration (this tensor also characterizes crystalline lattice distortion), and $\bar{f}^e, f^p$ are, in the general case, the non-symmetric second rank tensors.

The motion of the MCS is a rigid motion, and the motion relative to the MCS is a deformation motion: geometrically generalized nonlinear constitutive relations are formulated by an observer related to the MCS [65]. Relation (14) is a modification of the classical Kröner–Lee decomposition (1) via the explicit separation of the quasi-rigid motion r. Note that, in the general case, the tensor does not coincide with the orthogonal tensor r^e in the polar expansion $f^e = r^e \cdot u^e = v^e \cdot r^e$ in the case when the spin of the corresponding MCS is defined (2).

For analytical calculations, we also introduce the crystallographic coordinate system (CCS) $\mathrm{Oy^1y^2y^3}$ with basis q_i. Without loss of generality, we assume that, in the reference configuration, the CCS basis $q_i(0)$ is an orthonormal basis that coincides with k_i. Note that the lengths of the CCS basis vectors and the angles between them vary during deformation.

The constitutive laws are formulated in the context of the observer related to the MCS. This MCS is assumed to be linked to the symmetry axes of the material. Probably, the idea of the necessity of determining corotational derivatives with reference to the symmetric elements of the material (directors) was initially put forward by J. Mandel [77], yet, unfortunately, his study contains no specific relations for describing a crystallite lattice spin.

Based on the introduced multiplicative decomposition (14) and assuming that the elastic properties of the crystallite in MCS are constant, the constitutive elastoviscoplastic relation was formulated in terms of the unloaded configuration (unloading is carried out with $(\overline{f}^{e})^{-1}$, and the MCS remains fixed) [63,76]:

$$\overline{k} = \overline{\pi} : \overline{c}^{e}, \tag{15}$$

which includes the second Piola–Kirchhoff tensor $\overline{k}$, defined in the unloaded configuration, $\overline{k} = \overline{k}^{ij}\hat{k}_i \otimes \hat{k}_j = J(\overline{f}^{e})^{-1} \cdot \sigma \cdot (\overline{f}^{e})^{-T}$ and the four-rank elastic tensor $\overline{\pi} = \overline{\pi}^{ijmn}\hat{k}_i \otimes \hat{k}_j \otimes \hat{k}_m \otimes \hat{k}_n$, $\overline{c}^{e}$ is the elastic constituent of the right Cauchy–Green deformation tensor $\overline{c}^{e} = \overline{c}^{ij}\hat{k}_i \otimes \hat{k}_j = 1/2((\overline{f}^{e})^{T} \cdot \overline{f}^{e} - I)$, and I is the unit tensor. The accepted definition of the elastic tensor (its components are constant in the MCS) conforms to the principle of independence of constitutive relations from reference choice [21].

The plastic component of the deformation gradient f^{p} is determined using the viscoplastic or elastoplastic relations applied to describe the intragranular dislocation slip. In the model, as in other compared models, we use Relations (4)–(6).

An important component of the constitutive model is the kinetic relation for the motion of the MCS. As noted above, it seems reasonable to relate the mesoscale MCSs and the symmetry elements of the crystal lattice [65,76,77]. Plastic deformations occur due to the movement of dislocations and they do not cause symmetry distortion, and therefore r can be completely determined by the tensor f^{e}.

The elastic gradient $f^{e} = \overline{f}^{e} \cdot r$ contains both the quasi-rigid motion (rotation) of the MCS and the distortion of the lattice relative to it. Generally speaking, there are different variants of this relation. In order to overcome this challenge, it is necessary to accept a hypothesis for the definition of r. We consider here one of the options. Let a relation between CCS and MCS be as follows: (1) the axes Oy^1 and Op^1 coincide at every instant of deformation (the vector $\hat{k}_1$ is directed along the vector q_1); (2) the vector $\hat{k}_2$ is located in the plane Oy^1y^2 at each instant of deformation. In [65], the following relation for the MCS spin is presented:

$$\begin{aligned} \overline{\omega}_1 = \dot{r} \cdot r^{T} = \dot{\hat{k}}_i \otimes \hat{k}_i = \; & -(\hat{k}_2 \cdot l^{e} \cdot \hat{k}_1)\hat{k}_1 \otimes \hat{k}_2 + (\hat{k}_2 \cdot l^{e} \cdot \hat{k}_1)\hat{k}_2 \otimes \hat{k}_1 - \\ & -(\hat{k}_3 \cdot l^{e} \cdot \hat{k}_1)\hat{k}_1 \otimes \hat{k}_3 + (\hat{k}_3 \cdot l^{e} \cdot \hat{k}_1)\hat{k}_3 \otimes \hat{k}_1 - \\ & -(\hat{k}_3 \cdot l^{e} \cdot \hat{k}_2)\hat{k}_2 \otimes \hat{k}_3 + (\hat{k}_3 \cdot l^{e} \cdot \hat{k}_2)\hat{k}_3 \otimes \hat{k}_2. \end{aligned} \tag{16}$$

where $l^{e} = \dot{f}^{e} \cdot f^{e-1}$ is the elastic component of the velocity gradient. In [78], the authors described the way to define r in finite form taking into account the above relation between the CCS and the MCS and showed the compliance of the obtained result with that obtained by integrating (16).

Expression (16) can take a more compact form [79]:

$$\overline{\omega}_1 = I \times (\hat{k}_1 \otimes \hat{k}_2 \otimes \hat{k}_3 - \hat{k}_2 \otimes \hat{k}_1 \otimes \hat{k}_3 + \hat{k}_3 \otimes \hat{k}_1 \otimes \hat{k}_2) : l^{e} \tag{17}$$

where «×» denotes the vector product.

A model that involves kinematic Relations (4), (14), and (17) and constitutive law (15) is termed here the "LU1-model" (lattice unloading configuration type of model with spin 1).

Besides, the MCS can be related to other symmetry elements of the crystal lattice. As an alternative, we consider the case when the axes Oy^3 and Op^3 coincide at each instant of time (the vector $\hat{k}_3$ is directed along the vector q_3), the vector k_1 is positioned at each instant of deformation in the plane Oy^1y^3. Under these conditions, the spin can be calculated:

$$\overline{\omega}_2 = I \times (-\hat{k}_1 \otimes \hat{k}_3 \otimes \hat{k}_2 + \hat{k}_2 \otimes \hat{k}_3 \otimes \hat{k}_1 + \hat{k}_3 \otimes \hat{k}_1 \otimes \hat{k}_2) : l^{e}. \tag{18}$$

We call the modified version of the LU-model with spin (18) the "LU2-model".

Consider the issue regarding the difference between LU1-model and LU2-model.

We assume that, at equal kinematic effects $\boldsymbol{f}$, the current positions of the MCS related to the symmetry axes of the lattice in different ways can be described by different tensors $\boldsymbol{r}$ and $\boldsymbol{r}*$ (due to the smallness of elastic distortions, these tensors are similar). So, the deformation gradient can be written as $\boldsymbol{f} = \overline{\boldsymbol{f}}^e \cdot \boldsymbol{r} \cdot \boldsymbol{f}^p = \overline{\boldsymbol{f}}^e * \cdot \boldsymbol{r} * \cdot \boldsymbol{f}^p *$.

Based on the constitutive Relation (15), in the case when the orientation of the MCS is defined via the application of the tensor $\boldsymbol{r}$, the Cauchy stress are established as:

$$\sigma = 1/J\overline{\boldsymbol{f}}^e \cdot \left(\overline{\pi} : 1/2\left(\overline{\boldsymbol{f}}^{eT} \cdot \overline{\boldsymbol{f}}^e - \boldsymbol{I} \right) \right) \cdot \overline{\boldsymbol{f}}^{eT}, \tag{19}$$

where $\overline{\pi} = \overline{\pi}^{ijmn} \hat{\boldsymbol{k}}_i \hat{\boldsymbol{k}}_j \hat{\boldsymbol{k}}_m \hat{\boldsymbol{k}}_n = \overline{\pi}^{ijmn} (\boldsymbol{r} \cdot \boldsymbol{k}_i)(\boldsymbol{r} \cdot \boldsymbol{k}_j)(\boldsymbol{r} \cdot \boldsymbol{k}_m)(\boldsymbol{r} \cdot \boldsymbol{k}_n)$. Relation (19) is easy to rearrange [80] into the form:

$$\sigma = 1/J\boldsymbol{f}^e \cdot \left(\overline{\pi}^{ijmn} \boldsymbol{k}_i \otimes \boldsymbol{k}_j \otimes \boldsymbol{k}_m \otimes \boldsymbol{k}_n : 1/2\left(\boldsymbol{f}^{eT} \cdot \boldsymbol{f}^e - \boldsymbol{I} \right) \right) \cdot \boldsymbol{f}^{eT} \tag{20}$$

Analogously, when the position of the MCS position is defined via the application of the tensor $\boldsymbol{r}*$, the Cauchy stress can be obtained from the relation:

$$\sigma* = 1/J * \boldsymbol{f}^e * \cdot \left(\overline{\pi}^{ijmn} \boldsymbol{k}_i \otimes \boldsymbol{k}_j \otimes \boldsymbol{k}_m \otimes \boldsymbol{k}_n : 1/2\left(\boldsymbol{f}^{eT} * \cdot \boldsymbol{f}^e * - I \right) \right) \cdot \boldsymbol{f}^{eT} * \tag{21}$$

Comparative analysis of (20) and (21) shows that the difference in the definitions of stresses obtained using the LU1- and LU2-models can be attributed to the fact that $\boldsymbol{f}^e$ and $\boldsymbol{f}^e*$ are determined in different ways in case of inelastic deformation. As noted above, the definition for shear stress can be expressed as $\tau^{(k)} = \boldsymbol{k} : \boldsymbol{n}^{(k)} \otimes \boldsymbol{b}^{(k)}$. The unit vectors of the slip direction and the normal to the plane are found in terms of the actual configuration as $\boldsymbol{b}^{(k)} = \boldsymbol{f}^e \cdot \overset{o}{\boldsymbol{b}}{}^{(k)} / \| \boldsymbol{f}^e \cdot \overset{o}{\boldsymbol{b}}{}^{(k)} \|, \boldsymbol{n}^{(k)} = \boldsymbol{f}^{e-T} \cdot \overset{o}{\boldsymbol{n}}{}^{(k)} / \| \boldsymbol{f}^e \cdot \overset{o}{\boldsymbol{n}}{}^{(k)} \|$. It is evident that, at the elastic deformation stage and in a transition to plasticity (as $\boldsymbol{f}^e = \boldsymbol{f}^e*$), the crystallographic characteristics will be the same, and therefore the reasons for the differences between $\boldsymbol{f}^e$ and $\boldsymbol{f}^e*$ that may occur are absent.

Thus, the mesostresses obtained within the framework of both LU1- and LU2-models are equal. The final orientations of the lattices determined using the models are different, though due to the smallness of elastic distortions, they differ only slightly [80]. Since the LU1-and LU2-models give almost the same definitions of stresses, we will call them an "LU-model" (if the case in point concerns the definition of stresses).

It is easy to show that the constitutive relation of the LU-model is independent of the imposed rigid motion [21]. Actually, let's consider deformation gradient $\widetilde{\boldsymbol{f}} = \boldsymbol{O}^T \cdot \boldsymbol{f}$, where $\boldsymbol{O} = \boldsymbol{O}(t)(\boldsymbol{O}(0) = \boldsymbol{I})$ is the orthogonal tensor describing the rigid rotation added to the initial motion. For pure elastic deformation, we get $\widetilde{\boldsymbol{f}}^e = \boldsymbol{O}^T \cdot \boldsymbol{f}^e$, and the analysis of (20) yields $\sigma(\widetilde{\boldsymbol{f}}^e) = \boldsymbol{O}^T \cdot \sigma(\boldsymbol{f}^e) \cdot \boldsymbol{O}$. In view of $\widetilde{\sigma} = \boldsymbol{O}^T \cdot \sigma \cdot \boldsymbol{O}$, we have $\widetilde{\sigma} = \sigma(\widetilde{\boldsymbol{f}}^e)$. Thus, measures of the stress-strain state with an imposed rigid movement are related by exactly the same function. It means that the principle of independence of the constitutive relation from the imposed rigid motion is fulfilled. Hence, it follows that at the onset of the plastic flow $\widetilde{\boldsymbol{f}}^p = \boldsymbol{f}^p$ is correct, and, as a result, the principle of independence of the constitutive law from the imposed rigid motion (i.e., $\widetilde{\sigma} = \sigma(\widetilde{\boldsymbol{f}}^e)$ will also be fulfilled in the case of inelastic deformation as well.

By virtue of the fact that the used stress and strain measures are energy conjugate, the requirement for the absence of stress hysteresis and energy dissipation in the arbitrary closed cycles of elastic deformations is fulfilled automatically in the LU-model [63], as in the U-model.

In the general case, other physically justified rotation models (in particular, those capable of taking into account the interaction of defects in neighboring crystallites [81] or the contribution of grain boundary sliding [13]) can be incorporated into the proposed approach to determine the MCS spin.

3.2. Rate Form Relations in Terms of the Actual Configuration (LR-Model)

From the formulation of the LU-model with the constitutive law in finite Form (15) through the corotational differentiation [64,76], we can pass to the rate analogue for the observer in the MCS:

$$\overline{k}^{CR} = \overline{\pi} : \overline{c}^{eCR}, \tag{22}$$

where $\overline{k}^{CR} = \dot{\overline{k}} - \overline{\omega}_1 \cdot \overline{k} + \overline{k} \cdot \overline{\omega}_1$, $\overline{c}^{eCR} = \dot{\overline{c}}^{e} - \overline{\omega}_1 \cdot \overline{c}^e + \overline{c}^e \cdot \overline{\omega}_1$ (for implementation of this transition, we use the property of constancy of $\overline{\pi}$ for the observer in the MCS).

It was shown in [63,82] that, at $\overline{f}^e \approx \boldsymbol{I}$ which is characteristic of metals, the following formulation in the rate form can be derived in terms of the current configuration.

Since, at $\overline{f}^e \approx \boldsymbol{I}$, the unloaded and actual lattice configurations are close, the following estimate is valid for the velocity gradient:

$$\boldsymbol{l} \approx \dot{\overline{f}}^{e} \cdot \overline{f}^{e-1} + \overline{\omega}_1 + \boldsymbol{r} \cdot \dot{f}^{p} \cdot f^{p-1} \cdot \boldsymbol{r}^T = \overline{\omega}_1 + \dot{\overline{f}}^{e} \cdot \overline{f}^{e-1} + \sum_{k=1}^{K} \dot{\gamma}^{(k)} \overline{\boldsymbol{b}}^{(k)} \otimes \overline{\boldsymbol{n}}^{(k)}, \tag{23}$$

where $\overline{\boldsymbol{b}}^{(k)}, \overline{\boldsymbol{n}}^{(k)}$ are the vectors of the slip direction and the normal to the slip plane in the unloaded lattice configuration, i.e., they are in the actual position of the MCS. If the unloaded and actual lattice configurations are in close proximity, then we can assume $\overline{\boldsymbol{b}}^{(k)} = \boldsymbol{b}^{(k)}, \overline{\boldsymbol{n}}^{(k)} = \boldsymbol{n}^{(k)}$ for the latter relation and use $\sum_{k=1}^{K} \dot{\gamma}^{(k)} \boldsymbol{b}^{(k)} \otimes \boldsymbol{n}^{(k)}$ as an approximation for the inelastic strain rate.

According to (23), the clear motion decomposition is established in the rate form: the velocity gradient $\boldsymbol{l}$ is represented as a set of the MCS spin and the rate of elastic distortions and inelastic deformations determined by the observer related to this MCS.

Taking into account the described approximations, the constitutive law (22) is close to

$$\boldsymbol{k}^{cr} = \overline{\pi} : \left(\dot{\overline{f}}^{e} \cdot \overline{f}^{e-1} \right) = \overline{\pi} : \left(\boldsymbol{l} - \overline{\omega}_1 - \sum_{k=1}^{K} \dot{\gamma}^{(k)} \boldsymbol{b}^{(k)} \otimes \boldsymbol{n}^{(k)} \right) \tag{24}$$

where $\boldsymbol{k}$ is the weighted Kirchhoff stress tensor, and $\boldsymbol{k}^{cr} = \dot{\boldsymbol{k}} + \boldsymbol{k} \cdot \overline{\omega}_1 - \overline{\omega}_1 \cdot \boldsymbol{k}$ is its corotational derivative. Let us call the model involving the kinematic Relations (8), (17), and (23) and a constitutive law (24) the "LR1-model" (lattice rate type of model with spin 1). We also consider an analogous LR2-model in which Relation (18) is used for the spin. It was shown in [80] that the Cauchy stress determined using the formulation of this kind depends only slightly on the way they are related to specific material symmetry elements.

From the above discussion, we can draw a conclusion that the formulations of the U- and LU-models can provide the fulfillment of the conservative conditions for elastic deformation. On the other hand, for the study of technological processes of thermomechanical treatment with an a priori unknown configuration of the computational domain (including contact with the tool), the formulations in terms of the actual configuration in the rate form offer the most promise. Another advantage of such formulations is the possibility of an additive decomposition of the inelastic deformation rate into the contributions of different deformation mechanisms. This explains why the T-model is also popular.

The formulation based on the decomposition of motion with an explicit separation of the motion of the moving coordinate system (LU-model) makes possible a theoretically substantiated transition to a similar formulation in the rate form in terms of the actual configuration (LR-model). As will be illustrated below, this helps to easily show that all the models considered here give a similar response under equal influences.

4. Results and Discussion

In this section, the issue regarding a comparison of the formulations for crystal plasticity models described above is considered.

4.1. Analytical Comparison

Based on the Equation of motion (14), constitutive law (3) for the U-model can be written as:

$$J\boldsymbol{r}^{T}\cdot(\overline{\boldsymbol{f}}^{e})^{-1}\cdot\boldsymbol{\sigma}\cdot(\overline{\boldsymbol{f}}^{e})^{-T}\cdot\boldsymbol{r}=\overline{\pi}^{ijmn}\boldsymbol{k}_i\otimes\boldsymbol{k}_j\otimes\boldsymbol{k}_m\otimes\boldsymbol{k}_n:1/2(\boldsymbol{r}^{T}\cdot(\overline{\boldsymbol{f}}^{e})^{T}\cdot\overline{\boldsymbol{f}}^{e}\cdot\boldsymbol{r}-\boldsymbol{I}),\tag{25}$$

where $\boldsymbol{k}_i$ is the reference basis of the MCS related to the lattice symmetry.

The rearranged form of Relation (25) is:

$$J(\overline{\boldsymbol{f}}^{e})^{-1}\cdot\boldsymbol{\sigma}\cdot(\overline{\boldsymbol{f}}^{e})^{-T}=\overline{\pi}^{ijmn}(\boldsymbol{r}\cdot\boldsymbol{k}_i)\otimes(\boldsymbol{r}\cdot\boldsymbol{k}_j)\otimes(\boldsymbol{r}\cdot\boldsymbol{k}_m)\otimes(\boldsymbol{r}\cdot\boldsymbol{k}_n):1/2((\overline{\boldsymbol{f}}^{e})^{T}\cdot\overline{\boldsymbol{f}}^{e}-\boldsymbol{I}).\tag{26}$$

Relation (26) can be represented as:

$$J(\overline{\boldsymbol{f}}^{e})^{-1}\cdot\boldsymbol{\sigma}\cdot(\overline{\boldsymbol{f}}^{e})^{-T}=\overline{\pi}^{ijmn}\hat{\boldsymbol{k}}_i\otimes\hat{\boldsymbol{k}}_j\otimes\hat{\boldsymbol{k}}_m\otimes\hat{\boldsymbol{k}}_n:1/2((\overline{\boldsymbol{f}}^{e})^{T}\cdot\overline{\boldsymbol{f}}^{e}-\boldsymbol{I}).\tag{27}$$

Relation (27) corresponds to the constitutive law of the LU-model (15). Thus, the stresses in the U- and LU-models are equally determined. As shown in Section 3.1, the determined stresses are independent of the portion of the MCS motion in the deformation gradient. When we use models of this kind, it is important to keep in mind this distinction and verify separately texture description.

If, by analogy with the transition from the LU-model to the rate LR-model described in Section 3.2, we take for the U-model an MCS that moves with spin (2), then we get the rate formulation of the LR-model type, but with a corotational derivative of the Green–Naghdi type (the GN-model used, for instance, in [75,83,84]). In line with this assumption, the proposed model will give stresses close to those determined in the U-model.

In [75], the following relation between spins (2) and (9) is given:

$$\overline{\boldsymbol{\omega}}_{\boldsymbol{r}^e}=\overline{\boldsymbol{\omega}}_{T}-(\boldsymbol{B}:\boldsymbol{\sigma})\cdot\boldsymbol{d}^{p}+\boldsymbol{d}^{p}\cdot(\boldsymbol{B}:\boldsymbol{\sigma}),\tag{28}$$

where $\boldsymbol{B}$ is the four rank elastic compliance tensor, $\boldsymbol{d}^p$ is defined using Formula (8). According to (28), by virtue of the smallness of elastic deformations $(\boldsymbol{B}:\boldsymbol{\sigma})$, the use of spin Relations (2) and (9) in the mesolevel model will cause slightly different stresses to occur (in other words, the proximity of the T-model and the GN-model is their characteristic feature).

Hence, we can suggest that at small elastic distortions, all the models examined should give a close response (stresses). Table 1 presents information obtained when performing the analytical comparison of determination of stresses within the framework of different models.

In order to confirm the analytical results summarized in Table 1, we performed numerical calculations.

Table 1. Comparison of determination of stresses within the framework of different models (notation used here: «=»—precise coincidence of stresse obtained in the indicated models, «≈»—stresses are close; by "described" is meant the description of the essence of calculations with references to publications).

U	T	LU	LR1	LR2	GN
U	≈ (described in Section 4.1, follows from U ≈ GN ≈ T)	= (shown in Section 4.1)	≈ (shown in Section 4.1, described in Section 3.2, follows from U = LU ≈ LR1)	≈ (shown in Section 4.1, described in Section 3.2, follows from U = LU ≈ LR2)	≈ (described in Section 4.1, S by analogy with LU ≈ LR1)
T		≈ (described in Section 4.1, follows from LU = U ≈ GN ≈ T)	≈ (described in Section 4.1, follows from LR1 ≈ LU = U ≈ ≈ GN ≈ T)	≈ (described in Section 4.1, follows from LR2 ≈ LU = U ≈ ≈ GN ≈ T)	≈ described in Section 4.1)
LU			≈ (described in Section 3.2)	≈ (described in Section 3.2)	≈ (described in Section 4.1, follows from LU = U ≈ GN)
LR1				≈ (described in Section 3.2)	≈ (described in Section 4.1, follows from LR1 ≈ LU = = U ≈ GN)
LR2					≈ (described in Section 4.1, follows from LR2 ≈ LU = = U ≈ GN)

4.2. Illustrative Numerical Examples

We now turn to the consideration of the results obtained via the analysis of the application of kinematic relations and constitutive laws in the statistical constitutive model for simulating some kinematic loads imposed on the fcc polycrystal.

The representative volume included a sample of 343 crystallites, the initial orientations of which were distributed randomly according to a uniform law. The nominal properties of the polycrystal corresponded to those of copper. The mesolevel elastic property tensor contained the following independent components (constant for the observer in a rigid moving coordinate system linked with the lattice): $\pi_{1111} = 168.4$ GPa, $\pi_{1122} = 121.4$ GPa, $\pi_{1212} = 75.4$ GPa [85]; the viscoplastic relations included $\dot{\gamma}_0 = 0.001$ s^{-1}, $1/m = 0.012$; the hardening law parameters were $h_0 = 180$ MPa, $a = 2.25$, $\tau_{sat} = 148$ MPa, and the initial values of critical stresses for all slip systems were $\tau_c^{(k)}(0) = \tau_{c0} = 16$ MPa ($k = 1, \ldots, K$) [46,47]. Due to the significant nonlinearity of the systems of equations for all considered models, in particular, because of the presence of the Heaviside function in the viscoplastic law (5), which led to the necessity of discretization with a small time step to trace the activity of slip systems in crystallites, these equations were integrated using an explicit Euler method (in the calculations, the time step was 0.002 s). The results obtained with the aid of the LR1 model correspond satisfactorily to the experimental data for uniaxial compression and shear; the comparison of the results is given in [63].

We consider the affine deformations of the sample (corresponding to the representative volume of a polycrystal) subjected to uniformly distributed kinematic loads. Due to the uniform deformation, the radius vector of the material point at an arbitrary instant of time t is determined according to $\boldsymbol{r}(t) = \boldsymbol{f}(t) \cdot q^i \boldsymbol{p}_i$, where $\boldsymbol{f}(t)$ is the deformation gradient, q^i stands for the Lagrangian coordinates of the considered point, and $\boldsymbol{p}_i$ is the basis of the fixed laboratory coordinate system. The motion is determined by the deformation gradient (the choice of motion can be arbitrary):

$$f(t) = \begin{array}{l}\frac{1}{1+4h\sin(\omega t)}p_1p_1 + (1+4h\sin(\omega t))p_2p_2 + p_3p_3 + \\ +3h(1-cos(0.7\omega t))p_1p_2,\end{array} \quad (29)$$

where $\omega = 0.001\pi$, $h = 0.1$ is the constant parameter (deformation is considered within the time interval $t = [0, T]$, $T = 1000$ s). For illustration, the trajectory of the point with Lagrangian coordinates (0, L, 0) m is given in Figure 1.

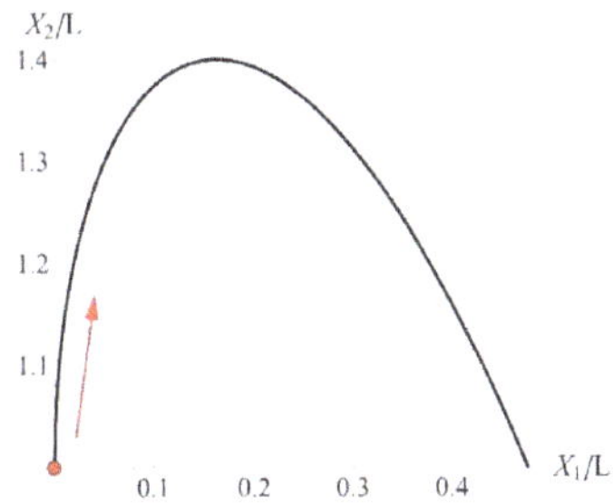

Figure 1. Trajectory of the point with Lagrangian coordinates (0, L, 0) for Motion (29) (in the plane OX_1X_2); the position of the point at the initial instant of time is defined by the coordinates $X_1/L = 0$, $X_2/L = 1$, $X_3 = 0$.

Figure 2 presents the time dependence of the Cauchy stress tensor components at a macrolevel (in the laboratory coordinate system) for the LU-model (Figure 2a, precise coincidence with the results of the U-model) and for LR1-model (Figure 2b).

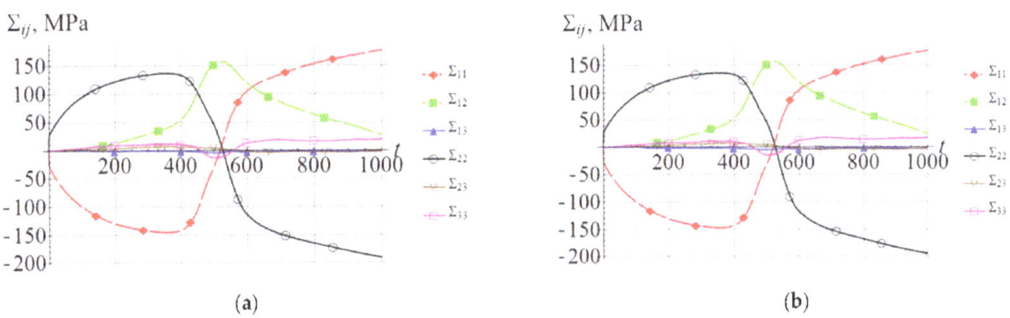

Figure 2. Time dependence of the Cauchy stress tensor components at a macrolevel (averaged mesostresses, $\boldsymbol{\Sigma} = \langle\sigma\rangle$) in the laboratory coordinate system for the fcc-crystal: (**a**) LU-model, U-model, (**b**) LR1-model.

We note the proximity of the results obtained (the curves plotted for the varying macrostress components are visually almost indistinguishable for all models examined in this study). For numerical assessment of the deviation in the results, we take the U-model as a base model and introduce the norm for comparing the results obtained using other models:

$$\Delta_G = \|\boldsymbol{\Sigma}_{t\in[0,T]}{}^{(U)} - \boldsymbol{\Sigma}_{t\in[0,T]}{}^{(G)}\|_{C^n_{L2}}, \quad (30)$$

where $\boldsymbol{\Sigma}_{t\in[0,T]}{}^{(U)}$ is the history of changes in stresses at $t = [0, 1000\ \text{s}]$, obtained via the use of the U-model, $\boldsymbol{\Sigma}_{t\in[0,T]}{}^{(G)}$ is the history of changes in stresses at $t = [0, 1000\ \text{s}]$ in case of the G-model (one of the models from Table 1). The norm is given based on the

Riemann integral $\|\mathbf{Y}_{t\in[0,T]}\|_{C^n_{L2}} = \left(\frac{1}{T}\int_0^T \mathbf{Y}(t) : \mathbf{Y}^T(t)dt\right)^{1/2}$ in the space C^n_{L2} continuous at $t \in [0,T]$ on the vector-function dimension n; in the calculations $n = 9$ (the number of the macrostress tensor component $\boldsymbol{\Sigma}$ that is equal to averaged mesostresses in crystallites).

Table 2 summarizes the results for deviations in the macrostresses obtained by the U-model from those determined by other models.

Table 2. Deviation of the history of changes in macrostresses obtained via the use of the U-model from the history of changes in macrostresses determined in the framework of other models.

Model	U	T	LU	LR1	LR2	GN
NormΔ_G, MPa	0	2.94	0	2.935	2.752	2.954
$\frac{\Delta_G}{\|\Sigma_{t\in[0,T]}{}^{(U)}\|_{C^n_{L2}}}$	0	0.013834	0	0.013813	0.012948	0.013903

(Zero deviations from Table 1 suggest that they do not exceed the order of the computer error when working with real numbers.).

Taking into account that the norms are integral over a sufficiently long period of time (T = 1000 s), the results summarized in Table 2 demonstrate the proximity of stresses obtained via the use of different formulations of mesolevel models, supporting thus the efficacy of theoretical statements given above (Table 1). Similar textures are obtained, which is indirectly confirmed by the proximity of macrostresses for the considered complex loading path. In [80], the issue of comparing the results of using some different spins (for other complex loading paths) was considered with an explicit analysis of the simulated lattice rotations.

Figure 3 shows the time dependence of the mesolevel stress components for a single crystallite (chosen arbitrarily). The initial orientations of the crystallographic and moving coordinate systems coincide and can be determined through the sequential rotation of the crystallographic system initially coincident with the coordinate system for the fcc-crystal around the Ox^1 axis with an angle ϕ_1= 4.078, around the Ox^2 axis with an angle ϕ_2= 0.0739, around the Ox^3 axis with an angle ϕ_3= 2.149 for the U-model (Figure 3a) and the LR1-model (Figure 3b).

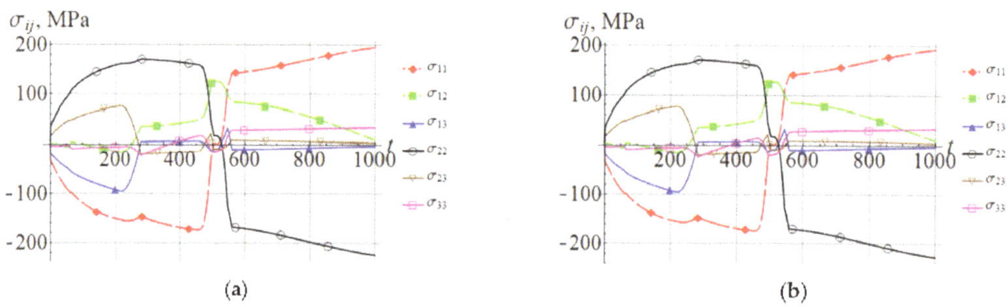

Figure 3. Time dependence of the Cauchy stress tensor components at a mesolevel in the laboratory coordinate system for crystallite No1: (**a**) LU-model, U-model, (**b**) LR2-model.

Analysis of the results shown in Figure 3 demonstrates that the mesostresses for the crystallite under study, which were obtained within the framework of the models mentioned above, are similar. Table 3 gives the norms for deviations in the mesostresses $\delta_G = \|\sigma_{t\in[0,T]}{}^{(U)} - \sigma_{t\in[0,T]}{}^{(G)}\|_{C^n_{L2}}$ for the considered models, which are similar to those for deviations in macrostresses (30).

Table 3. Deviation of the history of changes in mesostresses obtained via the use of the U-model for crystallite No1 from the history of changes in macrostresses determined within in the framework of other models.

Model	U	T	LU	LR1	LR2	GN
Norm δ_G, MPa	0	2.854	0	2.775	3.859	2.964
$\frac{\delta_G}{\|\sigma_{t\in[0,T]}{}^{(U)}\|_{C^n_{L^2}}}$	0	0.01145	0	0.011132	0.015481	0.011891

(Zero deviations from Table 1 suggest that they do not exceed the order of the computer error when working with real numbers.).

The results shown in Figure 3 and in Table 3 indicate the proximity of mesostresses. This is typical for the overwhelming number of crystallites (i.e., for almost all random orientations). In special cases, deviations of the mesostress components (difference modulus) may, at a certain instant of time, exceed 50 MPa. This can be attributed to some specific features in the motion of a representative point in the stress space near the yield surface [86] in transitions between the neighborhoods of the vertices. After transitions, in a great number of crystallites, the stresses become similar (there is a separate publication devoted to the study of these special cases). Due to the small fraction of such crystallites (37 crystallites out of 343) and averaging, the macroscale stresses obtained on the basis of the statistical constitutive model data using mesolevel models differ insignificantly (Table 2).

The differences in stresses obtained at the macrolevel are significantly lower than the experimental data scatter for samples from one production batch. Indeed, for each material, there is no complete identity of samples even from the same batch due to the presence of many random factors during manufacturing, which leads to differences in the microstructure (distributions of crystallite orientations, dislocation densities, etc.) and, as a consequence, to different experimental dependences of stresses on deformations. For example, such scatters for copper alloy samples are given in [87,88].

It should be noted here that the U-model and LU-model offer the most promise for modeling purely elastic behavior (e.g., elastic cyclic loading during exploitation) since the constitutive law is presented in the finite form, and there is no stress hysteresis in closed elastic cycles. Moreover, due to the use of energetically conjugate stress and strain measures, no energy is dissipated during the elastic deformation.

On the other hand, crystal plasticity models are primarily used to describe large inelastic deformations and material structure evolution. In this regard, formulations in terms of the actual configuration hold more promise because the physical meaning of the stress tensor is clearer in this case, and, consequently, the simulation of hardening using evolutionary equations is facilitated. These arguments lend additional support to using this formulation in case of the boundary value problem and in some advanced constitutive models at different deformation mechanisms. The theoretical predictions and the results of the numerical experiments indicate that the response produced by these models is similar to that obtained using the U-model. So, it turns out that the formulation in rate form would be most suitable for constructing advanced constitutive models (e.g., with complex hardening laws). The point is that the hypotheses adopted for constructing the evolutionary equations of the model via the use of the internal variables determined in the unloaded configuration can give more significant errors than the small deviations from the strict fulfillment of conservative conditions for the formulations written in terms of the actual configuration.

Geometrically nonlinear crystal plasticity models are applied to anisotropic crystallites, and therefore the decomposition of motion, used in rate formulations in terms of the current configuration, must be constructed by taking into account the symmetry of the material. In other words, the moving coordinate system determining the rigid motion in motion decomposition must be associated with the elements of material symmetry so that its symmetry properties can be taken into account correctly. Above, some options for constructing constitutive models with the introduction of such moving coordinate systems have been considered.

5. Conclusions

We have analyzed the most popular relations of crystal plasticity models in the context of geometric nonlinearity description and fulfillment of thermodynamic constraints in case of elastic deformation. These models are focused on the description of significant inelastic deformations and material structure evolution caused by these deformations. That is why the formulations in the rate form written in terms of the current configuration seem preferable to constructing advanced constitutive models able to take into account different deformation mechanisms and their complex interactions. The main advantages of these formulations over the formulation in the finite form are as follows:

- clear physical meaning of the stress tensor, which simplifies simulations with evolutionary hardening equations;
- use of the clear measure of the strain rate (the stretching tensor) and the possibility of an additive decomposition of the strain rate into contributions from various mechanisms;
- ease of their use in the rate formulation of the boundary value problem under varying, a priori unknown, contact boundary conditions.

The use of a corotational derivative in such a formulation means that the model is based on the decomposition of motion into the deformation motion and the rigid motion of a moving coordinate system, the rate of stress change relative to which is associated with the strain rate.

We have obtained the relations of the mesolevel model with an explicit separation of the motion of the moving coordinate system and elastic distortion of crystallites with respect to this system in the deformation gradient. The MCS is related to the elements of the material symmetry of the anisotropic crystallite. The proposed formalism with an explicit consideration of the MCS allows us to reasonably pass from the formulation in terms of the unloaded lattice configuration in the finite form (where the thermodynamic constraints for purely elastic deformation are strictly fulfilled) to the close formulation in rate form written in terms of the current configuration.

We have compared these relations with a popular crystal plasticity formulation, which makes it possible to establish their proximity to one another. The results of the performed analytical calculations show the equivalence or similarity (in the sense of the response determined under the same loads) of the formulations under consideration. These conclusions were supported by the results of the numerical calculations. It should be noted that the proposed approach aimed to determine the spin of the MCS also could include other physically justified rotation models, e.g., those taking into account the interaction of defects of neighboring crystallites or the contribution from the mechanism of grain boundary sliding. Such models cannot be introduced into other known approaches to constructing constitutive relations via the use of kinematic variables without explicit consideration of the material symmetric properties.

Author Contributions: Conceptualization, P.T. and A.S.; methodology, P.T. and A.S.; software, A.S.; validation, A.S., P.T., and N.K.; formal analysis, A.S. and N.K.; investigation, A.S., P.T., and N.K.; data curation, A.S. and N.K.; writing—original draft preparation, A.S.; writing—review and editing, P.T., A.S., and N.K.; visualization, A.S. and N.K.; supervision, P.T.; funding acquisition, N.K. All authors have read and agreed to the published version of the manuscript.

Funding: The study was carried out with financial support from the Ministry of Education and Science of the Russian Federation as part of the implementation of the national project "Science and Universities" (the state task fulfillment in the Laboratory of Multilevel Structural and Functional Materials Modeling).

Data Availability Statement: Not available.

Conflicts of Interest: The authors declare no conflict of interest.

References

1. McDowell, D.L. Internal State Variable Theory. In *Handbook of Materials Modeling*; Yip, S., Ed.; Springer: Dordrecht, The Netherlands, 2005; pp. 1151–1169.
2. Horstemeyer, M.F.; Bammann, D.J. Historical Review of Internal State Variable Theory for Inelasticity. *Int. J. Plast.* **2010**, *26*, 1310–1334. [CrossRef]
3. Trusov, P.; Ashikhmin, V.; Volegov, P.; Shveykin, A. Constitutive Relations and their Application to the Description of Microstructure Evolution. *Phys. Mesomech.* **2010**, *13*, 38–46. [CrossRef]
4. Maugin, G.A. The Saga of Internal Variables of State in Continuum Thermo-Mechanics (1893–2013). *Mech. Res. Commun.* **2015**, *69*, 79–86. [CrossRef]
5. Horstemeyer, M.F. Multiscale Modeling: A Review. In *Practical Aspects of Computational Chemistry: Methods, Concepts and Applications*; Leszczynski, J., Shukla, M.K., Eds.; Springer: Dordrecht, The Netherlands, 2009; pp. 87–135. ISBN 978-90-481-2687-3.
6. Diehl, M. Review and Outlook: Mechanical, Thermodynamic, and Kinetic Continuum Modeling of Metallic Materials at the Grain Scale. *MRS Commun.* **2017**, *7*, 735–746. [CrossRef]
7. Beyerlein, I.J.; Knezevic, M. Review of Microstructure and Micromechanism-Based Constitutive Modeling of Polycrystals with a Low-Symmetry Crystal Structure. *J. Mater. Res.* **2018**, *33*, 3711–3738. [CrossRef]
8. Roters, F.; Diehl, M.; Shanthraj, P.; Eisenlohr, P.; Reuber, C.; Wong, S.; Maiti, T.; Ebrahimi, A.; Hochrainer, T.; Fabritius, H.-O.; et al. DAMASK—The Düsseldorf Advanced Material Simulation Kit for Modeling Multi-Physics Crystal Plasticity, Thermal, and Damage Phenomena from the Single Crystal Up to the Component Scale. *Comput. Mater. Sci.* **2019**, *158*, 420–478. [CrossRef]
9. Trusov, P.V.; Shveykin, A.I. *Multilevel Models of Mono- and Polycrystalline Materials: Theory, Algorithms, Application Examples*; SB RAS: Novosibirsk, Russia, 2019. (In Russian)
10. Habraken, A.M. Modelling the Plastic Anisotropy of Metals. *Arch. Comput. Methods Eng.* **2004**, *11*, 3–96. [CrossRef]
11. Van Houtte, P. Crystal Plasticity Based Modelling of Deformation Textures. In Proceedings of the Microstructure and Texture in Steels, Jamshedpur, India, 5–7 February 2008; Haldar, A., Suwas, S., Bhattacharjee, D., Eds.; Springer: London, UK, 2009; pp. 209–224.
12. Zhang, K.; Holmedal, B.; Hopperstad, O.; Dumoulin, S.; Gawad, J.; Van Bael, A.; Van Houtte, P. Multi-Level Modelling of Mechanical Anisotropy of Commercial Pure Aluminium Plate: Crystal Plasticity Models, Advanced Yield Functions and Parameter Identification. *Int. J. Plast.* **2015**, *66*, 3–30. [CrossRef]
13. Shveykin, A.; Trusov, P.; Sharifullina, E. Statistical Crystal Plasticity Model Advanced for Grain Boundary Sliding Description. *Crystals* **2020**, *10*, 822. [CrossRef]
14. Lebensohn, R.A.; Castañeda, P.P.; Brenner, R.; Castelnau, O. Full-Field vs. Homogenization Methods to Predict Microstructure–Property Relations for Polycrystalline Materials. In *Computational Methods for Microstructure-Property Relationships*; Ghosh, S., Dimiduk, D., Eds.; Springer: Boston, MA, USA, 2011; pp. 393–441. ISBN 978-1-4419-0643-4.
15. McDowell, D.L. Viscoplasticity of Heterogeneous Metallic Materials. *Mater. Sci. Eng. R Rep.* **2008**, *62*, 67–123. [CrossRef]
16. Roters, F.; Eisenlohr, P.; Hantcherli, L.; Tjahjanto, D.D.; Bieler, T.; Raabe, D. Overview of Constitutive Laws, Kinematics, Homogenization and Multiscale Methods in Crystal Plasticity Finite-Element Modeling: Theory, Experiments, Applications. *Acta Mater.* **2010**, *58*, 1152–1211. [CrossRef]
17. Gurtin, M.E.; Reddy, B.D. Gradient Single-Crystal Plasticity within a Mises–Hill Framework Based on a New Formulation of Self- and Latent-Hardening. *J. Mech. Phys. Solids* **2014**, *68*, 134–160. [CrossRef]
18. Mayeur, J.; McDowell, D. A Comparison of Gurtin Type and Micropolar Theories of Generalized Single Crystal Plasticity. *Int. J. Plast.* **2014**, *57*, 29–51. [CrossRef]
19. McBride, A.; Bargmann, S.; Reddy, B.D. A Computational Investigation of a Model of Single-Crystal Gradient Thermoplasticity that Accounts for the Stored Energy of Cold Work and Thermal Annealing. *Comput. Mech.* **2015**, *55*, 755–769. [CrossRef]
20. Yalçinkaya, T.; Özdemir, I.; Simonovski, I. Micromechanical Modeling of Intrinsic and Specimen Size Effects in Microforming. *Int. J. Mater. Form.* **2018**, *11*, 729–741. [CrossRef]
21. Truesdell, C. *A First Course in Rational Continuum Mechanics*; Academic Press: London, UK, 1977.
22. Staroselsky, A.; Anand, L. Inelastic Deformation of Polycrystalline Face Centered Cubic Materials by Slip and Twinning. *J. Mech. Phys. Solids* **1998**, *46*, 671–673. [CrossRef]
23. Kalidindi, S. Modeling Anisotropic Strain Hardening and Deformation Textures in Low Stacking Fault Energy Fcc Metals. *Int. J. Plast.* **2001**, *17*, 837–860. [CrossRef]
24. Beyerlein, I.J.; Tomé, C.N. A Dislocation-Based Constitutive Law for Pure Zr Including Temperature Effects. *Int. J. Plast.* **2008**, *24*, 867–895. [CrossRef]
25. Beyerlein, I.J.; Mara, N.; Bhattacharyya, D.; Alexander, D.J.; Necker, C.T. Texture Evolution Via Combined Slip and Deformation Twinning in Rolled Silver–Copper Cast Eutectic Nanocomposite. *Int. J. Plast.* **2011**, *27*, 121–146. [CrossRef]
26. Cheng, J.; Ghosh, S. A crystal plasticity FE Model for Deformation with Twin Nucleation in Magnesium Alloys. *Int. J. Plast.* **2015**, *67*, 148–170. [CrossRef]
27. Zecevic, M.; Knezevic, M.; Beyerlein, I.J.; Tomé, C.N. An Elasto-Plastic Self-Consistent Model with Hardening Based on Dislocation Density, Twinning and De-Twinning: Application to Strain Path Changes in HCP Metals. *Mater. Sci. Eng. A* **2015**, *638*, 262–274. [CrossRef]

28. Fan, X.; Yang, H. Internal-State-Variable Based Self-Consistent Constitutive Modeling for Hot Working of Two-Phase Titanium Alloys Coupling Microstructure Evolution. *Int. J. Plast.* **2011**, *27*, 1833–1852. [CrossRef]
29. Yalcinkaya, T.; Brekelmans, W.; Geers, M. Deformation Patterning Driven by Rate Dependent Non-Convex Strain Gradient Plasticity. *J. Mech. Phys. Solids* **2011**, *59*, 1–17. [CrossRef]
30. Shveykin, A.; Sharifullina, E. Development of Multilevel Models Based on Crystal Plasticity: Description of Grain Boundary Sliding and Evolution of Grain Structure. *Nanomech. Sci. Technol. Int. J.* **2015**, *6*, 281–298. [CrossRef]
31. Toth, L.S.; Skrotzki, W.; Zhao, Y.; Pukenas, A.; Braun, C.; Birringer, R. Revealing Grain Boundary Sliding from Textures of a Deformed Nanocrystalline Pd–Au Alloy. *Materials* **2018**, *11*, 190. [CrossRef]
32. Zecevic, M.; Knezevic, M.; Mcwilliams, B.; Lebensohn, R.A. modeling of the thermo-mechanical response and texture evolution of we43 mg alloy in The Dynamic Recrystallization Regime Using a Viscoplastic Self-Consistent Formulation. *Int. J. Plast.* **2020**, *130*, 102705. [CrossRef]
33. Mellbin, Y.; Hallberg, H.; Ristinmaa, M. Recrystallization and Texture Evolution During Hot Rolling of Copper, Studied by a Multiscale Model Combining Crystal Plasticity and Vertex Models. *Model. Simul. Mater. Sci. Eng.* **2016**, *24*, 075004. [CrossRef]
34. Tang, T.; Zhou, G.; Li, Z.; Li, D.; Peng, L.; Peng, Y.; Wu, P.; Wang, H.; Lee, M.-G. A Polycrystal Plasticity Based Thermo-Mechanical-Dynamic Recrystallization Coupled Modeling Method and its Application to Light Weight Alloys. *Int. J. Plast.* **2019**, *116*, 159–191. [CrossRef]
35. Sarrazola, D.R.; Muñoz, D.P.; Bernacki, M. A New Numerical Framework for the Full Field Modeling of Dynamic Recrystallization in a CPFEM Context. *Comput. Mater. Sci.* **2020**, *179*, 109645. [CrossRef]
36. Taupin, V.; Chevy, J.; Fressengeas, C. Effects of Grain-To-Grain Interactions on Shear Strain Localization in Al–Cu–Li Rolled Sheets. *Int. J. Solids Struct.* **2016**, *99*, 71–81. [CrossRef]
37. Sokolov, A.S.; Trusov, P.V. A Two-Level Elasto-Viscoplastic Model: Application to the Analysis of the Crystal Anisotropy Influence. *J. Appl. Mech. Tech. Phys.* **2021**, *62*, 101–111. [CrossRef]
38. Kröner, E. General Continuum Theory of Dislocations and Proper Stresses. *Arch. Ration. Mech. Anal.* **1960**, *4*, 273–334. [CrossRef]
39. Lee, E.H.; Liu, D.T. Finite-Strain Elastic—Plastic Theory with Application to Plane-Wave Analysis. *J. Appl. Phys.* **1967**, *38*, 19–27. [CrossRef]
40. Lee, E.H. Elastic-Plastic Deformation at Finite Strains. *J. Appl. Mech.* **1969**, *36*, 35–40. [CrossRef]
41. Hutchinson, J.W. Bounds and Self-Consistent Estimates for Creep of Polycrystalline Materials. *Proc. R. Soc. London. Ser. A Math. Phys. Sci.* **1976**, *348*, 101–127. [CrossRef]
42. Asaro, R.J. Micromechanics of Crystals and Polycrystals. *Adv. Appl. Mech.* **1983**, *23*, 1–115. [CrossRef]
43. Asaro, R.J.; Needleman, A. Texture Development and Strain Hardening in Rate Dependent Polycrystals. *Acta Metall.* **1985**, *33*, 923–953. [CrossRef]
44. Gérard, C.; Cailletaud, G.; Bacroix, B. Modeling of Latent Hardening Produced by Complex Loading Paths in FCC Alloys. *Int. J. Plast.* **2013**, *42*, 194–212. [CrossRef]
45. Forest, S.; Rubin, M. A rate-Independent Crystal Plasticity Model with a Smooth Elastic–Plastic Transition and No Slip Indeterminacy. *Eur. J. Mech. A/Solids* **2016**, *55*, 278–288. [CrossRef]
46. Bronkhorst, C.A.; Kalidindi, S.R.; Anand, L. Polycrystalline Plasticity and the Evolution of Crystallographic Texture in FCC metals. *Philos. Trans. R. Soc. London Ser. A Phys. Eng. Sci.* **1992**, *341*, 443–477. [CrossRef]
47. Anand, L. Single-Crystal Elasto-Viscoplasticity: Application to Texture Evolution in Polycrystalline Metals at Large Strains. *Comput. Methods Appl. Mech. Eng.* **2004**, *193*, 5359–5383. [CrossRef]
48. Maresca, F.; Kouznetsova, V.; Geers, M. Reduced Crystal Plasticity for Materials with Constrained Slip Activity. *Mech. Mater.* **2016**, *92*, 198–210. [CrossRef]
49. Khadyko, M.; Dumoulin, S.; Cailletaud, G.; Hopperstad, O. Latent Hardening and Plastic Anisotropy Evolution in AA6060 Aluminium Alloy. *Int. J. Plast.* **2016**, *76*, 51–74. [CrossRef]
50. Grilli, N.; Janssens, K.; Nellessen, J.; Sandlöbes, S.; Raabe, D. Multiple Slip Dislocation Patterning in a Dislocation-Based Crystal Plasticity Finite Element Method. *Int. J. Plast.* **2018**, *100*, 104–121. [CrossRef]
51. Jeong, W.; Lee, C.-H.; Moon, J.; Jang, D.; Lee, M.-G. Grain Scale Representative Volume Element Simulation to Investigate the Effect of Crystal Orientation on Void Growth in Single and Multi-Crystals. *Metals* **2018**, *8*, 436. [CrossRef]
52. Sharma, L.; Peerlings, R.H.J.; Geers, M.G.D.; Roters, F. Microstructural Influences on Fracture at Prior Austenite Grain Boundaries in Dual-Phase Steels. *Materials* **2019**, *12*, 3687. [CrossRef] [PubMed]
53. Lieou, C.K.; Bronkhorst, C.A. Thermodynamic Theory of Crystal Plasticity: Formulation and Application to Polycrystal fcc Copper. *J. Mech. Phys. Solids* **2020**, *138*, 103905. [CrossRef]
54. Zhang, M.; Nguyen, K.; Segurado, J.; Montáns, F.J. A Multiplicative Finite Strain Crystal Plasticity Formulation based on Additive Elastic Corrector Rates: Theory and Numerical Implementation. *Int. J. Plast.* **2021**, *137*, 102899. [CrossRef]
55. Kalidindi, S.; Bronkhorst, C.; Anand, L. Crystallographic Texture Evolution in Bulk Deformation Processing of FCC Metals. *J. Mech. Phys. Solids* **1992**, *40*, 537–569. [CrossRef]
56. Knezevic, M.; Kalidindi, S.R. Crystal Plasticity Modeling of Microstructure Evolution and Mechanical Fields During Processing of Metals Using Spectral Databases. *JOM* **2017**, *69*, 830–838. [CrossRef]
57. Bittencourt, E. On the Effects of Hardening Models and Lattice Rotations in Strain Gradient Crystal Plasticity Simulations. *Int. J. Plast.* **2018**, *108*, 169–185. [CrossRef]

58. Van Houtte, P. Deformation Texture Prediction: From the Taylor Model to the Advanced Lamel Model. *Int. J. Plast.* **2005**, *21*, 589–624. [CrossRef]
59. Zecevic, M.; Knezevic, M. An Implicit Formulation of the Elasto-Plastic Self-Consistent Polycrystal Plasticity Model and its Implementation in Implicit Finite Elements. *Mech. Mater.* **2019**, *136*, 103065. [CrossRef]
60. Feng, Z.; Yoon, S.-Y.; Choi, J.-H.; Barrett, T.J.; Zecevic, M.; Barlat, F.; Knezevic, M. A Comparative Study between Elasto-Plastic Self-Consistent Crystal Plasticity and Anisotropic Yield Function with Distortional Hardening Formulations for Sheet Metal Forming. *Mech. Mater.* **2020**, *148*, 103422. [CrossRef]
61. Neil, C.; Wollmershauser, J.; Clausen, B.; Tomé, C.; Agnew, S. Modeling Lattice Strain Evolution at Finite Strains and Experimental Verification for Copper and Stainless Steel Using in Situ Neutron Diffraction. *Int. J. Plast.* **2010**, *26*, 1772–1791. [CrossRef]
62. Xiao, H.; Bruhns, O.T.; Meyers, A. The Integrability Criterion in Finite Elastoplasticity and its Constitutive Implications. *Acta Mech.* **2007**, *188*, 227–244. [CrossRef]
63. Trusov, P.V.; Shveykin, A.I.; Kondratev, N.S. Multilevel Metal Models: Formulation for Large Displacement Gradients. *Nanosci. Technol. Int. J.* **2017**, *8*, 133–166. [CrossRef]
64. Trusov, P.V.; Kondratev, N.S.; Shveykin, A.I. About Geometrically Nonlinear Constitutive Relations for Elastic Material. *PNRPU Mech. Bull.* **2015**, 182–200. [CrossRef]
65. Trusov, P.V.; Shveykin, A.I.; Yanz, A.Y. Motion Decomposition, Frame-Indifferent Derivatives, and Constitutive Relations at Large Displacement Gradients from the Viewpoint of Multilevel Modeling. *Phys. Mesomech.* **2017**, *20*, 357–376. [CrossRef]
66. Romanova, V.; Balokhonov, R.; Emelianova, E.; Sinyakova, E.; Kazachenok, M. Early Prediction of Macroscale Plastic Strain Localization in Titanium from Observation of Mesoscale Surface Roughening. *Int. J. Mech. Sci.* **2019**, *161-162*, 105047. [CrossRef]
67. Emelianova, E.S.; Romanova, V.A.; Balokhonov, R.R.; Pisarev, M.; Zinovieva, O.S. A Numerical Study of the Contribution of Different Slip Systems to the Deformation Response of Polycrystalline Titanium. *Phys. Mesomech.* **2021**, *24*, 166–177. [CrossRef]
68. Zaremba, S. Sur Une Forme Perfectionnée de La Théorie de La Relaxation. *Bull. Int. Acad. Sci. Cracovie.* **1903**, 595–614.
69. Jaumann, G. Geschlossenes System Physikalischer und Chemischer Differential-Gesetze. *Sitzungsber. Akad. Wiss. Wien Nat.-Naturwiss. Klasse. IIA* **1911**, *120*, 385–530.
70. Green, A.E.; Naghdi, P.M. A General Theory of an Elastic-Plastic Continuum. *Arch. Ration. Mech. Anal.* **1965**, *18*, 251–281. [CrossRef]
71. Xiao, H.; Bruhns, O.T.; Meyers, A. Hypo-Elasticity Model Based upon the Logarithmic Stress Rate. *J. Elast.* **1997**, *47*, 51–68. [CrossRef]
72. Xiao, H.; Bruhns, O.; Meyers, A. A Natural Generalization of Hypoelasticity and Eulerian Rate Type Formulation of Hyperelasticity. *J. Elast.* **1999**, *56*, 59–93. [CrossRef]
73. Jiao, Y.; Fish, J. Is an Additive Decomposition of a Rate of Deformation and Objective Stress Rates Passé? *Comput. Methods Appl. Mech. Eng.* **2017**, *327*, 196–225. [CrossRef]
74. Jiao, Y.; Fish, J. On the Equivalence Between the Multiplicative Hyper-Elasto-Plasticity and the Additive Hypo-Elasto-Plasticity Based on the Modified Kinetic Logarithmic Stress Rate. *Comput. Methods Appl. Mech. Eng.* **2018**, *340*, 824–863. [CrossRef]
75. Horstemeyer, M.F.; Potirniche, G.P.; Marin, E.B. Crystal Plasticity. In *Handbook of Materials Modeling*; Yip, S., Ed.; Springer: Dordrecht, The Netherlands, 2005; pp. 1133–1149.
76. Trusov, P.V.; Shveykin, A.I. On Motion Decomposition and Constitutive Relations in Geometrically Nonlinear Elastoviscoplasticity of Crystallites. *Phys. Mesomech.* **2017**, *20*, 377–391. [CrossRef]
77. Mandel, J. Equations Constitutives et Directeurs Dans les Milieux Plastiques et Viscoplastiques. *Int. J. Solids Struct.* **1973**, *9*, 725–740. [CrossRef]
78. Shveykin, A.I.; Trusov, P.V.; Kondratev, N.S. Multiplicative Representation of the Deformation Gradient Tensor in Geometrically Nonlinear Multilevel Constitutive Models. *Lobachevskii J. Math.* **2021**, *42*, 2047–2055. [CrossRef]
79. Ostapovich, K.V.; Trusov, P.V.; Yants, A.Y. Prediction of Crystallographic Texture Formation in Polycrystalline Samples under Severe Plastic Deformation Based on a Two-Level Statistical Elasto-Viscoplastic Model. *Phys. Mesomech.* **2021**, *24*, 225–236. [CrossRef]
80. Shveykin, A.; Trusov, P.V. Multilevel Models of Polycrystalline Metals: Comparison of Relations Describing the Crystallite Lattice Rotations. *Nanosci. Technol. Int. J.* **2019**, *10*, 1–20. [CrossRef]
81. Trusov, P.V.; Shveykin, A.I.; Nechaeva, E.S.; Volegov, P. Multilevel Models of Inelastic Deformation of Materials and their Application for Description of Internal Structure Evolution. *Phys. Mesomech.* **2012**, *15*, 155–175. [CrossRef]
82. Shveikin, A.I.; Trusov, P.V. Correlation between Geometrically Nonlinear Elastoviscoplastic Constitutive Relations Formulated in Terms of the Actual and Unloaded Configurations for Crystallites. *Phys. Mesomech.* **2018**, *21*, 193–202. [CrossRef]
83. Messner, M.; Beaudoin, A.; Dodds, R. Consistent Crystal Plasticity Kinematics and Linearization for the Implicit Finite Element Method. *Eng. Comput.* **2015**, *32*, 1526–1548. [CrossRef]
84. Ma, R.; Truster, T.J. FFT-Based Homogenization of Hypoelastic Plasticity at Finite Strains. *Comput. Methods Appl. Mech. Eng.* **2019**, *349*, 499–521. [CrossRef]
85. Harder, J. FEM-Simulation of the Hardening Behavior of FCC Single Crystals. *Acta Mech.* **2001**, *150*, 197–217. [CrossRef]
86. Rocks, U.; Canova, G.; Jonas, J. Yield Vectors in f.c.c. Crystals. *Acta Met.* **1983**, *31*, 1243–1252. [CrossRef]
87. Piao, M.; Huh, H.; Lee, I.; Park, L. Characterization of Hardening Behaviors of 4130 Steel, OFHC Copper, Ti6Al4V Alloy Considering Ultra-High Strain Rates and High Temperatures. *Int. J. Mech. Sci.* **2017**, *131–132*, 1117–1129. [CrossRef]
88. Zhang, Y.; Sun, H.; Volinsky, A.A.; Wang, B.; Tian, B.; Chai, Z.; Liu, Y.; Song, K. Small Y Addition Effects on Hot Deformation Behavior of Copper-Matrix Alloys. *Adv. Eng. Mater.* **2017**, *19*, 1700197. [CrossRef]

crystals

Article

Numerical Modeling and Simulations of Twinning-Induced Plasticity Using Crystal Plasticity Finite Element Method

Rashid Khan [1,*], Tasneem Pervez [2], Adel Alfozan [1], Sayyad Zahid Qamar [2] and Sumiya Mohsin [3]

[1] Mechanical & Industrial Engineering Department, College of Engineering, Imam Mohammad Ibn Saud Islamic University, Riyadh 11432, Saudi Arabia; afozan@imamu.edu.sa
[2] Mechanical & Industrial Engineering Department, College of Engineering, Sultan Qaboos University, Muscat 123, Oman; tasneem@squ.edu.om (T.P.); sayyad@squ.edu.om (S.Z.Q.)
[3] Mechanical Engineering Department, University Teklongi PETRONAS, Seri Iskandar 32610, Perak, Malaysia; mohsin_20001928@utp.edu.my
* Correspondence: rakhan@imamu.edu.sa

Abstract: In the current work, a fully implicit numerical integration scheme is developed for modeling twinning-induced plasticity using a crystal plasticity framework. Firstly, the constitutive formulation of a twin-based micromechanical model is presented to estimate the deformation behavior of steels with low stacking fault energy. Secondly, a numerical integration scheme is developed for discretizing constitutive equations through a fully implicit time integration scheme using the backward Euler method. A time sub-stepping algorithm and the two-norm convergence criterion are used to regulate time step size and stopping criterion. Afterward, a numerical scheme is implemented in finite element software ABAQUS as a user-defined material subroutine. Finally, finite element simulations are executed for observing the validity, performance, and limitations of the numerical scheme. It is observed that the simulation results are in good agreement with the experimental observations with a maximum error of 16% in the case of equivalent stress and strain. It is also found that the developed model is able to estimate well the deformation behavior, magnitude of slip and twin shear strains, and twin volume fraction of three different TWIP steels where the material point is subjected to tension and compression.

Keywords: slip; mechanical twinning; constitutive modeling; finite element simulations; deformation behavior; twin volume fraction

Citation: Khan, R.; Pervez, T.; Alfozan, A.; Qamar, S.Z.; Mohsin, S. Numerical Modeling and Simulations of Twinning-Induced Plasticity Using Crystal Plasticity Finite Element Method. *Crystals* **2022**, *12*, 930. https://doi.org/10.3390/cryst12070930

Academic Editor: Wojciech Polkowski

Received: 24 March 2022
Accepted: 26 May 2022
Published: 30 June 2022

1. Introduction

Advanced materials have a vital place along with other key technologies in the fourth industrial revolution (4IR) [1]. Technological developments and achievements depend (and will continue to depend) on the availability of advanced materials. In addition, advanced manufacturing techniques make it possible to produce a range of products, specifically for adverse and corrosive environments and cryogenic applications. In particular, manufacturing products with geometrically complex and enhanced properties becomes possible due to the development of advanced metallic alloys. However, regardless of the availability of the class of metallic alloys, several issues and limitations still restrict advancement in product development. According to the material point of view, one of the restrictions is the simultaneous requirement of high tensile strength and ductility. These properties become extremely crucial in large-deformation applications, such as superplasticity, sheet metal forming, cold rolling, and so on, where high strength with excellent formability is required to obtain highly deformed products. In traditional metallic alloys, these properties cannot be enhanced simultaneously; rather, improvement in one can only be achieved through the other's detriment. This long-lasting issue is resolved by developing innovative and advanced high-strength steels (AHSS) [2]. One of the prime characteristics of AHSS is an excellent balance between tensile strength and ductility. This makes AHSS an optimum choice in the automotive, aerospace, and oil and gas industries [3–5]. The classification of AHSS in three generations is based on the nature of the microstructure, phases, and composition [2,4]. The prominent members of the second and third generations are transformation and twinning-induced plasticity steels, respectively.

Twinning-induced plasticity (TWIP) steels have broad range of applications due to their promising combination of work hardening, formability, and tensile strength [6–8]. These characteristics make it a dominant candidate for many advanced material applications. The special amalgamation of TWIP steels' properties is achieved through controlled microstructure, and the fraction of primary and secondary phases [9]. The primary phase in these steels is retained austenite, while the secondary may include ferrite, martensite, and sometimes bainite. The main alloying element, which plays a significant role in enhancing the properties, is manganese [10]. The weight percentage of manganese varies, but it is normally greater than 15–20%. Due to the high percentage of manganese, the stacking fault energy (SFE) of TWIP steels is relatively lower [11], in the span of 20 to 40 mJ/m^2, than the other class of AHSS. The low magnitude of SFE favors activation of a secondary mode of plasticity, which is twinning [12]. The volume fraction of twinning governs the mechanical properties—more specifically, the strain hardening—of TWIP steels. Since twin regions behave as obstacles in dislocation glide, the dislocation mean free path may reduce. This eventually improves the strain hardening of TWIP steels [7,13,14]. Furthermore, in Fe–Mn–C grade, twin regions normally comprise a huge magnitude of sessile dislocations' density. The high magnitude of density results from twin nucleation and growth, which act as resilient inclusions that may hinder dislocation glide [15,16]. It can be stated that the excellent combination of strength and ductility in TWIP steels is primarily due to work hardening, which may be induced in a material due to the twinning mechanism. The secondary mode of plastic deformation, mechanical twinning, may occur in metals (in conjunction with slip), non-metals, and metallic glasses [6,7,17]. However, the fraction of twinning is highly dependent on the chemical composition of the alloy, as well as the physical conditions [18–21]. Moreover, stacking fault energy (SFE) plays an important role in defining the potential for the activation of twinning [22–24].

One of researchers' foremost challenges was to computationally couple two modes of plasticity, slip and twinning, in predicting thedeformation behavior of metals [25–27]. Among these obstacles, one was to computationally account for the huge number of twin orientations that may occur during the course of deformation [25,26]. Multiple solutions have been proposed to overcome this long-lasting issue. These mainly include Taylor's least work hypothesis [28] to minimize the orientation factor [29]; utilizing the statistical technique in determining the re-orientation of whole grain to account for the total number of grain orientations in computation [26]; and employing the condition of weighted grain orientations to avoid the generation of new grains [25]. Kalidindi et al. [27] presented another possible solution to incorporate mechanical twinning in crystal plasticity theory. It is based on the concept of multiplicative decomposition of the total deformation gradient into elastic and plastic, as initially proposed by Asaro and Rice [30]. Furthermore, plastic deformation gradient and strain hardening effects are defined according to slip and mechanical twinning. In addition, the rate of change of twin volume fraction depends on the resolved shear stress and twin resistance of potentially active twin systems, as in the definition of slip system hardening by Asaro and Rice [30]. In subsequent work, several attempts have been made to predict strain-hardening effects of mechanical twinning [31–33]. In these models, twin-related strain-hardening effects of α-titanium alloys are incorporated in crystal plasticity theory. The models have utilized the crystal plasticity finite element method to predict metallic alloys' deformation behavior, which undergoes slip and twinning.

The crystal plasticity finite element method (CPFEM) is frequently utilized to model mechanical twinning in shape-memory alloys [34–36], advanced high-strength steels [37–39], magnesium alloys [40], and high-Mn austenitic steels [19,41,42]. It has also found huge application in modeling and simulating the deformation and damage behaviors of transformation-induced plasticity (TRIP), twinning-induced plasticity (TWIP), and multiphase steels [43–45].

More specifically, Cui et al. [43] have presented a multi-mechanism and microstructure-based crystal plasticity model for estimating the shear deformation behavior of TRIP steel under cyclic load. In this, the simulation results show that the samples subjected to high strain magnitude exhibit stronger cyclic shear hardening and that activation of martensitic transformation promotes cyclic hardening. Furthermore, Qayyum et al. [44] have presented a novel physics-based crystal plasticity model with ductile damage criterion for predicting the damage behavior of austenite-based TRIP steel (X8CrMnNi16-6-6) with 10% zirconia particles. Among other conclusions, it was found that the energy absorbed by the zirconia particles in the course of deformation is comparatively high, and there is substantial stiffness degradation in the bulk material. These factors significantly influenced the composite material behavior. In a continuation of this work, Qayyum et al. [45] have utilized a similar physics-based crystal plasticity numerical modeling technique to create a semi-automatic virtual laboratory to analyze and create customized functional multiphase materials. The CPFEM has also successfully implemented in modeling the behavior of TWIP steels. More specifically, it has been implemented to estimate: elastic properties of single-crystal TWIP steel through nano-indentation [46]; deformation behavior, texture evolution, and earing mechanism in TWIP steels [47,48]; and combined effects of slip, mechanical twinning, and martensitic transformation on the overall behavior of high-Mn steels [49–51]. In particular, Madivala et al. [52] have investigated the strain-hardening and fracture behavior of high-manganese austenitic TWIP steel at temperatures ranging from 123 K to 773 K. It was observed that twinning becomes the dominant deformation mechanism at 298 K, and the twin fraction increases with temperature until a transition temperature of about 473 K. Beyond this, dislocation glide alone becomes dominant, instead of twinning and dislocation glide. Recently, Khan et al. [53] presented a micromechanical model of twinning-induced plasticity using crystal plasticity and thermodynamic frameworks. The deformation gradients resulting from crystallographic slip and mechanical twinning are modeled through the kinematic decomposition of the total deformation gradient. The constitutive formulation of dissipated energy and Helmholtz free energy and the driving potentials for inelastic deformation modes are represented through a thermodynamic framework. The deformation gradients resulting from crystallographic slip and mechanical twinning are modeled through the kinematic decomposition of the total deformation gradient. Finally, a numerical integration scheme is used to incorporate the constitutive formulation in commercial finite element code ABAQUS through the user-defined material subroutine. It was observed that when the material point in the single crystal is subjected to tension, twin deformation plays a dominant role, while the reverse is observed in compression. Furthermore, in both tension and compression, the variation in the volume fraction of twinned martensite is found in all crystallographic directions (i.e., [100], [110], and [111]), but with different magnitudes.

The current work is an extension of the micromechanical model presented by Khan et al. [53], where a novel numerical integration scheme is developed. In this, firstly, the constitutive equations of the micromechanical model of twinning-induced plasticity are presented. Secondly, a numerical integration scheme is utilized to update the constitutive equations using a fully implicit time integration procedure based on the backward Euler method. Thirdly, a three-level iterative scheme is developed to solve the coupled nonlinear system of equations through the Newton–Raphson method. In this, a $\mathcal{L}_2$ (two norm) convergence criterion is used to estimate the response of incremental values of state variables. A time sub-stepping algorithm is incorporated with the numerical integration scheme to improve the convergence and reduce computational time. The developed numerical scheme is then implemented as a user-defined material subroutine in the finite element software ABAQUS 6.14. The model is then validated through published experimental observations of TWIP steels. Finally, the deformation behavior of TWIP steel under different loading conditions is estimated through finite element simulations.

2. Constitutive Modeling

In this part, constitutive equations of the model developed by [53] are presented, where twinning-induced plasticity is incorporated with slip-based crystal plasticity theory. The constitutive equations include the formulations of: (i) multiplicative decomposition of the total deformation gradient into elastic and plastic parts, (ii) driving potentials for slip and twining, (iii) dissipated energy as a result of plastic flow, (iv) recovered energy due to elastic deformation, (v) plastic flow rule due to slip and twinning, and (vi) hardening law.

In the subsequent sections, standard symbols are employed for designating tensors and their operations. The tensor and vector quantities are, respectively, expressed through the capital and small bold letters. Fourth-, second-, and first-order tensors are symbolized as $\mathbb{C}$, $\mathbf{A}$, and $\mathbf{a}$, respectively. The notations $\mathbf{a} \otimes \mathbf{b}$ and $\mathbf{Aa}$ represent the dyadic product of vectors and the contraction of second-order tensor with vector, respectively. The mathematical operations of second-order tensors are illustrated as: (i) inner product: $\mathbf{AB}$, (ii) dyadic product: $\mathbf{A} \otimes \mathbf{B}$, and (iii) scalar product: $\mathbf{A}:\mathbf{B}$. The fourth- and second-order tensors' contraction is expressed as $\mathbb{C}:\mathbf{A}$. Any non-standardized notation will be defined explicitly.

2.1. Kinematic-Based Modeling

In all successive sections of mathematical modeling, only the final equations of each constitutive formulation are presented. For detailed derivations, the readers are advised to refer to [53,54].

Multiplicative Decomposition of Deformation Gradient

The decomposition of the total deformation gradient into elastic and plastic parts, as discussed by [55], can be represented as:

$$\mathbf{F} = \mathbf{F}^e\mathbf{F}^p, \tag{1}$$

where $\mathbf{F}$ is the total deformation gradient, while $\mathbf{F}^e$ and $\mathbf{F}^p$ are, respectively, elastic and plastic deformation gradients. The elastic part is further categorized in symmetric left stretch $\mathbf{V}^e$ and orthogonal rotation $\mathbf{R}^e$ tensors, $\mathbf{F}^e = \mathbf{V}^e\mathbf{R}^e$. The plastic deformation gradient incorporates crystallographic slip and mechanical twinning. The rotation and plastic deformation gradient tensors can be combined into a plastic rigid rotation tensor, $\mathbf{F}^*$, as discussed in [54]. The overall decomposition of deformation gradients can be represented by Equation (2) as:

$$\mathbf{F} = \mathbf{V}^e\mathbf{F}^* \;, \quad \text{where } \mathbf{F}^* = \mathbf{R}^e\mathbf{F}^p = \mathbf{R}^e\mathbf{F}^p_t\mathbf{F}^p_s \;. \tag{2}$$

In Equation (2), the slip and twinning parts of $\mathbf{F}^p$ are, respectively, represented as $\mathbf{F}^p_s$ and $\mathbf{F}^p_t$; $\mathbf{V}^e$ is the symmetric left stretch tensor; and $\mathbf{R}^e$ is an orthogonal rotation tensor. The elastic-plastic behavior of a material point involves slip and twin deformation modes, which can be elaborated through kinematic decomposition, as illustrated in Figure 1.

The undeformed and deformed configurations of a material point are represented by Ω_0 and Ω_f, respectively. In this, total deformation gradient is disintegrated into three intermediate configurations $\overline{\Omega}_{\mathrm{I}}$, $\check{\Omega}_{\mathrm{II}}$, and $\widetilde{\Omega}_{\mathrm{III}}$. The first and second intermediate configurations represent plastic deformation due to slip and mechanical twinning, respectively, while the third shows rigid body rotation. The deformed state, Ω_f, is projected from the third intermediate configuration through the stretch tensor $\mathbf{V}^e$. In the present model, the relaxed (third) intermediate configuration $\widetilde{\Omega}_{\mathrm{III}}$ is adopted for representing the constitutive equations. Theoretically, this configuration is obtained by elastically unloading, using $(\mathbf{V}^e)^{-1}$ without rotation, from the current (deformed) state to a stress-free configuration [54].

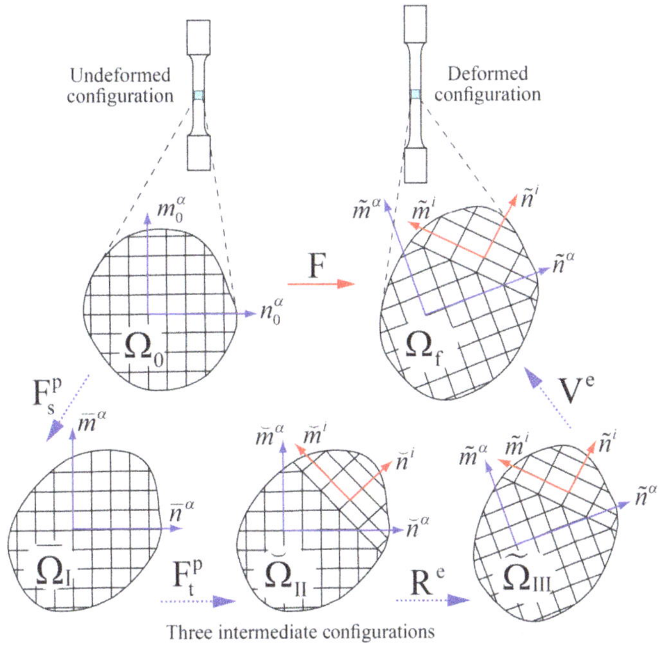

Figure 1. Kinematic decomposition of a material point into three intermediate configurations: $\bar{\Omega}_I$, $\check{\Omega}_{II}$, and $\tilde{\Omega}_{III}$.

Referring to the kinematic decomposition and definition of the velocity gradient, the velocity gradient in the current (deformed) configuration can be presented as:

$$\mathbf{L} = \dot{\mathbf{F}}\mathbf{F}^{-1} = \dot{\mathbf{V}}^e(\mathbf{V}^e)^{-1} + \mathbf{V}^e\widetilde{\mathbf{L}}^*(\mathbf{V}^e)^{-1}. \tag{3}$$

Here, $\widetilde{\mathbf{L}}^*$ is the velocity gradient in third intermediate configuration. By considering Equations (1) and (2), it can be written as:

$$\widetilde{\mathbf{L}}^* = \dot{\mathbf{F}}^*(\mathbf{F}^*)^{-1} = \dot{\mathbf{R}}^e(\mathbf{R}^e)^{-1} + \mathbf{R}^e\check{\mathbf{L}}^p(\mathbf{R}^e)^{-1}, \tag{4}$$

where $\check{\mathbf{L}}^p$, the plastic velocity gradient in second intermediate configuration $\check{\Omega}_{\mathrm{II}}$, incorporates plastic flow due to crystallographic slip and mechanical twinning in the constitutive model. Since the plastic deformation gradient can be divided into slip and twin contributions, $\mathbf{F}^p = \mathbf{F}^p_t\mathbf{F}^p_s$, it can be written as:

$$\check{\mathbf{L}}^p = \dot{\mathbf{F}}^p(\mathbf{F}^p)^{-1} = \bar{\mathbf{L}}^p_s + \mathbf{F}^p_s\check{\mathbf{L}}^p_t(\mathbf{F}^p_s)^{-1}. \tag{5}$$

In Equation (5), $\bar{\mathbf{L}}^p_s$ and $\check{\mathbf{L}}^p_t$ represent the plastic velocity gradients of slip and twinning, respectively. By using the definition of the velocity gradient, Equation (5) can be transformed as:

$$\check{\mathbf{L}}^p = \sum_{\alpha=1}^{N_s}\dot{\gamma}^\alpha\bar{\mathbf{S}}^\alpha + \mathbf{F}^p_s\left\{\sum_{i=1}^{N_t}\dot{\gamma}^i\check{\mathbf{S}}^i\right\}(\mathbf{F}^p_s)^{-1}, \tag{6}$$

where α represents the slip system's number as $(\alpha = 1, \ldots, N_s)$; N_s is the total number of slip systems; $\dot{\gamma}^\alpha$ is the plastic shear strain rate of α-slip system; $\bar{\mathbf{S}}^\alpha$ is the Schmid orientation tensor in first intermediate configuration (represented by slip direction, $\bar{\mathbf{m}}^\alpha$, and slip plane area normal vectors, $\bar{\mathbf{n}}^\alpha$, as $\bar{\mathbf{S}}^\alpha = \bar{\mathbf{m}}^\alpha \otimes \bar{\mathbf{n}}^\alpha$, where i denotes the twin system $(i = 1, \ldots, N_t)$, and N_t represents the entire number of twin systems); $\dot{\gamma}^i$ is the plastic shear strain rate of i-twin system; and $\check{\mathbf{S}}^i$ is the twin orientation tensor in second intermediate state (expressed through twin direction, $\check{\mathbf{m}}^i$, and twin plane area normal, $\check{\mathbf{n}}^i$, vectors as $\check{\mathbf{S}}^i = \check{\mathbf{m}}^i \otimes \check{\mathbf{n}}^i$). In the presented model, mechanical twinning is assumed to occur within the plastically deformed region by crystallographic slip. This leads to the inclusion of the volume fraction of each region (slip and twin) in the velocity gradient equation as folllows:

$$\check{\mathbf{L}}^p = \left(1 - \sum_{i=1}^{N_t} v^i\right) \sum_{\alpha=1}^{N_s} \dot{\gamma}^\alpha \bar{\mathbf{S}}^\alpha + \mathbf{F}_s^p \left\{ \sum_{i=1}^{N_t} v^i \dot{\gamma}^i \check{\mathbf{S}}^i \right\} (\mathbf{F}_s^p)^{-1} . \tag{7}$$

Here, v^i is the volume fraction of i-twin system in second configuration. Similarly, the plastic velocity gradient in third intermediate configuration can be expressed as:

$$\widetilde{\mathbf{L}}^* = \widetilde{\mathbf{\Theta}}^e + \mathbf{R}^e \left[\left(1 - \sum_{i=1}^{N_t} v^i\right) \sum_{\alpha=1}^{N_s} \dot{\gamma}^\alpha \bar{\mathbf{S}}^\alpha + \mathbf{F}_s^p \left\{ \sum_{i=1}^{N_t} v^i \dot{\gamma}^i \check{\mathbf{S}}^i \right\} (\mathbf{F}_s^p)^{-1} \right] (\mathbf{R}^e)^T , \tag{8}$$

where $\widetilde{\mathbf{\Theta}}^e (= \dot{\mathbf{R}}^e (\mathbf{R}^e)^T)$ is the spin tensor. The Schmid tensor— $\bar{\mathbf{S}}^\alpha$ from first and twin orientation tensor, $\check{\mathbf{S}}^i$ from second states—can be transformed to the third intermediate configuration through forward conversion procedure using rigid rotation tensor, to yield the following:

$$\widetilde{\mathbf{S}}^\alpha = \mathbf{R}^e \bar{\mathbf{S}}^\alpha (\mathbf{R}^e)^T , \tag{9}$$

$$\widetilde{\mathbf{S}}^i = \mathbf{R}^e \check{\mathbf{S}}^i (\mathbf{R}^e)^T . \tag{10}$$

The final form of velocity gradient in the third intermediate configuration can be expressed as:

$$\widetilde{\mathbf{L}}^* = \widetilde{\mathbf{\Theta}}^e + \left(1 - \sum_{i=1}^{N_t} v^i\right) \sum_{\alpha=1}^{N_s} \dot{\gamma}^\alpha \widetilde{\mathbf{S}}^\alpha + \mathbf{F}_s^p \left\{ \sum_{i=1}^{N_t} v^i \dot{\gamma}^i \widetilde{\mathbf{S}}^i \right\} (\mathbf{F}_s^p)^{-1} . \tag{11}$$

The symmetric part of the velocity gradient, as mentioned in Equation (3), is given as:

$$\mathbf{D} = \widetilde{\mathbf{D}}^e + \frac{1}{2} \left[(\mathbf{V}^e)^{-T} \widetilde{\mathbf{C}}^e \widetilde{\mathbf{L}}^* (\mathbf{V}^e)^{-1} + (\mathbf{V}^e)^{-T} (\widetilde{\mathbf{C}}^e \widetilde{\mathbf{L}}^*)^T (\mathbf{V}^e)^{-1} \right] , \tag{12}$$

where $\widetilde{\mathbf{D}}^e$ is the symmetric component of $\dot{\mathbf{V}}^e (\mathbf{V}^e)^{-1}$, and $\widetilde{\mathbf{C}}^e = (\mathbf{V}^e)(\mathbf{V}^e)^T = (\mathbf{V}^e)^2$ is the right Cauchy–Green deformation tensor in the third intermediate configuration. Equation (12) can also be represented as:

$$\mathbf{D} = \widetilde{\mathbf{D}}^e + (\mathbf{V}^e)^{-T} \widetilde{\mathbf{D}}^* (\mathbf{V}^e)^{-1} . \tag{13}$$

The symmetric part of $\widetilde{\mathbf{C}}^e \widetilde{\mathbf{L}}^*$ is given as:

$$\widetilde{\mathbf{D}}^* = \mathrm{sym}(\widetilde{\mathbf{C}}^e \widetilde{\mathbf{\Theta}}^e) + \mathbf{R}^e \bar{\mathbf{D}}^p (\mathbf{R}^e)^T , \tag{14}$$

where $\bar{\mathbf{D}}^p$ is the symmetric component of $\bar{\mathbf{C}}^e \bar{\mathbf{L}}^p$. Similarly, the skew-symmetric component of $\mathbf{L}$ can be represented as:

$$\mathbf{W} = \frac{1}{2} \left[\mathbf{L} - \mathbf{L}^T \right] = \widetilde{\mathbf{W}}^e + (\mathbf{V}^e)^{-T} \widetilde{\mathbf{W}}^* (\mathbf{V}^e)^{-1} , \tag{15}$$

where $\widetilde{\mathbf{W}}^e$ and $\widetilde{\mathbf{W}}^*$ are the skew-symmetric components of $\dot{\mathbf{V}}^e (\mathbf{V}^e)^{-1}$ and $\widetilde{\mathbf{C}}^e \widetilde{\mathbf{L}}^*$, respectively. The component $\widetilde{\mathbf{W}}^*$ can be derived as:

$$\widetilde{\mathbf{W}}^* = \mathrm{skew}(\widetilde{\mathbf{C}}^e \widetilde{\mathbf{\Theta}}^e) + \mathbf{R}^e \bar{\mathbf{W}}^p (\mathbf{R}^e)^T . \tag{16}$$

Here, $\bar{\mathbf{W}}^p$ is the anti-symmetric part of $\bar{\mathbf{C}}^e\bar{\mathbf{L}}^p$. The finite Green strain tensor in reference configuration (third intermediate state with reference to current configuration) can be represented in terms of the symmetric part of the velocity gradient through Equation (19).

A procedure of backward mapping, from current to third intermediate configurations, is adopted for developing an equation of the velocity gradient, $\widetilde{\mathbf{L}}$, that incorporates elastic stretch along with plasticity and rigid body rotation. This can be expressed through Equation (17) as:

$$\widetilde{\mathbf{L}} = (\mathbf{V}^e)^{-1}\mathbf{L}\mathbf{V}^e = (\mathbf{V}^e)^{-1}\dot{\mathbf{V}}^e + \widetilde{\mathbf{L}}^* \, . \tag{17}$$

After substituting the value of $\widetilde{\mathbf{L}}^*$ from Equation (11), Equation (17) becomes

$$\widetilde{\mathbf{L}} = (\mathbf{V}^e)^{-1}\dot{\mathbf{V}}^e + \widetilde{\mathbf{\Theta}}^e + \left[\left(1 - \sum_{i=1}^{N_t} v^i\right)\sum_{\alpha=1}^{N_s} \dot{\gamma}^\alpha \widetilde{\mathbf{S}}^\alpha + \mathbf{F}_s^p\left\{\sum_{i=1}^{N_t} v^i \dot{\gamma}^i \widetilde{\mathbf{S}}^i\right\}(\mathbf{F}_s^p)^{-1}\right] . \tag{18}$$

The finite Green strain tensor in reference configuration (third intermediate state with reference to current configuration) can be represented as:

$$\widetilde{\mathbf{E}}^e = \frac{1}{2}(\widetilde{\mathbf{C}}^e - \mathbf{I}) \, , \qquad \dot{\widetilde{\mathbf{E}}}^e = \frac{1}{2}\dot{\widetilde{\mathbf{C}}}^e . \tag{19}$$

Here, $\widetilde{\mathbf{C}}^e$ is the right Cauchy–Green strain tensor, and $\mathbf{I}$ is a second-order identity tensor. The 2nd Piola–Kirchhoff (PK2) stress, $\mathbf{T}^e$, in reference configuration (third intermediate configuration) can be represented in terms of finite Green strain tensor. The stresses in slipped and twinned regions can be written in the form of constitutive formulation as:

$$\mathbf{T}_s^e = \widetilde{\mathbb{C}}^s : \widetilde{\mathbf{E}}^e \, , \tag{20}$$

$$\mathbf{T}_t^e = \widetilde{\mathbb{C}}^t : \widetilde{\mathbf{E}}^e \, , \tag{21}$$

where $\mathbf{T}_s^e$ and $\mathbf{T}_t^e$ are the PK2 stress tensors, and $\widetilde{\mathbb{C}}^s$ and $\widetilde{\mathbb{C}}^t$ are the fourth-order elasticity tensors for slipped and twinned regions, respectively. In view of incorporating the effects of slip and twinning in the PK2 stress tensor, an equivalent form can be defined as:

$$\mathbf{T}^e = \widetilde{\mathbb{C}}^e : \widetilde{\mathbf{E}}^e \, . \tag{22}$$

Here, $\mathbf{T}^e$ is an equivalent PK2 stress, and $\widetilde{\mathbb{C}}^e$ is the equivalent elasticity tensor. An equivalent form of elasticity tensor can be expressed through volume averaging technique as:

$$\widetilde{\mathbb{C}}^e = \left(1 - \sum_{i=1}^{N_t} v^i\right)\widetilde{\mathbb{C}}^s + \sum_{i=1}^{N_t} v^i \widetilde{\mathbb{C}}^t \, . \tag{23}$$

Similarly, Cauchy stress can also be approximated on the basis of volume averaging technique as:

$$\mathbf{T} = \left(1 - \sum_{i=1}^{N_t} v^i\right)\mathbf{T}_s + \sum_{i=1}^{N_t} v^i \mathbf{T}_t . \tag{24}$$

2.2. *Kinetic-Based Modeling*

2.2.1. Dissipated Energy Formulation

The dissipated energy density (energy per unit reference volume) in the form of rate of change of entropy is estimated as:

$$\dot{E}_d = \boldsymbol{\sigma} : \dot{\mathbf{F}} + \rho_0 \theta \dot{\Pi}_m - \rho_0 \dot{e} - \nabla \vec{\mathbf{q}} \, , \tag{25}$$

where $\boldsymbol{\sigma}$ is PK1 (1st Piola–Kirchhoff) stress, $\dot{\mathbf{F}}$ is the rate of change of $\mathbf{F}$, ρ_0 is the density, θ is an absolute temperature, $\dot{\Pi}_m$ is the rate of change of entropy as a consequence of an external thermomechanical load, $\dot{e}$ is the rate of change of internal energy per unit mass, and $\vec{\mathbf{q}}$ is the heat flux due to temperature variation. The final form of the dissipated energy density rate is given as:

$$\dot{E}_d = \left(1 - \sum_{i=1}^{N_t} v^i\right) \sum_{\alpha=1}^{N_s} \left(\tau^\alpha + \Phi^\alpha - \rho_0 \frac{\partial \epsilon}{\partial \zeta} \Psi^\alpha\right) \dot{\gamma}^\alpha \\ + \sum_{i=1}^{N_t} \dot{v}^i \left(\tau^i + \Phi^i - \rho_0 \frac{\partial \epsilon}{\partial \zeta} \Psi^i\right) \gamma^i - (\theta) \nabla \Pi_q \,. \tag{26}$$

where τ^α and τ^i are the resolved shear stresses on α-slip and i-twin systems, respectively; Φ^α and Φ^i are the thermal equivalents of resolved shear stresses τ^α and τ^i, respectively; ϵ is the internal energy density; ζ is the crystal defect microstrain parameter; Ψ^α and Ψ^i are stress-like terms and functions of slip resistance of α-slip system and twin resistance of i-twin system, respectively; ∇ is the del operator; and Π_q is the entropy flux ($q = \Pi_q \theta$).

2.2.2. Helmholtz Free Energy Formulation

In the micromechanical model presented by [53], Helmholtz free energy (HFE) is evaluated through an additive decomposition of four forms of energies as:

$$E_h(\mathbf{F}^e, \theta, \zeta, v) = E_{hm}(\mathbf{F}^e) + E_{ht}(\theta) + E_{hd}(\zeta) + E_{hs}(v) \,, \tag{27}$$

where E_{hm}, E_{ht}, E_{hd}, and E_{hs} represent mechanical, thermal, crystal defect, and surface energy components of HFE. In this formulation, HFE depends on four state variables, which are elastic deformation gradient $\mathbf{F}^e$, absolute temperature θ, crystal defect microstrain parameter ζ, and twin martensite volume fraction v. In consideration of this, HFE can be derived as:

$$E_h = \frac{1}{\rho_0} \left\{ (\mathbf{F}^e \mathbf{F}^p) : (\widetilde{\mathbb{C}}^e : \widetilde{\mathbf{E}}^e \widetilde{\mathbf{E}}^e) \right\} (\mathbf{F}^p)^T : \left[\mathbf{F}^e + (\mathbf{F}^e)^T \right]^{-1} \\ + \theta \left[- h_e \ln\left(\frac{\theta}{\theta_r}\right) + \left(h_{eq} - \Pi^e_{m,0}\right) \right] + \frac{1}{2\rho_0} \varphi G^e \zeta^2 \\ + \frac{\chi}{\rho_0 l_0} \sum_{i=1}^{N_t} v^i \left(1 - \sum_{i=1}^{N_t} v^i\right) . \tag{28}$$

In Equation (28), h_e is an equivalent specific heat; θ_r is the reference temperature; $\Pi^e_{m,0}$ is the initial entropy density; φ is a dimensionless dislocation interaction parameter, which incorporates the effects of dislocations' mobility and their interactions in plasticity; G^e is the equivalent modulus of rigidity; l_0 is the length scale parameter; and χ is the interfacial energy per unit area.

2.2.3. Driving Potential Formulation

The driving potential (force) for inelastic deformation modes, slip and twinning, can be estimated through Equations (29) and (30), respectively.

$$\mathcal{G}^\alpha = \left(1 - \sum_{i=1}^{N_t} v^i\right) \left(\tau^\alpha + \Phi^\alpha - \varphi G^e \zeta \Psi^\alpha\right) . \tag{29}$$

$$\mathcal{P}^i = v^i \left(\tau^i + \Phi^i - \varphi G^e \zeta \Psi^i\right) . \tag{30}$$

2.3. *Material Flow Modeling*

The material flow due to the activity of α-slip systems can be estimated through a power function of the shear strain rate. [54,56] given as:

$$\dot{\gamma}^{\alpha} = \dot{\gamma}_0 \left| \frac{\tau^{\alpha}}{s_r^{\alpha}} \right|^{\frac{1}{m}} \operatorname{sign}(\tau^{\alpha}) . \quad (31)$$

In Equation (31), $\dot{\gamma}_0$ is the initial shear strain rate; τ^{α} is the shear stress on the α-slip system; s_r^{α} is slip resistance; and m is the strain rate sensitivity parameter. In a similar manner, material flow due to mechanical twinning is modeled through a nonlinear function, as discussed in [25–27,57]. Accordingly, the rate of change of the twinned-martensite volume fraction of i-twin system is given as:

$$\dot{v}^i = \frac{\dot{\gamma}_0}{\gamma^i} \left(\frac{\tau^i}{s_r^i} \right)^{1/m} , \quad (32)$$

where $\dot{\gamma}_0$ is the initial shear strain rate, γ^i is the shear strain of i-twin system, and s_r^i is the twin resistance of i-twin system.

2.3.1. Strain Hardening Rule

In the model presented by [53], a dislocation density-based hardening law is used to incorporate self, $\dot{s}_s^{\alpha}$, and latent, $\dot{s}_l^{\alpha}$, hardening contributions of α-slip systems as:

$$\dot{s}_r^{\alpha} = \dot{s}_s^{\alpha} + \dot{s}_l^{\alpha} = \sum_{\alpha=1}^{N_s} h_0^{\alpha} \left[1 - \left(\frac{s_r^{\alpha} - s_{r,0}^{\alpha}}{s_{r,S}^{\alpha} - s_{r,0}^{\alpha}} \right) \right] |\dot{\gamma}^{\alpha}| + \sum_{\kappa=1}^{N_s} h^{\alpha\kappa} |\dot{\gamma}^{\kappa}|. \quad (33)$$

In Equation (33), h_0^{α} and $s_{r,0}^{\alpha}$ are the initial values of hardening rate and strength of slip system, respectively; $s_{r,S}^{\alpha}$ is the saturation value of slip strength; $h^{\alpha\kappa}$ is an array of latent hardening values; and $\dot{\gamma}^{\kappa}$ is the shear strain rate of κ-slip system, where κ denotes a number of slip system except α ($\kappa = j,\ j = 1, \ldots, i-1, i+1, \ldots, N_s$). The hardening induced by mechanical twinning is incorporated in the model through Equation (34) as:

$$\dot{s}_t^i = h_{nc}^i \left(\sum_{i=1}^{N_t} v^i \right)^d \sum_{\mu_1=0}^{i} \gamma^i \dot{v}^{\mu_1} + h_{cp}^i \left(\sum_{i=1}^{N_t} v^i \right) \sum_{\mu_2=0}^{i} \gamma^i \dot{v}^{\mu_2} . \quad (34)$$

Here, h_{nc}^i and h_{cp}^i are the initial hardening rates of non-coplanar and coplanar twin systems; μ_1 ($\mu_1 \in i$) and μ_2 ($\mu_2 \in i$) represent the number of non-coplanar and coplanar twin systems, respectively; $\dot{v}^{\mu_1}$ and $\dot{v}^{\mu_2}$ are the rate of change of volume fractions for non-coplanar and coplanar twin systems, respectively; and d is a material parameter.

2.3.2. Microstrain Parameter

The microstrain, induced through crystal defects, in the case of slip and mechanical twinning can be estimated through Equations (35) and (36), respectively, as:

$$\dot{\zeta}_s = \frac{1}{\omega G^e N_s} \sum_{\alpha=1}^{N_s} \dot{s}_r^{\alpha} = \frac{1}{\omega G^e N_s} \sum_{\alpha=1}^{N_s} \sum_{\kappa=1}^{N_s} h^{\alpha\kappa} |\dot{\gamma}^{\kappa}| . \quad (35)$$

$$\dot{\zeta}_t = \frac{1}{\omega G^e N_t} \sum_{i=1}^{N_t} \left[\left\{ h_{nc}^i \left(\sum_{i=1}^{N_t} v^i \right)^d \sum_{\mu_1=0}^{i} \dot{v}^{\mu_1} + h_{cp}^i \left(\sum_{i=1}^{N_t} v^i \right) \sum_{\mu_2=0}^{i} \dot{v}^{\mu_2} \right\} \gamma^i \right] . \quad (36)$$

The stress-like parameters that are part of the driving potential equations for slip and twinning, Equations (29) and (30), are estimated as:

$$\Psi^{\alpha} = \frac{\left(1 - \sum_{i=1}^{N_t} v^i \right)^{-1}}{\omega G^e N_s} \sum_{\kappa=1}^{N_s} h^{\alpha\kappa} . \quad (37)$$

$$\Psi^i = \frac{\left(\sum_{i=1}^{N_t} \dot{v}^i\right)^{-1}}{\omega G^e N_t} \sum_{i=1}^{N_t} \left[h^i_{nc} \left(\sum_{i=1}^{N_t} v^i \right)^d \sum_{\mu_1=0}^{i} \dot{v}^{\mu_1} + h^i_{cp} \left(\sum_{i=1}^{N_t} v^i \right) \sum_{\mu_2=0}^{i} \dot{v}^{\mu_2} \right] . \tag{38}$$

3. Numerical Integration Scheme of Constitutive Equations

The development of the numerical integration scheme consists of the following:(i) identification of primary variables in constitutive formulation, (ii) discretization and numerical integration of equations in time domain, (iii) development of Newton–Raphson iterative scheme, and (iv) development of time sub-stepping algorithm to increase or decrease time step size, depending on the incremental values of primary variables.

3.1. Identification of Primary Variables

The constitutive equations of the slip- and twin-based crystal plasticity model are the first-order ordinary differential equations of the state variables mentioned in Equation (39), as follows:

$$\{\mathbf{T}^e,\ s^\alpha_r,\ s^i_t,\ \mathbf{R}^e,\ v^i\}\ , \tag{39}$$

where $\mathbf{T}^e$ represents second Piola–Kirchhoff stress tensor, $\{s^\alpha_r \mid \alpha = 1, \ldots, N_{sl}\}$ denotes slip resistance of α-slip system, $\{s^i_t \mid i = 1, \ldots, N_{tw}\}$ shows twin resistance of i-twin system, $\mathbf{R}^e$ is the rigid body rotation tensor, and $\{v^i \mid i = 1, \ldots, N_{tw}\}$ is twinned martensite volume fraction.

3.2. Discretization of Constitutive Equations

A time integration scheme is executed in sample coordinate axes through discretizing the deformation history in the time domain and subsequently numerically integrating constitutive equations for each time step. In order to define a deformation time history, the configurations of a material point are considered at time t_n and t_{n+1}. In this, a relation of $t_{n+1} = t_n + \Delta t$ is used, where t_n and t_{n+1} represent time at the start and end of the time step, respectively. Afterwards, the magnitudes of all variables are evaluated at t_n and t_{n+1}, and denoted with the subscripts n and $n+1$, respectively. The numerical time integration scheme is based on the following assumptions:

1. The total deformation gradient $\mathbf{F}_{n+1}$ and velocity gradient $\mathbf{L}_{n+1}$ are given.
2. The values of variables $\mathbf{T}^e_n$, $s^\alpha_{r,n}$, $s^i_{t,n}$, $\mathbf{R}^e_n$, and v^i_n at time t_n are known.
3. The initial values of time-independent slip (m^α_0, n^α_0) and twin (m^i_0, n^i_0) systems' vectors, elasticity tensors $\mathbb{C}^{sl}_0$ and $\mathbb{C}^{tw}_0$, initial crystallographic orientation of crystal $\mathbf{Q}_0$, initial kinetic flow rule (m and $\dot{\gamma}_0$), and hardening parameters $(h^\alpha_0, s^\alpha_{r,0}, s^\alpha_{r,S_0}, \dot{\gamma}_{S_0})$ are used as an input.

The output of the numerical integration scheme provides updated values of variables as: $\mathbf{T}^e_{n+1}$, $s^\alpha_{r,n+1}$, $s^i_{t,n+1}$, $\mathbf{R}^e_{n+1}$, and v^i_{n+1} at time t_{n+1}. The constitutive equations are discretized through fully implicit time integration procedure using backward Euler scheme. According to the kinematic formulation, an incremental form of Green strain tensor in terms of the symmetric part of the velocity gradient, Equation (25) in [53], can be written as:

$$\Delta\widetilde{\mathbf{E}}^e = \Delta t(\widetilde{\mathbf{D}} - \widetilde{\mathbf{D}}^*)\ . \tag{40}$$

Integration of Equation (40) using backward Euler scheme provides an updated value of Green strain tensor, as follows:

$$\widetilde{\mathbf{E}}^e_{n+1} = \widetilde{\mathbf{E}}^e_n + \Delta\widetilde{\mathbf{E}}^e_n = \widetilde{\mathbf{E}}^e_n + \Delta t(\widetilde{\mathbf{D}}_{n+1} - \widetilde{\mathbf{D}}^*_{n+1})\ . \tag{41}$$

where $\widetilde{\mathbf{D}}_{n+1}$ and $\widetilde{\mathbf{D}}^*_{n+1}$ are the symmetric parts of velocity gradients $\widetilde{\mathbf{L}}_{n+1}$ and $\widetilde{\mathbf{L}}^*_{n+1}$, respectively. The incremental values of these parameters can be illustrated as:

$$\widetilde{\mathbf{D}}_{n+1} = \frac{1}{2}[\widetilde{\mathbf{L}}_{n+1} + \widetilde{\mathbf{L}}^{\mathrm{T}}_{n+1}] = \mathrm{sym}(\widetilde{\mathbf{L}}_{n+1}) \; . \tag{42}$$

$$\widetilde{\mathbf{D}}^*_{n+1} = \mathrm{sym}(\widetilde{\mathbf{C}}^e_{n+1}\widetilde{\mathbf{\Theta}}^e_{n+1}) + \mathbf{R}^e_{n+1}\check{\mathbf{D}}^p_{n+1}(\mathbf{R}^e_{n+1})^T \; , \tag{43}$$

where an updated value of right Cauchy–Green deformation tensor in third intermediate configuration can be defined as:

$$\widetilde{\mathbf{C}}^e_{n+1} = \mathbf{V}^e_{n+1}(\mathbf{V}^e_{n+1})^{\mathrm{T}} \; , \tag{44}$$

Furthermore, an incremental magnitude of spin tensor $\widetilde{\mathbf{\Theta}}^e_{n+1}$ is represented as:

$$\widetilde{\mathbf{\Theta}}^e_{n+1} = \dot{\mathbf{R}}^e_{n+1}(\mathbf{R}^e_{n+1})^T = \Delta t \mathbf{R}^e_{n+1}(\mathbf{R}^e_{n+1})^T \; , \tag{45}$$

In addition, $\check{\mathbf{D}}^p_{n+1}$ is an updated symmetric part of $(\check{\mathbf{C}}^e_{n+1}\check{\mathbf{L}}^p_{n+1})$. Here, $\check{\mathbf{C}}^e_{n+1}$ is an incremental form of right Cauchy–Green deformation tensor in second intermediate configuration, which can be expressed as:

$$\check{\mathbf{C}}^e_{n+1} = (\mathbf{F}^e_{n+1})^T(\mathbf{F}^e_{n+1}) \; . \tag{46}$$

The incremental value of the plastic velocity gradient in second intermediate configuration can be defined as:

$$\begin{aligned} \check{\mathbf{L}}^p_{n+1} = & \left(1 - \sum\nolimits_{i=1}^{N_{tw}} v^i_{n+1}\right) \sum_{\alpha=1}^{N_{sl}} \Delta t \dot{\gamma}^\alpha_{n+1} \bar{\mathbf{S}}^\alpha_{n+1} \\ & + \mathbf{F}^p_{s,n+1}\left\{ \sum_{i=1}^{N_{tw}} v^i_{n+1} \Delta t \dot{\gamma}^i_{n+1} \check{\mathbf{S}}^i_{n+1} \right\} (\mathbf{F}^p_{s,n+1})^{-1} \; . \end{aligned} \tag{47}$$

where v^i_{n+1} represents an incremental change in the volume fraction of ith twin system, $\dot{\gamma}^\alpha_{n+1}$ is an updated value of the shear strain rate of α-slip system, $\dot{\gamma}^i_{n+1}$ is an updated value of shear strain induced by i-twin system, $\bar{\mathbf{S}}^\alpha_{n+1}$ and $\check{\mathbf{S}}^i_{n+1}$ are the updated magnitudes of Schmid and twin orientation tensors in second intermediate configuration, and $\mathbf{F}^p_{s,n+1}$ is an incremental change in the value of the plastic deformation gradient. Substitution of $\check{\mathbf{D}}^p_{n+1}$ in Equation (43) gives the following:

$$\begin{aligned} \widetilde{\mathbf{D}}^*_{n+1} = & \, \mathrm{sym}(\widetilde{\mathbf{C}}^e_{n+1}\widetilde{\mathbf{\Theta}}^e_{n+1}) \\ & + (\mathbf{R}^e_{n+1})\mathrm{sym}\left[\check{\mathbf{C}}^e_{n+1}\left\{ \left(1 - \sum\nolimits_{i=1}^{N_{tw}} v^i_{n+1}\right) \sum_{\alpha=1}^{N_{sl}} \Delta t \dot{\gamma}^\alpha_{n+1} \bar{\mathbf{S}}^\alpha \right.\right. \\ & \left.\left. + \mathbf{F}^p_{s,n+1}\left\{ \sum_{i=1}^{N_{tw}} v^i_{n+1} \Delta t \dot{\gamma}^i_{n+1} \check{\mathbf{S}}^i \right\} (\mathbf{F}^p_{s,n+1})^{-1} \right\} \right] (\mathbf{R}^e_{n+1})^T \; . \end{aligned} \tag{48}$$

Considering the effects of rigid body rotation tensor $\mathbf{R}^e_{n+1}$ on Schmid and twin orientation tensors, Equation (48) can also be written as:

$$\begin{aligned} \widetilde{\mathbf{D}}^*_{n+1} = & \, \mathrm{sym}(\widetilde{\mathbf{C}}^e_{n+1}\widetilde{\mathbf{\Theta}}^e_{n+1}) \\ & + \mathrm{sym}\left[\bar{\mathbf{C}}^e_{n+1}\left\{ \left(1 - \sum\nolimits_{i=1}^{N_{tw}} v^i_{n+1}\right) \sum_{\alpha=1}^{N_{sl}} \Delta t \dot{\gamma}^\alpha_{n+1} (\mathbf{R}^e_{n+1})(\bar{\mathbf{m}}^\alpha \otimes \bar{\mathbf{n}}^\alpha)(\mathbf{R}^e_{n+1})^T \right.\right. \\ & \left.\left. + \mathbf{F}^p_{s,n+1}\left\{ \sum_{i=1}^{N_{tw}} v^i_{n+1} \Delta t \dot{\gamma}^i_{n+1} (\mathbf{R}^e_{n+1})(\check{\mathbf{m}}^i \otimes \check{\mathbf{n}}^i)(\mathbf{R}^e_{n+1})^T \right\} (\mathbf{F}^p_{s,n+1})^{-1} \right\} \right] . \end{aligned} \tag{49}$$

The forward mapping approach for Schmid and twin orientation tensors can be implemented into Equation (49) as:

$$\begin{aligned}\left\{\mathbf{R}^e_{n+1}.(\bar{\mathbf{m}}^\alpha\otimes\bar{\mathbf{n}}^\alpha)\right\}.(\mathbf{R}^e_{n+1})^T &= \left\{\mathbf{R}^e_{n+1}.(\mathbf{Q}_0.\mathbf{m}^\alpha_0\otimes\mathbf{Q}_0.\mathbf{n}^\alpha_0)\right\}.(\mathbf{R}^e_{n+1})^T\\ &= \left\{\mathbf{R}^e_{n+1}.(\mathbf{Q}_0.\mathbf{m}^\alpha_0)\right\}\otimes\left\{\mathbf{R}^e_{n+1}.(\mathbf{Q}_0.\mathbf{n}^\alpha_0)\right\}.\end{aligned} \tag{50}$$

where $\mathbf{Q}_0$ is the initial rotation matrix that is used to transform crystal coordinates to sample coordinate systems through Euler angles, and $\mathbf{m}^\alpha_0$ and $\mathbf{n}^\alpha_0$ are the initial vectors representing α-slip system in reference configuration. It was stated previously that the rotation matrix (Euler angles) can be updated through a rigid body rotation tensor as follows: $\mathbf{Q}_{n+1} = \mathbf{R}^e_{n+1}.\mathbf{Q}_0$. The rotation matrix can also be used to transform Schmid orientation vectors from reference to first intermediate configuration as follows: $\bar{\mathbf{m}}^\alpha = \mathbf{Q}_0.\mathbf{m}^\alpha_0$ and $\bar{\mathbf{n}}^\alpha = \mathbf{Q}_0.\mathbf{n}^\alpha_0$. In this condition, Equation (50) can be rewritten as:

$$\left\{(\mathbf{R}^e_{n+1}).(\bar{\mathbf{m}}^\alpha\otimes\bar{\mathbf{n}}^\alpha)\right\}(\mathbf{R}^e_{n+1})^T = (\mathbf{Q}_{n+1}.\mathbf{m}^\alpha_0)\otimes(\mathbf{Q}_{n+1}.\mathbf{n}^\alpha_0) = \widetilde{\mathbf{m}}^\alpha_{n+1}\otimes\widetilde{\mathbf{n}}^\alpha_{n+1}\,. \tag{51}$$

In Equation (51), $\widetilde{\mathbf{m}}^\alpha_{n+1}$ and $\widetilde{\mathbf{n}}^\alpha_{n+1}$ are the slip incremental values of direction and area normal vectors of α-slip system in third intermediate configuration. It is clear from Equation (51) that the updated rotation matrix $\mathbf{Q}_{n+1}$ is used to transform Schmid orientation vectors from reference to third intermediate configuration. Similarly, twin orientation vectors are given as:

$$\left\{\mathbf{R}^e_{n+1}.(\check{\mathbf{m}}^i\otimes\check{\mathbf{n}}^i)\right\}.(\mathbf{R}^e_{n+1})^T = \widetilde{\mathbf{m}}^i_{n+1}\otimes\widetilde{\mathbf{n}}^i_{n+1}\,. \tag{52}$$

Substitution of Equations (51) and (52) in (49) provides

$$\begin{aligned}\widetilde{\mathbf{D}}^*_{n+1} &= \mathrm{sym}(\widetilde{\mathbf{C}}^e_{n+1}.\widetilde{\boldsymbol{\Theta}}^e_{n+1})\\ &+\mathrm{sym}\Bigg[\check{\mathbf{C}}^e_{n+1}.\Bigg\{\left(1-\textstyle\sum_{i=1}^{N_{tw}}v^i_{n+1}\right)\sum_{\alpha=1}^{N_{sl}}\Delta t\dot{\gamma}^\alpha_{n+1}\widetilde{\mathbf{S}}^\alpha_{n+1}\\ &+\mathbf{F}^p_{s,n+1}.\left\{\sum_{i=1}^{N_{tw}}v^i_{n+1}\Delta t\dot{\gamma}^i_{n+1}\widetilde{\mathbf{S}}^i_{n+1}\right\}(\mathbf{F}^p_{s,n+1})^{-1}\Bigg\}\Bigg].\end{aligned} \tag{53}$$

Here, $\widetilde{\mathbf{S}}^\alpha_{n+1} = \widetilde{\mathbf{m}}^\alpha_{n+1}\otimes\widetilde{\mathbf{n}}^\alpha_{n+1}$ and $\widetilde{\mathbf{S}}^i_{n+1} = \widetilde{\mathbf{m}}^i_{n+1}\otimes\widetilde{\mathbf{n}}^i_{n+1}$ are the updated Schmid and twin orientation tensors in third intermediate configuration. The updated values of shear strain for α-slip system and volume fraction of twinned martensite at time t_{n+1} can be defined by Equations (54) and (55), respectively, as:

$$\gamma^\alpha_{n+1} = \gamma^\alpha_n + \Delta\gamma^\alpha = \gamma^\alpha_n + \gamma_0\left|\frac{\tau^\alpha_{n+1}}{s^\alpha_{r,n+1}}\right|^{\frac{1}{m}}\sin(\tau^\alpha_{n+1})\,. \tag{54}$$

$$v^i_{n+1} = v^i_n + \Delta v^i = v^i_n + \frac{\gamma_0}{\gamma^i_{n+1}}\left|\frac{\tau^i_{n+1}}{s^i_{t,n+1}}\right|^{\frac{1}{m}}. \tag{55}$$

As discussed previously, the rigid body rotation tensor $\mathbf{R}^e_{n+1}$ updates the crystal orientation (Euler angles) matrix $\mathbf{Q}_{n+1}$, which can be used to transform the fourth order elasticity tensors for slip $\mathbb{C}^s_{n+1}$ and twinned $\widetilde{\mathbb{C}}^t_{n+1}$ regions to the third intermediate configuration as follows:

$$\widetilde{\mathbb{C}}^s_{n+1} = (\mathbf{Q}_{n+1}\otimes\mathbf{Q}_{n+1}):\widetilde{\mathbb{C}}^s_0:(\mathbf{Q}_{n+1}\otimes\mathbf{Q}_{n+1})^T\,. \tag{56}$$

$$\widetilde{\mathbb{C}}^t_{n+1} = (\mathbf{Q}_{n+1}\otimes\mathbf{Q}_{n+1}):\widetilde{\mathbb{C}}^t_0:(\mathbf{Q}_{n+1}\otimes\mathbf{Q}_{n+1})^T\,. \tag{57}$$

Any one of the elasticity tensors can be obtained in terms of another by using coordinate transformation rule defined as:

$$\widetilde{\mathbb{C}}^{t}_{ijkl} = \widetilde{\mathbb{C}}^{s}_{pqrs}\mathbf{P}_{ip}\mathbf{P}_{jq}\mathbf{P}_{kr}\mathbf{P}_{ls} = (\mathbf{P} \otimes \mathbf{P}) : \widetilde{\mathbb{C}}^{s} : (\mathbf{P} \otimes \mathbf{P})^{T} . \tag{58}$$

Here, $[\mathbf{P}]$ is the transformation matrix that relates lattice orientations in twinned and untwinned (slipped) regions. The components of transformation matrix $[\mathbf{P}]$ can be defined through Equation (59), given by [26], as:

$$\mathbf{P}_{jk} = 2n_j n_k - \delta_{jk} , \qquad j,k = 1,2,3 \tag{59}$$

where $\mathbf{n}$ is the area normal vector of the twin plane, and δ is the Kroneker delta. An equivalent elasticity tensor can be calculated as:

$$\widetilde{\mathbb{C}}^{e}_{n+1} = \left(1 - \sum_{i=1}^{N_t} v^{i}_{n+1}\right)\widetilde{\mathbb{C}}^{s}_{n+1} + \sum_{i=1}^{N_t} v^{i}_{n+1}\widetilde{\mathbb{C}}^{t}_{n+1} . \tag{60}$$

Furthermore, an updated form of equivalent second Piola–Kirchhoff stress tensor can be estimated as:

$$\mathbf{T}^{e}_{n+1} = \widetilde{\mathbb{C}}^{e}_{n+1} : \widetilde{\mathbf{E}}^{e}_{n+1} . \tag{61}$$

An evolution equation for rigid body rotation tensor $\mathbf{R}^{e}$ is integrated through the exponential map discussed by [58] as follows:

$$\mathbf{R}^{e}_{n+1} = \exp(\Delta t \widetilde{\mathbf{\Theta}}^{e}_{n+1}).\mathbf{R}^{e}_{n} , \tag{62}$$

where an updated value of the spin of lattice $\widetilde{\mathbf{\Theta}}^{e}_{n+1}$ can be estimated through the skew-symmetric component $\widetilde{\mathbf{W}}^{*}_{n+1}$ of the velocity gradient $\widetilde{\mathbf{L}}^{*}_{n+1}$ as:

$$\widetilde{\mathbf{W}}^{*}_{n+1} = \mathrm{skew}(\widetilde{\mathbf{C}}^{e}_{n+1}\widetilde{\mathbf{\Theta}}^{e}_{n+1}) + \sum_{\alpha=1}^{N_s} \Delta t \dot{\gamma}^{\alpha}_{n+1}\mathrm{skew}(\widetilde{\mathbf{C}}^{e}_{n+1}\widetilde{\mathbf{S}}^{\alpha}_{n+1}) . \tag{63}$$

By using backward mapping, the skew-symmetric component of velocity gradient $\widetilde{\mathbf{L}}_{n+1}$ can be estimated as:

$$\widetilde{\mathbf{W}}_{n+1} = \{(\mathbf{V}^{e}_{n+1})^{T}\mathbf{W}_{n+1}\}\mathbf{V}^{e}_{n+1} = (\mathbf{V}^{e})^{T}\mathrm{skew}\{(\mathbf{V}^{e}_{n+1})^{T}\{\Delta t \mathbf{V}^{e}_{n+1}\}\}\mathbf{V}^{e} + \widetilde{\mathbf{W}}^{*}_{n+1} . \tag{64}$$

Here, $\mathbf{W}_{n+1}$ is the updated skew-symmetric component of $\mathbf{L}_{n+1}$. Substitution of $\widetilde{\mathbf{W}}^{*}_{n+1}$ from Equation (63) in (64) provides

$$\begin{aligned}\mathrm{skew}(\widetilde{\mathbf{C}}^{e}_{n+1}\widetilde{\mathbf{\Theta}}^{e}_{n+1}) = {} & \widetilde{\mathbf{W}}_{n+1} - (\mathbf{V}^{e})^{T}\mathrm{skew}\{(\mathbf{V}^{e}_{n+1})^{T}\Delta t \mathbf{V}^{e}_{n+1}\}\mathbf{V}^{e} \\ & - \sum_{\alpha=1}^{N_s} \Delta t \dot{\gamma}^{\alpha}_{n+1}\mathrm{skew}(\widetilde{\mathbf{C}}^{e}_{n+1}\widetilde{\mathbf{S}}^{\alpha}_{n+1}) .\end{aligned} \tag{65}$$

For small elastic strain problems, the value of the right Cauchy–Green deformation tensor $\widetilde{\mathbf{C}}^{e}_{n+1}$ is typically small. In this case, Equation (65) can be written as:

$$\mathrm{skew}(\widetilde{\mathbf{\Theta}}^{e}_{n+1}) = \widetilde{\mathbf{W}}_{n+1} - (\mathbf{V}^{e})^{T}\mathrm{skew}\{(\mathbf{V}^{e}_{n+1})^{T}\Delta t \mathbf{V}^{e}_{n+1}\}\mathbf{V}^{e} - \sum_{\alpha=1}^{N_s} \Delta t \dot{\gamma}^{\alpha}_{n+1}\mathrm{skew}(\widetilde{\mathbf{S}}^{\alpha}_{n+1}) . \tag{66}$$

An incremental value of the slip resistance of α-slip system can be evaluated using a backward Euler scheme as follows:

$$s^{\alpha}_{r,n+1} = s^{\alpha}_{r,n} + \Delta s^{\alpha}_{r} = s^{\alpha}_{r,n} + \sum_{\alpha=1}^{N_s} h^{\alpha}_{0}\left[1 - \left(\frac{s^{\alpha}_{r,n+1} - s^{\alpha}_{r,0}}{s^{\alpha}_{r,S_{n+1}} - s^{\alpha}_{r,0}}\right)\right]\left|(\Delta t)\dot{\gamma}^{\alpha}_{n+1}\right| . \tag{67}$$

In the present model, slip resistances for all α-slip systems are considered to be similar. Therefore, Equation (67) can be modified as follows:

$$s_{r,n+1} = s_{r,n} + h_0 \left[1 - \left(\frac{s_{r,n+1} - s_{r,0}}{s_{r,S_{n+1}} - s_{r,0}} \right) \right] \sum_{\alpha=1}^{N_s} \left| (\Delta t) \dot{\gamma}_{n+1}^{\alpha} \right| , \quad (68)$$

In Equation (68), $s_{r,S_{n+1}}$ is an incremental saturation value of slip resistance, which can be calculated as:

$$s_{r,S_{n+1}} = s_{r,S_n} + \Delta s_{r,S} = s_{r,S_n} + s_{r,S_0} \left[\frac{\sum_{\beta}^{N_s} |\gamma_{n+1}^{\beta}|}{\gamma_{S_0}} \right]^{\frac{k\theta}{(G_e)_{n+1} b^3 Z}} + s_{r,p} \left(\sum_{\lambda=0}^{i} v_{n+1}^{\lambda} \right)^{a_1} , \quad (69)$$

where s_{r,S_0} is the initial value of saturation slip resistance, $\Delta\gamma_{S_0}$ is an incremental initial value of the slip system shear strain at the initial value of saturation slip resistance, a is the material parameter, $s_{r,p}$ is the material slip-hardening parameter, λ ($\lambda \in i$) represents the number of non-coplanar twin systems to slip system, and a_1 is a material parameter. In addition, the twin resistance can be estimated using Equation (34) as:

$$\begin{aligned} s_{t,n+1}^{i} = s_{t,n}^{i} + \Delta s_{t}^{i} = s_{t,n}^{i} + & \left[h_{nc}^{i} \left(\sum_{i=1}^{N_t} v_{n+1}^{i} \right)^{d} \sum_{\mu_1=0}^{i} \gamma_{n+1}^{i} \Delta t \dot{v}_{n+1}^{\mu_1} \right. \\ & \left. + h_{cp}^{i} \left(\sum_{i=1}^{N_t} v_{n+1}^{i} \right) \sum_{\mu_2=0}^{i} \gamma_{n+1}^{i} \Delta t \dot{v}_{n+1}^{\mu_2} \right] , \end{aligned} \quad (70)$$

The resistance of all twin systems are assumed to be identical. Therefore, Equation (70) can be expressed as:

$$\begin{aligned} s_{t,n+1} = s_{t,n} + & \left[h_{nc} \left(\sum_{i=1}^{N_t} v_{n+1}^{i} \right)^{d} \sum_{\mu_1=0}^{i} \gamma_{n+1}^{i} \Delta t \dot{v}_{n+1}^{\mu_1} \right. \\ & \left. + h_{cp} \left(\sum_{i=1}^{N_t} v_{n+1}^{i} \right) \sum_{\mu_2=0}^{i} \gamma_{n+1}^{i} \Delta t \dot{v}_{n+1}^{\mu_2} \right] . \end{aligned} \quad (71)$$

The updated driving potentials for slip and twin mode of deformations can be estimated using Equations (29) and (30), respectively as:

$$G_{s,n+1}^{\alpha} = G_{s,n}^{\alpha} + \Delta G_{s}^{\alpha} = G_{s,n}^{\alpha} + \left(1 - \sum_{i=1}^{N_t} v_{n+1}^{i} \right) \left(\tau_{n+1}^{\alpha} + \Phi^{\alpha} - \varphi G_{n+1}^{e} \zeta_{n+1} \Psi_{n+1}^{\alpha} \right) . \quad (72)$$

$$G_{t,n+1}^{i} = G_{t,n}^{i} + \Delta G_{t}^{i} = G_{t,n}^{i} + v_{n+1}^{i} \left(\tau_{n+1}^{i} + \Phi^{i} - \varphi G_{n+1}^{e} \zeta_{n+1} \Psi_{n+1}^{i} \right) . \quad (73)$$

The parameters Φ^{α} and Φ^{i} are the thermal analogues for α-slip and i-twin systems, respectively. For an isothermal process, these factors will remain constant during the deformation process. φ is a dimensionless dislocation parameter, and it is assumed to be constant throughout the deformation. An equivalent modulus of rigidity G_{n+1}^{e} can be estimated as:

$$G_{n+1}^{e} = \left(1 - \sum_{i=1}^{N_t} v_{n+1}^{i} \right) G_s + \sum_{i=1}^{N_t} v_{n+1}^{i} G_t . \quad (74)$$

The incremental forms of crystal defect microstrain parameters for the slip and twin modes of deformations can be calculated using Equations (35) and (36), respectively, as:

$$\zeta_{s,n+1} = \zeta_{s,n} + \Delta\zeta_s = \zeta_{s,n} + \frac{1}{\omega G_{n+1}^{e} N_s} \sum_{\alpha=1}^{N_s} \sum_{\beta=1}^{N_s} h_{n+1}^{\alpha\beta} |\gamma_{n+1}^{\beta}| . \quad (75)$$

$$\zeta_{t,n+1} = \zeta_{t,n} + \Delta\zeta_t = \zeta_{t,n} + \frac{1}{\omega G^e_{n+1} N_t} \left[\left\{ h_{nc} \left(\sum_{i=1}^{N_t} v^i_{n+1} \right)^d \sum_{\mu_1=0}^{i} (\Delta t) \dot{v}^{\mu_1}_{n+1} \right. \right. \tag{76}$$
$$\left. \left. + h_{cp} \left(\sum_{i=1}^{N_t} v^i_{n+1} \right) \sum_{\mu_2=0}^{i} (\Delta t) \dot{v}^{\mu_2}_{n+1} \right\} \sum_{i=1}^{N_t} \gamma^i_{n+1} \right] .$$

The updated stress-like functions Ψ^α_{n+1} and Ψ^i_{n+1} for slip and twin modes of deformation can be calculated using Equations (37) and (38), respectively, as:

$$\Psi^\alpha_{n+1} = \frac{\left(1 - \sum_{i=1}^{N_t} v^i_{n+1}\right)^{-1}}{\omega G^e_{n+1} N_s} \sum_{\beta=1}^{N_s} h^{\alpha\beta}_{n+1} . \tag{77}$$

$$\Psi^i_{n+1} = \frac{\left(\sum_{i=1}^{N_t} v^i_{n+1}\right)^{-1}}{\omega G^e_{n+1} N_t} \left[h_{nc} \left(\sum_{i=1}^{N_t} v^i_{n+1} \right)^d \sum_{\mu_1=0}^{i} v^{\mu_1}_{n+1} \right. \tag{78}$$
$$\left. + h_{cp} \left(\sum_{i=1}^{N_t} v^i_{n+1} \right) \sum_{\mu_2=0}^{i} v^{\mu_2}_{n+1} \right] .$$

Furthermore, the updated Cauchy stress tensor $\mathbf{T}_{n+1}$ can be calculated using second Piola–Kirchhoff stress tensor, $\mathbf{T}^e_{n+1}$, as:

$$\mathbf{T}_{n+1} = \mathbf{F}^e_{n+1} \left\{ \det \mathbf{F}^e_{n+1} . \mathbf{T}^e_{n+1} \right\} (\mathbf{F}^e_{n+1})^T . \tag{79}$$

3.3. Newton–Raphson Iterative Scheme

In the preceding section, a set of couple nonlinear algebraic Equations (61), (62), (68), (71) and (55) for the primary variables $\mathbf{T}^e_{n+1}$, $\mathbf{R}^e_{n+1}$, $s_{r,n+1}$, $s_{t,n+1}$, and v^i_{n+1}, respectively, were developed. In this, the updated form of these primary variables are calculated. A set of primary variables can be represented by a vector $\{p^v_i \mid i = 1, 2, \ldots, 5\}$, where

$$p^v_1 = \Delta \mathbf{T}^e , \; p^v_2 = \Delta s_r , \; p^v_3 = \Delta v^i , \; p^v_4 = \Delta \mathbf{R}^e , \; p^v_5 = \Delta s_t . \tag{80}$$

The primary variables are the main constituents (directly or indirectly) of the constitutive model. An elastic-plastic response of a system mainly governs by these variables. In order to obtain an overall response of a material, a set of five equations in terms of primary variables are constructed in a residual format. A residue of the system of equations can also be represented in a vector form as $\{\mathcal{R}_i \mid i = 1, 2, \ldots, 5\}$. The components of $\mathcal{R}_i$ are given through Equations (81)–(85) as:

$$\begin{aligned} \mathcal{R}_1 &= \hat{\mathcal{R}}_1(\mathbf{T}^e_{n+1}, s^\alpha_{r,n+1}, v^i_{n+1}, \mathbf{R}^e_{n+1}, s^i_{t,n+1}) \\ &= (\widetilde{\mathbb{C}}^e_{n+1})^{-1} : \mathbf{T}^e_{n+1} - \widetilde{\mathbf{E}}^e_{n+1} \\ &= (\widetilde{\mathbb{C}}^e_{n+1})^{-1} : \mathbf{T}^e_{n+1} - \widetilde{\mathbf{E}}^e_n - \Delta t (\widetilde{\mathbf{D}}_{n+1} - \widetilde{\mathbf{D}}^*_{n+1}) = 0 . \end{aligned} \tag{81}$$

$$\begin{aligned} \mathcal{R}_2 &= \hat{\mathcal{R}}_2(\mathbf{T}^e_{n+1}, s^\alpha_{r,n+1}, v^i_{n+1}, \mathbf{R}^e_{n+1}, s^i_{t,n+1}) \\ &= s^\alpha_{r,n+1} - s^\alpha_{r,n} - \sum_{\alpha=1}^{N_s} h^\alpha_0 \left[1 - \left(\frac{s^\alpha_{r,n+1} - s^\alpha_{r,0}}{s^\alpha_{r,S_{n+1}} - s^\alpha_{r,0}} \right) \right] \left| (\Delta t) \dot{\gamma}^\alpha_{n+1} \right| = 0 . \end{aligned} \tag{82}$$

$$\mathcal{R}_3 = \hat{\mathcal{R}}_3(\mathbf{T}^e_{n+1}, s^\alpha_{r,n+1}, v^i_{n+1}, \mathbf{R}^e_{n+1}, s^i_{t,n+1}) \\ = v^i_{n+1} - v^i_n - \frac{\gamma_0}{\gamma^i_{n+1}} \left| \frac{\tau^i_{n+1}}{s^i_{t,n+1}} \right|^{1/m} = 0\,. \tag{83}$$

$$\mathcal{R}_4 = \hat{\mathcal{R}}_4(\mathbf{T}^e_{n+1}, s^\alpha_{r,n+1}, v^i_{n+1}, \mathbf{R}^e_{n+1}, s^i_{t,n+1}) \\ = \mathbf{R}^e_{n+1} - \exp\{\Delta t \widetilde{\mathbf{\Theta}}^e_{n+1}\}.\mathbf{R}^e_n = 0\,. \tag{84}$$

$$\mathcal{R}_5 = \hat{\mathcal{R}}_5(\mathbf{T}^e_{n+1}, s^\alpha_{r,n+1}, v^i_{n+1}, \mathbf{R}^e_{n+1}, s^i_{t,n+1}) \\ = s^i_{t,n+1} - s^i_{t,n} - \left[h^i_{nc} \left(\sum_{i=1}^{N_t} v^i_{n+1} \right)^d \sum_{\mu_1=0}^{i} \gamma^i_{n+1}(\Delta t)\dot{v}^{\mu_1}_{n+1} \right. \\ \left. + h^i_{cp} \left(\sum_{i=1}^{N_t} v^i_{n+1} \right) \sum_{\mu_2=0}^{i} \gamma^i_{n+1}(\Delta t)\dot{v}^{\mu_2}_{n+1} \right] = 0\,. \tag{85}$$

In the next step, Equations (81) and (82) are solved using a full Newton–Raphson (N-R) method, since these two are implicit in nature; however, the rest are explicit. In the current work, a two-level iterative scheme, similar as that presented by [56], is used to obtain the values of primary variables. In the first level of iteration, the N-R method is used to solve Equation (81) for $\mathbf{T}^e_{n+1}$ by assuming the best possible values of the other primary variables $(\mathbf{R}^e_{n+1}, s^\alpha_{r,n+1}, s^i_{t,n+1}, v^i_{n+1})$. Once the updated value of the second Piola–Kirchhoff stress tensor is obtained, the second level of the iterative procedure is performed, which includes an N-R solution of the slip resistance $s^\alpha_{r,n+1}$ from Equation (82). This considers an updated value of $\mathbf{T}^e_{n+1}$ and the estimated values of $\mathbf{R}^e_{n+1}$, $s^i_{t,n+1}$, and v^i_{n+1}. Finally, updated values of the twinned martensite volume fraction v^i_{n+1}, rigid body rotation tensor $\mathbf{R}^e_{n+1}$, and twin resistance $s^i_{t,n+1}$ are calculated from Equations (83)–(85), respectively.

Convergence Criterion

The convergence criterion is required to terminate the iterative loop once the solution is assumed to be sufficiently accurate. The convergence criterion for the presented two-level iterative procedure is based on the variation of $\mathcal{L}_2$-norm for $\mathbf{T}^e_{n+1}$, and $s^\alpha_{r,n+1}$. If the $\mathcal{L}_2$-norm of the residuals is less than an imposed tolerance, then the incremental solution of the updated primary variables is converged (fully elastic). Otherwise, the trial value must be updated iteratively (based on the calculated value) until the residual satisfies the convergence criterion, given as:

$$||\mathcal{R}_{\text{trial}}||_2 < \text{Tol.} \tag{86}$$

In the current Newton–Raphson iterative scheme, iterations are carried out unless and until the variation in the $\mathcal{L}_2$-norm of the residuals for $\mathbf{T}^e_{n+1}$ and $s^\alpha_{r,n+1}$ satisfy the conditions in Equations (87) and (88), respectively, as follows:

$$||\Delta\mathbf{T}^e||_2 < (10^{-4})|\mathbf{T}^e|_{\text{trial}} \tag{87}$$

$$||\Delta s^\alpha_r||_2 < (10^{-4})|s^\alpha_r|_{\text{trial}} \tag{88}$$

In the current work, a time sub-stepping algorithm (TSSA) is used in the numerical integration scheme to control the time step size. The advantage of using time sub-stepping in an iterative procedure is to improve the convergence of a solution by reducing the time increment once needed. The TSSA must also be capable of increasing the time step size at a material point where convergence can easily be achieved to reduce computational time. The TSSA can reduce and increase the time step size based on the incremental variation. In this algorithm, if any of the convergence conditions, Equations (87) and (88), are not satisfied, then the N-R iterative scheme will call TSSA, which controls the time step size according to the incremental variation in the values of primary variables.

3.4. Time Sub-Stepping Algorithm

The proposed TSSA is based on the ratio of the maximum and desired incremental values of the primary variable. The ratio is represented by the parameter $\mathcal{K}$ as:

$$\mathcal{K} = \frac{\Delta\mathcal{A}_{\max}}{\Delta\mathcal{A}_x} , \tag{89}$$

where $\Delta\mathcal{A}_{\max}$ is the maximum value of $\Delta\mathcal{A}$ over all the crystals, all the integration points are in finite element mesh, and $\Delta\mathcal{A}_x$ is a desired incremental value of $\Delta\mathcal{A}$ in the numerical algorithm. The value of $\Delta\mathcal{A}$ is normally regulated by the computational performance and accuracy of the numerical integration procedure. In a fully implicit numerical scheme, the computational performance does not usually becomes a challenge, but the accuracy does. Therefore, the value of $\Delta\mathcal{A}$ is mainly controlled by the accuracy of the numerical solution. In accordance with the two-level iterative scheme, $\mathcal{A}$ is defined as a set of three primary variables $\mathcal{A} = \{\mathcal{A}_1\ \mathcal{A}_2\}$, where

$$\mathcal{A}_1 = \mathbf{T}^e\ ,\ \ \mathcal{A}_2 = s_r^\alpha\ \ . \tag{90}$$

In the proposed time sub-stepping algorithm, three ranges of values are defined for the parameter $\mathcal{K}$: (i) If the parameter $\mathcal{K}$ exceeds 1.25, the estimated value of $\mathcal{A}$ is rejected, and the new time increment (25% smaller than the previous) will be defined. This condition makes sure that the difference between the maximum and desired value of $\mathcal{A}$ will not reach beyond 25%, which could produce an inaccurate solution. (ii) If $\mathcal{K}$ lies in the range of 0.8 to 1.25, then the estimated solution is accepted, defining the new time step as $\Delta t_{n+1} = (\Delta t)_n/\mathcal{K}$. In this condition, the new time step size Δt_{n+1} is more or less identical to the previous Δt_n. (iii) If $\mathcal{K}$ is less than 0.8, then the estimated solution is assumed to be converged, and a new time step size is defined that is 25% larger than the previous. This condition ensures that the solution is well converged, so the time increment could be increased to reduce computational time. A summary of the time sub-stepping algorithm is given in Table 1. The numerical integration scheme for elastic-plastic deformation of a crystal based on crystal plasticity formulations is summarized in Figure 2.

Table 1. Time sub-stepping algorithm for Newton–Raphson Iterative Scheme.

1. Calculate the values of $\mathcal{K}$ for each component of vector $\mathcal{A}$

$$\mathcal{K} = \frac{\Delta\mathcal{A}_{\max}}{\Delta\mathcal{A}_x}$$

2. IF $\mathcal{K} > 1.25$
    THEN: Solution is rejected and define new time increment as:
    $$\Delta t_{n+1} = 0.75(\Delta t)_n$$
        GOTO N-R iterative algorithm
    ELSE GOTO step 3
3. IF $0.8 \leq \mathcal{K} \leq 1.25$
    THEN: Solution is accepted and define new time increment as:
    $$\Delta t_{n+1} = (\Delta t)_n/\mathcal{K}$$
        GOTO N-R iterative algorithm
    ELSE GOTO step 4
4. IF $\mathcal{K} < 0.8$
    THEN: Solution is accepted and define new time increment as:
    $$\Delta t_{n+1} = 1.25(\Delta t)_n$$
        GOTO N-R iterative algorithm
    ELSE END

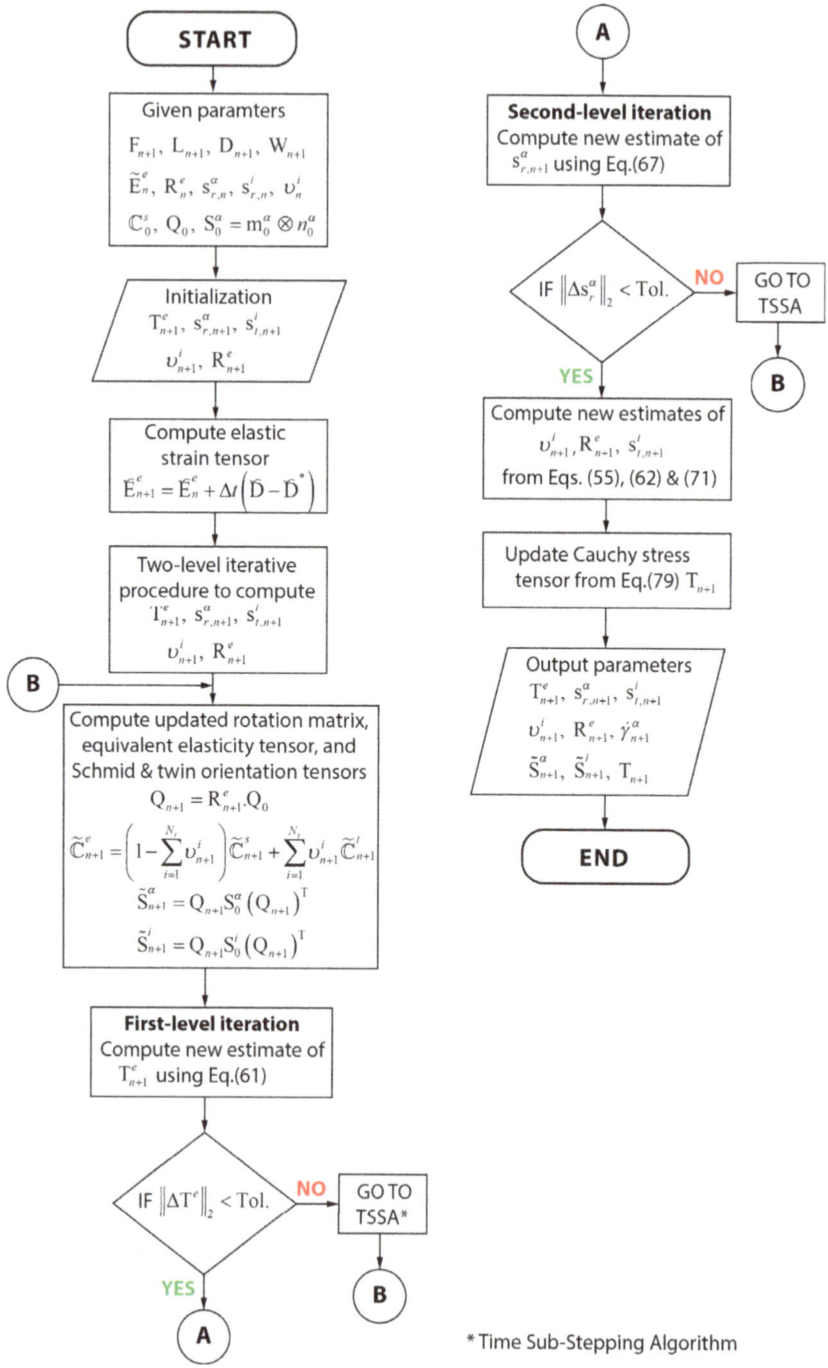

Figure 2. Numerical integration algorithm for crystal plasticity model.

3.5. Summary of Numerical Integration Algorithm

The presented numerical integration scheme for the crystal plasticity model is developed in a generalized framework. The integration scheme can also be used for a slip-based crystal plasticity model. In this case, the number of primary variables will reduce from five to three, that is, $\{\mathbf{T}^e,\ s_r^\alpha, v^i,\ s_t^i,\ \mathbf{R}^e,\}$ to $\{T^e,\ s_r^\alpha,\ \mathbf{R}^e\}$, by omitting twin resistance s_t^i and twinned martensite volume fraction v^i. However, the two-level Newton–Raphson iterative scheme remains two-level.

4. Finite Element Modeling

The numerical integration scheme of the twin-based crystal plasticity model is validated and further used to predict the deformation behavior of metals through finite element simulations. For this, numerical simulations are performed for single-crystal and polycrystal FCC-austenite subjected to biaxial and combined tension-shear loading using finite element software ABAQUS. The material model's constitutive equations are incorporated in finite element simulations through a user-defined material subroutine (UMAT). A material point in a single crystal of austenite is modeled through the eight-node brick element of unit side length with reduced integration (C3D8R). For polycrystalline simulations, each finite element represents 500 grains of random crystallographic texture. The random texture of grains, expressed in Euler angles, is developed using Kocks convention [59]. In addition, a weighted average procedure is utilized to estimate the cumulative response of polycrystalline austenite material. Furthermore, published experimental results of TWIP steel are referred to for validating the developed numerical model. Consequently, the deformation behavior of austenite-based TWIP steel, subjected to different loading conditions, is estimated and analyzed through finite element simulations.

4.1. Geometry and Boundary Conditions

In finite element simulations, two modes of loadings are considered: (i) uniaxial tension, and (ii) uniaxial compression. In uniaxial tension, a displacement of +0.15 mm is applied on a cube face, as shown in Figure 3a. The planar symmetric boundary condition is employed on three faces, while the two remaining surfaces are traction-free. Similar boundary conditions are adopted in uniaxial compression, except for a negative displacement of 0.15 mm on an element surface, as illustrated in Figure 3b. All loading conditions follow an analogous displacement rate, 1000 increments in a logical time bound of 0 to 1, of 1.5×10^{-4} mm/time. The effect of texture on the deformation pattern is incorporated through simulations of three crystallographic orientations, as illustrated in Figure 4. These are represented in two domains, that is, crystal and specimen coordinate systems, which are, respectively represented by the (e_1^c, e_2^c, e_3^c) and (e_1^s, e_2^s, e_3^s) axes. The crystal direction [100], corresponding to Euler angles $(\phi_1^1, \phi_2^1, \phi_3^1)$, is equal to $(-90^o, 0, 90^o)$ (see Figure 4a). Likewise, [110] and [111], corresponding to $(\phi_1^2, \phi_2^2, \phi_3^2)$ and $(\phi_1^3, \phi_2^3, \phi_3^3)$, are equivalent to $(-45^o, 0, 90^o)$ and $(-45^o, 35.26^o, 54.74^o)$, respectively, as represented in Figure 4b,c.

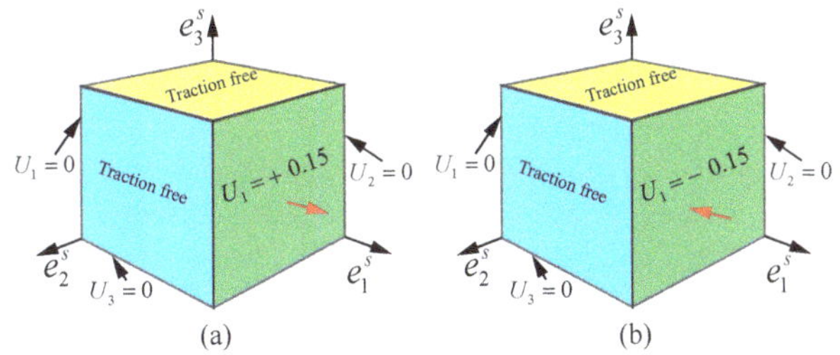

Figure 3. Finite element models subjected to (**a**) tension and (**b**) compression.

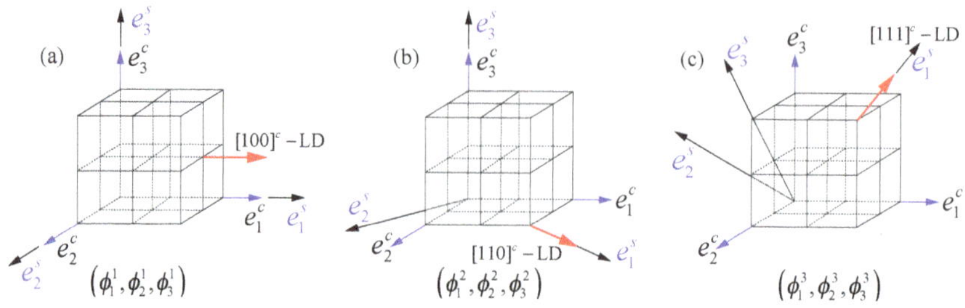

Figure 4. Loaded directions (LDs) in single crystal with respect to specimen orientation (**a**) [100]-LD, (**b**) [110]-LD, and (**c**) [111]-LD.

4.2. Modeling Parameters

In the present work, simulation parameters (material, hardening, and thermal) are considered as in [51,53,57], as summarized in Table 2.

Table 2. General modeling parameters of TWIP steels.

Type	Parameter
Material	
Moduli of rigidity	$G^s = 111.0$, $G^t = 98.4$ (GPa)
Bulk modulus	$K^a = 206.5$ (GPa)
Flow rule	
Initial shear strain rate	$\dot{\gamma}_0 = 0.001\ \text{s}^{-1}$
Hardening rule	
Initial hardening rate	$h_0^\alpha = 200$ (MPa)
Initial saturation value of slip resistance	$s^\alpha_{r,S_0} = 120$ (MPa)
Saturation slip resistance exponent	$a = 0.005$
Shear strain rate at saturation slip resistance	$\dot{\gamma}_{S_0} = 5 \times 10^{10}\ \text{s}^{-1}$
Latent hardening coefficient	$q^{\alpha\kappa} = 1.4$
Slip hardening parameter	$s_{r,p} = 350$ (MPa)
Boltzman's constant, Equation (80) [53]	$k = 1.38 \times 10^{-23}$ (J/K)
Absolute temperature, Equation (80) [53]	$\theta = 298$ (K)
Product of Burger's vector and material parameter, Equation (80) [53]	$b^3 B = 0.005$
Material parameter, Equation (80) [53]	$x = 0.5$
Initial hardening rate of non-coplanar twin systems	$h^i_{nc} = 800$ (MPa)
Initial hardening rate of coplanar twin systems	$h^i_{cp} = 8000$ (MPa)
Defect energy	
Crystal defect energy parameters	$\varphi = 10, \omega = 5$
Initial crystal defect energy	$\zeta_0 = 4.5 \times 10^{-4}\ \text{s}^{-1}$
Thermal energy	
Thermal analogous of resolved shear stresses	$\Phi^\alpha = 12, \Phi^i = 12$ (MPa)

5. Results and Discussion

In the current section, reported experimental investigations are utilized to validate the developed numerical integration scheme. Afterward, further finite element simulations of single-crystal and polycrystal austenite-based TWIP steel, subjected to uniaxial tension and compression, are executed and investigated. The prime purpose of these simulations is to test and verify the developed numerical scheme under different material deformation behaviors.

5.1. Model Validation

The developed numerical integration scheme is validated through the published experimental results of twinning-induced plasticity (TWIP) steels. In this, the uniaxial tensile test results, as reported in [60], of three TWIP steels with different chemical compositions are utilized. These steels are: TWIP 22% Mn-0.6% C, TWIP 30% Mn-0.5% C, and TWIP

29% Mn-0.8% C. In finite element simulations of uniaxial tension, displacement of 0.45 to 0.6 mm is applied on a surface of a cubic element with outward normal parallel to e_1^s axis, as shown in Figure 4. In uniaxial compression, displacement of 35.0×10^{-2} mm is applied along e_1^s vector on a surface with e_1^s as area normal vector. It is assumed that the cubic element of a material point is comprised of 500 grains with random texture. The orientation distribution function of grains is expressed in terms of Euler angles through the Kocks convention. The general modeling parameters of TWIP steels are summarized in Table 2, while parameters specific to TWIP steels 1, 2, and 3 are listed in Table 3.

The simulation results of uniaxial tension are in good agreement with the experimental observations of TWIP steels, as evident from Figure 5. However, it is noted that for TWIP 22% Mn-0.6% C and TWIP 29% Mn-0.8% C, the experimental results are under-predicted at higher strain (>0.35). For instance, the maximum absolute percent relative error ((experimental stress − simulation value)/experimental stress) in equivalent stress for TWIP 22% Mn-0.6% C at 0.43 equivalent strain is 11.63%. In TWIP 29% Mn-0.8% C, it is 7.44% at 0.45 strain. The deformation behavior of TWIP steels is solely dependent on the complex interactions of slip and twin planes, especially at higher strains. These interactions are the function of slip and twin planes' resistances and orientations. At higher strains, many slip and twin planes may activate and interact, which may cause a higher magnitude of strain hardening, as shown in the experimental results of Figure 5. This complicated phenomenon of higher strain hardening may not be fully encapsulated in numerical simulations. Nevertheless, the numerical model predicts experimental results of TWIP steels well. Therefore, it could be further used to simulate the deformation behavior of these steels, using similar modeling parameters, subjected to multiple types of loading conditions.

Table 3. Specific modeling parameters of TWIP 22% Mn-0.6% C, TWIP 30% Mn-0.5% C, and TWIP 29% Mn-0.8% C steels.

	TWIP 22% Mn-0.6% C
Type	**Parameter**
Material	
Elasticity tensor components	$E_{11}^a = 286.80$, $E_{12}^a = 166.40$, $E_{44}^a = 145.10$ (GPa)
Flow rule	
Rate sensitivity parameter	$m = 0.03$
Hardening rule	
Initial slip resistance	$s_{r,0}^\alpha = 70$ (MPa)
Initial twin resistance	$s_{r,0}^i = 80$ (MPa)
	TWIP 30% Mn-0.5% C
Type	**Parameter**
Material	
Elasticity tensor components	$E_{11}^a = 286.80$, $E_{12}^a = 166.40$, $E_{44}^a = 145.10$ (GPa)
Flow rule	
Rate sensitivity parameter	$m = 0.02$
Hardening rule	
Initial slip resistance	$s_{r,0}^\alpha = 80$ (MPa)
Initial twin resistance	$s_{r,0}^i = 80$ (MPa)
	TWIP 29% Mn-0.8% C
Type	**Parameter**
Material	
Elasticity tensor components	$E_{11}^a = 286.80$, $E_{12}^a = 166.40$, $E_{44}^a = 145.10$ (GPa)
Flow rule	
Rate sensitivity parameter	$m = 0.02$
Hardening rule	
Initial slip resistance	$s_{r,0}^\alpha = 90$ (MPa)
Initial twin resistance	$s_{r,0}^i = 120$ (MPa)

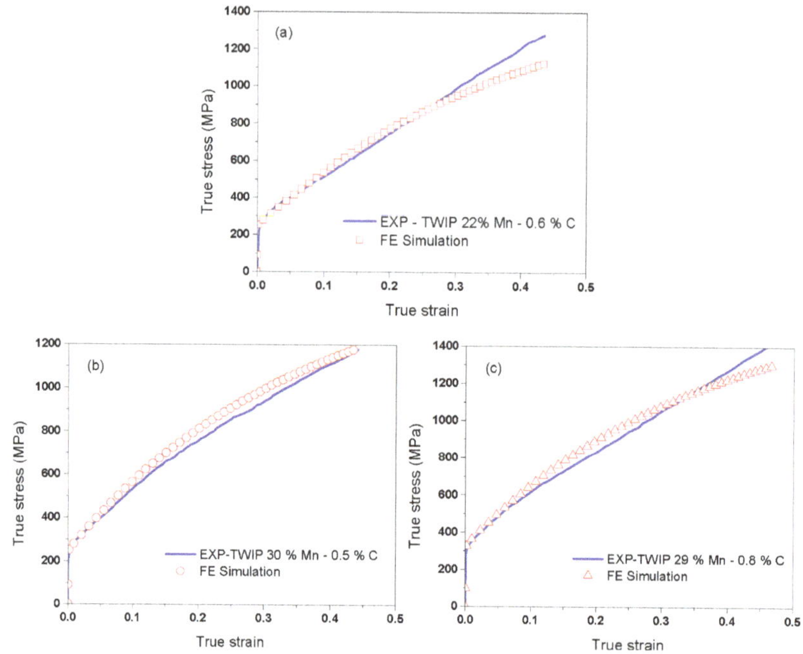

Figure 5. Comparison of experimental and simulation results of TWIP steels subjected to uniaxial tension: (**a**) TWIP 22% Mn-0.6% C, (**b**) TWIP 30% Mn-0.5% C, (**c**) TWIP 29% Mn-0.8% C.

5.2. Finite Element Simulations

After establishing the modeling parameters, further finite element simulations are performed to evaluate the deformation pattern of single-crystal and polycrystal TWIP steels subjected to uniaxial tension and compression. The other reasons for the simulations are to observe the effects of loading directions on the (i) activity of slip and twin-systems, (ii) magnitude of slip and twin shear strain, and (iii) volume fraction of the twinned region. In the subsequent sections, the variation of these parameters for three crystallographic directions ([100], [110], and [111]) in a single crystal of TWIP steels are analyzed. For polycrystal, the material point is represented by a set of five hundred randomly oriented grains. A detailed discussion about the choice of the optimum number of grains is presented in [53].

5.2.1. Slip and Twin Planes' Activity

In twinning- and transformation-induced plasticity steels, it would be worthwhile to investigate the contribution of slip and twinning in overall plastic strain. Moreover, their effects on deformation and hardening behaviors are equally essential. Therefore, the activity of slip and twin systems during the course of single crystals' deformation is evaluated through the current model in three crystallographic orientations, as illustrated in Figures 6 and 7. It is noted that slip and twin systems of austenite single crystal, as mentioned in Appendix A, are designated through numbers, as shown in these figures. For example, slip system 1 is the combination of slip plane normal, n^{α} and slip direction, and m_k^{α} vectors, where α and k are equal to 1, as shown in Table A1. In addition, vectors n^1 and m_2^1 form slip system 2. A similar convention is used for twin systems; see Table A2. The activity of slip and twin systems is assessed based on the ratios of resolved shear stress and slip or twin resistance. The slip or twin plane becomes active once these ratios are greater than 1.

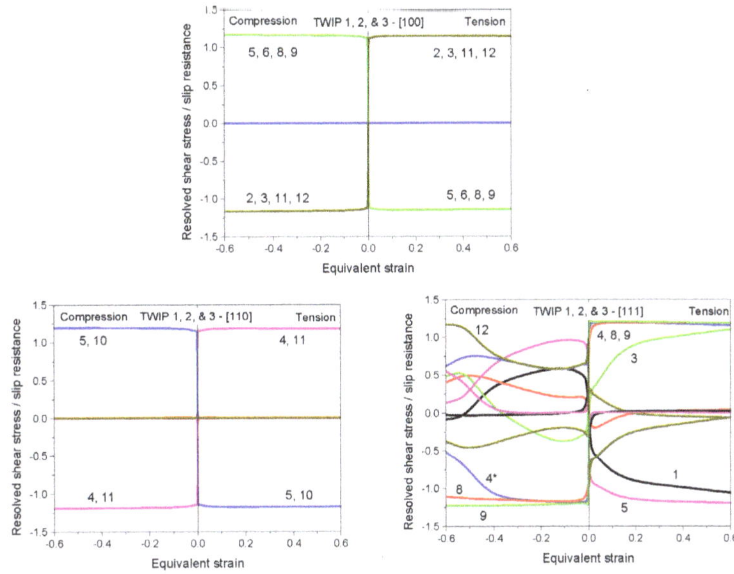

Figure 6. Activity of slip planes in single crystal of TWIP 1, TWIP 2, and TWIP 3 steels under tension and compression in three crystallographic directions.

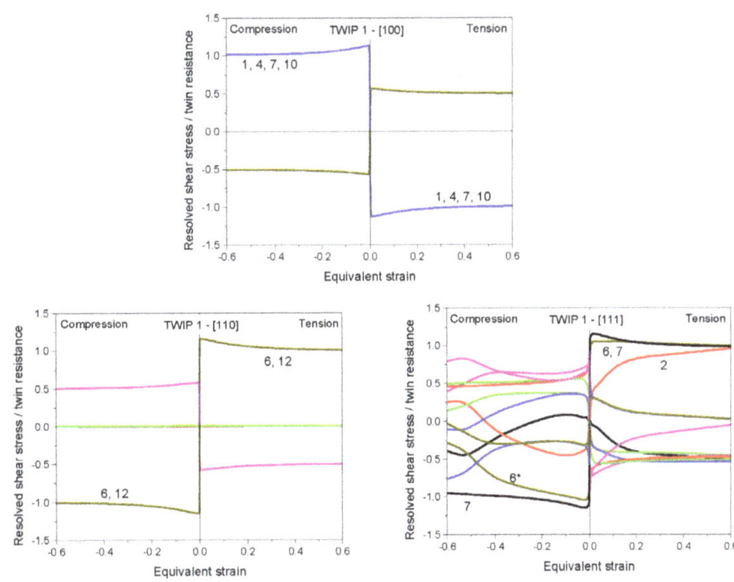

Figure 7. Activity of twin planes in single crystal of TWIP 1, TWIP 2, and TWIP 3 steels under tension and compression in three crystallographic directions.

It is evident from Figures 6 and 7 that all TWIP steels show similar activity of slip and twin systems, regardless of their varying compositions. Furthermore, a similar number of slip and twin systems are active while the crystal is subjected to tension and compression in the [100] and [110] directions. However, this is not the case in [111], where a higher number of systems show activity in tension (slip systems: 1, 3–5, 8, and 9; twin systems: 2, 6, and 7) than compression (slip systems: 4, 8, 9, and 12; twin systems: 6 and 7). It is also found that some slip and twin systems become activated at lower strain but deactivated at higher, or vice versa. This is probably due to the interaction among slip–slip, slip–twin, or twin–twin systems, and/or reorientation of the slip or twin planes during the course of deformation. These interactions are one of the main causes of material softening or hardening at the micro-level. Another important finding from Figures 6 and 7 is related to the contribution of the slip and twin modes in overall plasticity. As a whole, the contribution of slip is higher for all crystallographic directions. However, the highest level of twin activity is observed in the [100] direction.

5.2.2. Deformation and Hardening Behavior

The stress-strain response of single-crystal and polycrystal TWIP steels under tension and compression are represented, respectively, in Figures 8 and 9. It is noted that from now on, the conventions of TWIP 1, TWIP 2, and TWIP 3 are used for TWIP 22% Mn-0.6% C, TWIP 30% Mn-0.5% C, and TWIP 29% Mn-0.8% C, respectively. A prominent variation in deformation pattern, both in tension and compression, is seen in crystallographic orientations and polycrystal.

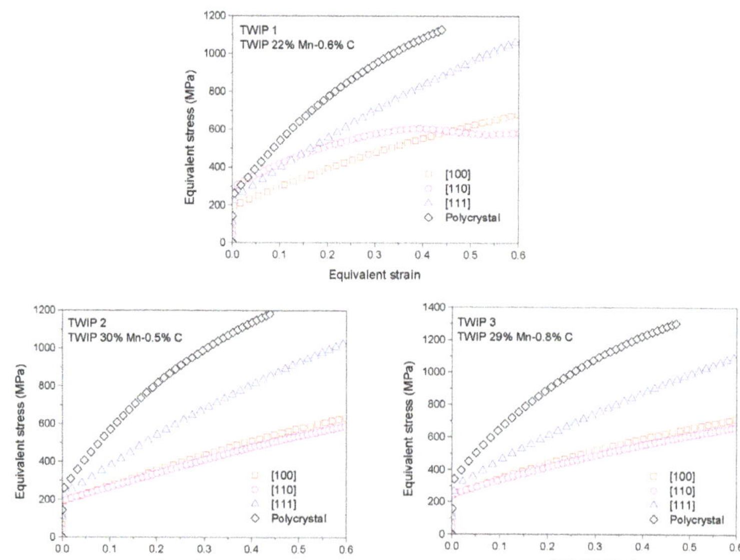

Figure 8. Deformation behavior of single-crystal and polycrystal TWIP steels subjected to tension.

It is noted in Figures 8 and 9 that orientations [100] and [110] show similar behavior in tension and compression, except for tension in TWIP 1. In this, a sample loaded in [110] initially shows hardening, and then softening after equivalent strain 0.4. The softening behavior may be induced due to the activation of mechanical twinning; however, this does not comply with twin systems' activity, as shown in Figure 6b, where a similar number of twin systems are active at strain 0.4. The other possible reasons may include: (i) reorientation of slip and/or twin systems that may enhance the overall shear strain rates of both modes, or (ii) variation in crystal defect energy as a result of dislocations' interaction. In all TWIP steels, crystals subjected to tensile load in the [111] direction present the largest magnitude of stress; however, in compression, a similar pattern is observed until the equivalent strain 0.4. After this, the [111] direction shows a lower magnitude of equivalent stress. Furthermore, it is also noted that all polycrystal TWIP steels represent a higher magnitude of stress in tension than compression. This may indicate a greater dominancy of slip and twin systems' reorientation (primary hardening) and interactions (latent hardening) in tension than compression. A quantitative comparison of stress magnitudes in single-crystal and polycrystal TWIP steels is presented in Table 4.

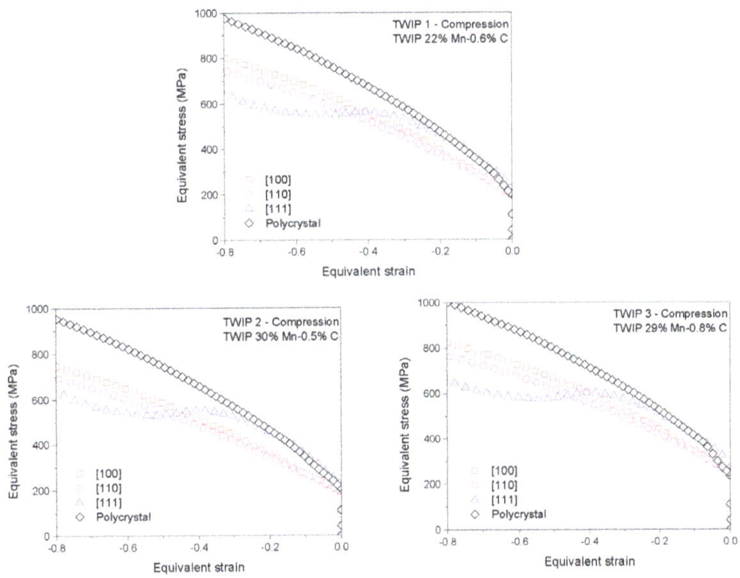

Figure 9. Deformation behavior of single-crystal and polycrystal TWIP steels subjected to compression.

Table 4. Comparison of equivalent stress at 0.4 equivalent strain of single-crystal and polycrystal TWIP steels under tension and compression.

	Equivalent Stress (MPa)							
	Tension				Compression			
	Poly	[100]	[110]	[111]	Poly	[100]	[110]	[111]
TWIP 1	1094	549	626	839	688	542	542	540
TWIP 2	1138	510	512	820	678	560	555	552
TWIP 3	1230	575	577	873	712	570	568	565

5.2.3. Slip and Twin Shear Strain

The magnitude of shear strain is another important parameter in crystal plasticity, especially if modes other than slip are also favorable. In addition, the slip and twin contribution in overall plasticity can only be quantitatively represented through shear strain magnitude, not by activity of slip and twin planes, as it is a qualitative measurement. In this regard, the magnitudes of shear strain in slip and twin modes are analyzed. For almost all TWIP steels, a linear variation of shear strain, under tension and compression, is observed in slip mode; however, nonlinearity is dominant in twin, as illustrated in Figures 10 and 11. Moreover, the magnitude of the shear strain in slip is far greater than twin in all steels' crystallographic directions, which refers to the dominancy of slip over twin mode in plasticity. In slip, all crystallographic directions represent, with few exceptions, a similar magnitude of shear strain under the same kind of loading; however, a significant variation is observed in twin mode. It is evident from Figure 11 that shear strain becomes nearly constant for all crystallographic directions of TWIP 1 and 2 after specific equivalent strain. On the contrary, shear strain's magnitude continuously increases in TWIP 3. This peculiar behavior could be an indication of latent hardening and planes' interaction effects. A quantifiable observation of shear strain in slip and twin modes is exhibited in Table 5.

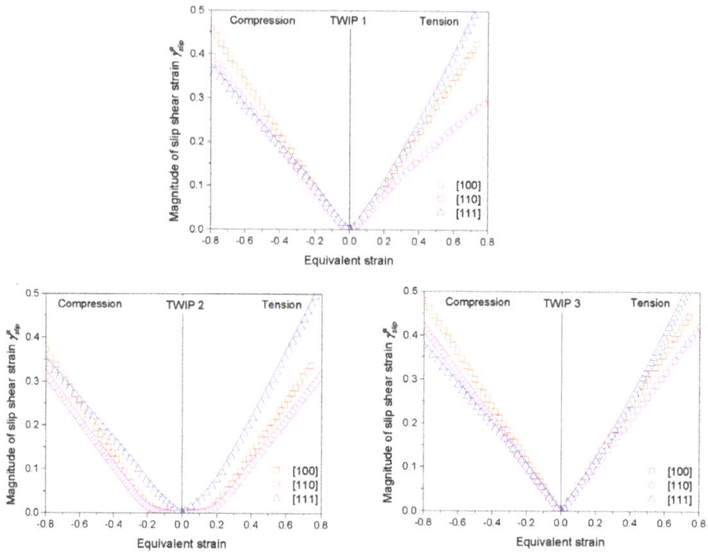

Figure 10. Magnitude of commulative slip shear strain of a single crystal of TWIP steels subjected to tension and compression.

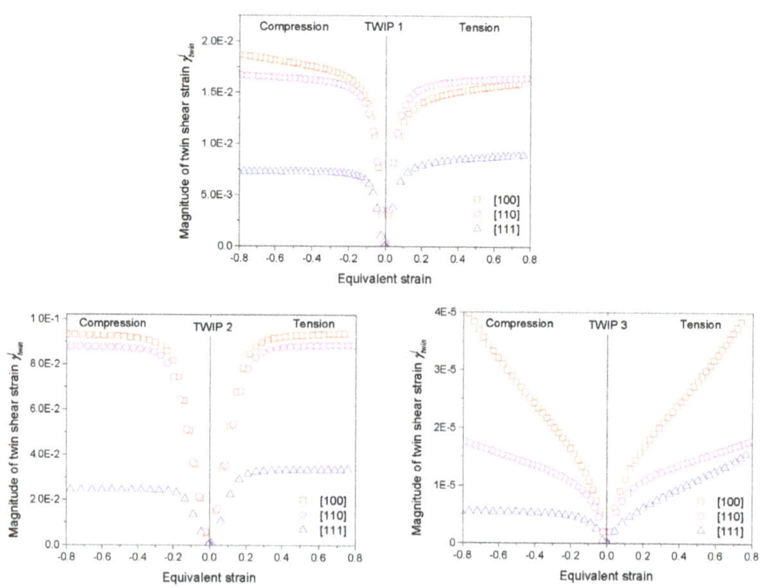

Figure 11. Magnitude of commulative twin shear strain of a single crystal of TWIP steels subjected to tension and compression.

Table 5. Shear strain's magnitude at 0.4 equivalent strain in slip and twin modes of TWIP 1, 2, and 3 steels under tension and compression.

		Twin Volume Fraction					
		Tension			Compression		
		[100]	[110]	[111]	[100]	[110]	[111]
Slip	TWIP 1	0.219	0.160	0.246	0.211	0.192	0.191
	TWIP 2	0.132	0.107	0.213	0.142	0.112	0.162
	TWIP 3	0.235	0.211	0.245	0.238	0.208	0.196
Twin	TWIP 1	0.015	0.016	0.008	0.017	0.016	0.007
	TWIP 2	0.091	0.087	0.033	0.091	0.086	0.024
	TWIP 3	2.4×10^{-5}	1.4×10^{-5}	9.8×10^{-5}	2.4×10^{-5}	1.3×10^{-5}	5.1×10^{-6}

An obvious observation from Table 5 is the highest magnitude of slip shear strain of TWIP 3 under tension and compression in all crystallographic directions. On the other hand, twin shear strain shows the uppermost values in TWIP 2 for all directions under both types of loading.

5.2.4. Twin Volume Fraction

The magnitude of shear strain alone does not present a complete quantitative estimation of twinning. A complete estimation requires the magnitude of the volume of twin in the overall plasticity. In view of this requirement, the twin volume fraction is estimated during the course of deformation, as represented through Figure 12. In conjunction with twin systems' activity, the twin volume fraction represents the ratio of active to total number of twin planes. A significant variation of twin volume is observed under tension and compression in all steels and crystallographic directions. As in the case of shear strain, the twin volume fraction becomes nearly steady for all crystallographic directions of TWIP 1 and 2, but not for TWIP 3, after a certain equivalent strain. The magnitude of twin volume fraction for TWIP steels is presented in Table 6. As is evident, the largest magnitude of twin volume is observed in the [100] crystallographic direction under tension and compression for all three TWIP steels. Overall, the twin volume fraction is not exceeded by 0.333 (33.33 %) in all directions and steels. This shows that the contribution of twinning in the plasticity of TWIP steels 1, 2, and 3 is limited to 33.33 % for a single crystal.

Table 6. Highest values of twin volume fraction for TWIP 1, 2, and 3 steels under tension and compression.

	Twin Volume Fraction					
	Tension			Compression		
	[100]	[110]	[111]	[100]	[110]	[111]
TWIP 1	0.333	0.166	0.249	0.333	0.166	0.166
TWIP 2	0.249	0.166	0.166	0.333	0.166	0.166
TWIP 3	0.249	0.166	0.166	0.200	0.110	0.080

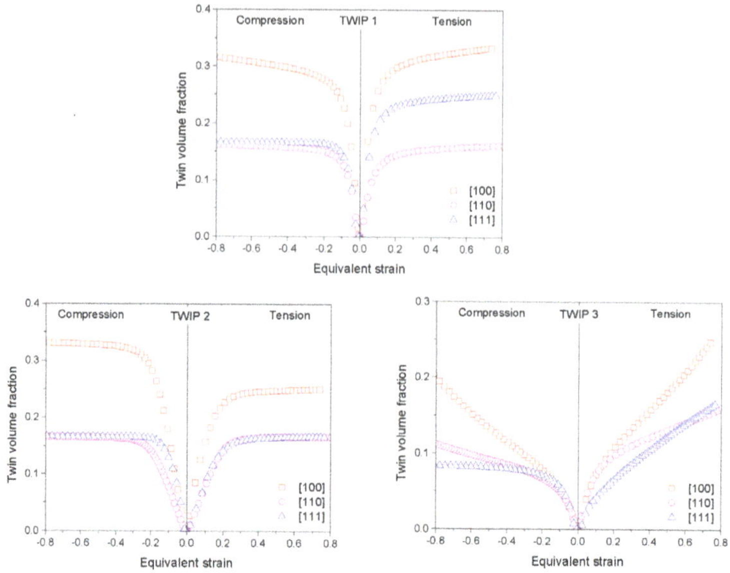

Figure 12. Twin volume fraction of a single crystal of TWIP steels subjected to tension and compression.

6. Conclusions

A numerical scheme is developed and implemented for modeling the elastic-plastic deformation behavior of twinning-induced plasticity steel, in which mechanical twinning contributes significantly to plasticity along with a crystallographic slip. Initially, a constitutive formulation of equations, reported in earlier work, is briefly discussed. Afterward, a numerical integration procedure is established by identifying primary variables, discretizing constitutive equations in the time domain, developing an iterative scheme, and introducing a time sub-stepping algorithm. A numerical integration scheme is then incorporated in finite element software ABAQUS through a user-defined material subroutine. Finite element models of single-crystal and polycrystal material points are developed and simulation results are compared with the published experimental observations. They are in close accord with the maximum error of 16.15% in TWIP steels for equivalent stress. However, the error becomes higher beyond 0.3 strain. This puts limitations on the current model to be implemented, with more accuracy, at high strain deformation. This must be further explored in future work. In addition, further simulations are executed to quantify slip and twin systems' activity, deformation behavior, shear strain pattern, and twin volume fraction of three TWIP steels subjected to uniaxial tension and compression. The slip and twin systems' activity shows that the slip contribution is higher for all crystallographic directions in all steels; however, the highest level of twin activity (number of active twin systems) is observed in the [100] direction. In addition, a higher magnitude of stress at 0.45 equivalent strain is observed in tension (1150 to 1300 MPa) than compression (780 to 850 MPa) in all polycrystal TWIP steels. It may indicate a more prominent role of slip and twin systems' reorientation and interactions in tension. Moreover, the twin shear strain becomes nearly constant for all crystallographic directions of TWIP 1 and 2 after 0.2 equivalent strain, but not for TWIP 3. It is also found that the fraction of twin volume is not surpassed by 33.33% for all directions and TWIP steels. The development and implementation of a fully implicit numerical integration scheme in modeling twinning-induced plasticity provide an effective platform to estimate the deformation behavior of TWIP steels. The current work can be enhanced to incorporate martensitic phase transformation and damage criterion in a coupled slip, twinning, and transformation-induced plasticity model.

Author Contributions: R.K.: conceptualization, methodology, investigation, and writing—original draft preparation; T.P.: supervision, and writing—review and editing; A.A.: validation and visualization; S.Z.Q.: supervision, formal analysis; S.M.: software and visualization. All authors have read and agreed to the published version of the manuscript.

Funding: This research received no external funding.

Acknowledgments: The authors acknowledge the support of Imam Mohammad Ibn Saud Islamic University for this research.

Conflicts of Interest: The authors declare no conflict of interest.

Appendix A. Slip and Twin Systems of Austenite Crystal

Face-centered cubic (FCC) austenite crystal's slip and twin systems are represented in terms of Miller indices and unit vectors in Tables A1 and A2, respectively.

Table A1. Slip systems of FCC crystal.

Slip Plane Normal n^α			Slip Direction m_k^α		
			α = 1–4, $k = 1, 2, 3$		
	Miller	Unit Vector		Miller	Unit Vector
n^1	(111)	$\left(\frac{1}{\sqrt{3}}, \frac{1}{\sqrt{3}}, \frac{1}{\sqrt{3}}\right)$	m_1^1	$[01\bar{1}]$	$\left(0, \frac{1}{\sqrt{2}}, \frac{-1}{\sqrt{2}}\right)$
			m_2^1	$[10\bar{1}]$	$\left(\frac{1}{\sqrt{2}}, 0, \frac{-1}{\sqrt{2}}\right)$
			m_3^1	$[1\bar{1}0]$	$\left(\frac{1}{\sqrt{2}}, \frac{-1}{\sqrt{2}}, 0\right)$
n^2	$(\bar{1}11)$	$\left(\frac{-1}{\sqrt{3}}, \frac{1}{\sqrt{3}}, \frac{1}{\sqrt{3}}\right)$	m_1^2	$[01\bar{1}]$	$\left(0, \frac{1}{\sqrt{2}}, \frac{-1}{\sqrt{2}}\right)$
			m_2^2	[101]	$\left(\frac{1}{\sqrt{2}}, 0, \frac{1}{\sqrt{2}}\right)$
			m_3^2	[110]	$\left(\frac{1}{\sqrt{2}}, \frac{1}{\sqrt{2}}, 0\right)$
n^3	$(\bar{1}\bar{1}1)$	$\left(\frac{-1}{\sqrt{3}}, \frac{-1}{\sqrt{3}}, \frac{1}{\sqrt{3}}\right)$	m_1^3	[011]	$\left(0, \frac{1}{\sqrt{2}}, \frac{1}{\sqrt{2}}\right)$
			m_2^3	[101]	$\left(\frac{1}{\sqrt{2}}, 0, \frac{1}{\sqrt{2}}\right)$
			m_3^3	$[1\bar{1}0]$	$\left(\frac{1}{\sqrt{2}}, \frac{-1}{\sqrt{2}}, 0\right)$
n^4	$(1\bar{1}1)$	$\left(\frac{1}{\sqrt{3}}, \frac{-1}{\sqrt{3}}, \frac{1}{\sqrt{3}}\right)$	m_1^4	[011]	$\left(0, \frac{1}{\sqrt{2}}, \frac{1}{\sqrt{2}}\right)$
			m_2^4	$[10\bar{1}]$	$\left(\frac{1}{\sqrt{2}}, 0, \frac{-1}{\sqrt{2}}\right)$
			m_3^4	[110]	$\left(\frac{1}{\sqrt{2}}, \frac{1}{\sqrt{2}}, 0\right)$

Table A2. Twin systems of FCC crystal.

Twin Plane Normal n^i				Twin Direction m_k^i $i = 1\text{–}4,\ k = 1,2,3$	
	Miller	Unit Vector		Miller	Unit Vector
n^1	(111)	$\left(\frac{1}{\sqrt{3}}, \frac{1}{\sqrt{3}}, \frac{1}{\sqrt{3}}\right)$	m_1^1	$[\bar{2}11]$	$\left(\frac{-2}{\sqrt{6}}, \frac{1}{\sqrt{6}}, \frac{1}{\sqrt{6}}\right)$
			m_2^1	$[1\bar{2}1]$	$\left(\frac{1}{\sqrt{6}}, \frac{-2}{\sqrt{6}}, \frac{1}{\sqrt{6}}\right)$
			m_3^1	$[11\bar{2}]$	$\left(\frac{1}{\sqrt{6}}, \frac{1}{\sqrt{6}}, \frac{-2}{\sqrt{6}}\right)$
n^2	$(\bar{1}11)$	$\left(\frac{-1}{\sqrt{3}}, \frac{1}{\sqrt{3}}, \frac{1}{\sqrt{3}}\right)$	m_1^2	$[211]$	$\left(\frac{2}{\sqrt{6}}, \frac{1}{\sqrt{6}}, \frac{1}{\sqrt{6}}\right)$
			m_2^2	$[\bar{1}\bar{2}1]$	$\left(\frac{-1}{\sqrt{6}}, \frac{-2}{\sqrt{6}}, \frac{1}{\sqrt{6}}\right)$
			m_3^2	$[\bar{1}1\bar{2}]$	$\left(\frac{-1}{\sqrt{6}}, \frac{1}{\sqrt{6}}, \frac{-2}{\sqrt{6}}\right)$
n^3	$(\bar{1}\bar{1}1)$	$\left(\frac{-1}{\sqrt{3}}, \frac{-1}{\sqrt{3}}, \frac{1}{\sqrt{3}}\right)$	m_1^3	$[2\bar{1}1]$	$\left(\frac{2}{\sqrt{6}}, \frac{-1}{\sqrt{6}}, \frac{1}{\sqrt{6}}\right)$
			m_2^3	$[\bar{1}21]$	$\left(\frac{-1}{\sqrt{6}}, \frac{2}{\sqrt{6}}, \frac{1}{\sqrt{6}}\right)$
			m_3^3	$[\bar{1}\bar{1}\bar{2}]$	$\left(\frac{-1}{\sqrt{6}}, \frac{-1}{\sqrt{6}}, \frac{-2}{\sqrt{6}}\right)$
n^4	$(1\bar{1}1)$	$\left(\frac{1}{\sqrt{3}}, \frac{-1}{\sqrt{3}}, \frac{1}{\sqrt{3}}\right)$	m_1^4	$[\bar{2}\bar{1}1]$	$\left(\frac{-2}{\sqrt{6}}, \frac{-1}{\sqrt{6}}, \frac{1}{\sqrt{6}}\right)$
			m_2^4	$[121]$	$\left(\frac{1}{\sqrt{6}}, \frac{2}{\sqrt{6}}, \frac{1}{\sqrt{6}}\right)$
			m_3^4	$[1\bar{1}\bar{2}]$	$\left(\frac{1}{\sqrt{6}}, \frac{-1}{\sqrt{6}}, \frac{-2}{\sqrt{6}}\right)$

References

1. Xu, M.; David, J.; Kim, S. The Fourth Industrial Revolution: Opportunities and Challenges. *Int. J. Financ. Res.* **2018**, *9*, 90. [CrossRef]
2. Demeri, M.Y. *Advanced High-Strength Steels: Science, Technology, and Applications*; ASM International: Almere, The Netherlands, 2013.
3. Kuziak, R.; Kawalla, R.; Waengler, S. Advanced High Strength Steels for Automotive Industry. *Arch. Civ. Mech. Eng.* **2008**, *8*, 103–117. [CrossRef]
4. Tamarelli, C.M. *The Evolving Use of Advanced High Strength Steels for Automotive Applications*; Technology Report; University of Michigan, USA, Materials Science and Engineering Department: Ann Arbor, MI, USA, 2011.
5. Al-Abri, O.S.; Pervez, T.; Qamar, S.Z.; Khan, R. On the performance analysis of AHSS with an application to SET technology—FEM simulations and experimental measurements. *Thin-Walled Struct.* **2016**, *101*, 58–74. [CrossRef]
6. Cooman, B.C.D.; Estrin, Y.; Kim, S.K. Twinning-induced plasticity (TWIP) steels. *Acta Mater.* **2018**, *142*, 283–362. [CrossRef]
7. Fan, J.; Qiao, J.W.; Wang, Z.H.; Rao, W.; Kang, G.Z. Twinning-induced plasticity (TWIP) and work hardening in Ti-based metallic glass matrix composites. *Sci. Rep.* **2017**, *7*, 1877. [CrossRef]
8. Grässel, O.; Krüger, L.; Frommeyer, G.; Meyer, L.W. High strength Fe-Mn-Al-Si TRIP/TWIP steels development: Properties & Application. *Int. J. Plast.* **2000**, *16*, 1391–1409.
9. Yang, G.; Kim, J.-K. An Overview of High Yield Strength Twinning-Induced Plasticity Steels. *Metals* **2021**, *11*, 124. [CrossRef]
10. Li, D.; Qian, L.; Wei, C.; Liu, S.; Zhang, F.; Meng, J. The role of Mn on twinning behavior and tensile properties of coarse- and fine-grained Fe–Mn–C twinning-induced plasticity steels. *Mater. Sci. Eng. A* **2020**, *789*, 139586. [CrossRef]
11. Pierce, D.; Jiménez, J.; Bentley, J.; Raabe, D.; Oskay, C.; Wittig, J. The influence of manganese content on the stacking fault and austenite-ϵ martensite interfacial energies in Fe-Mn-(Al-Si steels investigated by experiment and theory. *Acta Mater.* **2014**, *68*, 238–253. [CrossRef]
12. Lee, Y.K.; Lee, S.J.; Han, J. Critical assessment 19: Stacking fault energies of austenitic steels. *Mater. Sci. Technol.* **2016**, *32*, 1–8. [CrossRef]
13. Bouaziz, O.; Allain, S.; Scott, C.P.; Cugy, P.; Barbier, D. High manganese austenitic twinning induced plasticity steels: A review of the microstructure properties relationships. *Curr. Opin. Solid State Mater. Sci.* **2011**, *15*, 141–168. [CrossRef]
14. Zhi, H.; Zhang, C.; Antonov, S.; Yu, H.; Guo, T.; Su, Y. Investigations of dislocation-type evolution and strain hardening during mechanical twinning in Fe-22Mn-0.6C twinning-induced plasticity steel. *Acta Mater.* **2020**, *195*, 371–382. [CrossRef]
15. Idrissi, H.; Renard, K.; Ryel, L.; Schryvers, D.; Jacques, P.J. On the mechanism of twin formation in Fe–Mn–C TWIP steels. *Acta Mater.* **2010**, *58*, 2464–2476. [CrossRef]

16. Wang, S.; Liu, Z.; Wang, C.; Liu, J.; Liang, G.; Li, Q. Effects of Twin-Dislocation and Twin-Twin Interactions on the Strain Hardening Behavior of TWIP Steels. *J. Iron Steel Res. Int.* **2010**, *17*, 70–74. [CrossRef]
17. Pauly, S.; Gorantla, S.; Wang, G.; Kühn, U.; Eckert, J. Transformation-mediated ductility in CuZr-based bulk metallic glasses. *Nat. Mater.* **2010**, *9*, 473–477. [CrossRef]
18. Bouaziz, O.; Zurob, H.; Chehab, B.; Embury, J.D.; Allain, S.; Huang, M. Effect of chemical composition on work hardening of Fe—Mn—C TWIP steels. *Mater. Sci. Technol.* **2011**, *27*, 707–709. [CrossRef]
19. Peng, X.; Zhu, D.; Hu, Z.; Yi, W.; Liu, H.; Wang, M. Stacking fault energy and tensile deformation behavior of high-carbon twinning-induced plasticity steels: Effect of Cu addition. *Mater. Des.* **2013**, *45*, 518–523. [CrossRef]
20. Martin, S.; Wolf, S.; Martin, U.; Krüger, L.; Rafaja, D. Deformation Mechanisms in Austenitic TRIP/TWIP Steel as a Function of Temperature. *Metall. Mater. Trans.* **2016**, *47*, 49–58. [CrossRef]
21. Fei, L.; Weigang, Z.; Wenjiao, D. Stress-strain Response for Twinning-induced Plasticity Steel with Temperature. *Procedia Eng.* **2014**, *81*, 1330–1335.
22. Allain, S.; Chateau, J.P.; Bouaziz, O.; Migot, S.; Guelton, N. Correlations between the calculated stacking fault energy and the plasticity mechanisms in Fe-Mn-C alloys. *Mater. Sci. Eng.* **2004**, *387–389*, 158–162. [CrossRef]
23. Curtze, S.; Kuokkala, V.T.; Oikari, A.; Talonen, J.; Hänninen, H. Thermodynamic modeling of the stacking fault energy of austenitic steels. *Acta Mater.* **2011**, *59*, 1068–1076. [CrossRef]
24. Idrissi, H.; Ryelandt, L.; Veron, M.; Schryvers, D.; Jacques, P. Is there a relationship between the stacking fault character and the activated mode of plasticity of fe-mn-based austenitic steels? *Scr. Mater.* **2009**, *60*, 941–944. [CrossRef]
25. Tomé, C.N.; Lebensohn, R.A.; Kocks, U.F. A model for texture development dominated by deformation twinning: Application to zirconium alloys. *Acta Metall. Mater.* **1991**, *39*, 2667–2680. [CrossRef]
26. Houtte, P.V. Simulation of the rolling and shear texture of brass by the Taylor theory adapted for mechanical twinning. *Acta Metall.* **1978**, *26*, 591–604. [CrossRef]
27. Kalidindi, S.R. Incorporation of deformation twinning in crystal plasticity models. *J. Mech. Phys. Solids* **1998**, *46*, 267–290. [CrossRef]
28. Taylor, G.I. Plastic strain in metals. *J. Inst. Met.* **1938**, *62*, 307–324.
29. Chin, G.; Hosford, W.; Mendorf, D. Accommodation of constrained deformation in F. C. C. metals by slip and twinning. *Proc. R. Soc. Lond. A Math. Phys. Eng. Sci.* **1969**, *309*, 433–456.
30. Asaro, R.J.; Rice, J.R. Strain localization in ductile single crystals. *J. Mech. Phys. Solids* **1977**, *25*, 309–338. [CrossRef]
31. Salem, A.A.; Kalidindi, S.R.; Doherty, R.D. Strain hardening of titanium: Role of deformation twinning. *Acta Mater.* **2003**, *51*, 4225–4237. [CrossRef]
32. Salem, A.A.; Kalidindi, S.R.; Semiatin, S.L. Strain hardening due to deformation twinning in α-titanium: Constitutive relations and crystal-plasticity modeling. *Acta Mater.* **2005**, *53*, 3495–3502. [CrossRef]
33. Salem, A.A.; Kalidindi, S.R.; Doherty, R.D.; Semiatin, S.L. Strain hardening due to deformation twinning in α-titanium: Mechanisms. *Metall. Mater. Trans. A* **2006**, *37*, 259–268. [CrossRef]
34. Bhattacharya, K. Wedge-like microstructure in martensites. *Acta Metall. Et Mater.* **1991**, *39*, 2431–2444. [CrossRef]
35. Wang, J.; Sehitoglu, H. Modelling of martensite slip and twinning in NiTiHf shape memory alloys. *Philos. Mag.* **2014**, *94*, 2297–2317. [CrossRef]
36. Ostadrahimi, A.; Taheri-Behrooz, F. Analytical solution for twinning deformation effect of pre-strained shape memory effect beam-columns. *J. Intell. Mater. Syst. Struct.* **2019**, *30*, 2147–2165. [CrossRef]
37. Allain, S.; Chateau, J.P.; Bouaziz, O. A physical model of the twinning-induced plasticity effect in a high manganese austenitic steel. *Mater. Sci. Eng.* **2004**, *387–389*, 143–147. [CrossRef]
38. Bouaziz, O. Strain-hardening of twinning-induced plasticity steels. *Scr. Mater.* **2012**, *66*, 982–985. [CrossRef]
39. Lee, M.G.; Kim, S.J.; Han, H.N. Crystal plasticity finite element modeling of mechanically induced martensitic transformation (MIMT) in metastable austenite. *Int. J. Plast.* **2010**, *26*, 688–710. [CrossRef]
40. Khosravani, A.; Scott, J.; Miles, M.P.; Fullwood, D.; Adams, B.L.; Mishra, R.K. Twinning in magnesium alloy AZ31B under different strain paths at moderately elevated temperatures. *Int. J. Plast.* **2013**, *45*, 160–173. [CrossRef]
41. Shiekhelsouk, M.N.; Favier, V.; Inal, K.; Cherkaoui, M. Modelling the behaviour of polycrystalline austenitic steel with twinning-induced plasticity effect. *Int. J. Plast.* **2009**, *25*, 105–133. [CrossRef]
42. Khan, R.; Zahedi, F.I.; Siqqiui, A.K. Numerical modeling of twinning induced plasticity in austenite based advanced high strength steels. *Procedia Manuf.* **2016**, *5*, 772–786. [CrossRef]
43. Yang G.; Dayong A.; Fengbo H.; Liu, X.; Guozheng K.; Xu, Z. Multiple-mechanism and microstructure-based crystal plasticity modeling for cyclic shear deformation of TRIP steel. *Int. J. Mech. Sci.* **2022**, *222*, 107269.
44. Qayyum, F.; Guk, S.; Prahl, U. Studying the Damage Evolution and the Micro-Mechanical Response of X8CrMnNi16-6-6 TRIP Steel Matrix and 10% Zirconia Particle Composite Using a Calibrated Physics and Crystal-Plasticity-Based Numerical Simulation Model. *Crystals* **2021**, *11*, 759. [CrossRef]
45. Qayyum, F.; Guk, S.; Kawalla, R.; Prahl, U. On Attempting to Create a Virtual Laboratory for Application-Oriented Microstructural Optimization of Multi-Phase Materials. *Appl. Sci.* **2021**, *11*, 1506. [CrossRef]
46. Ching-Tun, P.; Mao, L.; Cheng, L.; Huijun, L. The determination of self hardening parameters of twinning induced plasticity steel via crystal plasticity modeling. *J. Comput. Theor. Nanosci.* **2015**, *12*, 2523–2530.

47. Li, Y.; Zhu, L.; Liu, Y.; Wei, Y.; Wu, Y.; Tang, D.; Mi, Z. On the strain hardening and texture evolution in high manganese steels: Experiments and numerical investigation. *J. Mech. Phys. Solids* **2013**, *61*, 2588–2604. [CrossRef]
48. Sun, C.; Wang, B.; Politis, D.J.; Wang, L.; Cai, Y.; Guo, X.; Guo, N. Prediction of earing in TWIP steel sheets based on coupled twinning crystal plasticity model. *Int. J. Adv. Manuf. Technol.* **2017**, *89*, 3037–3047. [CrossRef]
49. Wong, S.L.; Madivala, M.; Prahl, U.; Roters, F.; Raabe, D. A crystal plasticity model for twinning and transformation-induced plasticity. *Acta Mater.* **2016**, *118*, 140–151. [CrossRef]
50. Sun, C.; Guo, N.; Fu, M.; Wang, S. Modeling of slip, twinning and transformation induced plastic deformation for TWIP steel based on crystal plasticity. *Int. J. Plast.* **2016**, *76*, 186–212. [CrossRef]
51. Khan, R.; Pervez, T.; Qamar, S.Z. Modeling and simulations of transformation and twinning induced plasticity in advanced high strength austenitic steels. *Mech. Mater.* **2016**, *95*, 83–101. [CrossRef]
52. Manjunatha, M.; Alexander, S.; Su, L.W.; Franz, R.; Ulrich, P.; Wolfgang, B. Temperature dependent strain hardening and fracture behavior of TWIP steel. *Int. J. Plast.* **2018**, *104*, 80–103.
53. Khan, R.; Alfozan, A. Modeling of twinning-induced plasticity using crystal plasticity and thermodynamic framework. *Acta Mech.* **2019**, *230*, 2687–2715. [CrossRef]
54. Marin, E.B. *On the Formulation of a Crystal Plasticity Model*; Tech. Rep.; Sandia National Laboratories: Livermore, NM, USA, 2006.
55. Lee, E.H. Elastic-plastic deformation at finite strains. *J. Appl. Mech.* **1969**, *36*, 1–6. [CrossRef]
56. Kalidindi, S.R.; Bronkhorst, C.A.; Anand, L. Crystallographic texture evolution in bulk deformation processing of FCC metals. *J. Mech. Phys. Solids* **1992**, *40*, 537–569. [CrossRef]
57. Turteltaub, S.; Suiker, A.S.J. A multiscale thermomechanical model for cubic to tetragonal martensitic phase transformations. *Int. J. Solids Struct.* **2006**, *43*, 4509–4545. [CrossRef]
58. Simo, J.C.; Hughes, T.J.R. *Computational Inelasticity*; Springer: Berlin/Heidelberg, Germany, 1998; Volume 7.
59. Kocks, U.F.; Tome, C.N.; Wenk, H.R. *Texture and Anisotropy: Preferred Orientations in Polycrystals and Their Effect on Materials Properties*; Cambridge University Press: Cambridge, CA, USA, 1998.
60. Schmitt, J.H.; Iung, T. New developments of advanced high-strength steels for automotive applications. *Comptes Rendus Phys.* **2018**, *19*, 641–656. [CrossRef]

crystals

Editorial

Crystal Plasticity (Volume II)

Wojciech Polkowski

Łukasiewicz Research Network, Krakow Institute of Technology, Zakopiańska 73 Str., 30-418 Krakow, Poland; wojciech.polkowski@kit.lukasiewicz.gov.pl; Tel.: +48-12-26-18-115

Citation: Polkowski, W. Crystal Plasticity (Volume II). *Crystals* **2022**, *12*, 1344. https://doi.org/10.3390/cryst12101344

Received: 19 September 2022
Accepted: 22 September 2022
Published: 23 September 2022

1. Introduction

When we announced the first volume of a Special Issue dedicated to "Crystal Plasticity", we could not expect that a great collection of 25 excellent articles would be published [1]. Now, everyone is very welcome to use free access to read these articles at the link below:

https://www.mdpi.com/journal/crystals/special_issues/Crystal_Plasticity (accessed on 18 September 2022).

Our editorial efforts taken in the first volume provided us with a completely new collection of original, state-of-the-art research papers on both theoretical and experimental aspects of plastic deformation. Indeed, the wide spectrum of submitted papers allowed us to merge the most important topic areas of crystal plasticity—i.e., research on the theoretical modelling of dislocation mechanisms and lab-scale validation of materials' structural/mechanical responses to (semi-)industrial processing. Furthermore, both conventional (e.g., steels, nonferrous alloys) and novel (intermetallics, composites, and high-entropy alloys) materials were investigated. During the completion of the first volume, it was our honor to host well-recognized worldwide authorities, as well as young researchers and post-docs taking the "next-step" in their scientific careers. This versatility of contributing authors and topics has provided more proof for the high interest of the scientific community in revealing materials' behaviors from the atomic scale to macroscale under external loadings.

After closing the first volume, we had the feeling that there was still a lot of room for research in the field of crystal plasticity, and thus a lot of space for publishing activities ... Therefore, we had no doubts in announcing the second volume of a Special Issue on crystal plasticity. With the second volume, we aimed to continue our mission, which is still focused on providing theoretical and experimental research works, giving new insights and practical findings in the field of crystal plasticity-related topics.

So, how is the second volume on crystal plasticity? We can answer by paraphrasing a well-known song: "*Oops ... we did it again*".

Once again, a completely new set of 26 original works (including 22 research articles, 3 communications and 1 review) has been collected. As in the case of the first volume, here, a full spectrum of topics belonging to the field of crystal plasticity is represented, including both numerical simulations and experimental works.

By taking into account the investigated materials, the papers can be assigned to the following thematic groups:

- Steels and iron-based alloys [2–9];
- Non-ferrous alloys with fcc- (Ni- [10,11] and Cu-based [12–14]), or hcp crystal structure (Mg- [15,16] and Ti-based [17,18]). Other examples include Zirconium [19], Bi-Sn alloy [20] or polycarbonate resins [21];
- Multicomponent and high-entropy alloys [22–24];
- General theoretical studies on crystal plasticity [25–27].

I hope that the second volume of our Special Issue will be interesting for the scientific and academic communities, and that it will bring lot of inspiration for future research activities in the field of crystal plasticity.

Funding: This research received no external funding.

Acknowledgments: The contributions of all authors are gratefully acknowledged.

Conflicts of Interest: The author declares no conflict of interest.

References

1. Polkowski, W. Crystal Plasticity. *Crystals* **2021**, *11*, 44. [CrossRef]
2. Romanczuk-Ruszuk, E.; Nowik, K.; Sztorch, B. X-ray Line Profile Analysis of Austenitic Phase Transition and Morphology of Nickel-Free Fe-18Cr-18Mn Steel Powder Synthesized by Mechanical Alloying. *Crystals* **2022**, *12*, 1233. [CrossRef]
3. Liu, Q.; Tang, Z.; Yang, X.; He, Z.; Xue, H.; Zhuge, H. Mechanical Properties of Low Carbon Alloy Steel with Consideration of Prior Fatigue and Plastic Damages. *Crystals* **2022**, *12*, 967. [CrossRef]
4. Murali, A.P.; Ganesan, D.; Salunkhe, S.; Abouel Nasr, E.; Davim, J.P.; Hussein, H.M.A. Characterization of Microstructure and High Temperature Compressive Strength of Austenitic Stainless Steel (21-4N) through Powder Metallurgy Route. *Crystals* **2022**, *12*, 923. [CrossRef]
5. Lyu, H.; Ruimi, A. Understanding the Plastic Deformation of Gradient Interstitial Free (IF) Steel under Uniaxial Loading Using a Dislocation-Based Multiscale Approach. *Crystals* **2022**, *12*, 889. [CrossRef]
6. Katzarov, I.H.; Drenchev, L.B. Unveiling the Mechanisms of High-Temperature 1/2[111] Screw Dislocation Glide in Iron–Carbon Alloys. *Crystals* **2022**, *12*, 518. [CrossRef]
7. Bonifaz, E.A.; Mena, A.S. The Cooling Rate and Residual Stresses in an AISI 310 Laser Weld: A Meso-Scale Approach. *Crystals* **2022**, *12*, 502. [CrossRef]
8. Hussein, T.; Umar, M.; Qayyum, F.; Guk, S.; Prahl, U. Micromechanical Effect of Martensite Attributes on Forming Limits of Dual-Phase Steels Investigated by Crystal Plasticity-Based Numerical Simulations. *Crystals* **2022**, *12*, 155. [CrossRef]
9. Zhang, B.; Meng, L.; Ma, G.; Zhang, N.; Li, G.; Liu, K.; Zhong, S. Twinning Behavior in Cold-Rolling Ultra-Thin Grain-Oriented Silicon Steel. *Crystals* **2021**, *11*, 187. [CrossRef]
10. Engel, B.; Huth, M.; Hyde, C. Numerical Investigation into the Influence of Grain Orientation Distribution on the Local and Global Elastic-Plastic Behaviour of Polycrystalline Nickel-Based Superalloy INC-738 LC. *Crystals* **2022**, *12*, 100. [CrossRef]
11. Koneva, N.A.; Nikonenko, E.L.; Nikonenko, A.V.; Popova, N.A. Microstructural Changes in Ni-Al-Cr-Based Heat-Resistant Alloy with Re Addition. *Crystals* **2021**, *11*, 89. [CrossRef]
12. Wongsa-Ngam, J.; Noraphaiphipaksa, N.; Kanchanomai, C.; Langdon, T.G. Numerical Investigation of Plastic Strain Homogeneity during Equal-Channel Angular Pressing of a Cu-Zr Alloy. *Crystals* **2021**, *11*, 1505. [CrossRef]
13. Huang, W.; Pan, K.; Zhang, J.; Gong, Y. Strain Rate and Temperature Effects on Tensile Properties of Polycrystalline Cu_6Sn_5 by Molecular Dynamic Simulation. *Crystals* **2021**, *11*, 1415. [CrossRef]
14. Hsiao, S.-C.; Lin, S.-Y.; Chen, H.-J.; Hsieh, P.-Y.; Kuo, J.-C. Rolling Texture of Cu–30%Zn Alloy Using Taylor Model Based on Twinning and Coplanar Slip. *Crystals* **2021**, *11*, 1351. [CrossRef]
15. Gao, Y.; Wu, C.; Feng, W.; He, Y.; He, H.; Yang, J.; Chen, X. Effects of the Rare Earth Y on the Structural and Tensile Properties of Mg-based Alloy: A First-Principles Study. *Crystals* **2021**, *11*, 1003. [CrossRef]
16. Aljarrah, M.; Alnahas, J.; Alhartomi, M. Thermodynamic Modeling and Mechanical Properties of Mg-Zn-{Y, Ce} Alloys: Review. *Crystals* **2021**, *11*, 1592. [CrossRef]
17. Bai, F.; Zhu, Q.; Shen, J.; Lu, Z.; Zhang, L.; Ali, N.; Zhou, H.; Liu, X. Study on Phase Transformation Orientation Relationship of HCP-FCC during Rolling of High Purity Titanium. *Crystals* **2021**, *11*, 1164. [CrossRef]
18. Wang, Y.; Zhou, D.; Zhou, Y.; Sha, A.; Cheng, H.; Yan, Y. A Constitutive Relation Based on the Johnson–Cook Model for Ti-22Al-23Nb-2(Mo, Zr) Alloy at Elevated Temperature. *Crystals* **2021**, *11*, 754. [CrossRef]
19. Sedaghat, O.; Abdolvand, H. Strain-Gradient Crystal Plasticity Finite Element Modeling of Slip Band Formation in α-Zirconium. *Crystals* **2021**, *11*, 1382. [CrossRef]
20. Wang, C.-T.; Li, Z.; He, Y.; Wang, J.-T.; Langdon, T.G. Microstructural Evolution and Tensile Testing of a Bi–Sn (57/43) Alloy Processed by Tube High-Pressure Shearing. *Crystals* **2021**, *11*, 1229. [CrossRef]
21. Alsadi, J.; Ismail, R.; Trrad, I. An Integrative Simulation for Mixing Different Polycarbonate Grades with the Same Color: Experimental Analysis and Evaluations. *Crystals* **2022**, *12*, 423. [CrossRef]
22. Mhadhbi, M.; Polkowski, W. Synthesis and Characterization of Mechanically Alloyed Nanostructured (Ti,Cr) C Carbide for Cutting Tools Application. *Crystals* **2022**, *12*, 1280. [CrossRef]
23. Tseng, L.-W.; Chen, C.-H.; Tzeng, Y.-C.; Lee, P.-Y.; Lu, N.-H.; Chumlyakov, Y. Microstructure and Superelastic Properties of FeNiCoAlTi Single Crystals with the <100> Orientation under Tension. *Crystals* **2022**, *12*, 548. [CrossRef]
24. Tseng, L.-W.; Chen, C.-H.; Chen, W.-C.; Cheng, Y.; Lu, N.-H. Shape Memory Properties and Microstructure of New Iron-Based FeNiCoAlTiNb Shape Memory Alloys. *Crystals* **2021**, *11*, 1253. [CrossRef]

25. Zhou, H.; Wang, P.; Lu, S. Investigation on the Effects of Grain Boundary on Deformation Behavior of Bicrystalline Pillar by Crystal Plasticity Finite Element Method. *Crystals* **2021**, *11*, 923. [CrossRef]
26. Trusov, P.; Shveykin, A.; Kondratev, N. Some Issues on Crystal Plasticity Models Formulation: Motion Decomposition and Constitutive Law Variants. *Crystals* **2021**, *11*, 1392. [CrossRef]
27. Khan, R.; Pervez, T.; Alfozan, A.; Qamar, S.Z.; Mohsin, S. Numerical Modeling and Simulations of Twinning-Induced Plasticity Using Crystal Plasticity Finite Element Method. *Crystals* **2022**, *12*, 930. [CrossRef]

MDPI
St. Alban-Anlage 66
4052 Basel
Switzerland
Tel. +41 61 683 77 34
Fax +41 61 302 89 18
www.mdpi.com

Crystals Editorial Office
E-mail: crystals@mdpi.com
www.mdpi.com/journal/crystals

www.ingramcontent.com/pod-product-compliance
Lightning Source LLC
Chambersburg PA
CBHW041937260926
51850CB00018B/240
* 9 7 8 3 0 3 6 5 6 2 8 7 2 *